Prevention and Mitigation of Natural and Anthropogenic Hazards Due to Land Subsidence

Ninth International Symposium on Land Subsidence (NISOLS)

Proceedings of the International Association of Hydrological Sciences (PIAHS) Volume 372, 2015

Nagoya, Japan
15-19 November 2015

Editors:

K. Daito **D. Galloway**

ISBN: 978-1-5108-3908-3

Printed from e-media with permission by:

Curran Associates, Inc.
57 Morehouse Lane
Red Hook, NY 12571

Some format issues inherent in the e-media version may also appear in this print version.

Printed by Curran Associates, Inc. (2015)

International Association of Hydrological Sciences
Centre for Ecology and Hydrology
Wallingford, Oxfordshire OX10 8BB
United Kingdom

Phone: 441 491 692 442
Fax: 441 491 692 448

http://iahs.info/

Additional copies of this publication are available from:

Curran Associates, Inc.
57 Morehouse Lane
Red Hook, NY 12571 USA
Phone: 845-758-0400
Fax: 845-758-2633
Email: curran@proceedings.com
Web: www.proceedings.com

TABLE OF CONTENTS

RISK MANAGEMENT OF SUBSIDENCE RELATED HAZARDS

ANTHROPOGENIC LAND SUBSIDENCE IN COASTAL REGIONS

LAND SUBSIDENCE AND LIQUEFACTION, DUT TO EAST JAPAN GREAT EARTHQUAKE OF MARCH 11, 2011

GROUND FAILURE (FRACTURING, FAULT ACTIVATION, FISSURING)

LAND SURFACE DISPLACEMENT, MEASURING AND MONITORING

NUMERICAL MODELLING

SOCIAL, CULTURAL AND ECONOMIC INFLUENCE OF LAND SUBSIDENCE

WATER MANAGEMENT STRATEGIES

SUBSURFACE DEFORMATION DUE TO SHALE GAS PRODUCTION

Proc. IAHS, 372, 1–5, 2015
proc-iahs.net/372/1/2015/
doi:10.5194/piahs-372-1-2015

Comprehensive experimental study on prevention of land subsidence caused by dewatering in deep foundation pit with hanging waterproof curtain

T. L. Yang[1,2,3], X. X. Yan[1,2,3], H. M. Wang[1,2,3], X. L. Huang[1,2,3], and G. H. Zhan[1,2,3]

[1]Shanghai Institute of Geological Survey, Shanghai, 200072, China
[2]Key Laboratory of Land Subsidence Monitoring and Prevention,Ministry of Land and Resources, Shanghai, 200072, China
[3]Shanghai Engineering Research Center of Land Subsidence, Shanghai, 200072, China

Correspondence to: T. L. Yang (sigs_ytl@163.com)

Published: 12 November 2015

Abstract. Land subsidence caused by dewatering of deep foundations pit has currently become the focus of prevention and control of land subsidence in Shanghai. Because of the reliance on deep foundation dewatering pit projects, two comprehensive test sites were established to help prevent land subsidence. Through geological environmental monitoring during dewatering of a deep foundation pit, the analysis of the relation between artesian water level and soil subsidence, some basic features of land subsidence caused by dewatering of deep foundation pits are elucidated. The results provide a scientific basis for prevention and control of land subsidence caused by dewatering in deep foundation pits.

1 Introduction

The exploitation and utilization of the shallow subsurface in Shanghai keeps increasing with the demands of urban development deep foundation pits used in construction, leads to a series of environmental and geological problems (Yang and Gong, 2010; Wei et al., 2009). Among the problems, the most serious is the uneven subsidence caused by the compression of shallow strata, which arises from the dewatering of deep foundation pit engineering (Yang, 2010). Research on existing foundation pit engineering reveals that dewatering during excavation for the engineered hanging waterproof curtain has a great impact on the artesian water level at significant distances from the foundation pit (Yang et al., 2009). The large decline of artesian water level beyond the foundation pit leads directly to the uneven subsidence, which adversely affects the environment and important nearby structures. Furthermore, the subsidence is difficult to regulate and threatens the elevation safety in Shanghai (Yang et al., 2014).

Taking deep foundation pits of two construction projects in Shanghai as typical cases, we conducted a comprehensive dewatering-recharge test for prevention of land subsidence during the period of excavation and dewatering. The results can be applied elsewhere to prevent and manage subsidence caused by deep foundation pit engineering.

2 The testing design

2.1 Geological conditions of the testing field

The shallow strata of the testing field, which are comprised of clay, silt and sandy soil, are normal sedimentary strata. The objective of the foundation pit engineering is to decompress the first confined aquifer (the 7th soil layer). It is characterized by large depth, high water yield and permeability. According to the foundation pit retaining design, dewatering for decompression during the excavation is necessary. The underground continuous wall as a hanging waterproof curtain did not block the target aquifer – thus there is a directly hydraulic connection inside and outside the foundation pit.

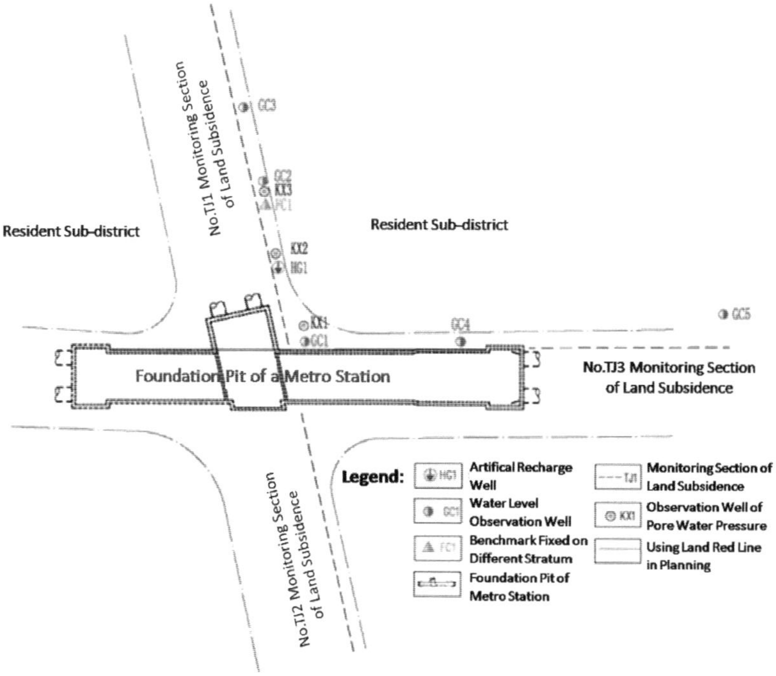

Figure 1. Layout chart of comprehensive testing site of land subsidence control in certain metro station foundation pit.

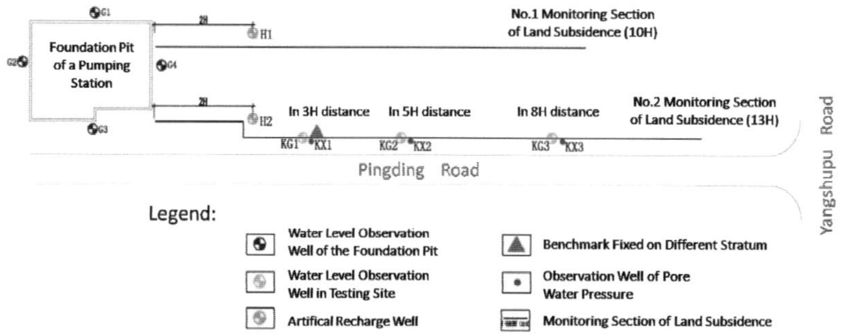

Figure 2. Layout chart of comprehensive testing site of land subsidence control in certain pumping station foundation pit.

2.2 Construction of the testing site

To monitor the geological and environmental effects caused by the pit excavation and dewatering, such as land subsidence and groundwater seepage, deep foundation pits of two construction projects in Shanghai were used as comprehensive testing sites to control engineering subsidence. The land subsidence monitoring section, groundwater observation wells for the confined aquifer, pore water pressure monitoring boreholes and extensometers for the layered subsidence of the soil were built inside the influence of the foundation pit and the dewatering (Yang et al., 2010). The length of monitoring section is more than 10 times the excavation depths (H) of the foundation pits. The interval between two monitoring points is 5 to 10 m. Observation wells for the confined aquifer and pore water pressure monitor holes and extensometers (benchmarks fixed on different stratum) for the layered subsidence of soil and recharge wells were arranged in the main monitoring sections of the foundation pits (Wang et al., 2012), shown in Figs. 1 and 2.

3 Comprehensive test

According to the test objective and the facilities at the testing sites, the comprehensive test included 2 parts, monitoring and recharge. The monitoring test included the monitoring

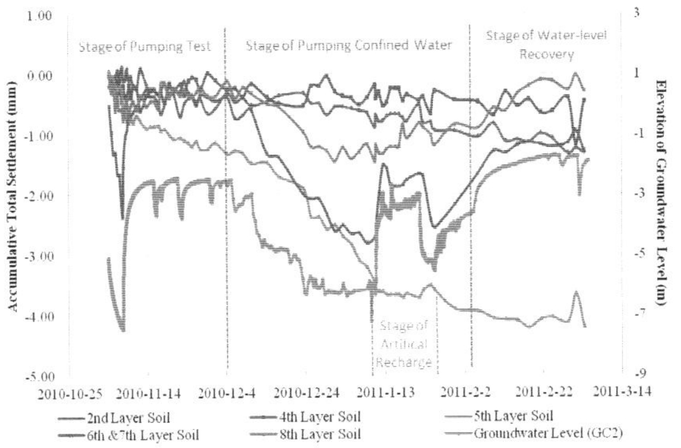

Figure 3. Duration graphical chart of soil laminar settlement and artesian head.

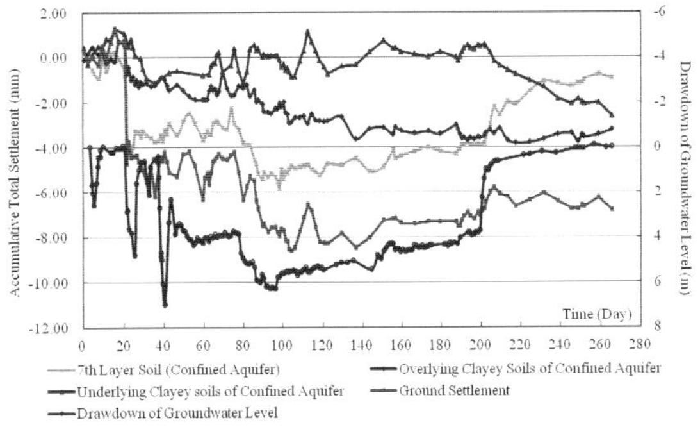

Figure 4. Duration graphical chart of soil laminar settlement and confined aquifer head drop depth at the point of 3 times the depth of foundation pit.

of the water level of the confined aquifer, the layered subsidence of the soil, the pore water pressure and the land subsidence. The monitoring spatial extent (10 times the excavation depth [H] of the foundation pit) covered most of the area influenced by dewatering. Through monitoring the groundwater seepage and the soil settlement during the dewatering period, the relation of the water-level change in the confined aquifer and the land subsidence caused by dewatering of the foundation pit was determined. The recharge test employed an artificial recharge well that penetrated the aquifer targeted for dewatering. A single-well pressurized recharge test was conducted to measure the effect of simultaneously recharging the shallow confined aquifer during the dewatering period on the control of land subsidence.

4 Comprehensive experimental analysis and study

The dewatering of the foundation pit during the construction period causes land subsidence in a wide area around the pit. Large scale water level declines in the confined aquifer that may cause environmental problems. About 70 m from the foundation pit, a group of extensometers, land subsidence monitor section and observation wells for the confined aquifer were set. The relationship between the water level in the confined aquifer and the soil displacement during the dewatering was studied.

4.1 Characteristics of confined aquifer level and different soil layer subsidence

A group of extensometers were arranged to measure the layered soil subsidence during the dewatering period. Figures 3 and 4 show the characteristic curves of subsidence of the soil layers. The results indicated that during the excavation and dewatering, with the change of water level, the characteristics of soil displacement were as follow:

1. During the initial stage of engineered dewatering, with the rapid decline of the confined aquifer water level, compression (settlement) is mainly focused on the soil layer of the confined aquifer. Furthermore, settlement gradually stabilized with stablization of the water level. Following engineered dewatering, the soil layer of the confined aquifer rebounded with recovery of the water level, while the overlying clayey soils sustained compression with gradual slow. At this time, settlement was mainly focused on the overlying clayey soil layer. In addition, during the engineered dewatering, compression of the underlying soil layer of the confined aquifer resulted in small settlement of that layer.

2. From the point of view of deformation characteristic caused by the engineered dewatering, initial settlement mainly showed elastic deformation of the confined aquifer, and overlying clayey soils were less affected. With the increase of pumping quantity and duration, elastic deformation of the confined aquifer was changed into elastic-plastic deformation of overlying clayey soils. After dewatering was over the soil layer of the confined aquifer rebounded, but compression occurred in the overlying clayey soils.

3. A short term of artificial recharge to the first confined aquifer was conducted during the dewatering period. The test result showed that the artificial recharge could increase the confined aquifer water level and reverse settlement of the confined aquifer soil layer, so that the land subsidence could be controlled.

4.2 Characteristics of confined aquifer level and land subsidence

Water level in the confined aquifer within the spatial extent of $10H$ around the foundation pit varied with distance from the

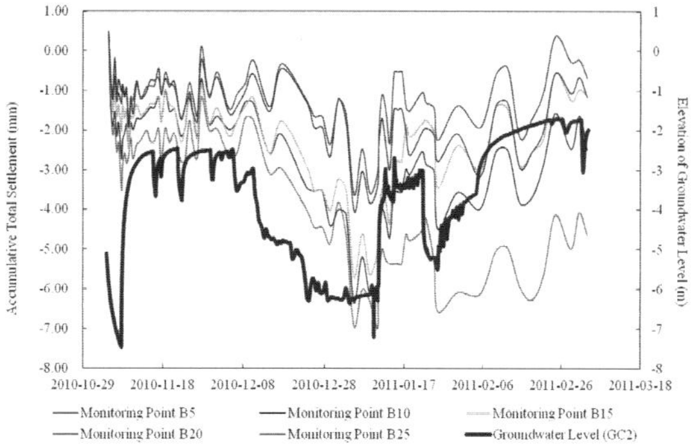

Figure 5. Duration graphical chart of land subsidence and artesian head in monitoring section.

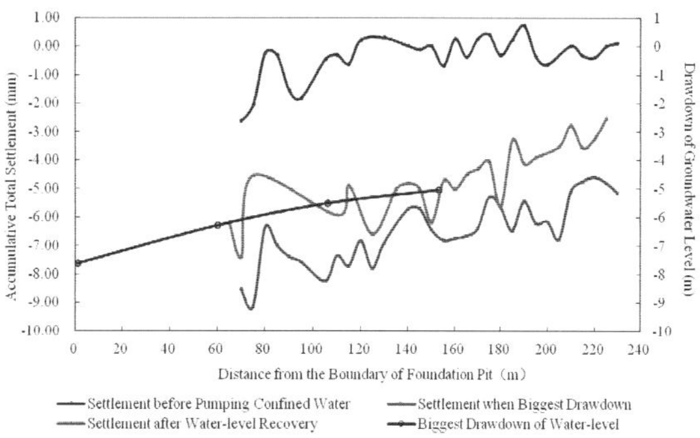

Figure 6. Graphical chart of spatio-temporal features of land subsidence and artesian head in monitoring section.

pit during the excavation and dewatering period. The corresponding subsidence and rebound were monitored. Figures 5 and 6 show the relationship between land subsidence and the water-level change or 3 monitor sections (Fig. 6). The results show the following characteristics:

1. The drawdown and recovery of the confined water level was directly linked with the land subsidence and rebound. The former was the triggering factor and the later the result.

2. The confined water level around the foundation pit dropped a lot during the dewatering period. The spatial extent of the confined aquifer water level influenced by dewatering was larger than $10H$, which corresponded with the land subsidence influence. The magnitude of subsidence and the water-level decline gradually decreased with distance from the foundation pit. The subsidence was positively associated with the variation of the confined aquifer water level.

3. The excavation and dewatering were the main reason why land subsidence occurred. Within $3H$ around the foundation pit, the land subsidence was caused by the superposition of excavation and dewatering, and the subsidence magnitude was very large. The subsidence profiles were spoon-shaped. The maximum cumulative subsidence occurred within the site of $0.5–1.0H$ from the foundation pit. Where the distance was larger than $3H$ from the foundation pit, the subsidence was caused mainly by dewatering. The subsidence magnitude there was relatively small, yet the scope was very large.

4. A short term of artificial recharge to the objective aquifer was conducted during the excavation and dewatering period. Within the influence range of recharging, soil rebound occurred and the subsidence magnitude was inversely proportion to the distance from the recharge well. So, the simultaneous artificial recharge to the confined aquifer during dewatering could help control land subsidence during the dewatering period.

5. After dewatering, the confined aquifer water level rose, but the subsidence kept developing. When the water level rose for a period of time, the land surface elevation recovered. Within $3H$ from the foundation pit, the final subsidence magnitude was large while the rebound value was small. Beyond $3H$ outside the scope from the foundation pit, the rebound magnitude was large while the final subsidence value was small.

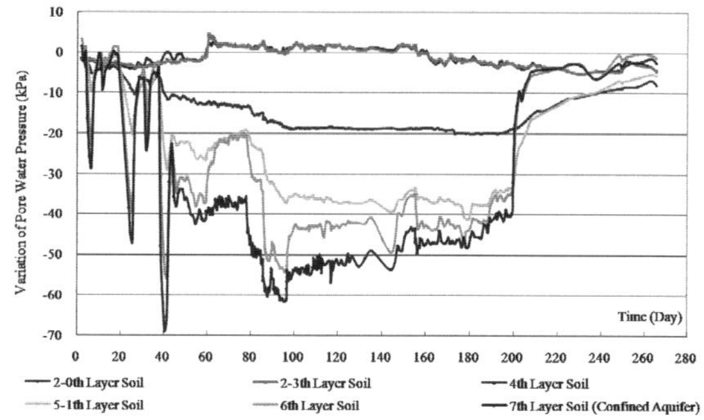

Figure 7. Duration graphical chart of pore water pressure at the point of 3 times the depth of foundation pit.

4.3 Characteristics of pore water-pressure change

As shown in Fig. 7, during the stage of pumping test and engineered dewatering, the water level (expressed as an equivalent pore water pressure in Fig. 7) in the confined aquifer fell. During this time there was an obvious decline of pore water pressure among the 5th and 6th soil layers overlying the confined aquifer. However, pore water pressure of the 4th soil layer and above was only slightly affected and relatively stable. Following dewatering of the foundation pit ,the water level of the confined aquifer rapidly recovered, but the pore water pressure of the overlying cohesive soil layers recovered slowly.

5 Conclusions

1. The dewatering of foundation pits using hanging waterproof curtains cause land subsidence.

2. The decline and recovery of the confined aquifer water level is directly linked with the land subsidence and rebound. The former is the triggering factor, and the later the result.

3. Monitoring results of extensometers showed that land subsidence caused by dewatering of the foundation pit was mainly comprised of the compression of the targeted aquifer for dewatering – the confined aquifer; and the neighboring soft soil layers. The compression of the confined aquifer was mainly elastic and was in accord with the variation of the confined aquifer water level while the neighboring soft soil layers was visco-elasto-plastic and was irreversible.

4. The spatial extent of land subsidence caused by dewatering of the foundation pit using the hanging waterproof curtain was larger than $10H$. The magnitude of subsidence and the decline of the confined aquifer water level gradually decreased with the increasing distance from the foundation pit. The subsidence was positively associated with the variation of the water level.

5. Within $3H$ around the foundation pit, the land subsidence was caused by the superposition of excavation and dewatering, and the subsidence value was large. The subsidence profile was spoon-shaped. The maximum cumulative subsidence occurred within a distance of $0.5–1.0H$ from the foundation pit. Beyond a distance of $3H$ from the foundation pit, the subsidence was caused mainly by the dewatering. The subsidence there was relatively small, yet the spatial extent was large.

6. A short term of artificial recharge to the confined aquifer was conducted during the excavation and dewatering period. Within the range of influence from recharging, soil rebound occurred and the magnitude was inversely proportion to the distance from the recharge well. So, simultaneously, artificially recharging the objective aquifer during its engineered dewatering could help control land subsidence during the dewatering period.

Acknowledgements. This work was supported by public welfare and scientific research subject projects funded by the Ministry of Land and Resources of China (201311045) and major scientific research subjects from the Science and Technology Commission of Shanghai Municipality (12231200700, 14DZ1207802).

References

Wang, J. X., Wu, Y. B., Zhang, X. S., Liu, Y., Yang, T. L., and Feng, B.: Filed experiments and numerical simulations of confined aquifer response to multi-cycle recharge-recovery process through a well, J. Hydrol., 464–465, 328–343, 2012.

Wei, Z. X., Wang, H. M., Wu, J. Z., Fang, Z. L., and Liu, G. B.: Land subsidence and its influences on urban security of Shanghai, Shanghai Geology, 30, 34–39, 2009.

Yang, T. L.: Analysis of the Land Subsidence Impact of Dewatering of Deep Foundation Pits, Shanghai Geology, 33, 41–44, 2010.

Yang, T. L. and Gong, S. L.: Microscopic analysis of the engineering geologic behavior of soft clay in Shanghai, China, Bull. Eng. Geol. Environ., 69, 607–615, 2010.

Yang, T. L., Yan, X. X., Wang, H. M., Xu, J. H., and Cheng, H. S.: Study on land subsidence induced by excavation engineering, Shanghai Geology, 30, 15–21, 2009.

Yang, T. L., Yan, X. X., Wang, H. M., Wu, J. Z, and Liu, J. B.: Experimental research of shallow aquifer pressure groundwater recharge based on dual control pattern of groundwater level and land subsidence, Shanghai Geology, 31, 12–17, 2010.

Yang, T. L., Wang, H. M., and Jiao, X.: Land subsidence zoning control in Shanghai, Shanghai Geology, 35, 105–109, 2014.

Proc. IAHS, 372, 7–11, 2015
proc-iahs.net/372/7/2015/
doi:10.5194/piahs-372-7-2015

Spatial-temporal variation of groundwater and land subsidence evolution in Beijing area

K. Lei[1], Y. Luo[1], B. Chen[2], M. Guo[3], G. Guo[1], Y. Yang[1], and R. Wang[1]

[1]Beijing Institute of Hydrogeology and Engineering Geology, Beijing 100195, China
[2]College of Resources Environment and Tourism, Capital Normal University, Beijing 100048, China
[3]Beijing Bureau of Geology and Mineral Exploration and Development, Beijing 100195, China

Correspondence to: K. Lei (279459260@qq.com)

Published: 12 November 2015

Abstract. Precipitation is the main recharge source of groundwater in the plain of Beijing, China. Rapid expansion of urbanization has resulted in increased built-up area and decreased amount of effective recharge of precipitation to groundwater, indirectly leading to the long-term over-exploitation of groundwater, and induced regional land subsidence. Based on the combination of meteorological data, groundwater level data, interferometric synthetic aperture radar (InSAR; specifically persistent scatterer interferometry, PSI), geographic information system (GIS) spatial analysis method and rainfall recharge theory, this paper presents a systematic analysis of spatial-temporal variation of groundwater level and land subsidence evolution. Results show that rainfall has been decreasing annually, while the exploitation of groundwater is increasing and the groundwater level is declining, which is has caused the formation and evolution of land subsidence. Seasonal and interannual variations exist in the evolution of land subsidence; the subsidence is uneven in both spatial and temporal distribution. In 2011, at the center of mapped subsidence the subsidence rate was greater than 120 mm a^{-1}. The results revealed good correlation between the spatial distribution of groundwater level declines and subsidence. The research results show that it is beneficial to measure the evolution of land subsidence to dynamic variations of groundwater levels by combining InSAR or PSI, groundwater-level data, and GIS. This apprpach provides improved information for environmental and hydrogeologic research and a scientific basis for regional land subsidence control.

1 Introduction

Land subsidence caused by groundwater extraction can be a severe geologic hazard, causing structural damages to infrastructure and environmental damages to land an water resources and can lead to a series of other hazards related to the formation of ground fissures (Galloway et al., 1998; Galloway and Hoffmann, 2007). Some studies attribute the leading cause of regional land subsidence in Beijing China to the overexploitation of groundwater (Chen, 2000). Large-scale and long-term overexploitation of groundwater leads to declining groundwater levels, the deformation(compaction) of clay sediments in the aquifer system, and land subsidence. As the capital of China, Beijing is a metropolis with serious water shortage problems (Lei et al., 2014). More than two-thirds of the water supply comes from groundwater. Long term overexploitation of groundwater has led to continuously declining water levels and plays an important role in the formation and development of regional land subsidence (Jia et al., 2007). In the center of mapped subsidence, the cumulative settlement is as much as to 1302 mm; and some areas are still subsiding at a high rate of 100–120 mm a^{-1}.

This study combines groundwater level data, the PSI technique, GIS spatial analysis methods and rainfall recharge theory to analyze spatial-temporal variations of the groundwater system and evolution of land subsidence in the Beijing plain area, China. The study also analyzes the spatial and temporal variability of rainfall groundwater recharge derived from precipitation. The relation between groundwater-level variations and the evolution of land subsidence is demonstrated.

2 Study area

Beijing (39°28′–41°05′ N, 115°25′–117°35′ E), located in the north of China, is divided into three parts: northern mountains, western mountains and south-eastern plains. The climate in Beijing is characterized by uneven seasonal precipitation, with over 70 % of the annual precipitation distributed between June and August. During recent decades, rainfall has been declining continuously attributed in part to the impact of global climate change; the average annual rainfall was 479.49 mm, equivalent to 84.2 % of the average annual rainfall for the period of record.

3 Methodologies

3.1 Estimation of rainfall recharge to ground water

Kriging was employed to spatially interpolate observed precipitation data from 20 national weather stations in Beijing. The principle of Kriging interpolation is based on:

$$Z_0^* = \sum_{i=1}^{n} \lambda_0^i Z_i, \tag{1}$$

where, Z_0^* is Kriging interpolation values; Z_i is observations of the position $\chi_i (i = 1, 2, \ldots, n)$. Groundwater recharge from rainfall was estimated by using

$$Q = 10^3 a \cdot X \cdot F, \tag{2}$$

where Q is rainfall infiltration recharge (m^3); a is the infiltration coefficient; X is average annual rainfall (mm); and F is infiltration area.

Landsat TM images of three periods (2000, 2006, 2009) covering the Beijing area were used to extract spatial-temporal information on the expansion of built-up areas by the supervised classification method. Then, loss in rainfall recharge(precipitation minus groundwater recharge derived from infiltration of precipitation) was calculated by GIS raster calculation using groundwater recharge from rainfall and the built-up area.

3.2 Deformation monitoring using PSI

Persistent scatterers (PSI) can effectively reduce the spatial and temporal decorrelation effects and reduce the error component caused by atmospheric delay versus traditional InSAR. Through error corrections on atmospheric, orbit and digital elevation model (DEM) phase components, accurate deformation rates can be extracted (Ferretti et al., 2001; Hooper, et al. 2004). The PSI method was used in this study:

$$\phi_{x,i} = \phi_{\text{def},x,i} + \phi_{\varepsilon,x,i} + \phi_{\text{atm},x,i} + \phi_{\text{orb},x,i} + \phi_{n,x,i}, \tag{3}$$

where, $\Phi_{\text{def},x,j}$ is the deformation phase of viewing direction; $\Phi_{\varepsilon,x,j}$ is the residual topographic phase introduced by errors in the external DEM; $\Phi_{\text{atm},x,j}$ is the phase difference of atmospheric delays between two satellite transits; $\Phi_{\text{orb},x,j}$ is the phase component of orbit error; $\Phi_{\text{n},x,j}$ is the noise component; x, j is row number, line number for pixels.

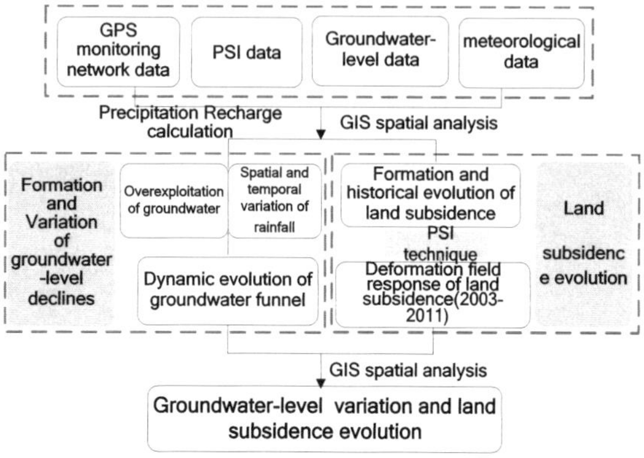

Figure 1. Framework diagram of the comprehensive analysis method.

3.3 Comprehensive analysis method

Combined with groundwater level data, meteorological data, PSI data and GIS were used to analyze groundwater level variations and the evolution land subsidence. The spatial and temporal variability of rainfall and its recharge to groundwater, is described in terms of the spatial and temporal correlations of groundwater level declines and land subsidence (Fig. 1).

4 Results and analysis

4.1 Spatial and temporal variability of precipitation and recharge to groundwater

4.1.1 Spatial and temporal variability of rainfall

Meteorological data was compiled from 20 weather stations for six typical periods (1997, 2000, 2003, 2006, 2009 and 2011) to develop time series contour maps of rainfall using the Kriging method (geostatistical tool of ArcGIS) (Fig. 2). The results show the spatial and temporal variation of rainfall in Beijing. In 1997, the rainfall was mainly concentrated in the northeast region and the average precipitation was 438 mm. In 2003, the rainfall was mainly concentrated in Yan-Qing and Fang-Shan where the average precipitation was 457 mm. In 2006, the rainfall was mainly concentrated in Ping-Gu and the average precipitation was 491 mm. In 2009–2011, there was a significant difference of precipitation between eastern areas and western areas. The precipitation in the central and eastern area was greater than 572 mm and distributed evenly.

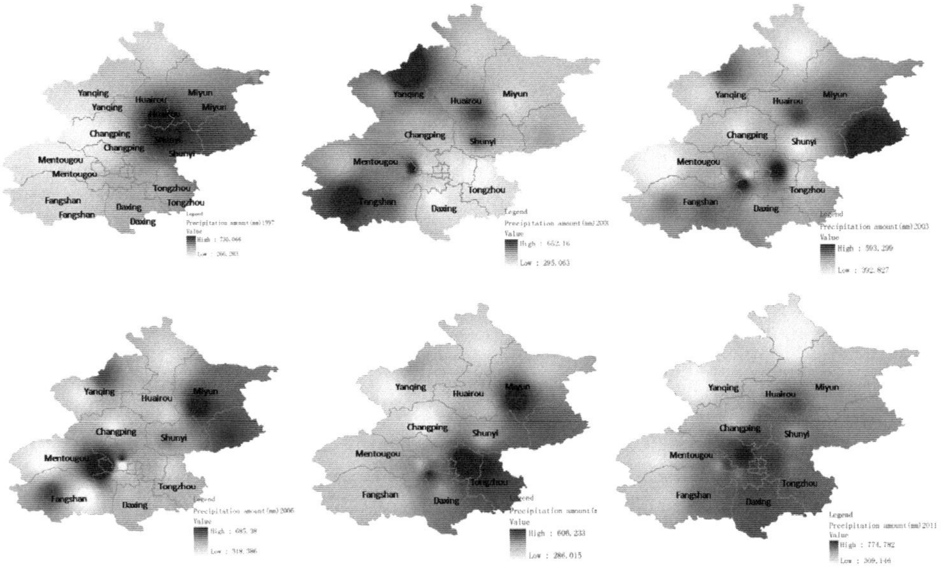

Figure 2. Spatial and temporal variability of rainfall (1997–2011).

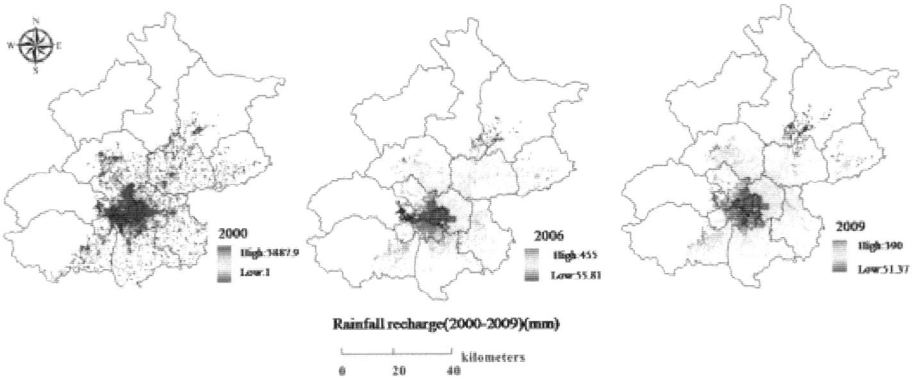

Figure 3. Spatial-temporal distribution of effective recharge to groundwater from rainfall (2000–2009).

4.1.2 Groundwater recharge from rainfall

With the decline of precipitation, groundwater recharge is decreasing considerably due to the growth of impervious areas, which result from the urbanization of Beijing.

Land use derived from Landsat TM images of three periods (2000, 2006, 2009) showed in Fig. 3 that the built-up area was 2246.94 km^2 in 2000, growing to 2344.97 km^2 in 2006, and growing further to 2767.22 km^2 in 2009 – a more than 20 % increase since 2000. With the expansion of impervious area (built-up area), and reduced groundwater recharge from rainfall led to the over-exploitation of groundwater.

Recharge from rainfall to the groundwater system was obtained by using GIS raster calculations and statistical analysis of groundwater recharge from rainfall accounting for the built-up area. In 2000, effective rainfall recharge was $18.81 \times 10^8\,m^3$, accounting for 48 % of the total annual rainfall ($39.41 \times 10^8\,m^3$), residual rainfall not contributing to groundwater recharge (loss) was 52 % in 2000, 61 % in 2006, and 62 % in 2009. The results suggest that with the increasing built-up area, the loss amount of rainfall recharge to groundwater was increasing. Two-thirds of the Beijing water supply is from groundwater. With urban area expansion, water requirements and groundwater demand and exploitation are increasing annually. Thus, groundwater levels have been declining, which leads to formation and evolution of subsidence.

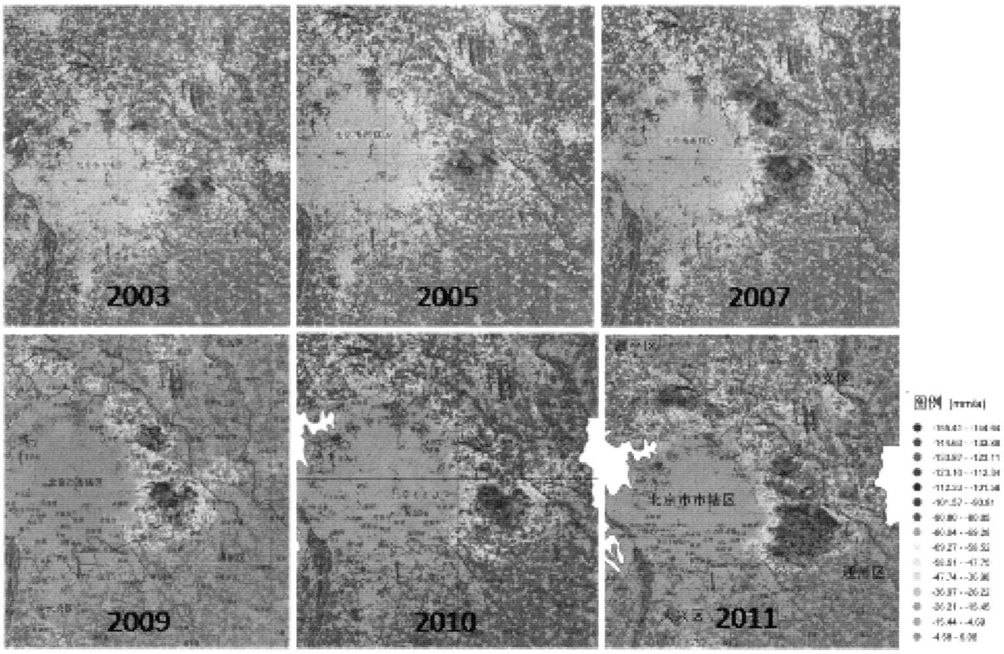

Figure 4. Land subsidence mapped using PSI technique (2003–2011).

4.2 Land subsidence evolution to resulting from variation of groundwater levels

4.2.1 Spatial-temporal distributions of land subsidence (2003–2011)

The time series land subsidence information was acquired using the InSAR PSI technique and the result was validated by leveling survey measurements.

According to the time-series PSI results, the subsidence expanded steadily during 2003–2011. In 2011, the cenrer of subsidence was subsiding 120 mm a^{-1}, and about 1087 km^2, mainly in the Changping, Shunyi, Chaoyang and Tongzhou Districts was subsiding more than 30 mm a^{-1} (Fig. 4).

4.2.2 Relationship between groundwater level variation and land subsidence evolution

The correlation between groundwater level variation and land subsidence evolution was analyzed by comparing groundwater level contoursfor select years during 2003–2011, with the average settlement rate obtained by the PSI technique during 2003–2011 (Fig. 5). Greater settlement rates are fairly well correlated with areas of greater groundwater-level declines-as indicated by the depth to water-level contours, mainly distributed in Tianzhu of Shunyi District, Chaoyang District and north-western Tongzhou District. The groundwater level was declining continuously with an average rate of 2.18 m a^{-1} and a maximum rate of 3.82 m a^{-1} in the area of maximum regional groundwater-level declines.

5 Conclusions

1. The paper analyzed the spatial and temporal variability of rainfall and its recharge to groundwater during 2000–2009. With declining precipitation, groundwater recharge was decreasing heavily due to reduced precipitation and to the growth of impervious areas. Meanwhile, water requirements and groundwater exploitation were increasing annually. Groundwater level has been declining, which has led to formation and evolution of a regional cone of depression in the groundwater system and land subsidence.

2. The land subsidence (during 2003–2011) was measured and mapped using PSI. In 2011, the settlement rate at the center of subsidence was larger than 120 mm a^{-1}, 1087 km^2 was subsiding at a rate larger than 30 mm a^{-1}.

3. From the comparison between variation of groundwater levels and subsidence, the results suggested that the spatial distribution of groundwater-level declines was fairly well correlated with the mapped subsidence, but not entirely.

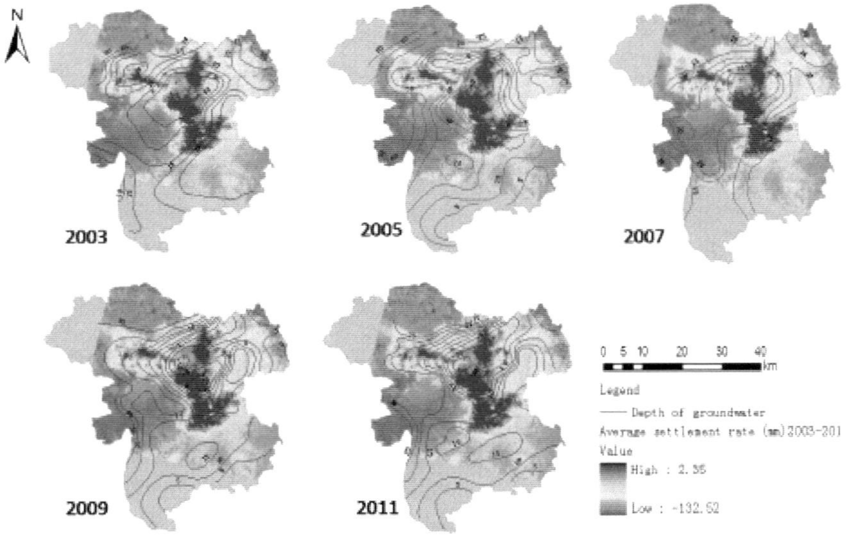

Figure 5. Groundwater contours (depth to water) and land subsidence.

Acknowledgements. This research was funded by the Beijing Municipal Science and Technology Project (Z131100005613022).

References

Chen, C. X.: Thinking on Land Subsidence Caused By Groundwater Exploitation, Hydrogeology and Engineering Geology, 27, 45–60, 2000.

Ferretti, A., Prati, C., and Rocca, F.: Permanent scatterers in SAR interferometry, IEEE T. Geosci. Remote, 39, 8–20, 2001.

Galloway, D. L. and Hoffmann, J.: The application of satellite differential SAR interferometry-derived ground displacements in hydrogeology, Hydrogeol. J., 15, 133–154, 2007.

Galloway, D. L., Hudnut, K. W., Ingebritsen, S. E., Phillips, S. P., Peltzer, G., Rogez, F., and Rosen, P. A.: Detection of aquifer-system compaction and land subsidence using interferometric synthetic aperture radar, Antelope Valley, Mojave Desert, California, Water Resour. Res., 34, 2573–2585, 1998.

Hooper, A., Zebker, H., Segall, P., and Kampes, B.: A New Method for Measuring Information on Volcanoes and Other Natural Terrains Using InSAR Persistent Scatterers, Geophys. Res. Lett., 31, 1–5, 2004.

Jia, S. M., Wang, H. G., Zhao, S. S., and Luo, Y.: A Tentative Study of the Mechanism of Land Subsidence in Beijing, City Geology, 2, 20–26, 2007.

Lei, K. C., Chen, B. B., Jia, S. M., Wang, S. F., and Luo, Y.: Formation and Genetic Mechanism of Land Subsidence Based on PS-InSAR Technology in Beijing, Spectrosc. Spect. Anal., 34, 2185–2189, 2014.

Proc. IAHS, 372, 13–15, 2015
proc-iahs.net/372/13/2015/
doi:10.5194/piahs-372-13-2015

Open Access
PIAHS

Investigation of alternative mechanisms of aquifer-system compaction and land subsidence in Shanghai

Y. Yuan, Y.-S. Xu, S.-L. Shen, and N. Zhang

State Key Laboratory of Ocean Engineering, Department of Civil Engineering,
Shanghai Jiao Tong University, Shanghai, China

Correspondence to: S.-L. Shen (slshen@sjtu.edu.cn)

Published: 12 November 2015

Abstract. Alternative subsidence mechanisms related to groundwater extration and surface loading of built infrastructure in Shanghai have been evaluated to explain continued subsidence despite the controlled reduction of groundwater extraction. The traditional theory of aquifer-system compaction embodied in the aquitard drainage model cannot fully explain this phenomenon. Two possible alternative mechanisms were studied previously: (i) surface loading attributed to urban construction; (ii) creep occuring in sandy deposits of aquifers. These mechanisms could not fully explain the observed subsidence. Two additional alternative mechanistic models are proposed that involve principles of load transfer considering the temporally and spatially redistributed stresses associated with groundwater extraction: (i) the Cosserat continuum mechanism, considering shear force on the permeable coarse-grained matrial in the aquifer due to hydraulic gradients in the aquifer; and (ii) internal erosion of fine-grained (clay and silt) particles within the aquifer. Initial results based on simulations incorporating Cosserat mechanics look promising.

1 Introduction

The aquifer system in the Yangtze River Delta deposits in Shanghai comprise marine and estuarine sediments with alternating, sub-horizontal strata of permeable, coarse-grained (mostly sand with some silt and clay fraction) strata constituting the aquifers and soft (relatively more compressible), low-permeablity fine-grained (predominately clay and silt) strata constituting the aquitards. The aquitards may be either laterally extensive where they act as confining units in the aquifer system, or interbeded within the aquifers. Generally, when groundwater pumping occurs, groundwater levels decline, causing consolidation of the aquifer system leading to land subsidence. Most of the subsidence in Shanghai has been attributed to the consolidation of the aquitards within the aquifer system.

Recently new subsidence occurring in Shanghai and elsewhere with different geological and hydrogeological conditions, cannot be adequately interpreted using the traditional theory of aquifer-system compaction. Groundwater withdrawal in Shanghai has been controlled, but subsidence has continued since the 1990s. Xu et al. (2009) showed that with the development of urban construction, land subsidence has been poorly correlated with the volume of groundwater withdrawal, but has been well correlated with groundwater drawdown. The groundwater drawdowns have been attributed to the groundwater damming effect of deep foundations for high-rise buildings and other civilian infrastructure. In recent years, field observations show that the percentage of subsidence due to deformation occurring in the aquifers has increased, and that this deformation lags the change in groundwater level. The objective of this paper is to summarize the alternative approaches to illustrate the mechanism of the aforementioned phenomena.

2 New phenomena of land subsidence

The thickness of soft deposits of Shanghai is generally 300 m with a maximum of 400 m. There is one phreatic aquifer group (Aq0) and five confined aquifers (AqI to AqV). The

Published by Copernicus Publications on behalf of the International Association of Hydrological Sciences.

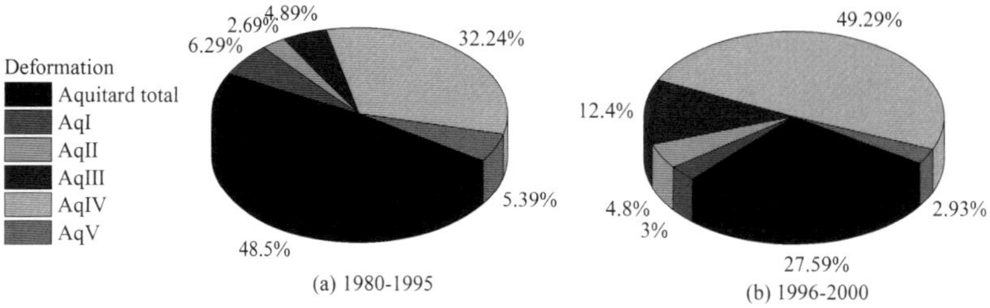

Figure 1. Pie chart of deformation of each aquifer in Shanghai (Data from Wei et al., 2010a).

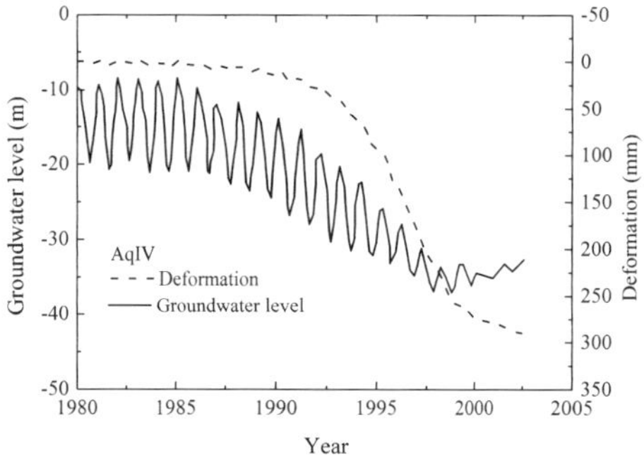

Figure 2. Groundwater level and deformation of AqIV in Yangpu district, Shanghai (Data from Wei et al., 2010b).

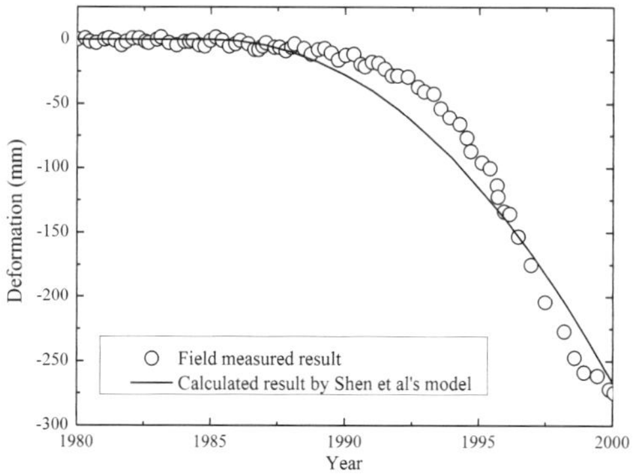

Figure 3. Comparison of simulated deformation using Cosserat mechanics to measured deformation of AqIV at Laodong Park (modified from Shen et al., 2013).

aquifers are separated by six aquitards (AdI to AdVI) (Xu et al., 2009). Historical land subsidence in Shanghai, which is mainly due to groundwater withdrawal from AqII to AqV, has been officially recorded since 1921. Land subsidence was controlled in the 1970s after some measurements were taken. The subsidence rate in the urban center increased again under the same conditions of groundwater withdrawal in the late 1980s. The average subsidence rate during 1996–2005 was four times of the rate during 1972–1988. Figure 1 shows the percentage of the deformation of each aquifer and all of the aquitards to the total land subsidence during 1980–2000. The increasing percentage of subsidence caused by deformation of aquifers during 1996–2000 is attributed mostly to deformation occurring in AqIV. Fig. 2 shows the variations of deformation and groundwater level of AqIV in Yangpu District, which is located in the urban center of Shanghai. As shown in Fig. 2, deformation of AqIV lags the recovery of groundwater level.

3 Mechanism of land subsidence

3.1 Urbanization-induced land subsidence

Much infrastructures has been constructed in Shanghai since the 1990s. Studies of recent accelerated subsidence associated with urbanization point to two main causes: (i) the additional stresses caused by the building loads; (ii) the reduction of recharged groundwater volume due to the cut-off effect imposed by underground structures penetrating the shallow portion of the aquifer system on groundwater flow. The cut-off effect influences only the deformation of the upper strata overlying AqII. It cannot explain the continuous deformation mechanism of the deeper aquifers (e.g. AqIV).

3.2 Deformation of the sand

The compaction of the aquitards due to the drawdown of the groundwater level has been considered the main factor causing land subsidence. However, situ data collected over the past twenty years indicate that deformation of the aquifers cannot be ignored and has become significant.

It was proposed to be the impact of the silt particles in the sand strata. Wang et al. (2010) concluded sand creep to be mainly responsible for the land subsidence. But the sand creep in aquifer could not create such scale of aquifer compaction. Thus, the sand creep theory can be used to illustrate the new phenomenon of the land subsidence to a degree. Some researchers proposed that deformation creep of sand just partly explained the magnitude and timing of the subsidence, that the deformation and subsidence lagged the recovery of the groundwater level (Zhang et al., 2007).

3.3 Research based on Cosserat mechanics

Cai (2004) proposed that sand in aquifers would yield when the shear force was high enough. Groundwater pumping from the aquifer results in a large hydraulic gradient in the aquifer. This hydraulic gradient will cause an unbalanced shear stress in the aquifer. Cosserat continuum mechanics was proposed to describe the deformation of a continuous medium under shear stress, which is different from Cauchy mechanics (Budhu and Adiyaman, 2010). Shen et al. (2013) analysed the land subsidence in Shanghai based on Cosserat mechanics, as shown in Fig. 3. The result indicated the model based on Cosserat mechanics simulated the measured data well.

3.4 Internal erosion of fine particles in sand

An alternative mechanism causing subsidence in Shanghai is proposed here. Large quantities of silt and clay particles dispersed within the matrix pores of the aquifers in Shanghai (Wei et al., 2010a, b). The impact of these particles on the deformation of the aquifers is very small when the particles are static (Zhang et al., 2007). The hydraulic gradients induced by groundwater pumping can lead to shear stress in the aquifers and seepage stress in the silt and clay particles. The hydraulic gradients would lead to movement of the silt and clay particles in the pores of the aquifer, namely internal erosion of the particles. The internal erosion would result in the redistribution of land subsidence in both space and time. So it is important to investigate the mechanism of the internal erosion of the fine-grained particles and its impact on land subsidence in Shanghai.

4 Conclusions

Recently land subsidence in Shanghai is attributed to deformation occurring in the aquifers (eapecially Aq04) that continues and lags drawdowns. The traditional theory of aquifer-system compaction embodied in the aquitard drainage model cannot reasonably explain these phenomena. Research to investigate potential alternative mechanisms is addressing cut-off of groundwater flow caused by urban construction, sand creep, shear stress on soils induced by hydraulic gradients, and internal erosion of the fine-grained particles dispersed in the aquifers.

Acknowledgements. The research work described herein was funded by the National Nature Science Foundation of China (Grant No. 41472252) and partially funded by the National Basic Research Program of China (973 Program: 2015CB057806). These financial supports are gratefully acknowledged.

References

Budhu, M. and Adiyaman, I. B.: Mechanics of land subsidence due to groundwater pumping, Int. J. Numer. Anal. Met., 34, 1459–1478, 2010.

Cai, Z. Y.: The deformation behavior of sand, Yellow River Conservancy Press, China, 172 pp. 2004.

Shen, S. L., Ma, L., Xu, Y. S., and Yin, Z. Y.: Interpretation of increased deformation rate in aquifer IV due to groundwater pumping in Shanghai, Can. Geotech. J., 50, 1129–1142, 2013.

Wang, F., Miao, L. C., and Lv, W. H.: Study on the new deformation characteristics of the pumped aquifers in Su-Xi-Chang area, China, Geoenvironmental Engineering and Geotechnics (GSP 204), edited by: He, Q. and Shen, S. L., ASCE Press, Reston, Virginia, 57–62, 2010.

Wei, Z. X., Zhai, G. Y., and Yan, X. X.: Shanghai Urban Geology, Geology Press, Beijing, China, 123 pp., 2010a.

Wei, Z. X., Zhai, G. Y., and Yan, X. X.: Shanghai Urban Geology, Geology Press, Beijing, China, 374 pp., 2010b.

Xu, Y. S., Shen, S. L., and Du, Y. J.: Geological and hydrogeological environment in Shanghai with geohazards to construction and maintenance of infrastructures, Eng. Geol., 109, 241–254, 2009.

Zhang, Y., Xue, Y. Q., Wu, J. C., and Ye, S. J.: Stress-strain measurements of deforming aquifer systems that underlie Shanghai, China, Environmental and Engineering Geoscience, 13, 217–228, 2007.

Proc. IAHS, 372, 17–21, 2015
proc-iahs.net/372/17/2015/
doi:10.5194/piahs-372-17-2015

Open Access
PIAHS

Groundwater-abstraction induced land subsidence and groundwater regulation in the North China Plain

H. Guo[1], L. Wang[2], G. Cheng[1], and Z. Zhang[1]

[1]China Institute of Geo-Environment Monitoring, Beijing, 100081, China
[2]Beijing Geo-Environmental Monitoring Station, Beijing, 100195, China

Correspondence to: H. Guo (hpguo79@gmail.com)

Published: 12 November 2015

Abstract. Land subsidence can be induced when various factors such as geological, and hydrogeological conditions and intensive groundwater abstraction combine. The development and utilization of groundwater in the North China Plain (NCP) bring great benefits, and at the same time have led to a series of environmental and geological problems accompanying groundwater-level declines and land subsidence. Subsidence occurs commonly in the NCP and analyses show that multi-layer aquifer systems with deep confined aquifers and thick compressible clay layers are the key geological and hydrogeological conditions responsible for its development in this region. Groundwater overdraft results in aquifer-system compaction, resulting in subsidence. A calibrated, transient groundwater-flow numerical model of the Beijing plain portion of the NCP was developed using MODFLOW. According to available water supply and demand in Beijing plain, several groundwater regulation scenarios were designed. These different regulation scenarios were simulated with the groundwater model, and assessed using a multi-criteria fuzzy pattern recognition model. This approach is proven to be very useful for scientific analysis of sustainable development and utilization of groundwater resources. The evaluation results show that sustainable development of groundwater resources may be achieved in Beijing plain when various measures such as control of groundwater abstraction and increase of artificial recharge combine favourably.

1 Introduction

Land subsidence can be defined as the sinking of the ground surface with respect to surrounding terrain or sea level. Land subsidence is the result of consolidation of subsurface strata caused by natural causes such as tectonic motion or man-induced causes such as the withdrawal of groundwater, oil and gas (Abidin et al., 2013; Zeitoun and Wakshal, 2013; Galgaro et al., 2014; Papadaki, 2014). Due to increased groundwater abstractions to supply rapid industrial and urban development, subsidence is prevalent in the NCP (Fig. 1), and critically impacts sustainable economic and social development. The first occurrence of land subsidence in the NCP took place in Tianjin in the 1920s, and by the 1960s it became a severe hazard in this city. Since the 1980s, the land affected by subsidence extended from cities to rural areas. Ground cracks or fissures are observed in areas where large differential subsidence occurs. Commonly subsidence in the NCP is slow, accumulative, and irreversible. Economic losses due to subsidence in the NCP have exceeded Yuan 200 billion (Ye and He, 2006). Due to groundwater overexploitation, the maximum subsidence reached up to 1.1 m in Beijing plain during the period 1955–2007 (Zhang et al., 2014). To control subsidence in the Beijing plain, reducing groundwater abstraction and increasing artificial recharge will be important goals for groundwater resources management.

This paper presents a brief summary of the subsidence mechanism in the NCP. A transient 3-D groundwater flow model of the Beijing plain is used to simulate 10 proposed future groundwater regulation scenarios. A multi-criteria fuzzy pattern recognition method is used to evaluate the simulated scenarios based on a set of specified groundwater management goals and constraints. Evaluation of the proposed regulation scenarios is discussed and one scenario is recommended.

Published by Copernicus Publications on behalf of the International Association of Hydrological Sciences.

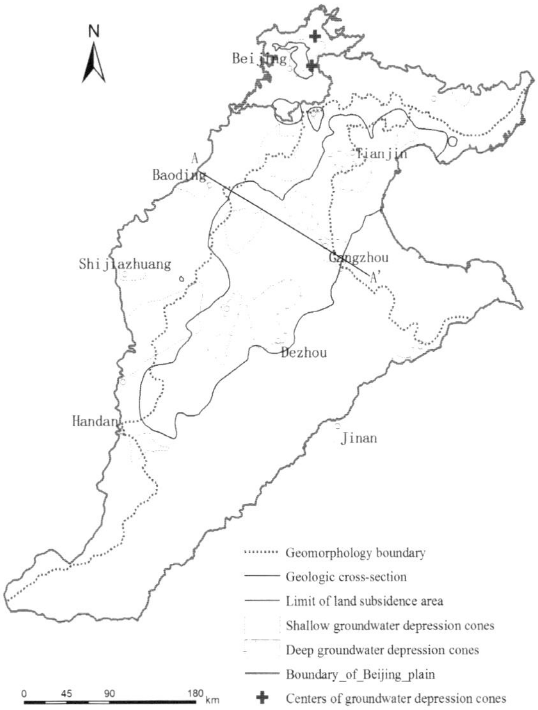

Figure 1. Distribution of the depression cones of shallow and deep groundwater, and major land subsidence area in the NCP.

2 Regional hydrogeology

The NCP is located in the eastern part of China, has a total area of about 140 000 km^2. This region comprises the plain area of Beijing, Tianjin and Hebei Province and the plain area of Henan and Shandong provinces to the north of the Yellow River, which is a typical plain landscape with elevation less than 100 m a.s.l. The NCP comprises three distinct hydrogeological settings within the Quaternary sediments (Fig. 1): the piedmont plain and associated major alluvial fans, the alluvial plains with many abandoned river channels, and the coastal plain strip around the margin of the Bohai Sea (Foster et al., 2004). The main stratigraphy of this region consists of unconsolidated Quaternary sediments with thickness ranging from 200 m to more than 600 m, comprising unconsolidated pebble, gravel, sand, silt, and clay. Groundwater occurs principally in the pores of the unconsolidated Quaternary sediments. From the piedmont plain to alluvial and coastal plains in the middle-east portion of the NCP, the aquifer systems typically transistion from a single aquifer of sandy gravel to multiple aquifers of sand separated by silt or clay layers (Fig. 2).

3 Influencing factors of land subsidence

The factors affecting subsidence in the NCP are complex. Geologically, the aquifer system comprises aquifers and aquitards in the multi-layered Quaternary sediments, consisting of layers of principally coarse-grained materials, mainly unconsolidated sands, as well as layers of principally fine-grained deposits, which are predominantly clays and silts that have medium to high compressibility. The alternating sequences of sand and compressible clay layers are very susceptible to deformation due to increases in effective stress accompanying groundwater-level declines. Excessive withdrawal of underground fluids such as groundwater, oil, and gas is the important driving condition for subsidence. Shallow groundwater abstraction may be the main driving condition for subsidence in the piedmont area and associated alluvial fans (e.g. Beijing and Handan). Figure 1 shows that most serious subsidence areas in the NCP lie in the alluvial plain and in the coastal strip around the margin of the Bohai Bay, which are well correlated with the distribution of depression cones of deep groundwater. In the alluvial and coastal plains, subsidence is commonly caused by withdrawal of deep groundwater, but in a few areas where large volumes of shallow groundwater, and oil and gas are extracted, subsidence is also observed.

4 Groundwater regulation in Beijing plain

A transient, calibrated regional groundwater flow model was developed using MODFLOW for a simulation period of 1995–2011. The model consists of 9 model layers: 5 aquifers separated by 4 aquitards. The top aquifer is unconfined and the underlying four aquifers are semi-confined.

4.1 Formulation of regulation scenarios

During the past 30 years, overexploitation of the groundwater resource has led to severe land subsidence, which seriously affects economically and socially sustainable development in Beijing municipality. Some actions such as artificial recharge and reduction of groundwater abstraction could be taken to overcome the subsidence crisis. The major regulation measures include: reduce or stop the abstraction from emergency groundwater well fields, reduce the abstraction for irrigation and from well fields in the downtown area and self-supplied water supply wells, artificially increase groundwater recharge to aquifers in alluvial fans of rivers. Groundwater demand will increase with the growth of population. Ten groundwater regulation scenarios (Table 1) were designed based on the water demand and the capacity of water supply in Beijing plain. In scenario 1, the total groundwater abstraction of $2600 \times 10^6\ m^3\ a^{-1}$ would remain unchanged as before. The groundwater flow model described above was used to simulate effects of proposed regulation scenarios. The sce-

Table 1. Groundwater regulation scenarios (10^6 m^3 a^{-1}).

Abstraction and recharge	Case 1	Case 2	Case 3	Case 4	Case 5	Case 6	Case 7	Case 8	Case 9	Case 10
Domestic and industrial	847	508	847	508	847	847	508	847	847	847
Agriculture	1023	1023	1023	1023	1023	1023	818	1023	1023	1023
Well fields	354	354	354	501	364	331	354	402	354	354
Emergency water source	291		291			204				291
Self-water supply wells	113	113	113				113			113
Artificial recharge			405			150			215	520
Total abstraction	2628	1999	2628	2033	2234	2415	1795	2270	2234	2628

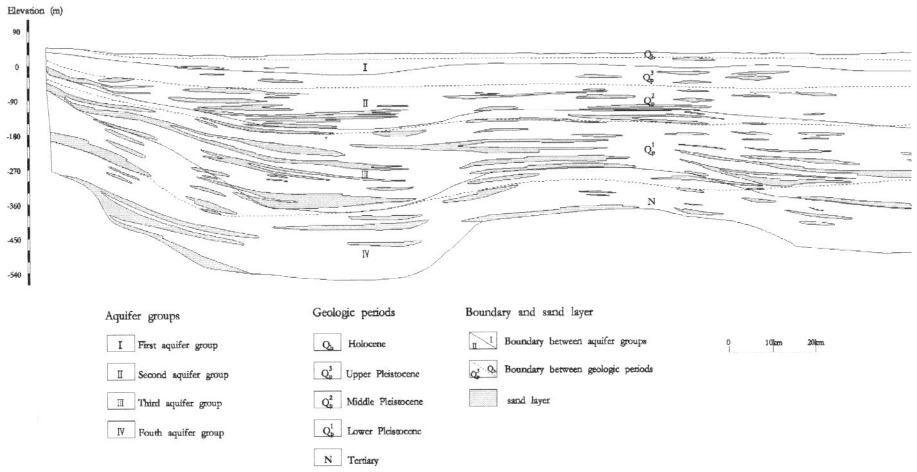

Figure 2. Hydrogeological cross section A-A' (location shown in Fig. 1).

narios were simulated for the 50 year period from 2012 to 2061.

4.2 Multi-criteria fuzzy pattern recognition (MFPR) approach for quantitative estimation of regulation scenarios

Assume $\boldsymbol{X} = (x_1, x_2, \ldots, x_n)$ be a finite set of n groundwater regulation scenarios. Each regulation scenario can be valued by m quantitative factors, and then the decision-making matrix can be denoted as

$$\boldsymbol{X} = \left(x_{ij}\right)_{m \times n} \tag{1}$$

where x_{ij} is the value of the ith quantitative factor for the jth regulation scenarios.

Generally, there exist three types of estimation factors: larger-the-better, smaller-the-better and nominal-the-better by characteristics. These three types of estimation factors can, respectively, be normalized by the following equations (Chen, 2002):

$$r_{ij} = \frac{x_{ij} - \min_j x_{ij}}{\max_j x_{ij} - \min_j x_{ij}}, \ i \tag{2}$$

$$r_{ij} = \frac{\max_j x_{ij} - x_{ij}}{\max_j x_{ij} - \min_j x_{ij}}, \ i \tag{3}$$

$$r_{ij} = 1 - \frac{\left|x_{ij} - x_i\right|}{\max_j \left|x_{ij} - x_i\right|}, \ i \tag{4}$$

where r_{ij} is the relative membership degree of the jth scenario regarding the ith factor, $\max_j x_{ij}$ and $\min_j x_{ij}$ are the maximum and minimum value regarding the ith factor, respectively; and x_i is the optimum target value of the ith factor.

Each groundwater regulation scenario can be evaluated by c levels, where the relative membership degree of the "superior" level is assigned to be 1 and relative membership degree of "inferior" level is 0. With $c = 2$, the relative membership matrix of each groundwater regulation scenario belonging to each level can be established (Chen, 2002):

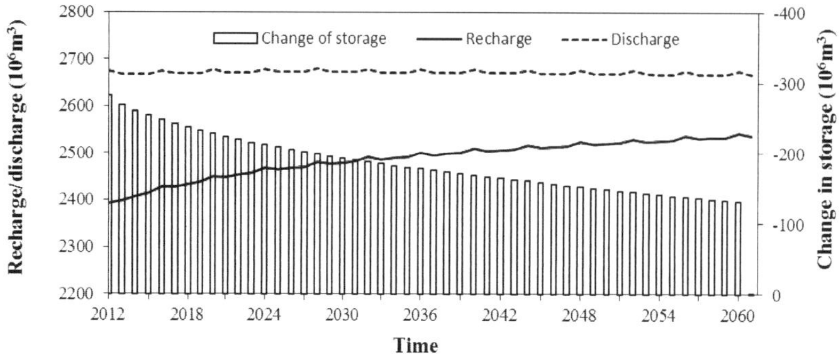

Figure 3. Change of groundwater balance under scenario 1.

$$u_{hj} = \left(\sum_{k=1}^{2} \frac{\sum_{i=1}^{m} \left[\boldsymbol{w}_i \left(r_{ij} - s_{ih} \right) \right]^2}{\sum_{i=1}^{m} \left[\boldsymbol{w}_i \left(r_{ij} - s_{ik} \right) \right]^2} \right)^{-1} \tag{5}$$

where u_{hj} is the relative membership grade of the jth scenario belonging to the hth ranking, $j = 1, 2, \ldots, n$; $h = 1$–2; and $\boldsymbol{w}_i$ is the weight vector of factor set.

The ranking feature of each groundwater regulation scenario is,

$$\boldsymbol{H} = (1, 2, \ldots, c) \cdot \left(u_{hj} \right) = (H_1, H_2, \ldots, H_n) \tag{6}$$

where H_j is ranking feature value of the jth scenario. Ranking feature values for the groundwater development scenarios are ordered lowest to highest representing best to worst, respectively.

5 Quantitative estimation of regulation scenarios

Seven main factors are chosen to quantatively estimate different groundwater regulation scenarios, which include amount of groundwater exploitation, evaporation from unconfined aquifer, groundwater outflows to rivers, regional average groundwater-level depth below the ground, drawdowns in depression cone centers and the ratio of groundwater storage to total recharge. The amount of groundwater exploitation needed to meet the requirements for everyday life, city construction, agriculture and industry, and sustain groundwater levels above critical thresholds should be balanced so that severe land subsidence cannot be induced. Regional average groundwater levels should be controlled in a reasonable range so that the ecosystems dependent on groundwater can be sustained and nurtured. Sufficient groundwater discharge to rivers is required to sustain baseflow for riverine ecosystems. Evaporation from unconfined aquier is unproductive and should be reduced as much as possible. The ratio of groundwater storage to total recharge ought to keep in a reasonable range.

The ranking feature value vector for the 10 regulation scenarios (1 : 10) subject to the set of criteria described above was obtained from Eq. (6):

$$\boldsymbol{H} = (1.906\ 1.174\ 1.140\ 1.174\ 1.187\ 1.137\ 1.140\ 1.199\ 1.128\ 1.119).$$

Scenarios 9 and 10 clearly have the lowest ranking feature values and are therefore considered as the best choices for managing the groundwater resource in the Beijing plain. Scenario 1 has the highest ranking feature value representing the worst choice for managing the groundwater resource. From a water balance perspective, the groundwater system in all scenarios approach a new equilibrium at the end of the 50 year simulation period, except scenario 1 where groundwater levels continue to decline significantly (Fig. 3), indicating poor near-future sustainability of the groundwater resources. The simulation results indicate that two major groundwater depression cones are developed, which are located in the southeast of Beijing city (Fig. 1).

In scenario 9, production from emergency well fields in the plain and most deep self-supply wells would be stopped, and reclaimed wastewater and imported surface water would be artificially recharged to aquifers in alluvial fans of major rivers. The annual total artificial recharge reaches $215 \times 10^6\ m^3\ a^{-1}$ (Table 1). The total reduction of annual abstraction rates amounts to $394 \times 10^6\ m^3\ a^{-1}$ compared with scenario 1. Implementing scenario 9 would result in immediate, rapid recovery of groundwater levels, especially at the depression cone centers. As a result, the continuous decline in groundwater levels and subsidence would be greatly mitigated. Figure 4 shows the change of groundwater balance under scenario 9. Here, groundwater recharge exceeds abstraction, and groundwater storage is gradually increasing as the change in groundwater flowing out of storage is decreasing. Sustainable groundwater resources development may be achieved after many decades under this scenario.

In scenario 10, sustainable groundwater resources development may be achieved sooner. However, in the Bejing plain with limited water supply, it would be difficult to ensure

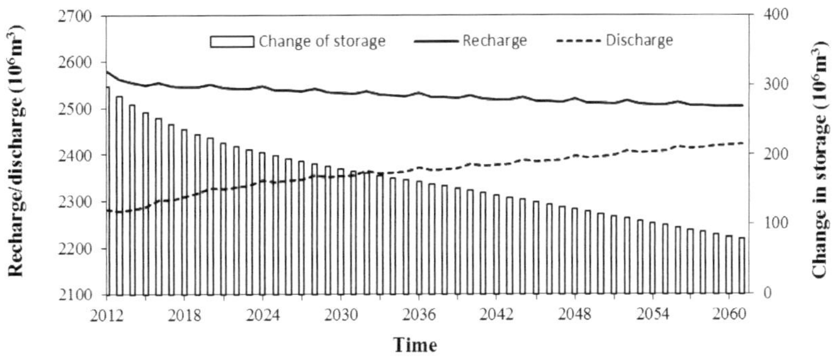

Figure 4. Change of groundwater balance under scenario 9.

an alternative water source of $520 \times 10^6\,m^3\,a^{-1}$ for artificial recharge specified in this scenario. In scenario 9, one of the best evaluated scenarios, a reasonable combination of artificial recharge and reduced abstraction would be implemented, which is more feasible and easier to implement. Thus, scenario 9 may be the best groundwater regulation scenario for the Beijing plain.

6 Summary and conclusions

The main factors leading to land subsidence in the NCP are represented by geological conditions such as the presence of highly compressible, thick, unconsolidated sediments and dynamic conditions of the groundwater system, e.g. groundwater abstraction. The best method to control subsidence is to recover groundwater levels especially in the known subsidence areas. Beijing is facing challenges in managing scarce water resources and controling subsidence. A multi-criteria fuzzy pattern recognition (MFPR) approach is introduced and applied to quantitatively evaluate 10 proposed groundwater regulation scenarios simulated using a regional groundwater flow model for the Beijing plain. The evaluation model will be helpful for determining the most reasonable and feasible scenarios for sustainable development and utilization of groundwater resources.

Currently, in the NCP The areal extent and accumulative magnitude of subsidence is increasing though some controlling measures have been taken. Subsidence in the NCP has a direct relationship with groundwater exploitation, so rationally reducing and controlling groundwater abstraction should be carried out to decrease groundwater production. However, because both population and economic activities are heavily dependent upon the groundwater resource, it is important to seek a balance between controlling subsidence and ensuring adequate water supply so that the NCP can be developed in a sustainable and harmonious manner.

Acknowledgements. This work was supported by the China Geological Survey (No. 12120113011700), and the National Basic Research Program of China (973 Program, 2010CB428806).

References

Abidin, H. Z., Gumilar, I., Andreas, H., Murdohardono, D., and Fukuda, Y.: On causes and impacts of land subsidence in Bandung Basin, Indonesia, Environ. Earth Sci., 68, 1545–1553, 2013.

Chen, S. Y.: Optimal Fuzzy Recognition Theory and Application in Complex Water Resources System, JiLin University Press, Jilin, China, 2002.

Foster, S., Garduno, H., Evant, R., Olson, D., Tian, Y., Zhang, W., and Han, Z.: Quaternary Aquifer of the North China Plain – assessing and achieving groundwater resource sustainability, Hydrogeol. J., 12, 81–93, 2004.

Galgaro, A., Boaga, J., and Rocca, M.: HVSR technique as tool for thermal-basin characterization: a field example in N-E Italy, Environ. Earth Sci., 71, 4433–4446, 2014.

Papadaki, E. S.: Monitoring subsidence at Messara basin using radar interferometry, Environ. Earth Sci., 72, 1965–1977, 2014.

Ye, X. B. and He, Q. C.: Estimation of economic loss due to land subsidence in the North China Plain, China Land Press, Beijing, 2006.

Zeitoun, D. G. and Wakshal, E.: Land Subsidence Analysis in Urban Areas: the Bangkok Metropolitan Area Case Study, Springer Environmental Science and Engineering, Springer, Dordrecht, 2013.

Zhang, Y. Q., Gong, H. L., Gu, Z. Q., Wang, R., Li, X. J., and Zhao, W. J.: Characterization of land subsidence induced by groundwater withdrawals in the plain of Beijing city, China, Hydrogeol. J., 22, 397–409, 2014.

Proc. IAHS, 372, 23–27, 2015
proc-iahs.net/372/23/2015/
doi:10.5194/piahs-372-23-2015

Open Access
PIAHS

Land subsidence in the San Joaquin Valley, California, USA, 2007–2014

M. Sneed and J. T. Brandt

US Geological Survey, 6000 J Street, Placer Hall, Sacramento, CA 95819, USA

Correspondence to: M. Sneed (micsneed@usgs.gov)

Published: 12 November 2015

Abstract. Rapid land subsidence was recently measured using multiple methods in two areas of the San Joaquin Valley (SJV): between Merced and Fresno (El Nido), and between Fresno and Bakersfield (Pixley). Recent land-use changes and diminished surface-water availability have led to increased groundwater pumping, groundwater-level declines, and land subsidence. Differential land subsidence has reduced the flow capacity of water-conveyance systems in these areas, exacerbating flood hazards and affecting the delivery of irrigation water.

Vertical land-surface changes during 2007–2014 were determined by using Interferometric Synthetic Aperture Radar (InSAR), Continuous Global Positioning System (CGPS), and extensometer data. Results of the InSAR analysis indicate that about 7600 km^2 subsided 50–540 mm during 2008–2010; CGPS and extensometer data indicate that these rates continued or accelerated through December 2014. The maximum InSAR-measured rate of 270 mm yr^{-1} occurred in the El Nido area, and is among the largest rates ever measured in the SJV. In the Pixley area, the maximum InSAR-measured rate during 2008–2010 was 90 mm yr^{-1}. Groundwater was an important part of the water supply in both areas, and pumping increased when land use changed or when surface water was less available. This increased pumping caused groundwater-level declines to near or below historical lows during the drought periods 2007–2009 and 2012–present.

Long-term groundwater-level and land-subsidence monitoring in the SJV is critical for understanding the interconnection of land use, groundwater levels, and subsidence, and evaluating management strategies that help mitigate subsidence hazards to infrastructure while optimizing water supplies.

1 Introduction and background

The extensive withdrawal of groundwater from the unconsolidated deposits of the San Joaquin Valley (SJV), California has caused widespread land subsidence–locally reaching 9 m by 1981 (Ireland, 1986). Long-term groundwater-level declines can result in a vast one-time release of "water of compaction" from compacting silt and clay layers in the aquifer system, which causes land subsidence (Galloway et al., 1999). Land subsidence in the SJV from groundwater pumping began in the mid-1920s (Poland et al., 1975; Bertoldi et al., 1991; Galloway et al., 1999), and by 1970, about half of the SJV, or about 13 500 km^2, had subsided more than 0.3 m (Poland et al., 1975).

Surface-water imports from the Delta-Mendota Canal (DMC) since the early 1950s and the California Aqueduct since the early 1970s resulted in decreased groundwater pumping in some parts of the valley, which was accompanied by a steady recovery of water levels and a reduced rate of aquifer-system compaction in some areas (Ireland, 1986). During the droughts of 1976–1977 and 1987–1992, diminished availability of surface water prompted increased pumping of groundwater to meet irrigation demands. This increased groundwater pumping resulted in water-level declines and periods of renewed compaction. Following each of these droughts, recovery to pre-drought water levels was rapid and compaction virtually ceased (Swanson, 1998; Galloway et al., 1999). Similarly, during the more recent droughts of 2007–2009, and 2012–present, groundwater pumping has increased in some parts of the valley. Groundwater levels declined during these periods in response to increased pumping, approaching or surpassing historical low levels, which reinitiated compaction.

Groundwater pumping that resulted in renewed aquifer-system compaction and land subsidence caused serious operational, maintenance, and construction-design problems for the California Aqueduct, the DMC, and other water-delivery and flood-control canals in the SJV. Subsidence has reduced the flow capacity and freeboard of several canals that deliver irrigation water to farmers and transport floodwater out of the valley; structural damages have already required millions of dollars' worth of repairs, and more repairs are expected in the future (Bob Martin, San Luis and Delta-Mendota Water Authority, and Chris White, Central California Irrigation District, personal communication, 2010). Even small amounts of subsidence in critical locations, especially where canal gradients are small, can impact canal operations. On the DMC between the canal intakes and San Luis Reservoir, where less than 15 mm of subsidence was measured during 2007–2010 (Sneed et al., 2013), a 5-day window of opportunity to recharge the Reservoir in spring 2014 fell short because of reduced flow capacity (Bob Martin, San Luis and Delta-Mendota Water Authority, personal communication, 2014).

The objective of this paper is to describe the location, extent, and magnitude of land subsidence in the SJV during 2007–2014, which includes both drought and non-drought periods (http://www.ncdc.noaa.gov/cag/, assessed 18 April 2015). The SJV is a broad alluviated structural trough constituting the southern two-thirds of the Central Valley of California, that is a substantial source of the nut, fruit, and vegetable supply for the United States (Faunt, 2009).

2 Land subsidence and groundwater levels

Interferometric Synthetic Aperture Radar (InSAR), continuous Global Positioning System (CGPS), and extensometer data were used to determine the location, extent, and magnitude of aquifer-system compaction and resultant land subsidence. Analysis of interferograms generated from synthetic aperture radar images from the European Space Agency's ENVISAT satellite and the Japan Aerospace Exploration Agency's ALOS satellite acquired between 2008 and 2010 indicated 50–540 mm of subsidence in two large agriculturally-dominated areas in the SJV. One area is centered near the town of El Nido (2100 km^2) and the other near the town of Pixley (5500 km^2; Fig. 1). The period 2008–2010 is shown in Fig. 1 because interferograms covering the entire study area were generated for this period only. Because suitable InSAR data were not available for 2010–2014, CGPS data collected during 2007–2014 were used to generate land subsidence time series, which confirmed the InSAR-derived rates and generally indicated that these rates continued or accelerated through 2014 (Fig. 2). Extensometer data collected at four sites during 2012–2014 (data were not available for 2007–2011) were used to generate aquifer-system compaction time series at locations along the major canals, and generally indicated higher compaction rates during the growing season of 2014, the third consecutive year of drought, than for the previous two growing seasons (Fig. 3). To help explain the variability in location and magnitude of land subsidence, computed subsidence and compaction were compared with water-level measurements retrieved from US Geological Survey and California Department of Water Resources databases (Figs. 2 and 3).

3 El Nido

The largest subsidence magnitude in the SJV during 2007–2014 was measured near El Nido. The interferograms are the only measurements that captured the maximum magnitudes because the CGPS stations and extensometers are located on the periphery of the most rapidly subsiding area (Fig. 1); however, InSAR data were only available for 2008–2010. The interferograms indicated a local maximum of about 540 mm during January 2008–January 2010, or 270 mm yr^{-1}, which is among the highest rates ever measured in the SJV. The maximum subsidence measured at nearby CGPS station P303 was about 50 mm during that same time period, indicating a large subsidence gradient between the two locations (Fig. 1). Assuming the same rate of subsidence occurred during 2007–2014 as occurred during 2008–2010 at the local subsidence maximum near El Nido, about 2 m of subsidence may have occurred during 2007–2014.

Data from the three CGPS stations and two extensometers near the periphery of the El Nido subsidence area show seasonally variable subsidence and compaction rates (including uplift as elastic rebound during the fall and winter), but different characters over longer periods of time. Vertical displacement at P307 and P303 indicated subsidence at fairly consistent rates during and between drought periods (Fig. 2a). These fairly consistent subsidence rates indicate that these areas continued to pump groundwater despite climatic variations (possibly due to lack of surface water availability); residual compaction also may be a factor. Vertical displacement at P304, however, indicated that most subsidence occurred during drought periods and very little occurred between drought periods (Fig. 2a). This suggests that this area received other sources of water, most likely surface water, when it was available between drought periods, and also that residual compaction was not very important in this area. Data from the extensometers 12S/12E-16H2 and 14S/13E-11D6 were available only during the recent drought period, so comparison to a non-drought period was not possible. CGPS and extensometer data indicated subsidence and compaction rate increases during 2014, the third year of drought (Figs. 2a and 3a). In parts of the El Nido subsidence area, where the planting of permanent crops (vineyards and

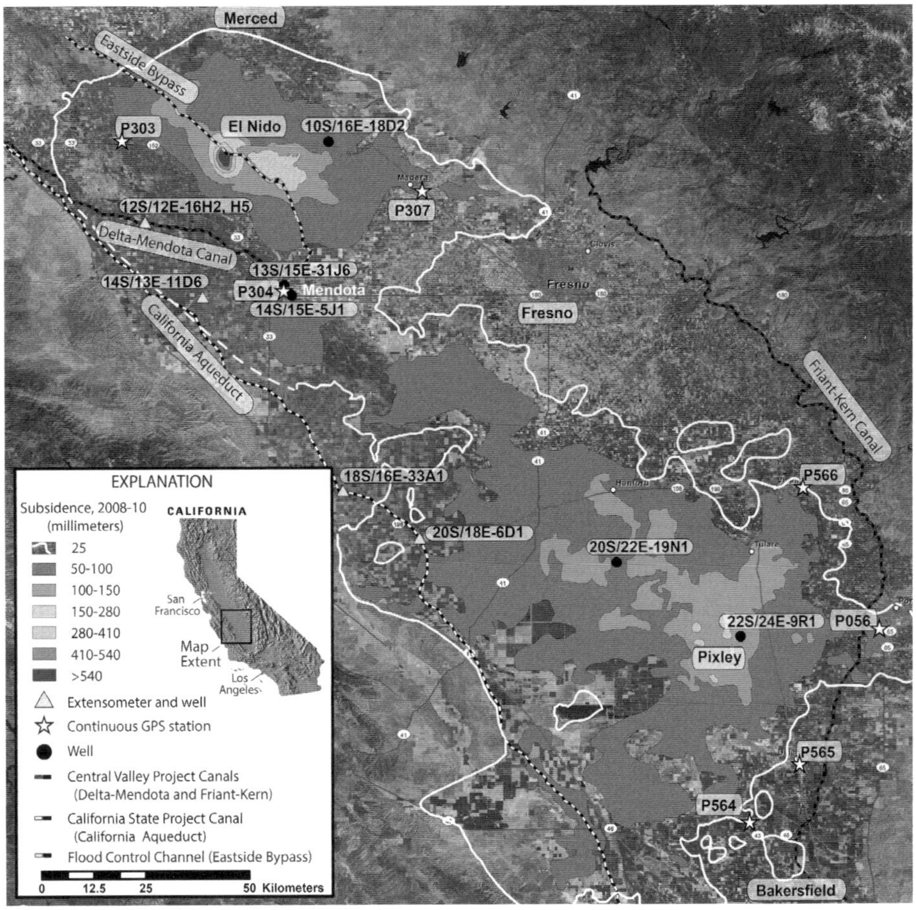

Figure 1. Map showing regions of subsidence derived from interferograms for 2008–2010, and locations of major canals, extensometers, Continuous GPS stations, and wells.

orchards) has increased, groundwater was either the primary source of water or groundwater pumping increased when surface-water availability was reduced, and groundwater levels declined to near or below historical lows during 2007–2010 and 2012–2014 (Figs. 2a and 3a). The correlation between high rates of compaction or land subsidence and water levels near or below historical lows indicates that the preconsolidation stress likely was exceeded; if so, the subsidence likely is mostly permanent.

4 Pixley

The Pixley subsidence area is larger than the El Nido subsidence area, but subsided at a lower rate during 2007–2014. Similar to the El Nido area, the interferograms provided the only measurements that captured the maximum magnitudes because the CGPS stations and extensometers are located on the periphery of the most rapidly subsiding area (Fig. 1); however, InSAR data were only available for 2007–2010. The interferograms indicated a maximum subsidence of about 180 mm during January 2008–January 2010, whereas the maximum measured subsidence at nearby CGPS station P056 (40 km distant) was about 65 mm during that same time period (Figs. 1 and 2b). If the same rate of subsidence occurred during 2007–2014 as occurred during 2008–2010 at the local maximum near Pixley, then about 0.7 m of subsidence may have occurred during 2007–2014. Data from the four CGPS stations and two extensometers near the periphery of the Pixley subsidence area show seasonally variable subsidence and compaction rates (including uplift as elastic rebound during the fall and winter), but different characters over longer periods of time. Vertical displacement at P564 and P565 indicated that most subsidence occurred during drought periods and very little occurred between drought periods (Fig. 2b). This suggests that this area received other sources of water, most likely surface water, when it was avail-

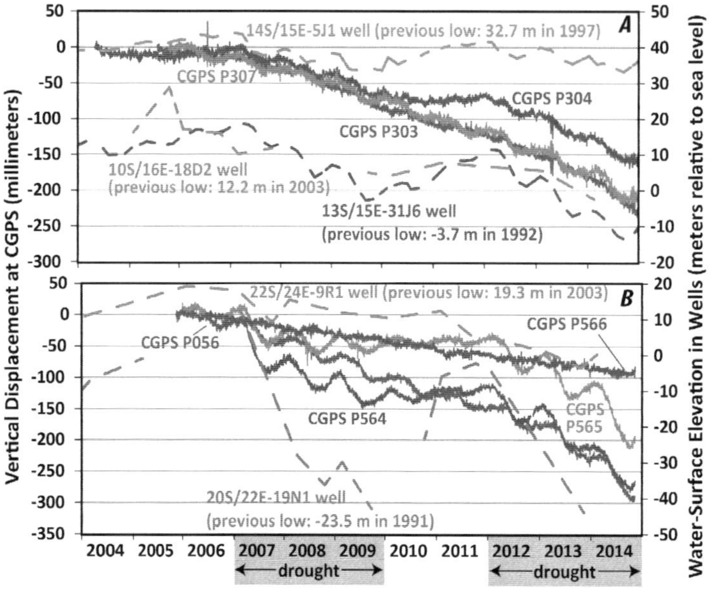

Figure 2. Graphs showing vertical displacement at selected CGPS stations and water-surface elevation in selected wells for 2004–2014 near **(a)**. the El Nido subsidence area and **(b)**. the Pixley subsidence area. (See Fig. 1 for CGPS and well locations.)

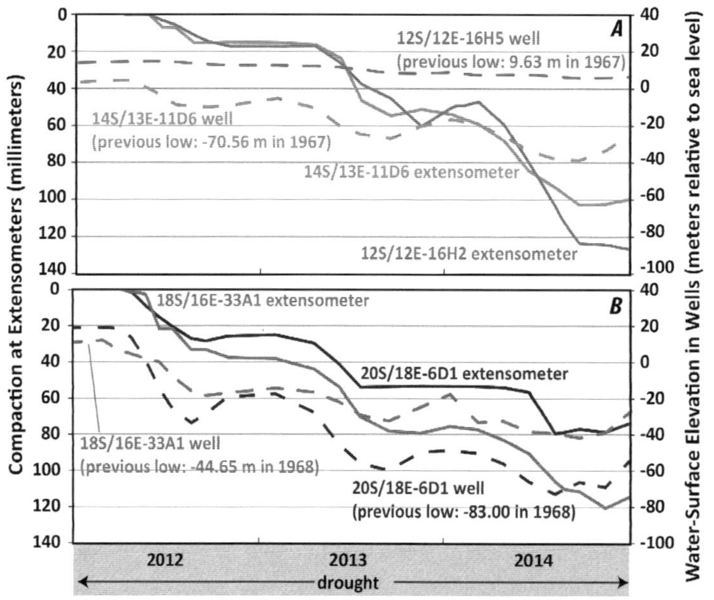

Figure 3. Graphs showing compaction and water-surface elevation at selected extensometers and associated wells for 2012–2014 near **(a)**. the El Nido subsidence area and **(b)**. the Pixley subsidence area. (See Fig. 1 for extensometer and well locations.)

able between drought periods, and also that residual compaction was not very important in this area. Vertical displacement at P056 and P566 indicated subsidence at fairly consistent rates during and between drought periods (Fig. 2b). These fairly consistent subsidence rates indicate that these areas continued to use groundwater despite climatic variations (possibly due to limited surface water availability); residual compaction also may be a factor. Data from extensometers

18S/16E-33A1 and 20S/18E-6D1 were available only during the recent drought period, so comparison to a non-drought period was not possible. CGPS and extensometer data indicated subsidence and compaction rate increases during 2014, the third year of drought (Figs. 2b and 3b). In the Pixley area, groundwater pumping continued or increased when surface-water availability was reduced, and groundwater levels declined to near or below historical lows during 2007–2010 and 2012–2014 (Figs. 2b and 3b). The correlation between high rates of compaction or land subsidence and water levels near or below historical lows indicates that the preconsolidation stress likely was exceeded; if so, the subsidence likely is mostly permanent.

5 Summary and conclusions

Groundwater and surface water are generally used conjunctively in the SJV (Faunt, 2009). During recent drought periods (2007–2009 and 2012–present), groundwater pumping increased in areas where surface-water deliveries were curtailed; in response, groundwater levels declined. However, in areas where surface-water deliveries were normally an absent or minor component of the water supply, pumping was fairly steady during drought and non-drought periods; accordingly, groundwater levels declined at fairly consistent rates regardless of climatic conditions. Groundwater levels in both water-supply-source scenarios declined to levels approaching or surpassing historical low levels, which caused aquifer-system compaction and land subsidence that likely is mostly permanent.

Land use in some parts of the SJV has trended toward the planting of permanent crops (vineyards and orchards) at the expense of non-permanent land uses such as rangeland or row crops. This may have the effect of "demand hardening", which refers to the need for stable water supplies to irrigate crops that cannot be fallowed. As land use and surface-water availability continue to vary in the SJV, long-term groundwater-level and land-subsidence monitoring is critical because continued groundwater use in excess of recharge, which the historical record indicates is likely, would result in additional water-level declines and associated subsidence. Such long-term data can be used to better understand the interconnection of land use, groundwater levels, and subsidence, and to enable the evaluation of management strategies to mitigate subsidence hazards to infrastructure while optimizing water supplies. This knowledge will be critical for successful implementation of California's recent legislation aimed toward sustainable groundwater use without damage from land subsidence.

References

Bertoldi, G. L., Johnston, R. H., and Evenson, K. D.: Ground water in the Central Valley, California–A summary report, US Geological Survey Professional Paper 1401-A, 1991.

Faunt, C. C. (Ed.): Groundwater availability of the central valley aquifer, California, US Geological Survey Professional Paper, 1766, 2009.

Galloway, D. L., Jones, D. R., and Ingebritsen, S. E.: Land subsidence in the United States, US Geological Survey Circular 1182, 1999.

Ireland, R. L.: Land subsidence in the San Joaquin Valley, California, as of 1983, US Geological Survey Water-Resources Investigations Report 85–4196, 1986.

Poland, J. F., Lofgren, B. E., Ireland, R. L., and Pugh, A. G.: Land subsidence in the San Joaquin Valley, California, as of 1972, US Geological Survey Professional Paper 437-H, 1975.

Sneed, M., Brandt, J., and Solt, M.: Land subsidence along the Delta-Mendota Canal in the northern part of the San Joaquin Valley, California, 2003–2010, US Geological Survey Scientific Investigations Report 2013–5142, 2013.

Swanson, A. A.: Land subsidence in the San Joaquin Valley, updated to 1995, in: Borchers, J.W., ed., Land subsidence case studies and current research, Proceedings of the Joseph F. Poland Symposium on Land Subsidence, Sacramento, Calif., 4–5 October 1995, Association of Engineering Geologists, Special Publication no. 8, 75–79, 1998.

Proc. IAHS, 372, 29–32, 2015
proc-iahs.net/372/29/2015/
doi:10.5194/piahs-372-29-2015

Open Access PIAHS

Evaluation of the potential of InSAR time series to study the spatio-temporal evolution of piezometric levels in the Madrid aquifer

M. Béjar-Pizarro[1,2,3], P. Ezquerro Martín[1,3,5], G. Herrera[1,2,3,4], R. Tomás[2,3,6], C. Guardiola-Albert[1,7], J. M. Ruiz Hernández[7], M. Marchamalo Sacristán[1,3,5], and R. Martínez Marín[3,5]

[1]Geohazards InSAR laboratory and Modeling group (InSARlab), Geoscience research department, Geological Survey of Spain (IGME), Alenza 1, 28003 Madrid, Spain
[2]Unidad Asociada de investigación IGME-UA de movimientos del terreno mediante interferometría radar (UNIRAD), Universidad de Alicante, P.O. Box 99, 03080 Alicante, Spain
[3]Grupo Español de Trabajo en Subsidencia del Terreno (SUBTER) from the UNESCO, P.O. Box 99, 03080 Alicante, Spain
[4]Earth Observation and Geohazards Expert Group (EOEG), EuroGeoSurveys, The Geological Surveys of Europe, 36-38, Rue Joseph II, 1000 Brussels, Belgium
[5]Technical University of Madrid. Laboratorio de Topografía y Geomática, ETSI Caminos, Canales y Puertos C/Profesor Aranguren s/n, 28040 Madrid, Spain
[6]Departamento de Ingeniería Civil, Escuela Politécnica Superior, Universidad de Alicante, P.O. Box 99, 03080 Alicante, Spain
[7]Geoscience research department, Geological Survey of Spain (IGME), Alenza 1, 28003 Madrid, Spain

Correspondence to: M. Béjar-Pizarro (m.bejar@igme.es)

Published: 12 November 2015

Abstract. The Tertiary detritic aquifer of Madrid (TDAM), with an average thickness of 1500 m and a heterogeneous, anisotropic structure, supplies water to Madrid, the most populated city of Spain (3.2 million inhabitants in the metropolitan area). Besides its complex structure, a previous work focused in the north-northwest of Madrid city showed that the aquifer behaves quasi elastically trough extraction/recovery cycles and ground uplifting during recovery periods compensates most of the ground subsidence measured during previous extraction periods (Ezquerro et al., 2014). Therefore, the relationship between ground deformation and groundwater level through time can be simulated using simple elastic models. In this work, we model the temporal evolution of the piezometric level in 19 wells of the TDAM in the period 1997–2010. Using InSAR and piezometric time series spanning the studied period, we first estimate the elastic storage coefficient (S_{ke}) for every well. Both, the S_{ke} of each well and the average S_{ke} of all wells, are used to predict hydraulic heads at the different well locations during the study period and compared against the measured hydraulic heads, leading to very similar errors when using the S_{ke} of each well and the average S_{ke} of all wells: 14 and 16 % on average respectively. This result suggests that an average S_{ke} can be used to estimate piezometric level variations in all the points where ground deformation has been measured by InSAR, thus allowing production of piezometric level maps for the different extraction/recovery cycles in the TDAM.

Published by Copernicus Publications on behalf of the International Association of Hydrological Sciences.

1 Introduction

Piezometric maps are one of the most useful hydrogeological tools to manage aquifers, providing information about groundwater flow direction, interactions between groundwater and surface water and identifying groundwater recharge zones (e.g. Todd and Larry, 2005). These maps are built using discrete water level measurements in wells that are usually sparsely distributed, leading to accurate point measurements but low spatial resolution contour maps (Fasbender et al., 2008).

Here we assess the potential of InSAR data to build piezometric maps with high spatial resolution of the Tertiary detritic aquifer of Madrid (TDAM). Using InSAR-derived displacements and piezometric time series spanning 1997–2010, we first estimate the elastic storage coefficient (S_{ke}) for 19 well locations and also an average S_{ke} of all wells. We then use these coefficients to model hydraulic heads at the different well locations and compare them with in situ measurements. Finally, the implications of our results for the construction of piezometric maps and limitations of the approach are discussed.

2 Study area

The Madrid Metropolitan area is located above a triangular shape basin, the Madrid Basin, consisting of a tectonic depression filled with continental deposits of Tertiary age. The detritic facies of these deposits, located in the northwestern part of the basin, define the TDAM. This is a single, heterogeneous and anisotropic aquifer with an average thickness of 1500 m and altitudes ranging from 650 to 800 m a.s.l. (Martínez-Alfaro, 1977; Yélamos and Villarroya Gil, 2007).

3 Data used

3.1 InSAR data

We use the InSAR time series processed in Ezquerro et al. (2014), covering the period April 1992 to September 2010. These data were processed with the PSP-IFSAR technique (Costantini et al., 2008) to calculate the radar line-of-sight (LOS) displacement temporal evolution of each Persistent Scatterer (PS). See Ezquerro et al. (2014) for more details about the InSAR processing. Figure 1 shows one of the resulting deformation maps. The main deformation signal corresponds to the Fuencarral and Pozuelo Well Fields (labelled as Well Fields 1 and 2 respectively in Fig. 1), The deformation pattern follows a $\sim$ N45° E direction. Maximum deformation occurs at the centre of the well fields, reaching 80 mm of cumulated displacement during 1992–2010.

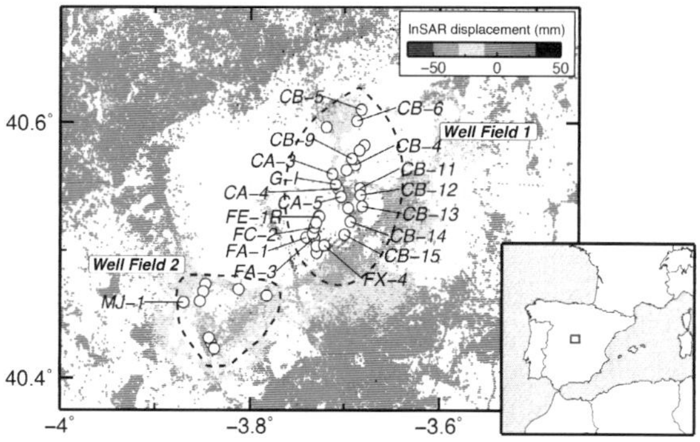

Figure 1. Location map and data distribution over the study region. InSAR deformation corresponding to an extraction episode from January 1999 to April 2000 is shown in the background. Black dashed lines outline the two well fields in the area.

3.2 Piezometric time series

We have used the water level data available for the period 1997–2010 for 19 wells distributed over the two wells fields in the study area (Fig. 1). The average depth of these wells varies between 300 and 700 m below the surface. Since 1997, these well fields have undergone three cycles of water extraction, coinciding with drought periods, with dates: February 1999–March 2000, March 2002–December 2002, April 2005–November 2006 (Fig. 2). After each extraction period there is a recovery phase of different duration. Elevation of water level during these extraction/recovery cycles varies between 390 and 600 m a.s.l.

4 Models

4.1 Estimates of elastic storage coefficients

The elastic storage coefficient (S_{ke}) can be computed by means of the graphical methodology proposed by Riley (1969), which consists of the determination of the slope of the stress–strain curve branch. Using this method, we determined the S_{ke} at well locations comparing water levels and InSAR-derived ground displacements (Fig. 2). The resulting elastic storage coefficient estimates ranged from 2.61×10^{-4} at well CB-5 in the northern part of Well Field 1 to 7.09×10^{-4} at well CB-15 in the southern part of Well Field 1, with an average S_{ke} of 4.48×10^{-4}.

4.2 Models of hydraulic head

The S_{ke} values estimated in the previous section were used to predict hydraulic head at specific well locations during the period 1997–2010. Figure 3 shows the comparison of estimated and measured hydraulic heads at two well locations. The error between observations and each of these models ranges between 10 and 28 % among the 19 wells. We also

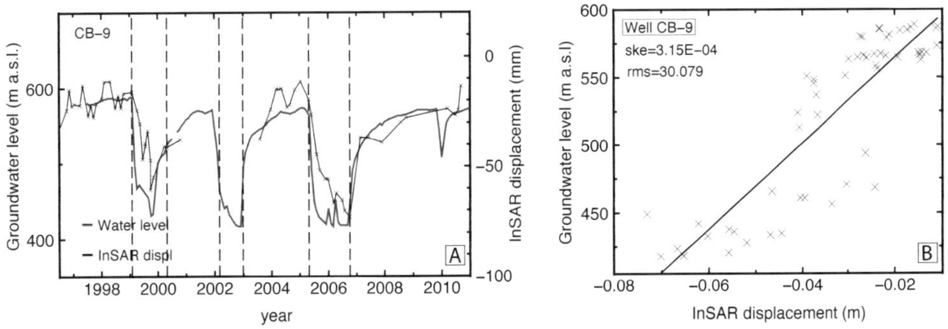

Figure 2. Estimation of skeletal elastic storage coefficient from stress displacement analysis for well CB-9 (see location in Fig. 1). On the left, InSAR displacement and piezometric time series On the right, stress-strain diagram. Vertical dashed lines represent dates limiting water extraction cycles.

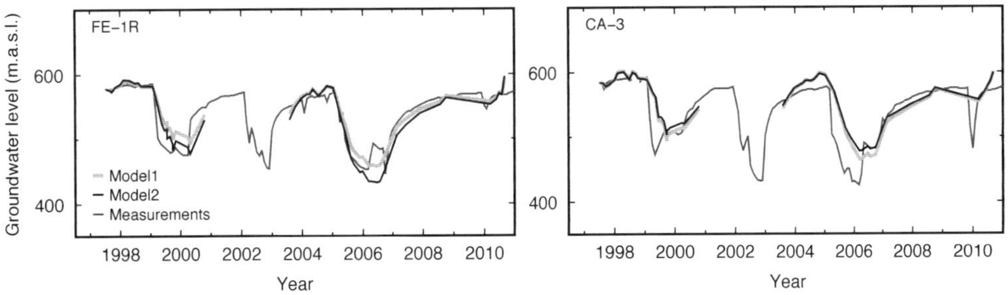

Figure 3. Estimated and measured hydraulic head at two well locations (FE-1R and CA-3). Red line represents measured hydraulic head, gray line represents estimated hydraulic head using the elastic storage coefficient S_{ke} of each well and black line represents estimated hydraulic head using the average elastic storage coefficient S_{ke} computed from all 19 wells labelled in Fig. 1.

tested a model of hydraulic heads using the average S_{ke} for all wells, which resulted in Model 2 in Fig. 3. The error using this model ranges between 10 and 24 % among the 19 wells. Therefore, both Models 1 and 2 produce very similar errors (14 and 16 % on average among the 19 wells, respectively)

5 Conclusions

Our results indicate that an average S_{ke} can be used to estimate piezometric level variations in the 19 wells analysed here. The next step would be to use the complete InSAR dataset to extend this analysis to the complete region covered by InSAR measurements (an area of 40 km × 40 km around the two Well Fields producing the observed deformation), thus allowing to produce piezometric level maps for the different extraction/recovery cycles in the TDAM. These maps will help identify and monitor vulnerable areas of the TDAM and should help mitigate future geological and geotechnical risks induced by the ongoing ground deformation.

This approach however has some limitations: (i) there is a densely vegetated area with no deformation data in the middle of our deformation region (Fig. 1). The use of geostatistical techniques would allow interpolating deformation in these regions; (ii) a piezometric initial map is necessary to add/subtract the water level differences obtained in the models. The uncertainties of this map would be incorporated to the InSAR-derived piezometric maps. All these limitations should be taken into account and appropriately assessed when producing the final piezometric maps, clearly indicating the associated error and reliability of each region.

Acknowledgements. This work is supported by the Spanish Ministry of Economy and Competitiveness and EU FEDER funds under projects TEC2011-28201-C02-02, TIN2014-55413-C2-2-P and ESP2013-47780-C2-2-R, by the Ministry of Education, Culture and Sport trough the project PRX14/00100 and by the project 15224/PI/10 from the Regional Agency of Science and Technology in Murcia.

References

Costantini, M., Falco, S., Malvarosa, F., and Minati, F.: A new method for identification and analysis of persistent scatterers in series of SAR images, in: Geoscience and Remote Sensing Symposium, IGARSS 2008, Boston, MA, 7–11 July 2008, IEEE International, 2, II-449–II-452, doi:10.1109/IGARSS.2008.4779025, 2008.

Ezquerro, P., Herrera, G., Marchamalo, M., Tomás, R., Béjar-Pizarro, M., and Martínez, R.: A quasi-elastic aquifer deformational behavior: Madrid aquifer case study, J. Hydrol., 519, 1192–1204, doi:10.1016/j.jhydrol.2014.08.040, 2014.

Fasbender, D., Peeters, L., Bogaert, P., and Dassargues, A.: Bayesian data fusion applied to water table spatial mapping, Water Resour. Res., 44, W12422, doi:10.1029/2008WR006921, 2008.

Martínez-Alfaro, P. E.: Historia del abastecimiento de aguas a Madrid. El papel de las aguas subterraneas, Anales del Instituto de Estudios Madrileños, 14, 29–51, 1977.

Todd, D. K. and Larry, W. M.: Groundwater hydrology edition, Wiley, New Jersey, 2005.

Riley, F. S.: Analysis of borehole extensometer data from central California, in: Land subsidence: Proceedings of the Tokyo Symposium: Tokyo, International Association of Hydrological Sciences, September 1969, edited by: L. J., Tison, IAHS Publication 88, 423–431, 1969

Yélamos, J. G. and Villarroya Gil, F.: El acuífero terciario detrítico de Madrid: pasado, posibilidades actuales y retos pendientes, Enseñanzas de las Ciencias de la Tierra, 15.3, 317–324, 2007.

Proc. IAHS, 372, 33–38, 2015
proc-iahs.net/372/33/2015/
doi:10.5194/piahs-372-33-2015

Open Access
PIAHS

Land subsidence caused by a single water extraction well and rapid water infiltration

I. Martinez-Noguez and R. Hinkelmann

Water Resources Management and Modeling of Hydrosystems, TU Berlin, Berlin, Germany

Correspondence to: I. Martinez-Noguez (isaac.martinez@wahyd.tu-berlin.de)

Published: 12 November 2015

Abstract. Nowadays several parts of the world suffer from land subsidence. This setting of the earth surface occurs due to different factors such as earth quakes, mining activities, and gas, oil and water withdrawal. This research presents a numerical study of the influence of land subsidence caused by a single water extraction well and rapid water infiltration into structural soil discontinuities. The numerical simulation of the infiltration was based on a two-phase flow-model for porous media, and for the deformation a Mohr–Coulomb model was used.

A two-layered system with a fault zone is presented. First a single water extraction well is simulated producing a cone-shaped (conical) water level depletion, which can cause land subsidence. Land Subsidence can be further increased if a hydrological barrier as a result of a discontinuity, exists. After water extraction a water column is applied on the top boundary for one hours in order to represent a strong storm which produces rapid water infiltration through the discontinuity as well as soil deformation. Both events are analysed and compared in order to characterize deformation of both elements and to get a better understanding of the land subsidence and new fracture formations.

1 Introduction

Land subsidence processes can occur due to extraction of minerals in mining galleries, tunnels construction, fluids extraction (water, oil or gas) from natural reserves, decrease of groundwater level during prolonged extraction, natural land dissolution, compaction of soil materials or tectonic activity. In modern times, cities have grown rapidly and likewise have the agricultural and industrial activities. In order to satisfy the growing demands for the vital liquid, for which the superficial water bodies have not been sufficient, groundwater has been exploited to such an extent that its extraction has outpaced groundwater recharge. Especially in arid and semi-arid zones this has produced a deficit leading to a rapid decline of the groundwater table. In heterogeneous strata the fast decline of the groundwater table leads to a differential adjustment and in some cases to a formation of cracks and fractures on the surface of the ground. The valley of Queretaro city, Mexico (Fig. 1, right), is an example of active land subsidence. Since the 1970s, with the rapid growth of the city, the water demand increased rapidly and as a result the urban area is affected by differential compaction and formation of a reticular system of faults and fractures. Many of them appear on the surface and have caused economical damages in the last 40 years.

Groundwater overexploitation causing water table decline in unconfined aquifers leads to deformation of the porous media. This is mainly controlled by the increase in effective stress in the soil mass, reducing pore volume and, hence, the aquifer will compresse causing subsidence (Rojas et al., 2002; Martinez et al., 2013) and may cause the generation of earth fissures at land surface. Fault and fractures can influence groundwater flow in aquifer and aquitard layers (Carreón-Freyre et al., 2005). Structural discontinuities can perform as hydrological barriers. Such a barrier can result in different water levels on either side of the fault and as a result differential land subsidence occurs across the fault on the surface of the earth. Furthermore, it has been shown that fast water infiltration through a pre-existing fracture zone into a soil system could perform as a mechanism for fracturing and triggering of land subsidence (Martinez et al., 2013). Groundwater depletion occurs at scales ranging from a single well to regional aquifer systems. The extents of the resulting effects depend on several factors including pumpage and natural dis-

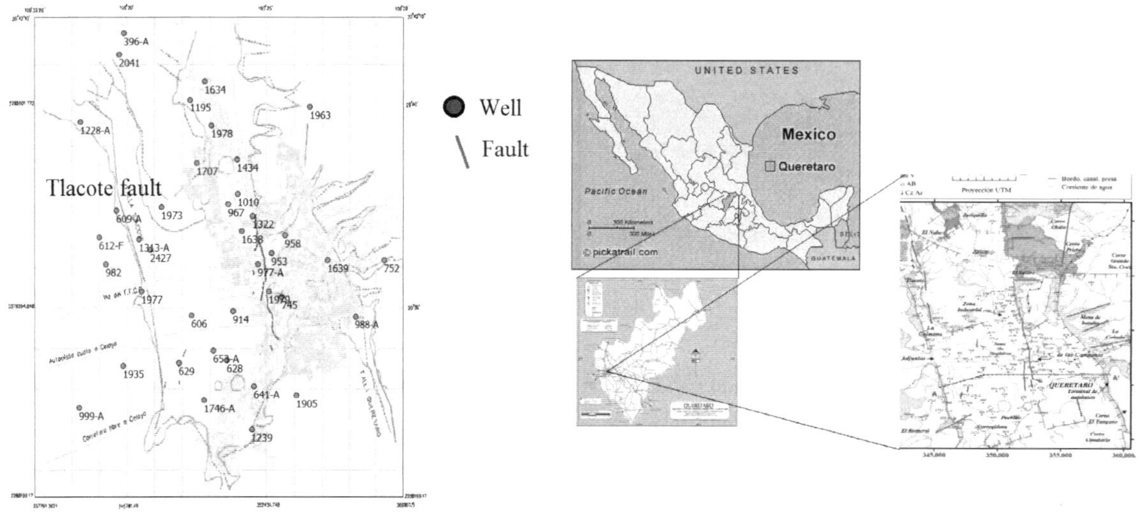

Figure 1. Queretaro, urban area, well distribution and faults and fractures location after Pacheco (2007); State of Querétaro within Mexico.

charge rates, physical properties of the aquifer, and natural and human-induced recharge rates.

Figure 1 left shows the urban area of Queretaro city (light green), faults and fractures (red line) and some of the active wells in the city (circles). In many of the cases the wells at both sides of the faults show a considerable difference of piezometric levels. An example of different water levels across a foult are given by, the static levels of the wells number 612-F at the left side of the fault Tlacote and the well number 1313-A of the year 2011 are 153.9 and 140 m respectively, which is a difference almost of 14 mm. The same behaviour exists also in wells number 1322 and 1638 located at the right and left sides of the fault Epigmenio Gonzales, respectively, where the water depths in 2011 were 116 and 129.5 m, respectively (Table 1).

The hydrogeological system is complex because of the heterogeneity of the soil layers, the basement, the fracture system and the well system. Beside the complexity of the system, another problem is the lack of data. In the literature there are several examples of analysis of subsidence produced by one single well, for example Ziaie and Rahnama (2007) presented in a study the relationship between classical soil parameters and subsidence driven for a pumping single well. Also Martinez et al. (2013) demostrated that fast water infiltration could be a mechanism for fracture generation and for triggering land subsidence. In this research both elements are analysed.

The aim of this research is to simulate the dynamic groundwater table of a homogeneous aquifer with a fault zone which has been caused by a single water extraction well and fast water infiltration after a strong rainfall. Although only idealized hydrogeological systems are modelled, the results help us to understand the land subsidence intensity resulting from both events, extraction and rapid infiltration and the interrelation of them.

Table 1. Water depths of some wells in Queretaro in 2011.

	Well no.	Water depth 2011 (m)
Fault Tlacote		
right side	2427	133.1
right side	1313-A	140
left side	612-F	153.9
Fault E. Gonzales		
right side	1322	116
left side	1638	129.5
right side	953	132.2

2 Model concepts

2.1 Two-phase flow in porous media

If two fluid phases are not or only slightly miscible into each other, a two-phase flow model concept for porous medium can be applied (Hinkelmann, 2005). The continuity equation must be fulfilled for each phase α, one for the liquid phase w (water) and one for the gas phase n (air):

$$\frac{\partial (S_\alpha \varphi_\alpha \rho_\alpha)}{\partial t} + \mathrm{div}\left(\rho_\alpha \underline{v}_\alpha\right) - \rho_\alpha q_\alpha = 0. \quad (1)$$

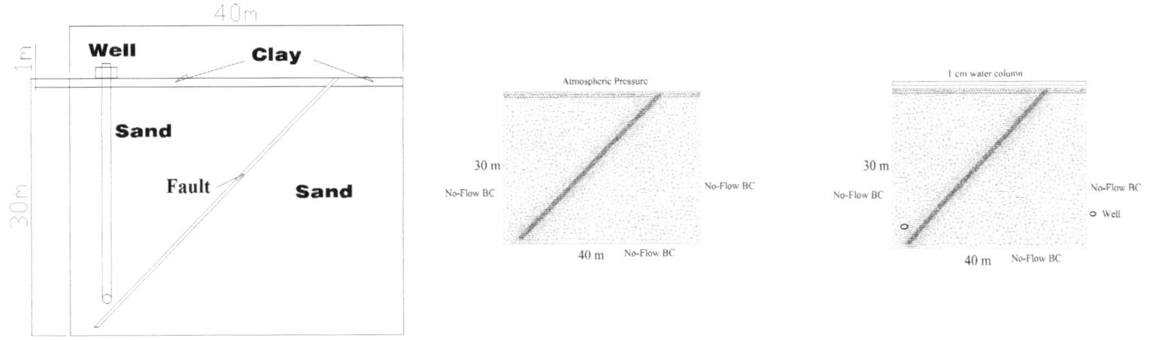

Figure 2. Idealised system (left). Unstructured grid of a 40 m × 30 m system with higher resolution around the fault zone (middle); Complete 1 cm water column on the top boundary (right).

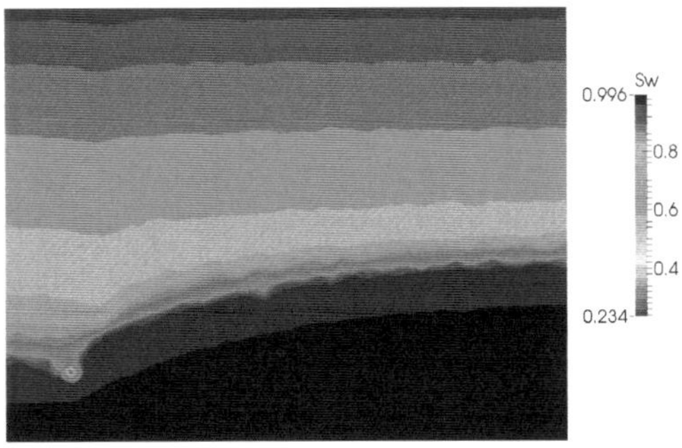

Figure 3. Water saturation after water extraction of a single system without fracture.

Figure 4. Water saturation in the case of a single well extraction with a fault zone.

Here, φ_α is the void space filled with phase α, S_α the saturation, ρ_α the density, $\underline{v}_\alpha$ the filter or Darcy velocity and q_α a sink or source of the phase.

To describe the Darcy velocity of each phase, the generalized form of the Darcy law can be used:

$$\underline{v}_\alpha = -\frac{k_{r\alpha}}{\mu_\alpha}\mathbf{K}\cdot(\text{grad}\, p_\alpha - \rho_\alpha g), \tag{2}$$

where $k_{r\alpha}$ is the relative permeability, $\mathbf{K}$ the tensor of intrinsic permeability, μ_α the dynamic viscosity, p_α the pressure and g the gravity. $\frac{k_{r\alpha}}{\mu_\alpha} = \lambda_\alpha$ is referred to as the mobility of phase α.

The relative permeability for liquid phase w and gas n phase are computed with the Brooks–Corey (Brooks and Corey, 1964) relationship:

$$k_{rw} = S_e^{\frac{2+3\lambda}{\lambda}}, \tag{3}$$

$$k_{rn} = (1 - S_e)^2\left(1 - S_e^{\frac{2+\lambda}{\lambda}}\right). \tag{4}$$

The parameter λ characterizes the grain-size distribution. A small value describes a single grain-size material, while a larger value indicates highly non-uniform material. S_e stands for the effective (water) saturation defined as:

$$S_e = \frac{(S_w - S_{wr})}{(1 - S_{nr} - S_{wr})}. \tag{5}$$

The pore volume is completely filled with the wetting and the non-wetting phase saturation, S_w, S_n:

$$S_w + S_n = 1. \tag{6}$$

At the interface between both, the difference between the phase pressure of gas p_n and liquid phase p_w is called capillary pressure.

$$p_c = p_n - p_w \tag{7}$$

The authors would like to mention that a Richards model concept also would have been suitable for the simulations. Further information about the model is found e.g. in Hinkelmann (2005).

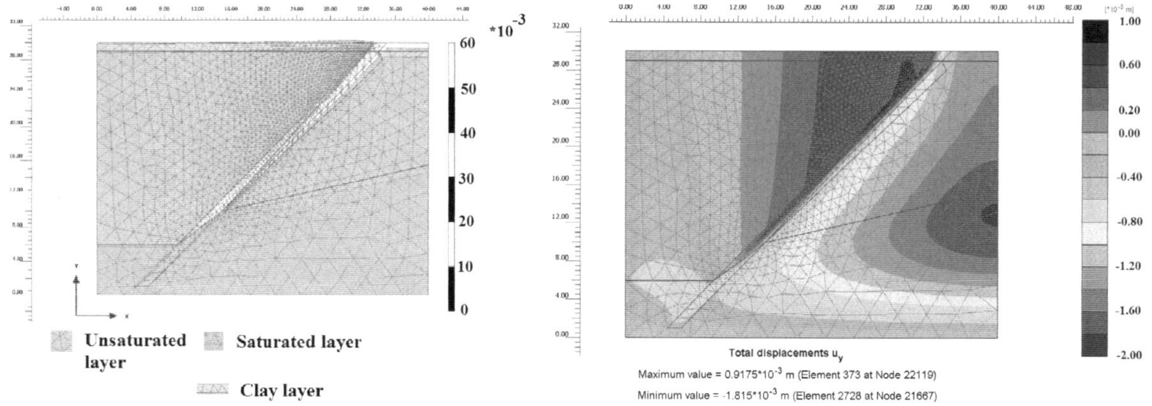

Figure 5. Deformed mesh after water extraction (left). Maximum total vertical displacement after water extraction (right).

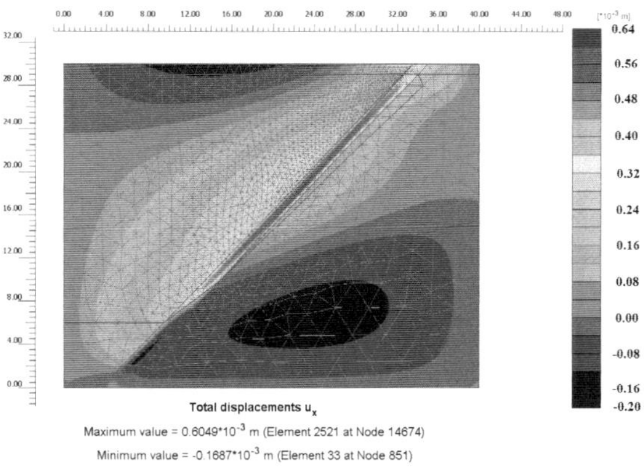

Figure 6. Maximum total horizontal displacement after water extraction.

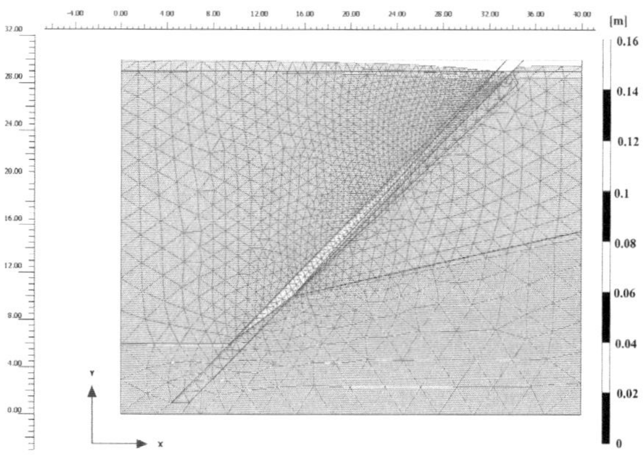

Figure 7. Deformed mesh after fast water infiltration.

2.2 Linear-elastic perfectly plastic model with Mohr–Coulomb failure criterion: deformation

Plasticity refers to irreversible strains. As soil behaviour is highly non-linear and irreversible, a linear-elastic perfectly plastic Mohr–Coulomb model has been applied (Waterman et al., 2004). This model involves five input parameters, i.e. Young's modulus E, Poisson's ratio υ, friction angle φ, dilatancy angle ψ and cohesion C. For simplification, in this investigation the dilatancy angle is taken equal to zero.

Material models for soil and rock are generally expressed as a relationship between infinitesimal increments of effective stress ("effective stress rates") and infinitesimal increments of strains ("strain rates"). This relationship is expressed in Hooke's law:

$$\underline{\dot{\sigma}}' = \underline{\underline{\mathbf{M}}}\underline{\dot{\varepsilon}},$$

where $\underline{\underline{\mathbf{M}}}$ is a material stiffness matrix.

This model decomposed strains and strain rates into elastic and plastic parts:

$$\underline{\varepsilon} = \underline{\varepsilon}^{\mathrm{e}} + \underline{\varepsilon}^{\mathrm{p}} \qquad \underline{\varepsilon} = \underline{\dot{\varepsilon}}^{\mathrm{e}} + \underline{\dot{\varepsilon}}^{\mathrm{p}}. \tag{8}$$

If Eq. (8) is inserted into the Hook's law Eq. (8) leads to:

$$\underline{\dot{\sigma}}' = \underline{\underline{\mathbf{D}}}^{\mathrm{e}}\underline{\dot{\varepsilon}}^{\mathrm{e}} = \underline{\underline{\mathbf{D}}}^{\mathrm{e}}\left(\dot{\varepsilon} + \dot{\varepsilon}^{\mathrm{p}}\right) \tag{9}$$

$\underline{\underline{\mathbf{D}}}^{\mathrm{e}}$ is the elastic material stiffness matrix $\underline{\dot{\sigma}}'$ effective stress rates.

A plastic potential function g is introduced. The plastic strain rates are written as:

$$\underline{\dot{\varepsilon}}^{\mathrm{p}} = \lambda \frac{\partial g}{\partial \underline{\sigma}'}, \tag{10}$$

in which λ is the plastic multiplier. For elastic behaviour λ is zero and for plastic behaviour λ is positive.

These model concepts are implemented in the tool PLAXIS. Further information about the model is found in Waterman et al. (2004).

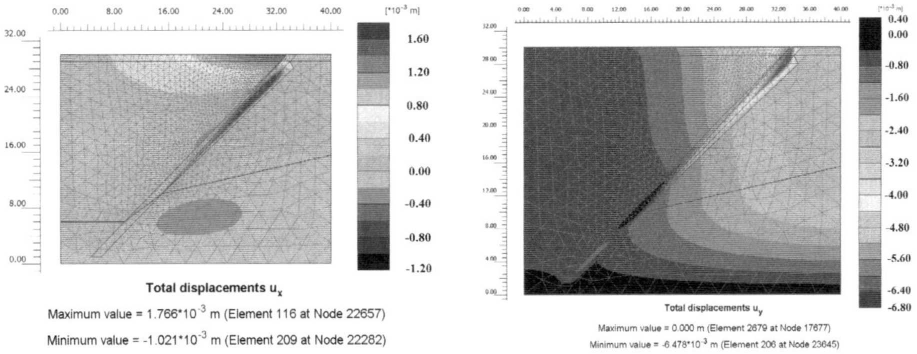

Figure 8. Maximum total vertical displacement after fast water infiltration (left). Maximal total horizontal displacement after fast water infiltration (right).

3 Idealized system and parameter

A rectangular system of 40 m length and 30 m depth was simulated. A 2-D steady-state model of a two-layered system with a clay layer on the top, a sand layer underneath and a 1 m fault zone was set up. A fracture zone was set up with a 45° inclination with respect to the horizontal (Fig. 2, left). In this study the fault zone was idealized as a porous medium damage band with higher permeability (blue in Fig. 2, middle) compared to the sand and the clay layers (green and red, respectively, in Fig. 2, middle). An unstructured grid was generated with 9798 cells with a higher resolution around the fault zone as shown in Fig. 2 middle. The boundary condition (BC) at the top of the system is a Dirichlet BC with a fixed atmospheric pressure. At the bottom, left and right of the domain a fixed Neumann no flow BC was chosen. For all simulations, as an initial condition, full water saturation was considered.

The flow parameters used in the numerical simulation for the sand, clay and fracture zone are described in Table 2.

Table 2. Soil parameters for two-phase flow modeling.

		Sand	Clay	Fault zone
S_{wr}	[–]	0.15	0.15	0.05
S_{ar}	[–]	0.01	0.01	0.01
K	[–]	5×10^{-11}	3.5×10^{-13}	Variable
λ	[–]	2.3	3.5	2.2
P_d	[Pa]	700	2400	200
φ	[–]	0.37	0.43	0.32

4 Results

4.1 Reference case

Firstly a system without fracture was modelled as a reference to be compare later with the results of the other setups. Figure 3 shows the normal water depletion cone that appears around a well extraction.

4.2 Water withdrawal of a single well

With difference to the reference case, in the first step of simulation, the water extraction of a single well near a high permeable fault zone, a barrier effect is produced, i.e. the water level at both sides of the fault zone are different. The results of water saturation of the system with a 45° high permeability fault zone and a single well extraction (Fig. 4) show a difference of approximately 3.5 m on both sides of the fault zone.

This barrier produces an irregular soil deformation, especially on the top of the surface (Fig. 5, left). The results of the simulation show a maximum total vertical displacement of approximately 1.8×10^{-3} m (Fig. 5, right) and approximately a 0.6×10^{-3} m maximum total horizontal displacement (Fig. 6).

4.3 Fast water infiltration

To simulate an accumulation of water caused by a runoff after a strong rainfall a 1 cm water column on the complete surface was applied. The BC at the top of the system is a Dirichtlet BC with a fixed water pressure $p_n = p_{atm} + \rho g h$ with $h = 1$ cm simulating an accumulation of a 1 cm water column (Fig. 2, right).

As shown in Fig. 7, which highlights the deformed mesh after fast water infiltration, the maximal displacements take place in the upper part of the fault zone.

The results show that after 30 min of fast water infiltration through the high permeable fault zone a maximum horizontal displacement of approximately 1.8×10^{-3} m and a vertical displacement of 6.8×10^{-3} m take place (Fig. 8).

5 Conclusions

The present study is a conceptual model of land subsidence produced water extraction from a single well and fast water infiltration through a fault zone.

A two-layered system consisting of a sand layer and a small clay layer on the top with an inclined high permeable fault zone was investigated.

The differences of the water levels on both sides of the fault zone produced by the single well water extraction lead to different settlements on both sides. However, there is a greater impact on the vertical than on the horizontal displacement. The fast water infiltration impacts the displacements in both directions, vertical and horizontal, and the horizontal displacement can cause plastic deformation and also could produce new fracture formations.

If the fault zone is narrower, for example 40 cm, the infiltration through the fault zone is not significant and the water spreads out more into the partially saturated zone because of capillarity forces and gravity (Martinez et al., 2013). This phenomenon increased the deformation of the soil layers, specially the horizontal one. The results also show that a fault zone with high permeability could act as hydrological barrier. This phenomenon will be studied in more detail in future work.

References

Brooks, R. H. and Corey, A. T.: Hydraulic properties of porous media, in: Hydrology papers, Vol. 3, Colorado State University, Fort Collins, CO, 1964.

Carreón-Freyre, D., Cerca, M., Luna-González, L., and Gámez-González F. J.: Influencia de la estratigrafía y estructura geológica en el flujo de agua subterránea del Valle de Querétaro, Revista Mexicana de Ciencias Geológicas, 22, 1–18, 2005.

Hinkelmann, R.: Efficient numerical methods and information processing techniques for modeling hydro-and environmental systems, in: Lecture Notes in Applied and Computational Mechanics, Vol. 21, Springer, Berlin, Heidelberg, 2005.

Martinez, I., Hinkelmann, R., and Savidis, S.: Fast water infiltration: a mechanism for fracture formation during land subsidence, Hydrogeol. J., 21, 761–771, doi:10.1007/s10040-013-0971-6, 2013.

Pacheco, J.: Modelo de subsidencia del Valle de Querétaro y predicción de agrietamientos superficiales, PhD thesis, UNAM Mexico, 2007.

Rojas, E., Arzate, J., and Arroyo, M.: A method to predict the group fissuring and faulting caused by regional groundwater decline, Eng. Geol., 65, 245–260, 2002.

Waterman, D., Al-Khoury, R., Bakker, J. K., Bonnier, P. G., Burd, H. J., Soltys, G., Vermeer, P. A.: PLAXIS, 2D, Version 8, Manual, PLAXIS, Delft, The Netherlands, available at: http://www.plaxis.nl/shop/135/info/manuals, last access: 8 March 2014, 2004

Ziaie, A. and Rahnama, M. B.: Prediction of single well land subsidence due to ground water drainage, Int. J. Agric. Res., 2, 349–358, 2007.

Proc. IAHS, 372, 39–44, 2015
proc-iahs.net/372/39/2015/
doi:10.5194/piahs-372-39-2015

Open Access
PIAHS

New information on regional subsidence and soil fracturing in Mexico City Valley

G. Auvinet, E. Méndez-Sánchez, and M. Juárez-Camarena
Instituto de Ingeniería, UNAM, Mexico
Correspondence to: G. Auvinet (gauvinetg@iingen.unam.mx)

Published: 12 November 2015

Abstract. In this paper, updated information about regional subsidence in Mexico City downtown area is presented. Data obtained by R. Gayol in 1891, are compared with information obtained recently from surveys using the reference points of Sistema de Aguas de la Ciudad de México (2008) and on the elevation of a cloud of points on the ground surface determined using Light Detection and Ranging (LiDAR) technology. In addition, this paper provides an overview of recent data obtained from systematic studies focused on understanding soil fracturing associated with regional land subsidence and mapping of areas susceptible to cracking in Mexico City Valley.

1 Introduction

Mexico Valley lacustrine subsoil has an exceptionally high compressibility and low resistance. Additionally, a regional subsidence phenomenon is affecting the urban area since the early Twentieth Century. One consequence of this phenomenon is the generation of cracks in the soil in many places. Both problems, regional subsidence and soil fracturing represent a risk for the stability of buildings and affect the urban infrastructure.

The demographic development of Mexico City has created an accelerated demand for services, especially supply of drinking water. One of the cheapest ways to respond to this demand has been the exploitation of the aquifer underneath the urban area by pumping water from deep wells. This activity has produced a regional subsidence phenomenon and the cracking of the soil in the lacustrine and alluvial-lacustrine areas of Mexico City. Due to the high cost of other water-supply alternatives, it is expected that extraction of water from the local aquifer will continue for many years.

The regional subsidence in Mexico City has severe consequences. It affects the drainage system, transport infrastructure, foundations of buildings and generates serious risks to the population, since it induces other problems such as flooding of low areas. Therefore, although the regional subsidence is an ancient phenomenon, its study and analysis remain a priority nowadays, inasmuch as it has not been possible to control its basic cause, which frequently leads to adopt emergency solutions.

More and more frequently, cracks appear in the soil of Mexico City causing alarm among the population and damaging buildings. Therefore, since 2005, the Geocomputing Laboratory group of the Geotechnical Section of Instituto de Ingeniería, UNAM in collaboration with the Mexican Society for Geotechnical Engineering and with the support of municipal authorities has undertaken a systematic study of the phenomenon of soil cracking. The occurrence of cracking may result from any condition that causes important tension stresses in the soil (Auvinet, 2008) and the occurrence of cracking has different causes, including contraction of the lacustrine clays by drying, existence of tension stresses associated with buildings weight, hydraulic fracturing in areas of flooding, etc. However, the most important and destructive cracks are a direct consequence of regional subsidence that occurs in Mexico City as a result of pumping water from the aquifer.

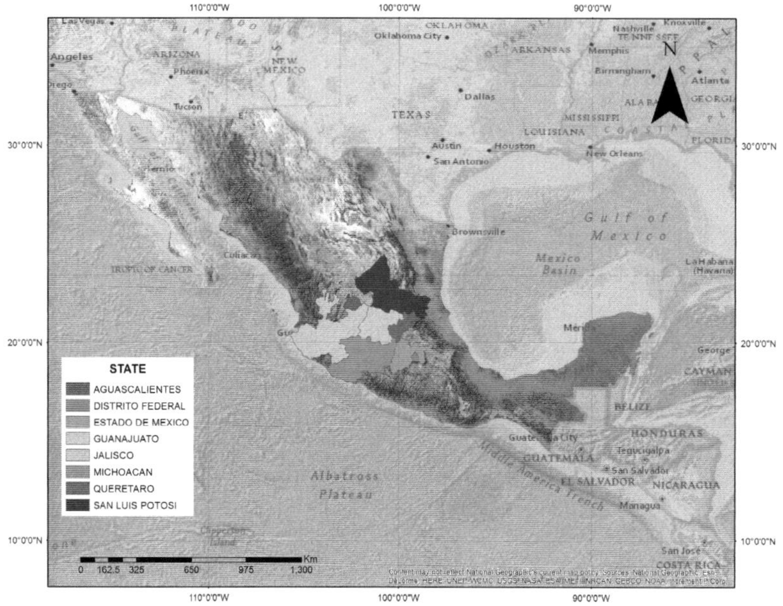

Figure 1. Spatial distribution of the phenomenon of soil cracking in Mexico.

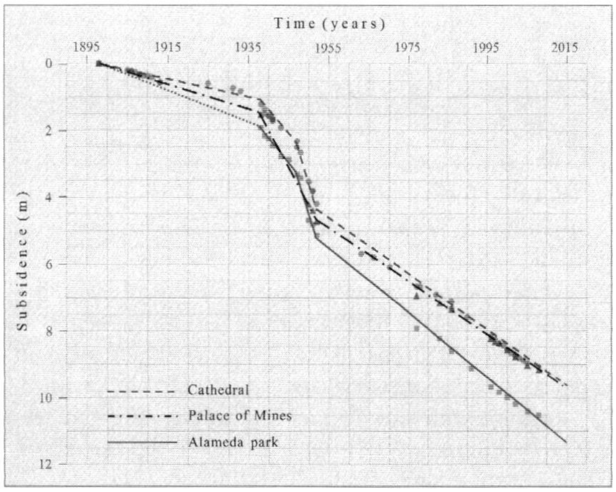

Figure 2. Evolution of subsidence for period 1898–2015.

2 Previous studies

The regional subsidence of Mexico basin was studied in the late nineteenth century by Téllez Pizarro (1899). In 1925, Roberto Gayol informed the Society of Engineers and Architects of Mexico that Mexico City was subsiding and that the probable cause was the "disturbance that the drainage of subsoil water was producing, in the bottom of Mexico basin". Gayol based his assertion on some surveys in Mexico City downtown and in the Texcoco Lake (1891).

Between 1920 and 1930, José A. Cuevas gave his support to the ideas of Gayol and asked Nabor Carrillo to study the influence of the pumping from water wells on subsidence. Carrillo (1948) explained the subsidence using newly developed techniques of Soil Mechanics and established that the cause of this phenomenon was the consolidation of clays due to the increase in effective stress caused by the drawdown of interstitial water pressure.

In 1952, the systematic study of subsoil and the first piezometric measurements made by Sandoval, Hiriart and Marsal, corroborated the findings of Nabor Carrillo. Also very meaningful were the investigations by Zeevaert and the periodical surveys and piezometric measurements performed since 1953 by Hydrological Commission of Mexico Basin Valley (CHCVM, SRH) and subsequently by the Water System of Mexico City (SACMEX).

Soil cracking is not exclusive of Mexico City, this phenomenon has spread to other states of the country (Estado de Mexico, Querétaro, Morelia, Silao, Aguascalientes, between others). In Fig. 1, the states of the country where the cracking phenomenon has been detected are indicated.

In Mexico City, due to increasing subsidence, soil fissuring that previously only occurred in the dry zone of former-Lake of Texcoco has extended to the transition geotechnical zone of the city (0 to 20 m of lacustrine clay). This can be attributed to differential settlements between zones of soft and hard soils, damaging buildings and urban services. The problem of soil cracking in Mexico City Valley is of large magnitude and will require continued attention in the future.

3 Subsidence evaluation

From the available historical documents, it has been possible to reconstruct the history of subsidence of the old downtown area of Mexico City. Analyzing graphs in Fig. 2 and in particular, those corresponding to the Cathedral and Alameda Central, it can be concluded that the history of subsidence has gone through four stages during the twentieth century and in the first decade of the current century while, in the Palace of Mines, the subsidence history has gone through only three stages. These stages correspond to different rates of water extraction.

Figure 3a shows the shaded relief model (SRM) built from topographical data obtained in 1891 by Gayol. The model corresponds to the area of the old "traza" (center) of Mexico City. The so-called "corographic" map of Federal District by Manuel Fernandez Leal (1899) was superimposed on the model. Figure 3b shows the shaded relief model (SRM) constructed from a dense cloud of points obtained by scanning using the Light Detection and Ranging (LiDAR) technique by the National Institute of Geography Statistics and Computing (INEGI, 2010), corresponding to the old central area of Mexico City. Figure 4 shows the remnants of the buildings of the ceremonial precinct of the Mexica Empire and the trace of the Historic center of Mexico City superimposed on a shaded relief model, obtained by INEGI (2010).

Figure 5 shows an assessment of the magnitude of regional subsidence, both in the west-east direction along Tacuba Avenue and in the North-South direction along José María Pino Suárez Avenue, in the historical center of Mexico City for the 1891–2010 period (119 years).

4 Soil cracking associated to regional subsidence

Cracks originating in areas of abrupt transition between compressible and hard soils, are a direct consequence of the regional subsidence of Mexico Valley as a result of pumping water from the local aquifer.

These cracks are very destructive; they present a step towards the area of greatest settlement and are generally parallel to the elevation contours at the foot of the mountains, hills or outcrops of rock or rigid materials (Fig. 6).

The cracking phenomenon associated with regional subsidence in areas of abrupt transition takes a particular importance in the area surrounding Santa Catarina range, Peñón Viejo-Peñón Márquez hill, Guerrero hill, Peñón de los Baños hill, Chimalhuacán hill, Xico hill and in the northern limit of the Chichinautzin range. Figure 7 shows the principal zones where soil is most suceptible to cracking as defined from the gradient of variation of clay layers thickness and recorded differential settlements.

Photographs of Fig. 8 show the destructive capacity of soil cracking associated with regional subsidence.

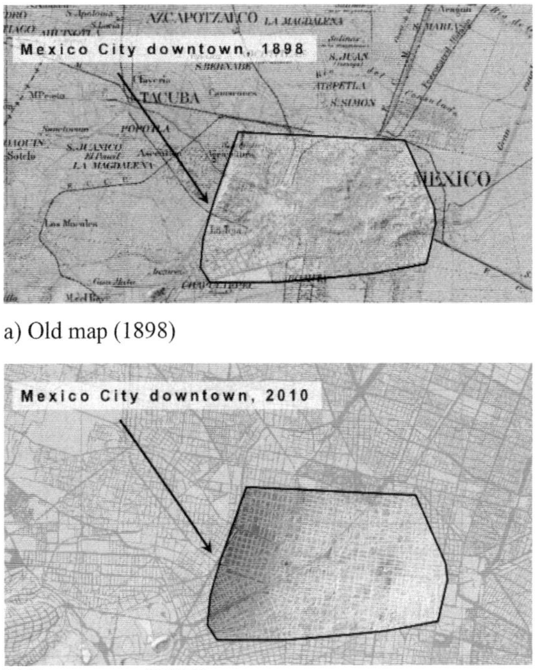

a) Old map (1898)

b) Recent map (2010)

Figure 3. Maps of Mexico City downtown area over a shading relief model.

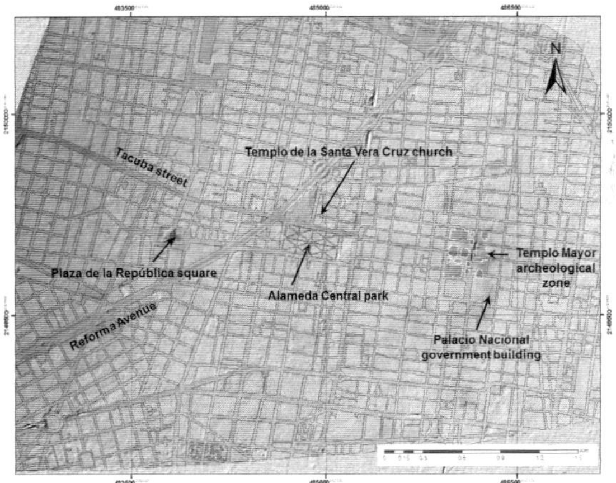

Figure 4. Mexica constructions remnants in Mexico City downtown area over a raster model obtained from LiDAR data.

5 Conclusions

The effort undertaken by different groups and in particular by the Geocomputing Laboratory of the Engineering Institute, UNAM to achieve a satisfactory evaluation of the phenomenon of subsidence, as well as other aspects of geotechnical problems such as soil fissuring has given useful and

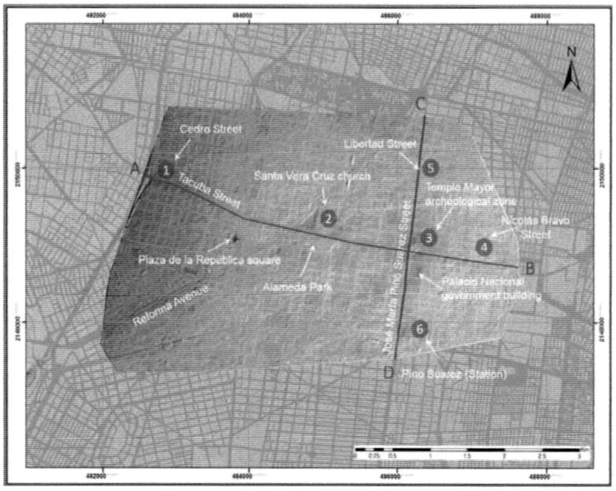

a) Location of profiles

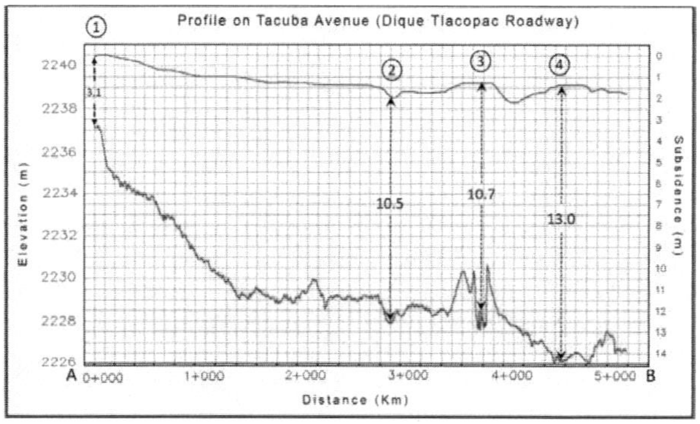

b) Profile along Tacuba Avenue (A-B)

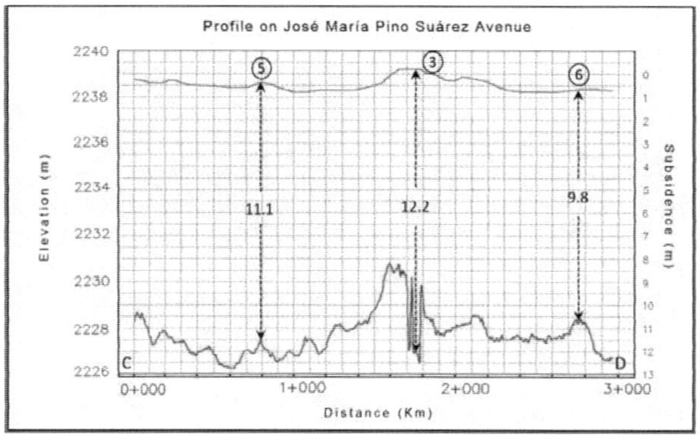

c) Profile along José María Pino Suárez Avenue (C-D)

Surface Profile, 1891 | Surface Profile, 2010

Figure 5. Location map and topographic profiles along Tacuba Avenue and J. M. Pino Suárez Avenue.

Figure 6. Crack with vertical step associated with the subsidence in abrupt transition zone.

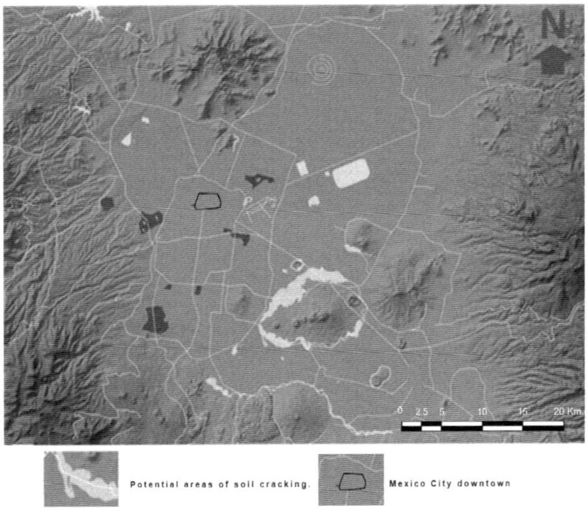

Figure 7. Main zones of Mexico City Valley subject to surface cracking (yellow strips).

Figure 8. Effects of differential soil subsidence and cracks on constructions.

promising results, but this is only the first stage of a huge job to be performed consistently in the future.

In a short period, it has been possible to update and expand the database of sites where cracks are found and to categorize them according to their mechanism of generation and/or propagation; further efforts will be necessary to define and extend the techniques to control the cracking phenomenon and mitigate its harmful effects.

Acknowledgements. The authors acknowledge the support of different institutions: Mapoteca Manuel Orozco y Berra, Palacio de Minería library, CONACYT, DGAPA-UNAM, IPN, GDF, CNA SACMEX and other institutions and agencies for the realization of this work. The valuable contributions from companies and geotechnical consultants who kindly gave to the authors access to their collections of subsoil explorations are also recognized.

References

Auvinet, G.: Agrietamiento de suelos, Proceedings, XXIV Reunión Nacional de Mecánica de Suelos, special volume, 299–313, Mexico, 2008.

Carrillo, N.: Influence of Artesian Wells on the Sinking of Mexico City, Proceedings of 2th Congress of International Soil Mechanics and Foundation Engineering, Vol. VI, 156–159, Rotterdam, 1948.

Fernandez Leal, M.: Carta Corográfica del Distrito Federal, Secretaría de Fomento, Official map, Mexico, 1899.

Gayol, R.: Plano General de las atarjeas conforme al proyecto de desagüe y saneamiento de la Ciudad de México, official map, Mexico, 1891.

Gayol, R.: A study of disturbances caused by draining out water from the subsoil of the Valley of Mexico, First part, Revista Mexicana de Ingenieria y Arquitectura, Vol. III, No. 3, 96–132; Second part, Revista Mexicana de Ingenieria y Arquitectura, Vol. III, No. 8, 507–559, Mexico, 1925 (in Spanish).

Instituto Nacional de Estadística, Geografía e Informática (INEGI): Data of LiDAR technology in Mexico City, data bases, Mexico, 2010.

Sistema de Aguas de la Ciudad de México: Set of topographic data of Mexico City, from 1977 to 2008, Mexico, 2008.

Téllez-Pizarro, A.: Apuntes acerca de los cimientos de los edificios en la ciudad de México, Reprinted in the commemorative volume of the XXV anniversary of the Mexican Society for Soil Mechanics (1982), Mexico, 1899.

Proc. IAHS, 372, 45–48, 2015
proc-iahs.net/372/45/2015/
doi:10.5194/piahs-372-45-2015

Open Access
PIAHS

Application of multi-sensor advanced DInSAR analysis to severe land subsidence recognition: Alto Guadalentín Basin (Spain)

R. Bonì[1], G. Herrera[2,3,4], C. Meisina[1], D. Notti[1,10], M. Béjar-Pizarro[2,3], F. Zucca[1], P. J. González[5], M. Palano[6], R. Tomás[7,3], J. Fernández[8], J. A. Fernández-Merodo[2,3], J. Mulas[2,3], R. Aragón[2,3], C. Guardiola-Albert[2], and O. Mora[9]

[1]Department of Earth and Environmental Science, University of Pavia, Via Ferrata 1, 27100 Pavia, Italy
[2]Geohazards InSAR laboratory and modeling group, Instituto Geológico y Minero de España (IGME), C/. Alenza 1, 28003 Madrid, Spain
[3]Unidad Asociada de investigación IGME-UA de movimientos del terreno mediante interferometría radar (UNIRAD), Universidad de Alicante, P.O. Box 99, 03080 Alicante, Spain
[4]Earth Observation and Geohazards Expert Group (EOEG), EuroGeoSurveys, the Geological Surveys of Europe, 36–38, Rue Joseph II, 1000 Brussels, Belgium
[5]Institute of Geophysics and Tectonics, School of Earth and Environment, University of Leeds, Leeds, LS2 9JT, UK
[6]INGV Sezione di Catania, Piazza Roma 2, 95123 Catania, Italy
[7]Universidad de Alicante, Dpto. de Ingeniería Civil, Escuela Politécnica Superior de Alicante, P.O. Box 99, 03080 Alicante, Spain
[8]Instituto de Geociencias, UCM-CSIC, Plaza de Ciencias 3, Ciudad Universitaria, 28040 Madrid, Spain
[9]Altamira Information, C/. Còrsega 381-387, 08037 Barcelona, Spain
[10]Departamento de Geodinámica, Universidad de Granada, Avenida Fontenueva, Granada, Spain

Correspondence to: R. Bonì (roberta.boni01@universitadipavia.it)

Published: 12 November 2015

Abstract. Multi-sensor advanced DInSAR analyses have been performed and compared with two GPS station measurements, in order to evaluate the land subsidence evolution in a 20-year period, in the Alto Guadalentín Basin where the highest rate of man-induced subsidence (> 10 cm yr^{-1}) of Europe had been detected. The control mechanisms have been examined comparing the advanced DInSAR data with conditioning and triggering factors (i.e. isobaths of Plio-Quaternary deposits, soft soil thickness and piezometric level).

1 Introduction

Since the 1960s, the Alto Guadalentín Basin (south-eastern Spain), where the city of Lorca is located, has been affected by severe changes of the groundwater level.

The objectives of this work are (i) to quantify the cumulated displacements over a 20-year period with multi-sensor advanced DInSAR data, and (ii) to investigate the control mechanisms that could explain this land motion.

2 Study area

The Alto Guadalentín Basin is located in south-eastern Spain, and it includes an aquifer system that covers an area of about 277 km^2, constituted by Plio-Quaternary detrital and alluvial material including clays, sands and conglomerates; Miocene detritical with conglomerate and sand deposits; and locally Triassic carbonate rocks. The Mesozoic metamorphic rocks represent the lower impermeable limit. Since 1960 the agriculture development, triggered uncontrolled aquifer sys-

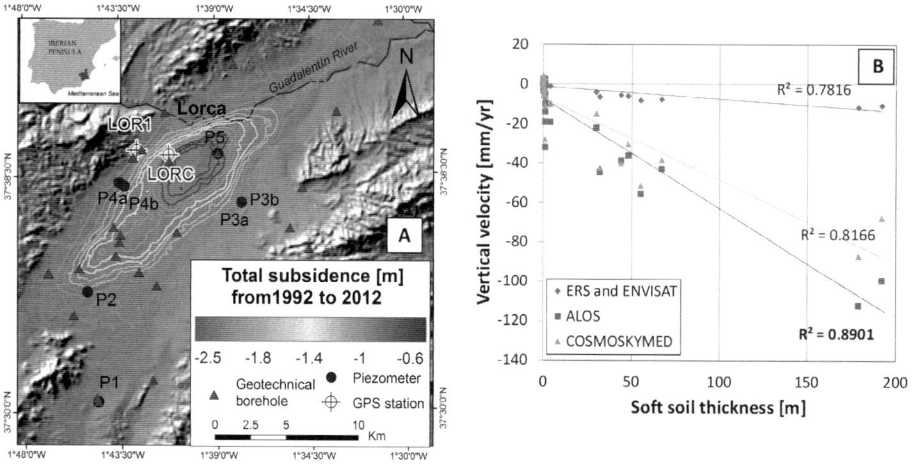

Figure 1. **(a)** Cumulated displacements (m) of the Alto Guadalentín Basin from 1992 to 2012. **(b)** Soft soil thickness calculated for each geotechnical boreholes and vertical velocity detected by the different sensors in the proximity of the boreholes. R^2 stand for Pearson correlation coefficient.

tem exploitation that led it to declare temporarily overexploited in 1987 (CHS, 2006).

3 SAR dataset and processing details

A large set of C-band SAR data (ERS-1, ERS-2 and ENVISAT), L-band SAR data (ALOS PALSAR) and X-band SAR data (COSMO-SkyMed) has been employed to reconstruct the temporal and spatial subsidence evolution from 1992 to 2012 that affected the Alto Guadalentín Basin. ERS and ENVISAT-ASAR datasets, covering the 1992–2001 and the 2003–2007 time intervals, respectively, were processed by adopting the Small Baseline approach (Berardino et al., 2002). The ALOS PALSAR dataset from 2007 to 2010 and the COSMO-SkyMed dataset from 2011 to 2012 were processed using the DIAPASON interferometric software and the SPN software (Arnaud et al., 2003) using the Persistent Scatterer approach.

4 Results

4.1 Spatio-temporal evolution

Note that since the multi-sensors displacement data using different incidence angles (ERS-1/2 and ENVISAT incidence angle permits to detect 92 % of the vertical displacements, whereas ALOS and COSMO-SkyMed satellites measure 83 and 75 % of the vertical displacements, respectively) in order to homogenize the geometrical distortions introduced by the different acquisition angles of the different sensors, both the average velocity and the cumulated surface displacements in the LOS obtained for each dataset have been projected along the vertical direction. In addition, the DInSAR displacement accuracies have been computed at local scale, by comparing them against Global Position System (GPS) measurements. Using continuous records of displacement detected by two GPS stations (LOR1 and LORC) located in a stable and subsiding area, respectively (see the location in Fig. 1a). Average absolute error of 4.6 ± 4 mm for the ALOS data and of 4.8 ± 3.5 mm for the COSMO-SkyMed data has been detected, confirming the high reliability of the employed DInSAR data. Considering the vertical cumulated surface displacements of the different datasets, the total subsidence from 1992 to 2012 reach up to 2.5 m with an average subsidence of 1.8 m over an area of 14.8 km^2, located on the central sector of the Alto Guadalentín Basin (Fig. 1a). Subsidence areal extent exhibits a SW-NE elliptical shape parallel to the valley direction.

4.2 Control mechanisms

The relationship between the evolution of the groundwater levels and land subsidence is controlled by several factors (geological, hydrogeological and geotechnical factors). These factors can have a direct role in the trigger of subsidence (piezometric level) and others have a conditioning role (lithology and thickness of deformable soil).

4.2.1 Conditioning factors

The isobaths map produced by Cerón and Pulido-Bosch (1996), using information from electrical geophysics surveys, represents the variation of the thickness of the Plio-Quaternary deposits. A spatial analysis has been performed in order to determine the subsidence related to each isobaths unit. The result of this comparison demonstrates that the

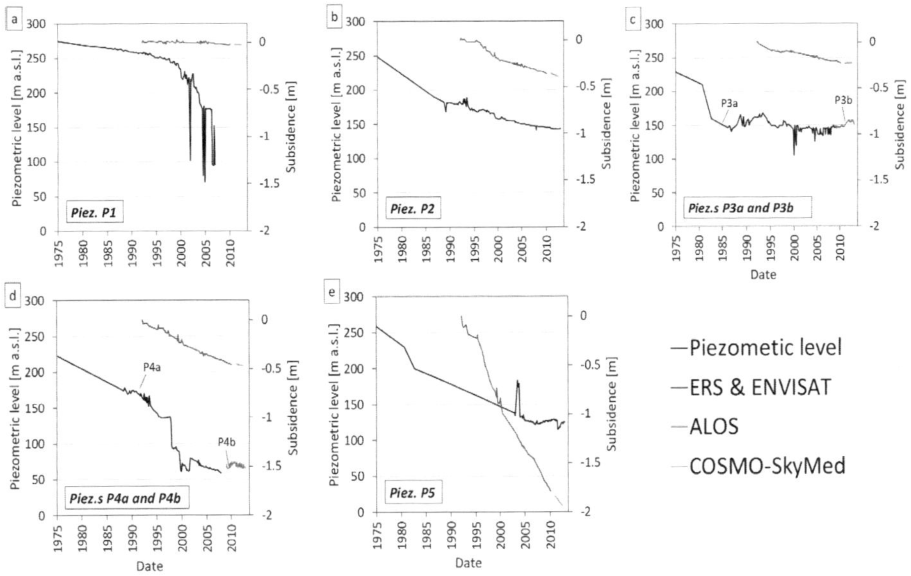

Figure 2. Comparison of the water level variations from 1975 to 2012 (black and pink line) at the different piezometers located on Fig. 1a, and the vertical displacements detected by the satellite sensors (blue, red and green lines).

maximum subsidence does not coincide with the higher value of thickness of Plio-Quaternary filling, so there are other control mechanisms. The Plio-Quaternary deposits contain discontinuous compressible layers and these layers have been selected to analyse the soft soil distribution. Using geotechnical borehole information, soft soils thicker than 100 m has been detected in the central part of the basin, whereas towards the southwest and bordering areas of the basin thinner soft soils are found. The relationship of soft soil thickness with subsidence rates estimated from the different satellites is straightforward, i.e. maximum subsidence rates are coincident with the thickest soft soils and contrarily, smaller subsidence rates are coincident with the thinnest soft soils found in the bordering areas (Fig. 1b).

4.2.2 Triggering factor

A detailed time series analysis of subsidence and ground water evolution is presented for the available (seven) piezometers during the 20-year period (Fig. 2). No correlation is observed at piezometer P1, located in the southern zone, where the highest groundwater drawdown (162 m) from 1992 to 2007 corresponds to 13 cm of cumulated subsidence in the period 1992–2012. This is probably due to the lesser amounts of soft soils in the southern border of the basin. Certain correlation is observed in the western part of the basin (piezometers P2, P4a and P4b). The piezometer P2 displays a steady lowering of the piezometric level from 1989 to 2013 (Fig. 2b) that correlates with the steady subsidence rate. Contrarily piezometers P4a-b, which are jointly analyzed due to their proximity and characteristics (Fig. 1), exhibit a high ground water level variability that is not in agreement with the estimated subsidence linear rate (Fig. 2d). In the eastern part of the basin, piezometers P3a-b display a water level evolution that is in agreement with subsidence accelerations and decelerations. In the central part of the basin where the thickest soft soils are found (around 190 m), the time series of piezometer P5 reveals a groundwater level recovery from 2007 to 2013 that does not correspond with measured subsidence steady rates (Fig. 2e). This fact reveals an inelastic, unrecoverable and permanent deformation of this aquifer as previously reported by González and Fernández (2011) and Rigo et al. (2013), which has been triggered by the groundwater exploitation started in the early 60s. Overall, a clear correlation between groundwater level drawdown and subsidence accelerations or decelerations is not observed.

5 Conclusions

This research shows that Advanced DInSAR multi-sensor data obtained from a large set of SAR images are effective in analysing the subsidence spatio-temporal evolution in a 20-year period over a wide area (277 km^2). Moreover, the analysis of the subsidence trend improves the understanding of the apparent response time between the increase of water withdrawals starting from 1960 and the measured subsidence rates. From the analyses we can conclude that the areas where soft soils is thickest, magnitudes of subsidence are greatest, and that the thickness of soft soils plays an important role in the time delay between water-level depletion and ground-surface displacement.

Acknowledgements. This work is financially supported by the DORIS project (Ground Deformation Risk Scenarios: an Advanced Assessment Service) funded by the EC-GMES-FP7 initiative (Grant Agreement 423 no. 242212). ALOS PALSAR images were provided by the project JAXA-1209. Part of this work is supported by the Spanish Government under project TEC2011-28201-C02-02 and TIN2014-55413-C2-2-P and by the project 15224/PI/10 from the Regional Agency of Science and Technology in Murcia. Additional funding was obtained from the Spanish Research Program through the projects AYA2010-17448, ESP2013-47780-C2-1-R and ESP2013-47780-C2-2-Rand by the Ministry of Education, Culture and Sport through the project PRX14/00100.

References

Arnaud, A., Adam, N., Hanssen, R., Inglada, J., Duro, J., Closa, J., and Eineder, M.: ASAR ERS interferometric phase continuity. International Geoscience and Remote Sensing Symposium, 21–25 July 2003, Toulouse, France, 2003.

Berardino, P., Fornaro, G., Lanari, R., and Sansosti, E.: A new algorithm for surface deformation monitoring based on small baseline differential SAR interferograms, IEEE T. Geosci. Remote, 40, 2375–2383, 2002.

Cerón, J. C. and Pulido-Bosch, A.: Groundwater problems resulting from CO_2 pollution and overexploitation in Alto Guadalentín aquifer (Murcia, Spain), Environ. Geol., 28, 223–228, doi:10.1007/s002540050096, 1996.

CHS: Plan especial ante situaciones de alerta y eventual sequia en la cuenca del Segura: Confederacion hidrografica del Segura, Capitulo 3: Recursos y demandas, p. 5, https://www.chsegura.es/export/descargas/cuenca/sequias/pes/docsdescarga/Memoria.pdf, 2006.

González, P. J. and Fernández, J.: Drought-driven transient aquifer compaction imaged using multitemporal satellite radar interferometry, Geology, 39, 551–554, doi:10.1130/G31900.1, 2011.

Rigo, A., Béjar-Pizarro, M., and Martínez-Díaz, J.: Monitoring of Guadalentín valley (southern Spain) through a fast SAR Interferometry method, J. Appl. Geophys., 91, 39–48, 2013.

Proc. IAHS, 372, 49–52, 2015
proc-iahs.net/372/49/2015/
doi:10.5194/piahs-372-49-2015

The deformation behavior of soil mass in the subsidence region of Beijing, China

F. Tian[1], J.-R. Liu[1], Y. Luo[1,2], L. Zhu[3], Y. Yang[1,4], and Y. Zhou[1,4]

[1]Beijing Institute of Hydrogeology and Engineering Geology, Beijing, China
[2]College Of Earth Science, University of Chinese Academy Of Sciences, Beijing, China
[3]College of Resources Environment and Tourism, Capital Normal University, Beijing, China
[4]School of Earth and Space Sciences, Peking University, Beijing, China

Correspondence to: F. Tian (tfh-001@163.com)

Published: 12 November 2015

Abstract. Land subsidence induced by excessive groundwater withdrawal has been a major environmental and geological problem in the Beijing plain area. The monitoring network of land subsidence in Beijing has been established since 2002 and has covered the entire plain area by the end of 2008. Based on data from extensometers and groundwater observation wells, this paper establishes curves of variations over time for both soil mass deformation and water levels and the relationship between soil mass deformation and water level. In addition, an analysis of deformation behavior is carried out for soil mass with various lithologies at different depths depending on the corresponding water level. Finally, the deformation behavior of soil mass is generalized into five categories. The conclusions include: (i) the current rate of deformation of the shallow soil mass is slowing, and most of the mid-deep and deep soil mass continue to compress at a more rapid speed; (ii) the sand strata behaves elastically, while the clay soil mass at different depths is usually characterized by elastic-plastic and creep deformation, which can be considered as visco-elastoplastic.

1 Introduction

Serious land subsidence problems due to excessive groundwater withdrawal have emerged in more than 150 cities all over the word (Hu et al., 2004; Tomás et al., 2010). Groundwater is the main supply source of public water in Beijing. In the past few decades, groundwater supply has accounted for approximately two-thirds of the total water supply in Beijing (Zhang et al., 2008). At the same time, some environmental geological problems arising from excessive groundwater exploitation, especially land subsidence and ground fissures, have attracted much attention. A land subsidence monitoring system in Beijing began to be constructed in 2002. In 2008, the system began collecting data from extensometers, leveling networks, InSAR and GPS surveys. The monitoring results show that by the end of 2012, the region has experienced a cumulative subsidence more than 50 mm and has affected about 4300 km^2, accounting for about 68 % of the Beijing plain area.

Many stress-strain analyses for the aquifer-system were carried out based on borehole extensometers (Burbey, 2001; Zhang et al., 2006). This paper presents a comprehensive analysis of data from the extensometers to reveal the deformation behavior of soil mass in the subsidence prone area of Beijing.

2 Deformation behaviors of soil mass at different depth

Research has shown that temporal relation patterns exist between land subsidence and groundwater withdrawal. The primary subsidence layers have gradually shifted from shallow to deeper formations (with depth more than 100 m) coinciding with the increase of the mining depth of groundwater.

F1, F2 and F3 are three subsidence stations which are equipped with a series of extensometers and, pore pressure transducers yielding groundwater level observations at different depths (Fig. 1). Generally, the Quaternary unconsoli-

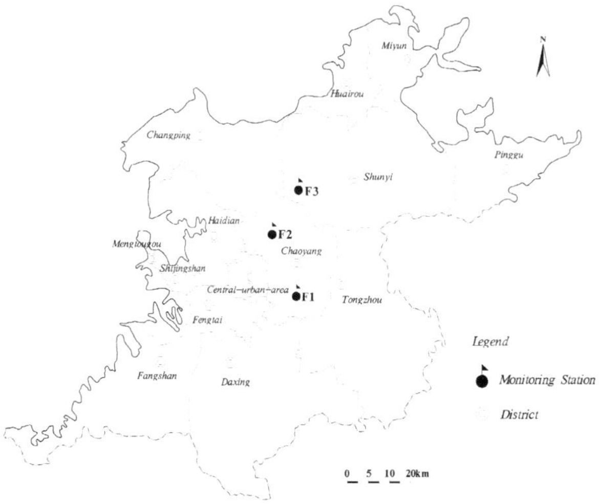

Figure 1. Location of three monitoring stations (F1, F2 and F3) for land subsidence in Beijing.

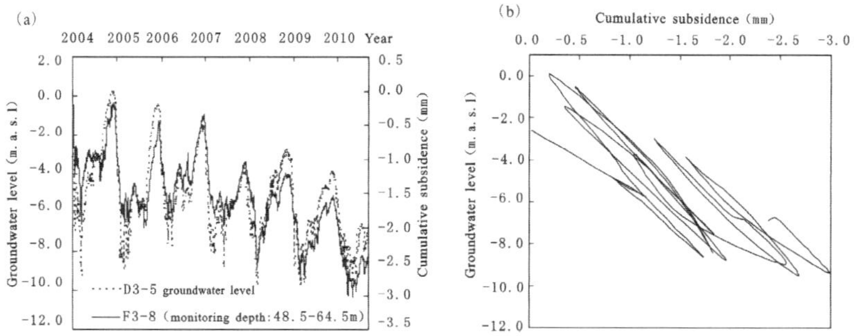

Figure 2. Deformation behavior of sand strata vs. its corresponding groundwater level at F3-8. **(a)** Groundwater level and deformation over time, **(b)** groundwater level vs. deformation hysteresis loops.

dated strata in the Beijing plain area can be divided into three compressible layers. The deformation behaviors of soil mass at different depth are explored by using stress-strain curves, obtained from the subsidence and the piezometric head data.

2.1 Sand layer

The F3-8 extensometer reveals a confined aquifer composed of fine sand and medium sand. The data shows corresponding uplift, when piezometric heads increase (Fig. 2a). The stress-deformation loops reflecting the elastic deformation feature (Fig. 2b). The annual deformation is less than 0.5 mm.

2.2 The first compressible layer

The F3-10 extensometer reveals that the shallow strata are mainly composed of silt. The annual average groundwater level fluctuates slightly (Fig. 3a). This layer subsides continuously and the deformation rate slows with time. Figure 3b shows that the compression continues whether the changes in the annual groundwater level in the phreatic aquifer. This layer is dominated by silt and behaves plastically with time-dependent creep deformation. The stress-deformation loops after 2008, showing that the deformation is elastic.

2.3 The second compressible layer

The F1-3 extensometer collects deformation data from the second compressible layer which is composed of interbedded clay and sand layers. The cumulative thickness of the clay layers is similar to that of the sand layers. The variation in groundwater levels implies that the effective stress is continuously increasing (Fig. 4a). Figure 4b shows that the layer compress when the groundwater level declines, however, no uplift is observed in the compression record when water levels increase. This layer shows plastic deformation with greater residual subsidence, time-dependent creep deformation and a significant visco-elasto-platic behavior.

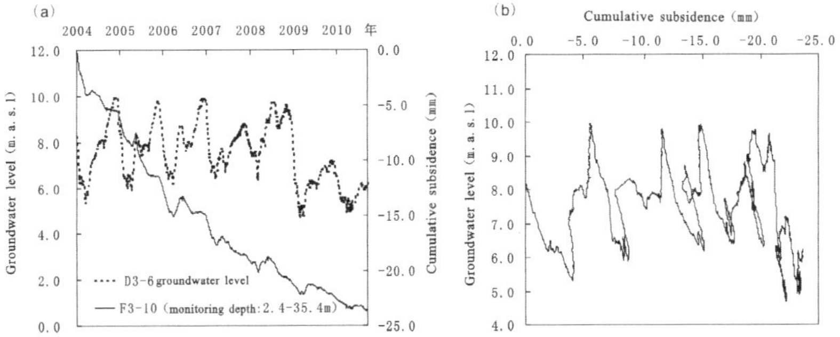

Figure 3. Deformation behavior of shallow soil mass vs. its corresponding groundwater level at layered mark F3-10. **(a)** Groundwater level and deformation over time, **(b)** groundwater level vs. deformation hysteresis loops.

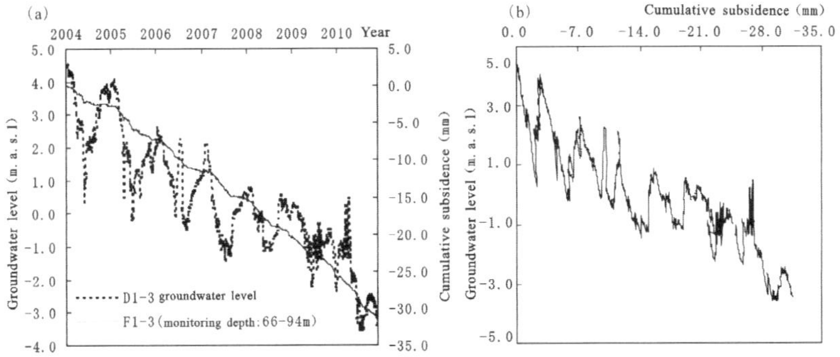

Figure 4. Deformation behavior of the mid-deep soil mass vs. its corresponding groundwater level at layered mark F1-3. **(a)** Groundwater level and deformation over time, **(b)** groundwater level vs. deformation hysteresis loops.

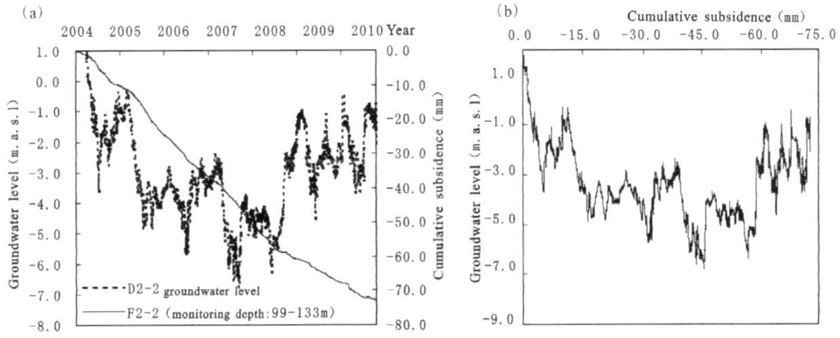

Figure 5. Deformation behavior of the deep soil mass vs. its corresponding groundwater level at layered mark F2-2. **(a)** Groundwater level and deformation over time, **(b)** groundwater level vs. deformation hysteresis loops.

2.4 The third compressible layer

The F2-2 extensometer monitors a thick clay layer in the upper part accounting for 75 % of the total thickness of this layer. The groundwater level declined from 2004 to August 2008 and then increased. Initially, the layer continuously compressed regardless of whether the groundwater level increased or declined, and subsequent to 2008 the compression rate decreased (Fig. 5a). The deformation is in the plastic range and creep deformation reveals an obvious lag effect (Fig. 5b). The compression shows obvious lag effect with the changes in groundwater level.

The F3-3 extensometer monitors the deformation of the layer composed of thick silts and clays with a cumulative thickness accounting for 80 % of this layer. The groundwater level of this layer continuously declines and behaves plasti-

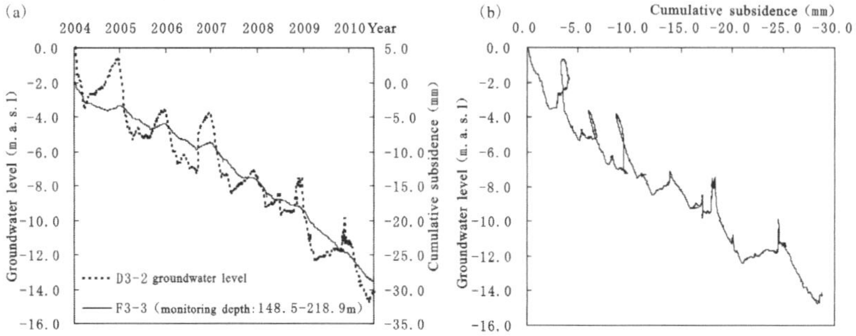

Figure 6. Deformation of the deep soil mass vs. its corresponding groundwater level at layered mark F3-3. **(a)** Groundwater level and deformation over time, **(b)** groundwater level vs. deformation hysteresis loops.

cally, while creep is largely elastic with visco-elasto-platic features (Fig. 6).

The above reveals the different deformation patterns associated with different groundwater levels at different depths. The deformation occurs mainly in the second compressible layer. In recent years, the discharge from the deep aquifers drastically increased, and the groundwater level has subsequently declined. The deformation of the third compressible layer accounts for the greater quantity of the total land subsidence in the region.

3 Conclusions

Five deformation patterns exist for different depths and groundwater change patterns in land subsidence region of Beijing. The sandy layer shows elastic deformation. The clay layers at different depths show elastic, plastic and creep deformations with visco-elasto-platic features. The deformation is controlled by the thickness of the compressible deposit, lithology and the physical and mechanical properties when the pattern of groundwater level change is provided. The deformation of the first compressible layer accounts for least amount of total land subsidence. Most of the deformation occurs in the second and third compressible layers.

Acknowledgements. This study was supported by the project of land subsidence monitoring operation in Beijing and the authors would also like to acknowledge colleagues for their contribution to this paper.

References

Burbey, T. J.: Stress-strain analyses for aquifer-system characterization, Ground Water, 39, 128–136, 2001.

Hu, R. L., Yue, Z. Q., Wang, L. C., and Wang, S. J.: Review on current status and challenging issues of land subsidence in China, Eng. Geol., 76, 65–77, 2004.

Tomás, R., Herrera, G., Delgado, J., Lopez-Sanchez, J. M., Mallorquí, J. J., and Mulas, J.: A ground subsidence study based on DInSAR data: Calibration of soil parameters and subsidence prediction in Murcia City (Spain), Eng. Geol., 111, 19–30, 2010.

Zhang, A. J., Ye, Ch., Li, Y., and Xie, Z. H.: Beijing Groundwater, China Land Press, Beijing, 2008 (in Chinese).

Zhang, Y., Xue, Y. Q., Ye, S. J., Wu, J. C., and Li, Q. F.: Analysis of deformation of sand strata and land subsidence based on modes of groundwater level changes in Shanghai City, The Chinese Journal of Geological Hazard and Control, 17, 103–108, 2006 (in Chinese with English abstract).

Proc. IAHS, 372, 53–57, 2015
proc-iahs.net/372/53/2015/
doi:10.5194/piahs-372-53-2015

Open Access
PIAHS

Groundwater management based on monitoring of land subsidence and groundwater levels in the Kanto Groundwater Basin, Central Japan

K. Furuno[1], **A. Kagawa**[2], **O. Kazaoka**[2], **T. Kusuda**[3], **and H. Nirei**[4]

[1]IUGS-GEM Japan branch, 6-41-8, Kotehashidai, Hanamigawa-ku, Chiba 262-0005, Japan
[2]Research Institute of Environmental Geology, Chiba, Japan
[3]Chiba Environmental Foundation, Chiba, Japan
[4]The Geo-pollution Control Agency, Chiba, Japan

Correspondence to: K. Furuno (kuniofurunojp@gmail.com)

Published: 12 November 2015

Abstract. Over 40 million people live on and exploit the groundwater resources of the Kanto Plain. The Plain encompasses metropolitan Tokyo and much of Chiba Prefecture. Useable groundwater extends to the base of the Kanto Plain, some 2500 to 3000 m below sea level. Much of the Kanto Plain surface is at sea level. By the early 1970s, with increasing urbanization and industrial expansion, local overdraft of groundwater resources caused major ground subsidence and damage to commercial and residential structures as well as to local and regional infrastructure. Parts of the lowlands around Tokyo subsided to 4.0 m below sea level; particularly affected were the suburbs of Funabashi and Gyotoku in western Chiba. In the southern Kanto Plain, regulations, mainly by local government and later by regional agencies, led to installation of about 500 monitoring wells and almost 5000 bench marks by the 1990's. Many of them are still working with new monitoring system. Long-term monitoring is important. The monitoring systems are costly, but the resulting data provide continuous measurement of the "health" of the Kanto Groundwater Basin, and thus permit sustainable use of the groundwater resource.

1 Introduction

The Kanto plain is also called "the Kanto fore-arc submarine basin" or "the paleo-Kanto-submarine basin", based on surrounding geologic and physiographic features such as Nasu Volcanic Zone, Fuji Volcanic Zone, Japan trench and Izu-Ogasawara Trench (Nirei et al., 1990). Over 40 million people live on the Kanto plain which includes the Tokyo metropolis. The waterfront area around Tokyo Bay mostly includes both alluvium and man-made strata such as reclaimed areas and is densely covered with houses and factories. Sediments of the Kanto basin are 2500–3000 m thick and are of Miocene to Holocene age. The basin is also called the Kanto groundwater basin from the standpoint of the production of groundwater, natural gas and iodine which are contained in fossil sea water. Uncontrolled use of groundwater has caused land subsidence. To ensure effective use of groundwater, a monitoring system for their management has been developed. The system has about 500 self-recording observation wells, almost 5000 bench marks (leveled every year). Over-pumping of the groundwater (fresh water and brine groundwater) resulted in a serious land subsidence problem in various areas, especially alluvial plains and reclaimed area so that it was inevitable to control the groundwater use to solve the problem (Aihara et al., 1969). Consequently, systematic observations of pumping up volumes, groundwater level changes and land subsidence records for the effective groundwater use have been carried out for preventing the land subsidence from a geological point of view (Research Committee on Land subsidence Prevention in Southern Kanto District, 1974).

Published by Copernicus Publications on behalf of the International Association of Hydrological Sciences.

2 Kanto groundwater basin and underground fluid resources

The Kanto Paleo-submarine basin can be called the Kanto groundwater basin from the standpoint of the distribution and the fluidity of the underground fluid resources (Nirei and Furuno, 1986). The bottom of the groundwater basin corresponds to the base of Kanto tectonic basin, which is situated at 3000–2500 m below the surface. The sediments of the Kazusa sub-groundwater basin in the basin contains brine groundwater (fossil sea water) including natural gas and iodine (Figs. 1, 2). The equivalent layer of the Kazusa Group that extended to the western and northern part of the Kanto plain is composed of alternating beds of coarse- to fine-grained deposits is suitable for groundwater. The lower part of the Shimosa sub-groundwater basin contains groundwater colored with humid material from the northern to central part of the Boso peninsula, and the colored groundwater is unsuitable for drinking. In the upper part of the Shimosa sub-groundwater basin, most of the aquifer layers contain fresh water; this basin had the largest pumpage volume of water in the Kanto groundwater basin. However, the more pumping up volume increased, the lower the groundwater level became in the Shimosa upper sub-groundwater basin. Which is also resulted in lowering of groundwater level even in the alluvial deposits as well as subsiding of ground surface. It is now widely confirmed that the amount of ground subsidence is affected by the thickness of alluvial deposits (Research Institute of Environmental Geology, 1979).

3 Status of underground fluid resource use

Natural gas and groundwater have been produced as the underground fluid resources in the Kanto groundwater basin. The former is mainly derived from the Kazusa sub-groundwater basin. Annual production of natural gas was 520 million Nm^3 in 1980, 460 million Nm^3 in 2003 and 440 million Nm^3 in 2011. The natural gas produced from the Kazusa sub-groundwater basin was contained in fossil sea water of the Kazusa group. Accordingly, it was inevitable to pump up the fossil sea water when releasing natural gas, which frequently caused a land subsidence problem. Therefore, the pumping is now limited to the area of hills and the Pacific coast in the southern Kanto, a considerable distance from the metropolitan area. Recently, the groundwater is mainly pumped up at the marginal area of the Shimosa lower sub-groundwater basin and the upper Shimosa sub-groundwater basin. Annual pumped volume was 3.4 million $m^3\,day^{-1}$ in 1978 and 3.1 million $m^3\,day^{-1}$ in 2003 (Prefectural Governors' Committee for Land subsidence Prevention in Kanto District, 1983, 2008) and 2.8 million $m^3\,day^{-1}$ in 2015 (Ministry of Environment, 2015). These data were obtained in the limited area where reporting of pumped volume is required of groundwater users by regulation for preventing land subsi-

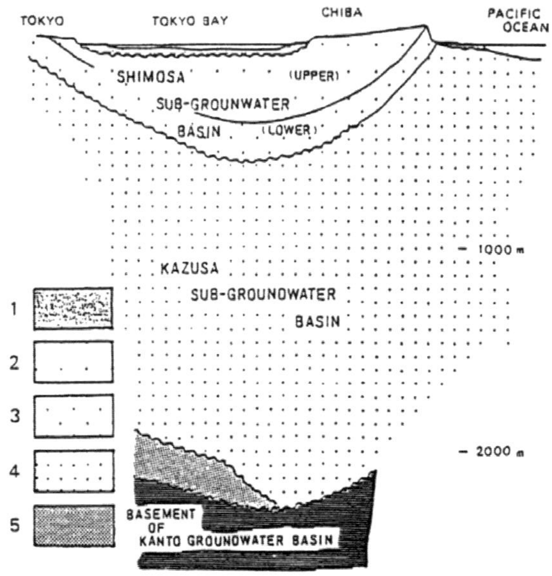

Figure 1. Schematic section of the Kanto groundwater basin: 1: alluvial deposits, 2: upper part of the Shimosa Group, 3: lower part of the Shimosa Group, 4: Kazusa Group, 5: Miura Group.

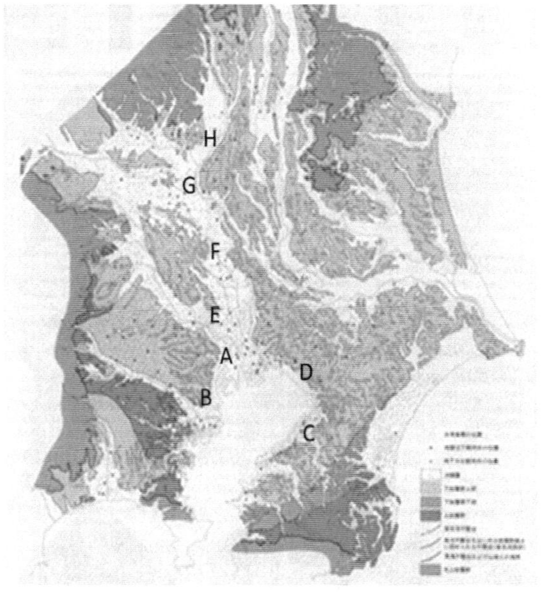

Figure 2. Kanto groundwater basin and the distribution of monitoring wells and benchmarks. Large dots: monitoring wells, small dots: benchmarks. A–E: monitoring wells which data were used in Fig. 3.

dence. This area covers about half of the Kanto groundwater basin. Consequently, the actual pumping volume is estimated to have been twice as much as the values mentioned above. The groundwater has been used for (1) drinking water supply, (2) industries, (3) agriculture (4) buildings and (5) other uses. The used volume increases generally according as drinking water supply, industries and so on increase in scale. However the used volume shows a little difference based on each autonomy on the Kanto groundwater basin and the scale of the volume used by the agriculture and the building is different at each prefectural government. The volume of other uses is the smallest amount of the volume which is the fifth order. The fact that the drinking water supply make up the largest amount of the ground water use may be the consequence of the concept that necessary of groundwater use is for the benefit of inhabitants (Shibasaki, 1976; Research Group for Water Balance, 1976).

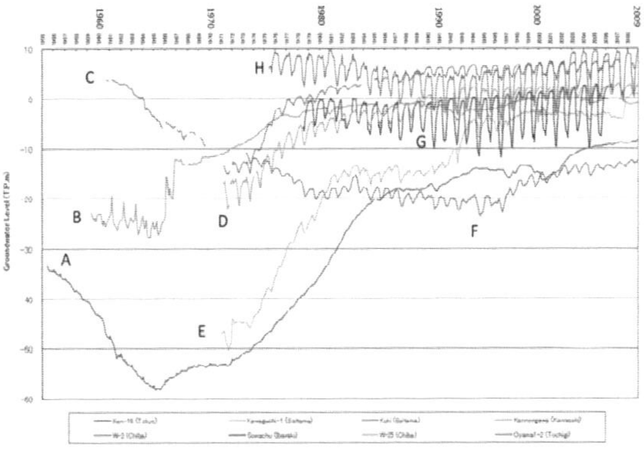

Figure 3. Historical change of the groundwater level in the Kanto groundwater basin. A: Ken-15 (Tokyo), B: Kannongawa (Kawasaki), C: W-2 (Chiba), D: W-25 (Chiba), E: Kawaguchi-1 (Saitama), F: Kuki (Saitama), G: Sowachu (Ibaraki), H: Oyama1–2 (Tochigi).

4 Control of pumping volume and change of ground level and land subsidence

Groundwater and brine groundwater were pumped up disorderly by companies, farmers and local governments for industry, agriculture, drinking water supply and production of natural gas and iodine before 1960s. However, the regulation by national government began in the middle 1960s and local governments began from the 1970s as progress of the subsidence became remarkable in order to prevent land subsidence.

Each local government has attempted to regulate the pumping volume of the groundwater in order to prevent land subsidence from 1970s. For example, users are required to report the pumping volume compulsorily. Criteria of depth and diameter of discharge pipes were set up (Prefectural Governors' Committee for Land subsidence Prevention in Kanto District, 1983). An agreement to control the pumping of fossil sea water containing natural gas was also made between the Chiba prefecture administration and each gas production company. The condition of groundwater levels in the whole Kanto groundwater basin has recently been clarified. The number of observation wells was 375 in total as of 1982 (Furuno et al., 1983), 459 as of 1991 and about 500 as of 2010s. The amount of land subsidence has been measured by yearly precise leveling. The number of benchmarks attained 4880 and the total length of the leveling, 7363.6 km as of 1983, 4945 benchmarks and 7387.6 km as of 1986, about 5000 benchmarks and 7500 km as of 2010s, The groundwater management in the shimosa sub-groundwater basin is described as below.

5 Control of groundwater over-pumping

In the Keihin industrial zone (coastal zone of Tokyo, Yokohama and Kawasaki) located in the south western part of the Kanto groundwater basin, the land subsidence caused by over-pumping of groundwater were recognized in early 1940s. As a result, industrial aqueducts were established to replace the groundwater pumping from the coastal zone of Kawasaki and Yokohama. In the industrial zone, the pumping was controlled in the 1950s by regulations on the depth of wells and the diameter of discharge pipes. In the industrial zone of Tokyo and its suburbs, the pumping volume was controlled in the 1960s by two laws concerning industrial water and pumping at building sites. However, the countermeasures for land subsidence were often loosely applied by groundwater users, especially the industrial users. In the 1970s, with the rapid growth of the Japanese economy, land subsidence have been recognized as the most important problems. In such a social background, not only the national government enacted the strict criteria for the regulation concerning the industrial water and the pumping up law for building use, but also the prefectural governments in the southern part of the Kanto groundwater basin strictly applied these regulations. As a result, the groundwater level and the land subsidence stopped gradually. It was even recognized in certain regions that the groundwater level rose. However, excessive regulations produce another problem in some areas. Tokyo Station and Ueno Station in Tokyo set up deep underground platforms before early 1960s when groundwater level was low. Groundwater level rose from late 1960s and buoyancy acted on the building. Those stations had to set up anchors and weights as a countermeasure of buildings rising by buoyancy. Rising of groundwater level in man-made strata and Holocene deposit increase the risk of liquefaction-fluidization. We need to keep proper groundwater level.

6 Change of groundwater level

It was in the early 1960s that the groundwater level fell to 60 m below sea level in the coastal industrial zone of Tokyo metropolis. After that, the recovery of the groundwater level was monitored according to the pumping up control. In the 1970s the regulations were strictly adhered to in the southern Kanto groundwater basin. Accordingly, in the years from 1975 to 1980, the groundwater level recovered 30–40 m in the area where groundwater level lowered. Thus the groundwater level showed a recovering tendency in the lowest groundwater level area. The area where groundwater level dropped 20–30 m below sea level has expanded. In other words, a center of groundwater depression moved to the northern part because the pumping control in the southern Kanto groundwater basin is applied strictly. Figure 3 shows historical groundwater change of the monitoring wells and Fig. 4 shows groundwater level of Kanto groundwater basin (Shimosa sub-groundwater) in 2011 based on the monitoring wells. As Fig. 4 shows, groundwater level is not so low as the former 60 but 20 m below sea level. The lowest groundwater level is stable around 10–20 m below sea level recently. We should think about the use of appropriate groundwater based on monitoring. Drinking water should have top priority.

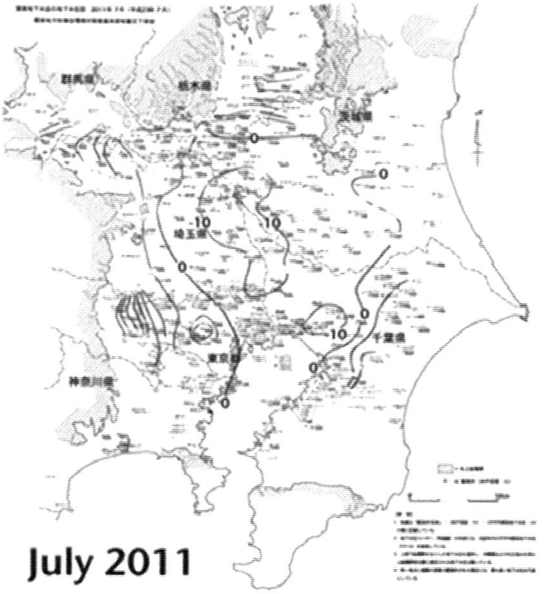

Figure 4. Confined groundwater level in the Kanto groundwater basin (especially the Shiosa sub-groundwater basin). Lowest contour line shows 10 m below sea level.

7 Change of land subsidence

The land surface has been sinking to below sea level in the downtown Tokyo area for mixed industry and residential uses (Shitamachi) since 1910s. Coincidentally, the areas showing ground level lower than mean sea level and high water level of ordinary spring tide of Tokyo Bay were widely scattered. Recently, however, these phenomena have been recognized in alluvial area not only in Shitamachi of Tokyo, but also in suburbs of Tokyo such as Kawasaki and Yokohama in Kanagawa Prefecture, Funabashi, Gyotoku and Urayasu on western Chiba Prefecture. This land subsidence has been stabilized by the regulation of the pumping. The land subsidence was stopped in certain areas and the ground surface slightly uplifted by the recovery of the groundwater level. On the other hand, in the area from central to northern part of the Kanto groundwater basin, land subsidence has begun to sink as a result of over-pumping. That is, land subsidence area is moving northward. Land subsidence before 1970s spread lowland area around Tokyo Bay and Kujyukuri in Boso Peninsula which faced the Pacific Ocean (Fig. 5).

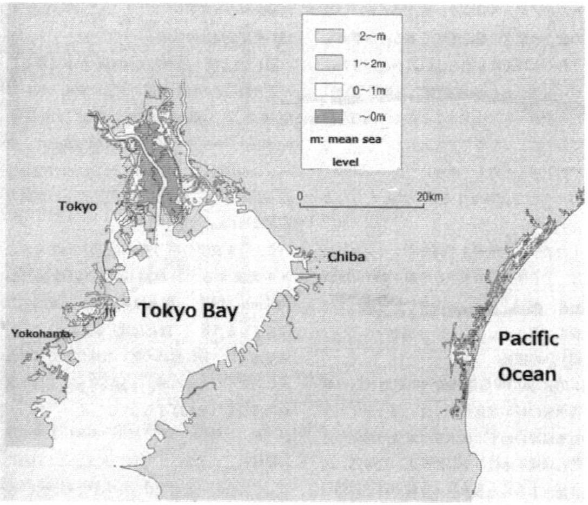

Figure 5. Lowland area around Tokyo Bay and Kujyukuri which faced the Pacific Ocean.

8 Conclusion

It is necessary to establish a more effective monitoring system for continuous observation of the groundwater pumping, groundwater level and land subsidence, and to decide the pumping volume in consideration of the change of the groundwater level and the ground movement with the law of dynamic equilibrium between man and nature due to the relation among human groups. In other words, the groundwater resources can be used while the health condition of the groundwater basin is examined by periodical health checks. The long-term monitoring of land subsidence and groundwater is important for the sustainable use of the groundwater in the groundwater basin.

Acknowledgements. We sincerely appreciate members of Research Institute of Environmental Geology, Chiba and members of the local governments of Kanto district who have developed a monitoring system and kept on recording land subsidence and groundwater level.

References

Aihara, S., Ugata, H., Miyazawa, K., and Tanaka, U.: Problems on groundwater control in Japan, in: Land subsidence, IAHS-UNESCO, 2, 635–644, 1969.

Furuno, K., Nemoto, K., Takanashi, Y., Yada, T., Takizawa, H., Oshima, K., Saotome, W., and Nirei, H.: The Monitoring of land subsidence and groundwater head in the Kanto groundwater basin, Bulletin of Chiba Prefectural Research Institute for Environmental Pollution, 15, 99–108, 1983.

Ministry of Environment: Review of land subsidence, available at: http://www.env.go.jp/water/jiban/chinka.html (last access: 20 September 2015), 2015 (in Japanese).

Nirei, H. and Furuno, K.: Development of Quaternary Resources and Environmental Protection: Status of Underground fluid resources use in the Kanto groundwater basin, in: Recent Progress of Quaternary Research in Japan, edited by: Kaizuka, S., Aso, M., Endo, K., et al., National Committee for Quaternary Research in Japan, Tokyo, 11, 71–80, 1986.

Nirei, H., Kusuda, T., Suzuki, K., Kamura, K., Furuno, K., Hara, Y., Satoh, K., and Kazaoka, O.: The 1987 East Off Chiba Prefecture Earthquake and It's Hazards, Memoirs of the Geological Society of Japan, 35, 31–46, 1990.

Prefectural Governors' Committee for land subsidence Prevention Kanto District: Report of groundwater volume pumping up, 228 pp., 1983 (in Japanese).

Prefectural Governors' Committee for land subsidence Prevention Kanto District: Report of groundwater volume pumping up, 124 pp., 2008 (in Japanese).

Research Committee on Land subsidence in Southern Kanto District: Report of research on land subsidence in Southern Kanto District, 297 pp., 1974 (in Japanese).

Research Group for Water Balance: Groundwater basin management, Tokai University Press, Tokyo, 242 pp., 1976 (in Japanese).

Research Institute of Environmental Geology: The Data Analysis on the Urayasu Land Subsidence Observation Well System, Research Institute of Environmental Geology, Chiba, Project Research Report No. 4, 86 pp., 1979. (in Japanese)

Shibasaki, T.: Carrying capacity of groundwater development and conservation, Groundwater basin management, Tokai University Press, Tokyo, 47–75, 1976 (in Japanese).

Proc. IAHS, 372, 59–62, 2015
proc-iahs.net/372/59/2015/
doi:10.5194/piahs-372-59-2015

Subsidence characterization and modeling for engineered facilities in Arizona, USA

M. L. Rucker, K. C. Fergason, and B. B. Panda

Amec Foster Wheeler, Phoenix, Arizona, USA

Correspondence to: M. L. Rucker (michael.rucker@amecfw.com)

Published: 12 November 2015

Abstract. Several engineered facilities located on deep alluvial basins in southern Arizona, including flood retention structures (FRS) and a coal ash disposal facility, have been impacted by up to as much as 1.8 m of differential land subsidence and associated earth fissuring. Compressible basin alluvium depths are as deep as about 300 m, and historic groundwater level declines due to pumping range from 60 to more than 100 m at these facilities. Addressing earth fissure-inducing ground strain has required alluvium modulus characterization to support finite element modeling. The authors have developed Percolation Theory-based methodologies to use effective stress and generalized geo-material types to estimate alluvium modulus as a function of alluvium lithology, depth and groundwater level. Alluvial material modulus behavior may be characterized as high modulus gravel-dominated, low modulus sand-dominated, or very low modulus fines-dominated (silts and clays) alluvium. Applied at specific aquifer stress points, such as significant pumping wells, this parameter characterization and quantification facilitates subsidence magnitude modeling at its' sources. Modeled subsidence is then propagated over time across the basin from the source(s) using a time delay exponential decay function similar to the soil mechanics consolidation coefficient, only applied laterally. This approach has expanded subsidence modeling capabilities on scales of engineered facilities of less than 2 to more than 15 km.

1 Introduction

Subsidence has been analyzed at engineered facilities in Arizona for over two decades (Keaton et al., 1998; Rucker et al., 2008). Prior to developing the current Percolation Theory (PT)-based techniques, finite element model parameter studies were used to best-fit modeled subsidence profiles to available InSAR-, historic survey-, and survey-derived elevation records versus design elevation subsidence records. Alluvium material classification included well logs (ranging from commonly poor-quality historical qualitative descriptions to high-quality quantitative borehole geophysical logging) and surface geophysical measurements (especially resistivity) to supplement published data (AMEC, 2009).

2 Modeling alluvium as percolating systems

Basin alluvium can be modeled as networks of interconnected but highly dissimilar (heterogeneous) materials, such as interfingered lenses of finer alluvium (behavior dominated by silts and clays) and coarser alluvium (behavior dominated by sands and gravels). Connectivity of alluvium aquifer systems consisting of fines-dominated, sand-dominated or gravel-dominated units can be quantified through concepts of PT. For characterizing subsidence, three alluvium parameters are addressed using PT.

- Modulus is a function of alluvium density (porosity) as controlled by effective stress or original depositional energy. Fines-dominated, sand-dominated and gravel-dominated alluvium units tend to have very low, low and high modulus, respectively.
- Resistivity relates to the connectivity of the fines (silts and clays) portion of the alluvium and indicates alluvium modulus behavior and permeability characteristics. In fresh water aquifer conditions, fines-dominated, sand-dominated, and gravel dominated alluvium units typically have resistivities of less than 10 ohm-meters, 10 to about 30 ohm-meters and greater than 30 ohm-meters, respectively.

Published by Copernicus Publications on behalf of the International Association of Hydrological Sciences.

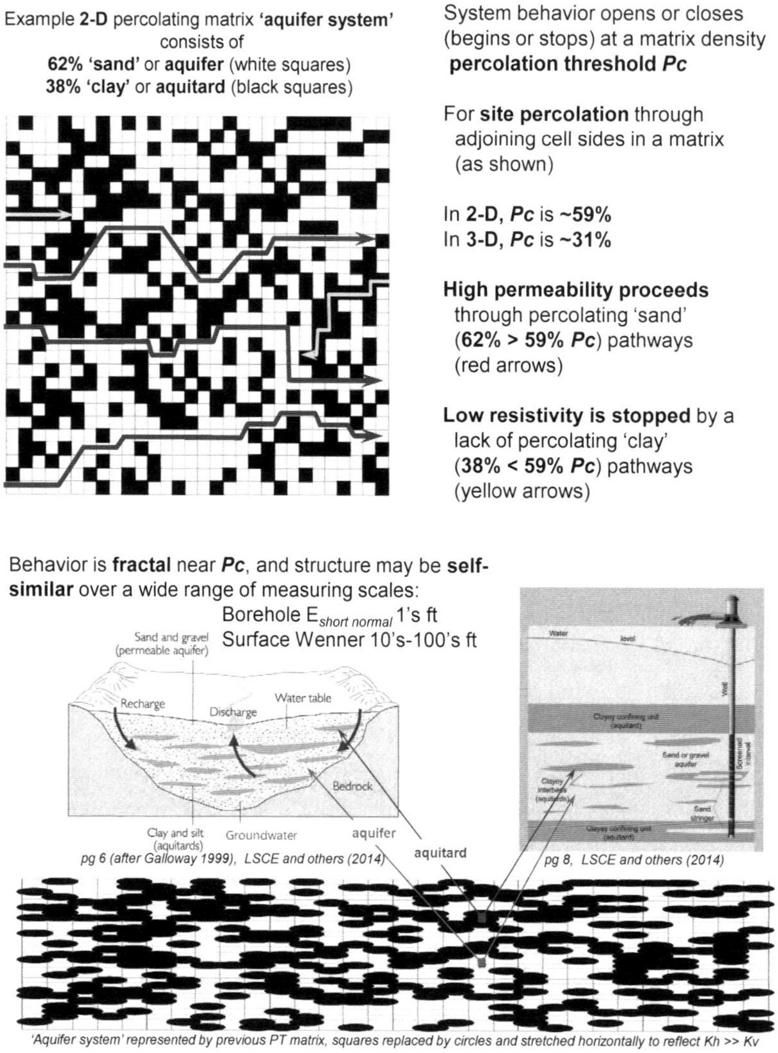

Figure 1. Basic PT (Sukop et al., 2002) site-percolation model (Fig. 1a top) is modified to address basin alluvium connectivity and behaviour (Fig. 1b bottom).

- Permeability relates to the connectivity of the coarse (sands and gravels) portion of the alluvium to indicate time delay and potential presence of a "water of compaction" confined aquifer condition. Fines-dominated, sand-dominated and gravel-dominated alluvium units typically have very low permeability (with potential water of compaction conditions), moderate permeability and high permeability, respectively.

Concepts of permeability and resistivity connectivity in an alluvial PT system (exhibiting significant randomness) are visualized in Fig. 1. Connectedness in the random matrix occurs as like squares that share a side "connect" into a continuous pathway. When a continuous pathway of connected squares across the matrix is present, then the matrix is a "percolating network". The percolation threshold (Pc) is the density of the matrix (probability in a random system) at which percolation begins. Percolating network permeable flow (from left to right), and lack of a percolating network low resistivity pathway are demonstrated in the 2-D matrix example (Fig. 1a). The same fraction of clay (38 % in the 2-D example) in a 3-D network (Pc of 31 %) would exhibit network connectivity, and the 3-D matrix would have a moderately low resistivity.

These concepts are applied to characterize heterogeneous basin alluvium (in 3-D) as shown in the 2-D representation

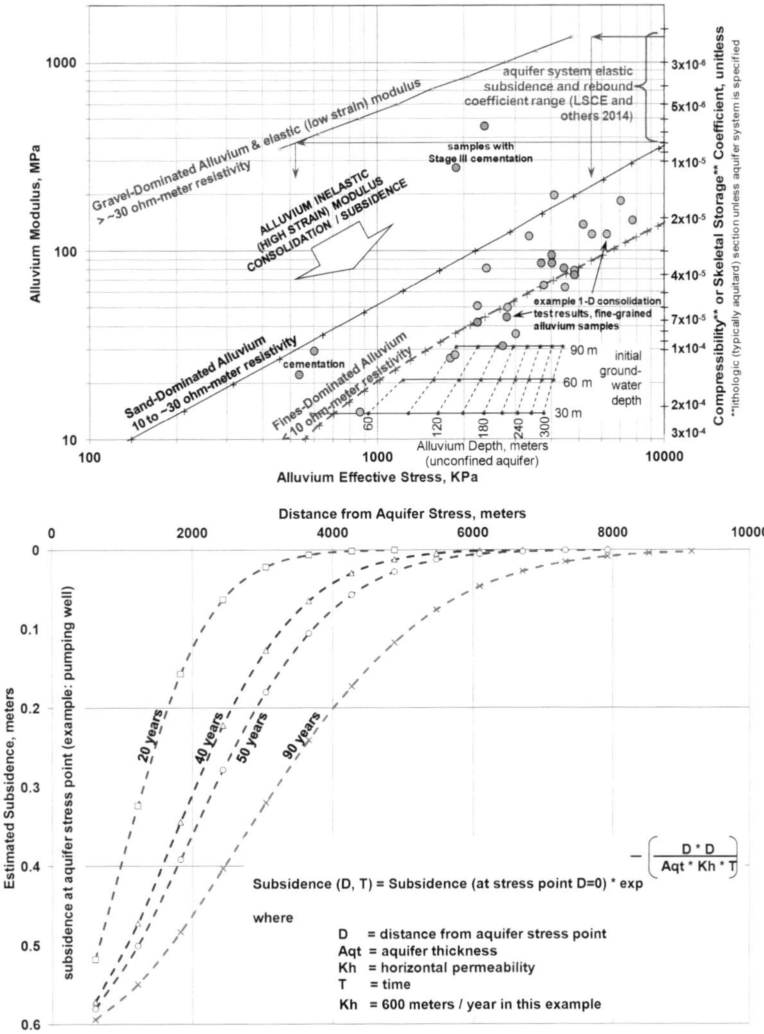

Figure 2. Subsidence modulus parameters are related to compressibility over a range of effective stress in Fig. 2a (top). Example test results are from deep wells in the east Salt River Valley, Arizona. Lateral subsidence propagation over time is shown in Fig. 2b (bottom).

in Fig. 1b. Fines-dominated alluvium and sands-dominated alluvium are most likely to contribute to significant subsidence. Sand-dominated alluvium contains significant quantities (greater than about 31 % to less than about 69 %) of fines in a connected permeable sand network to provide drainage paths for those fines; sand-dominated alluvium exhibits high subsidence due to the network of moderate to high permeability material surface areas draining adjacent fine-grained materials. Gravel-dominated alluvium is likely to contribute only minor subsidence. Boundaries between these alluvium lithologies may strongly correlate with 3-D Pc thresholds defining the edges of fines-dominated to sand-dominated and sand-dominated to gravel-dominated modulus behaviors.

3 Subsidence estimation using PT-based modulus concepts

The objective of this PT-based subsidence modeling process is estimation of the modulus profile through the compressible alluvium. For materials consisting of a matrix of particles and porosity, PT provides means to define modulus as a threshold and power-law relationship to porosity (Sahimi, 1994), where modulus increases exponentially as porosity decreases. A PT approach to estimate elastic (low strain) modulus as a function of porosity (or density) in geomaterials is presented in Rucker (2008). A relationship of inelastic (high strain) modulus relative to elastic modulus is modeled as a power-law (Rucker, 2008) anchored at very low modulus by sand (Seed and Idriss, 1970) and at very high modulus by rock (van Heerden, 1987). Finally, an approach to relate inelastic modulus to effective stress is presented in AMEC (2009). Starting with a material at low density (high porosity), a modeled incremental reduction of porosity becomes a modeled strain. From that strain, and an average of the modulus at initial and incremented porosity, a stress may be calculated. This process is incremented over a range of decreasing porosity, so that density and modulus increase as a function of the accumulated applied stress. Resulting estimated initial consolidation high strain modulus trends for alluvium as a function of effective stress, and thus depth and groundwater levels, are presented in Fig. 2a. Lateral propagation of subsidence over time, where horizontal permeability is the primary alluvium material parameter, is demonstrated in Fig. 2b.

4 Comparison of USGS and PT approaches to subsidence characterization

The PT approach appears to be consistent with USGS concepts (Helm, 1975) since compaction parameters (compressibility or skeletal storage coefficients, typically determined using compaction extensometers) are an inverse relation to PT-based modulus as shown in Fig. 2a. An example of this consistency is the range of alluvium elastic modulus and alluvium effective stress compared with the range of aquifer system elastic subsidence and rebound coefficient (purple arrows in Fig. 2a) as reported by LSCE, Borchers and Carpenter (2014). Individual laboratory-scale, relatively undisturbed soil samples (fine-grained sediments) obtained for subsidence studies (circle symbols in Fig. 2a), tend to behave as and relate best to the fines-dominated modulus trend. The fines-dominated modulus trend is about half the modulus of the sand-dominated modulus trend; an alluvium matrix including a mix of inelastic fines material and elastic coarse material (Fig. 1b) reflects an overall sand-dominated modulus utilizing the PT approach.

5 Results and conclusions

Reasonable estimates of basin alluvium modulus facilitate calculation of subsidence. Modeled results have had typical estimate errors of about 10 % with standard deviations of about 30 to 40 % compared to available project historic survey point data. Once historical subsidence has been modeled, rational future predictions and projections have been made to inform long-term design criteria for facility rehabilitation.

Acknowledgements. The strong support and encouragement of the Flood Control District of Maricopa County, Arizona and the Arizona Electric Power Cooperative, Inc., was essential in the development and application of the new concepts and methods described in this paper.

References

AMEC E and E, Inc.: White Paper, in Procedural Documents for Land Subsidence and Earth Fissure Appraisals, for Flood Control District of Maricopa County, Phoenix, AZ, available at: ftp://ftp.mcdot.maricopa.gov/anonymous/FCD (last access: 22 October 2015), 2009.

Galloway, D. L., Jones, D. R., and Ingebritsen, S. E.: Land subsidence in the United States, US Geological Survey Circular, Reston, Virginia, 1182, 177 pp., 1999.

Helm, D. C.: One-dimensional simulation of aquifer system compaction near Pixley, California – 1. Constant parameters, Water Resour. Res., 11, 465–478, 1975.

Keaton, J. R., Rucker, M. L., and Cheng, S.: Geomechanical Analysis of an Earth Fissure Induced by Ground-Water Withdrawal for Design of a Proposed Ash and Sludge Impoundment, Southeastern Arizona, Land Subsidence Case Studies and Current Research: Proceedings of the Joseph F. Poland Symposium on Land Subsidence, 4–5 October 1995, Spec. Publ. No. 8, AEG, Star Publishing Co., Belmont/CA, 217–226, 1998.

LSCE: Luhdorff & Scalmanini Consulting Engineers, Borchers, J. W. and Carpenter, M.: Luhdorff and Scalmanini Consulting Engineers (LSCE), Land Subsidence from Groundwater Use in California, Full Report of Findings, California Water Foundation, 9–10, April 2014.

Rucker, M. L.: Estimating in-situ geo-material mass density, modulus and unconfined compressive strength from field seismic velocity measurements. Highway Geophysics – Nondestructive Evaluation Conference, Charlotte, NC, 1–4 December, available at: ftp://ftp.mcdot.maricopa.gov/anonymous/FCD (last access: 22 October 2015), 2008.

Rucker, M. L., Greenslade, M. D., Weeks, R. E., Fergason, K. C., and Panda, B. B.: Geophysical and Remote Sensing Characterization to Mitigate McMicken Dam, in: GeoCongress 2008: Geosustainability and Geohazard Mitigation (GSP 178), edited by: Reddy, K. R., Khire, M. V., and Alshawabkeh, A. N., 207–214, American Society of Civil Engineers, doi:10.1061/40971(310)26, 2008.

Sahimi, M.: Applications of Percolation Theory, Taylor & Francis, Bristol, Pennsylvania, 181–186, 1994.

Seed, H. B. and Idriss, I. M.: Soil Moduli and Damping Factors for Dynamic Response Analysis, Report No. EERC 70–10, Earthquake Engineering Research Center, University of California, Berkeley, California, p. 19, 1970.

Sukop, M. C., Van Duk, G., Perfect, E., and Van Loon, W. K. P.: Percolation thresholds in 2-dimensional prefractal models of porous media, Transport in Porous Media, Vol. 48, 187–202, Kluver Academic Press, the Netherlands, 2002.

van Heerden, W. L.: General Relations Between Static and Dynamic Moduli of Rocks, Int. J. Rock Mech. & Geomech. Abstr., 24, 381–385, 1987.

Proc. IAHS, 372, 63–68, 2015
proc-iahs.net/372/63/2015/
doi:10.5194/piahs-372-63-2015

Modelling ground rupture due to groundwater withdrawal: applications to test cases in China and Mexico

A. Franceschini[1], P. Teatini[1], C. Janna[1], M. Ferronato[1], G. Gambolati[1], S. Ye[2], and D. Carreón-Freyre[3]

[1]Department ICEA, University of Padova, Padova, Italy
[2]School of Earth Sciences and Engineering, Nanjing University, Nanjing, China
[3]Laboratorio de Mecanica de Geosistemas, Mexican National University, Queretaro, Mexico

Correspondence to: A. Franceschini (franc90@dmsa.unipd.it)

Published: 12 November 2015

Abstract. The stress variation induced by aquifer overdraft in sedimentary basins with shallow bedrock may cause rupture in the form of pre-existing fault activation or earth fissure generation. The process is causing major detrimental effects on a many areas in China and Mexico. Ruptures yield discontinuity in both displacement and stress field that classic continuous finite element (FE) models cannot address. Interface finite elements (IE), typically used in contact mechanics, may be of great help and are implemented herein to simulate the fault geomechanical behaviour. Two main approaches, i.e. Penalty and Lagrangian, are developed to enforce the contact condition on the element interface. The incorporation of IE incorporation into a three-dimensional (3-D) FE geomechanical simulator shows that the Lagrangian approach is numerically more robust and stable than the Penalty, thus providing more reliable solutions. Furthermore, the use of a Newton-Raphson scheme to deal with the non-linear elasto-plastic fault behaviour allows for quadratic convergence. The FE – IE model is applied to investigate the likely ground rupture in realistic 3-D geologic settings. The case studies are representative of the City of Wuxi in the Jiangsu Province (China), and of the City of Queretaro, Mexico, where significant land subsidence has been accompanied by the generation of several earth fissures jeopardizing the stability and integrity of the overland structures and infrastructure.

1 Introduction

The exploitation of subsurface resources, including freshwater aquifers, involves various environmental problems. One of these is land subsidence. Recently, attention has been directed to the activation of pre-existing regional faults, as well as the possible generation of new fractures, caused by groundwater withdrawals. Indeed, the activation of faults can trigger or induce a seismic activity (González et al., 2012). Another problem related to the fault activation is the creation of preferential pathways for fluid leakage. This problem can be also relevant in the case of storage of hydrocarbons or carbon dioxide in the subsoil. Activation of faults and the generation of fissures have a direct effect on the land surface and may cause unacceptable damage to engineered structures and disruption to infrastructures. This is especially problematic in densely populated areas, such as in China (Wang et al., 2007), Mexico (Carreón-Freyre, 2010; Carreón-Freyre and Cerca, 2006), as well as in many semiarid sedimentary basins worldwide (e.g., Azat and Shahram, 2010). A different but somewhat related problem is the hydraulic fracturing or so-called "fracking" technique, which involves the intentional generation of underground fractures in the rock by injecting fluids at high pressure to create preferential pathways that facilitate the extraction of hydrocarbons. A reliable mathematical model to predict the consequences of subsurface fluid withdrawal must take into account the presence of discontinuities, if any, in the porous medium.

At present, the existing numerical-mathematical models to simulate the discontinuities in the porous medium are based on two techniques: (i) use of a continuous model where the fault is characterized by physical properties different from those of the hosting medium (Rutqvist et al., 2007); (ii) introduction of a discontinuity surface, which behaves as an internal boundary and can be present or absent depending on the stress state. At a numerical level, in the context of the Finite Element Method, the first approach implies a finer discretization of the neighbourhood of the fault, substantially to simulate a layer of different material, while the second requires the introduction of Interface Finite Elements. The first method is simpler, as it involves no real discontinuities, but the difficulties and limitations rest on the proper characterization of a continuous medium that simulates a discontinuity. With Interface Finite Elements (Goodman et al., 1968; Ferronato et al., 2008), there are two approaches to prescribe the necessary mechanical-geometric constraint conditions, that are the non-penetration of solid bodies and the stress state consistent with some failure criterion, e.g., Mohr-Coulomb (Labuz et al., 2012). The first method, referred to as penalty (Bathe, 2006), consists of the introduction of very stiff springs between the faces of the rupture, which impose the condition of non-penetration until they break from overcoming a stress limit. From a mathematical point of view, this method is not exact, because the elastic springs have displacements for any non-zero stress value, while, in nature, in case of a closed fracture the relative displacements are exactly zero. Furthermore, numerically, this method causes a strong ill-conditioning of the stiffness matrix (Ferronato et al., 2012), because of the introduction of the penalty coefficients. However, mainly for the ease of implementation, this approach is widely used. Alternatively the constraint conditions can be imposed by using Lagrange multipliers (Bathe, 2006; Simo et al., 2000), namely in an analytically exact way. The Lagrange multipliers, which physically represent the contact stresses, are additional unknowns, so that the problem is numerically more complex. The increase of the computational cost however, is offset by a more robust convergence in all non-linear steps and a more stable numerical behaviour.

With a Lagrange approach, it is of paramount importance, in case of a rupture sliding, to account for the relative orientation of the shear stress vector with respect to the relative displacement. As the Mohr-Coulomb failure criterion is planar, there is no information on three-dimensionality. A way to compute such an orientation is based on an application of the criterion of "Maximum Plastic Dissipation" (Wriggers, 2006), which uniquely provides the direction of the shear stress as a function of the relative displacement so as to maximize the friction work. By implementing this criterion in the derivation of the variational formulation, we get a numerical scheme that converges quadratically.

2 Numerical modelling of fault mechanics

A ground rupture is a discontinuity in a porous medium, whose behaviour is governed by the combination of stress and geometry of the close surroundings. To model such behaviour, we choose a failure criterion that prescribes the rupture activation. The shear stress limit τ_{L} is given by (Labuz et al., 2012):

$$\tau_{\mathrm{L}} = c - \sigma_{\mathrm{n}} \tan \phi \tag{1}$$

where c is the cohesion and σ_{n} the normal stress acting on the fault, taken to be negative in compression. Equation (1) defines the modulus but not the direction of the limit shear stress. According to the criterion of "Maximum Plastic Dissipation", the maximization of the friction work $W_{\mathrm{f}} = \boldsymbol{\tau}_{\mathrm{L}}^T \boldsymbol{u}_{\mathrm{r}}$, with $\boldsymbol{u}_{\mathrm{r}}$ the relative tangential sliding and $\boldsymbol{\tau}_{\mathrm{L}}$ the limit shear stress, yields:

$$\boldsymbol{\tau}_{\mathrm{L}} = \tau_{\mathrm{L}} \frac{\boldsymbol{u}_{\mathrm{r}}}{\|\boldsymbol{u}_{\mathrm{r}}\|_2} \tag{2}$$

Notice that boldface characters denote vectors. In case of a fault opening, i.e. $\sigma_{\mathrm{n}} > 0$, the two sides of the rupture become completely free to move with no loading.

To attain equilibrium, the minimization of the functional representing the total potential energy produces the principle of virtual work:

$$\delta \Pi = \delta W_{\mathrm{i}} - \delta W_{\mathrm{e}} \tag{3}$$

where the internal virtual work reads $\delta W_{\mathrm{i}} = \int_{\Omega} \delta \boldsymbol{\varepsilon}^T \hat{\boldsymbol{\sigma}} \mathrm{d}V$, with $\hat{\boldsymbol{\sigma}}$ the total stress, and the external virtual work is $\delta W_{\mathrm{e}} = \int_{\Omega} \delta \boldsymbol{u}^T \boldsymbol{b} \mathrm{d}V + \int_{\delta\Omega} \delta \boldsymbol{u}^T \boldsymbol{t} \mathrm{d}S$, with $\boldsymbol{b}$ and $\boldsymbol{t}$ the volume and surface forces, respectively. The contribution provided by the work of the discontinuity must be added to Eq. (3). The total rupture area is denoted by Γ, while the sliding portion, where Eqs. (1) and (2) hold, is indicated with $\bar{\Gamma}$. If opened, the rupture is an unloaded boundary with no contributions to the virtual work. Otherwise, the virtual work δW_{f} reads:

$$\delta W_{\mathrm{f}} = \int_{\Gamma \setminus \bar{\Gamma}} \delta \boldsymbol{u}^T \boldsymbol{\lambda} \mathrm{d}S + \int_{\Gamma \setminus \bar{\Gamma}} \delta \boldsymbol{\lambda}^T \boldsymbol{u} \mathrm{d}S + \int_{\bar{\Gamma}} \delta \boldsymbol{u}_{\mathrm{r}}^T \boldsymbol{\tau}_{\mathrm{L}} \mathrm{d}S \tag{4}$$

As long as the discontinuity is close, both stresses and displacements are independent variables and can have virtual

Table 1. Geomechanical properties used for the two test cases. E: Young modulus; ν: Poisson's ratio; e: void ratio.

City	Material	E (MPa)	ν	e
Queretaro City	Sand	15	0.25	0.78
	Clay	15	0.25	0.78
	Rock	4900	0.25	0.78
Wuxi	Sand	12	0.30	0.77
	Clay	8	0.30	0.77
	Rock	1200	0.30	0.77

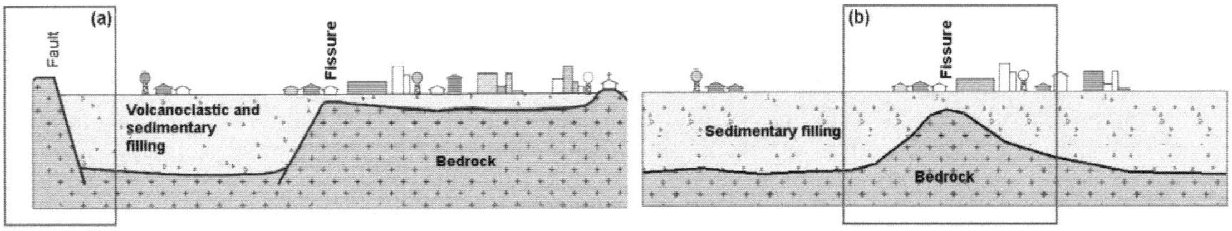

Figure 1. Sketches of the typical geological setting ground rupture in Queretaro City (**a**) and Wuxi, China (**b**). The simulated configurations are highlighted by the red boxes.

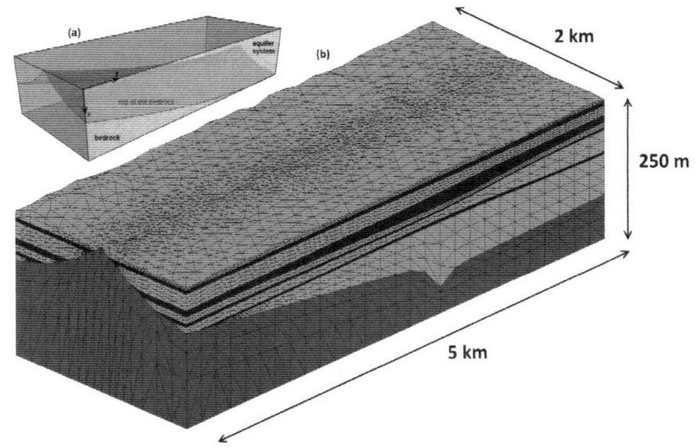

Figure 2. Test case 1: perspective views of the geological setting (**a**) and the 3-D FE-IE grid (**b**). The materials are highlighted in various colours: rock in red, aquifers in blue, and aquitards in green.

variations. Conversely, in case of sliding, only the displacement is an independent variable, as the shear stress is fully defined. In this work, we adopt a one-way coupled approach, so the pressure is a priori known and can be treated as an external force, separating it from the term of total stress, according to Terzaghi's principle (Bishop, 1959), that is $\hat{\boldsymbol{\sigma}} = \boldsymbol{\sigma} - \alpha p \boldsymbol{i}$, where $\boldsymbol{\sigma}$ are the effective stress, α is the Biot coefficient, p is the pressure and $\boldsymbol{i}$ is the vector form of the Kronecker delta. So we may write the complete conditional formulation as:

$$\int_{\Omega} \delta\boldsymbol{\varepsilon}^T \boldsymbol{\sigma} \mathrm{d}V + \int_{\Gamma\backslash\overline{\Gamma}} \delta\boldsymbol{u}^T \boldsymbol{\lambda} \mathrm{d}S + \int_{\Gamma\backslash\overline{\Gamma}} \delta\boldsymbol{\lambda}^T \boldsymbol{u} \mathrm{d}S + \int_{\overline{\Gamma}} \delta\boldsymbol{u}_{\mathrm{r}}^T \boldsymbol{\tau}_{\mathrm{L}} \mathrm{d}S = \int_{\Omega} \delta\boldsymbol{\varepsilon}^T \alpha p \boldsymbol{i} \mathrm{d}V + \int_{\Omega} \delta\boldsymbol{u}^T \boldsymbol{b} \mathrm{d}V + \int_{\partial\Omega} \delta\boldsymbol{u}^T \boldsymbol{t} \mathrm{d}S \qquad (5)$$

2.1 Numerical model

The displacements and stresses are approximated by the functions belonging to two different finite Hilbert spaces that, according to the Finite Elements Method, have a dimension equal to the nodal grid number of the entire domain and of the surface of the discontinuity, respectively. In general, we can write:

$$\boldsymbol{u}^h(\boldsymbol{x}) = \mathbf{N}_u(\boldsymbol{x})\boldsymbol{u}, \quad \boldsymbol{\lambda}^h(\boldsymbol{x}) = \mathbf{N}_\lambda(\boldsymbol{x})\boldsymbol{\lambda} \qquad (6)$$

With the assumption of small displacements, deformations are calculated as $\boldsymbol{\varepsilon}^h = \mathbf{B}\boldsymbol{u}^h = \mathbf{L}\mathbf{N}\boldsymbol{u}$, where $\mathbf{L}$ is the symbolic matrix of differential operators. Denoting by $\mathbf{D}$ the tangent stiffness matrix, the variation of the effective stress is $\mathrm{d}\boldsymbol{\sigma}^h = \mathbf{D}\mathrm{d}\boldsymbol{\varepsilon}^h$. Moreover, we have to introduce the relative displacements between the two rupture faces, $\boldsymbol{u}_{\mathrm{r}}^h = \mathbf{S}\boldsymbol{u}$, where the matrix $\mathbf{S}$ properly maps the nodes of the global mesh. It is also necessary to express the stress on the discontinuity in a local reference system, using an appropriate rotation matrix $\mathbf{R}$, which binds the global system to the local one. Thus, the discretized form of the Eq. (2) becomes:

$$\boldsymbol{\tau}_{\mathrm{L}}^h = (c - \sigma_{\mathrm{n}} \tan\phi) \frac{\mathbf{R}\mathbf{N}_u \mathbf{S}\boldsymbol{u}}{\sqrt{(\mathbf{R}\mathbf{N}_u \mathbf{S}\boldsymbol{u})^T \mathbf{R}\mathbf{N}_u \mathbf{S}\boldsymbol{u}}} \qquad (7)$$

In Eq. (7), the normal stress to the rupture surface can be written as $\sigma_{\mathrm{n}} = \boldsymbol{n}^T\boldsymbol{\lambda}$, where $\boldsymbol{n}$ is the unit vector normal to the discontinuity plane. The functional Eq. (5) is now written as:

$$\begin{aligned} &\delta\boldsymbol{u}^T \int_{\Omega} \mathbf{B}\boldsymbol{\sigma}^h \mathrm{d}V + \delta\boldsymbol{u}^T \int_{\Gamma\backslash\overline{\Gamma}} \mathbf{N}_u^T \mathbf{N}_\lambda \mathbf{R}^T \boldsymbol{\lambda} \mathrm{d}S + \delta\boldsymbol{\lambda}^T \\ &\int_{\Gamma\backslash\overline{\Gamma}} \mathbf{R}\mathbf{N}_\lambda^T \mathbf{N}_u \boldsymbol{u} \mathrm{d}S + \delta\boldsymbol{u}^T \int_{\overline{\Gamma}} \left(c - \boldsymbol{n}^T\boldsymbol{\lambda}\tan\phi\right) \frac{\mathbf{H}\boldsymbol{u}}{\sqrt{\boldsymbol{u}^T\mathbf{H}\boldsymbol{u}}} \mathrm{d}S \\ &= \delta\boldsymbol{u}^T \int_{\Omega} \mathbf{B}^T \alpha p \boldsymbol{i} \mathrm{d}V + \delta\boldsymbol{u}^T \int_{\Omega} \mathbf{N}_u^T \boldsymbol{b} \mathrm{d}V \\ &+ \delta\boldsymbol{u}^T \int_{\partial\Omega} \mathbf{N}_u^T \boldsymbol{t} \mathrm{d}S \end{aligned} \qquad (8)$$

Equation (8) must hold for every virtual displacement and stress. So, two non-linear equations are obtained. To solve this system, we use the Newton-Raphson scheme. Computing the Jacobian, we obtain a non-symmetric saddle point matrix. The non-symmetry is caused by the contribution due to the principle of "Maximum Plastic Dissipation" which introduces a component of virtual work where the displacements are independent but the stresses are dependent variables.

3 FE-IE model applications

The algorithm presented above has been implemented and tested. We use linear interpolation functions for displace-

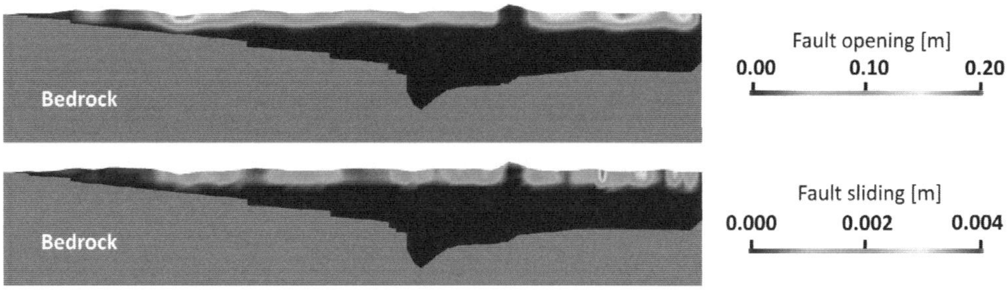

Figure 3. Test case 1: opening and sliding of the ground rupture after 10 years of water withdrawal.

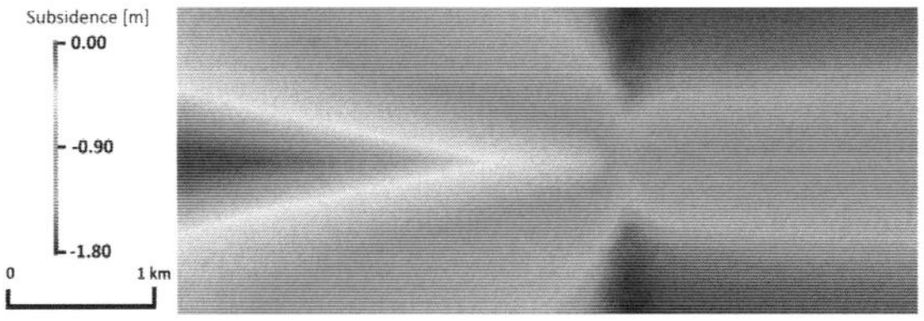

Figure 4. Test case 1: land subsidence after 10 years of water withdrawal.

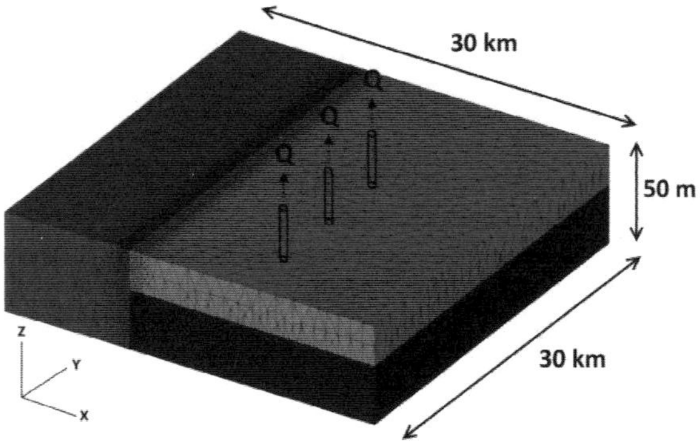

Figure 5. Test case 2: 3-D mesh with the materials highlighted in various colours. Rock is red, aquifer is blue, and clay is green. The fault divides rock from the sedimentary clay-aquifer system.

ments and constant piecewise functions for the Lagrangian multipliers. This choice was performed for compatibility reasons with the stresses of the 3-D FE grid. The discretization relies on the classic tetrahedral FE and the fault is a union of tetrahedra faces.

We address two applications: (i) the first concerns earth fissuring in Wuxi, near Shanghai, China, (Wang et al., 2007) caused by groundwater withdrawal from a shallow aquifer, with shallow confined bed; (ii) the second case concerns the City of Queretaro, Mexico (Carreón-Freyre and Cerca, 2006), placed above a graben structure with a volcanoclastic and sedimentary filling (stiff clay, silt and sandy materials) where several earth fissures have been generated by groundwater withdrawal; A sketch of the hydrogeological setting of the two test cases is shown in Fig. 1. For each case, we present the essential geometry and material properties, the history of pumping and the results, with special attention to the behaviour of both subsidence and rupture generation. Preliminary simulations with a simplified stress history are carried out.

3.1 Test case 1: Wuxi

In China, earth fissuring related to groundwater pumping has occurred since the 1970s. To cope with the rapid economic development large volumes of groundwater were withdrawal in the Wuxi area, Jiangsu Province, resulting in extensive land subsidence (Shi et al., 2007) and fissure development (Wang et al., 2007). Earth fissures cause serious damages, including cracking of buildings and failure in underground pipelines, with huge economic losses. The geologic setting, which is characterized by an undulating shaped, relatively shallow rocky paleo-basement covered by Quaternary compressible sedimentary deposits from the Yangtze River, strongly enhances the risk of fissure development.

Geological information has been used to develop the static model. A rock ridge, characterized by a depth between few meters and about 100 m from the ground surface is buried below the Quaternary sedimentary sequence (Fig. 2a). The large land subsidence measured in the area between the 1980s and 2000s, has been accompanied by the development of a fissure in correspondence with the tip of the subsurface ridge. The geometry of the main sandy and clayey layers forming the multi-aquifer system has been derived from a number of boreholes drilled down to the bedrock. The target of the study is to simulate the development and propagation of the ground rupture, using a realistic history of groundwater extraction.

The domain extends 2 km × 5 km in the horizontal plane and from the land surface down to 250 m depth in the vertical direction. A refined mesh consisting of 23 303 nodes and 121 942 tetrahedral elements is used to accurately represent the actual lithostratigraphic configuration (Fig. 2b). A num-

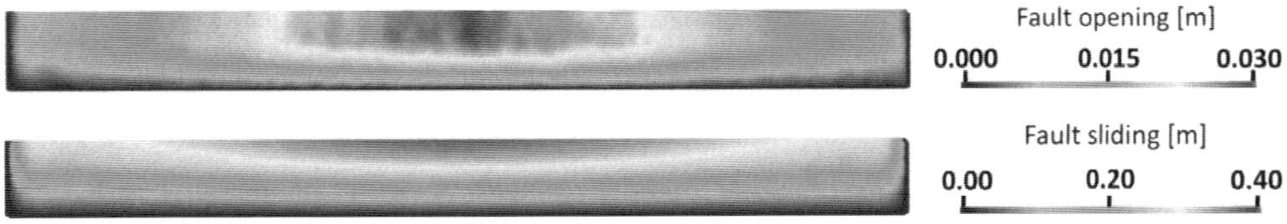

Figure 6. Test case 2: opening and sliding on the fault after 10 years of water extraction.

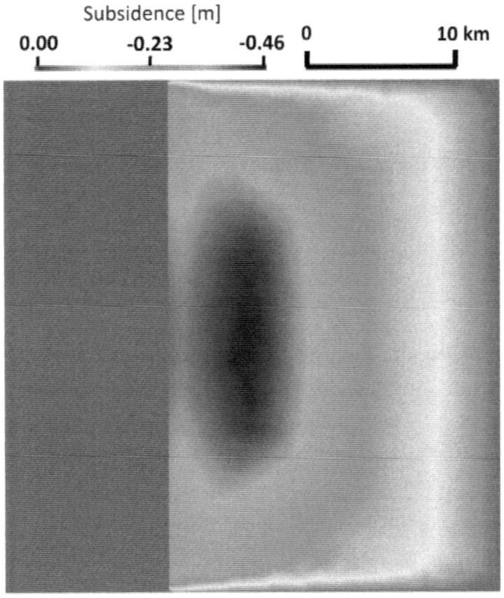

Figure 7. Test case 2: land subsidence after 10 years of water extraction.

ber of 1876 IE are introduced into the 3-D FE grid along the trace of the ridge tip, extending from the land surface to the bedrock top. The IE are characterized by a friction angle $\phi = 30°$ and zero cohesion. The geomechanical properties of the bedrock, aquifers, and aquitards are provided in Table 1.

A uniform pressure decline is prescribed in the sandy layers. The drawdown increases linearly from 0 to 20 m over 10 years. To account for a likely delay of pressure propagation in the clay layers, a half pressure change (i.e., 10 m at the end of the simulation period) is prescribed in the aquitards. No pressure change propagates in the bedrock.

The numerical results confirm the development of an earth fissure along the ridge tip. Due to the quasi-symmetric geological setting, the rupture is characterized by an opening and a negligible sliding. At the end of the simulation period the maximum opening amounts to a couple of decimeters (Fig. 3). Figure 4 provides the corresponding land subsidence distribution. The basement depth of burial significantly affects both the rupture opening and land subsidence.

3.2 Test case 2: Queretaro City

Subsidence related groundwater withdrawal and water-level decline has been documented in Queretaro City since the 1970s (Trejo-Moedano and Martinez-Baini, 1991). Local fractures in the fluvio-lacustrine sediments have been reported. The intensity of fracturing has increased and caused numerous problems to urban infrastructure. In the simplified case addressed in this study, an extensive shallow aquifer is covered with a undersaturated and stiff clayey sequence and confined on one side by a hard rock mass. The rock mass is separated from the aquifer by a regional pre-existing vertical fault. The clayey sediments have a thickness of 20 m, while the aquifer extends for 30 m in depth. The mesh covers a 30 km-side square area, one third of which is occupied by rock. In the z direction, the mesh extends from the free surface down to 50 m. Pumping takes place from the aquifer through three wells, as shown in Fig. 5, yielding a maximum pressure drawdown of about 20 m. In 10 years, a linearly increasing flow rate is implemented, up to the value of $300\,\mathrm{L\,s^{-1}}$ per well. The geomechanical parameters are provided in Table 1. Permeability of clayey sediments and rock is $K = 10^{-10}\,\mathrm{m\,s^{-1}}$, while permeability of aquifer is $K = 10^{-4}\,\mathrm{m\,s^{-1}}$. The fault has zero cohesion and friction angle $\phi = 30°$. The FE mesh consists of and 297 971 tetrahedra with 57 050 nodes. The fault is discretized by 8074 IE.

The model results provided by the IE and FE are shown in Figs. 6 and 7, respectively. In this geological setting the fault mainly slides. It can be noted that the fault opens negligibly (less than 3 cm), while the sliding in the fault plane approaches 40 cm, a value of the same order of magnitude as the maximum recorded subsidence, as can be seen in Fig. 7. This shows how important the presence of the fault is for the exact prediction of the land subsidence. The greatest calculated displacements along the fault are in the upper central part, i.e., in areas where the influence of the wells is largest and compaction of the underlying layers is the greatest. The displacements are quite regular and representative of the real order of magnitude.

4 Conclusions

The main subject of this investigation is how to simulate a geological discontinuity in a continuous medium. We used the Interface Finite Element method with a Lagrangian approach. At variance with existing algorithms, we introduce the principle of "Maximum Dissipation Plastic" in the formulation of the functional to be minimized. The solution is generally very smooth, as can be seen from the results discussed above.

The algorithm was implemented and preliminarly tested in two cases: (i) an aquifer with a shallow undulating bedrock representative of the geological setting in Wuxi, China. and (ii) a laterally rock confined aquifer covered with stiff clayey sediments, representative of the typical geological setting in Queretaro City In both cases the algorithm is robust and provides numerically stable and physically plausible solutions. Further improvements will be carried out for both the case studies, with the implementation of a more realistic evolution of the pressure head the calibration of FE-IE models on available measurements, and the development of future scenarios accounting for the expected management of the subsurface water resources.

Acknowledgements. The work has been developed within the IGCP Project 641 (M3EF3 – Deformation and fissuring caused by exploitation of subsurface fluids) funded by UNESCO.

References

Azat, E. and Shahram, S.: Land subsidence and fissuring due to ground water withdrawal in Yazd-Ardakan Basin, Central Iran, World Academy of Science, Engin. Technol., 71, 535–538, 2010.

Bathe, K.-J.: Finite Element Procedures. Prentice Hall, Upper Saddle River, New Jersey, 2006.

Bishop, A. W.: The principle of effective stress, Teknisk Ukeblad, 106, 859–863, 1959.

Carreón-Freyre, D.: Land subsidence processes and associated ground fracturing in Central Mexico, in: Land Subsidence, Associated Hazards and the Role of Natural Resources Development, edited by: Carreon-Freyre D., Cerca, M., and Galloway, D., IAHR Publ. no. 339, 149–157, 2010.

Carreón-Freyre, D. and Cerca, M.: Delineating the near-surface geometry of the fracture system affecting the valley of Queretaro, Mexico: Correlation of GPR signatures and physical properties of sediments, Near Surface Geophysics, EAGE (European Assoc. of Geoscientists and Engineers), 4, 49–55, 2006.

Ferronato, M., Gambolati, G., Janna, C., and Teatini P.: Numerical modelling of regional faults in land subsidence prediction above gas/oil reservoirs, Int. J. Numer. Anal. Meth. Geomech., 32, 633–657, 2008.

Ferronato, M., Janna, C., and Pini, G.: Parallel solution to ill-conditioned FE geomechanical problems, Int. J. Numer. Anal. Meth. Geomech., 36, 422–437, 2012.

González, P. J., Tiampo, K. F., Palano, M., Cannavó, F., and Fernándex, J.: The 2011 Lorca earthquake slip distribution controlled by groundwater crustal unloading, Nature Geosci., 5, 821–825, 2012.

Goodman, R. E., Taylor, R. L., and Breeke T. L.: A model for the mechanics of jointed rock, J. Soil Mech. Foundation Div. (ASCE), 94, 637–659, 1968.

Labuz, J. F. and Zang, A.: Mohr–Coulomb Failure Criterion, Rock Mech. Rock Eng., 45, 975–979, 2012.

Rutqvist, J., Birkholzer, J. T., and Tsang, C.-F.: Coupled reservoir-geomechanical analysis of the potential for tensile and shear failure associated with CO_2 injection in multilayered reservoir-caprock systems, Int. J. Rock. Mech. Mining Sci., 45, 132–143, 2007.

Shi, X., Xue, Y., Ye, S., Wu, J.-C., Zhang, Y., and Jun, Y.: Characterization of land subsidence induced by groundwater withdrawals in Su-Xi-Chang area, China, Environ. Geol., 52, 27–40, 2007.

Simo, J. C. and Hughes, T. J. R.: Computational Inelasticity, Springer, New York, 2nd Edn., 2000.

Trejo-Moedano, A. and Martinez-Baini, A.: Soils cracking in the Querétaro zone, Proc. Symp. "Agrietamientos de suelos", Sociedad Mexicana de Mecánica de Suelos, México, 67–74, 1991 (in Spanish).

Wang, G. Y., You, G., Shi, B., Yu, J., Li, H. Y., and Zong, K. H.: Earth fissures triggered by groundwater withdrawal and coupled by geological structures in Jiangsu Province, China, Environ. Earth Sci., 57, 1047–1054, 2009.

Wriggers, P.: Computational Contact Mechanics. Springer, Heidelberg, 2nd Edn., 2006.

Proc. IAHS, 372, 69–72, 2015
proc-iahs.net/372/69/2015/
doi:10.5194/piahs-372-69-2015

Open Access
PIAHS

Modeling of earth fissures caused by land subsidence due to groundwater withdrawal

B. B. Panda, M. L. Rucker, and K. C. Fergason

Amec Foster Wheeler, Phoenix, Arizona, USA

Correspondence to: B. B. Panda (bibhuti.panda@amecfw.com)

Published: 12 November 2015

Abstract. Land subsidence and earth fissures are phenomena related to groundwater withdrawal in a sedimentary basin. If the rock basement or basin lithology is irregular, both vertical and horizontal displacements can be induced due to differential settlement and tensile stresses appearing in the soil mass. If the differential settlement is of sufficient magnitude, earth fissuring can occur within tensile zones. The magnitudes of compaction and fissure geometry are closely related to the thickness and skeletal compressibility of fine-grained sediments within the aquifer system. Land subsidence and earth fissuring were modeled by employing a two-dimensional (2-D) coupled seepage and stress-strain finite element analysis. The basin bedrock geometry, lithological variation, measurements of surface displacements, and changes in hydraulic head were the critical input parameter for the subsidence modeling. Simulation results indicate that strain had exceeded the approximate threshold for fissure formation of 0.02 to 0.06 % in the area of the identified fissures. The numerical model was used to predict future subsidence and potential earth fissures for flood control structures within the metro Phoenix area.

1 Introduction

Land subsidence due to groundwater withdrawal in alluvial basins in the southwest region of USA is a process of differential compaction of deep sediments. The effect of groundwater pumping on ground deformation in south central Arizona was investigated by many researchers (Boling et al., 1980). The application of interferometric synthetic aperture radar (InSAR) to characterize the distribution and rate of land subsidence provided invaluable detail into the probable positioning of tensile ground stresses and earth fissuring (AMEC, 2003). The differential rates and magnitudes of subsidence over relatively short distances, cause horizontal strains due to tensile stresses and eventually can become sufficient to cause earth fissuring. Jachens and Holzer (1982) concluded that most fissuring occurred at horizontal tensile strains in the range of 0.02 to 0.06 %. Numerical models have been developed by the authors (Panda, 2007; AMEC, 2003, 2013) to assimilate both InSAR and historic subsidence data for prediction of subsidence and earth fissure due to groundwater withdrawal.

2 Theory

The change in pore pressure in any soil layer due to withdrawal of water with time is related to the change in void ratio by the following equation:

$$p = \frac{e}{a_v} \quad (1)$$

where p = change in pore pressure, e = change in void ratio, and a_v = coefficient of compressibility (related to skeletal compressibility).

The vertical settlement of any soil depends on the void ratio and thickness of the soil layer.

$$s = e2H \quad (2)$$

where s = vertical settlement at time t and $2H$ = total vertical thickness of the medium.

The amount of compaction and fissure location are closely related to the thickness and skeletal compressibility of fine-grained sediments within the aquifer system.

Published by Copernicus Publications on behalf of the International Association of Hydrological Sciences.

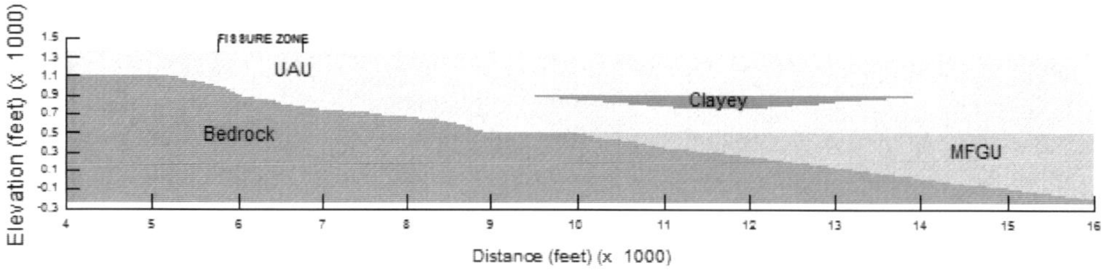

Figure 1. Finite Element Model Geometry along McMicken Dam.

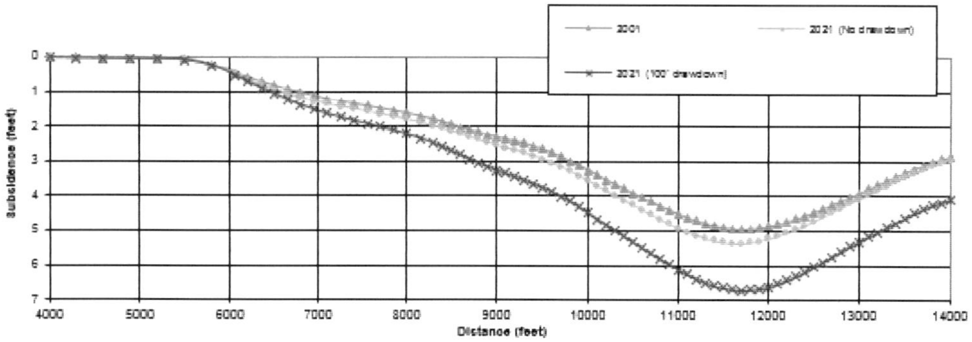

Figure 2. Prediction of vertical subsidence along the McMicken Dam.

3 Modeling concepts

A fully coupled analysis requires that both the stress-deformation and seepage dissipation equations be solved simultaneously. This coupling was achieved by using two finite element-based computer programs, SEEP/W and SIGMA/W (Geo-Slope, 2012). Modeling of the earth fissure for two dimensional (2-D) geometry is performed by calculating the tensile strain along the surface of any linear structure. The tensile strain is developed mainly due to differential settlement of the ground surface. Input data for the modeling effort consisted of an array of interrelated geometric and numerical components, including alluvial lithology, hydrological data and assigned mechanical properties of the alluvial materials. The numerical model was calibrated using InSAR and survey data. The input parameters for the 2-D models were estimated from the 1-D model based on percolation theory approach (AMEC, 2013).

4 Case history

4.1 McMicken Dam, Phoenix, Arizona

McMicken Dam is part of a system of flood retention and conveyance facilities that protect the metropolitan area of Phoenix, Arizona. The 9-mile long, earthen embankment of McMicken Dam was constructed in 1955 along the alluvial flank of the White Tank Mountains. McMicken Dam has experienced over 5 feet of differential subsidence due to groundwater pumping in the deep basin sediments of the Phoenix aquifer (AMEC, 2003). The vertical subsidence along the dam profile is modeled using the finite element method. The conceptual model geometry used in finite element model along the dam is presented in Fig. 1. Two different future drawdown scenarios were assumed for subsidence prediction. In a first scenario, no drawdown was assumed for 20 years while 100 feet drawdown in 20 years was assumed for a second scenario. The predicted vertical subsidence along the dam profile is presented in Fig. 2. The modeled horizontal strain along the dam profile is presented in Fig. 3. The modeled horizontal strain in excess of the threshold 0.02 % strain was estimated at the location of known earth fissures. The model also shows another two locations along the profiles, where the tensile strain is approaching or at the threshold value. These two locations may be regarded as zones with a greater potential for the development of earth fissure.

4.2 Powerline floodway, Mesa, Arizona

Subsidence and earth fissuring in the vicinity of Powerline floodway, has been investigated and numerical models were developed (AMEC, 2013) to predict future subsidence and earth fissure potential along the Powerline floodway. An

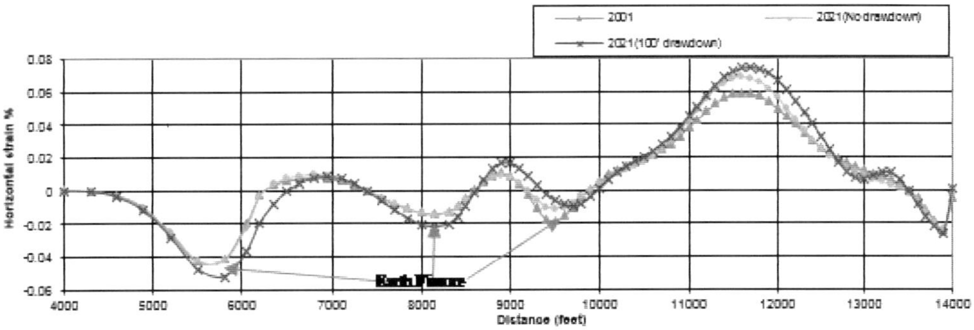

Figure 3. Prediction of horizontal strain and location of potential earth fissure zone along the McMicken Dam.

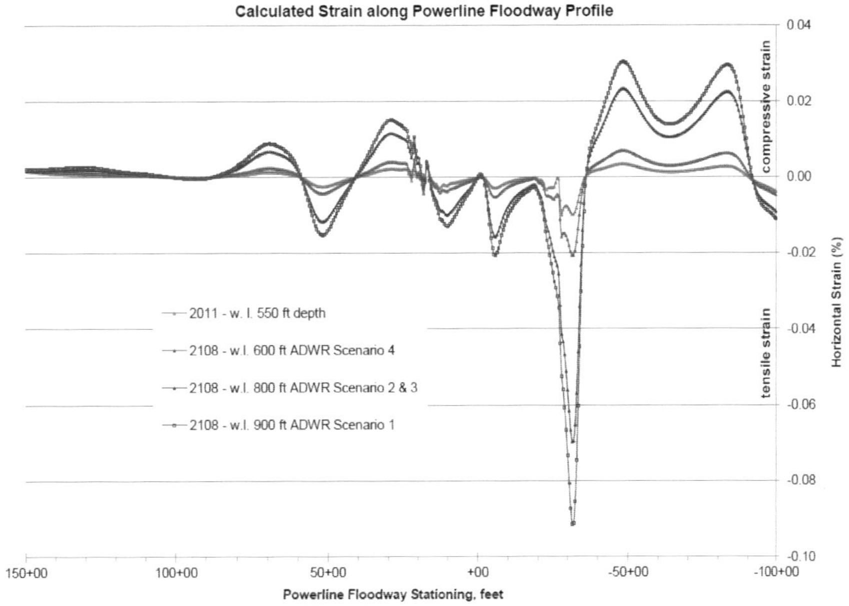

Figure 4. Prediction of horizontal strain along the Powerline floodway structures.

earth fissure observed along the dam was calibrated through numerical modeling (AMEC, 2013). The calibrated model was used to predict the future earthfissure potential along the floodway based on various ground water pumping scenarios. The estimated horizontal strain from the finite element model simulation along the powerline floodway is presented in Fig. 4. Although the simulated current ground strains have not reached the threshold limit of 0.02, future strains under various groundwater pumping scenarios clearly shows the potentail for future development of earth fissures along the floodway.

5 Conclusions and discussions

The earth fissures resulting from ground subsidence due to the withdrawal of groundwater were modeled using coupled flow and deformation finite element models. The numerical model simulated the past ground subsidence and observed earth fissure activity along the flood control retention structures in Arizona. The numerical simulation demonstratted the strong influence of the depth of bedrock and the distribution of alluvial lithology on the subsidence and location of earth fissures. It is recommended that the ground subsidence and earth fissure activity due to future groundwater pumping in the area should be consindered in the design of flood control structures. The numerical model was used as a suitable tool to estimate future subsidence and earthf issure potential based on future groundwater pumping scenarios. The identification of potential earth fissure zones based on finite element horizontal strain estimation also assists in the planning of the earth fissure monitoring system for the structures.

References

AMEC Earth & Environmental, Inc. (AMEC): Earth fissure investigation report, McMicken Dam, Work Assignment Nos. 4 & 5, Contract FCD 2000C006, Maricopa County, 60 pp., 11 April 2003.

AMEC Environmental 7 Infrastructures, Inc. (AMEC): Final design level subsidence and earthfissure risk zoning report Powerline, Vineyard Road and Rittenhouse Flood retarding Structures, prepared for Flood Control District of Maricopa County Contract FCD 2011C005 Work Assignment No. 12, AMEC Job No. 17-2013-4045, 421 pp., December 2013.

Boling, J. K., Carpenter, M. C., Johnson, N. M., and Davis, S. N.: Measurements, prediction and hazard evaluation of earth Fissures and subsidence, South Central Arizona, OWRT Project No. A-092-Ariz, The University of Arizona, Tucson, 46 pp., 1980.

Geo-Slope International Ltd.: Finite element stress and deformation analysis, Computer Program SIGMA/W, Calgary, Alberta, Canada, 2012.

Jachens, R. C. and Holzer, T. L.: Differential Compaction mechanism for earth fissures near Casa Grande, Arizona, Geol. Soc. Am. Bull., 93, 998–1012, 1982.

Panda, B. B.: Prediction of Ground Deformation due to Groundwater Withdrawal using Numerical Simulations, Proceedings of First Sri Lankan Geotechnical society International conference on Soil and Rock engineering, Columb, 8 pp., 6–11 August 2007.

Proc. IAHS, 372, 73–76, 2015
proc-iahs.net/372/73/2015/
doi:10.5194/piahs-372-73-2015

Assessing the potential of the multi-aquifer subsurface of the Mekong Delta (Vietnam) for land subsidence due to groundwater extraction

P. S. J. Minderhoud[1,2], G. Erkens[2,1], V. H. Pham[1,2,3], B. T. Vuong[3], and E. Stouthamer[1]

[1]Department of Physical Geography, Utrecht University, Utrecht, the Netherlands
[2]Department of Subsurface and Groundwater Systems, Deltares Research Institute, Utrecht, the Netherlands
[3]Division of Water Resources Planning and Investigation for the South of Vietnam (DWRPIS), Ho Chi Minh City, Vietnam

Correspondence to: P. S. J. Minderhoud (p.s.j.minderhoud@uu.nl)

Published: 12 November 2015

Abstract. Land subsidence rates of $\sim$ 1–4 cm yr^{-1} are measured in the low-lying Vietnamese Mekong Delta. These relatively high subsidence rates are attributed to groundwater extraction, which has increased drastically over the past decades due to growing domestic, agricultural and industrial demands. As a result, hydraulic heads in aquifers are dropping, on average 0.3–0.7 m yr^{-1}. There is an urgent need to go from measurements to predictions in order to test possible future groundwater management scenarios and to reduce the increase of flood risk, salt water intrusion and, on the longer term, prevent the delta from drowning. In this study, we aim to assess the subsidence potential of the multi-aquifer subsurface of the Mekong delta due to groundwater extraction. The first step is to gain a thorough understanding of the complex sedimentary architecture of the heterogeneous subsurface. Combined with the related geotechnical properties, the subsurface build-up determines the subsidence potential. Here, we present our approach to develop a 3-D geo-hydrological model based on lithological borehole data, geophysical sedimentary properties, palaeogeography and conceptual models of delta evolution.

1 Introduction

The Vietnamese Mekong delta is experiencing strong intensifications of agricultural and aquaculture practices and cities in the delta are growing fast (Renaud and Kuenzer, 2012). Pressure on the existing natural resources is high and the demand for fresh water is steadily increasing. On daily basis over 2 million m^3 of groundwater is extracted from the upper 500 m of the multi-aquifer subsurface. Aquifer drawdown occurs at rates of 0.3–0.7 m yr^{-1} (Wagner et al., 2012). Such declines in hydraulic head can trigger high compaction in the drained subsurface sediments, causing land subsidence at the surface (e.g. Galloway and Burbey, 2011). However, the actual impact of groundwater extraction remains unclear, because: (1) subsidence rates measured at the surface are caused by multiple drivers, not just groundwater extraction, and (2) studies on subsidence in the Mekong delta are scarce and reliable ground-based, time series of land elevation are hard to find, if they at all exist. At present, InSAR data is the only source of surface elevation data available in time series, allowing the estimation of subsidence rates (Fig. 1).

Analysis of InSAR data measured during the period 2006–2010 reveals subsidence rates of $\sim$ 1–4 cm yr^{-1} within the Mekong delta, with an average rate of 1.6 cm yr^{-1} (Erban et al., 2014). These rates are high, especially when considering that the majority of the delta is elevated less than 2 m above mean sea level and already prone to flooding. Subsidence poses a real threat for the Mekong delta. Even more, if also absolute sea-level rise of $\sim$ 3 mm yr^{-1} is taken into account (Church et al., 2013). Subsidence and absolute sea level rise determine *relative* sea level rise, the rise of mean sea level experienced from the land over time, increasing salt water intrusion and flood risks. Furthermore, subsidence is causing damage to buildings and infrastructure. There is an urgent need to come up with adequate groundwater quality and extraction regulation to secure a sustainable future for

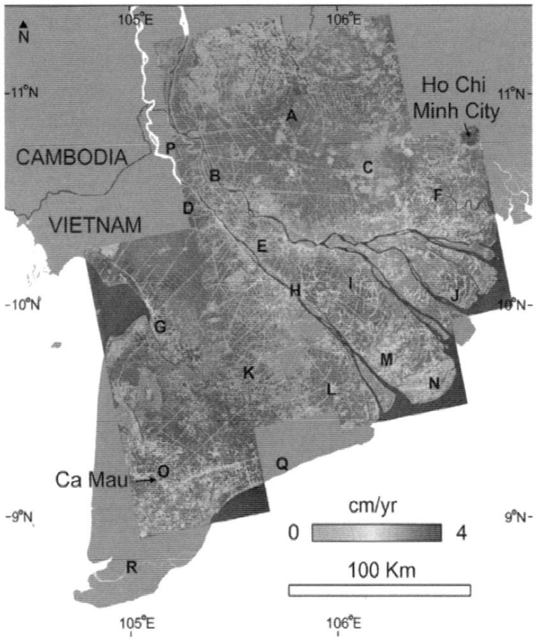

Figure 1. InSAR-based subsidence rates measured between 2006 and 2010 in the Mekong Delta. Locations of monitoring wells are marked by letters. Data ©JAXA, METI 2011 (Erban et al., 2014).

the delta. Decision-making requires insights in the spatial extent and magnitude of groundwater extraction as driver of subsidence in the delta. Therefore, we aim to assess the subsidence potential of the upper 500 m of the subsurface due to groundwater extraction. The first step is to get a better understanding of the complex sedimentary architecture of the heterogeneous subsurface of the Mekong delta. In this paper we present our approach to develop a new 3-D subsurface model, which enables groundwater and subsidence modelling, based on thorough analysis of the complex multi-aquifer subsurface. Ultimately, this model can be used to create meaningful predictions and evaluate (ground)water management scenarios for policy-makers.

2 From monitoring to predicting subsidence

Total measured subsidence at the earth surface, e.g. by InSAR analysis, is the sum of subsidence resulting from all natural and human-induced drivers acting in an area (Fig. 2). This cumulative signal needs to be unraveled to determine the contribution of each driver and to go from measuring to predicting subsidence. We distinguish between shallow and deep drivers and processes of subsidence, because in the shallow phreatic aquifer other drivers and processes are dominant (e.g. drop in phreatic water table and consequent oxidation), than in the deeper confined aquifer-aquitard system (e.g. groundwater extraction causing consolidation). The boundary between shallow and deep processes depends on the local lithology and is situated at the top of the upper most aquitard. Total subsidence is the sum of all shallow and deep subsidence rates, resulting from all processes forced by all drivers of subsidence; the subsidence balance equation (Fig. 2). For the Mekong delta the subsidence balance equation is still unsolved and impacts of different drivers and processes are unknown. Erban et al. (2014) identified groundwater extraction as a potential main driver of subsidence, using 1-D consolidation calculations with hydraulic head time series at several locations in the delta. We will further explore the potential of groundwater extraction on subsidence for the Mekong delta in a step towards solving the total subsidence balance.

3 Research approach

Our first step to assess the subsidence potential of the multi-aquifer system of the Mekong delta due to groundwater extraction, is to gain a profound understanding of the subsurface sedimentary architecture. The aquifer-aquitard architecture, the geotechnical properties and the characteristics and occurrence of interbeds will determine the potential for subsidence. Present models of the subsurface are coarse and do not have the level of detail required for accurate subsidence modelling (DWRPIS, 2010). Therefore we develop a 3-D lithological subsurface model based on a higher lithological data density and combining additional sources of subsurface data (Fig. 3). Furthermore, concepts of delta evolution and palaeogeographic development of the Mekong delta are constructed and provide valuable information on the spatial occurrence of palaeo-depositional environments. This allows to reconstruct the complex nature of the subsurface build-up through time. These insights are used to interpolate the lithological borehole data into a 3-D lithological subsurface model, in addition to standard interpolation methods, like linear interpolation of borehole data at point locations. Through this approach we create a more detailed and, from a geological point of view, more realistic representation of the complex multi-aquifer subsurface than present in current available models. Moreover, this approach enables exploration of aquifer characteristics such as aquifer connectivity and sediment properties. Subsequently, this lithological subsurface model is converted into a geo-hydrological model and transferred to a groundwater flow software package (e.g. MODFLOW), including a subsidence module. This enables hydrological and subsidence modelling of the Mekong delta at regional and local scales and assessing the subsidence potential of the subsurface by groundwater extraction. Finally, this model can be applied to create meaningful quantitative predictions of subsidence driven by groundwater extraction and evaluate groundwater management scenarios.

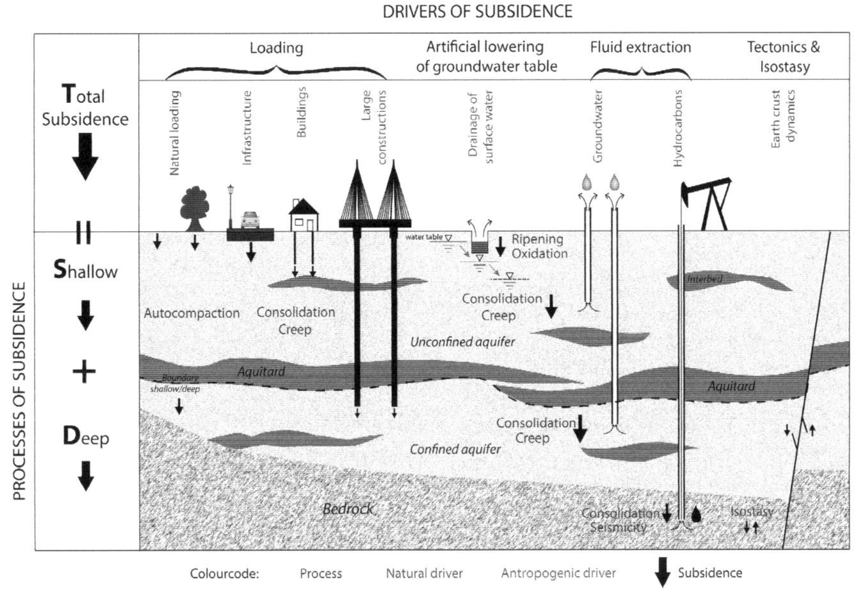

Figure 2. Schematization of the main subsidence drivers and processes within the upper (phreatic) aquifer and deeper (confined) aquifer(s). Natural and anthropogenic drivers are distinguished. The subsidence balance equation is given on the left side, being the total sum of all shallow and deep subsidence rates.

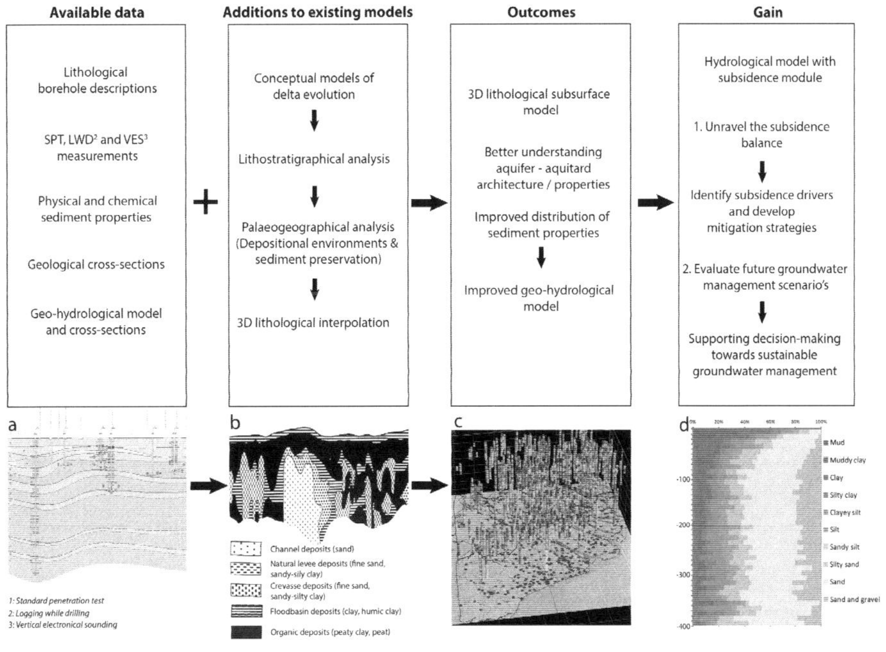

Figure 3. Workflow to develop the 3-D lithological subsurface model of the Mekong Delta, improve the geo-hydrological model and analyze the potential for subsidence driven by groundwater extraction. Steps graphically depicted below the workflow. **(a)** Cross-section of current geo-hydrological model. **(b)** Example of a cross-section including geomorphological elements as a result from lithological and palaeogeographical analyses (modified after Bierkens and Weerts, 1994). **(c)** 3-D view of the new available lithological borehole data in the Mekong delta up to 500 m deep. **(d)** Depth distribution of lithological classes for the Mekong delta. Data: ©Division of Water Resources Planning and Investigation of South Vietnam (DWRPIS).

4 Conclusions and future outlook

Surface measurements by InSAR analysis show relatively high subsidence rates of $\sim$ 1–4 cm yr^{-1} for the Mekong delta, causing increased flooding risk, salinization and damage to buildings and infrastructure. Groundwater extraction is hypothesized to be a main driver of subsidence, but the contribution of other possible human-induced drivers, like loading by buildings and infrastructure and drainage have not been quantified yet. The development of a new lithological model of the Mekong delta based on subsurface data and incorporating palaeogeographical and delta evolutional insights is the first step towards resolving the subsidence balance. The lithological model, converted into a 3-D geo-hydrological model, enables calculations of groundwater dynamics and resulting subsidence, using groundwater flow software including a subsidence module. This allows assessing the subsidence potential of the multi-aquifer system of the Mekong delta caused by groundwater extraction. The model can furthermore be used to create meaningful predictions of future subsidence rates using different delta management scenarios, supporting sustainable decision-making in groundwater management to limit the total amount and rates of subsidence

Acknowledgements. The Division of Water Resources Planning and Investigation for the South of Vietnam (DWRPIS), Ho Chi Minh city, Vietnam is thanked for providing subsurface and hydrological data for this research. This paper is part of a PhD research carried out by P. S. J. Minderhoud at the Department of Physical Geography, Utrecht University, The Netherlands. The PhD project is funded by NWO-WOTRO (W 07.69.105), Deltares and TNO-Geological Survey of the Netherlands.

References

Bierkens, M. F. P. and Weerts, H. J. T.: Application of indicator simulation to modelling the lithological properties of a complex confining layer, Geoderma, 62, 265–284, doi:10.1016/0016-7061(94)90040, 1994.

Church, J. A., Clark, P. U., Cazenave, A., Gregory, J. M., Jevrejeva, S., Levermann, A., Merrifield, M. A., Milne, G. A., Nerem, R. S., Nunn, P. D., Payne, A. J., Pfeffer, W. T., Stammer, D., and Unnikrishnan, A. S.: Sea Level Change, in: Climate Change 2013: The Physical Science Basis. Contribution of Working Group I to the Fifth Assessment Report of the Intergovernmental Panel on Climate Change, edited by: Stocker, T. F., Qin, D., Plattner, G.-K., Tignor, M., Allen, S. K., Boschung, J., Nauels, A., Xia, Y., Bex, V., and Midgley, P. M., Cambridge University Press, Cambridge, United Kingdom and New York, NY, USA, 2013.

DWRPIS (Division for Water Resources Planning and Investigation for the South): Report on the results of the National Groundwater Monitoring Network for Nam Bo Plain, Division of Water Resources and Planning Investigation of Viet Nam, unpublished, 2010.

Erban, L. E., Gorelick, S. M., and Zebker, H. A.: Groundwater extraction, land subsidence, and sea-level rise in the Mekong Delta, Vietnam, Environ. Res. Lett., 9, 1–6, doi:10.1088/1748-9326/9/8/084010, 2014.

Galloway, D. L. and Burbey, T. J.: Review: Regional land subsidence accompanying groundwater extraction, Hydrogeol. J., 19, 1459–1486, doi:10.1007/s10040-011-0775-5, 2011

Renaud, F. G. and Kuenzer, C.: The Mekong Delta System, Interdisciplinary analyses of a river delta, Springer, Dordrecht, Netherlands, 463 pp., doi:10.1007/978-94-007-3962-8, 2012.

Wagner, F., Tran, V. B., and Renaud, F. G.: Groundwater Resources in the Mekong Delta: Availability, Utilization and Risks, in: The Mekong Delta System, edited by: Renaud, F. G. and Kuenzer, C., Springer Environmental Science and Engineering, Springer Science+Business Media, Dordrecht, Netherlands, Chapter 7, 201–220, doi:10.1007/978-94-007-3962-8_7, 2012.

Proc. IAHS, 372, 77–82, 2015
proc-iahs.net/372/77/2015/
doi:10.5194/piahs-372-77-2015

Open Access PIAHS

Combination with precise leveling and PSInSAR observations to quantify pumping-induced land subsidence in central Taiwan

C. H. Lu[1], C. F. Ni[1], C. P. Chang[2,3], J. Y. Yen[4], and W. C. Hung[5]

[1]Graduate Institute of Applied Geology, National Central University, Zhongli District, Taoyuan City, Taiwan
[2]Center for Space and Remote Sensing Research, National Central University, Zhongli District, Taoyuan City, Taiwan
[3]Institute of Geophysics, National Central University, Zhongli District, Taoyuan City, Taiwan
[4]Department of Natural Resources and Environmental Studies, National Dong Hwa University, Shoufeng, Hualien, Taiwan
[5]Green Environmental Engineering Consultant Co. Ltd., Hsinchu, Taiwan

Correspondence to: C. H. Lu (fox52600@gmail.com)

Published: 12 November 2015

Abstract. Choushui River Fluvial Plain (CRFP) is located in the western central Taiwan, where the geomaterials are composed of alluvial deposits. Because the CRFP area receives highly variable rainfall in wet and dry seasons, the groundwater becomes the main resource of residential water. The precise leveling monitoring from 1970s indicated that the coastal areas of CRFP had been threatened by serious pumping-induced land subsidence. On the basis of relatively accurate measurements of precise leveling measurements, we used cokriging technique to incorporate a number of InSAR images to quantify the surface deformation in CRFP. More specifically, the well-developed Persistent Scatterer InSAR (PSI) was employed to process 34 Envisat images (2005–2008) and the results of PSI was then used for improving the spatial resolution of data from precise leveling. The results of cokriging estimation indicate whether the rate or the area of the land subsidence slows down gradually from 2005 to 2008. The subsidence in the northern part of CRFP was influenced by the groundwater decline in aquifer III, and the southern part was influenced by groundwater decline in aquifer II and III. The cokriging estimation was also comparable with continuous GPS data, and their correlation coefficient is 0.9603 and the root mean square is 10.56 $mm\,yr^{-1}$.

1 Introduction

Groundwater resource management is a notable issue due to growing population and rapid urban development, especially in areas where surface water is limited or shortage. Many cities, such as Shanghai, Las Vegas, and Bangkok had faced serious land subsidence because of unlimited groundwater withdrawal. To identify the land subsidence range effectively the geodetic techniques with different scales record the behaviors of surface deformation. Precise leveling, Global Positioning System (GPS) and Interferometric Synthetic Aperture Radar (InSAR) have been the most powerful techniques in geodesy (Galloway et al., 1998; Massonnet and Feigl, 1998; Galloway and Burbey, 2011; Hung et al., 2010).

In the central Taiwan, Choushui River fluvial plain (CRFP) faced the serious problem of land subsidence by poignant groundwater drawdown since 1970's. Central Geological Survey (CGS) has investigated several hydrologic and geologic studies. Thereafter, Water Resources Agency (WRA) have been collecting geodetic data such as precise leveling, GPS and compaction monitoring well (CMW) to monitoring the rate of land subsidence in CRFP (Central Geological Survey, 1999; Liu et al., 2001, Liu et al., 2004; Hwang et al., 2008). Hung et al. (2011) apply the Persistent Scatter InSAR (PSI), which was developed by Hooper et al. (2004,

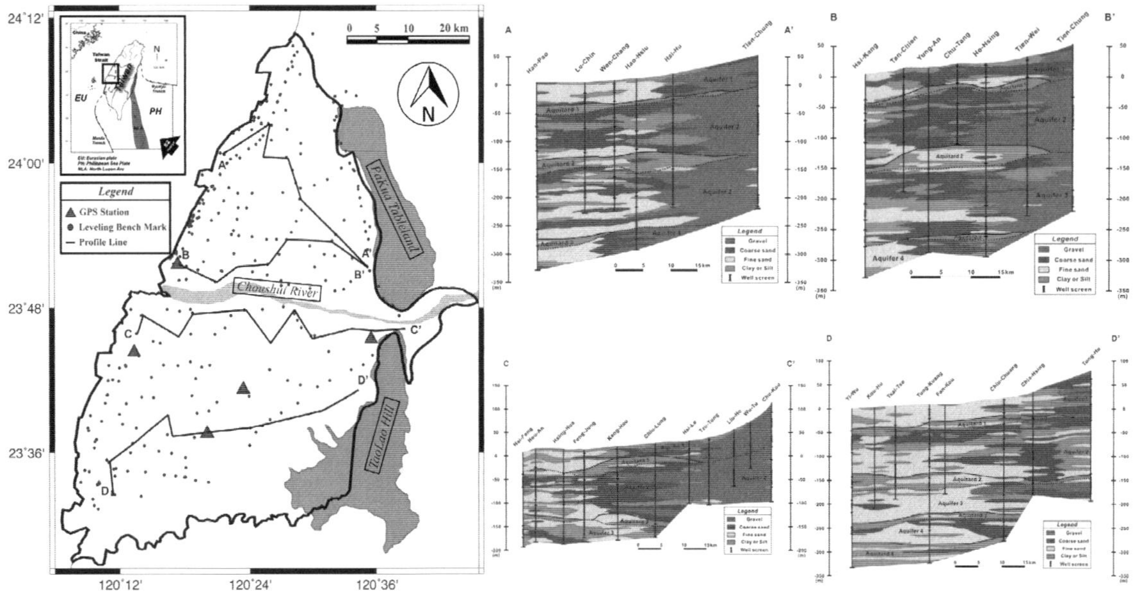

Figure 1. The study area. The black circle is the leveling benchmarks. The red triangle is continuous GPS station. The black lines indicate the location of hydrogeological profile. The right side of figure shows the hydrogeological profiles. The hydrogeological profiles were modified from Central Geological Survey (1999).

2007) and Hooper (2008), for identifying PS pixels and extracting more than 100 pixels km^{-2} in CRFP from 2006 to 2008. In this paper, the main aim is to integrate precise leveling data and PSI result with geostatistical method, variogram and cokriging, for retaining the accuracy of precise leveling and the spatial resolution of PSI and estimating the surface deformation, then comparing cokriging estimation with the difference of yearly groundwater level, at last verifying the cokriging estimation with the observed GPS data from 2005 to 2008 in CRFP.

2 Background of study area

2.1 Geological setting

Taiwan is located at the boundary between the Philippine Sea Plate (PSP) and Eurasian Plate (EUP), where the still ongoing collision started about 5–7 Ma. The PSP move toward the northwest to EUP at a rate of 82 $\mathrm{mm\,yr}^{-1}$. Choushui River Fluvial Plain (CRFP) is one part of the coastal plain in the western Taiwan. This area covers about 2000 km^2 and the boundary is enclosed by Pakua tableland and Taolao hill to the east, Taiwan Strait to the west, Wu River to the north and Putzu River to the south. Choushui River is the longest river in Taiwan and flows cross Central Range, Western Foothill and Coastal Plain then in to Taiwan Strait. The range of the sedimentary thickness is from 750 to 3000 m and the grain size become finer from the east to the west. The materials of the sedimentary contain clay, fine sand, coarse sand, and gravel. According the drilling core in the depth of 300 m, the subsurface hydrogeology can be divided into eight overlapping sequences. There are four marine sequences (aquitard I–IV) and four non-marine sequences (aquifer I–IV). The right side of Fig. 1 shows four hydrogeological profiles in CRFP (Central Geological Survey, 1999; Liu et al., 2001; Liu et al., 2004).

2.2 Observation data

2.2.1 Precise leveling

The land subsiding surveys determined by precise leveling in CRFP has recorded since 1975. These data acquired from Water Resource Agency (WRA) are based on the first or second order standard of leveling procedures proposed by Ministry of the Interior (MOI). The specifications demand that any loop misclosure should be below 3 mm $\sqrt{K}$, where K is the distance between two neighboring benchmarks in km. In this study 298 precise leveling benchmarks were be used from 2005 to 2008 (Fig. 1).

2.2.2 Persistent Scatterers InSAR (PSI)

Due to the area of CRFP covers two SAR image modes (Track 232, Frame 3123 and Track 232, Frame 3141), 34 Envisat radar images (16 images from Track 232, Frame 3123; 18 images from Track 232, Frame: 3141) from European

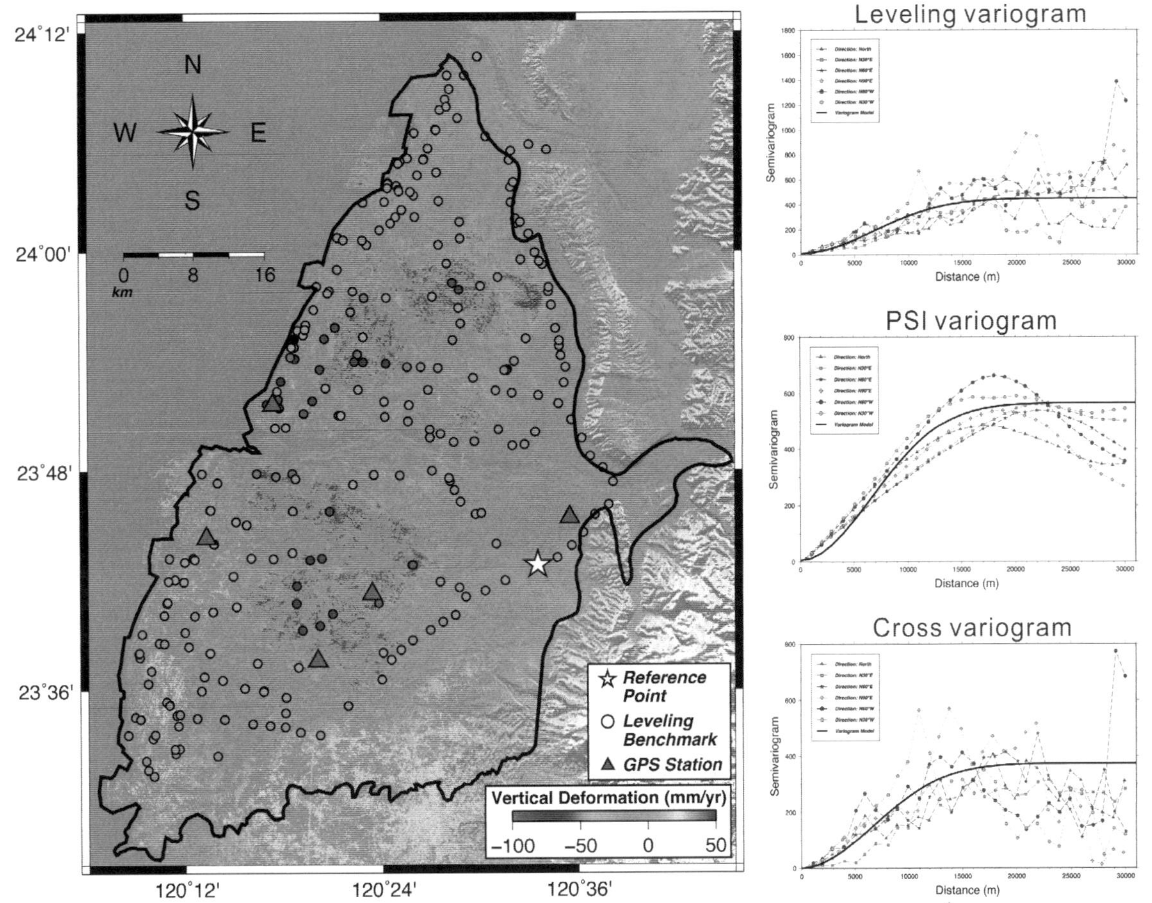

Figure 2. The left side of the figure is the vertical deformation of precise leveling and PSI between 2005 and 2008. The cold color means the land subsidence. The right side of the figure from top to bottom are precise leveling variogram, PSI variogram and cross-variogram between precise leveling and PSI. Every variogram has 6 directions and a Gaussian model.

Space Agency (ESA) were used in this study. Envisat radar images were taken from May 2005 to September 2008. The PSI was processed with StaMPS/MTI (Hooper et al., 2007; Hooper, 2008). The topographic effect was removed by the DEM from NASA's SRTM mission and the precise orbit parameters from Delft Institute for Earth-Oriented Space Research were used to correct the orbit errors.

2.2.3 Groundwater level

Since 1992 WRA installed hydrologic monitoring well evenly, and dug the depths ranging from 200 to 300 m in CRFP. According to the hydrogeology profiles (Fig. 1) CRFP can be divided into four aquifers in depth of 300 m. In this study we collect 98 monitoring well screens in the aquifer I, 97 screens in the aquifer II, 44 screens in the aquifer III, and 21 screens in the aquifer IV. The yearly average of groundwater level from 2005 to 2008 was interpolated by using kriging method.

3 Methodology

The grid of 9867 active cells is set up in the study area and the size of each cell is 500 m × 500 m. Two geostatistical methods, variogram and cokriging, are adopted for analyzing and combining the precise leveling and PSI result. The FORTRAN codes gamv and cokb3d in GSLIB (Deutsch and Journel, 1998) make use of generating experimental variograms and cokriging respectively.

3.1 Variogram analysis

An experimental semivariogram, $\gamma'(h)$, is defined as half of the average squared difference between two values of an uni-

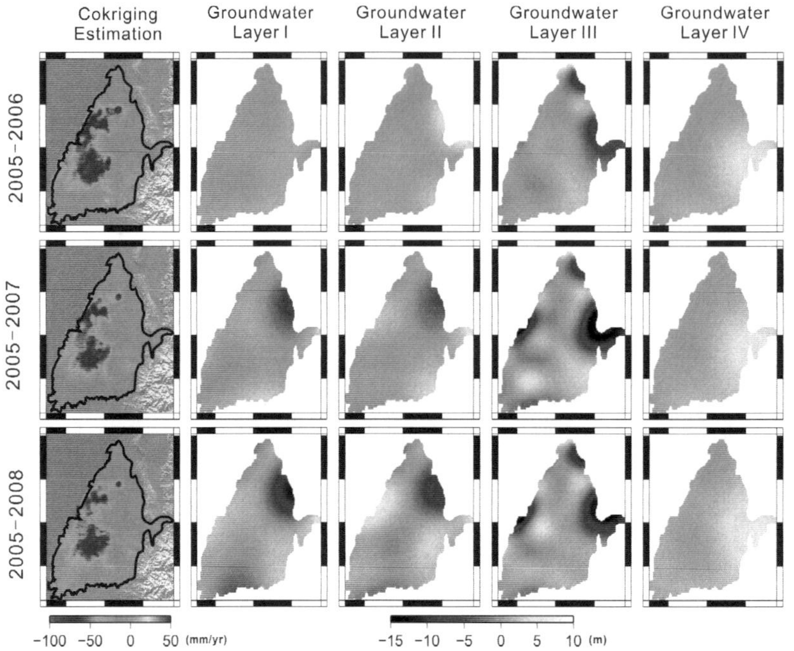

Figure 3. First column is the cokriging estimation in the three intervals (2005–2006, 2005–2007 and 2005–2008). The rest columns are the difference of yearly average of groundwater level in four aquifers.

variate approximately separated by vector $\boldsymbol{h}$ and can be written as:

$$\gamma'(h) = \frac{1}{2N(h)}\left\{\sum_{i=1}^{N(h)}[z(u_i+h)-z(u_i)]^2\right\}. \quad (1)$$

where $N(h)$ is the number of pairs; $z(u_i)$ and $z(u_i+h)$ are the value of precise leveling or PSI at the location u_i and u_i+h of the pair i; and the vector $\boldsymbol{h}$ can be specified with particular directions and lag distance.

3.2 Cokriging interpolation

The cokriging estimate is a linear combination of both primary and secondary data value and minimizes the variance of the estimation error by exploiting the cross-correlation between two variables (Isaaks and Srivastava, 1989). The precise leveling is set to the primary variable, $L(u_{a_1})$, the PSI results is sets to the secondary variable, $P(u'_{a_2})$, and the cokring estimation, $Z_{\mathrm{COK}}(u)$, can be written as:

$$Z_{\mathrm{COK}}(u) = \sum_{a_1=1}^{n_1}\lambda_{a_1}(u)\cdot L(u_{a_1}) + \sum_{a_2=1}^{n_2}\lambda'_{a_2}(u)\cdot P(u'_{a_2}), \quad (2)$$

where λ_{a_1} are the weights applied to the n_1 precise leveling data, u_{a_1} are the n_1 locations of precise leveling benchmarks, λ'_{a_2} are the weights applied to the n_2 PSI results, u'_{a_2} are the n_2 locations of PSI pixel.

4 Results and discussion

4.1 Variogram analysis

15 Envisat interferograms from Track 232, Frame 3123 and 17 Envisat interferogram form Track 232, Frame 3141 were used to estimate the displacement of the radar line-of-sight (LOS). At the same period 298 precise leveling data recorded the vertical deformation in CRFP. The left side of Fig. 2 shows the vertical displacement rate of precise leveling and PSI from 2005 to 2008. The cold color represents the land subsidence relative to the reference point and the warm color mean surface uplift. The most subsidence rate of PSI result almost reaches 60 mm yr^{-1} and the most subsidence rate of precise leveling data reaches 80 mm yr^{-1} in the center of southern CRFP. The lag distance of 1 km was used to analyze the semivariograms and cross semiveriogram in 6 horizontal directions (right side of Fig. 2). The red triangle is north direction, the green square is N–30° E direction, the blue star is N–60° E, the yellow diamond is N–90° E, the purple circle is N–60° W, the gray hexagon is N–30° W, and the black solid line is the omnidirectional semivariogram fitted by Gaussian model.

4.2 Cokriging estimation and groundwater level

The cokriging estimations in the three intervals display on first column of Fig. 3. The negative values of the rate (cold

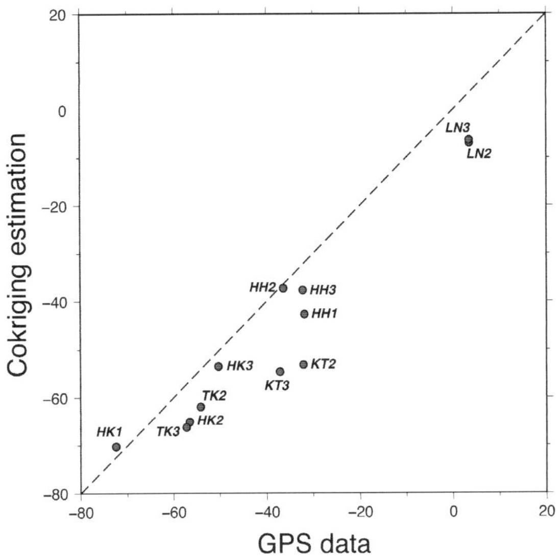

Figure 4. The comparison of cokriging estimation with GPS data. The code name of GPS station is written in text.

color) mean the area of land subsidence. In 2005–2006 the highest rate of land subsidence reaches $-99.98\,\mathrm{mm\,yr^{-1}}$, in 2005–2007 is $-93.17\,\mathrm{mm\,yr^{-1}}$, and in 2005–2008 is $80.97\,\mathrm{mm\,yr^{-1}}$. There are $440\,\mathrm{km^2}$ (1760 cells) where the subsidence rate is higher than $-50\,\mathrm{mm\,yr^{-1}}$ in 2005–2006, $340\,\mathrm{km^2}$ (1360 cells) in 2005–2007, and $316.5\,\mathrm{km^2}$ (1266 cells) in 2005–2008. Whether the rate or the area of the land subsidence slows down gradually from 2005 to 2008.

In general the subsidence in the northern part of CRFP was influenced by groundwater decline in aquifer III, and the southern part was influenced by groundwater decline in aquifer II and III. In 2005–2006 the land subsidence is most serious however the groundwater decline is slightest. It reveals that the elastic compaction would not be a major factor to induce the land subsidence during this interval. In 2005–2007 and 2005–2008, the regional groundwater decline corresponds the same location of land subsidence. Especially in aquifer III the depth is about 175–300 m, and the regional decline of groundwater level distributes in CRFP obviously. It may reflect the pumping-water location or behavior in aquifer III. The geo-material near eastern CRFP contains the gravel higher than 90 % and the permeability of the gravel is between 1.0×10^{-4} and 9.9×10^{-4} (Central Geological Survey, 1999). It gave a good path for groundwater flow and supply. Therefor even the groundwater level near Pakua Tableland draws down almost 15 m, the surface deformation reveals slight.

The five continuous CPS stations in CRFP were acquired from WRA. Two stations, Hsi-Kang (HK) and Hsin-Hsing (HH) recorded the vertical displacement from 2005 to 2008 and the others stations, Tu-Ku (TK), Lin-Nei (LN) and Ko-Tso (KT) recorded the vertical displacement from 2006 to 2008. Figure 4 shows the direct comparisons of cokriging estimation with GPS data. The correlation coefficient between the cokriging estimation and GPS data within the same cell is 0.9603 and the RMS is $10.56\,\mathrm{mm\,yr^{-1}}$. Thus the cokriging estimation is a reliable data for detecting the land subsidence in this study.

To sum above results, the mechanism of land subsidence in CRFP is quite complication. The models not only include elastic and inelastic compaction behaviors but also should consider the characteristics of different aquifers.

This research was partially supported by the Ministry of Science and Technology of the Republic of China (contracts MOST 103-2221-E-008 -049-MY3 and MOST 102-2116-M-008-010).

References

Central Geological Survey: Project of groundwater monitoring network in Taiwan during first stage – research report of Choushui River alluvial fan, Water Resources Bureau press, Taipei County, 383 pp., 1999.

Deutsch, C. V. and Journel, A. G.: Geostastistical software library and user's guide, Oxford University Press, Oxford, UK, New York, USA, 369 pp., 1998.

Galloway, D. L. and Burbey, T. J.: Review: Regional land subsidence accompanying groundwater extraction, Hydrogeol. J., 19, 1459–1486, doi:10.1007/s10040-011-0775-5, 2011.

Galloway, D. L., Hudnut, K., Ingebritsen, S., Phillips, S., Peltzer, G., Rogez, F., and Rosen, P.: Detection of aquifer system compaction and land subsidence using interferometric synthetic aperture radar, Antelope Valley, Mojave Desert, California, Water Resour. Res., 34, 2573–2585, 1998.

Hooper, A.: A multi-temporal InSAR method incorporating both persistent scatterer and small baseline approaches, Geophys. Res. Lett., 35, L16302, doi:10.1029/2008GL034654, 2008.

Hooper, A., Zebker, H., Segall, P., and Kampes, B.: A new method for measuring deformation on volcanoes and other natural terrains using InSAR persistent scatterers, Geophys. Res. Lett., 31, 1–5, 2004.

Hooper, A., Segall, P., and Zebker, H.: Persistent scatterer interferometric synthetic aperture radar for crustal deformation analysis, with application to Volcán Alcedo, Galápagos, J. Geophys. Res., 11, B07407, doi:10.1029/2006JB004763, 2007.

Hung, W. C., Hwang, C., Chang, C. P., Yen, J. Y., Liu, C. H., and Yang, W. H.: Monitoring severe aquifer-system compaction and land subsidence in Taiwan using multiple sensors: Yunlin, the southern Choushui River Alluvial Fan, Environ. Earth Sci., 59, 1535–1548, 2010.

Hung, W. C., Hwang, C., Chen, Y. A., Chang, C. P., Yen, J. Y., Hooper, A., and Yang, C. Y.: Surface deformation from persistent scatterers SAR interferometry and fusion with leveling data: A case study over the Choushui River Alluvial Fan, Taiwan, Remote Sens. Environ., 115, 957–967, 2011.

Hwang, C., Hung, W. C., and Liu, C. H.: Results of geodetic and geotechnical monitoring of subsidence for Taiwan High Speed Rail operation, Nat. Hazards, 47, 1–16, doi:10.1007/s11069-007-9211-5, 2008.

Isaaks, E. H. and Srivastava, R. M.: An introduction to applied geostatistics, Oxford University press, Oxford, UK, 561 pp., 1989.
Liu, C. W., Lin, W. S., Shang, C., and Liu, S. H.: The effect of clay dehydration on land subsidence in the Yun-Lin coastal area, Taiwan, Environ. Geol., 40, 518–527, 2001.
Liu, C. H., Pan, Y. W., Liao, J. J., Huang, C. T., and Ouyang, S.: Characterization of land subsidence in the Choshui River alluvial fan, Taiwan, Environ. Geol., 45, 1154–1166, 2004.
Massonnet, D. and Feigl, K.: Radar interferometry and its application to changes in the Earth's surface, Rev. Geophys., 36, 441–500, 1998.

Proc. IAHS, 372, 83–87, 2015
proc-iahs.net/372/83/2015/
doi:10.5194/piahs-372-83-2015

Open Access
PIAHS

Towards a global land subsidence map

G. Erkens[1,2] and E. H. Sutanudjaja[1,2]

[1]Deltares Research Institute, Utrecht, the Netherlands
[2]Utrecht University, Utrecht, the Netherlands

Correspondence to: G. Erkens (gilles.erkens@deltares.nl)

Published: 12 November 2015

Abstract. Land subsidence is a global problem, but a global land subsidence map is not available yet. Such map is crucial to raise global awareness of land subsidence, as land subsidence causes extensive damage (probably in the order of billions of dollars annually). With the global land subsidence map relative sea level rise predictions may be improved, contributing to global flood risk calculations.

In this paper, we discuss the approach and progress we have made so far in making a global land subsidence map. Initial results will be presented and discussed, and we give an outlook on the work needed to derive a global land subsidence map.

1 Introduction

Although its impact on flood risk is locally outranging the impact of absolute sea level rise, over the last decades land subsidence retrieved much less attention in terms of research. One of the reasons for this is the unknown extent of land subsidence around the world, in contrast to absolute sea level rise. For the latter, there are models that predict global absolute sea level rise (Church et al., 2013). For land subsidence there are many examples around the world, but a comprehensive picture of the global extent is currently unavailable. This originates from the heterogeneity of the land subsidence signal: an array of potential drivers and the heterogeneous subsurface hydrology and geology, makes land subsidence mainly a local phenomenon on a global scale.

A map showing the extent of land subsidence around the world, the global land subsidence map, would be very useful. It would raise awareness among scientist and policymakers alike. The land subsidence rates could be used – together with regional absolute sea level rise predictions – to estimate relative sea level rise in coastal areas. This serves as input for flood risk calculations. Lastly, a global land subsidence map would not only locate current land subsidence hotspots but also help to identify future sinking areas under predicted socio-economic development scenarios.

The maps that come close to a global land subsidence map are those based on a collection of globally spread case studies. However, these have two major limitations:

i. They are static and show the current land subsidence rates, but not necessarily at a single moment in time. They also have no predictive power.

ii. They are biased towards well-studied areas. Large cities or classic study sites in western countries show up most prominently.

Considering these limitations, we set out to produce a global land subsidence map that is derived from numerical model calculations. In this way, we are able to introduce a temporal component showing both historical and predicted future land subsidence under different development scenarios.

In this paper, we discuss the approach and progress we have made so far. Initial results will be presented and discussed, and we give an outlook on the work needed to derive a global land subsidence map.

2 The approach

Land subsidence is caused my many different drivers, both natural and anthropogenic-induced. Ideally, a global land subsidence map would cover all known drivers, but for sake of simplicity, in first instance we restricted ourselves

Published by Copernicus Publications on behalf of the International Association of Hydrological Sciences.

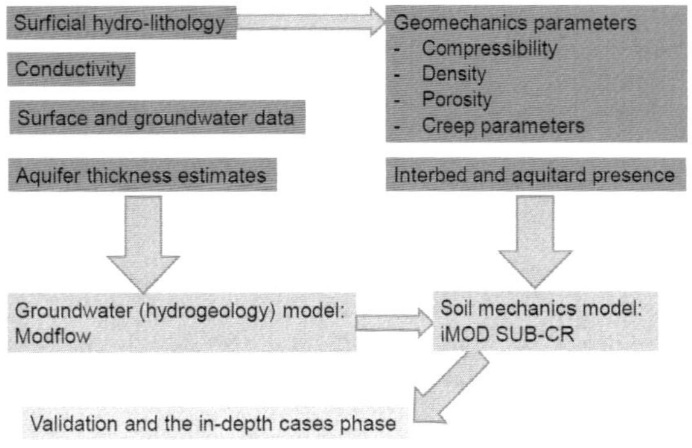

Figure 1. The proposed model suite to derive a global land subsidence map consists of three coupled models: the surface water model, the ground water model and a soil mechanics model.

to one driver. One of the most prominent causes for land subsidence is aquifer compaction as a result of excessive groundwater extraction for domestic, agricultural and industrial use (Galloway and Burbey, 2011). For instance, the Vietnamese Mekong Delta sinks on average 1.6 cm yr^{-1}, attributed to groundwater extraction (Erban et al., 2014). Crucially, in many coastal mega-cities, land subsidence is accelerated by ongoing urbanization and related increased demands for (ground) water. In Jakarta land subsidence is up to 20 cm yr^{-1} (Abidin et al., 2008). With ongoing economic development in the near future the expectation is that groundwater abstraction and, consequently, land subsidence rates, will accelerate and areas affected will expand as well as. Therefore, in this paper, we chose to focus on aquifer abstraction induced land subsidence and present our numerical approach on constructing a global land subsidence map.

Figure 1 illustrates different the model components used in this study. The integrated global hydrological and water resources model, PCR-GLOBWB (Van Beek et al., 2011; Sutanudjaja et al., 2015; Van Beek, 2008; Van Beek and Bierkens, 2009), serves as the starting point. The model is fed by global meteorological forcing data (i.e. precipitation, temperature and reference potential evaporation) and parameterized based on only global datasets (for an extensive list, see e.g. Sutanudjaja et al., 2011). PCR-GLOBWB simulates daily river discharge and groundwater recharge, as well as surface water and groundwater abstraction rates. The latter are estimated internally within the model based on the simulation of their availabilities and water demands for irrigation and other sectors (De Graaf et al., 2014; Wada et al., 2014). The daily output of PCR-GLOBWB would then be aggregated to the monthly resolution and used to force the MODFLOW groundwater model (McDonald and Harbaugh, 1988; Sutanudjaja et al., 2011, 2014) resolving spatio-temporal groundwater head dynamics, incorporating the simulated groundwater abstraction of PCR-GLOBWB. Subsequently, the simulated monthly groundwater head changes are fed into a land subsidence module, iMOD-SUB-CR (Bakr, 2015), which is an extension of the subsidence and aquifer-system compaction package (SUB-WT) of MODFLOW (Leake and Galloway, 2007). For this study all aforementioned models are simulated at the spatial resolution of 5 arc minutes (approximately 10 km at the equator).

3 First results

In this study we perform all aforementioned model simulations for the period 1958–2010. Our results are focused on the simulation results of PCR-GLOBWB that are fed in the MODFLOW model, particularly the rates of groundwater depletion, i.e. the abstraction of groundwater stores that are not being replenished by groundwater recharge. More specifically, we present and summarize the groundwater depletion rates in urban areas.

3.1 First attempt: depletion in urban area missed

Figure 2 shows first results for the island of Java, Indonesia. This is an area that is well known for its extensive land subsidence as a result ground water extraction. The map shows the depletion of ground water as calculated by the PCR-GLOBWB hydrology model. Most of the areas that show depletion are irrigation areas, where the model suggests that groundwater is excessively used. Large urban areas, such as Jakarta, with well-known depletion of groundwater and related land subsidence, are completely missed.

This misconception is attributed to the previous assumption in the PCR-GLOBWB model that simplifies that the fraction of water demands that is satisfied by groundwater resource (GW_{frac}) as a ratio between simulated baseflow and river discharge. Agricultural irrigation results in a high water demand, and because the river discharge in those areas is dominated by baseflow, the model assumes the use of groundwater. While in the urban area of Jakarta where river flows are not baseflow dominant, water demands are lower and as a result, surface water availability as simulated for Jakarta is sufficient for the domestic and industrial water demand. However, in reality, the urban areas in Java rely on ground water as a clean and reliable source of water. In turn, it is foremost surface water that is used for irrigation in Java.

3.2 Second attempt: including water resources uses

To improve the model, we included (global) databases on the relative contribution of ground water and surface water to the total water usage, for both urban agglomerations (McDonald et al., 2014: Dataset of water infrastructure) and agricultural areas (Siebert et al., 2010: Groundwater use for irrigation). Figure 3 shows the result of the second attempt in which the contribution of actual groundwater and surface water is included. The fraction of water demands that is satisfied by groundwater resource (GW_{frac}) is prescribed based on the

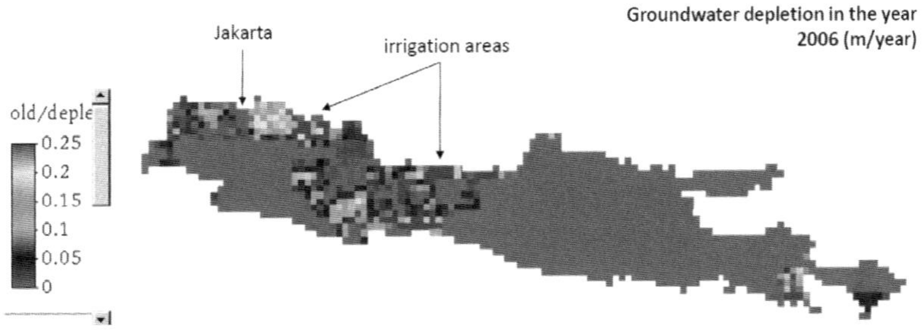

Figure 2. Screen shot of initial depletion ($m\,yr^{-1}$) results as calculated by the global hydrology model for the island of Java, Indonesia. It shows the depletion of the groundwater as a result of abstraction for irrigation of agricultural areas, but misses known depleted urban areas, such as Jakarta.

Table 1. Calculated depletion ($m\,yr^{-1}$ for the period 2000–2010) for a selection of cities that are known to subside.

Country	City	Depletion ($m\,yr^{-1}$)
United States	San Jose	3.27
Mexico	Mexico City	1.30
United States	Tucson	1.29
United States	San Francisco	1.25
Iran	Tehran	1.23
United States	Sacramento	1.18

aforementioned inventories. This outcome resembles much closer the expected outcome: dominant depletion in the urban areas.

In Fig. 4, the results of the same model run for the South China Sea area are depicted (depletion in $m\,yr^{-1}$). This matches known areas of subsidence and groundwater extraction, such as the western half of Taiwan, the cities of Hong Kong, Shanghai, Qingdao and Beijing. It also captured the North China Plain, which is an agricultural area with strongly depleted aquifers. These results give confidence in the validity of the model.

3.3 A global city database of groundwater abstractions and discussion

Based on the model calculations, we are now able provide depletion per year for different urban areas. Table 1 lists some examples of calculated depletion rates for a selected number of cities for the period 2000–2010.

Most of the cities that are known to have depletion of the aquifers are found on the list. Some cities, such as Ho-Chi-Minh City (Vietnam) or Albuquerque (New Mexico, USA) are, however, missed. One of the possible explanations for this is the unknown contribution of irrigation. This may be overestimated in urban areas, which suffer from land surface sealing as a result of building. The rates as calculated by the model are on the lower end of reported depletion rates. This may be the consequence of the scale that we use: depletion rates are averaged over a 10×10 km gridcel. Groundwater extraction and depletion, specifically for industrial sites, may however be much more local than that. Another interesting outcome is the prominent position of the city of San Jose (USA) with $3.27\,m\,yr^{-1}$ depletion calculated for the 2000–2010 period (Table 1). In reality, depletion in the San Jose area has essentially stopped after dramatic subsidence forced the local government to reduce groundwater abstraction (Ingebritsen and Jones, 1999). Surface water is now being imported from elsewhere and the imported water is partly used for recharging the aquifer systems. The model however, still calculates the consumption of ground water in San Jose, as in the model local surface water availability is insufficient.

4 Towards a global land subsidence map

The PCR-GLOBWB outcomes will be used to force a two-layer global Modflow model, consisting of an upper unit with low permeability and a deeper confined aquifer system. The groundwater model will yield hydraulic heads. A decrease in hydraulic head will not automatically imply that subsidence will occur. If the deposits consist of sand, or are (over)consolidated, the hydraulic pressure decrease and effective stress increase will not lead to inelastic volume reduction. There is enough global information on the surface geology (e.g. Dürr, et al., 2005) to be able to focus only on those areas that are susceptible to subside. The required subsurface information to parameterize the geotechnical model is however unavailable on a global scale at this time, but will be approached by using different scenarios of subsurface build-up.

The outcomes will be compared to measured or modeled land level lowering in well-known case study areas, such as Jakarta and the Vietnamese Mekong Delta. The final map will include also future land subsidence rates under different development scenarios for the entire earth, and includes a sensitivity test for different subsurface build-up. The entire map will be used as input for a global flood risk model. This

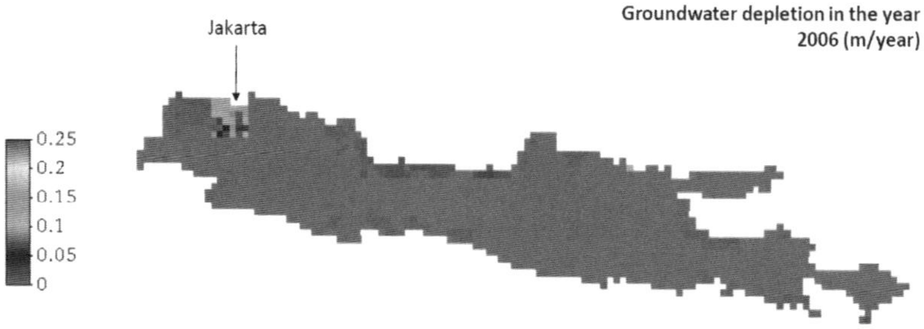

Figure 3. Screen shot of revised depletion (m yr^{-1}) results as calculated by the global hydrology model for the island of Java, Indonesia. It shows the depletion of the groundwater as a result of ground water abstraction in the urban area of Jakarta. The calculated depletion in the irrigation areas is now minor.

Figure 4. Screen shot of depletion (m yr^{-1}) results as calculated by the global hydrology model for the east coast of China and Taiwan. The cities of Hong Kong, Shanghai, Qingdao and Beijing can be recognised. The large area in the north is the North China Plain, which is an agricultural area.

will be done within the concept of a dynamic DEM (digital elevation model). Within this concept, the currently available height information (DEM) for a certain area is updated with the cumulative subsidence (or uplift) over a certain period of time. The predicted global land subsidence effects will be included in flood hazard and storm surge assessments within a global flood risk model.

Acknowledgements. The authors thank their colleagues Marijn Kuijper (Deltares Research Institute), Hessel Winsemius (Deltares Research Institute), Inge de Graaf (Utrecht University) and Marc Bierkens (Utrecht University, Deltares Research Institute) for data, support and discussion. This paper is a part of the World Resources Institute project Flood Risk and Intervention Assessment for Global Cities funded by the Netherlands Ministry of Infrastructure and Environment and the Netherlands Ministry of Foreign Affairs.

References

Abidin, H. Z., Andreas, H., Djaja, R., Darmawan, D., and Gamal, M.: Land subsidence characteristics of Jakarta between 1997 and 2005, as estimated using GPS surveys, GPS Solut., 12, 23–32, 2008.

Bakr, M.: Influence of Groundwater Management on Land Subsidence in Deltas A Case Study of Jakarta (Indonesia), Water Resour. Manag., 29, 1541–1555, doi:10.1007/s11269-014-0893-7, 2015.

Church, J. A., Clark, P. U., Cazenave, A., Gregory, J. M., Jevrejeva, S., Levermann, A., Merrifield, M. A., Milne, G. A., Nerem, R. S., Nunn, P. D., Payne, A. J., Pfeffer, W. T., Stammer, D., and Unnikrishnan, A. S.: Sea Level Change, in: Climate Change 2013: The Physical Science Basis. Contribution of Working Group I to the Fifth Assessment Report of the Intergovernmental Panel on Climate Change, edited by: Stocker, T. F., Qin, D., Plattner, G.-K., Tignor, M., Allen, S. K., Boschung, J., Nauels, A., Xia, Y., Bex, V., and Midgley, P. M., Cambridge University Press, Cambridge, United Kingdom and New York, NY, USA, 2013.

De Graaf, I. E. M., van Beek, L. P. H., Wada, Y., and Bierkens, M. F. P.: Dynamic attribution of global water demand to surface water and groundwater resources: Effects of abstractions and return flows on river discharges, Adv. Water Resour., 64, 21–33, 2014.

Dürr, H. H., Meybeck, M., and Dürr, S. H.: Lithologic composition of the Earth's continental surfaces derived from a new digital map emphasizing riverine material transfer, Global Biogeochem. Cy., 19, GB4S10, doi:10.1029/2005GB002515, 2005.

Erban, L. E., Gorelick, S. M., and Zebker, H. A.: Groundwater extraction, land subsidence, and sea-level rise in the Mekong Delta, Vietnam, Environ. Res. Lett., 9, 084010, doi:10.1088/1748-9326/9/8/084010, 2014.

Galloway, D. L. and Burbey, T. J.: Review: Regional land subsidence accompanying groundwater extraction, Hydrogeol. J., 19, 1459–1486, 2011.

Ingebritsen, S. E. and Jones, D. R.: Santa Clara Valley, California – A case of arrested subsidence, in: Land Subsidence in the United States by Galloway, D., Jones, D. R., and Ingebritsen, S. E., U.S. Geological Survey Circular 1182, 177 pp., 15–22, 1999.

Leake, S. A. and Galloway, D. L.: MODFLOW ground-water model – User guide to the Subsidence and Aquifer-System Compaction Package (SUB-WT) for water-table aquifers, U.S. Geological Survey, Techniques and Methods, 6-A23, 42 pp., 2007.

McDonald, M. G. and Harbaugh, A. W.: A modular three-dimensional finite-difference ground-water flow model: Techniques of Water-Resources Investigations of the United States Geological Survey, Book 6, Chapter A1, 586 pp., 1988.

McDonald, R. I., Weber, K., Padowski, J., Flörke, M., Schneider, C., Green, P. A., Gleeson, T., Eckman, S., Lehner, B., Balk, D., Boucher, T., Grill, G., and Montgomery, M.: Water on an urban planet: Urbanization and the reach of urban water infrastructure, Global Environ. Chang., 27, 96–105, 2014.

Siebert, S., Burke, J., Faures, J. M., Frenken, K., Hoogeveen, J., Döll, P., and Portmann, F. T.: Groundwater use for irrigation – a global inventory, Hydrol. Earth Syst. Sci., 14, 1863–1880, doi:10.5194/hess-14-1863-2010, 2010.

Sutanudjaja, E. H., van Beek, L. P. H., de Jong, S. M., van Geer, F. C., and Bierkens, M. F. P.: Large-scale groundwater modeling using global datasets: a test case for the Rhine-Meuse basin, Hydrol. Earth Syst. Sci., 15, 2913–2935, doi:10.5194/hess-15-2913-2011, 2011.

Sutanudjaja, E. H., van Beek, L. P. H., de Jong, S. M., van Geer, F. C., and Bierkens, M. F. P.: Calibrating a large-extent high-resolution coupled groundwater-land surface model using soil moisture and discharge data, Water Resour. Res., 50, 687–705, 2014.

Sutanudjaja, E. H., van Beek, L. P. H., Bosmans, J. H. C., Drost, N., de Graaf, I. E. M., de Jong, K., Peßenteiner, S., Straatsma, M. W., Wada, Y., Wanders, N., Wisser, D., and Bierkens, M. F. P.: PCR-GLOBWB 2.0: a 5 arc-minute global hydrological and water resources model, Geosci. Model Dev. Discuss., in preparation, 2015.

Van Beek, L. P. H.: Forcing PCR-GLOBWB with CRU data, [online] available at: http://vanbeek.geo.uu.nl/suppinfo/vanbeek2008.pdf (last access: 2 June 2015), 2008.

Van Beek, L. P. H. and Bierkens, M. F. P.: The Global Hydrological Model PCR-GLOBWB: Conceptualization, Parameterization and Verification, [online] available at: http://vanbeek.geo.uu.nl/suppinfo/vanbeekbierkens2009.pdf (last access: 2 June 2015), 2009.

Van Beek, L. P. H., Wada, Y., and Bierkens, M. F. P.: Global monthly water stress: 1. Water balance and water availability, Water Resour. Res., 47, W07517, doi:10.1029/2010WR009791, 2011.

Wada, Y., Wisser, D., and Bierkens, M. F. P.: Global modeling of withdrawal, allocation and consumptive use of surface water and groundwater resources, Earth Syst. Dynam., 5, 15–40, doi:10.5194/esd-5-15-2014, 2014.

Proc. IAHS, 372, 89–93, 2015
proc-iahs.net/372/89/2015/
doi:10.5194/piahs-372-89-2015

Open Access
PIAHS

Surface deformation on the west portion of the Chapala lake basin: uncertainties and facts

M. Hernandez-Marin[1], J. Pacheco-Martinez[1], J. A. Ortiz-Lozano[1], G. Araiza-Garaygordobil[1], and A. Ramirez-Cortes[2]

[1]Departamento de Geotecnia e Hidráulica, Universidad Autónoma de Aguascalientes, Aguascalientes, Mexico
[2]Doctorado en Ciencias de los Ámbitos Antrópicos, Universidad Autónoma de Aguascalientes, Aguascalientes, México

Correspondence to: M. Hernandez-Marin (mhernandez@correo.uaa.mx)

Published: 12 November 2015

Abstract. In this study we investigate different aspects of land subsidence and ground failures occurring in the west portion of Chapala lake basin. Currently, surface discontinuities seem to be associated with subsiding bowls. In an effort to understand some of the conditioning factors to surface deformation, two sounding cores from the upper sequence (11 m depth) were extracted for analyzing physical and mechanical properties. The upper subsoil showed a predominant silty composition and several lenses of pumice pyroclastic sand. Despite the relative predominance of fine soil, the subsoil shows mechanical properties with low clay content, variable water content, low plasticity and variable compressibility index, amongst some others. Some of these properties seem to be influenced by the sandy pyroclastic lenses, therefore, a potential source of the ground failure could be heterogeneities in the upper soil.

1 Introduction

Mexico is one of the most affected countries by anthropogenic subsidence. Here, conditioning factors such as topography, hydrostratigraphy, depositional environments, and others are combined with large periods of groundwater withdrawal, resulting in surface deformation in the form of subsidence (vertical deformation), or faulting and fissuring (horizontal or combined horizontal and vertical deformation). The zone known as the Transmexican Volcanic Belt is perhaps the most affected area in the country, mostly due to the post-depositional processes resulting in fine sediments that potentially contain smectitic clays, usually derived from volcanic materials deposited in lacustrine and currently urbanized valleys such as Mexico City, Queretaro, Chapala, and others.

In this work, new findings regarding surface deformation on the west portion of the Chapala lake basin are analyzed and discussed, and previous measured land subsidence previously measured and mapped faults are recalled. Novel information includes the description and geotechnical characterization of sediments from two cores of 11 m from of the upper sedimentary sequence. Additionally, uncertainties regarding the factors that may condition or trigger surface deformation in this particular area are discussed.

2 Conditions of Chapala lake basin

The lake of Chapala and its potential flooding area is bounded by ranges. The elevation difference between the top of the mountains and the lacustrine plain is close to 500 m. This difference is more significant along the western portion of the basin. Due to the potential for tourism on the lake shore, several urban centers around the lake have experienced significant growth in the past few decades, one of them is the community of Jocotepec on the west with close to 37 700 inhabitants (INEGI, 2010). The increasing population in all communities around the lake has triggered the demand of water to satisfy agricultural and municipal needs in the last decades. Usually people from the valley prefer groundwater because is cleaner and more accessible than the water from the lake.

In the basin of Chapala lake, as most inside the Transmexican Volcanic Belt, tectonism, volcanism and sedimen-

Published by Copernicus Publications on behalf of the International Association of Hydrological Sciences.

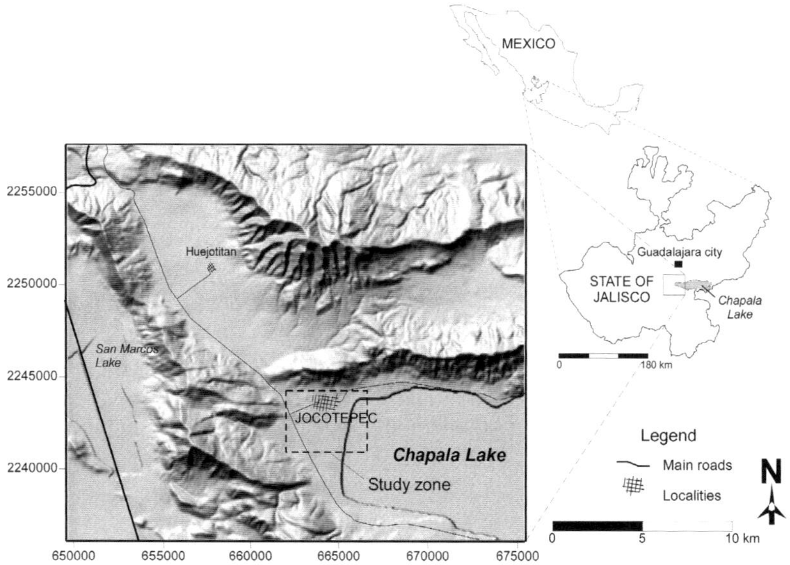

Figure 1. Location of the study area in the west part of the Chapala lake. From Hernandez-Marin et al. (2014).

tation occurred simultaneously during tertiary an quaternary ages. Regionally, this basin is seismically active with recent shallow earthquakes. This basin presents its longest side in an east-to-west direction, which is collinear to the main regional fault system. The ranges bounding the lake plain are mainly volcanic and plutonic, formed by basalts, rhyolite and pyroclastic material (Rosas-Elguera and Urrutia-Fucugauchi, 1998); and the sedimentary sequence of the plain presents alluvial, lacustrine and volcanic deposits (ash) including sands, silts and clays (Zarate-del-Valle and Simoneit, 2005). Figure 1 depicts the location and main characteristics of the study area.

Land subsidence in this part of Mexico is recent since the first reports of surface deformation are from the early 90's, however, it was not documented until 2007 by Castillo-Aja and Valdivia-Ornelas (2007). Locally, the first cores for physically analyzing sediment samples were used to determine mineralogy of the fine sediments and the rate of sedimentation (Fernex et al., 2001). Through the soil samples from these cores it was observed that the main minerals of the clay portion are halloysite, hectorite and illite, and the sand grains are mainly composed of vitreous volcanic ash (pumice) (Fernex et al., 2001).

3 Monitoring campaign of surface deformation

Land subsidence and surface faulting have recently been a recurrent problem in the urban area of Jocotepec and suburbs. The problems associated to ground deformation include damage to buildings, roads and other constructions. In 2012, as part of a project to quantify subsidence, identify ground failures and determine damage to infrastructures, it was found that for a period of 8 months (April–November), a maximum deformation of 7.16 cm was measured with a calculated rate of 0.89 cm month^{-1}. Additionally, two zones of the surveyed area showed uplifting close to 2 cm. Four alignments of surface faults with preferential direction SW-NE are affecting several constructions (Hernandez-Marin et al., 2014). Land subsidence, fault alignments and other characteristics are shown in Fig. 2.

4 Sampling and analyses

In order to determine the stratigraphy of the upper sequence (< 11 m) inside the subsiding zone, and determine its main geotechnical properties, two cores of 11 m depth were sampled. Analyses included stratigraphy determination, sediment description and geotechnical testing in accordance with ASTM standards.

4.1 Stratigraphy of the analyzed sequence

Observations from samples in both cores indicate a predominance of green-olive fine sediments with minor layers of brown fine sediments and alternating gray-light and gray-dark pyroclastic sand. The thickness of the sandy layers is variable in both cores, in core 1 (northernmost) is more than 2 m, but in the core 2 (drilled at 287 m to the southeast from the core 1) the thickness is close to 20 cm only. Figure 3 shows the observed stratigraphy with descriptions of the sediments from the sampled cores.

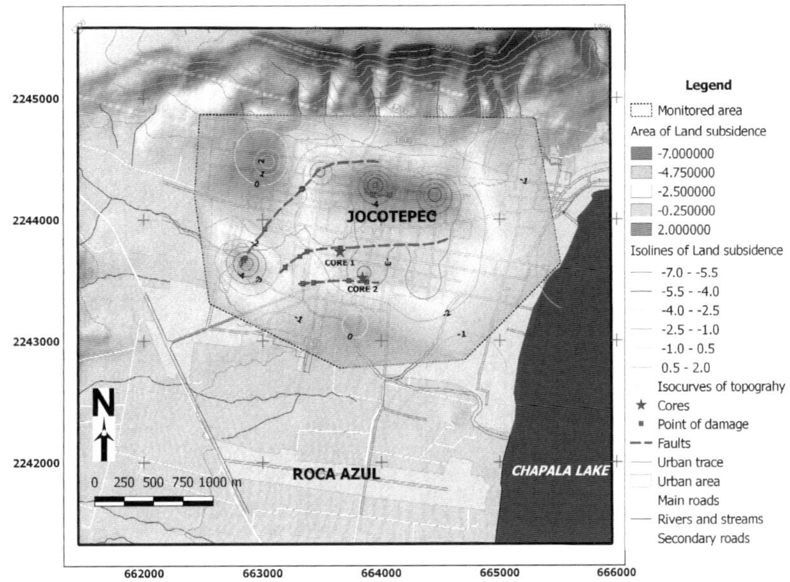

Figure 2. General characteristics of the study zone at west Chapala lake basin. Subsidence registered and fault alignments observed in 2012 are present. Also the two cores for geotechnical analysis of the subsoil are indicated. Units of area and isolines of land subsidence are in centimeters.

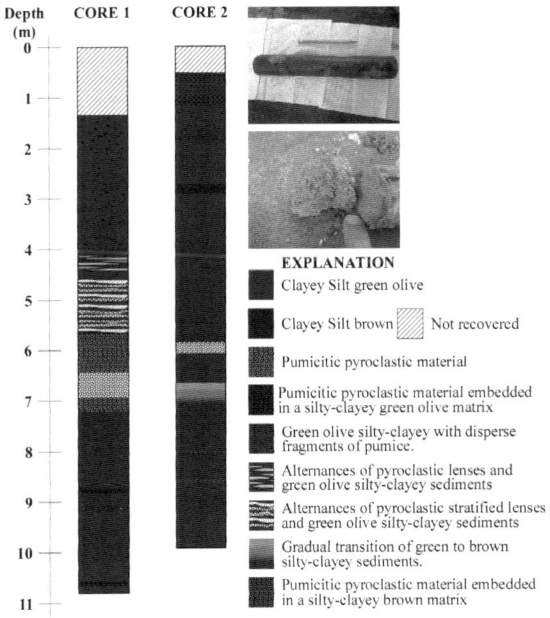

Figure 3. Described stratigraphy of both cores. Sample photographs of the predominant green-olive sediments, along with an example of a pumice sandy layer are shown.

4.2 Grain-size, water content, consistency limits and compressibility

Even though the sequence appears to be macroscopically dominated by fine sediments, the grain-size of sediments passing sieve # 10 (≤ 2 mm) using the hydrometer technique, demonstrated the presence of fine sands, particularly in core 1. As observed in Fig. 4 the clay content is very low in core 1, in comparison with the high content found at some depths in core 2, were proportion is as high as 37 % between 1 and 2 m (Fig. 5). Gravimetric water content is low compared with soil samples from other volcano-lacustrine basins such as Mexico City. Maximum water content is close to 215 %, although most of the water content measurements are no higher than 150 %, (in Chalco basin, adjacent to Mexico City, for instance, water content can reach 350 %, Hernandez-Marin, 2003). Consistency limits are low with liquid limits lower than 50 %. Finally, to obtain the compressibility properties, consolidation test were performed from unaltered samples. In general, compressibility index is lower than 0.2 (only one determination was close to 0.4), as shown in Figs. 4 and 5.

5 Uncertainties and discussion

Many questions remain unanswered regarding the process of subsidence in the locality of Jocotepec and in general in the Chapala lake basin. Some of these questions are for instance: is the groundwater depletion the unique process causing land subsidence? As demonstrated in Hernandez-Marin

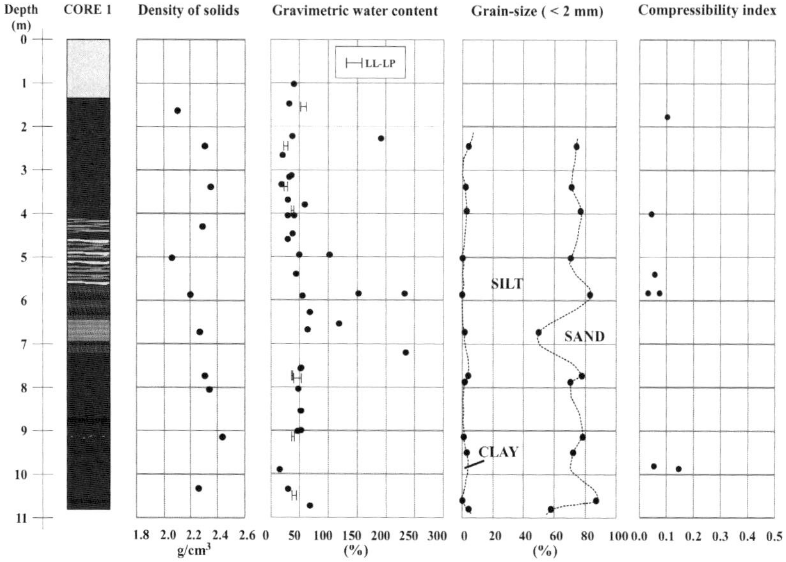

Figure 4. Core one and some of its geotechnical properties. LL-LP represents the values of liquid and plastic Atteberg limits.

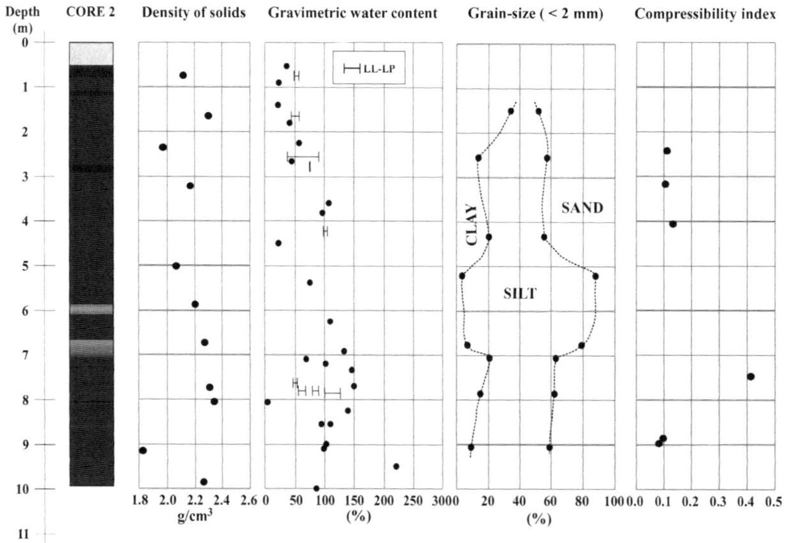

Figure 5. Core two and some of its geotechnical properties. LL-LP represents the values of liquid and plastic Atteberg limits.

et al. (2014), subsidence bowls do not coincide with groundwater cones of depression, even though some cones of depression are remarkable due to intensive pumping for irrigation at the south of the study area. Even though groundwater levels are shallow in most of the area, since as observed during drilling of cores, the groundwater table was no deeper than 2 m. Another question is regarding the tectonic component: is this stress contributing to surface deformation? The zone is tectonically active and is regionally dominated by a triple junction rift–rift–rift (Pacheco et al., 1999). The last registered earthquake occurred in 17 May 2012, of magnitude 4.4 and the epicenter was located 6 km from the study zone, as reported by the National Seismological Service, indicating tectonic activity and potential stressing. Fault alignments tend to follow isolines of subsidence, as presented in Fig. 2, however that tendency is only suggestive; therefore, the real relationship between faults and subsidence is still under investigation. The origin of faults is also unknown.

A first hypothesis is that differential subsidence due to heterogeneities in the lacustrine sequence is the possible cause. However, future work must corroborate this hypothesis.

6 Conclusions

New information regarding the geotechnical properties of the upper sequence of the west portion of Chapala lake basin is presented and discussed, and the previous study to quantify subsidence and delineate ground failures is revisited. Two cores for the geotechnical analysis of sediments show predominance of silt and fine sand, whereas geotechnical properties such as water content, consistency limits, and compressibility index are low compared with samples from other volcano-lacustrine basins such as Mexico basin. Differences in the stratigraphy, in particular those related to pumice layers, less compressible than silty and clayey layers, are a potential cause to differential subsidence and surface faulting.

Acknowledgements. This work was financed by the Autonomous University of Aguascalientes with the internal project: PIIC13-4N. The authors are also thankful for the reviewers comments, which improved the quality of this manuscript.

References

Castillo-Aja, M. R. and Valdivia-Ornelas, L.: Amenazas por agrietamiento en el Valle de Tesistán [Risks due to cracking in the valley of Tesistan], Universidad de Guadalajara, 2007

Fernex, F., Zarate-del-Valle, P., Ramirez-Sanchez, H., Michaud, F., Parron, C., Dalmasso, J., Barci-Funel, G., and Guzman-Arroyo, M.: Sedimentation rates in Lake Chapala (western Mexico): possible active tectonic control, Chem. Geol., 177, 213–228, 2001.

Hernandez-Marin, M.: Estudio del comportamiento hidromecánico de sedimentos vulcanolacustres durante el proceso de consolidación [Study of the hydromechanic behavior of the vulcanolacustrine sediments during the process of consolidation], Master Dissertation, National Autonomous University of Mexico, Mexico, 2003.

Hernandez-Marín, M., Pacheco-Martínez, J., Ramírez-Cortés, A., Burbey, T. J., Ortíz-Lozano, J. A., Zermeño-De-Leon, M. E., Guinzberg-Velmont, J., and Pinto-Aceves, G.: Evaluation and analysis of surface deformation in west Chapala basin, central Mexico, Environ. Earth Sci., 72, 1491–1501, 2014.

INEGI: Censo de población y vivienda [Census of population and housing], Mexico, 2010.

Pacheco, J. F., Montera-Gutierrez, C. A., Delgado, H., Singh, S. K., Valenzuela, W., Shapiro, N. M., Santoyo, M. A., Hurtado, A., Barron, R., and Gutierrez-Miguel, E.: Tectonic significance of an earthquake in the Zacoalco half-graben, Jalisco, Mexico, J. S. Am. Earth Sci., 12, 557–565, 1999.

Rosas-Elguera, J. G. and Urrutia-Fucugauchi, J.: Tectonic control of the volcano-sedimentary sequence of the Chapala graben, western Mexico, Int. Geol. Rev., 40, 350–362, 1998.

Zarate-del-Valle, P. and Simoneit, B.: La generación de petróleo hidrotermal en sedimentos del Lago Chapala y su relación con la actividad geotérmica del rift Citala en el estado de Jalisco, México [Generation of hidrotermal oil into the Chapala Lake sediments and its relationship with the geotermal activity of the Citala rift in the state of Jalisco, Mexico], Revista Mexicana de Ciencias Geológicas, 22, 358–370, 2005.

Proc. IAHS, 372, 95–99, 2015
proc-iahs.net/372/95/2015/
doi:10.5194/piahs-372-95-2015

Factor analysis on land subsidence in the Nobi Plain, southwest Japan

A. Kouda[1], K. Nagata[2], and T. Sato[3]
[1]Teikoku International Corporation, 2-8 Hashimoto-cho, Gifu, Japan
[2]Geospatial Information Authority of Japan, 2-5-1 Sannomaru, Naka-ku, Nagoya, Japan
[3]Dept. of Civil Engineering, Gifu University, 1-1 Yanagido, Gifu, Japan
Correspondence to: A. Kouda (kouda@teikoku-eng.co.jp)

Published: 12 November 2015

Abstract. The land subsidence of the Nobi Plain largely ceased with the commencement of pumping regulations beginning in 1975. However, a small amount of land subsidence, less than one centimeter, is still observed at the Delta zone in the southwest part of the Plain. The authors attempted to investigate the cause of the small amount of existing subsidence. The alluvial clay layer deposits are more than 15–20 m thick and the withdrawal is small. The decrease of the yearly average groundwater level has not been confirmed.

On the other hand, the seasonal change in the groundwater level is clearly observed. This investigation focuses on the seasonal change of groundwater level each year and its relation to the thickness of alluvial clay layer. The Delta zone was divided into several cells and a multiple regression analysis was applied to the seasonal change of the groundwater level and the thickness of alluvial clay layer of the cell. The study concluded that the small amount of land subsidence was caused of drawdown of piezometric head of groundwater every year during the summer.

1 Introduction

Three large rivers (Fig. 1), the Kiso R. to the east, the Nagara R. in the middle and the Ibi R. in the west, flow it into the Ise Bay which is located at the bottom of Fig. 1.

The area of the plain is almost 1300 km^2 and the population reached almost 6 million in 2010 which is a 16 % increases from 2005. Average precipitation has been recorded as 1543 mm year−1 since 1961. Groundwater withdrawal was recorded as 1 251 million m^3 year−1 in 1976, which breaks down to 743 million m^3 for industry, 228 million m^3 for drinking and 156 million m^3 for agricultural use.

Large amounts of land subsidence have occurred due to excess groundwater use (Fig. 2). Groundwater use has been regulated since 1975. As a result the piezometric heads have gradually increased and the rate of land subsidence has gradually decreased. However, small amounts of land subsidence have continued in the Delta zone of the southwest part since the 1980s (Fig. 3). Groundwater use in the Delta zone is small. However, 15–20 m of soft alluvial clay occur in this area.

In this paper, we conducted a factor analysis using the potential factors and the induction factor for the Delta zone to elucidate the cause of the small amount of ongoing land subsidence.

2 Multiple regression analysis of land subsidence

2.1 Features of land subsidence

Since 1989 the land subsidence area shows good agreement with the area of the alluvial clay layer with more than 15 m thickness (Fig. 4). Piezometric groundwater levels severely drop more than 2 m in the summer in the northern part of the area, A decrease of piezometric head in the Delta zone and is estimated to be about 0.2–2.0 m each year during the summer (Fig. 5). Figure 5 indicates that the area of greatest cumulative land subsidence does not necessarily align with that of groundwater level declines each year during the summer. Special attention is placed on the relationship between piezometric head decline and thickness of the alluvial clay layer and cumulative land subsidence amount.

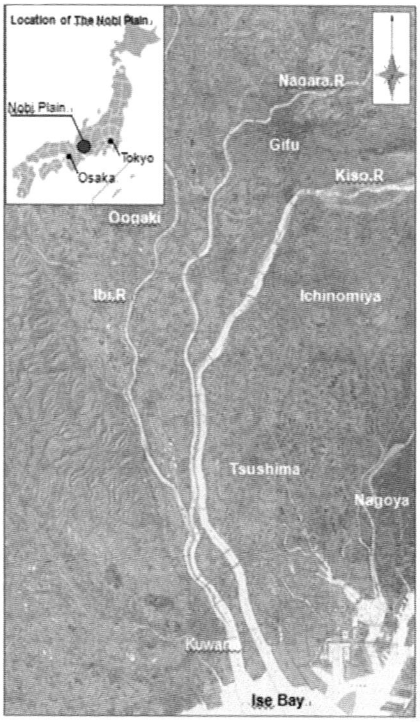

Figure 1. Reserch area and major rivers.

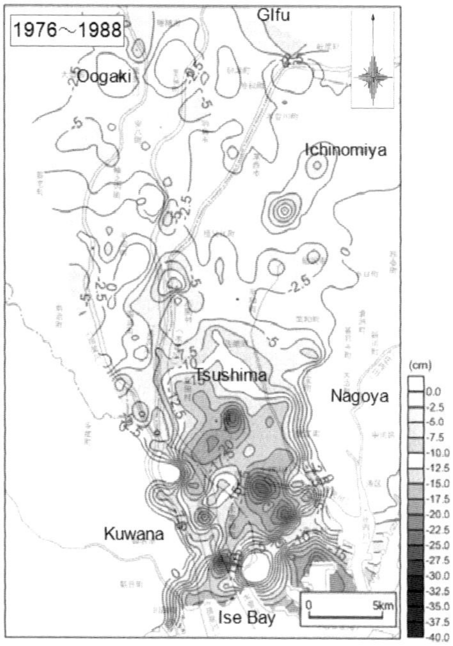

Figure 2. Cumulative land subsidence amount in the Nobi Plain (1976–1988).

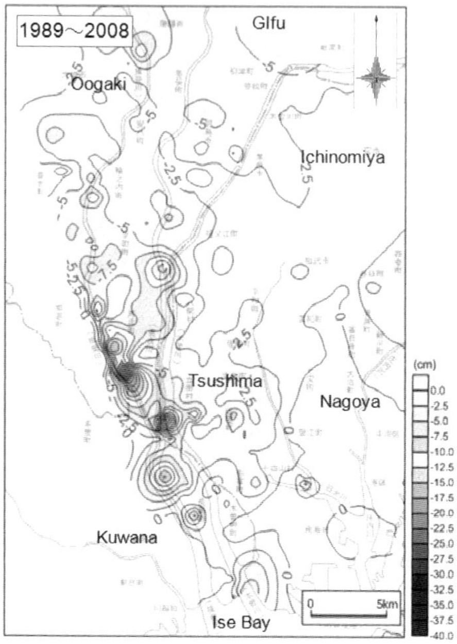

Figure 3. Cumulative land subsidence amount in the Nobi Plain (1989–2008).

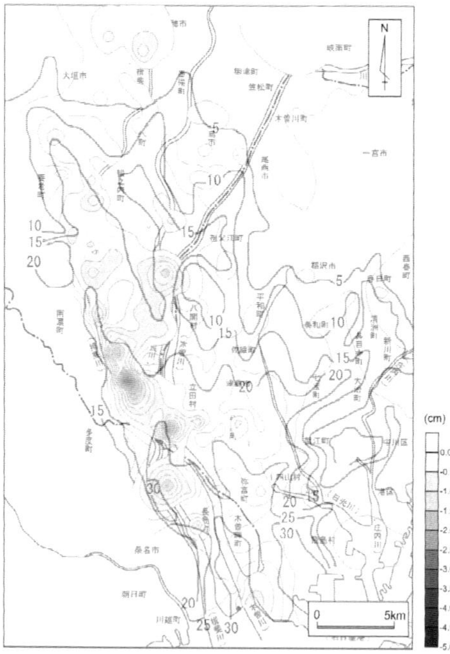

Figure 4. The thickness of alluvial clay layer and land subsidence area (m).

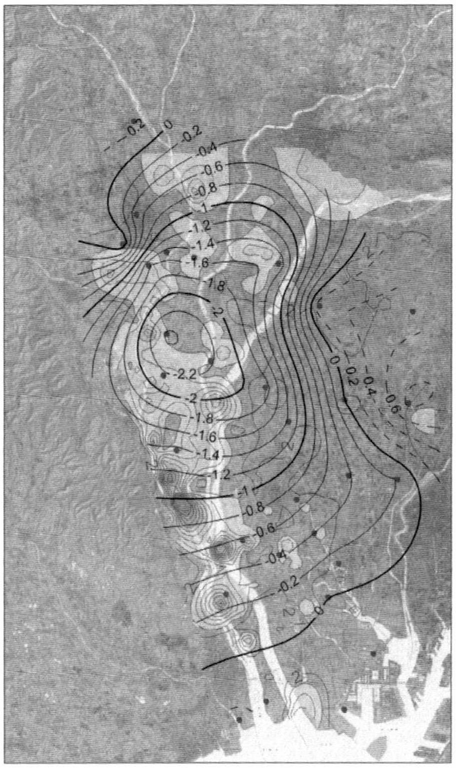

Figure 5. Decrease of piezometric head of groundwater in summer and land subsidence area (2007).

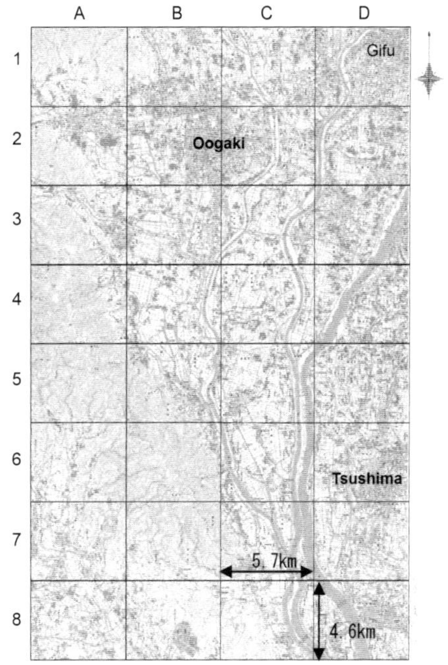

Figure 6. Analysis area.

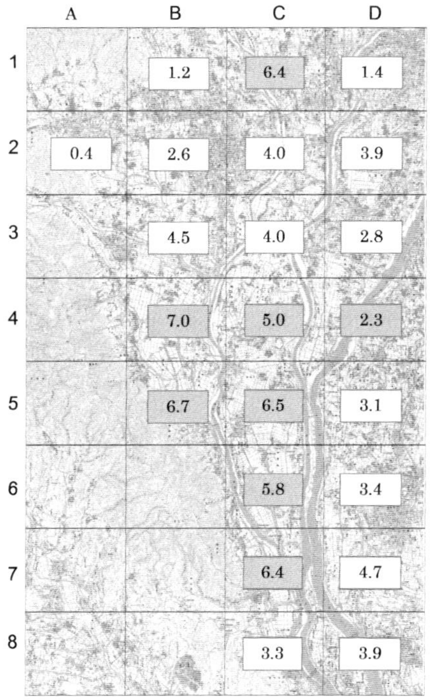

Figure 7. Average Cumulative value of land subsidence (1989–2006).

2.2 Model of the regression analysis

The research area is divided into 32 cells each with a length of 5.7 km in the east-west direction and 4.6 km in the north-south direction (Fig. 6). The measured values at the 307 benchmarks are applied to the calculation of the cumulative land subsidence of the cells. The results during 17 years of analysis, from 1989 to 2006, are shown in Fig. 7. The thickness of the alluvial clay layer is shown in Fig. 8. Seasonal variation of piezometric head of groundwater was calculated from differences in the yearly average groundwater level from 1989 to 2006 from the 28 observation wells. The amount of seasonal variation in each cell, ΔP is determined from the isolines of the seasonal variation, which is derived from the difference of the yearly average value of the piezometric head of groundwater during the summer (from June to September) and the winter (from October to May). The amount of ΔP is obtained from the difference between the yearly average value of piezometric head of groundwater in the winter and the yearly average value of piezometric head of groundwater in the summer. Positive values represent the decline in the piezometric head during the summer season (Fig. 9).

	A	B	C	D
1		7.6	9.7	9.1
2	1.1	11.4	14.3	12.2
3		19.9	20.4	18.3
4		26.3	20.7	15.7
5		21.5	26.0	14.4
6			26.5	20.5
7			22.6	23.4
8			26.9	28.7

Figure 8. Average thickness of alluvial clay layer.

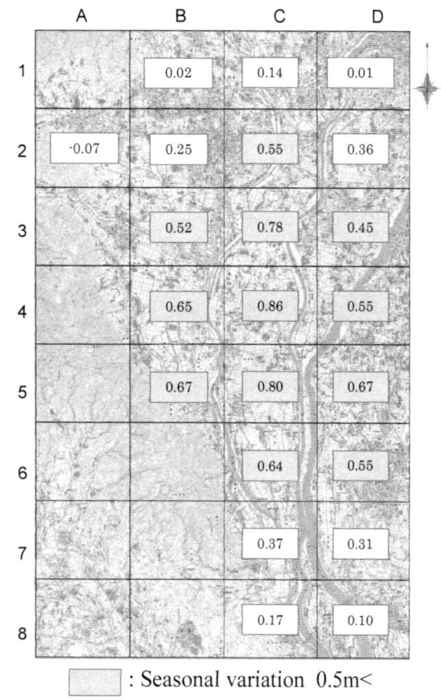

Figure 9. Seasonal variation of average value of groundwater level (1989–2006).

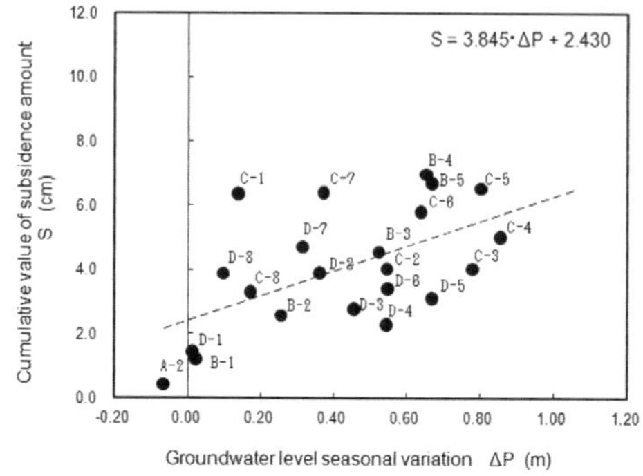

Figure 10. Cumulative value of land subsidence and seasonal variation groundwater level.

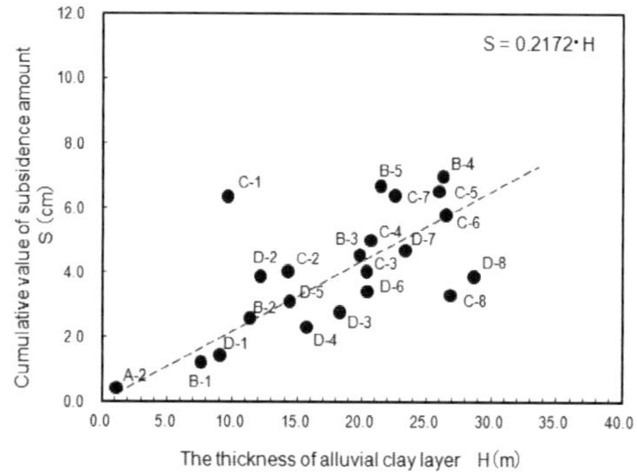

Figure 11. Cumulative value of land subsidence and thickness of the alluvial clay layer.

2.3 Multiple regression analysis results

The relationship between the amount of seasonal variation of groundwater level (ΔP) and the cumulative land subsidence (S) is shown in Fig. 10. Calculated values vary widely and the correlation coefficient is low, $r = 0.566$. One of the tendencies is that cumulative land subsidence is proportional to the increase of the amount of seasonal variation of groundwater level.

The relationship between the thickness of the alluvial clay layer (H) and the cumulative land subsidence (S) is shown in Fig. 11. The correlation coefficient is $r = 0.650$. Cumulative land subsidence is largely proportional to the increase of the thickness of the alluvial clay layer.

Multiple regression analysis provides the influence of (H) and (ΔP). The relationship between the measured and the calculated values of the cumulative land subsidence is shown in Fig. 12. The multiple correlation coefficient improves to $R = 0.704$.

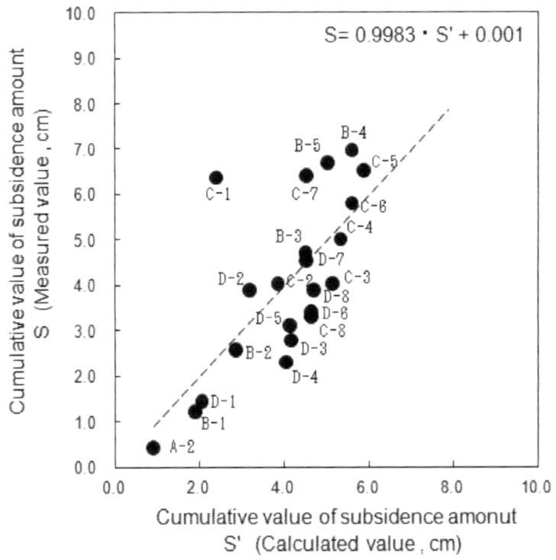

Figure 12. Comparison of Measured value and Calculated value of land subsidence.

3 Discussions and conclusions

Land subsidence occurs every year during the summer in the region where the thickness of soft alluvial clays deposits are greatest. The results of the multiple regression analysis suggests that the seasonal variation of the piezometric head is a suitable measure for describing the inducing factor of land subsidence,which is less than one centimeter per year in this area.

Consolidation curves under cyclical loading are below the normal consolidation line (NCL). It will takes a significant amount of time to reach the final land subsidence, which is greater than the NCL of the same level of an effective mean stress. Piezometric heads decline each year during the summer, which coincides with the cyclical loading curves in consolidation theory for alluvial clay layers.

This represents the main cause of continuous land subsidence over the long term at this site.

References

Gifu Prefecture: Report of Measures summary promote research for land subsidence revention, etc., 70–80, 2008.

Yasuhara, K. and Anderson, K. H.: Recompression of normally consolidated clay after cyclic loading, Soils and Foundations, 31, 83-94, 1991.

Land subsidence research meeting of tokai three Prefecture: Land subsidence and groundwater in Nobi Plain, 59–76, 1985.

Proc. IAHS, 372, 101–104, 2015
proc-iahs.net/372/101/2015/
doi:10.5194/piahs-372-101-2015

Open Access
PIAHS

Frequency anomaly of groundwater level before major earthquakes in Taiwan

T.-K. Yeh[1], C.-H. Chen[2], C.-H. Wang[3], and S. Wen[2]

[1]National Taipei University, New Taipei, Taiwan
[2]National Chung Cheng University, Chiayi, Taiwan
[3]Academia Sinica, Taipei, Taiwan

Correspondence to: T.-K. Yeh (bigsteel@mail.ntpu.edu.tw)

Published: 12 November 2015

Abstract. Anomalous decreases on water levels were observed in 78 % of wells in central Taiwan about 250 days before the Chi-Chi earthquake whose magnitude is 7.6 on 20 September 1999. Variations in groundwater levels measured on anomalous wells from 1 August 1997 to 19 September 1999, were transferred into the frequency domain to unveil frequency characteristics. Analytical results show that amplitudes at the frequency band between 0.02 and 0.04 day^{-1} generally maintained at low stage and were apparently enhanced a few weeks before the Chi-Chi earthquake. Variations of amplitude at this particular frequency band were further examined along with other Taiwan earthquakes whose magnitude is larger than 6 from 1 August 1997 to 31 December 2009. Features of the enhanced amplitudes at the frequency band are consistently observed prior to the other two earthquakes during the 12.5-year study period. Result confirms that abnormal rise and fall changes in groundwater level yield an agreement with forward and backward surface displacements around epicenter prior to the Chi-Chi earthquake.

1 Introduction

Taiwan is an island located at the western margin of the Pacific Ocean with about 400 km in length and 150 km in width (Fig. 1). The convergent plate interaction between the Philippine Sea plate and the Eurasia plate has raised Taiwan and formed the central mountain with an altitude up to about 3952 m since around 6 million years ago. Annual precipitation in Taiwan is about 2500 mm on average based on the long-term records of Water Resources Agency, but rainfall is mainly concentrated in the wet season (May–October, about 78 % of the annual total; also see Fig. 2) due to high contributions from Meiyu and typhoons. Rainfall retention for effective water resource management is very important in Taiwan because uneven distribution of precipitation is very prominent both in temporal and spatial modes. When water consumption has rapidly increased by population growth and economic development after 1980, a great quantity of groundwater has been used to make up the insufficiency of surface water. Consequently, land subsidence happens following rapid depletion of groundwater resources in the western and southwestern regions of Taiwan from the past decades, due to excessive extraction and slow recharge (Liu et al., 2004).

The most notorious land subsidence region with an active subsiding area of over 600 km^2 and a maximum subsiding rate up to 10 $cm\,yr^{-1}$ is located at the Choshuichi Alluvial Fan of central Taiwan (Chen et al., 2010). To effectively utilize groundwater resource and control land subsidence along coastal areas, where pumping is intensive and recharge is very slow, 54 evenly distributed hydrologic stations that record hourly changes in water levels using piezometers were installed at depths ranged from 24 to 306 m in the Choshuichi Alluvial Fan during 1992–1997 (Fig. 1). The Choshuichi Alluvial Fan, with an area of about 1800 km^2, can be divided into three aquifers for a depth of 250 m according to hydrogeological surveys in subsurface (Chen and Yuan, 1999). Each station may have one to five screens situated in different wells for fully observing groundwater level changes from shallow to deep aquifers.

Published by Copernicus Publications on behalf of the International Association of Hydrological Sciences.

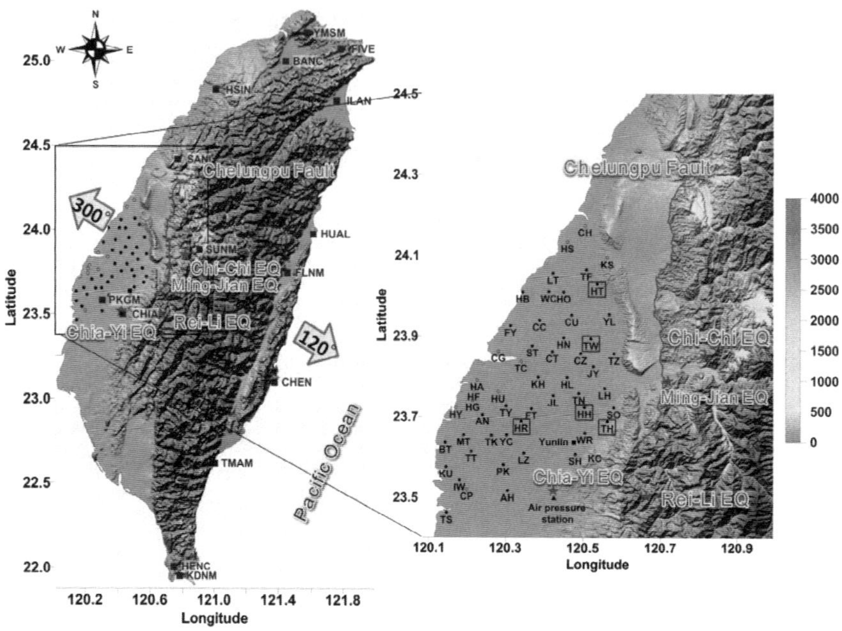

Figure 1. Locations of 4 earthquakes, 54 groundwater wells and 15 GPS stations in Taiwan.

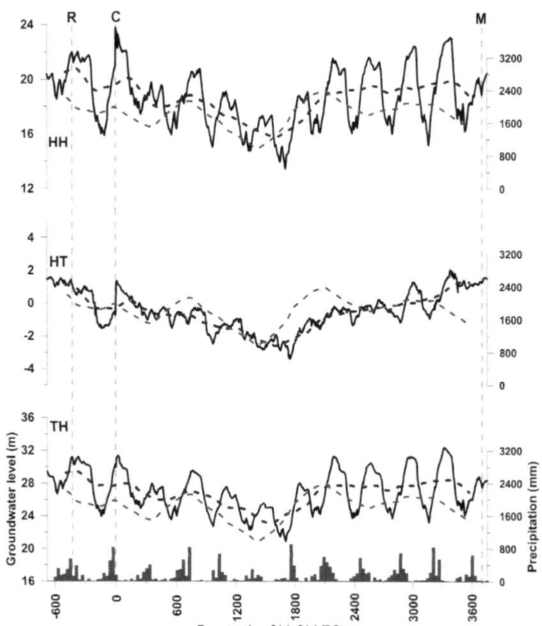

Figure 2. Variations in groundwater levels at HH, HT and TH stations from 1 August 1997 to 31 December 2009. The red line shows the average annual precipitation and the blue line displays the groundwater levels.

In this study, effects of the barometric pressure and Earth's tide on groundwater levels at 54 monitoring wells in the Choshuichi Alluvial Fan are mitigated first to examine any unusual variation associated with the Chi-Chi earthquake in the normally temporal domain. Notably, time-series water levels in the period of 2.5-year right before the Chi-Chi earthquake are utilized to reveal frequency characteristics associated with the Chi-Chi earthquake via the Hilbert–Huang Transform (HHT) and retrieve suspect signals buried in the temporal domain (Huang et al., 1998; Huang and Wu, 2008). After the earthquake-related frequency band have been determined from records before the Chi-Chi earthquake, amplitude at the particular frequency bands, which has deduced from 12.5-year data from 1 August 1997 to 31 December 2009 in groundwater stations of Huhsi (HH), Huatang (HT) and Tungho (TH), are further examined to check whether the identified anomalous amplitude can be repeatedly observed for other two major earthquakes. In addition, stress variations deduced by the surface displacements from the GPS data are compared with abnormal changes in groundwater level during the Chi-Chi earthquake to examine possible physical mechanisms.

2 Long-term variations of groundwater level in central Taiwan

It is well known that precipitation is highly related to climatic changes and is the major factor dominating the variations of groundwater levels. Before retrieving earthquake-related signals from groundwater levels, long-term changes (3–5 years) were considered first. Figure 2 shows variations in groundwater level at the HH, HT and TH stations and records in precipitation at the Yunlin station (120.476° E, 23.636° N) between 1 August 1997 and 31 December 2009. Precipitation effects on groundwater level can be viewed as two distinct features during this 12.5-year period. The first feature, which comprises many short (3–5 years) cycles lying on the generally decreasing trend in the first half and then a rising tendency in the latter half, yields an agreement between 1-year running average in annual precipitation (red line) and groundwater levels (blue line) at the HH, HT and TH sites.

The second one is apparent annual cycles that dominate in groundwater levels, particularly for those of the HH and TH sites. Annual cycles of the groundwater levels in Taiwan are caused by the apparent shift between wet and dry seasons in one year. For stations that are located relatively far away from recharge areas, their annual changes in the time-series data are comparatively weak (such as the HT shown in Fig. 2). It is worth mentioning that an unusual drop can be clearly identified from records at 78 % (= 42/54) stations (including HT, HH and TH) about 250 days prior to the Chi-Chi earthquake. The HT site had this obviously unusual drop standing out in the temporal domain due to weak annual changes. By contrast, for sites of HH and TH, the abnormal drops would be masked by in-phase annual variations in groundwater levels in the time-series record. This feature vividly illustrates the primary drawbacks in retrieving the abnormal drops of groundwater levels via the temporal records.

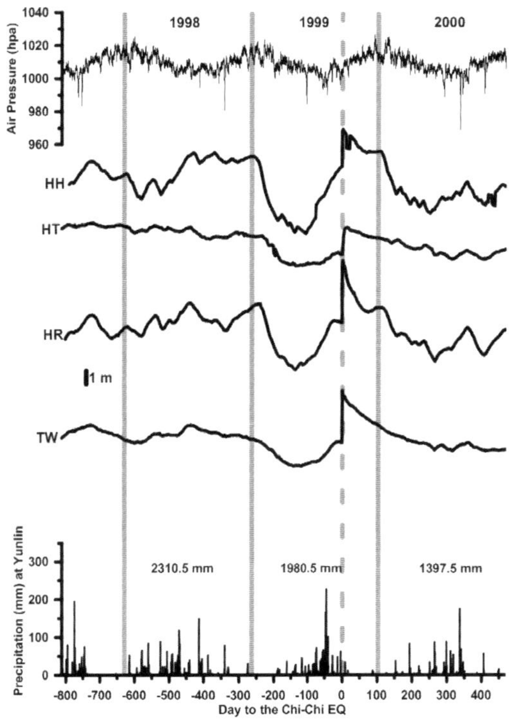

Figure 3. Variations of air pressure, groundwater level at HH, HT, HR and TW sites as well as annual precipitations about 800 days before and 450 days after the Chi-Chi earthquake.

3 Removal of short-term effects on groundwater level

Since the unusual drop with duration of ~ 250 days should not relate with long-term climatic changes, variations of atmospheric pressure, Earth's tides, precipitation and human extraction, which are considered to be the major factors affecting groundwater levels, are further taken into account. We examine these short-term effects (i.e. atmospheric pressure, Earth's tides, precipitation and human extraction) using groundwater data with a time span between 1 August 1997 and 31 December 2009 to adequately cover the unusual drop. The upper panel in Fig. 3 illustrates the variations of air pressure measured at the southern site of the study area. It is clear that annual fluctuations are dominated in the air pressure records. Low and high air pressures were regularly observed in summer and winter seasons, respectively. Effects of the atmospheric pressure on the groundwater level ranged between −0.2 and 0.2 m were then utilized in data correction. The Earth's tides with the frequencies of 1 and 2 day^{-1} could be easily eliminated while hourly data were down-sampled as a daily record. Corrected groundwater level data could thus be obtained while responses of air pressure and the Earth's tide were removed.

The second (from top) and bottom plots in Fig. 3 present variations in corrected groundwater data at the HH, HT, Honglung (HR) and Tienwei (TW) sites and precipitation at the Yunlin station (120.476° E, 23.636° N) from 1 August 1997 to 31 December 2000, respectively. Unusual decreases of approximately 2–4 m in corrected water levels can be clearly identified in these groundwater wells about 150 days before the Chi-Chi earthquake. We further examined records of 54 monitoring wells in the Choshuichi Alluvial Fan and found that similar patterns of unusual decrease are observed at 78 % wells, primarily distributed near the Chelungpu fault. Unusual decreases by artificial water pumping can be eliminated as the cause because temporal duration of these anomalous decreases exceeded 200 days and the anomalous wells are widely distributed in most areas of the Choshuichi Alluvial fan. Drought can be excluded as well, because the annual precipitation in 1999 accounted to 80 % of Taiwan's annual average. The annual precipitation in 1999 (1980.5 mm) is higher than that in 2000 (1397.5 mm) and lower than 1998 (2310.5 mm). If drought had played a major factor, the same pattern of low groundwater levels would have been repeated in 2000. Thus, possibility of the unusual drops in the Choshuichi Alluvial Fan can be ascribed to the Chi-Chi earthquake in both temporal and spatial domains, after factors from the long-term and short-term effects have been removed. We further survey the 12.5-year time-series data; however, no similar drop-like variations can be repeatedly observed for other two major earthquakes in the temporal domain.

4 Discussion and conclusions

The debate regarding pre-earthquake groundwater anomalies is primarily due to the limited number of observation wells over a wide area and the anomalous patterns for different strong earthquakes, which are often very hard to repeat and define. In this study, we have shown that an anomalous decrease in groundwater levels was observed in effect-corrected records at the 78 % (= 42/54) of the wells widely distributed across the Choshuichi Alluvial Fan nearby the Chelungpu fault a few months before the Chi-Chi earthquake. Thus, groundwater level has the potential to reflect the tectonic stress accumulation prior to the occurrence of the Chi-Chi earthquakes if the monitoring wells are densely distributed and close to the epicenter, as demonstrated in Fig. 4. Apart from the large signals, some relatively-small peaks of enhanced amplitude ratios can be observed at HT and/or TH in Fig. 4, but cannot be simultaneously observed at the other two stations, implying that they are local and small-magnitude events resulting limited peaks.

Critical earthquake-related signals are often hidden behind substantial and common effects with various frequencies. For the stand-out of anomalous patterns, our study shows that it is difficult to retrieve and/or understand these vital earthquake-related signals without removing these common effects via an adequate filter. The seismo-groundwater anomalies in the Choshuichi Alluvial Fan in Taiwan could be clearly identified for all three strong earthquakes ($M > 6$) because amplitudes at the frequency band between 0.02 and 0.04 day^{-1} were consistently enhanced in groundwater level records of many monitoring wells in the frequency domain. In short, this new method sheds the light for the precursory-seeking research of strong earthquakes, but needs further work to refine the technique and test its applicability for the forthcoming strong earthquakes. Naturally, the method presented in this work has also limitation and prerequisite. For example, an adequately dense monitoring network has to be established first. Another one is that it cannot apply to the earthquake right after a strong one, such in the case of Chia-Yi earthquake discussed in this study. Nonetheless, if this method is well integrated with other geophysical observations, such as geomagnetic anomalies and crustal displacements, variation in groundwater levels and electromagnetic anomalies could be quantified and/or estimated by using volume changes via possibly physical mechanisms. These integrated results will be highly promising to understand underlying mechanisms of major earthquakes and to build a strong earthquake forecast system in the future.

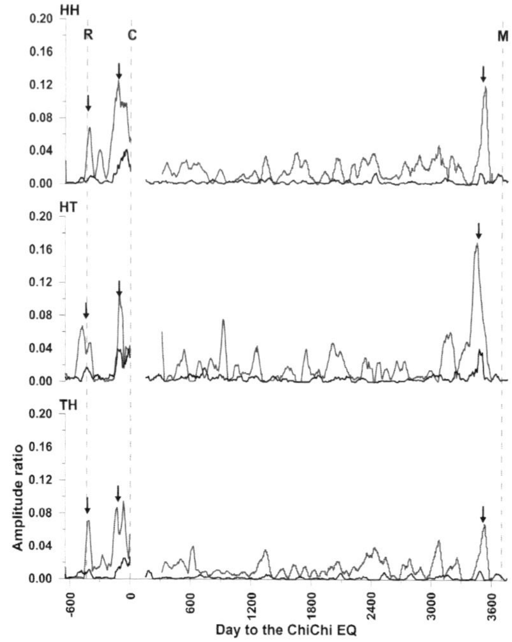

Figure 4. The amplitude ratios of groundwater levels at HH, HT and TH stations from 1 August 1997 to 31 Dececember 2009.

Acknowledgements. The authors would like to thank the Ministry of Science and Technology of the Republic of China, Taiwan, for financially supporting this research under Contract No. MOST 103-2119-M-305-001.

References

Chen, C. H., Wang, C. H., Hsu, Y. J., Yu, S. B., and Kuo, L. C.: Correlation between groundwater level and altitude variations in land subsidence area of the Choshuichi Alluvial Fan, Taiwan, Eng. Geol., 115, 122–131, 2010.

Chen, W. F. and Yuan, P. A.: Preliminary study on sedimentary environments of Choshui fan-delta, Journal of the Geological Society of China, 42, 269–288, 1999.

Huang, N. E. and Wu, Z.: A review on Hilbert–Huang transform: method and its applications to geophysical studies, Rev. Geophys., 46, RG2006, doi:10.1029/2007RG000228, 2008.

Huang, N. E., Shen, Z., Long, S. R., Wu, M. C., Shih, H. H., Zheng, Q., Yen, N. C., Tung, C. C., and Liu, H. H.: The empirical mode decomposition and the Hilbert spectrum for nonlinear and non-stationary time series analysis, P. Roy. Soc. Lond. A Mat., 454, 903–995, 1998.

Liu, C. H., Pan, Y. W., Liao, J. J., and Hung, W. C.: Estimating coefficients of volume compressibility from compression of strata and piezometric changes in a multiaquifer system in west Taiwan, Eng. Geol., 75, 33–47, 2004

Proc. IAHS, 372, 105–109, 2015
proc-iahs.net/372/105/2015/
doi:10.5194/piahs-372-105-2015

Open Access
PIAHS

Fuzzy-logic assessment of failure hazard in pipelines due to mining activity

A. A. Malinowska and R. Hejmanowski

AGH University of Science and Technology, Cracow, Poland

Correspondence to: A. A. Malinowska (amalin@agh.edu.pl)

Published: 12 November 2015

Abstract. The present research is aimed at a critical analysis of a method presently used for evaluating failure hazard in linear objects in mining areas. A fuzzy model of failure hazard of a linear object was created on the basis of the experience gathered so far. The rules of Mamdani fuzzy model have been used in the analyses. Finally the scaled model was integrated with a Geographic Information System (GIS), which was used to evaluate failure hazard in a water pipeline in a mining area.

1 Introduction

Water pipelines localized in active mining areas are particularly hazarded with factors which may potentially evoke additional strains. Previous work shows that such factors include: seismic tremors, landslides, and continuous and discontinuous deformations of terrain due to underground extraction (Kalisz, 2007; Knothe, 1953; Kowalski and Kwiatek, 1995; Kwiatek and Mokrosz, 1996; Mendec et al., 1996; Mokrosz and Zawora, 1997; Szadziul, unpublished data; Talesnick and Baker, 2008; Zhao et al., 2005). Therefore, the failure-prone zones hazarding the pipelines, need to be managed because of the potential risk they bring. Depending on the significance of a given object, various evaluation methods are used for assessing the risk in particular situations. In the case of objects, whose failure may threaten the general safety and well-being of people, the applied methods are both time-consuming and very costly. Among the most frequent methods of the strain modeling is Finite Element Method (FEM; Zhao et al., 2005). Another approach is applicable when the linear objects are numerous and their failure may be only noxious for the users. In this case the applied estimation methods allow for a quick but approximated estimation of hazard. Unfortunately, approximated methods are inaccurate, and still time consuming. Therefore an attempt to work out a fuzzy model for assessing hazards caused by continuous deformations acting on linear objects was made. This model is integrated with GIS which significantly accelerates and simplifies the evaluation and reporting about the zones potentially threatened by underground extraction operations.

2 Background

2.1 Pipeline hazard estimation – approximated point method

In Poland, the strength of water pipelines in mining areas is evaluated with the point method. This method was created in the 1950s and has been commonly applied by the coal mine industry. This method offers an approximated evaluation of failure hazard of pipelines and consists of three stages:

1. resistance evaluation of water pipeline (expressed as resistance category (Table 1);

2. prediction of continuous strains hazard with Knothe function theory (expressed as terrain category) (Knothe, 1953);

3. comparison of terrain category with pipeline resistance category.

The resistance of the pipelines is assessed on the basis of four risk factors, and each of them is ascribed a certain number of points. On this basis the pipeline is given its category of resistance to horizontal strains. The higher is the number of points, the lower is the resistance category. A low resistance category means that a linear object is vulnerable and

Table 1. Resistance classification of water pipelines – point method.

Hazard factor		Number of points			
Matter	PE	0–10			
	Cast iron or steel	10–15			
	PCV	15–20			
	Asbestos	20–30			
Compensation	Pipe sleeve	0–10			
	Compensators	10–20			
	No compensators	20–30			
Type and number of connections	< 1 connection $100\,m^{-1}$	0–10			
	1–3 connection $100\,m^{-1}$	10–20			
	3–5 connection $100\,m^{-1}$	20–30			
	> 5 connection $100\,m^{-1}$	30–40			
Technical condition	Very good	0–10			
	Good	10–20			
	Acceptable	20–30			
	Poor	30–40			
Number of points	0–24		25–48	49–80	> 80
Resistance category	4		3	2	1
Acceptable horizontal strains [$mm\,m^{-1}$]	9.0		6.0	3.0	1.5

prone to surface strains. In this case the probability of a failure occurrence is higher.

The terrain hazard is defined on the basis of predicted horizontal strains and tilts (Knothe, 1953). Then the terrain, where continuous deformations are possible, is categorized. Six categories have been distinguished depending on the intensity of predicted maximal horizontal strain ε max:

1. 0 (ε max $\in$ 0–0.3 $mm\,m^{-1}$);
2. I (ε max $\in$ 0.3–1.5 $mm\,m^{-1}$);
3. II (ε max $\in$ 1.5–3.0 $mm\,m^{-1}$);
4. III (ε max $\in$ 3.0–6.0 $mm\,m^{-1}$);
5. IV (ε max $\in$ 6.0–9.0 $mm\,m^{-1}$);
6. V (ε max > 9.0 $mm\,m^{-1}$).

The terrain categories are compared with pipeline resistance categories and on this basis the ultimate assessment is made.

2.2 Estimation of continuous deformation hazard in linear objects – shortcomings

The method used in Poland for evaluating hazard with continuous deformations in linear objects is very simplified. The assessment lies in finding objects where, the resistance category is exceeded by the terrain category (Fig. 1) (Szadziul, unpublished data).

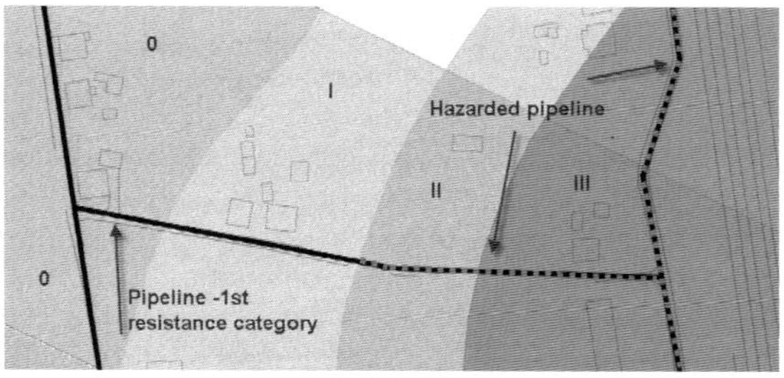

Figure 1. Example of the course of a linear object through terrains of belonging to different terrain categories (0, I, II, and III represent different terrain categories, with III being the most hazardous).

The practical application of this method has been exemplified below. A water pipeline belonging to the first resistance category is deposited in a ground of category 0–III (Fig. 1). Horizontal strains of terrain in that zone range between 0.1 $mm\,m^{-1}$ (0 terrain category) to 4.5 $mm\,m^{-1}$ (III terrain category). The pipeline can withstand horizontal strains under 1.5 $mm\,m^{-1}$ (Table 1), therefore is hazarded in the terrain belonging to the II and III terrain category.

The major shortcoming of the presently applied method is discretisation of the described hazard. Moreover, the resistance of pipelines is defined in a very general way. Two sections of water pipeline (5 m from each other) can be evaluated as hazarded and not hazarded. Such discretisation creates serious problems to the organization responsible for safety in

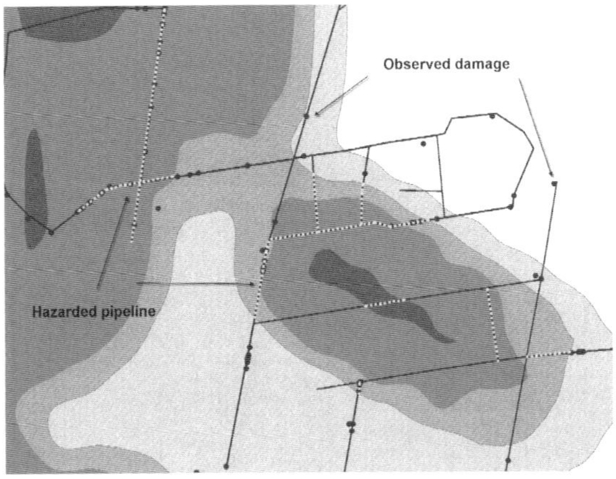

Figure 2. Hazard evaluation of water pipeline with point method (yellow dotted line: endangered part of the pipeline, red dot: observed damage).

mining areas, which have to make decisions about possible hazards and replacements of particular pipeline sections. As visualized in the presented example, the congruence between actual failures and places of potential hazard is low (Fig. 2).

Another problem lies in the lack of direct connection between terrain categories and pipeline resistance categories. Moreover, the present methodology is burdened with high uncertainty stemming from a number of factors including: incomplete information about the mining-geology conditions, incorrect modeling of continuous deformations, subjective evaluation of pipeline resistance; and, incomplete information about pipelines. This prompted the current study to identify other solutions, which would increase the efficiency of evaluation of pipelines hazard in areas staying under constant deformations of terrain.

3 Modeling of pipeline hazard with fuzzy logic

The analyses were focused on assuming additional risk factor for the assessment of pipeline hazard and developing a fuzzy-logic model based on risk factors.

Fuzzy logic has been used in analyses since the 1960s (Zadeh, 1965). Initially it was used in industry for image processing, complex processes control, and computer-aided decision making. In the subsequent years attempts were undertaken to implement fuzzy theory in social systems, economy and even medicine. In environmental sciences fuzzy theories were used for evaluating hazards which could be generated by selected natural elements (Adriaenssensa et al., 2009; Bojorquez-Tapia et al., 2002; Busch and Maas, 2006; Gheorghe et al., 2000; Liu et al., 2005), e.g. for evaluating the landslide hazard (Lee, 2007), and evaluation of environmental elements in the aspect of their degradation (e.g., hazard evaluation of a river ecosystem, Ioannidou et al., 2003). It should be emphasized that in a majority of cases the fuzzy analyses were supported by GIS tools. The present work represent the first time fuzzy logic has been used for assessing hazard assessment in pipelines placed in mining areas.

Table 2. Rule base.

			HSM/HSR					
			0	I	II	III	IV	V
			L	VL	M	H	VH	EH
V	4	L	La	La	La	VL	L	M
	3	VL	La	La	VL	L	M	A
	2	M	La	VL	L	M	A	H
	1	H	VL	L	M	A	H	VH

The first stage of creating a reasoning fuzzy system was determining input and output variables of the model. Basing on the studies by Malinowska et al. (unpublished data) two variables regarding the linear objects hazard were defined: predicted maximum horizontal strains (HSM $\in (0,9)$) and the difference between predicted major strains in two directions (HSR $\in (0,18)$. Another linguistic variable was the sensitivity of linear objects (V), expressed as resistance points ($V \in (0,100)$. The failure hazard of a linear object expressed in the point scale ($R \in (0,100)$) was the output variable. This variable can be used for determining the failure hazard of a pipeline in points. The space in which these variables were analyzed was assumed on the basis of extreme values observed in the mining areas in Poland. Then the linguistic values were defined for each of the variables (Table 1): very low (VL), low (L), medium (M), high (H), very high (VH), extremely high (EH).

Pipeline vulnerability was described by 4 linguistic variables: very low (VL), low (L), medium (M), high (H), very high (VH).

Hazard of pipeline failure was characterized by 7 linguistic variables: lack (La), very low (VL), low (L), medium (M), appreciable (A), high (H), very high (VH).

Each linguistic value has a defined fuzzy set described by a characteristic function. The information which could describe the characteristic function is very limited, therefore authors assumed a triangle shape of the membership function for fuzzy sets. The membership functions were modeled based on the following assumptions:

1. characteristic points of the membership function are defined on the basis of limits of terrain category and resistance category,

2. membership functions meet the unity condition.

Then, a rule base was defined on the following assumptions (Table 2).

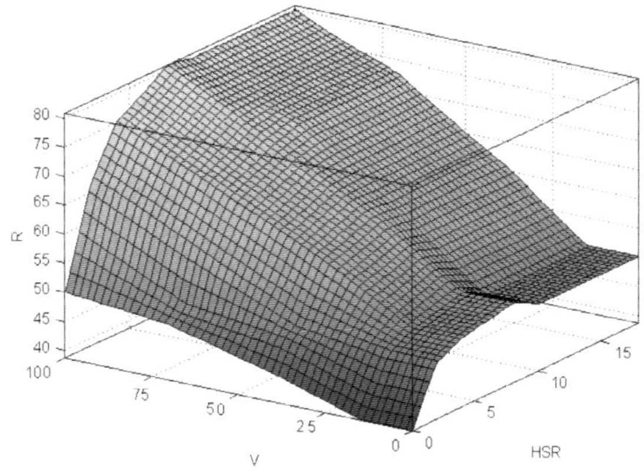

Figure 3. Surface of fuzzy model (V vs. HSR).

1. If the terrain category is higher than the resistance category by two or more grades, the failure hazard is serious.
2. If the terrain category is equal to the resistance category, the failure hazard is very low.

Fuzzy modelling was realized with the use of the Mandami dependence. The degree to which the premises in the inference block have been met was evaluated with the product operator (PROD). The successively activated premises allowed for determining the degree to which conclusions have been met. The last stage of fuzzy reasoning lied in determining the output function of rule base conclusions. The output membership function was defined on the basis of membership functions of particular conclusions of rules in the accumulation process. The accumulation in the model was performed for the SUM operator. The output membership function obtained in the course of inference was used for determining a sharp output value representing this set in the most reliable manner. Defuzzification was performed with the gravity center method. On this basis a surface was created, allowing for the evaluation of pipeline hazard with continuous strains (Fig. 3).

4 Model application

Modeling of pipeline failure hazard is supported by GIS programming. This is of special significance in predominantly urbanized areas, where the water network is extensive. It should be stressed that a sound evaluation of the failure hazard with GIS can be done only when reliable data about the resistance of the pipeline and the expected deformations which may occur in the study area are available (Fig. 4).

In the proposed solution based on a fuzzy model hazard in pipelines can be determined in a continuous manner, for each segment of the pipeline (the length to be defined arbitrarily). The pipeline can be ascribed a point value of failure risk ($R \in (0,100)$). The pipeline sections marked in intense colors are more endangered with failures (Fig. 5). The congruence between actual failures and places of potential hazard is high.

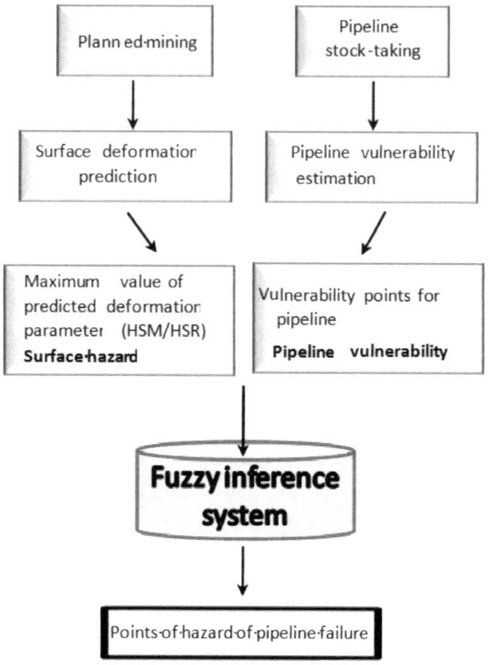

Figure 4. Pipeline damage risk assessment algorithm.

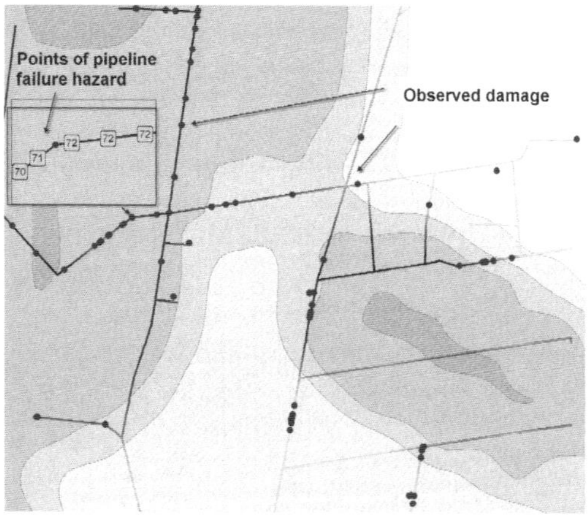

Figure 5. Evaluation of failure hazard of a pipeline performed with a fuzzy model (yellow dotted line: endangered part of the pipeline, red dot: observed damage, line with the blue intensive color-predicted high risk of pipeline damage).

Organizations responsible for water pipelines are free to define the limits of risk points, for which the hazard is very high. The proposed model can be used for evaluating hazards in other linear objects, such as gas pipelines, when parameters have been adapted to the local conditions.

5 Summary

In a fuzzy-set based model the uncertainty resulting from the subjective evaluation of experts or incomplete data can be accounted for. The applied solution increases the efficiency of estimation of a water pipeline hazard. Moreover, thanks to the application of the inference block, the hazard can be presented as a continuous variable, which seems to be very advantageous compared to other methods. The comparison of modeling results with actual observations of damaged pipelines revealed a considerably higher accuracy using this new method. In addition, the presented method of a failure hazard evaluation for water pipelines seems to be readily and easily integrated with geographic information systems.

Acknowledgements. The research reported in this paper has been supported by a grant from the National Science Centre no. 2011/01/D/ST10/06958.

References

Adriaenssensa, V., Baetsb, B. D., Goethalsa, P. L. M., and Pauwa, N. D.: Fuzzy rule-based models for decision support in ecosystem management, Sci. Total Environ., 319, 1–12, 2009.

Bojorquez-Tapia, L. A., Juarez, L., and Cruz-Bello, G.: Integrating Fuzzy Logic, Optimization, and GIS for Ecological Impact Assessments, Environ. Manag., 30, 418–433, 2002.

Busch, W. and Maas, K.: Remarks to the risk assessment for abandoned mine sites, Acta Montanistica Slovaca Ročník 12, 3, 340–348, 2006.

Gheorghe, A. V., Mrock, R., and Kröger, W.: Risk assessment of regional systems, Reliab. Eng. Syst. Safe., 70, 141–156, 2000.

Ioannidou, I. A, Paraskevopoulos, S., and Tzionas, P.: Fuzzy modeling of Interactions among Environmental Stressors in the Ecosystem of Lake Koronia, Greece, Environ. Manag., 32, 624–638, 2003.

Kalisz, P.: The impact of the mining on the pipeline resiliency, Zeszyty Naukowe Politechnika Śląska S. Górnictwo, 278, 191–200, 2007 (in Polish).

Knothe, S.: Equation of the profile of finitely formed subsidence trough, Archiwum Górnictwa i Hutnictwa, 1(1), 1953 (in Polish).

Kowalski, A. and Kwiatek, J.: Surface and surface structure object protection on the mining area, Przegląd Górniczy, 4, 11–19, 1995 (in Polish).

Kwiatek, J. and Mokrosz, R.: Gas networks on the mining areas, WUG, 3, 1996 (in Polish).

Lee, S.: Application and verification of fuzzy algebraic operators to landslide susceptibility mapping, Environ. Geol., 52, 615–623, 2007.

Liu, H., Chen, W., and Kang, Z.: Fuzzy Multiple Attribute Decision Making for Evaluating Aggregate Risk in Green Manufacturing, J. Tsinghua Sci. Technol., 10, 627–632, 2005.

Mendec, J., Kliszczewicz, B., and Wytrychowska, M.: Surface and surface structure object protection on the mining area, Rules for the water pipeline protection against the effects of underground mining, Glówny Instytut Górnictwa, Katowice, Poland, 53 pp., 1996 (in Polish).

Mokrosz, R. and Zawora, J.: The impact of the mining on the buried pipeline, Materiały konferencyjne: Ochrona powierzchni i o biektów budowlanych przed szkodami górniczymi, Glówny Instytut Górnictwa, Katowice, Poland, 20-28, 1997 (in Polish).

Talesnick, M. and Baker, R.: Failure of flexible pipe with a concrete liner, Eng. Fail. Anal., 5, 247–259, 2008.

Zhao, W., Nassar, R., and Hall, D.: Design and reliability of pipeline rehabilitation liners, Tunn. Undergr. Sp. Tech., 20, 203–212, 2005.

Zadeh, L. A.: Fuzzy sets, Inform. Control, 38, 1–14, 1965.

Proc. IAHS, 372, 111–113, 2015
proc-iahs.net/372/111/2015/
doi:10.5194/piahs-372-111-2015

Superior coexistence: systematicALLY regulatING land subsidence BASED on set pair theory

Y. Chen[1] and S.-L. Gong[2]

[1]School of Foreign Languages, Tongji University, Shanghai 200092, China
[2]Center for Land Subsidence of China Geological Survey, Shanghai 200072, China

Correspondence to: S.-L. Gong (gong_shiliang@126.com)

Published: 12 November 2015

Abstract. Anthropogenic land subsidence is an environmental side effect of exploring and using natural resources in the process of economic development. The key points of the system for controlling land subsidence include cooperation and superior coexistence while the economy develops, exploring and using natural resources, and geological environmental safety. Using the theory and method of set pair analysis (SPA), this article anatomises the factors, effects, and transformation of land subsidence. Based on the principle of superior coexistence, this paper promotes a technical approach to the system for controlling land subsidence, in order to improve the prevention and control of geological hazards.

1 Introduction

Land subsidence is associated with the consequences of economic development and the use of groundwater and land resources, particularly in the process of industrialisation and urbanisation. It is an environmental, geological problem and an important type of geological disaster, which occurs widely in cities and urban areas. It becomes a critical factor that influences geological environmental quality, and confines regional economic and social sustainable development.

Based on the principles of mutual connection and superior coexistence in the course of economic and social development, as well as exploiting resources and ecological environmental protection, this article discusses a working plan and a technical approach for systematically regulating land subsidence using set pair analysis (SPA).

2 SPA for land subsidence

SPA is an important theory that systematically analyses mutual connections and deals with uncertainty problems (Zhao, 1996). The core of this theory is contradiction analysis to examine unity and opposites. The basic unit of contradictory unity is called a "set pair". Uncertainty exists objectively and generally in all sorts of things and systems; it is an inherent attribute and embodies movement and development. The two objects of one set pair have a relatively certain connection of contradictory unity. Differences exist in the uncertainties of transition status and incomplete information, whose process of change tends toward identity or opposite. By studying the unity, difference, contrast connection (identity or opposite), variability, and transformation of the "set pair", we can understand the rules for systematic development; furthermore, it is possible to analyse and deal with various, complex actual problems (Zhao, 2000).

Due to numerous restriction factors, as well as environmental and social influences, and a wide range of prevention and control methods, land subsidence is a complex system that contains many contradictions between the aspects of unity and system, between interconnectedness and situational changes. Uncertainty remains prevalent in the internal mechanisms, of land subsidence as well as its processes, feedback on control, and many other aspects due to the system's complexity, as well as the state of dominant and limitations. As a result, SPA could be a reference for research and prevention, and controlling land subsidence.

The essential "set pair" of land subsidence consists of the interactions between people and the land, which are interrelated and interdependent, restrict and influence each other, and constitute the system of the inner core. Land subsidence is influenced by natural factors more closely associated with economic engineering activities such as human effects (Galloway, 2014; Gambolati, 2014). The geological environment contains natural resources on which humans rely for survival and development, and shows identity. In addition, the geological environment is the main victim of geological problems and disasters; therefore, it shows the opposite. Its mechanism, process, and form of mutual influence show difference and uncertainty; in other words, it fully embodies the characteristics of identity or the opposite.

The key to analysing land subsidence is the identical or opposite situation and influence on system characteristics. The development of land subsidence is a process of decreasing the consistency degree and increasing the antagonism degree. The actual performance is the gradual loss of resources, environmental carrying capacity, increasingly obvious latent disasters, and mutations caused by slow quantitative change. Land subsidence turns into an environmental geological problem and disaster, and then, it triggers a security crisis (Gong, 2003). Its situational changes are closely related to the rise of industrialisation, urbanisation, and development.

For example, river deltas, littoral zones, plains, inland basin areas, and other sectors with natural geographical advantages often become the centres of economic activity and urban agglomeration, where the geological environmental system is relatively weak. Regarding the background of highly intense development and utilising natural resources in geological environments, land subsidence (produced by superposing the mutual coupling of inner and outer factors) leads to land subsidence constantly progressing and being extended.

The development and use of land resources is the same as the development and use of groundwater resources, which can also lead to land subsidence. The main performance is the intensive use of land, particularly in urban areas. Since urban land resources are rare, land utilisation and three-dimensional municipal development become the inevitable choice; this makes the city expand outwardly to build up all kinds of high-level or high-rise buildings, underground space development, and utilisation on a large scale.

The permanent ground elevation loss and its subsidence difference will seriously threaten city flood prevention and security for all kinds of important projects. Land subsidence is an important part of the urban disaster system, and becomes a constraint factor affecting environmental security and sustainable economic and social development. This is an extrusive performance form of interaction between the two aspects of the "set pair" form, and causes a trend of situational change.

Ultimately, preventing and controlling land subsidence aims to keep the geological environmental system stable and secure, to make full use of underground water, land, and other natural resources, and reduce the risk of geological disasters.

3 Superior coexistence: systemic control of land subsidence

Superior coexistence includes the two aspects of the superior state and coexistence. The superior state is the safe direction of the object, and characterises the living conditions for sustainable development of the body's behaviour, and causes security to reach a more profound goal, which is development and security (Yu, 2005).

Superior state co-existence is a philosophy of safety, a new way of thinking based on the humanities and social sciences. The concept exists for purposes of natural science research, especially in terms of preventing geological hazards and controlling the engineering system, which undoubtedly leads to enlightenment and guidance. Geological disaster prevention and environmental security are also an important part of the field of non-traditional security. Superior state co-existence is a new way of thinking in terms of environmental safety to prevent and control geological disasters.

Regarding the regulation of the land subsidence system, based on SPA and superior state co-existence, in order to realise the three elements (mutual improvement and harmonious coexistence among economic and social development, the resources of development and utilisation, and ecological environmental protection) we should pay attention to the basic ideas of systemic and dialectical nature. We should also attach great importance to preventing disasters, and creating countermeasures for the dynamic integration of rigid control and flexible adjustment.

If we regard the geological environmental resource attributes as the identity of economic and social development, the degree is characterised by identity degree. We regarded the geological environmental disaster attributes as the contrariety of economic and social development. Its degree is characterised by a contrary degree. The geological environment development and utilisation attributes were regarded as the otherness of economic and social development. Its intensity is characterised by the difference degree. To ensure that the geological environment is in a safe state and cannot lead to geological disasters, we must exploit and utilise the resources effectively. In other words, that should be less than the recoverable resources and the development intensity of the disasters caused. Consequently, the geological environmental system appears in the same healthy and stable state. In addition, we can avoid or reduce the reverse state easily in order to induce geological disasters.

Preventing and controlling land subsidence requires implementing economic disaster reduction, which is synergy. Its outstanding characteristic is "negative changed into positive, changing from positive and negative to positive". Based on the influential effect of "negative" benefits caused by disaster prevention, we were able to reduce disaster losses, while reducing losses part was deemed a positive benefit. Using the dialectical standpoint of the SPA identity-diversity-oppose relation, we decreased the antagonism degree or accordingly increased the consistency degree. Therefore, we need to improve the identity of the system situation, which we called stability. The implementation of disaster prevention measures is helpful to effectively reduce uncertainty, as well as to promote an increased consistency degree and lowered antagonism degree. Comprehensive benefits obviously appeared.

As a whole, regarding the technology level of regulating and controlling the land subsidence system, how to better coordinate between urban development and environmental quality is the key content for selecting subsidence control measures and its implementation process. From the perspective of subsidence control, it is obvious that we should eliminate the various factors of subsidence. Based on the actual situation of urban development, we will obtain the greatest economic and social benefits, but consume as few resources as possible. At the same time, land subsidence also has a certain inevitability, and controlling land subsidence is the goal of subsidence prevention and control. We are allowed to have a certain amount of safety guaranteeing subsidence, on the basis of reasonable scientific allocations, and make full use of limited natural resources. This should also be the starting point to solve the dilemma of resource utilisation and environmental quality.

4 Conclusion

Land subsidence seriously restricted the social economy's sustainable development. Land subsidence prevention should emphasise economic development and resource utilisation, with the superior state co-existence of geological environment. According to academic thought and SPA, in order to comprehensively understand the unity of opposites, which contains many contradictions, and the system's uncertainty of function and influence, we must pay attention to the status, properties, and catastrophe process of the geological environmental system. We must also organically integrate long-term strategies for land subsidence prevention and treatment and practical technical measures. This will effectively promote harmony between humans and the environment, and promote security and development for humans and society.

References

Galloway, D. L.: The current situations and developing trend of international land subsidence research, Shanghai Land Resour., 35, 1–8, doi:10.3969/j.issn.2095-1329.2014.02.001, 2014.

Gambolati, G.: The simulation and forecast of land subsidence, Shanghai Land Resour., 35, 1–11, doi:10.3969/j.issn.2095-1329.2014.04.001, 2014.

Gong, S. L.: Set pair theory norms for reasonable exploitation of geological environment. Journal of Geological Hazards and Environment Preservation, 14, 34–38, doi:10.3969/j.issn.1006-4362.2003.02.008, 2003.

Yu, X. F.: Superior co-existence: A new concept in security philosophy. Journal of Zhejiang University (Humanities and Social Sciences), 35, 5–12, doi:10.3785/j.issn.1008-942X.2005.02.001, 2005.

Zhao, K. Q.: Set pair analysis: A new theory, method and application for uncertainty, Syst. Eng., 14, 18–23, 72, 1996.

Zhao, K. Q.: Set pair analysis and its application, Hangzhou, Zhejiang Scientific and Technical Publishers, 2000.

Proc. IAHS, 372, 115–120, 2015
proc-iahs.net/372/115/2015/
doi:10.5194/piahs-372-115-2015

Open Access
PIAHS

Study on the risk and impacts of land subsidence in Jakarta

H. Z. Abidin[1], H. Andreas[1], I. Gumilar[1], and J. J. Brinkman[2]

[1]Geodesy Research Group, Faculty of Earth Science and Technology, Institute of Technology Bandung, Bandung 40132, Indonesia
[2]Deltares, Delft, Netherlands

Correspondence to: H. Z. Abidin (hzabidin@gmail.com)

Published: 12 November 2015

Abstract. Jakarta is the capital city of Indonesia located in the west-northern coast of Java island, within a deltaic plain and passes by 13 natural and artificial rivers. This megapolitan has a population of about 10.2 million people inhabiting an area of about 660 km^2, with relatively rapid urban development. It has been reported for many years that several places in Jakarta are subsiding at different rates. The main causative factors of land subsidence in Jakarta are most probably excessive groundwater extraction, load of constructions (i.e., settlement of high compressibility soil), and natural consolidation of alluvial soil. Land subsidence in Jakarta has been studied using leveling surveys, GPS surveys, InSAR and Geometric-Historic techniques. The results obtained from leveling surveys, GPS surveys and InSAR technique over the period between 1974 and 2010 show that land subsidence in Jakarta has spatial and temporal variations with typical rates of about 3–10 cm year^{-1}. Rapid urban development, relatively young alluvium soil, and relatively weak mitigation and adapatation initiatives, are risk increasing factors of land subsidence in Jakarta. The subsidence impacts can be seen already in the field in forms of cracking and damage of housing, buildings and infrastructure; wider expansion of (riverine and coastal) flooding areas, malfunction of drainage system, changes in river canal and drain flow systems and increased inland sea water intrusion. These impacts can be categorized into infrastructural, environmental, economic and social impacts. The risk and impacts of land subsidence in Jakarta and their related aspects are discussed in this paper.

1 Introduction

Land subsidence is a natural-anthropogenic hazard affecting quite many large urban areas (cities) in the world, including Jakarta, the capital city in Indonesia. It is located on the lowland area in the northern coast of West Java, centered at the coordinates of about 6°15′ S and +106°50′ E (see Fig. 1), within a deltaic plain and passes by 13 natural and two canals. This megapolitan has a population of about 10.1 million people in 2014, inhabiting an area of about 662 km^2, with relatively rapid urban development. Topographically, the area of Jakarta has slopes ranging between 0 and 2° in the northern and central parts, between 0 and 5° in the southern part, and its southern-most area has an altitude of about 50 m above mean sea level.

Land subsidence is a well known phenomena affecting Jakarta, and it has been observed using several geodetic techniques and has been widely reported for many years (Rismianto and Mak, 1993; Murdohardono and Sudarsono, 1998; Purnomo et al., 1999; Rajiyowiryono, 1999; Abidin et al., 2001, 2004, 2008, 2010, 2011, 2013; Koudogbo et al., 2012; Ng et al., 2012; Chaussard et al., 2013). According to those studies, land subsidence in Jakarta has spatial and temporal variations with typical rates of about 3–10 cm year^{-1}.

Land subsidence impacts in Jakarta can be seen already in the field in forms of cracking and damage of housing, buildings and infrastructure; wider expansion of (riverine and coastal) flooding areas, malfunction of drainage system, changes in river canal and drain flow systems and in-

Published by Copernicus Publications on behalf of the International Association of Hydrological Sciences.

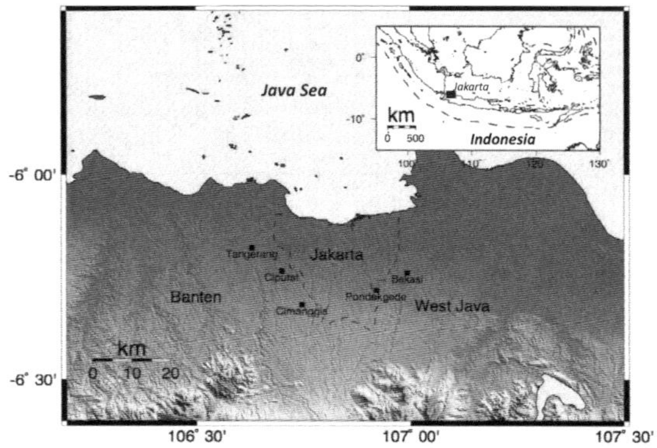

Figure 1. Location of Jakarta, the capital city of Indonesia.

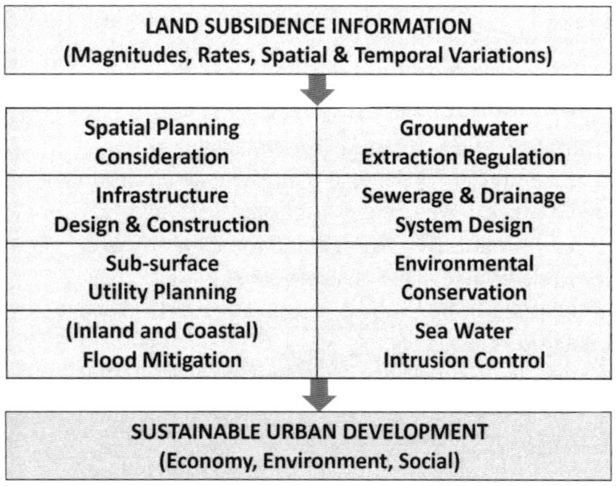

Figure 2. Importance of land subsidence information for urban development.

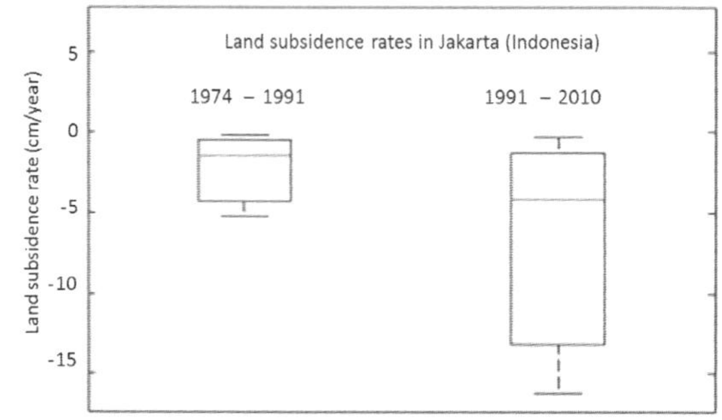

Figure 3. Box plots of land subsidence rates in Jakarta during the period of 1974–1991 (left) and 1991–2010 (right). Subsidence rate unit is cm $year^{-1}$.

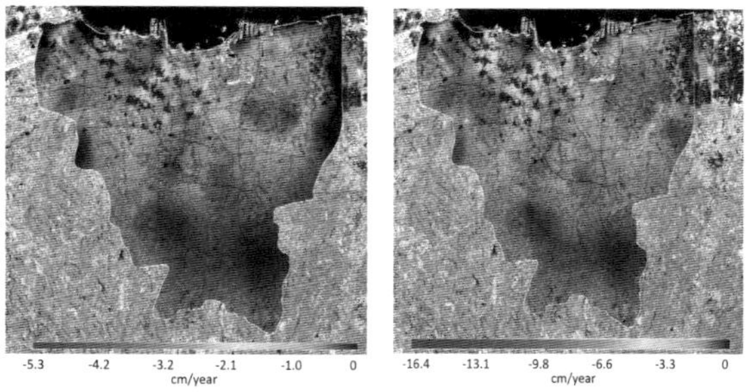

Figure 4. Rates of land subsidence in Jakarta during the period of 1974–1991 (left) and 1991–2010 (right); courtesy of Irwan Gumilar of Geodesy Research Group of ITB.

creased inland sea water intrusion. In general, these impacts can be categorized into infrastructural, environmental, economic and social impacts.

Since information of land subsidence characteristics is important for sustainable urban development of Jakarta (see Fig. 2), then understanding the risk and impacts of land subsidence in Jakarta and their related aspects is very important and strategic, and land subsidence charactristics should be continuously monitored.

2 Land subsidence characteristics in Jakarta

Land subsidence phenomena in Jakarta has actually been observed using several geodetic methods, such as Leveling survey, GPS survey, InSAR, Microgravity survey, and Geometric-Historic method. In general, the observed land subsidence in Jakarta has spatial and temporal variations, with the typical rates of about 3–10 cm $year^{-1}$ (see Figs. 3, 4). These subsidence rates are obtained from the integration of results obtained by the Leveling surveys (1974–1997), GPS surveys (1997–2010), and InSAR technique (2006–2010). In this period of about 37 years (1974–2010), the total subsidence of up to about 4 m have been observed in several locations in Jakarta. The observed land subsidences along the coastal areas of Jakarta are relatively have larger rates than the inland areas. Figures 3 and 4 also show the existence of spatial variation in subsidence rates along the coastal zone of Jakarta, in which the central and western parts have relatively larger rates. These larger rates of subsidence are mainly due to higher volumes of groundwater extraction, combined with relatively younger alluvium soil composition. More information on land subsidence characteristics in Jakarta can be seen in Abidin et al. (2001, 2004, 2008, 2010, 2011, 2013, 2014, 2015a, b), Ng et al. (2012), Chaussard et al. (2013).

The studies from Murdohardono and Sudarsono (1998), Rismianto and Mak (1993), Harsolumakso (2001), and Hutasoit (2001) suggested that the possible causative factors of land subsidence in Jakarta are most probably: excessive groundwater extraction, load of constructions (i.e., settlement of high compressibility soil), natural consolidation of alluvial soil, and tectonic activities. Land subsidence in certain area is usually caused by combination of those factors. Considering the spatial variation of land subsidence rates in Jakarta area, then it can be expected that the contribution of each factor on land subsidence at each location also will has spatial variation. In Jakarta, tectonic activities seem to be the least dominant factor, while excessive groundwater extraction is considered to be one of dominant factor for causing land subsidence.

3 Land subsidence risk in Jakarta

Considering the aforementioned causative factors of land subsidence in Jakarta, actually there are three main risk increasing factors of land subsidence in Jakarta that can be considered, namely: relatively young alluvium soil in which Jakarta is located, rapid urban development, and relatively weak imposement of land subsidence mitigation and adaptation measures (see Fig. 5).

Jakarta is located in a lowland coastal areas, composed by relatively young and soft alluvium soil. According to Rimbaman and Suparan (1999), there are five main landforms of Jakarta, namely: alluvial landforms (southern part), landforms of marine-origin (northern part adjacent to the coastline), beach ridge landforms (northwest and northeast parts), swamp and mangrove swamp landforms (coastal fringe), and former channels (perpendicular to the coastline). There are also 13 rivers and 2 canals flowing across Jakarta from its southern part and have estuaries in Java sea along an approximately 35 km of coastline. Jakarta basin, according to Yong et al. (1995), its consisted of a 200–300 m thick sequence of quaternary deposits which overlies tertiary sediments, in which the top sequence is thought to be the base of the groundwater basin. The quaternary sequence itself can be further subdivided into three major units, which, in ascending order are: a sequence of pleistocene marine and nonmarine sediments, a late pleistocene volcanic fan deposit, and holocene marine and floodplain deposits.

The above geological and geomorphological conditions make Jakarta has a higher risk (prone) toward land subsidence phenomena, especially in cases of excessive groundwater extraction and massive loading from buildings and infrastructures.

Rapid urban development in Jakarta also contributes to increasing risk of land subsidence occurences in Jakarta (Abidin et al., 2011). In this case, the increases in built-up areas, population, economic and industrial activities, will then increase groundwater extraction and also buildings and infrastructure loadings; which usually in turn lead to land subsidence phenomena. The relatively very rapid urban development of Jakarta as a megapolitan city is mainly in the sectors of industry, trade, transportation, housing, hotel and apartment, and many others (Firman, 1999, 2004; Hudalah et al., 2013); and they have introduced several negative environmental problems (Firman and Dharmapatni, 1994; Hudalah and Firman, 2012), such as: extensive conversion of agricultural areas into residential and industrial areas, significant disturbance to ecological and hydrological functions of the upland of Jakarta area and river catchment areas, and increase in groundwater extraction due to development of industrial activities and the high population increase. Coastal area of Jakarta which is mainly composed by relatively young and soft alluvium soil, has also experienced extensive urban development, such as establishment of sea port, coastal resort, golf course, residential areas, industries, apartments, malls, hotels, and commercials and office buildings. Coastal reclamation has also been conducted to accomodate more coastal development initiatives. These extensive urban development activities will have contribution into increasing land subsidence risk in several places in Jakarta (Abidin et al., 2011).

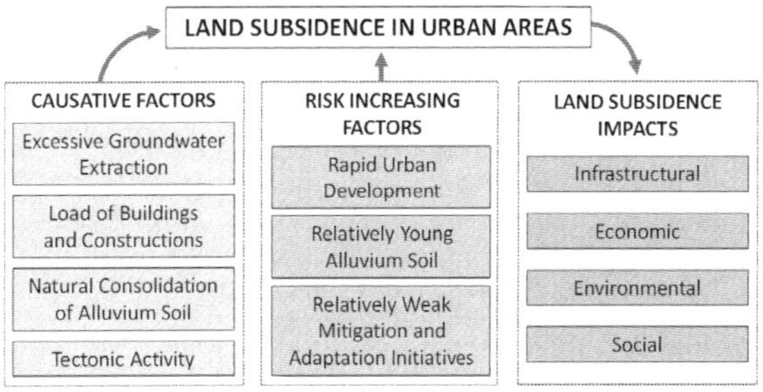

Figure 5. The causative factors, risk increasing factors and impacts of land subsidence in urban areas of Indonesia.

The other risk increasing factor of land subsidence in Jakarta is a relatively weak mitigation and adaptation initiatives (see Table 1) being implemented in Jakarta. At present, land subsidence hazard is not yet properly considered in urban development and spatial planning, groundwater extraction regulation system, and building codes in Jakarta. Although the regulation has been introduced to limit groundwater extraction in several subsidence prone areas in Jakarta; however, its enforcement is still need to be proven.

4 Land subsidence impacts in Jakarta

In general, the impacts of land subsidence in Jakarta can be seen in the field in various forms as shown in Fig. 6, such as cracking of permanent constructions and roads, tilting of houses and buildings, "sinking" of houses and buildings, changes in river canal and drain flow systems, wider expansion of coastal and/or inland flooding areas, and increased inland sea water intrusion. These subsidence impacts can be categorized into infrastructure, environmental, economic, and social impacts (see Table 2). Most of these impacts are indirectly generated due to land subsidence in the areas, and several of them are directly caused by subsidence. More over, these subsidence impacts have also rela-

Table 1. Examples of mitigation and adaptation initiatives for decreasing land subsidence risk in urban areas.

Causative factors	Mitigation	Adaptation
Excessive groundwater extraction	– Limit or prohibit groundwater extraction in subsidence prone areas. – Enforcement of strict regulation and punishment for groundwater extraction in subsidence prone areas.	– Increase the surface groundwater supply and resources. – Continuous monitoring of subsidence characteristics.
Load of buildings and infrastructures	– Urban development planning takes into account the land subsidence characteristics in the areas. – Urban development rate in subsidence prone areas is properly controlled.	– Implementation of special building codes for land subsidence prone areas. – Continuous monitoring of subsidence characteristics.
Natural consolidatium of alluvium soil	None	– Implementation of subsidence-adaptive urban developmen and spatial planning.
Tectonic activities		– Continuous monitoring of subsidence characteristics.

Table 2. Characteristics of land subsidence impacts; after Abidin et al. (2015c).

No.	Category	Representation of impact	Level of impact
1.	Infrastructural	cracking of permanent constructions and roads	direct
		tilting of houses and buildings	direct
		"sinking" of houses and buildings	direct
		breaking of underground pipelines and utilities	direct
		malfunction of sewerage and drainage system	indirect
		deterioration in function of building and infrastructures	indirect
2.	Environmental	changes in river canal and drain flow systems frequent coastal flooding wider expansion of flooding areas inundated areas and infrastructures increased inland sea water intrusion deterioration in quality of environmental condition	indirect
3.	Economic	increase in maintenance cost of infrastructure decrease in land and property values abandoned buildings and facilities disruption to economic activities	indirect
4.	Social	deterioration in quality of living environment and life (e.g. health and sanitation condition) disruprion to daily activities of people	indirect

tion among each other, and its connection system is simplified by Fig. 7.

The potential losses due to land subsidence in urban areas such as Jakarta are actually quite significant (Ward et al., 2011; Viets, 2010). Related infrastructural, social and environmental costs due to direct and indirect impacts of land subsidence are economically quite significant, and can not be under estimated in sustainable urban development. In this regard for example, the planning, development and maintenance costs of building and infrastructures in the affected areas are usually much higher than the normal situation. The collateral impact of coastal subsidence in Jakarta, in the form of coastal flooding during high tides is also quite damaging (Abidin et al., 2011, 2015b). This repeated coastal flooding in several areas along the coast, will deteriorates the structure and function of building and infrastructures, badly influences the quality of living environment and life (e.g. health and sanitation condition), and also disrupts economic and social activities in the affected areas. Inland subsidence should also has an impact on inland flooding phenomena in Jakarta (Texier, 2008), since it will theoretically lead to expanded coverage and deeper water depth of flooded (inundated) areas

Figure 6. Examples of representation in the field of land subsidence impacts in Jakarta.

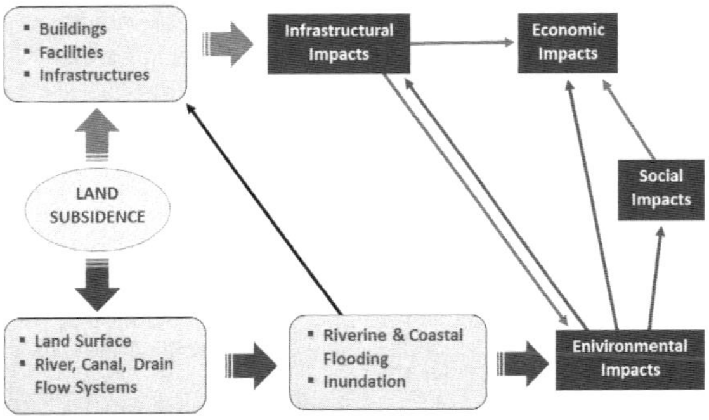

Figure 7. Land subsidence impacts in urban areas and its connection system, from Abidin et al. (2015c).

(Abidin et al., 2015b, c). The losses due to inland flooding in Jakarta are also significant since it affects many economic related activities in the city and surrounding regions.

5 Closing remarks

Land subsidence is a natural-anthropogenic hazard affecting Jakarta, with typical rates of about 3–10 cm year^{-1}. Its direct and indirect impacts can be easily seen nowadays in the field. Without good and effective mitigation and adaptation initiatives, land subsidence hazard in Jakarta can have more disastrous impacts in the future. In this case, the infrastructural, economic, environmental and social impacts of land subsidence can be quite a heavy burden for sustainable development of Jakarta.

At present, detail characteristics and mechanisms of land subsidence in Jakarta, both in spatial and temporal domains, are still not yet fully established. Contribution of each causative factors on observed land subsidence in certain areas still needs to be estimated and modelled. In this regard, the implementation of continuous subsidence monitoring system (e.g. GNSS CORS system) in Jakarta is necessary. Since, there is quite a strong linkage between geological and geomorphological condition, and rapid urban development in Jakarta with land subsidence characteristics, then subsidence mitigation and adaptation initiatives should be integrated in urban development program and spatial planning of Jakarta.

Finally, it should be pointed out that based on preliminary studies (Abidien et al., 2015b), it is found that there is some spatial correlation between land subsidence affected areas with flooded (inundated) areas in Jakarta. However, since the exact relation mechanism between the two phenomena is not yet established; then more quantitative characteristics of this correlation cannot be estimated. In order to establish a quantitative relation between land subsidence and flooding phenomena in Jakarta, then the following activities should be conducted, namely: detail mapping of the spatial and temporal rates and impacts of land subsidence, detail mapping of flooded (inundated) area during the flooding events, and detail flood risk modelling for Jakarta.

Acknowledgements. Land subsidence study in Jakarta using space geodetic techniques has been conducted by the Geodesy Research group of ITB since 1997, by using several research grants from Ministry of Science, Technology, and Higher Education of Indonesia, from the Provincial Government of Jakarta, and from ITB Research programs. The GPS surveys were conducted by the Geodesy Research Group of ITB, the Geospatial Agency of Indonesia, and the staffs and students from the Department of Geodesy and Geomatics Engineering of ITB.

References

Abidin, H. Z., Djaja, R., Darmawan, D., Hadi, S., Akbar, A., Rajiyowiryono, H., Sudibyo, Y., Meilano, I., Kusuma, M. A., Kahar, J., and Subarya, C.: Land Subsidence of Jakarta (Indonesia) and its Geodetic-Based Monitoring System, Nat. Hazards, 23, 365–387, doi:10.1023/A:1011144602064, 2001.

Abidin, H. Z., Djaja, R., Andreas, H., Gamal, M., Hirose, K., and Maruyama, Y.: Capabilities and Constraints of Geodetic Techniques for Monitoring Land Subsidence in the Urban Areas of Indonesia, Geomatics Research Australia, 81, 45–58, 2004.

Abidin, H. Z., Andreas, H., Djaja, R., Darmawan, D., and Gamal, M.: Land subsidence characteristics of Jakarta between 1997 and 2005, as estimated using GPS surveys, GPS Solutions, Springer Berlin/Heidelberg, 12, 23–32, 2008.

Abidin, H. Z., Andreas, H., Gamal, M., Gumilar, I., Napitupulu, M., Fukuda, Y., Deguchi, T., Maruyama, Y., and Riawan, E.: Land

Subsidence Characteristics of the Jakarta Basin (Indonesia) and its Relation with Groundwater Extraxtion and Sea Level Rise, in: Groundwater Response to Changing Climate, IAH Selected Papers on Hydrogeology No. 16, edited by: Taniguchi, M. and Holman, I. P., CRC Press, London, Chapter 10, 113–130, 2010.

Abidin, H. Z., Andreas, H., Gumilar, I., Fukuda, Y., Pohan, Y. E., and Deguchi, T.: Land subsidence of Jakarta (Indonesia) and its relation with urban development, Nat. Hazards, 59, 1753–1771, 2011.

Abidin, H. Z., Andreas, H., Gumilar, I., Sidiq, T. P., and Fukuda, Y.: On the Roles of Geospatial Information for Risk Assessment of Land Subsidence in Urban Areas of Indonesia, in: Intelligent Systems for Crisis Management, Lecture Notes in Geoinformation and Cartography, edited by: Zlatanova, S., Peters, R., Dilo, A., and Scholten, H., Springer-Verlag Berlin Heidelberg, 277–288, 2013.

Abidin, H. Z., Andreas, H., Gumilar, I., and Yuwono, B. D.: Performance of the Geometric-Historic Method for Estimating Land Subsidence in Urban Areas of Indonesia, Proceedings of the XXV FIG Congress, TS 10B Session – Earth Geodynamics and Monitoring 2, 16–21 June 2014, Kuala Lumpur, Malaysia, 2014.

Abidin, H. Z., Andreas, H., Gumilar, I., Yuwono, B. D., Murdohardono, D., and Supriyadi, S.: On integration of geodetic observation results for assessment of land subsidence hazard risk in urban areas of Indonesia, in: International Association of Geodesy Symposia Series, Springer Berlin Heidelberg, 143, 1–8, 2015a.

Abidin, H. Z., Andreas, H., Gumilar, I., and Wibowo, I. R. R.: On correlation between urban development, land subsidence and flooding phenomena in Jakarta, Proc. IAHS, 370, 15–20, doi:10.5194/piahs-370-15-2015, 2015.

Abidin, H. Z., Andreas, H., Gumilar, I., Sidiq, T. P., and Gamal, M.: Environmental Impacts of Land Subsidence in Urban Areas of Indonesia, in: Proceedings of the FIG Working Week 2015, TS 3 – Positioning and Measurement, Sofia, Bulgaria, Paper no. 7568, 17–21 May 2015, 2015c.

Chaussard, E., Amelung, F., Abidin, H. Z., and Hong, S.-H.: Sinking cities in Indonesia: ALOS PALSAR detects rapid subsidence due to groundwater and gas extraction, Remote Sens. Environ., 128, 150–161, 2013.

Firman, T.: From Global City to City of Crisis: Jakarta Metropolitan Region Under Economic Turmoil, Habitat Int., 23, 447–466, 1999.

Firman, T.: New town development in Jakarta Metropolitan Region: a perspective of spatial segregation, Habitat Int., 28, 349–368, 2004.

Firman, T. and Dharmapatni, I. A. I.: The challenges to suistanaible development in Jakarta metropolitan region, Habitat Int., 18, 79–94, 1994.

Harsolumakso, A. H.: Struktur Geologi dan Daerah Genangan, Buletin Geologi, 33, 29–45, 2001.

Hudalah, D. and Firman, T.: Beyond property: Industrial estates and post-suburban transformation in Jakarta Metropolitan Region, Cities, 29, 40–48, 2012.

Hudalah, D., Viantari, D., Firman, T., and Woltjer, J.: Industrial land development and manufacturing deconcentration in Greater Jakarta, Urban Geogr., 34, 950–971, 2013.

Hutasoit, L. M.: Kemungkinan Hubungan antara Kompaksi Alamiah Dengan Daerah Genangan Air di DKI Jakarta, Buletin Geologi, 33, 21–28, 2001.

Koudogbo, F. N., Duro, J., Arnaud, A., Bally, P., Abidin, H. Z., and Andreas, H.: Combined X- and L-band PSI analyses for assessment of land subsidence in Jakarta, Proc. SPIE, Remote Sensing for Agriculture, Ecosystems, and Hydrology XIV, 8531, 853107, doi:10.1117/12.974821, 2012.

Murdohardono, D. and Sudarsono, U.: Land subsidence monitoring system in Jakarta, Proceedings of Symposium on Japan-Indonesia IDNDR Project: Volcanology, Tectonics, Flood and Sediment Hazards, 21–23 September 1998, Bandung, Indonesia, 243–256, 1998.

Ng, A. H.-M., Ge, L., Li, X., Abidin, H. Z., Andreas, H., and Zhang, K.: Mapping land subsidence in Jakarta, Indonesia using persistent scatterer interferometry (PSI) technique with ALOS PALSAR, Int. J. Appl. Earth Obs., 18, 232–242, 2012.

Purnomo, H., Murdohardono, D., and Pindratno, H.: Land Subsidence Study in Jakarta, Proceedings of Indonesian Association of Geologists, 30 November–1 December 1999, Jakarta, Volume IV: Development in Engineering, Environment, and Numerical Geology, 53–72, 1999.

Rajiyowiryono, H.: Groundwater and Landsubsidence Monitoring along the North Coastal Plain of Java Island, CCOP Newsletter, 24, 19 pp., 1999.

Rimbaman and Suparan, P.: Geomorphology, in: Coastplan Jakarta Bay Project, Coastal Environmental Geology of the Jakarta Reclamation Project and Adjacent Areas, Jakarta, Bangkok, CCOP COASTPLAN Case Study Report No. 2, 21–25, 1999.

Rismianto, D. and Mak, W.: Environmental aspects of groundwater extraction in DKI Jakarta: Changing views, Proceedings of the 22nd Annual Convention of the Indonesian Association of Geologists, 6–9 December 1993, Bandung, Indonesia, 327–345, 1993.

Texier, P.: Floods in Jakarta: when the extreme reveals daily structural constraints and mismanagement, Disaster Prevention and Management, 17, 358–372, 2008.

Viets, V. F.: Environmental and Economic Effects of Subsidence, Publication of Lawrence Berkeley National Laboratory, LBNL Paper LBL-8615, 251 pp., available at: http://escholarship.org/uc/item/1sb4c8vh, last access: 20 October 2015, 2010.

Ward, P. J., Marfai, M. A., Yulianto, F., Hizbaron, D. R., and Aerts, J. C. J. H.: Coastal inundation and damage exposure estimation: a case study for Jakarta, Nat. Hazards, 56, 899–916, 2011.

Yong, R. N., Turcott, E., and Maathuis, H.: Groundwater extraction-induced land subsidence prediction: Bangkok and Jakarta case studies, Proceedings of the Fifth International Symposium on Land Subsidence, IAHS Publication no. 234, 89–97, 1995.

Proc. IAHS, 372, 121–127, 2015
proc-iahs.net/372/121/2015/
doi:10.5194/piahs-372-121-2015

Application of InSAR and gravimetric surveys for developing construction codes in zones of land subsidence induced by groundwater extraction: case study of Aguascalientes, Mexico

J. Pacheco-Martínez[1,a], S. Wdowinski[2], E. Cabral-Cano[3], M. Hernández-Marín[4], J. A. Ortiz-Lozano[1], T. Oliver-Cabrera[2], D. Solano-Rojas[2], and E. Havazli[2]

[1]Departamento de Construcción y Estructuras, Universidad Autónoma de Aguascalientes, Aguascalientes, Mexico
[2]Department of Marine Geosciences, University of Miami, Miami, Florida, USA
[3]Unidad de Geomagnetismo y Exploración, Universidad Nacional Autónoma de Mexico, D.F., Mexico, USA
[4]Departamento de Geotécnia e Hidráulica, Universidad Autónoma de Aguascalientes, Aguascalientes, Mexico
[a]now at: Department of Marine Geosciences, University of Miami, Miami, USA

Correspondence to: J. Pacheco-Martínez (jesus.pacheco@edu.uaa.mx)

Published: 12 November 2015

Abstract. Interferometric Synthetic Aperture Radar (InSAR) has become a valuable tool for surface deformation monitoring, including land subsidence associated with groundwater extraction. Another useful tools for studying Earth's surface processes are geophysical methods such as Gravimetry. In this work we present the application of InSAR analysis and gravimetric surveying to generate valuable information for risk management related to land subsidence and surface faulting. Subsidence of the city of Aguascalientes, Mexico is presented as study case. Aguascalientes local governments have addressed land subsidence issues by including new requirements for new constructions projects in the State Urban Construction Code. Nevertheless, the resulting zoning proposed in the code is still subjective and not clearly defined. Our work based on gravimetric and InSAR surveys is aimed for improving the subsidence hazard zoning proposed in the State Urban Code in a more comprehensive way. The study includes a 2007–2011 ALOS InSAR time-series analysis of the Aguascalientes valley, an interpretation of the compete Bouguer gravimetric anomaly of the Aguascalientes urban area, and the application of time series and gravimetric anomaly maps for improve the subsidence hazard zoning of Aguascalientes City.

1 Introduction

The city of Aguascalientes (725 000 inhabitants), as well as other medium-sized cities (with a total of 275 000 inhabitants), is located within the Aguascalientes Valley, 430 km NW of Mexico City (Fig. 1). The growing population and increased agricultural and industrial activities along the valley have resulted in an intensive groundwater extraction, leading to develop land subsidence since earlies 1970's (Pacheco-Martínez et al., 2013).

In Aguascalientes City, surface faulting was first observed in the late of 1970's, when a group of houses developed cracks and fissures in walls and floors. Because all damaged properties were aligned along a linear feature, several studies concluded that the damage was caused by an active geological fault (Aranda-Gómez, 1989; Lermo et al., 1996).

From the discovery of this fault in the late 1970's until present, new surface faults have developed along the entire valley affecting the urban areas of Cosio, San Francisco del Rincón, San Francisco de los Romos, Jesús María and Aguascalientes City, as well as other smaller towns and rural communities (Fig. 1). As a consequence of this process, several legal disputes concerning land and housing selling affected by faulting were presented in local courts. Thus, sur-

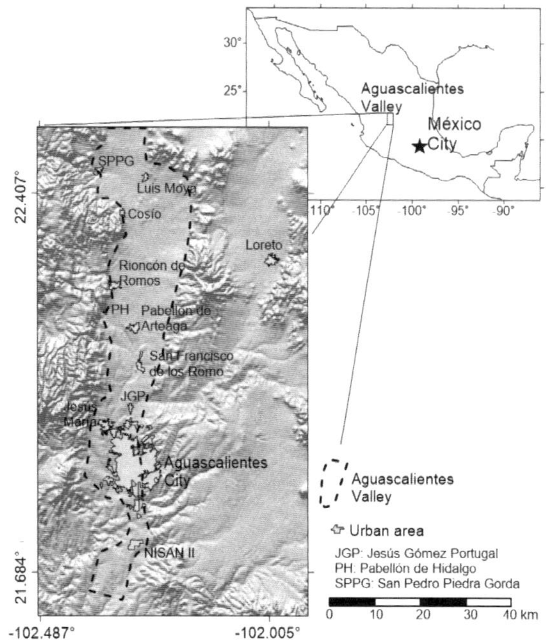

Figure 1. Location of Aguascalientes Valley. The names of minor cities and small towns are provided in the map legend.

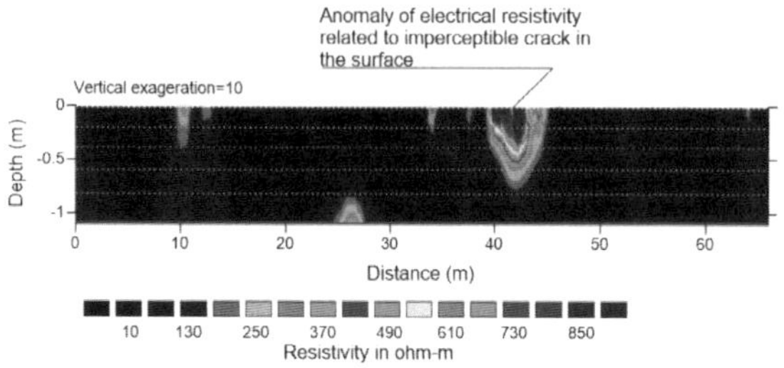

Figure 2. An example of resistivity anomaly related to a hidden or incipient surface fault surveyed in Aguascalientes City (After Pacheco-Martínez et al., 2013). Location of profile is shown in Fig. 3.

face faulting became a major concern for the local government, specially for those bureaus responsible for providing regulatory construction codes and for bureaus dealing with planning and growing of urban infrastructure.

In this work, we present geophysical data and geodetic observations, which provide information to improve the risk zoning defined in the current construction code of Aguascalientes. These observations include ALOS InSAR time-series the Aguascalientes valley, and gravimetric measurements in the city of Aguascalientes.

2 Land subsidence and related faulting in Aguascalientes City

The first reports of surface faults in the city of Aguascalientes were due to damage that a surface fault caused in several buildings in the late 1970's (Lermo et al., 1996; Aranda-Gómez, 1989), but there were no estimates about the subsidence magnitude at that time. However, currently the affected area by subsidence is around 942 km^2 with rates reaching 10 $cm\,yr^{-1}$ as shown in Sect. 4.

Ground faulting within the city is a great concern for local population, because of the structural damage that it induces to buildings, as well as to public infrastructure (Pacheco-Martínez et al., 2011; Romero-Navarro et al., 2010). Table 1 shows that there are at least 208 ground faults locations, which affect over 1860 properties (SIFAGG, 2015).

A new and bigger concern to the local government is the damage to heritage buildings in the city caused by ground faulting. In recent years, three historical buildings in the valley have been affected by surface faulting; two of them, the Cathedral of Aguascalientes and the Museum of the Insurgency are listed in the national catalogue of historic buildings.

3 Current geotechnical practices and the Urban Code for the Aguascalientes State

The local government responded to the ground faulting problem by implementing several actions, including the recommendations and requirements for new construction projects which have been incorporated in the Urban Code for the Aguascalientes State. The most relevant statements in the code concerning ground failure damage prevention are:

1. Certification of proficient professionals in geology or geophysics, who have to prove a background in ground faulting studies in order to be certified.
2. A requirement for municipalities to prepare and publish ground faulting maps.
3. Proposal of a zoning risk based on the shortest distance to mapped ground faults.
4. A statement that requires constructors to present geophysical or geological studies prepared by the certificated geologist or geophysics in order to obtain a construction license.

Some of the weaknesses of the Urban Code concerning the ground faulting are:

1. The code does not recognize the phenomenon of regional subsidence of the valley. The ordinance is focused solely to prevent the effects of previously mapped surface faults.

Table 1. Ground faults length in municipalities and number of affected properties. AGS: Aguascalientes, JM: Jesús María, COS: Cosio, PAB: Pabellón, RIN: Rincón de Romos, SFCO: San Fancisco de los Romo.

	AGS	JM	COS	PAB	RIN	SFCO	TOTAL
Number of mapped surface fault locations	66	49	15	37	28	13	208
Accumulated length of ground faults (km)	83.575	60.473	30.620	43.534	47.275	25.231	290.678
Number of damaged properties	1438	183	35	86	35	88	1865

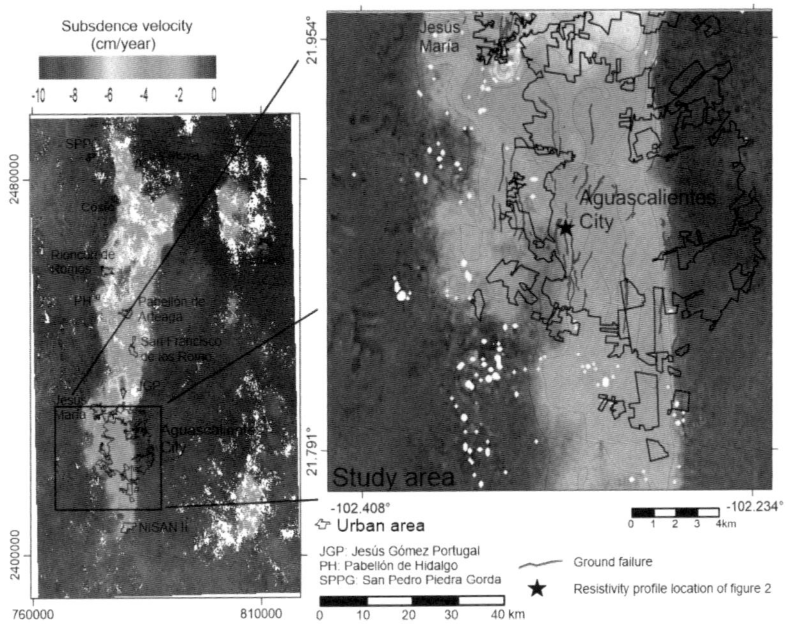

Figure 3. Subsidence velocity map of Aguascalientes Valley as determine from InSAR time series analysis of 2007–2011 ALOS data.

2. The proposed zoning in the code only points out the potential seismic risks to buildings related to the ground fault motion. The zoning does not include considerations about the hazard related to future ground fault evolution nor about the hazard associated to the differential subsidence in progress. The code recognizes three risk zones:

Zone I. High risk: terrain located within a 5 m wide buffer zone (or that determined by a municipality technician) along the axis of the ground failure. The code states that this zone has a high risk for urban development because the zone is under the influence of seismic activity and differential settlements caused by the motion of ground faults.

Zone II. Medium risk: terrain located within a 200-m wide buffer zone on each side from the zone I. The code states that in zone II the buildings or structures are still under the influence of seismic activity, which are originated along ground faults.

Zone III. All the terrain outside of zone I and II. The code states that the constructions in this zone are not influenced by ground faults.

3. The Urban Code does not clearly define what type of geological or geophysical study must be carried out depending the risk zone.

4. The code compels constructors to present geological and geophysical studies for any new constructions authorization, even if new constructions are outside of the subsiding area, for example at the eastern part of Aguascalientes City (see Fig. 3) where neither subsidence nor ground failure have been documented.

Regarding to the ground fault assessments, all studies that certificated geologist and geophysics develop for construction permissions focus on providing information that can lead to determine if the studied area is affected by ground faulting. In affirmative case, the studies have to include the determination of the area affected by the fault.

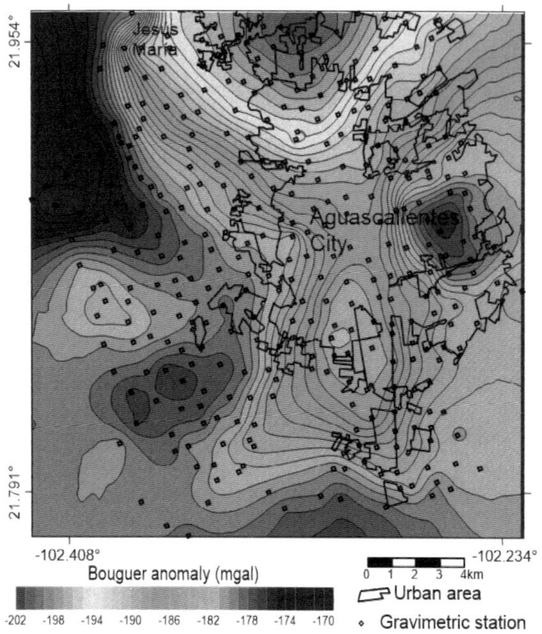

Figure 4. Complete Bouguer anomaly of the Aguascalientes city and surroundings.

The more difficult case is detecting faults at an early stage of formation. In this stage, faults may not be recognizable on the surface because either offset is very small and can be hard to detect by the erosion or by "in situ" pre-construction activities. To overcome this issue, measurements of electrical resistivity profiles of subsoil has been used with relative success to detect ground faulting in an early stage. Fractured subsoil under favorable conditions for the technique application (homogeneous stratigraphy and undisturbed soil) appears as an anomaly of high resistivity (Fig. 2).

However, under unfavourable conditions, such as a subsoil with a complex stratigraphy, deposits of anthropogenic fillings, or in zones with strong lateral water content variations, the interpretation of electrical resistivity data can be ambiguous. In such cases, geophysical measurements are complemented by trench excavation in order to directly explore where resistivity anomalies are observed. Exposed trenches help to verify the existence of incipient ground failures.

4 The role of InSAR and gravimetric methods in subsidence studies

InSAR techniques have been used successfully to study land subsiding areas (Ketelaar, 2009; Galloway and Burbey, 2011). The large-scale coverage of the technique and the relatively wide-spread availability of InSAR imagery have fostered its application for land subsidence studies. In some studies, InSAR was used to detect the extent of subsiding areas and subsidence rates (Chaussard et al., 2014; Galloway et al., 1998; Bell et al., 2008). Some authors have reported the use of InSAR for hazard evaluation of ground failures related to subsidence (Cabral-Cano et al., 2010, 2011).

The large-scale coverage of SAR data enabled to demonstrate that the large spatial extent of land subsidence is controlled by geologic structures and stratigraphy of the subsiding area (Amelung et al., 1999; Cigna et al., 2012). These studies concluded that groundwater level decrease is the triggering factor of the subsidence process, whereas the presence of deformable sediments determines its occurrence.

In areas where the process of subsidence began long ago, the decrease in groundwater level has taken place along the entire aquifer system, suggesting that subsidence has affected the deformable sediments in greater or lesser extent. In these circumstances, a subsidence map should show the limits of the area which have potential to develop land subsidence. The map should also show those areas in which subsidence is not likely to develop.

In the case of Aguascalientes Valley, the deformation-prone aquifer system is composed of an alluvial sediments sequence with abundant silty sand and gravel contained within a regional graben; thus the limits of the subsiding-prone area are also the limits of the graben. Those limits are clearly identifiable in the subsidence map of Aguascalientes Valley (Fig. 3).

The velocity map of Fig. 3 was elaborated from the processing of 34 SAR images acquired by the ALOS satellite from 2007 to 2011. Images cover two frames (420, 430) from the ascending track 191. Imagery was processed with the differential InSAR technique (D-InSAR) to obtain 28 interferograms and the small baseline (SB) to generate the InSAR time-series (Rosen et al., 2000; Berardino et al., 2002; Lanari et al., 2004). We use for processing ROI_PACK software developed by NASA's Jet Propulsion Laboratory (Rosen et al., 2004). Topographic correction was applied according to Fattahi and Amelung (2013). Pixels with temporal coherence lesser than 0.7 were dismissed for elaborating the velocity map.

In Mexico, the subsurface geology of many subsiding areas consists of sequences of unconsolidated granular strata lying over a rigid bedrock, usually volcanic or carbonated rocks. Consequently, the density contrast between these units is significant, and very favorable for gravimetric analysis. Gravimetry provide in this way valuable information about the configuration of the bedrock in sedimentary basins (Pacheco-Martínez et al., 2006, 2010; Yutsis et al., 2014).

Furthermore, some studies have shown that in geological settings susceptible to develop subsidence by groundwater extraction, there is an inverse correlation between sediment thickness and the gravimetric anomaly (Jachens and Holzer, 1979; Pacheco-Martínez et al., 2006, 2010). In other words, the higher values of the gravimetric anomaly correlate with

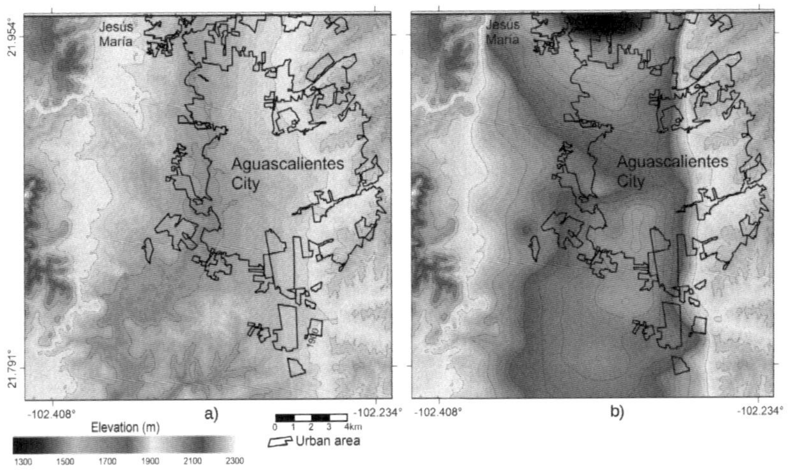

Figure 5. Surface (**a**) and bedrock (**b**) topography in the Aguascalientes City study area.

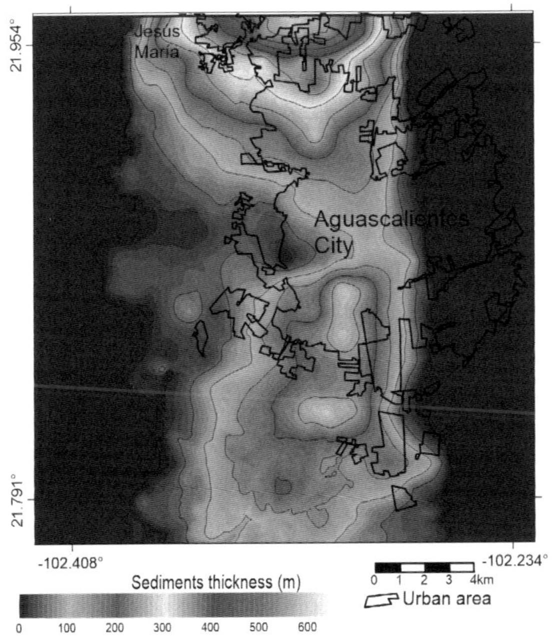

Figure 6. Map of sediments thickness in Aguascalientes City.

small sediment thicknesses. Inversely, lower anomaly values correspond to thicker sedimentary sequences.

Figure 4 shows the complete Bouguer's anomaly of Aguascalientes City subsidence area, which was elaborated from a surveying of 339 ground-based gravimetric measurements. Gravity field data were surveyed using a CG5 gravimeter from Scintrex. Standard procedures for fieldworks and data processing where followed (Telford et al., 1990) . The gravity data were inverted according to Singh and Guptasarma (2001) using PyGMI software developed by Cole (2015) in order to determine a model of the rocky stratum (Fig. 5). The Rocky stratum model was restrained in several points in which rocky stratum depth is known through lithology logs wells and outcrops in both sides of the graben. Finally, an isopach map (Fig. 6), which is a sediment thickness variations map, was calculated from the difference in elevations of the rocky stratum and the terrain surface.

5 Subsidence hazard zoning map elaboration

The hazard level related to subsidence depends on the existence of deformable sediments. It depends also on the advance degree of the subsidence process, whether superficial sinking and ground failure have been observed. Hence, three hazard factors were considered in order to complete a hazard factors array, from which the hazard level was calculated according to the existence of such factors (Table 2).

The zoning map (Fig. 7) was prepared according to a spatial analysis of the hazard factors presented in Table 2. Each hazard zone and the suggested requirements for geotechnical studies needed to evaluate the existence of ground failure in the terrain of new projects are described bellow from low to high level.

Zone 4 non-existent hazard: neither subsidence nor ground failure have been detected. Furthermore deformable filling thicknesses are negligible in this zone. Standard geotechnical studies should be enough.

Zone 3 low hazard: neither subsidence nor ground failure have been detected. Nevertheless the isopach map shows thicknesses of deformable sediments which can be significant, then subsidence could develop in the future. Standard geotechnical studies plus a superficial geological study

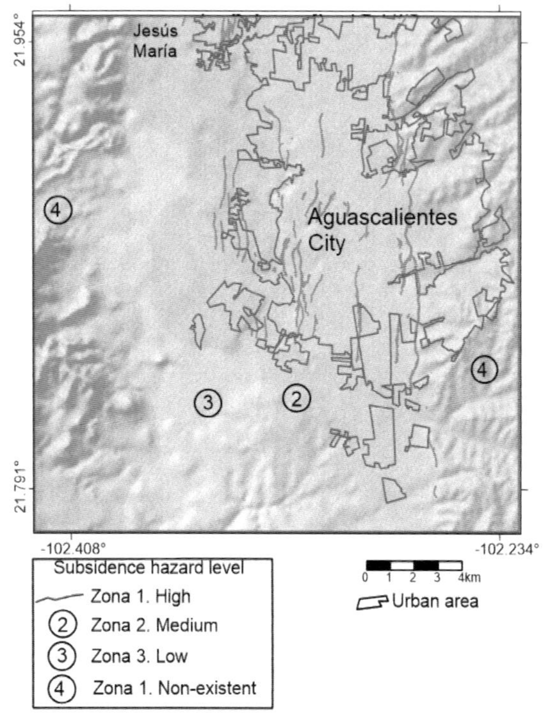

Figure 7. Subsidence hazard zoning map of Aguascalientes City.

searching for evidences of pre-subsidence cracks should be enough.

Zone 2 medium hazard: subsidence is developing, but ground failure has not been detected. Same requirements of level 3 plus a geophysical study to find out incipient ground failure.

Zone 1 high hazard: this zone corresponds to the existent ground failure traces. Same requirements of level 2 plus a detailed study to determine the trace width over which differential subsidence could cause damage on new constructions.

6 Conclusions

The zoning map of subsidence and surface faulting prone area is an important element in risk management of land subsidence and associated surface faults. In this vein, Gravimetric measurements and InSAR-derived velocity maps provide valuable information for its generation. The first allows determining sediment thickness prone to deform, and the second one helps determining accurately where subsidence is present.

The combination of both information sources, along with reliable subsurface geological information, allows us to determine in a comprehensible and sustainable way the subsidence hazard map for Aguascalientes. This map will allow government agencies responsible for urban growth planning to better define the needed technical studies according to the hazard level of the zone in which new constructions are planned.

Due to the dynamic nature of the subsidence, the hazard map have to be updated with new information generated through the geotechnical studies required for construction permissions of new projects. New SAR-images acquisitions and their processing to obtain updated velocity maps will allow regular updates of the hazard map.

Table 2. Hazard factor array and resulting hazard levels. H: High level, M: Medium level, L: Low level, N: non-existent level.

	Hazard levels			
Hazard factors	H	M	L	N
Deformable sediments existence	Yes	Yes	Yes	No
Detected surface sinkings	Yes	Yes	No	No
Ground failure precense	Yes	No	No	No

Acknowledgements. J. Pacheco-Martínez thanks CONACYT for financial support: "Investigación apoyada por el CONACYT". S. Wdowinski and E. Cabral-Cano acknowledge support through NASA-ROSES grant NNX12AQ08G. E. Cabral-Cano is supported by UNAM-PAPIIT projects IN104213-2, and IN109315-3. This material is partly based on data provided by the UNAVCO Facility with support from the National Science Foundation (NSF) under award EAR-1338091.

References

Amelung, F., Galloway, D. L., Bell, J. W., Zebker, H. A., and Laczniac, R. J.: Sensing the Ups and Downs of Las Vegas: InSAR Revelas Structural Control of Land Subsidence and Aquifer-System Deformation, Geology, 27, 483–486, 1999.

Aranda-Gómez, J. J.: Geología Preliminar Del Graben de Aguascalientes, Revista Mexicana de Ciencias Geológicas, 8, 22–32, 1989.

Bell, J. W., Amelung, F., Ferretti, A., Bianchi, M., and Novali, F.: Permanent Scatterer InSAR Reveals Seasonal and Long-Term Aquifer-System Response to Groundwater Pumping and Artificial Recharge, Water Resour. Res., 44, W02407, doi:10.1029/2007WR006152, 2008.

Berardino, P., Fornaro, G., Lanari, R., and Sansosti, E.: A New Algorithm for Surface Deformation Monitoring Based on Small Baseline Differential SAR Interferograms, IEEE T. Geosci. Remote, 40, 2375–2383, 2002.

Cabral-Cano, E., Osmanoglu, B., Dixon, T., Wdowinski, S., Demets, C., Cigna, F., and Díaz-Molina, O.: Subsidence and Fault Hazard Maps Using PSI and Permanent GPS Networks in Central Mexico, Proceedings of Eighth International Symposium on Land Subsidence, 255–59, IAHS Red Books, Querétaro, Mexico: IAHS Press, Publication no. 339, 2010.

Cabral-Cano, E., Díaz-Molina, O., and Delgado-Granados, H.: Subsidencia Y Sus Mapas de Peligro: Un Ejemplo En El Área Nororiental de La Zona Metropolitana de La Ciudad de Mexico, Boletín de La Sociedad Geológica Mexicana, 63, 53–60, 2011.

Chaussard, E., Wdowinski, S., Cabral-Cano, E., and Amelung, F.: Land Subsidence in Central Mexico Detected by ALOS InSAR Time-Series, Remote Sens. Environ., 140, 94–106, 2014.

Cigna, F., Osmanoğlu, B., Cabral-Cano, E., Dixon, T. H., Ávila-Olivera, J. A., Garduño-Monroy, V. H., DeMets, C., and Wdowinski, S.: Monitoring Land Subsidence and Its Induced Geological Hazard with Synthetic Aperture Radar Interferometry: A Case Study in Morelia, Mexico, Remote Sens. Urban Environ., 117, 146–161, 2012.

Cole, P.: PyGMI-Python Geophysical Modeling and Interpretation (version 2.2.6). South Africa: Council for Geosciences, Geological Survey of South Africa, available at: http://patrick-cole.github.io/pygmi/index.html, 2015.

Fattahi, H. and Amelung, F.: DEM Error Correction in InSAR Time Series, IEEE T. Geosci. Remote, 51, 4249–4259, 2013.

Galloway, D. L. and Burbey, T. J.: Review: Regional Land Subsidence Accompanying Groundwater Extraction, Hydrogeol. J., 19, 1459–1486, 2011.

Galloway, D. L., Hudnut, K. W., Ingebritsen, S. E., Phillips, S. P., Peltzer, G., Rogez, F., and Rosen, P. A.: Detection of Aquifer System Compaction and Land Subsidence Using Interferometric Synthetic Aperture Radar, Antelope Valley, Mojave Desert, California, Water Resour. Res., 34, 2573–2585, 1998.

Jachens, R. C. and Holzer, T. L.: Geophysical Investigation of Ground Failure Related to Ground Water Withdrawal-Picacho Basin, Arizona, Ground Water, 17, 574–585, 1979.

Ketelaar, V. B. H.: Satellite Radar Interferometry – Subsidence Monitoring Techniques, Springer, 2009.

Lanari, R., Mora, O., Manunta, M., Mallorqui, J. J., Berardino, P., and Sansosti, E.: A Small-Baseline Approach for Investigating Deformations on Full-Resolution Differential SAR Interferograms, IEEE T. Geosci. Remote, 42, 1377–1386, 2004.

Lermo, J., Nieto-Obregón, J., and Zermeño, M.: Faults and Fractures in the Valley of Aguascalientes. Preliminary Microzonification, Proceedings of the Eleventh World Conference on Earthquake Engineering, Acapulco Mexico, 1996.

Pacheco-Martínez, J., Arzate-Flores, J., Rojas, E., Arroyo, M., Yutsis, V., and Ochoa, G.: Delimitation of Ground Failure Zones due to Land Subsidence Using Gravity Data and Finite Element Modeling in the Querétaro Valley, Mexico, Eng. Geol., 84, 143–160, 2006.

Pacheco-Martínez, J., Arzate-Flores, J. A., López-Doncel, R., Barboza-Gudiño, R., Mata-Segura, J. L., Del-Rosal-Pardo, A., and Aranda-Gómez, J. J.: Zoning Map of Ground Failure Risk due to Land Susbidence of San Luís Potosí, Mexico, Proceedings of Eighth International Symposium on Land Subsidence, 179–184, Querétaro, Mexico: IAHS Press, Publication no. 339, 2010.

Pacheco-Martínez, J., Zermeño-De-León, M. E., Ortiz-Lozano, J. A., Solís-Pinto, A., Romero-Navarro, M. A., Aguilar-Valdés, F. J., and Fuente-López, J. A.: Soil cracks related to land subsidence. The main geotechnical hazard affecting to construction in Aguascalientes City, Proceedings of the 14th Pan-American Conference on Soil Mechanics and Geotechnical Engineering, 8, 1–8, Toronto, Canada, 2011.

Pacheco-Martínez, J., Hernandez-Marín, M., Burbey, T. J., González-Cervantes, N., Ortíz-Lozano, J. A., Zermeño-De-Leon, M. E., and Solís-Pinto, A.: Land Subsidence and Ground Failure Associated to Groundwater Exploitation in the Aguascalientes Valley, Mexico, Eng. Geol., 164, 172–186, 2013.

Romero-Navarro, M. A., Pacheco-Martínez, J., Ortiz-Lozano, J. A., Zermeño-De-León, M. E., Araiza-Garaygordobil, G., and Mendoza-Otero, E.: Land Subsidence in Aguascalientes Valley, Mexico. Historical Review and Present Situation, Proceedings of Eighth International Symposium on Land Subsidence, 207–9, IAHS Red Books, Querétaro, Mexico: IAHS Press, Publication no. 339, 2010.

Rosen, P. A., Hensley, S., Joughin, I. R., Li, F. K., Madsen, S. N., Rodríguez, E., and Goldstein, R. M.: Synthetic Aperure Radar Interferometry, Proceedings of the IEEE, 88, 333–382, 2000.

Rosen, P. A., Hensley, S., and Peltzer, G.: Updated Repeated Orbit Interferometry Package Released, Eos, Transactions American Geophysical Union, 85, 47, 2004.

SIFAGG: Sistema de Información de Fallas Geológicas Y Grietas on Line: http://www.aguascalientes.gob.mx/sop/sifagg/web/mapa.asp, Cartografía Digital Interactiva, Aguascalientes, Mexico: Secretaría de Obras Públicas del Estado de Aguascalientes, 2015.

Singh, B. and Guptasarma, D.: New Method for Fast Computation of Gravity and Magnetic Anomalies from Arbitrary Polyhedral, Geophysics, 66, 521–526, 2001.

Telford, W. M., Geldart, L. P., and Sheriff, R. E.: Applied Geophysics, New York, USA: Cambridge University Press, 1990.

Yutsis, V., Aranda-Gómez, J. J., Arzate-Flores, J. A., Bohnel, H., Pacheco Martínez, J., and López-Loera, H.: Maar Geophysics: Valle de Santiago Study, Proceedings of The 5th International MAAR Conference, 32–33, Querétaro, Mexico, 2014.

Proc. IAHS, 372, 129–139, 2015
proc-iahs.net/372/129/2015/
doi:10.5194/piahs-372-129-2015

Production induced subsidence and seismicity in the Groningen gas field – can it be managed?

J. A. de Waal, A. G. Muntendam-Bos, and J. P. A. Roest

State Supervision of Mines, The Hague, the Netherlands

Correspondence to: J. A. de Waal (j.a.dewaal@minez.nl)

Published: 12 November 2015

Abstract. Reliable prediction of the induced subsidence resulting from gas production is important for a near sea level country like the Netherlands. Without the protection of dunes, dikes and pumping, large parts of the country would be flooded. The predicted sea-level rise from global warming increases the challenge to design proper mitigation measures. Water management problems from gas production induced subsidence can be prevented if measures to counter its adverse effects are taken timely. This requires reliable subsidence predictions, which is a major challenge. Since the 1960's a number of large, multi-decade gas production projects were started in the Netherlands. Extensive, well-documented subsidence prediction and monitoring technologies were applied. Nevertheless predicted subsidence at the end of the Groningen field production period (for the centre of the bowl) went from 100 cm in 1971 to 77 cm in 1973 and then to 30 cm in 1977. In 1984 the prediction went up again to 65 cm, down to 36 cm in 1990 and then via 38 cm (1995) and 42 cm (2005) to 47 cm in 2010 and 49 cm in 2013. Such changes can have large implications for the planning of water management measures.

Until 1991, when the first event was registered, production induced seismicity was not observed nor expected for the Groningen field. Thereafter the number of observed events rose from 5 to 10 per year during the 1990's to well over a hundred in 2013. The anticipated maximum likely magnitude rose from an initial value of less than 3.0 to a value of 3.3 in 1993 and then to 3.9 in 2006. The strongest tremor to date occurred near the village of Huizinge in August 2012. It had a magnitude of 3.6, caused significant damage and triggered the regulator into an independent investigation. Late 2012 it became clear that significantly larger magnitudes cannot be excluded and that values up to magnitude 5.0 cannot be ruled out. As a consequence the regulator advised early 2013 to lower Groningen gas production by as much and as fast as realistically possible. Before taking such a decision, the Minister of Economic Affairs requested further studies. The results became available early 2014 and led to the government decision to lower gas production in the earthquake prone central area of the field by 80 % for the next three years. In addition further investigations and a program to strengthen houses and infrastructure were started.

Important lessons have been learned from the studies carried out to date. It is now realised that uncertainties in predicted subsidence and seismicity are much larger than previously recognised. Compaction, subsidence and seismicity are strongly interlinked and relate in a non-linear way to production and pressure drop. The latest studies by the operator suggest that seismic hazard in Groningen is largely determined by tremors with magnitudes between 4.5 and 5.0 even at an annual probability of occurrence of less than 1 %. And that subsidence in 2080 in the centre of the bowl could be anywhere between 50 and 70 cm. Initial evaluations by the regulator indicate similar numbers and suggest that the present seismic risk is comparable to Dutch flooding risks.

Different models and parameters can be used to describe the subsidence and seismicity observed so far. The choice of compaction and seismicity models and their parameters has a large impact on the calculated future subsidence (rates), seismic activity and on the predicted response to changes in gas production. In addition there are considerable uncertainties in the ground motions resulting from an earthquake of a given magnitude and in the expected response of buildings and infrastructure. As a result uncertainties in subsidence and seismicity become very large for periods more than three to five years into the future. To counter this a control loop based on interactive modelling, measurements and repeated calibration will be used. Over the coming years, the effect of the production reduction in the centre of the field on subsidence and seismicity will be studied in detail in an effort to improve understanding and thereby reduce prediction uncertainties. First indications are that the reduction has led to a drop in subsidence rate and seismicity within a period of a few months. This suggests that the system can be controlled and regulated. If this is the case, the integrated loop of predicting, monitoring and updating in combination with mitigation measures can be applied to keep subsidence (rate) and induced seismicity within limits. To be able to do so, the operator has extended the field-monitoring network. It now includes PS-InSAR and GPS stations for semi-permanent subsidence monitoring in addition to a traditional network of levelling benchmarks. For the seismic monitoring 60 shallow (200 m) borehole seismometers, 60 + accelerometers and two permanent downhole seismic arrays at reservoir level will be added. Scenario's spanning the range of parameter and model uncertainties will be generated to calculate possible subsidence and seismicity outcomes. The probability of each scenario will be updated over time through confrontation with the measurements as they become available. At regular intervals the subsidence prediction and the seismic risk will be re-evaluated. Further mitigation measures, possibly including further production measures will need to be taken if probabilities indicate unacceptable risks.

1 Introduction

By end 2013 production induced subsidence above the centre of the Groningen gas field in the Netherlands amounted to some 33 cm with subsidence increasing at a rate of some 7–8 mm per year (van Thienen-Visser and Breunese, 2015a; Pijpers, 2014a). Such an amount of subsidence requires the design and implementation of proper mitigation measures in a near sea-level country like the Netherlands (de Waal et al., 2012). The country would largely flood without the presence of dykes and active pumping and the predicted sea-level rise from global warming increases the challenge.

As reported for a number of other oil and gas fields (Lee, 1979; Merle et al., 1976; de Waal and Smits, 1986; Hettema et al., 2002; Mallman and Zoback, 2007; Kosloff and Scott, 1980) subsidence rate in Groningen was initially significantly lower than predicted on the basis of laboratory rock-mechanical measurements carried out on samples taken from the reservoir. At a later stage the subsidence rate accelerated, coming closer to expected values and in line with the notion of "delayed subsidence" in (Hettema et al., 2002). While the subsidence accelerated, earthquakes started to occur. Over time these have increased in numbers and strength (Fig. 1). The strongest tremor so far with a magnitude of 3.6 occurred near the village of Huizinge in 2012, causing damage to thousands of homes and raising anxiety among those living in the area (Dost and Kraaijpoel, 2012). The mechanism behind the earthquakes is understood to be (differential) compaction at reservoir level reactivating existing faults that have become critically stressed as a result of production induced stress changes. Better understanding of the physics of reservoir compaction and fault reactivation is important to predict the further evolution of the Groningen subsidence and seismicity and to assess if and how the resulting seismic hazards and risks can be managed, e.g. through production measures such as changes in production rates, off-take pattern, nitrogen injection etc. and/or by executing a strengthening program for houses and infrastructure.

After discussing some general aspects of the Groningen field and its production history, the evolution of subsidence and seismicity since the start of production will be discussed in more detail with emphasis on the lessons that have been learned so far. Thereafter the recent developments will be discussed that followed the 2012 insight that risks from reservoir compaction induced seismicity in Groningen might be significantly higher then previously realised (Muntendam-Bos and de Waal, 2013).

2 Field properties and production history

The Groningen gas field was discovered in 1959 in the Lower Permian Rotliegend at a depth of some 3000 m by exploration well Slochteren-1 near the Dutch village of Kolham (van Hulten, 2009). The structure containing the Groningen reservoir is a NNW-SSE trending high formed by normal faulting during the Late Jurassic to early Cretaceous (Stäuble and Milius, 1970; Whaley, 2009). The high has an overall horst geometry, lying between the Eems Graben to the east and the Lauwerszee Trough to the west. The reservoir is covered by Late Permian Zechstein carbonate, anhydrite and halite evaporates which provide an excellent seal. The field covers an area of some 900 km^2 and is located below a relatively densely populated area with some 250 000 houses including several urban centres. The average thickness of the gas-bearing sandstone interval is some 100 m. The field has excellent reservoir properties with porosity averaging around 17 % and permeabilities in the hundreds of mD (Mijnlieff and Geluk, 2011; van Ojik et al., 2011). Some 1800 larger and smaller faults have been mapped based on the available 3-D seismic using ant tracking on the 3-D seismic cube (NAM, 2013).

History matching and pressure measurements in observation wells indicate all of the faults to be non-sealing with the exception of a number of larger faults at the field boundaries. The original volume of recoverable Groningen gas reserves is estimated at some 2800 BCM (100 Tcf), which makes it

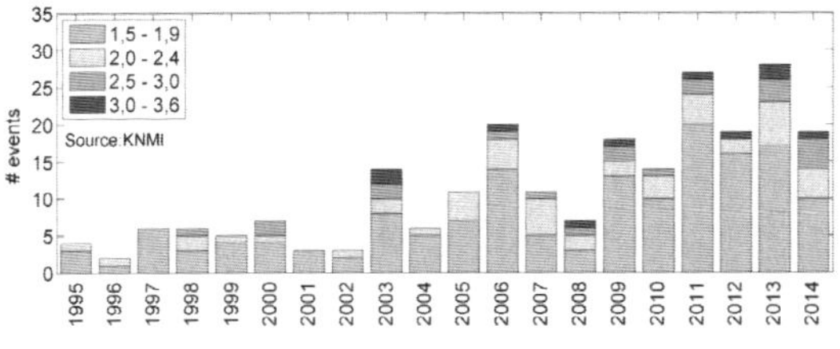

Figure 1. Groningen seismicity ($M > 1.5$) vs. time.

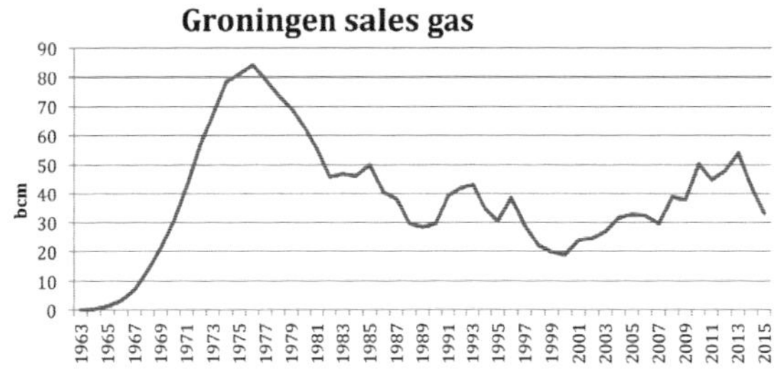

Figure 2. Groningen gas production (BCM/year) vs. time.

one of the larger gas fields worldwide. The gas is produced through more than 300 wells distributed over 22 production clusters. It contains slightly more then 15 % nitrogen. Gas appliances in the Netherlands and surrounding countries are optimised for this. As a result Groningen gas cannot be replaced by gas with a different composition without enough time to replace these appliances economically. An option, which also works short-term, is to mix higher methane content gas from other sources with nitrogen. For a substantial replacement volume this is not a small task given the amounts of nitrogen required. Groningen production levels over time are shown in Fig. 2.

During the early seventies of the previous century the field was produced at high rates as it was anticipated that the value of the gas would be eroded away by cheap nuclear power. When it was realised that this would not happen, the production philosophy was changed drastically and Groningen became the regional swing producer. Preference was given to production from smaller onshore and offshore fields, the development of which is stimulated by tax measures (Small Fields Policy). The policy was so successful that the annual Groningen production dropped from values above 90 BCM in 1976 to as low as 20 BCM by 2000. Thereafter production from the "Small Fields" started to decline with Groningen reaching a new annual high of 54 BCM in 2013. Presently more then two-third of the recoverable gas has been produced and first stage compression has been introduced. Over time the capacity of the Groningen field to act as the regional swing producer will decline and this role will be taken over gradually by large underground gas storage facilities.

3 Subsidence

3.1 Early days

Already during the early stages of field development it was realised that the gas production from the Groningen field could lead to a significant amount of surface subsidence. Detailed rock-mechanical measurement and modelling studies were executed (NAM, 1973) leading in 1973 to a predicted maximum subsidence of 100 cm at the end of field life, then expected around 2020. The prediction was based on an analytical linear elastic nucleus of strain model, supported by calculations with a linear elastic finite element numerical model. The use of a linear elastic model seemed justified by the results of a large number of zero pore pressure laboratory compaction measurements carried out on core samples taken from the reservoir. An extensive monitoring network was installed around the same time. It involved a significant extension of the existing local geodetic surface network, monitoring of near surface layer compaction (unrelated to gas production) and a number of dedicated down-hole reservoir compaction monitoring wells with radioactive bullets shot into the formation (NAM, 1973).

By 1973 it had become clear that both reservoir compaction and surface subsidence were occurring at a much lower rate then had been predicted. This remained the case also after applying a Biot correction factor to account for the effect of the field pore pressure, bringing the 1971 prediction down to 77 cm in 1973. In 1976 the predicted maximum subsidence at the end of field life was brought down to 30 cm by calibrating a linear elastic model against the 1977 levelling data (Schoonbeek, 1976). The reason for the large difference between the laboratory measured and the field observed rock compressibility was not clear. It was speculated to be caused by core damage from drilling, transport and laboratory procedures applied to the core samples prior to testing. A number of studies were started to investigate this and other potential explanations.

3.2 The rate type compaction model

The additional studies looked at the relation between reservoir compaction and surface subsidence, core disturbance, the validity of the effective stress concept, the in-situ stress state, possible previous deeper burial, pressure lag (between different permeability layers) and loading rate (de Waal, 1986). The study included an extensive re-analysis of the available data from Groningen and from other fields worldwide with significant subsidence, as well as rock-mechanical laboratory experiments on the effects of stress state, pore pressure, previous deeper burial, loading path, core disturbance and (changes in) loading rate. Based on the outcomes, the cause for the discrepancy in Groningen was postulated to be the large increase in loading rate that occurs in the reservoir at the start of depletion. Such an increase in loading rate cannot be easily simulated in laboratory experiments and even if this could be done, the rates at which laboratory experiments can be carried out remain much higher than those occurring in the field over geological times. Validating this explanation for field application is therefore problematic. The newly developed rate-type compaction model explained the difference between the initial field observed and the lab-derived compressibility for Groningen and explained the observed reduced and/or delayed subsidence above a number of other depletion-drive oil and gas reservoirs (de Waal, 1986; de Waal and Smits, 1988).

The physical explanation for the laboratory observed rate sensitivity was postulated to be time-dependent friction at the sliding mineral contact points (Dieterich, 1978, 1994) between (assemblies of cemented) grains or on microscopic fractures responsible for the non-elastic component in the compaction. Using a simple model incorporating the time dependent friction equation proposed by Dieterich, results in a differential equation popular in soil mechanics to describe the known rate dependent compaction behaviour of sands and soils at lower stress levels (Bjerrum, 1967; Kolymbas, 1977; Vermeer and Neher, 1999).

The use of the new model for predicting subsidence in Groningen was accepted in 1986. Subsidence at the end of field life was predicted to be around 65 cm with a one-sigma uncertainty of 10 cm.

3.3 Back to linear models

Based on differences between the observed development of the subsidence above the Groningen field and predictions based on the rate type compaction model, the Groningen field operator in 1990 requested a review of the various subsidence predictions by two independent MIT advisors of high international standing (Toksöz and Walsh, 1990). The reviewers advised NAM to discount the prediction of Groningen subsidence on the basis of the rate type compaction model as they did not expect that loading or strain rate could have a significant effect on the deformation of the consolidated reservoir rock in the Groningen field. They advised the operator to return to the use of a linear elastic model calibrated against the field observed subsidence, leading to a new prediction for the ultimate maximum subsidence of 36 cm. Further adjustments increased the prediction to 38 cm in 1995 and to 42 cm in 2005 to honour results of new surface levelling data. By 2005 it had become clear again that a linear model could not describe the field behaviour. The operator introduced a bi-linear compaction model in which the uniaxial compressibility increases after a given amount of pressure drop after which it stays at that higher value. The pivot point and the compressibility before and after the pivot point were fitted to honour new levelling survey measurements as they came in over time. Using this approach, the predicted maximum subsidence at the end of field life was increased to 47 cm in 2010, including 2 cm of additional subsidence from the use of a lower gas abandonment pressure.

3.4 Back to non-linear models

By 2012 other Rotliegend gas reservoirs in the Netherlands showed continuing subsidence at a more or less constant gas pressure at the end of the production period. A bi-linear compaction model cannot explain such behaviour. For the time being NAM decided to switch to a time decay model, similar to a model proposed earlier by (Houtenbos, 2007) to describe the subsidence behaviour of compacting reservoirs. Mossop (Mossop, 2012) postulated a non-uniform heavy tailed permeability distribution resulting in a non-uniform pressure distribution within the bulk reservoir rock to be the explanation.

At the request of the regulator a new investigation into the physical background of the observed subsidence delays was commenced taking the permeability distribution explanation as only one of a number of possibilities. In addition salt flow, bottom and lateral aquifer inflow, intrinsic sandstone rate sensitivity, core disturbance, stress state and stress path are again being investigated.

One of the alternatives presently under consideration is a modified version of the rate-type compaction model abandoned for consolidated sandstones in 1990. The new formulation (Pruiksma et al., 2015) combines the features of the original rate type model with the soil mechanics isotach concept that enables transitions between different loading rates. The new isotach formulation combines a direct (linear elastic) strain component and a creep component, which elegantly explains the moderate loading rate dependence of the isotach compressibility. This way a number of limitations of the original rate type model have been resolved (de Waal et al., 2015), be it at the cost of an additional material property representing the linear elastic component of the deformation. The fit between predictions based on the modified isotach model and the measured subsidence in Groningen is better than that obtained with the time-decay model or the bi-linear model (van Thienen-Visser and Breunese, 2015).

The predicted impact of drastic changes in production rate during later stages of the production period is very different for the different compaction models. For the bi-linear model the predicted effect on compaction and subsidence is instantaneous, while for the isotach model the response is more gradual over a period of up to a year. The time-decay model predicts that it will take 5 to 10 years for the rock to respond to significant changes in production rate.

4 Induced seismicity

4.1 History

Until 1991 no induced seismicity was registered in the Groningen field, nor was it expected. The independent study by MIT experts (Toksöz and Walsh, 1990) estimated the probability of an induced seismic event with a magnitude above 3.0 at less then 10 % for the next 50 years. If such an event would occur at all, they expected it to happen on the sealing faults at the field edges, not in the field centre. History proved different. For a number of other fields in the Netherlands induced seismicity had been linked to gas production a few years earlier (BOA, 1993). While subsidence in Groningen was accelerating, induced seismicity started to occur. The first production-induced earthquake with a local magnitude M_L of 2.4 was recorded at Middelstum in 1991. The frequency and magnitude of the events thereafter continued to increase over time as shown in Fig. 1. The local seismic network was upgraded a number or times and is complete for magnitudes above 1.5 since 1995. To date, close to a thousand gas production induced earthquakes have been registered in Groningen. Most events have been of a small magnitude ($M_L < 1.5$), while by mid 2015 some 250 earthquakes had magnitudes $M_L \geq 1.5$. In contrast to early expectation, the tremors do not occur on the sealing faults at the field boundaries. They occur mainly in more central areas where cumulative compaction and the density of faults with unfavourable offsets are highest. The mechanism behind the earthquakes is now generally considered to be (differential) compaction at reservoir level reactivating offset faults as originally proposed by (Roest and Kuilman, 1994).

Initially doubts on the relation between gas production and observed seismicity continued, even after an official investigation in 1993 (known as the BOA study) concluded that the two were linked. Based on an analysis of the seismicity, initial estimates were made of the maximum probable magnitude that could be expected to occur. The 1993 BOA study came to an initial estimated value of 3.3. KNMI later increased this estimate to 3.3–3.5 in 1995 (de Crook et al., 1995), to 3.7 in 2006 (de Crook et al., 1998) and then to 3.9 in 2006 (van Eck et al., 2006). In (de Lange et al., 2011) it is estimated that induced seismic events with magnitudes below 3.9 would not result in severe structural damage to buildings and hence the associated risk was deemed acceptable until things changed by end 2012. As of 2008 State Supervision of Mines (the Dutch regulator) became more and more concerned about the increasing frequency of the induced seismicity and the increasing magnitudes. TNO was asked to carry out a study into possible explanations (Muntendam-Bos, 2009). The conclusions and recommendations of this study led to a request to the operator by end 2009 for further research into possible mechanisms and feasible remedies, the results of which needed to be included in the required 2013 update of the Groningen winningsplan.

4.2 New insights

By mid 2012 the larger than predicted annual frequency of earthquakes with a magnitude equal or above 3.0 further increased the concerns of State Supervision of Mines (SSM) and intensified discussions with the operator. On 16 August 2012 an induced seismic event with a magnitude of 3.6 occurred near the village of Huizinge. It was the highest magnitude event registered in Groningen to that date, causing damage to thousands of homes and raising anxiety among those living on or near the field (Dost and Kraaijpoel, 2012). It triggered SSM into commencing its own investigation, including a new statistical analysis of all the available Groningen seismicity data (Muntendam-Bos and de Waal, 2013). The results were worrying. The SSM analysis indicated a continuous and systematic increase in the number of events and their magnitude (Fig. 3). The same figure suggests that smaller magnitude events might well have started earlier, without being detected given the limitations of the local seismic network at the time. The SSM analysis also showed that an upper magnitude bound – generally deemed to be 3.9 – could not be derived from a statistical analysis of the Groningen seismicity while other methodologies to arrive at such an upper estimate could not exclude values up to at least a magnitude 5 (Muntendam-Bos and de Waal, 2013). The results were confirmed by studies carried out by the operator (NAM, 2013, Bourne et al., 2014). In addition the SSM analysis suggested that increases and decreases in the annual number of seismic events could be linked to increases and decreases in the annual gas production rate with a delay of 6 to 12 months (Fig. 4). Based on the results of its investigation, SSM judged the seismic risk in Groningen as "high" and advised early January 2013 to reduce gas production by as much and as fast as realistically possible.

The SSM advice was followed by a year of concerted data acquisition and further studies, while production continued and reached its highest annual level in nearly 20 years (54 BCM). The results of the studies became available by end 2013 and suggested that the local seismic activity rate in Groningen increases exponentially with cumulative reservoir compaction (NAM, 2013). If correct it implies that production reductions in areas with large compaction are most effective to reduce future seismicity. Based on this insight the earlier SSM advice became more focussed: stop gas production in the central area of the Groningen field - where the largest amount of reservoir compaction takes place - for at least three years (Staatstoezicht, 2014). The ministerial decision, based on this advice, was to reduce the production in the centre of the field by 80 %. It was implemented on 17 January 2014. Initially the operator was allowed to increase production in areas outside the centre of the field to partly compensate for the production reduction in the centre. These production increases were capped in December 2014 after early observations suggested they resulted in increases in seismicity in these areas outside the field centre. It was realised from the start that the implemented production measures would probably only work temporarily (if at all) as pressure depletion in the centre of the field will resume after a few years given the continued production elsewhere. The large-scale data acquisition and study program was therefore

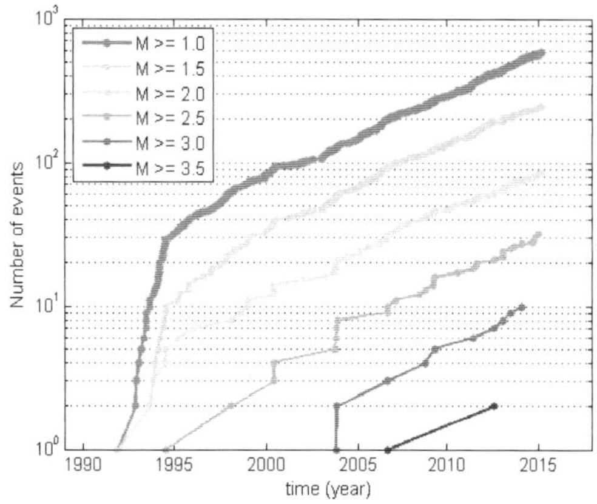

Figure 3. SSM analysis of the Groningen induced seismicity.

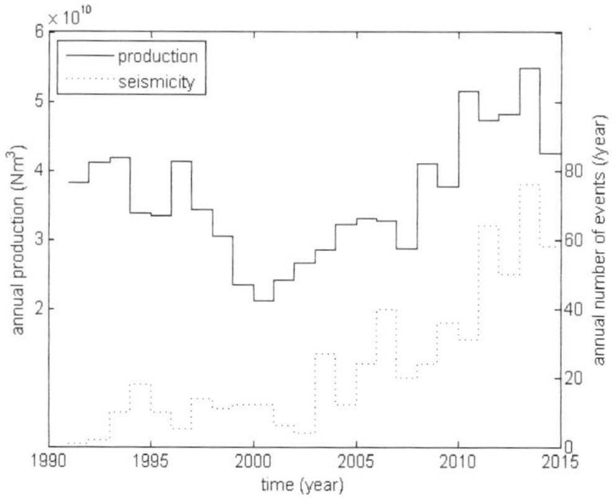

Figure 4. Groningen annual production and annual induced seismicity ($M > 1.0$).

continued and extended and a program to strengthen houses and infrastructure was started.

5 Recent developments

5.1 Field response

The main question related to the January 2014 production measures were how effective they would be to reduce the seismicity and on what time scale. Time is required for the pressure signal resulting from the shut-in at the central clusters to spread through the reservoir. Given the Groningen reservoir properties it takes some three months for that signal to reach a distance of 3 km (van Thienen-Visser and Breunese, 2015). As (differential) compaction is the engine behind the induced seismicity, the time scale on which compaction responds to stress changes is a second major factor. The time decay model predicts a typical response time of five to seven years. Linear and bi-linear compaction models predict an instant response. The rate type isotach model predicts part of the response to be immediate and part of it to happen over the next 6 to 12 months. In addition there is a delay from the time it takes for the pressure effect from the production reduction to diffuse from the wells into the reservoir.

Another factor influencing the response is the seismological model linking seismicity to compaction. On the one hand it can cause additional delays as a result of rate and state friction on the fault surfaces where the seismicity is generated. On the other hand the postulated exponential rise of seismic activity rate in combination with compaction creep would dampen the response significantly. Given the epistemic uncertainties around the compaction model and the seismological model in combination with the absence of empirical data, it was very uncertain how the seismicity would respond and a "wait and see" approach was taken. To be able to do so, the operator has extended the field monitoring to include PS-InSAR and a network of differential GPS stations for semi-permanent subsidence monitoring. For the seismic monitoring 60 shallow (200 m) borehole seismometers, 60 + accelerometers and two temporally installed downhole seismic arrays at reservoir level in the centre of the field are now in place. The latter will be replaced by permanent installations shortly.

Conclusions on the seismic response to the reduced production in the central area of the field within a period of a year are a challenge given the statistical character of the seismicity and the challenge to accurately measure compaction rate changes over relatively short periods. Nevertheless results to date look promising. A double-double difference analysis of the GPS stations in the centre and those at the edge of the field carried out by Statistics Netherlands (Pijpers, 2014a) shows a clear break in subsidence rate around April 2014 as shown in Fig. 5. Seismic event density maps over the Groningen field for three consecutive periods of a year are discussed in (van Thienen-Visser, 2015b). A clear change in the pattern of seismic events is visible during the third period with seismicity reducing significantly in the central area of the field after April 2014 (Fig. 6). An independent second study by Statistics Netherlands (Pijpers, 2014b) confirms the statistical significance of the changes in seismic activity. Although not yet scientifically conclusive at a 99 % confidence level these results support decision making on remediating measures to reduce future seismic risk as they suggest that the seismicity can be controlled and regulated at relatively short notice.

5.2 Risk

Early 2013 SSM estimated the seismic risk level in Groningen as "high", based on the realisation that events with a magnitude well above 3.9 could not be excluded. Based on SSM's analysis an assumed upper magnitude limit of 5.0 re-

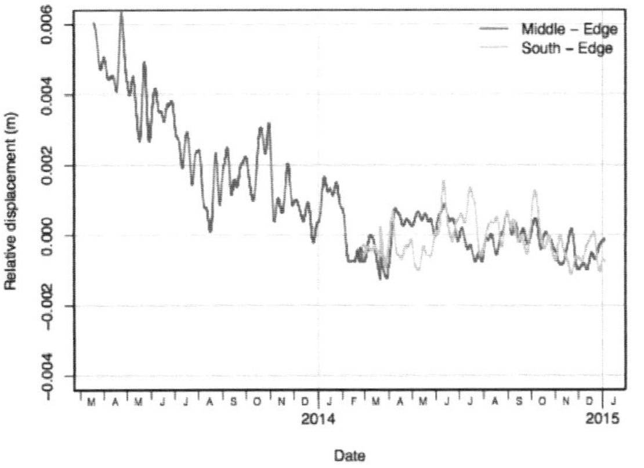

Figure 5. Analyses of GPS data showing break in subsidence rate around April 2014.

sulted in an annual probability of 3 % for an earthquake with a magnitude 4.5 or higher, a level at which serious damage cannot be excluded as houses in the Netherlands are not built to sustain seismic ground motions. By end 2013 a more detailed risk analysis was made for the central area of the field (Staatstoezicht, 2013; Muntendam et al., 2013). Based on a probabilistic analysis of the ground motions that can occur and taking into account the fragility of the local housing stock, data made available by the operator was translated into annual LPR (local personal risk) numbers and a Group risk F-N (frequency-number) plot. Results show that the seismic risk levels in Groningen are considerable and comparable to the highest flooding risk levels in the Netherlands (Fig. 7). Subsequent work by a dedicated impact assessment expert committee during 2014 estimated the number of houses in Groningen exceeding an LPR of 10–4 to be between 40 000 and 90 000 (Stuurgroep NPR, 2015). Supporting a widely accepted minimum LPR level of 10–5, the committee recommends strengthening the housing stock to 10–4, assuming actual LPR levels will come down when more refined and less conservative calculations become available. A final norm needs to be developed taking into account acceptable seismic risk, security of supply issues and financial consequences.

5.3 Measurement and control loop

Target is the cyclic development of a well organised risk management plan with clear objectives, well defined risk norms and control measures with proven effectiveness. To start, scenario's spanning the range of parameter and model uncertainties will be generated to calculate possible subsidence and seismicity outcomes. The probability of each scenario will be updated over time through confrontation with measurements as they become available. At regular intervals the subsidence prediction and the seismic risk will be re-evaluated. Further mitigation measures, possibly including

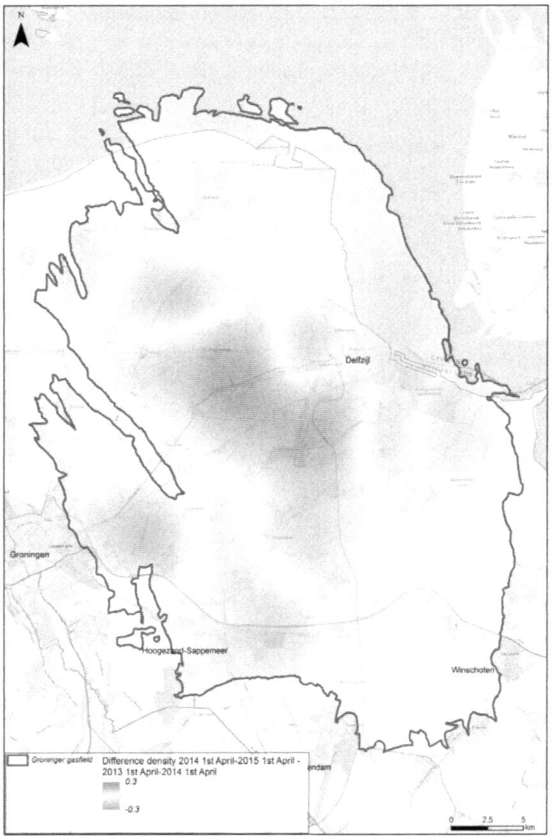

Figure 6. Increases and decreases in seismic event density after production reductions (from van Thienen-Visser et al, 2015b).

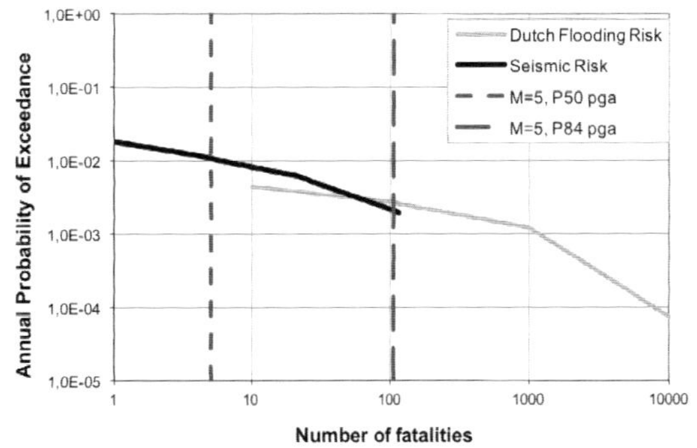

Figure 7. Calculated seismic Group Risk Loppersum area (from Muntendam et al., 2015).

further production measures and further building strengthening measures will need to be taken if probabilities indicate unacceptable risks. A possible framework for a measurement and control loop enabling adjustments when seismicity metrics indicate a risk of exceeding acceptable risk is shown in Fig. 8. To apply the concept it needs to be part of a "Seismic Risk Management Protocol" that should contain the following elements:

1. agreed metrics to express the seismic risk;
2. a norm for the amount of seismic risk considered acceptable considering societal issues, security of supply and financial consequences, translated into an enforceable, operational criterion;
3. a probabilistic assessment of the seismic risk (taking into account all uncertainties):
   a. as a function of location and time;
   b. under different production scenarios;
   c. taking into account the effect of strengthening programs;
4. an agreed procedure, including regular analysis of proper signal parameters like GPS subsidence rate and seismic event density.to update the predicted probabilistic seismic risk as new data is gathered;
5. a measurement plan;
6. agreed 'SMART" measures (production/pressure-maintenance/strengthening) to adjust the seismic risk using a multi-objective optimisation (minimising seismic risk, while maximising security of supply and state income);
7. an independent audit system to ensure compliance and to verify the technical integrity of the underlying work.

6 On-going and future work

The on-going large study and data acquisition effort by the operator is continued and extended to arrive at a better understanding of the gas production induced reservoir compaction, surface subsidence and induced seismicity in Groningen. Aim is to reduce the large epistemic and aleatory uncertainties, to arrive at better monitoring and prediction of subsidence and seismicity, to better estimate seismic risk as a function of location and time and to assess the possibility to implement a measurement and control loop to contain seismic risk.

Regular surface levelling campaigns continue, complemented by semi-continuous PS-InSAR measurements and a network of permanent GPS stations distributed over the field. The existing network of shallow seismic monitoring wells is being upgraded to some 60 deep (100–200 m) boreholes spread over the field. The expectation is that this results in a completeness level of 0.5 by end 2015, leading to a tenfold increase in the detectable number or seismic events and a threefold reduction of the aleatory uncertainty in the seismicity activity level. In addition some 60 + accelerometers and two permanent downhole seismic arrays at reservoir level are being installed. The accelerometers will help to reduce the uncertainties in the GMPE (ground motion prediction equation) relating earthquake magnitude to surface acceleration.

Extensive statistical (trend) analyses will be carried out to investigate the effects of compaction (rate) and compaction distribution on the spatial and temporal changes in seismic activity rate and b value (determining the ratio between weak and strong events). A similar analysis will be carried out to investigate the effect of production rate and production distribution. The improved detection limit of the extended seismic network is expected to be very beneficial for this type of work. The outcomes of the various analyses will be used to validate the present (activity rate) seismological model developed by the operator and to introduce model improvements if necessary. Potential examples are the incorporation of (non-linear) compaction rate dependence and the incorporation of a critical stressing rate above which the seismic activity rate increases rapidly (Llenos et al., 2009; Toda et al., 2002).

Better static and dynamic subsurface models are being constructed to model reservoir property distributions (e.g. porosity, permeability, faults, compressibility). An example is the latest structural model for which more than 1800 faults have been mapped at reservoir level using ant tracking on the 3-D seismic cube (NAM, 2013). Another example is the detailed characterisation of the near surface layers to a depth of several hundred meters, important for seismic wave attenuation calculations. The subsurface models will be updated as new data becomes available while the operator and TNO have started on inversion of the surface subsidence data to better determine the amount and distribution of the downhole reservoir compaction.

The operator is combining the static, dynamic and geomechanical reservoir models with a seismological model, a ground motion predication equation, a database of building types and locations and with fragility curves for the various building types to arrive at probabilistic (Monte Carlo) predictions of seismic hazard and seismic risk under a number of different production scenarios and housing reinforcement programs (Bourne et al., 2014).

7 Lessons

A number of important lessons can be learnt from the history of subsidence and induced seismicity in Groningen. First it is clear that epistemic uncertainties were largely underestimated from the start and during a long period thereafter. This is illustrated by the large changes over time in the predicted maximum subsidence at the end of field life in the centre of the Groningen field. It illustrates that initial uncertainties, prior to the start of production, are perhaps as large as a fac-

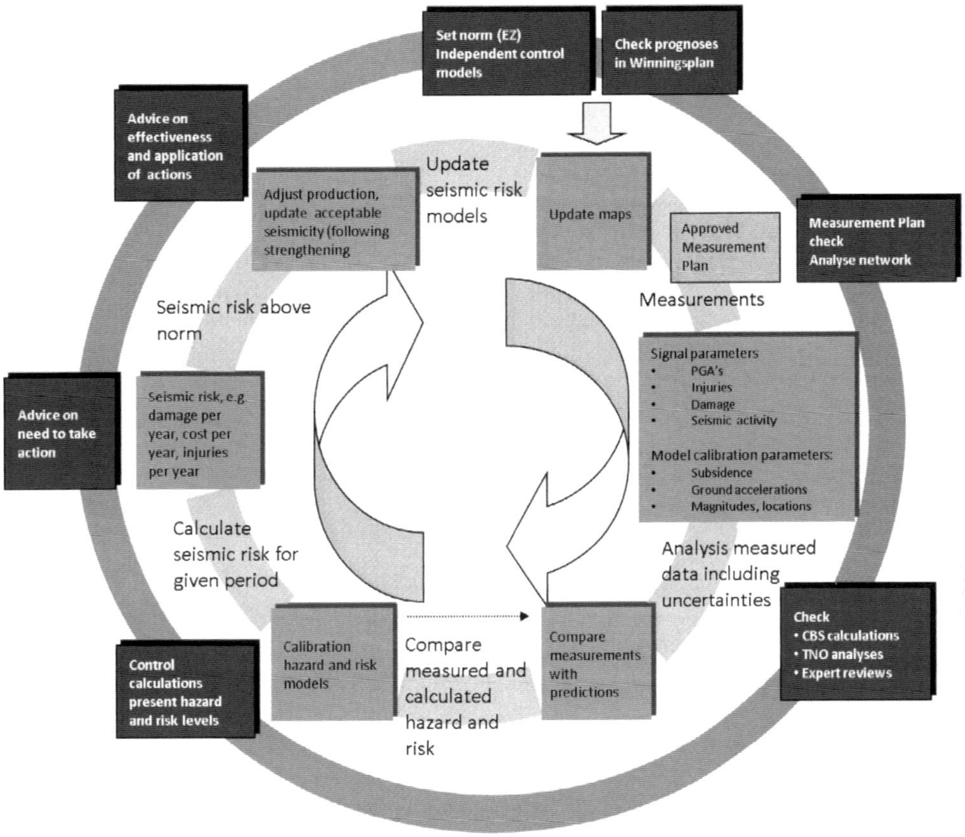

Figure 8. A possible seismic risk "measurement and control" loop.

tor of two (up or down) relative to the predicted expectation case. This has meanwhile been confirmed for a number of other fields in the Netherlands (de Waal et al., 2012). Major potential contributors are uncertainties in subsurface models, unknown influx of bottom and lateral aquifers, core damage, non-linear rock compressibility, the difference in loading rate between laboratory and field, the very large change in loading rate at the start of production, the in-situ stress state, reservoir burial history, salt flow, the relation between subsidence and compaction etc. Rather than predicting a single number it is therefore much better to provide a range of possible outcomes, only reducing the range when certain scenarios become too unlikely given field measurements (Nepveu et al., 2010). And to keep the range of possible outcomes, as it narrows over time, in line with the range that the area can sustain e.g. by timely taking appropriate mitigation measures where and when required.

Secondly it seems that non-linear compaction is the norm rather than the exception with the rate type isotach compaction model as a plausible explanation for the transition zone that follows the large increase in loading rate at the start of the production period. The effects are well known from Soil Mechanics where a long time geological loading manifests itself as an apparent over consolidation from previous deeper burial (which it is not). This also implies that a certain degree of additional subsidence can be expected as a result of creep after the production period has ended. Subsidence and seismicity response time following the significant production adjustment in January 2014 seem in the order of a couple of months. Although not rigidly proven at a 95 % confidence level, it does seem to favour the rate type isotach model over the time decay model which uses a single 5–7 year response time to explain the initial subsidence delay after the start of production.

Also for induced seismicity, history proved very different from the original expectations. At the start of the production period induced seismicity was not expected and when it happened it took many years before its linkage to the gas production was widely accepted, an good illustration of the strength of paradigms. Contrary too initial expectation the seismicity has focussed in the central area of the field where the faults are not sealing. Also initial expectations on the ability to predict an upper bound on the likely magnitude to occur proved wrong and it is now realised that there is significant seismic risk that needs to be contained through a measurement and control loop including production measures and a housing and infra-structure strengthening programs.

The response time of a few months of subsidence rate and seismic activity rate to the production measures that have been taken suggests that the risk can be managed by adjusting production rates. To sustain a reduced seismicity level over longer periods of time, field-wide production measures will need to be taken as pressure decline in the centre of the field will re-start within a few years given the continued production from production clusters outside of the centre of the field. It is still unclear if production reductions will only smear the risk out over time (film-rate effect) or if real reductions in the total number of earthquakes can be expected. As a minimum, production reduction will buy time for further study and to strengthen buildings and infrastructure.

Acknowledgements. The authors want to thank the management of State Supervision of Mines for permission to publish this paper. We also thank staff at TNO, KNMI, CBS (Statistics Netherlands) and NAM for stimulating discussions and support although they may not agree with all of the interpretations and conclusions of this paper. We are particularly grateful for the contributions and support provided by Karin van Thienen-Visser and Jaap Breunese at TNO and by Jan van Herk at State Supervision of Mines.

References

Bjerrum, L.: Engineering geology of Norwegian normally consolidated marine clays as related to settlements of buildings, in: Geotechnique, Seventh Rankine Lecture 17, 81–118, 1967.

BOA (Begeleidingscommissie Onderzoek Aardbevingen): Eindrapport multidisciplinair onderzoek naar de relatie tussen gaswinning en aardbevingen in Noord-Nederland, Koninklijk Nederlands Meteorologisch Instituut, Afdeling Seismologie, De Bilt, 76 pp., 1993 (in Dutch).

Bourne, S. J., Oates, S. J., Elk, J. van, and Doornhof, D.: A seismological model for earthquakes induced by fluid extraction from a subsurface reservoir, J. Geophys. Res. Solid Earth, 119, 8991–9015, doi:10.1002/2014JB011663, 2014.

de Crook, Th., Dost, B., and Haak, H. W.: Analyse van het seismische risico in Noord Nederland, KNMI TR-168, de Bilt, 30 pp., 1995 (in Dutch).

de Crook, Th., Dost, B., and Haak, H. W.: Seismisch risico in Noord-Nederland, KNMI, 1998 (in Dutch).

de Lange, G., Oostrom, N. G. C. van, Dortland, S., Borsje, H., and de Richemont, S. A. J.: Gebouwschade Loppersum, Deltaris 1202097-000, 84 pp., 2011 (in Dutch).

de Waal, J. A.: On the rate type compaction behaviour of sandstone reservoir rock, PhD dissertation, Delft University of Technology, 166 pp., 1986.

de Waal, J. A. and Smits, R. M. M.: Prediction of reservoir compaction and surface subsidence: field application of a new model, SPE Formation Evaluation 3, 347–356, 1988.

de Waal, J. A., Roest, J. P. A., Fokker, P. A., Kroon, I. C., Breunese, J. N., Muntendam-Bos, A. G., Oost, A. P., and van Wirdum, G.: The effective subsidence capacity concept: How to assure that subsidence in the Wadden Sea remains within defined limits, in: Geologie en Mijnbouw Neth. J. of Geoscience, 91, 385–399, 2012.

de Waal, J. A., van Thienen-Visser, K., and Pruiksma, J. P.: Rate type isotach compaction of consolidated sandstone, 49th US Rock mechanics/Geomechanics symposium (ARMA), San Francisco, 28 June–1 July, ARMA 15-436, 2015.

Dieterich, J. H.: Time-dependent friction and the mechanics of stick-slip, Pageoph., 116, 790–806, 1978.

Dieterich, J. H.: A constitutive law for rate of earthquake production and its application to earthquake clustering, J. Geoph. Res. 99, 2601–2618, 1994.

Dost, B. and Kraaijpoel, D.: The August 16, 2012 earthquake near Huizinge (Groningen), KNMI publication TR 16/1/2013, 26, 2012.

Hettema, M., Papamichos, E., and Schutjens, P.: Subsidence delay: field observations and analysis, Oil and Gas Science and Technology, in: Revue dIFP, Energies Nouvelles, 57, 443–458, 2002.

Houtenbos, A. P. E. M.: Subsidence and gas production: an empirical relation, SubsvsProdFinal.doc, 1–8, available at: http://bodemdaling.houtenbos.org/resources (last access: 5 November 2015), 2007.

Kolymbas, D.: A rate-dependent constitutive equation for soils, Mech. R. Comms., 4, 367–372, 1977.

Kosloff, D. and Scott, R. F.: Finite element simulation of Wilmington oil field subsidence: II Nonlinear Modelling, Tectonoph., 70, 59–183, 1980.

Lee, L. L.: Subsidence earthquake at a California oil field, in: Evaluation and prediction of subsidence, edited by: Saxena, S. K., New York, ASCE, 549–564, 1979.

Llenos, A. L., McGuire, J. J., and Ogata, Y.: Modelling seismic swarms triggered by aseismic transients, Earth Planet. Sci. Lett., 281, 59–69, 2009.

Mallman, E. P. and Zoback, M. D.: Subsidence in the Louisiana coastal zone due to hydrocarbon production, J. of Coastal Res. SI, 50, 443–449, 2007.

Merle, H., Kentie, C., van Opstal, G., and Schneider, G.: The Bachaquero study – a composite analysis of the behavior of a compaction drive/solution gas drive reservoir, J. of Petr. Techn., 27, 1107–1115, 1976.

Mijnlieff, H. F. and Geluk, M.: Palaeotopography-governed sediment distribution-a new predictive model for the Permian Upper Rotliegend in the Dutch sector of the Southern Permian Basin, in: The Permian Rotliegend of the Netherlands, edited by: Grötsch, J. and Gaupp, R., SEPM Special Publication No. 98, 147–159, ISBN: 978-1-56576-300-5, 2011.

Mossop, A.: An explanation for anomalous time dependent subsidence, in: Proc. 46th US Rock Mech. Symp., 24-27 June, Chicago, 60–68, 2012.

Muntendam-Bos, A. G.: Geïnduceerde trillingen in of nabij het Groningen gasveld, een aanzet tot een geomechanische verklaring, available at: http://www.nlog.nl/nl/reserves/reserves/AGE09-10100_RapportAardtrillingenGroningenveld_PUBLIEK_NLOG.pdf (last access: 5 November 2015), 2009 (in Dutch).

Muntendam-Bos, A. G. and de Waal, J. A.: Reassessment of the probability of higher magnitude earthquakes in the Groningen gas field: technical report, available at: http://www.researchgate.net/publication/271764011_Reassessment_of_the_probability_of_higher_magnitude_earthquakes_in_the_Groningen_gas_field (last access: 5 November 2015), February 2013.

Muntendam-Bos, A. G., Roest, J. P. A., and de Waal, J. A.: A guideline for gas production induced seismic risk, The Leading Edge, special issue Injection Induced Seismicity, 672–677, 2015.

NAM: The analysis of surface subsidence resulting from gas production in the Groningen area, the Netherlands, Trans. Royal Dutch Soc. of Geol. and Mining Eng., edited by: Nederlandse Aardolie Maatschappij B. V., 28, 1973.

NAM: Wijziging winningsplan Groningen 2013, inclusief technische bijlage Groningen winningsplan 2013, Versie 29, November 2013 (partly in Dutch).

Nepveu, M., Kroon, I. C., and Fokker, P. A.: Hoisting a Red Flag: An Early Warning System for Exceeding Subsidence Limits, Math Geosci. 42, 187–198, doi:10.1007s11004-009-9252-2, 2010.

Pijpers, F.: Phase0 report1, Significance of trend changes in ground subsidence in Groningen, Statistics Netherlands Scientific paper, CBS, 14 pp., 2014a.

Pijpers, F.: Phase0 report2, Significance of trend changes in tremor rates in Groningen, Statistics Netherlands Scientific paper, CBS, 18 pp., 2014b.

Pruiksma, J. P., Breunese, J. N., van Thienen-Visser, K., and de Waal, J. A.: Isotach formulation of the rate type compaction model for sandstone, Int. J. Rock Mech. Min., 78, 127–132, doi:10.1016/j.ijrmms.2015.06.002, 2015.

Roest, J. P. A. and Kuilman, W.: Geomechanical Analysis of Small Earthquakes at the Eleveld gas Reservoir, in: Proc. Eurock '94, Balkema, Rotterdam, 573–580, 1994.

Schoonbeek, J. B.: Land subsidence as a result of natural gas extraction in the province of Groningen, SPE European Spring Meeting, 8–9 April, Amsterdam, the Netherlands, 20 pp., SPE 5751, 1976.

Staatstoezicht op de Mijnen: Risico Analyse Aardgasbevingen Groningen, available at: https://www.sodm.nl/sites/default/files/redactie/RisicoanalyseaardgasbevingenGroningen.pdf (last access: 5 November 2015), Den Haag, December 2013 (in Dutch).

Staatstoezicht op de Mijnen: Advies winningsplan 2013/Meet- en Monitoringplan NAM Groningen gasveld, available at: https://www.sodm.nl/sites/default/files/redactie/AdviesSodMwinningsplanGroningen2013.pdf (last access: 5 November 2015), 2014 (in Dutch).

Stäuble, A. J. and Milius, G.: Geology of the Groningen gas field, Netherlands Am. Ass. Petrol. Geol., Mem., 14, 359–369, 1970.

Stuurgroep NPR: Impact assessment Nederlandse Praktijk Richtlijn aardbevingsbestendig bouwen, 8 January 2015 (in Dutch).

TNO: Toetsing van de bodemdalingsprognoses en seismische hazard ten gevolge van gaswinning van het Groningen veld, R11953, Eindrapport, Utrecht, 218 pp., 2013 (in Dutch).

Toda, S., Stein, R. S., and Sagiya, T.: Evidence from the AD 2000 Izu islands swarm that stressing rate governs seismicity, Nature, 418, 58–61, 2002.

Toksöz, M. N. and Walsh, J. B.: An evaluation of the subsidence program directed by Nederlandse Aardolie Maatschappij, NAM EP90-20043, 51 pp., 1990.

van Eck, T., Goutbeek, F. H., Haak, H. W., and Dost, B.: Seismic hazard due to small-magnitude, shallow-source, induced earthquakes in the Netherlands, Eng. Geol., 87, 105–121, doi:10.1016/j.enggeo.2006.06.005, 2006.

van Hulten, F. F. N.: Brief history of petroleum exploration in the Netherlands, in: Symposium fifty years of petroleum exploration in the Netherlands after the Groningen discovery, EBN, TNO, Geo-Energy and PGK, 15–16 January 2009, Utrecht, 2009.

van Ojik, K., Böhm, A. R., Cremer, H., Geluk, M. C., De Jong, M. G. G., Mijnlieff, H. F., and Djin Nio, S.: The rationale for an integrated stratigraphic framework of the Upper Rotliegend depositional system, in: The Permian Rotliegend of the Netherlands, edited by: Grötsch, J. and Gaupp, R., SEPM Special Publication No. 98, 37–48, ISBN: 978-1-56576-300-5, 2011.

van Thienen-Visser, K. and Breunese, J.: Induced seismicity of the Groningen gas field: history and recent developments, The Leading Edge, special issue Injection Induced Seismicity, 665–671, 2015a.

van Thienen-Visser, K., Fokker, P., Nepveu, M., Sijacic, D., Hettelaar, J., and van Kempen, B.: Recent developments of the seismicity of the Groningen field in 2015, TNO R10755, 2015b.

Vermeer, P. and Neher, H.: A soft soil model that accounts for creep, in: Beyond 2000 in computational geotechnics-10 years of Plaxis international, Balkema, Rotterdam, ISBN: 90 5809 040 X, 1999.

Whaley, J.: The Groningen gasfield, GEO ExPro Magazine, Vol. 6–4, available at: http://www.geoexpro.com/articles/2009/04/the-groningen-gas-field (last access: 5 November 2015), 2009.

Proc. IAHS, 372, 141–144, 2015
proc-iahs.net/372/141/2015/
doi:10.5194/piahs-372-141-2015

Open Access
PIAHS

Full-scale experimental and numerical study about structural behaviour of a thin-walled cold-formed steel building affected by ground settlements due to land subsidence

J. A. Ortiz, L. A. Hernández, M. Hernández, J. Pacheco, M. E. Zermeño, and R. Salinas

Autonomous University of Aguascalientes, Aguascalientes, México

Correspondence to: J. A. Ortiz (jose.ortiz.lozano@gmail.com)

Published: 12 November 2015

Abstract. Land subsidence due to ground water withdrawal is a problem in many places around the world (Poland, 1984). This causes differential ground settlements that affect masonry structures, because these structural materials do not exhibit an adequate performance beyond a certain level of angular distortion. This work presents the experimental and numerical results about a study regarding the performance of a full-scale thin-walled cold-formed steel building affected by ground differential settlements due to land subsidence. The experimental stage consisted in the construction of a test-building to be subjected to differential settlements in laboratory. The numerical stage consisted in performing a numerical non-linear static pull-down analysis simulating the differential ground settlements of the test-building. The results show that the structural performance of the tested building was very suitable in terms of ductility.

1 Introduction

Ground failure associated with land subsidence is the primary geotechnical hazard in several cities in central Mexico (Arroyo et al., 2004), that has caused enormous economic loss by damaging public and private properties, including dwellings. The damage caused by ground failures is variable and considerable, even though the local government is aware of this problem and has undertaken steps to mitigate ground failure-related damages. Active fracture zones associated with land subsidence induce displacements in buildings in the form of differential settlements, which cause damage to structural systems in dwellings, especially those based on masonry which is the principal construction method used in one or two-story dwellings,. The most common and visible damage occurs in walls which fails by shear force because of their low capacity to absorb angular distortion.

Cold-formed steel sections have instability problems because of the effects of certain buckling modes due to the slenderness of the section (Anapayan and Mahendran, 2012). One of the principal uses for this type of material is for structural frames in the form of modules, which are most commonly assembled using elements with edge-stiffened flanges and elements with simple flanges with different dimensions and gages. The majority of studies on cold-formed steel structures have been primarily focused on their performance under seismic events, e.g., the performance under lateral loads using non-linear static push-over analysis, as described in the specifications provided by the Applied Technology Center (ATC-40, 1996). However, few studies have focused on the case of cold-formed steel structures affected by ground settlement due to land subsidence phenomenon; therefore, conducting research on this issue is a priority for zones affected by land subsidence.

Because of its mechanical properties, the structural performance of thin-walled cold-formed steel structures could allow large displacements to be absorbed without failing. In this report, the results of an experimental and numerical study, with the objective of assessing the performance of a full-scale thin-walled cold-formed steel building under angular distortion and simulating the differential ground settlements due to land subsidence phenomenon, is presented.

Figure 1. Interior view of the cold-formed steel structure.

This test building is a new construction system to be used in houses and the main objective is to know the structural performance for land subsidence.

2 Experimental and numerical methodology

2.1 The test building subjected to vertical displacements

A full scale thin-walled cold-formed steel one-story building with 8 bearing walls of 2.5 m high was built. Figure 1 shows interior view of the housing which was structured with simple channel section 350T125-33 elements (tracks) on the top and bottom parts of the frame and vertical stiffened channel section 350S162-33 elements (studs). The center to center distance between the studs was 400 mm. Used as a sheathing material; a high-density expanded polystyrene panel with a thickness of 75 mm was inserted between the studs. The connections between the studs and the tracks were made using #8 flat head self-drilling screws with a longitude of 20 mm and applied to each joint; thus, four screws were used. To fasten the wall frames to the ground, "hold-down" type anchors at a right angle, constructed of steel plate A-36 of 4-mm thickness, were put in each bottom end of the frame; the anchors fastened the frame using 14 self-drilling screws #10 that were 38 mm in longitude, and a steel screw A-307 Gr. B with a diameter of 16 mm was used to anchor the frame to the ground. A double action hydraulic actuator with a 160 kN capacity connected to the ground was necessary to apply the monotonic vertical pull-down load on the bottom end of the frame, which was gradually increased to reach a 140 mm target vertical displacement. Measuring instruments were placed on the points of interest to evaluate the displacements for a total of 20 points distributed uniformly on the area of each frame. The average rate of load application was 12.5 mm min^{-1}.

Tests on lateral load in thin walled cold-formed steel wall frames were conducted to obtain information related to the mechanical performance of the wall in terms of the elastic rotational stiffness of the system to calibrate the parameters of the finite element models.

2.2 Static non-linear pull-down analysis

The methodology followed in this study consisted on performing a non-linear pull-down analysis to a full scale thin-walled cold-formed steel one-story building affected by vertical displacement and simulating the effects produced by land subsidence, which develops gradually in buildings over the years.

The pull-down analysis of a structure can be considered when one of its supports suffers a vertical displacement, generally downwards. The results for this type of analysis are similar to the results that occur when a static non-linear push-over is used; the only difference is the direction in which the displacements are evaluated. In the push-over analysis, the horizontal displacements are assessed; in the pull-down analysis, the focus is on the vertical displacements, which can be generated by different causes. An important parameter in this type of vertical displacement (settlement) is its rate, which depends on the landslide type or other phenomenon that affects the structure (Negelescu and Foerster, 2010). A calibration of the model analysed was conducted based on all of the variables of influence, such as the aspect ratio, the gauge of the studs and tracks, the distance between the studs, the thickness of the sheathing, the elastic rotational stiffness and the mechanical properties of the materials involved, e.g., the modulus of elasticity and shear modulus. Afterwards several numerical simulations were performed using finite element software (SAP 2000, 2008) to verify that the numerical structural behaviour was equal to that of the experimental tests. Subsequently, the numerical pull-down analysis was performed on the virtual structural model by applying a gradual vertical monotonic incremental vertical displacement on the support (control joint), which simulated a differential settlement of the ground due to the land subsidence phenomenon.

The angular distortion is the ratio between the displacement and the longitude of the analysed structural frame; neither the studs nor the wall frames supported any vertical or horizontal load. The behaviour of the virtual model was assessed by obtaining the values of the moment dependent on the angular distortion for each step in the application of incremental displacement beyond the linear regime and adopting the following methodology:

1. Building the virtual model of cold-formed steel without any type of sheathing in the non-linear analysis SAP 2000 (2008) software based on the geometry and number of elements (studs and tracks).

2. Assigning properties: mechanical parameters of the materials (steel and sheathing), geometrical properties of the cross-section of cold-formed steel elements, applied loads and states of the load.

Figure 2. North Wall at the experimental test.

3. Calibrating the virtual model to reproduce the structural behaviour with lateral loads based on experimental tests conducted for the expanded polystyrene system, which are based on elastic rotational stiffness values, elasticity and shear modulus.
4. Assigning moment-rotation curves for distortional and local buckling on the ends of each stud to simulate plastic hinges (Ayhan and Schafer, 2002)
5. Configuring parameters for static non-linear pull-down analysis: maximum displacement, control joint and number of steps.
6. Running non-linear analysis on the virtual model and processing the results.

3 Results and analysis

3.1 Experimental results

The experimental phase of housing prototype was carried out in three stages vertical displacement applied in the central part of the rear wall located at the North end of the building. Figure 2 show the North wall during the experimental test. The cumulative total displacement was around 150 mm. The affected area by the pull-down effect corresponds to the West, Central, East and North walls. Figure 3 shows the structure and the location of the North and Central walls. Elements with the most significant damage were the Central and North walls. In the first phase of displacement (50 mm) damage and deformations due local buckling in bottom tracks were observed. The more affected tracks at this stage correspond with those located in the north and central walls, that connect at the point of application of displacements. In the second phase of displacement (100 mm accumulated), the polystyrene sheathing begins to detach from the metal structure without being broken. The diagonal bracing which at minimum vertical movements begin to bear and transmit tensile and compression forces. At this stage bending stresses are greatly enhanced by generating excessive deformations in bottom tracks. This condition of stresses, coupled with the presence of stiffening elements such as studs and connecting plates, causes plastic hings mainly in those segments of walls that connect to the bearing wall (Fig. 4). At the final stage (150 mm accumulated) some structural elements reached their maximum allowable load. Bottom tracks start to break at vulnerables areas near to bolted connections. Polystyrene sheathing completely breaks in top and bottom ends of walls affected and most of the connectors between slab and walls completely fail.

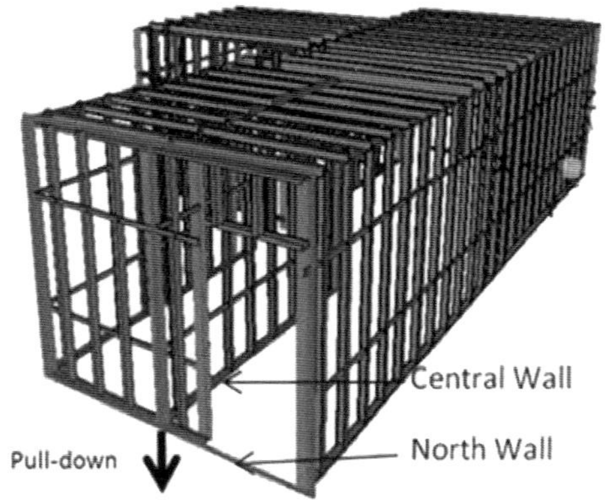

Figure 3. Structuring of housing and location of North and Central walls (SAP 2000, 2008).

3.2 Numerical results

The parameter used to evaluate the performance of studs in affected walls is the relative moment value (M/M_p), which measures the ability of the element to withstand bending moments in relation to the distortion applied in each of the walls. According with this was observed that in the case of the north wall, the stud 416 located at the point of application of displacement, distortion reached a value of 12 to 58 % of its bending capacity. In the case of Central wall stud 528 reached 98 % of capacity to a distortion value only 5 %. The stud 531 achieves 100 % capacity for a distortion of about 7 %. The rest of the studs on each of the walls studied showed lower values than those listed in this paragraph. According to the numerical analysis we can see that the behavior of structural elements is excellent as it allows high levels of distortion, without elements reach fail.

4 Conclusions

The cold-formed steel structure with polystyrene sheathing presents a greater flexibility in comparison with other materials like OSB wood, calcium silicate, and gypsum because the

Figure 4. Excessive deformations in bottom track and fail of polystyrene.

polystyrene allows significant displacements (vertical settlement expressed as angular distortion) without suffering excessive damage; this result has been verified from a numerical standpoint by determining the relative moment. This construction system has an excellent behaviour when considering flexibility. Lastly, using structures based on cold-formed steel wall frames with polystyrene sheathing would be suitable to reduce damages and guarantee structural safety in housing structures constructed in zones affected by ground settlement due to land subsidence. Nowadays this construction system is adequate to apply it in one or two-story buildings. On the other hand, due its characteristics is possible to use with different materials in order to improve the structural performance.

References

Anapayan, T. and Mahendran, M.: Improved design rules for hollow flange sections subject to lateral distortional buckling, Thin Wall. Struct., 50, 128–140, 2012.

Arroyo, M. G., Zermeño, E., Castañeda, J. A.: El agrietamiento en Aguascalientes, causas y efectos, primera edición; Universidad Autónoma de Aguascalientes, Aguascalientes, 2004.

ATC-40: Seismic evaluation and retrofit of concrete buildings, Seismic Safety Commission, State of California, Sacramento, California, Developed by the Applied Technology Council, 1–316 (SSC Report no. 96-01), 1996.

Ayhan, D. and Schafer, B.: Moment-rotation characterization of cold-formed steel beams, Advancing Cold-Formed Steel Earthq Eng. (Research Report), 02, 1–33, 2002.

Negulescu, C. and Foerster, E.: Parametric studies and quantitative assessment of the vulnerability of a RC frame building exposed to differential settlements, Nat. Hazards Earth Syst. Sci., 10, 1781–1792, doi:10.5194/nhess-10-1781-2010, 2010.

Poland, J. F.: Guidebook to studies of land subsidence due to ground-water withdrawal, prepared for the International Hydrological Program, Working Group 8.4 UNESCO, 1–303, 1984.

SAP 2000: Structural Analysis Program Advanced version 12.0.0, Berkeley, Ca, Computers and Structures, Inc., 2008.

Proc. IAHS, 372, 145–150, 2015
proc-iahs.net/372/145/2015/
doi:10.5194/piahs-372-145-2015

Land subsidence risk assessment and protection in mined-out regions

A. Zhao[1,2] **and A. Tang**[2,1]

[1]School of Civil Engineering, Harbin Institute of Technology, Harbin, China
[2]School of Civil Engineering, Heilongjiang University, Harbin, China

Correspondence to: A. Tang (tangap@hit.edu.cn)

Published: 12 November 2015

Abstract. Land subsidence due to underground mining is an important hazard that causes large damages and threatens to social and economic activities. The China government has started a national project to estimate the risk of land subsidence in the main coal production provinces, such as Heilongjiang, Anhui and Shanxi Provinces. Herein, the investigation methods for land subsidence identification were reported, some types of land settlement are summarized, and some successful engineering measures to mitigate the subsidence are discussed. A Geographical Information System (GIS) for land subsidence risk assessment is developed and is based on site investigations and numerrical simulation of the subsidence process. In this system, maps of mining intensity and risk ranks are developed.

1 Introduction

In China, land subsidence in mined-out regions is an important hazard for sustainable economic and social development, particularly in coal mine areas. By 2003, land subsidence attributed to mining production and groundwater pumping with a cumulative subsidence larger than 200 mm affected an area of more than 79 000 km^2 (Xue et al., 2005; Wu et al., 2008; Zhang et al., 2014) in mainland China. The subsidence damaged mine, building and mining infrastructure,caused the deaths of many miners, and damaged the affected lands and their their land uses. In these affected areas, agricultural production and industry had to be curtailed, and a large number of the native inhabitants became homeless and had to move from the subsidence regions. Thus, the land subsidence in mined-out regions caused vast economic losses and posed great risks for local social, agricultural and industrial developments. Since 2000, the China government has implemented a national project to real-time monitor and assess the hazard and risk of the mined-out regions, and the core of this project is land subsidence risk assessment and prevention. After 15 years, the project has made great progress. Highly effective methods to mitigate subsidence have been implemented in the main coal producing regions, although the project plan has not completely solved these problems.

Many specialists have done a lot of research on the land subsidence mechanism and how to estimate the subsidence effects and asses the safety of structures built on the sinking areas. In the beginning of 19th century, the vertical-line theory (VLT) and Gnoto-normal line theory (GNLT) had been used to calculate land subsidence induced by minning in Belgium (Klatesch, 1983), VLT and GNLT proposed the equation to calculate the subsidence: $w = m\cos\alpha$, here, w is the subsidence, m is the thickness of mine strata, and α is the dip angle of mine strata. In the middle of the 19th century, many more theories were developed to predict land subsidence and explore its mechanism, such as bisector theory (Jlcinsky, 1876–1884), round arch theory (Fayol, 1885), strip-belt zone theory (House, 1897), cantilever beam theory (Halbnam, 1903), and influence line theory (Schimizx, 1932), among others (Liu, 1965). Many scholars have studied how land subsidence affects the safety of mining engineering, and a number of scientific publications on subsidence emanating from European countries and proposed several alternative methods (Tugrul et al., 2013). In the 1970s, Jones and Spencer (1977) discussed how mining subsidence damged highway structure in USA. Since the 1950s, Chi-

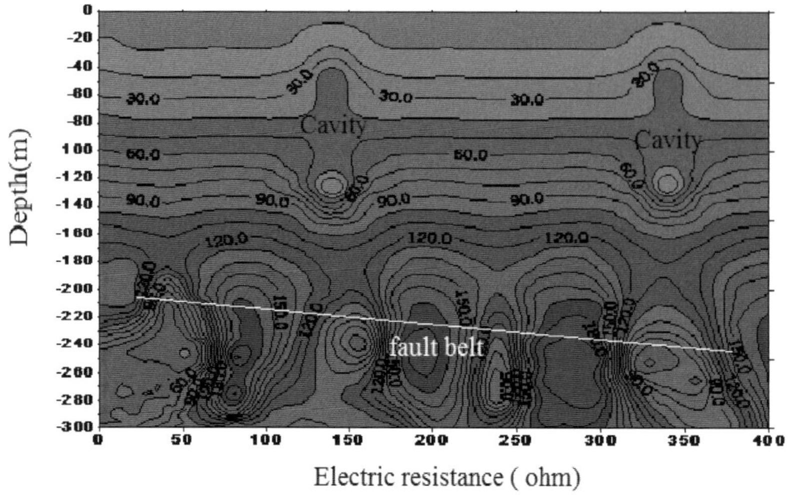

Figure 1. Electromagnetic exploring for cavity identification.

nese scholars have studied the land subsidence mechanism and protective methods, and many new theories and practical engineering measures have been referred (He, 1992).

In this paper, a comprehensive risk assessment framework and investigation methods for land subsidence identification are introduced, several typical subsidence mechanisms are summarized, and some successful engineering measures to mitigate the subsidence are discussed. A GIS for land subsidence risk assessment is developed and is based on site investigations and numerical simulation of subsidence process. In this system, maps of mining intensity and risk ranks are developed.

2 Identification and types of land subsidence in mined-out regions

In mining regions, because land subsidence is closely related to the underground excavation, it is commonly used to explore the underground cavity. In order to discover these cavities, geophysical exploration techniques have been mostly applied, such as induced polarization electrical method, 3-D seismic exploration, magnetic and electromagnetic exploration, ground penetrating radar, light detection and ranging (LIDAR), geophysical computerized tomography, electrical prospecting, and gravity methods. Generally, all of these methods have been successfully applied in identifying mined-out cavities within less than 300 m depth below land surface. In China, the first three methods have been most applied. Figure 1 shows a typical result for electromagnetic exploration of cavities.

The traditional theories divide subsiding materials above the mine into three belts or zones or areas: the bending belt, the fault and cracking belt, and the collapsing belt (from the

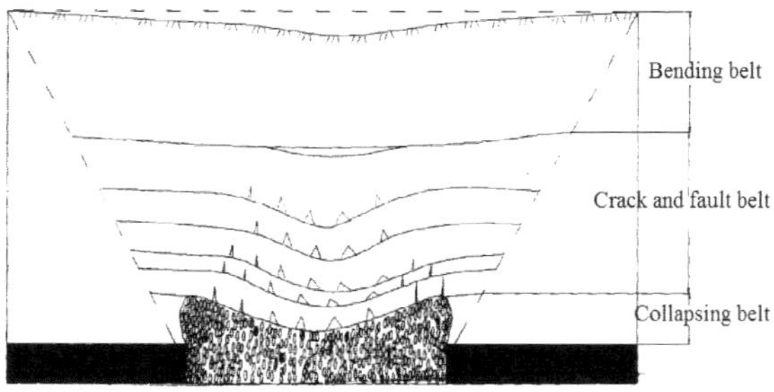

Figure 2. "Three belt" model of land subsidence types in mined-out region.

surface to the underground cavity). In the bending belt, the soil and rock are deformed but for the most parts are not mechanically failed, there are few fractures developed mainly in the vertical direction, the strata have little differential displacement and subsidence with no strata separation cracks. The thickness of the bending belt depends on the overlying rock characteristics and depth of mined-out cavity (Fig. 2). If the rock is soft and the cavity is deep, the bending deformation is more widely distributed. For the fault and crack belt, there are many cracks and faults of both horizontal and vertical directions in the rocks, and layer separation cracks are usually observed. The number and width of the rocks' cracks and faults are greater and inversely related to the distance between land surface and the cavity. In the top layer of this crack belt, the orientation of the cracks and faults is vertical toward the cavity, but in the bottom layer, the orientation is horizontal, and some strata are separated, and easily slip toward the cavity. In the collapsing belt, the rock is completely damaged, and partly or completely fill the mined-out cavities

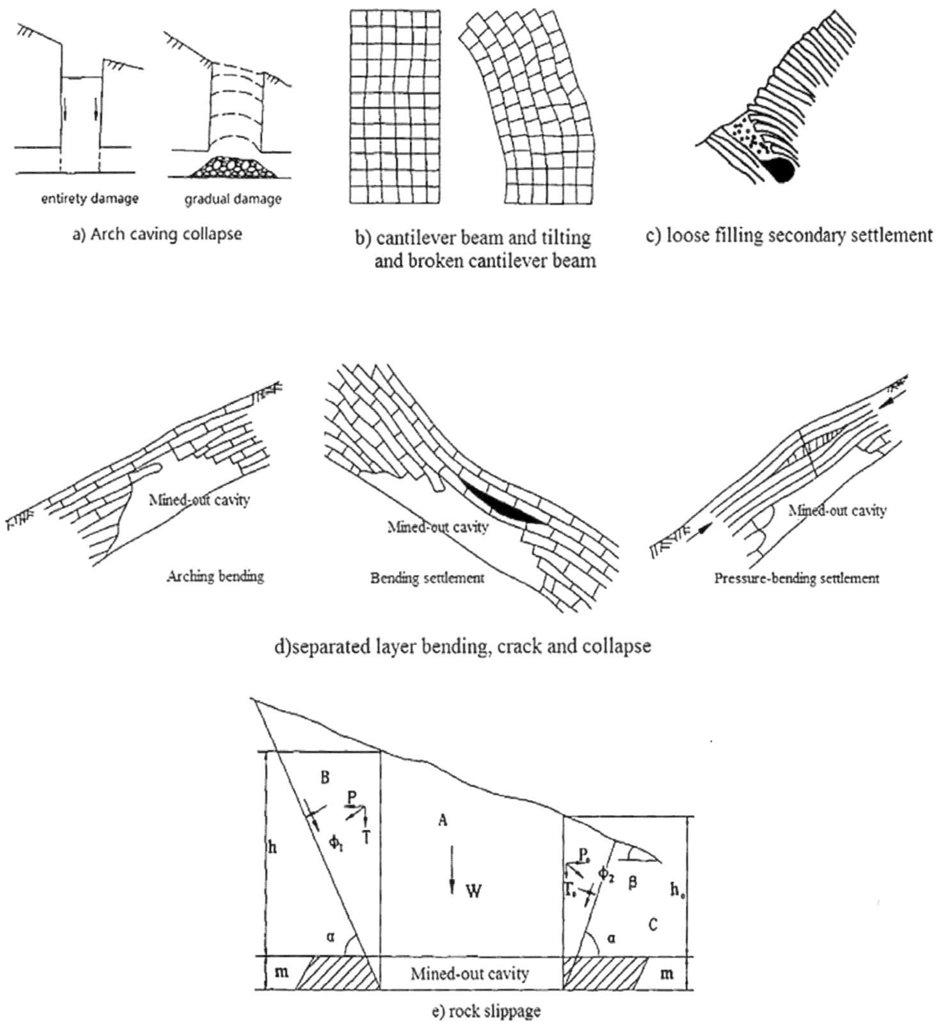

Figure 3. Land subsidence patterns in mined-out region.

depending on the rock types and thickness of mined strata. The height of the collapsing zone can be calculated as:

$$h = \frac{m}{(k-1)\cos a}, \tag{1}$$

where h is the height of the collapsing zone, m is the thickness of the mining stratum, k is the hulking coefficient of the cracked rock (Table 1), and α is the inclination angle of the stratum. This traditional theory describes the rock deformation characteristics, but does not describe the land subsidence mechanism. Therefore, based on this theory alone, it is difficult or not suitable to design or plan reliable protective measures against land subsidence induced by different rock damage mechanisms. Since the 1950s, many important theories of subsidence processes in mining regions were developed by Chinese scientists, such as the rheological theory of the overburden strata in mining fields (Liu, 1995; Ran and Gu, 1998), nonlinear smoothing finite element theory and large deformation theory (He and Peng, 1994; He and Zhao, 1995), probability influence function theory (He, 1992), stochastic and fuzzy model theories for land subsidence prediction (Liu, 1965; Li et al., 2010), chaotic support vector machine theory (Li et al., 2008), and the thin-layer plate roof caving model theory (Zhang and Cao, 2015). According to the theoretical, numerical and in situ observations, mining-related land subsidence in China was classified into seven main types: (1) arch caving collapse (Fig. 3a), (2) cantilever beam caving of natural rock strata, including tilting and broken strata (Fig. 3b), loose filling caving secondary settlement (Fig. 3c), (4) separated layer bending, crack and

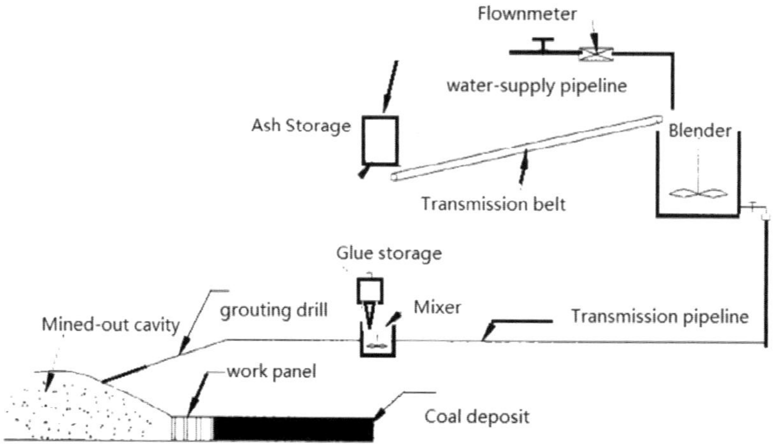

Figure 4. Simplified flowchart of a popular grouting filling method used in China.

Table 1. Hulking coefficient values of cracked rock (k).

Rock Type	Initial crack status	Residual status (compacted)
sand	1.05–1.15	1.01–1.03
Clay	< 1.2	1.03–1.07
Cracked coal	< 1.2	1.05
Clayed shale	1.40	1.10
Sandy shale	1.60–1.80	1.10–1.15
Hard sandstone	1.50–1.80	
Soft rock		1.02
Middle-hardness rock		1.025
Hard rock		1.03

collapse (Fig. 3d), (5) rock slippage (Fig. 3e), (6) rock softening and creep due to flooding and groundwater penetration, (7) collapsing due to overburden load, dynamic loads, and groundwater-level variation from natural or man-made causes (e.g. climatic variation, flooding and groundwater extraction).

3 Main engineering treatment methods for subsidence protection

Reclamation and protection of the areas affected by land subsidence are the important tasks for achieving sustainable development in mining regions. The Chinese government has paid much more attention to this since the 1950s, and during the recent past two decades, vast investments have been made to treat land subsidence. Based on the reclamation and protection practices, several principal engineering treatment methods have shown some success.

3.1 Grouting filling method

For a jointed rock mass and cavity, the grouting filling method (GFM) has been a highly effective method to control land subsidence. In our practice, the principle GFMs have included goaf-stowing with cement materials (GSC), ash grouting (AG), and low density slurry (LDS) methods. The following parameters must be considered to obtain satisfactory results with GFMs: grouting pressure, injection velocity, fluid type and density, crack density of rock mass, buried depth of cavity, volume of the collapsing belt, hydrology, construction technique and equipment. The most popular GFM in China is illustrated in Fig. 4. Here, the main filling material is ash from a power plant. The glue is a mixture of aluminium-alumine, inorganic materials and additives. Its relative density is 3.0 and its density is $1.2\,\mathrm{t\,m^{-3}}$, and it has a volumetric water content of 90 %. After mixing with water, the glue mixture can have a bearing capacity of 0.5–1.0 MPa in one hour. Typically, the mass fraction of the component of grouting materials are ash (50–70 %), glue (5 %), and water (25–45 %).

3.2 Anchor and anti-slide pile supporting system

If the cavity is buried at less than 100 m depth and is overlain by hard rock mass, the anchor supporting system (ASS) or anti-slide pile supporting system (APSS) can be used to mitigate land subsidence. However, ASS and APSS are more effective in preventing the lateral deformation of the cavity. In China, ASS and APSS are usually applied to treat the subsidence in metal (gold, iron and cobber etc.) mining regions. For ASS and APSS, the important parameters include the anchorage depths of piles and anchors, pile and anchor length and spacing, pile cross section, and ratio of height and width, and stability and strength of the supporting systems. From 1995, new ASS and APSS technique have been

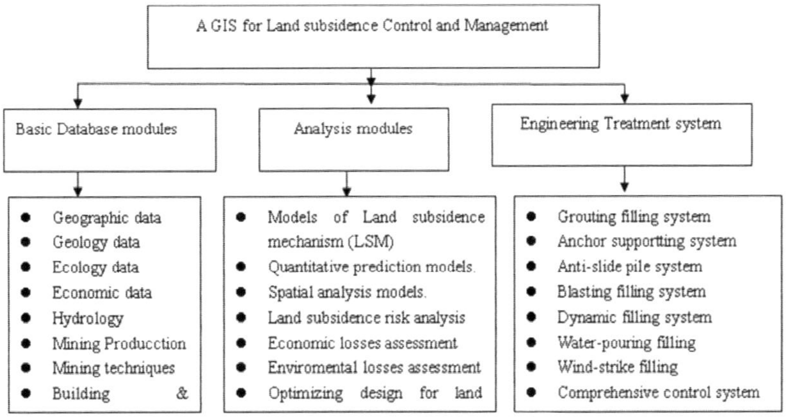

Figure 5. A GIS focussed land subsidence management system for in mined-out regions.

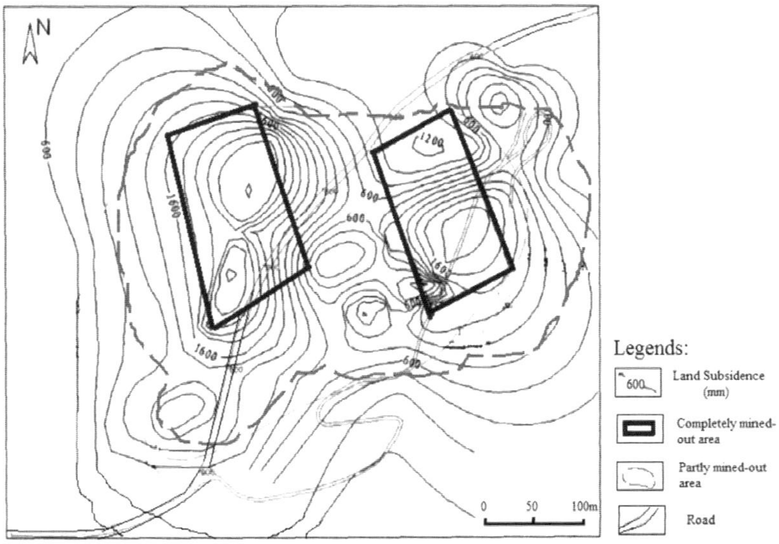

Figure 6. One of the typical maps of mining intensity and risk ranks based on GIS.

continually applied in coal mining regions, like Heilongjiang province, Shanxi province, Inner-Mongolia province and Anhui province.

3.3 Comprehensive treatment method

Dynamic compaction, blasting and filling, water-pouring filling (WPF) and wind-strike filling methods are used to control the subsidence in mined-out regions in China. For a large and deep subsidence area with little population, the blasting and filling method is often used, and is capable of rapidly controlling subsidence for relatively short durations. Where acceptable environmental conditions permit, the reclaimed land use is typically recreation and agriculture. Dynamic compaction is applied to treat the shallow land subsidence, particular surface-based transportation system. WPF is often applied to prevent the deep mined-out cavity from subsiding, in those cases, it is necessary where there is rich and easily extracted groundwater. However, subsidence control with WPF method often takes a long time before it is optimally effective due to long-term creep deformation and softening with soil-rock-water interaction. Wind-strike-filling is also used to treat shallow mined-out cavities where fine sand and clay are easily available.

4 A GIS for land subsidence management in mined-out regions

Managing land subsidence in mined-out regions is a complex. It entails multi-disciplinary knowledge of the geology, hydrology, geography, ecology, soil and rock mechanics, engineering, management, and risk assessment. A powerful data management and spatial analysis platform, GIS has been applied as a development tool for land subsidence management systems in China since 2000. Presently, almost all of land subsidence management systems for mining areas are based on GIS. The general management framework, flowchart and function for these systems are described in Fig. 5. One of the typical maps of mining intensity and risk ranks based on GIS is illustrated in Fig. 6. For a GIS system based on land subsidence risk assessment, prediction and prevention, these sub-systems are often at least included inside them: (1) subsidence mechanism analysis system; (2) subsidence risk assessment,mitigation and warning system; (3) subsidence spatial distribution estimation system; (4) engineering techniques to controlling and treatment system; (5) daily management system; and (6) emergency response system.

5 Summary and conclusions

Land subsidence in mining regions is an important manmad hazard in China and throughout the world that poses large risks for land use and environmental protection. Despite the successful application of engineering treatment methods based in part on traditional theories of subsidence processes related to underground mining, more research on subsidence processes in mined-out regions is needed particularly with respect to application of modern monitoring techniques, and the development of numerical simulation and computational methods based on refined process models. As the result of new advance in mining engineering and the growing demand for energy and mineral products, mining depths will increase. Land subsidence induced by underground mining will face new challenges, related to the increased mining depths and the potential reactivation of previously treated land subsidence. Therefore, reliable new methodologies to forecast and prevent land subsidence, for example by reducing the disturbing effect on the rock mass, must be developed. Additionally, laws and ordinances related to restricting subsidence induced by mining must been made and strictly executed, otherwise, it is possible that land subsidence risk will rapidly increase for the need to high-speed economic development of China.

References

He, G.: Mining Land subsidence, China University of Mining and Technology Press, 1992.

He, M. and Peng, T.: The problem of large deformation in soft-rock engineering and practical analysis of floor-heaving, J. China Univ. Mining Technol., 4, 86–91, 1994.

He, M. and Zhao, J.: Mining theory and practice under building, China University of Mining and Technology Press, 1995.

Jones, C. J and Spencer, W. J.: The implication of mining subsidence for modem highway structure, UK, Wales: Proceedings of Large Ground Tunnel Movement and Structure, University of Wales, Cardiff, UK, Halstead Press, New York, 515–526, 1977.

Karmis, M., Agioutantis, Z., and Jarosz, A.: Subsidence prediction techniques in the United States: a state-of-the-art review, Min. Res. Eng., 3, 197–210, 1990.

Klatesch, H.: Mining subsidence engineering, Springer Verlag Press, Berlin, Germany, 1983.

Li, W. X., Liu, L., and Dai, L. F.: Fuzzy probability measure (FPM) based non-symmetric membership function: engineering examples of ground subsidence due to underground mining, Eng. Appl. Artif. Intel., 23, 420–431, 2010.

Li, W. X., Gao, C. Y., Yin, X., Li, J.-F., Qi, D.-L., and Ren, J.-C.: A visco-elastic theoretical model for analysis of dynamic ground subsidence due to deep underground mining, Appl. Math. Model., 39, 5495–5506, doi:10.1016/j.apm.2015.01.003, 2015.

Li, X., Wang, Q., Yao, J., and Zhao, G.: Chaotic time series prediction for surrounding rock's deformation of deep mine lanes in soft rock, J. Centre South University of Technology, 15, 224–229,2008.

Liu, B. C.: The basic laws of land surface movement in coal mining, China Industry Press, Beijing, China, 1965.

Liu, T.: Influence of mining activities on mine rockmass and control engineering, J. China Coal Soc., 20, 1–5, 1995.

Ran, Q. and Gu, X.: A coupled model for land subsidence computation with consideration of rheological property, Chinese J. Geol. Hazard Control, 9, 99–103, 1998.

Tugrul, U., Hakan, A., and Ozgur, Y.: An integrated approach for the prediction of subsidence for coal mining basins, Eng. Geol., 166, 186–203, 2013.

Wu, J. C., Shi, X. Q., Xue, Y. Q., Zhang, Y., Wei, Z. X., and Yu, J.: The development and control of the land subsidence in the Yangtze Delta China, Environ. Geol., 55, 1725–1735, 2008.

Xue, Y., Zhang, Y., Ye, S. J., Wu, J., and Li, Q.: Land subsidence in China, Environ. Geol., 48, 713–720, 2005.

Zhang, B. and Cao, Sh.: Study on first caving fracture mechanism of overlying roof rock in steep thick coal seam, Int. J. Mining Sci. Technol., 25, 133–138, 2015.

Zhang, Y. Q., Gong, H. L., Gu, Z. Q., Wang, R., Li, X. J., and Zhao, W.: Characterization of land subsidence induced by groundwater withdrawals in the plain of Beijing city, China, Hydrogeology, 22, 397–409, 2014.

Proc. IAHS, 372, 151–155, 2015
proc-iahs.net/372/151/2015/
doi:10.5194/piahs-372-151-2015

Subsidence at the "Trébol" of Quito, Ecuador: an indicator for future disasters?

T. Toulkeridis[1], D. Simón Baile[1], F. Rodríguez[1], R. Salazar Martínez[1], N. Arias Jiménez[2], and D. Carreon Freyre[3]

[1]Universidad de las Fuerzas Armadas ESPE, Sangolquí, Ecuador
[2]Empresa Metropolitana de Alcantarrillado y Agua Potable de Quito, Quito, Ecuador
[3]Universidad Nacional Autónoma de México, Queretaro, Mexico

Correspondence to: T. Toulkeridis (ttoulkeridis@espe.edu.ec)

Published: 12 November 2015

Abstract. A sinkhole of great proportions was produced in one of the most trafficked zones of Quito. Constructed in the late sixties, this area is of high importance in solving the traffic jams of the capital city. The sinkhole called "El Trebol" started to be generated in the form of a crater, reached finally dimensions of approximately 120 m in diameter and some 40 m of depth, where at its base the river Machangara appeared. The generation of this sinkhole paralyzed the traffic of the south-central part of the city for the following weeks and therefore the state of emergency was declared. Soon the cause of the sinkhole was encountered being the result of the lack of monitoring of the older subterranean sewer system where for a length of some 20 m the concrete tunnel that canalized the flow of the river collapsed generating the disaster. The collapse of this tunnel resulted from the presence of a high amount of trash floating through the tunnel and scratching its top part until the concrete was worn away leaving behind the sinkhole and the fear of recurrence in populated areas. The financial aspects of direct and indirect damage are emphasized.

1 Introduction to the event and its causes

On Monday, 31 March 2008, around 14:00 LT, a giant subsidence in form of a sinkhole has been generated in a site called "El Trebol", where 25 000 m^3 of earthy material disappeared. This area is the most important southern inter-connector of the city to the Valley "De los Chillos" in the eastern end of Quito, but serves as well to connect the southern to the northern side of the city, where an extremely high amount of vehicles transit daily. This site was constructed four decades before its collapse. The sinkhole started in form of a crater with a diameter of approximately 30 m, amplifying its size constantly due to the instability of the slopes, which being wet and saturated with water of the high precipitations in that area of those days prior to the event. Further presence of more subterranean water has been noticed in the southern slope of the sinkhole, when the diameter reached some 120 m in diameter and a depth of 40 m, determined from the top part down to the previously covered river. That river was detected after the visible collapse of the concrete channeling which leads the river Machangara for a distance of some 20 m. Fortunately, no victims where reported. The result of this subsidence has been a traffic collapse and the declaration of the state of emergency of the Municipality of Quito. This event forced the authorities of Quito first to find out the causes of this subsidence and afterwards to search for further areas with similar problems and vulnerabilities in order to avoid future disasters where potentially people could be involved.

Three hypothesis were taken into consideration as the most potential cause for this disaster: (a) extreme discharge, (b) erosional process and chemical activity through time and finally (c) activity o subterranean waters, of which it resulted to be a combination of all three causes in different proportions.

Table 1. INAMHI's Meteorological Automatic Stations closest to El Trébol area.

Code	Station name	Latitude	Longitude	Altitude (m)
M003	Izobamba	−0.366089	−78.555061	3085.00
M024	Iñaquito	−0.175000	−78.485278	2789.12
M002	La Tola	−0.231667	−78.370333	2503.00

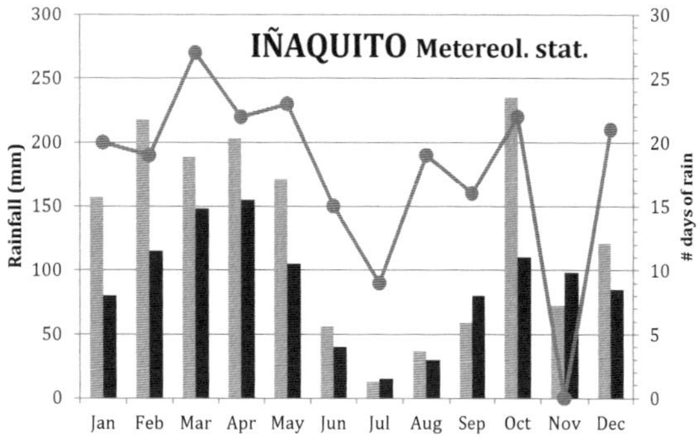

Figure 1. Total monthly rainfall, average multiannual rainfall (of the available data) number of days with rain per month (higher than 0.1 mm) of the closest meteorological station to El Trebol being Iñaquito. From National Institute of Meteorology and Hydrology (2010).

2 Hydrological characteristics

The hydrographic network of the South Valley of Quito is classified as dendritic, having the Machángara as its main river course. The Machángara rises in the steep foothills of volcano Atacazo and crosses the valley from south-west to north-east running parallel to the basin until it reaches the Trébol area (Panecillo) where it takes a turn to the East, gets deeper and flows to the valleys of Cumbayá and Tumbaco. Another important drainage is the so-called "Quebrada Grande" that has its source in the northwest foothills of volcano Atacazo and for a strecht it runs parallel to the Machángara river until it becomes its tributary. When it exits to the valley, the same river caudal varies between 3 $m^3 s^{-1}$ during dry season and 170 $m^3 s^{-1}$ during the wet season. Furthermore, the Machángara river is the main sewage receiver of Quito.

Regarding rainfall, the distribution along the year shows the prevalence of two periods with abundant rainfall being these, February–May and October–November respectively, (EMAAP-Q, 2006; Table 1, Fig. 1). On the other hand, as for the multiannual distribution of rainfall it is important to highlight that the wet season between January and March 2008 was considered the strongest of the last 20 years with accumulated rainfall during the first 3 months of 2008 being much higher that the recorded during any other heavy rainy seasons of 1989, 1993 and 2000 (Salazar et al., 2009). In particular, a peak of 21.2 L m^{-2} was registered on 31 March 2008 during a heavy rainfall that lasted 4 h from 1 to 5 p.m. (INAMHI, 2008; El Comercio, 2008; Salazar et al., 2009). Furthermore, within the whole year 2008 and coinciding in the three meteorological stations closest to "El Trébol", March recorded continuous raining since it was the month with the highest number of days of rain, with 27 and 29 out of 31 days in Iñaquito and Izobamba respectively (Fig. 1.)

3 Historical background of the area

The site where El Trébol was constructed is about 300–600 m. downstream of the confluence of Jerusalem and El Tejar streams, with Machángara River. Today, the channeled stream of Jerusalem is the Boulevard 24 de Mayo Avenue and its extension to the old bus station, current Qumandá Park. The channeled stream of El Tejar, after passing below the historic center, merges and flows with the stream of Manosalvas down in the low neighborhood of San Juan, in the area that is the center of mass transit transfer, now known as La Marin, before reaching Machángara River.

As a result of uncontrolled growth of the city, this natural drainage system was conducted on sewer channels below the construction of roads for vehicular traffic, and historically have been recorded 12 sinkholes (Fig. 2a) of different magnitudes, defined as "declines or collapses of roadway in the filler material streams, caused by faulty sewers". The responsibility gets under the fill type initially performed with debris and garbage, as well as the age, type of construction and degree of deterioration of the physical work of the channels, sewers and collectors (Pewter, 1986; Fig. 2b). The location of these sinkholes is defined about the old bed of the streams filled so we can assume that on its way continue to occur, especially in periods of daily rainfall above the historical average. In the area of EL Trébol, the first serious collapse was registered on 31 March 2008 and took ten years after the river and fill channels were made for the construction of the infrastructure to solve the problems of the traffic jams. There was a second sink lower rates on 10 January 2014 (Fig. 2c).

4 Direct and indirect financial damage

The day after sinkhole, the Ecuadorean Government created a line of credit of USD 60 million to help the city and start the reconstruction (La Hora, 2008). At the same day, the Quito's Major and the City's Council created an emergency fund of USD 200 000. This emergency fund was increased to about USD 1 million.

The day after drainage vault collapsed, rebuilding process started with a team of 5 hydraulic excavators, 5 power shovels, 10 roll-off trucks, 4 system equipment of mobile industrial lighting. In the rebuilding process 210 workers teams were at the site 24/7 (Explored, 2009). The reconstruction of

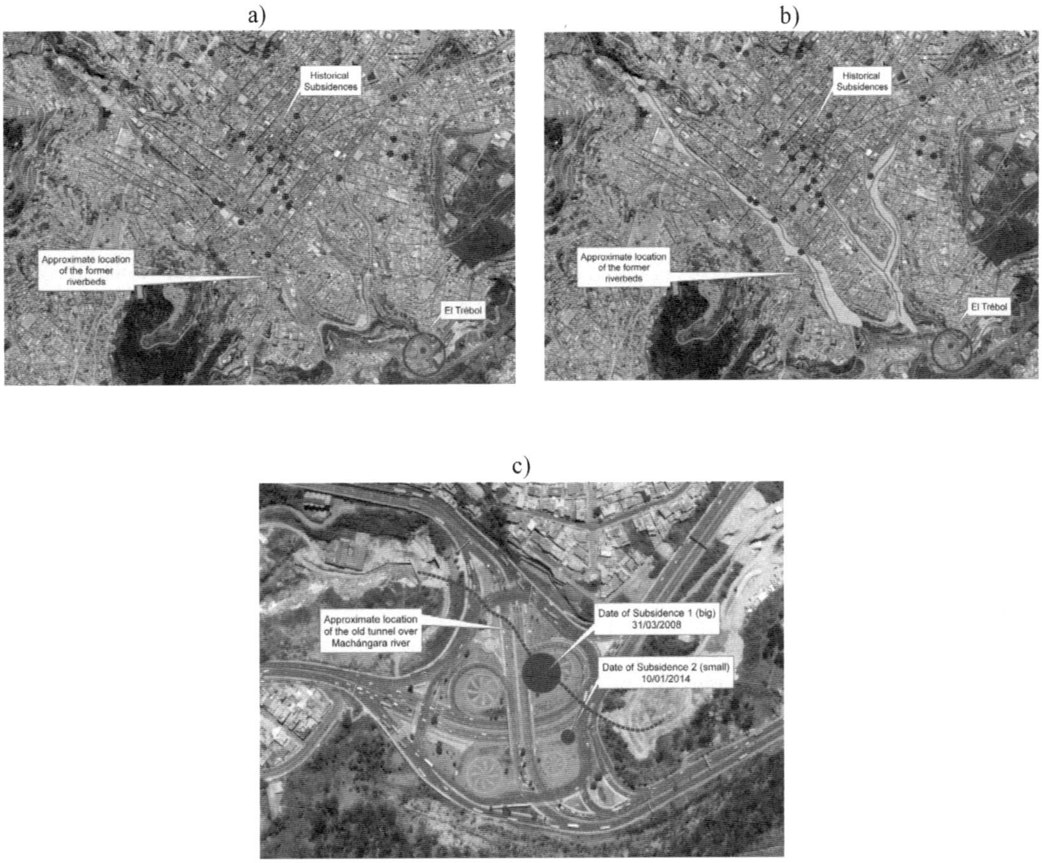

Figure 2. (a–c) Historical subsidences at Machángara river, El Trébol sector.

the cloverleaf interchange took 8 months to full traffic recovery, yet the entire reconstruction of the site took 22 months (Explored, 2010). El Trebol total reconstruction cost reached over USD 13 million, yet it did not required to use the full line of credit; details of expenditures are presented in Table 2.

However, this cost did not reflect the real cost because it does not take in account the externalities implicit with the sinkhole. Externalities associated to how users were affected by the sinkhole and cloverleaf interchange reconstruction. For instance, local authorities closed schools during the first week after the event, cost of students losing classes are not included neither teachers income lost, who were in per hour contracts.

We analyzed those costs of users who were affected by and were not compensated for. Because of lack of official information, we concentrated our efforts to estimate the cost of losing time during the reconstruction of the cloverleaf interchange and drainage vault, as well as the additional cost in gasoline of users of this crucial cloverleaf interchange.

Table 2. List of the costs of the reconstruction in USD (Source: MDMQ, 2015).

New tunnel	7 575 872.21
Rental of equipment	352 622.93
Land movement	151 526.15
Labor	801 546.35
Construction materials	435 495.03
Services	219 061.16
Other costs	31 236.78
EMOP	2 000 000.00
Vida para Quito	2 000 000.00
Total	13 567 360.61

In case of additional cost in gasoline, there were 80 000 of vehicles circulate and use El Trebol every day, in addition of 400 inter-parish buses of public transportation (La Hora, 2008). In order to estimate the value lost by users, we concentrated in private transportation under the assumption that a car-owner who uses his car to go to his job and back home

Table 3. Vehicle type and different gasoline prices in USD.

Vehicle	Gasoline price	
type	USD gallon^{-1}	Total cost*
17 456	2.00	11 210 941.44
62 544	1.48	40 168 258.56
Total gasoline		51 379 200.00

* Adding USD 0.07

Table 4. Cost category expressed in delay of 25 and 35 min.

Cost category	25 min	35 min
CPVTOTAL	34 185 200.00	47 859 280.00
CPSTOTAL	15 668 216.67	21 935 503.33
CPAvVITOTAL	7 121 916.67	9 970 683.33
CPúTOTAL	11 395 066,67	15 953 093.33
Total cost	68 370 400.00	95 718 560.00

fills his car gas tank once a week. This assumption seems reasonable for most of car owners. Yet, we did not include public transportation because we do not know how many times a transportation unit fills bus gas tank in a week. Based on public estimations, we used a value 0.07 additional dollar that an owner has to pay extra to fill his gas tank. It is reasonable also to assume that a car owner spends USD 20 week^{-1} filling his car tank. These USD 0.07 seems a low bound, but still reasonable. Then, we multiply the value of one gallon of gasoline adding these USD 0.07.

Based on AIHE statistics (AIHE, 2015), we know that 21 % of the car owners use a "super premium" gasoline which costs USD 2.00 gallon^{-1} and 78 % uses "extra" gasoline with a price of USD 1.48 gallon^{-1}. Based on the number of vehicles which circulated at that time "El Trebol" every day, 80 000 cars day^{-1}, we can say that approximately 17 000 cars used "super premium" gasoline and over 62 000 used "extra" gasoline type. We estimated that additional cost in gasoline for private car owners was USD 85 million for those 8 months of traffic problems at "El Trebol" (Table 3).

We did not include capital depreciation, even now, it is public knowledge that keep a car running while waiting depreciate its value faster than normal conditions. We did not include both public transportation and car owners who own a diesel engine car.

Concerning the cost of time lost (an opportunity cost), we estimate its value from per hour salary multiply for the additional time that users had to spend during the reconstruction process. We considered that a time between 25 up to 35 min is reasonable to believe users lost during their travel to workplaces or going back their homes. A user lost of USD 2.2–3.12 h^{-1} seems to be reasonable. This value is multiplied by the total time of site reconstruction. Since user came from different directions, we estimate the lost value separately. According to media reports, 48 000 came private own vehicles from "Los Chillos" Valley (CPVTOTAL), 22 000 came from southern part of Quito (CPSTOTAL), 10 000 were coming toward the valley or southern part of the city (CPAvVITOTAL) and 400 units of public transportation (CPúTOTAL). Regarding public transportation, we assume that each unit was carrying 40 passengers each trip, which it is a low bound because rush hour, these units can be a full capacity (around 72 passengers).

Table 5. Cost category.

Cost category	Amount (USD)	Ratio
Reconstruction cost	13 567 360.61	
Additional gasoline cost	51 379 200.00	0.264063290
Opportunity cost 25 min	68 370 400.00	0.198439100
Opportunity cost 35 min	95 718 560.00	0.141742214

Finally we estimate users' opportunity cost multiplying per hour lost times the time of "El Trebol" reconstruction, which was total of 8 months. The users' opportunity cost for each category is presented in Table 4.

Users lost a considerable amount of time when reconstruction took place, adding all users (the aggregate value) it turned out that the real cost of the sinkhole increases significantly. As the table shows, under the assumption that users lost only 25 min, the opportunity cost reaches over USD 68 million during the 8 months of reconstruction, and under the assumption that user lost up to 35 min, the opportunity cost reaches over USD 95 million. As a result the real cost (under 25 min assumption) reached more than USD 133 million, and under 35 min assumption reached more than USD 160 million as real cost (Table 5).

5 Conclusions and recommendations

As result of this event, we consider as a priority to take corresponding actions to prevent future collapses. Taking in consideration the alignment of the actual and past sinkholes, these alignments need to be reinforced in order to avoid future disasters in that area. As demonstrated in our study, the real costs of damages are much higher in the indirect damage of such sinkhole events rather in the reconstruction of the disaster site itself. Unfortunately, the enforcement of the potential subsidence areas did not take place yet, as demonstrated by a new sinkhole in 2014 in a zone where the vulnerability has been previously emphasized.

References

AIHE: Estadísticas, Asociación de la Industria Hidrocarburífera del Ecuador, Quito, Asociación de la Industria hidrocarburífera del Ecuador (AIHE), Quito, Ecuador, available at: http://www.aihe.org.ec/index.php?option=com_content&view=article&id=122:estadisticas&catid=67&Itemid=142, 2015.

Cowen, T. (Ed.): Public Goods and Market Failures, Transaction Publishers, New Brunswick, NJ, USA, 1992.

Explored: Obra vial desaparece en la capital después de 40 años, Explored Noticias Ecuador publicado en 1 enero 2009, http://www.explored.com.ec/noticias-ecuador/nuevo-colector-ya-funciona-en-el-trebol-437565.html, 2009.

National Institute of Meteorology and Hydrology (INAHMI): Meteorological yearbook, Nr. 48, Quito, Ecuador, 2010.

La Hora: El Trebol se cayó en Quito, Diario La Hora Nacional, sección País, Domingo 6 de abril de 2008, available at: http://www.lahora.com.ec/index.php/noticias/show/703827/1/El_Tr/%C3/%A9bol_se_cay/%C3/%B3_en_Quito.html#.VT4MtJ_NHw, 2008.

Municipio del Distrito Metropolitano de Quito (MDMQ): Balance de los Estudios Urbanos (1985–2005) de la Coperación IRD-Municpio de Quito, in: Movilidad, elementos esenciales y riesgos en el Distrito Metropolitano de Quito, edited by: Demoraes, F., Co-edición MDMQ-IFEA-IRD, Quito, 2005.

O'Rourke, T. D. and Crespo, E.: Geotechnical properties of cemented volcanic soil, J. Geotech. Eng., 114, 1126–1147, 1988.

Peltre, P.: Quebradas y riesgos naturales en Quito, in: Riesgos Naturales en Quito, edited by: Pierre, P., Corporación Editora Nacional, Quito, Ecuador, 45–89, 1989.

Proc. IAHS, 372, 157–161, 2015
proc-iahs.net/372/157/2015/
doi:10.5194/piahs-372-157-2015

Has land subsidence changed the flood hazard potential? A case example from the Kujukuri Plain, Chiba Prefecture, Japan

H. L. Chen[1], Y. Ito[2], M. Sawamukai[3], T. Su[2], and T. Tokunaga[2]

[1]School of Environmental Science and Engineering, Zhejiang Gongshang University, Hangzhou, 310018, China
[2]Department of Environment Systems, School of Frontier Sciences, The University of Tokyo, Tokyo, 277-8563, Japan
[3]VisionTech Inc., Tsukuba-city, Ibaraki, 305-0045, Japan

Correspondence to: H. L. Chen (hualichen57@126.com)

Published: 12 November 2015

Abstract. Coastal areas are subject to flood hazards because of their topographic features, social development and related human activities. The Kujukuri Plain, Chiba Prefecture, Japan, is located nearby the Tokyo metropolitan area and it faces to the Pacific Ocean. In the Kujukuri Plain, widespread occurrence of land subsidence has been caused by exploitation of groundwater, extraction of natural gas dissolved in brine, and natural consolidation of the Holocene and landfill deposits. The locations of land subsidence include areas near the coast, and it may increase the flood hazard potential. Hence, it is very important to evaluate flood hazard potential by taking into account the temporal change of land elevation caused by land subsidence, and to prepare hazard maps for protecting the surface environment and for developing an appropriate land-use plan. In this study, flood hazard assessments at three different times, i.e., 1970, 2004, and 2013 are implemented by using a flood hazard model based on Multicriteria Decision Analysis with Geographical Information System techniques. The model incorporates six factors: elevation, depression area, river system, ratio of impermeable area, detention ponds, and precipitation. Main data sources used are 10 m resolution topography data, airborne laser scanning data, leveling data, Landsat-TM data, two 1 : 30 000 scale river watershed maps, and precipitation data from observation stations around the study area and Radar data. The hazard assessment maps for each time are obtained by using an algorithm that combines factors with weighted linear combinations. The assignment of the weight/rank values and their analysis are realized by the application of the Analytic Hierarchy Process method. This study is a preliminary work to investigate flood hazards on the Kujukuri Plain. A flood model will be developed to simulate more detailed change of the flood hazard influenced by land subsidence.

1 Introduction

Flooding in coastal areas is one of the most devastating natural hazards, which causes considerable property damage and personal injury (Doornkamp, 1998; Hunt and Watkiss, 2011). Different from other areas, coastal lowland areas can be influenced by both natural process and anthropogenic activities which make the area at higher risk of floods than others. At the same time, there is a growing awareness of anthropogenic impact and positive/negative feedbacks on flooding (IPCC, 2013). Rapid expansion in human settlements increases imperviousness which in turn increases peak discharge and surface runoff. Widespread land subsidence caused by excessive groundwater withdrawal or engineering activities further amplifies the harmful effects of flooding by modifying the ground elevation, stream slopes and flow pathways. On the other hand, some activities such as construction of detention ponds can buffer the flow to reduce the risk of flooding. Factors related to flood hazards are changing in spatial and temporal scales which increase the difficulties of accurate flood hazard assessment.

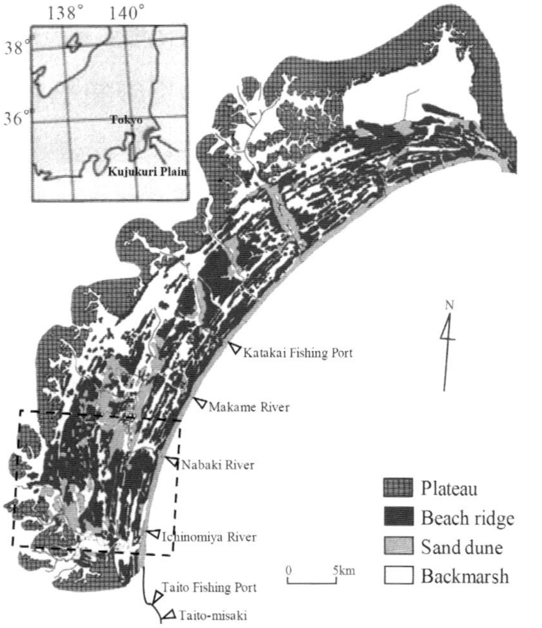

Figure 1. Geomorphological map of the Kujukuri Plain (Obanawa et al., 2010). Square marked by dashed line indicates the study area.

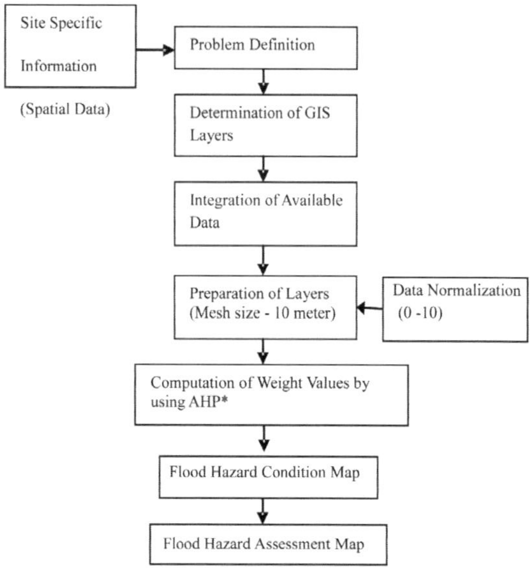

Figure 2. Flow chart for making the flood hazard map. * AHP – Analytical Hierarchy Process.

The Kujukuri Plain (Fig. 1), Chiba Prefecture, Japan, suffers from flood hazards. Many natural processes and anthropogenic activities have changed the surface environment of the Plain, such as agricultural development, urban and industrial development, change of drainage patterns, and deposition and/or erosion of river beds. Due to the exploitation of groundwater, extraction of natural gas dissolved in brine, and natural consolidation of the Holocene and landfill deposits, widespread land subsidence has occurred in the Plain (Environmental and Community Affairs Department, Chiba Prefecture 1970–2007). All of these factors could more or less increase the risk of flooding in the Kujukuri Plain. Therefore, constructing flood hazard potential maps for different periods is meaningful to study the change of the spatial and temporal patterns of the hazard potential for discussing future land-use planning of the Kujukuri Plain.

2 Methodology

A semi-quantitative method was selected to investigate the flood hazard assessment due to the scarcity of good quality data in the study area (Chen et al., 2015). Figure 2 shows the steps used in this study including data gathering, Geographic information system (GIS) manipulation, and Multicriteria Decision Analysis. The main data used are airborne laser scanning data, leveling data, Landsat-TM data, two river watershed maps with 1 : 30 000 scale, and radar precipitation data. Based on these data, five layers expressing condition factors were constructed, i.e., river system, elevation, depression area, ratio of impermeable area, and detention ponds. Precipitation was the main factor to trigger floods. A layer for the triggering factor, i.e., precipitation distribution, was created from radar precipitation data. The next step was to assign weights to each layer. The Analytic Hierarchy Process method (Saaty, 1980) was applied to assign the weight values (Table 1). The pixel value of the condition and assessment map was calculated by overlapping the factor layers.

3 Analysis and results

As shown in Fig. 3, each layer was created for three periods, i.e., year 1970, 2004 and 2013. The river system data for 2004 and 2013 were similar because no significant change was found between two periods. Data quality of year 1970 were not good compared with those of year 2004 and 2013, especially the elevation and the precipitation data. For detention ponds layer, there was no such structure for year 1970. So, the condition and assessment maps were calculated only for year 2004 and 2013 (Fig. 4). Here, the higher values indicate the higher likelihood of floods. In the flood hazard assessment map, the studied area was divided into five classes by using the quantiles method which created nearly equal ratios of the coverage in each class, i.e., very high, high, moderate, low, and very low hazards potential. The assessment map (Fig. 4) shows that the southeastern, the northeastern and the central parts of the study area are expected to have highest flood hazard potential as a consequence of the combination of high precipitation with lower elevation difference

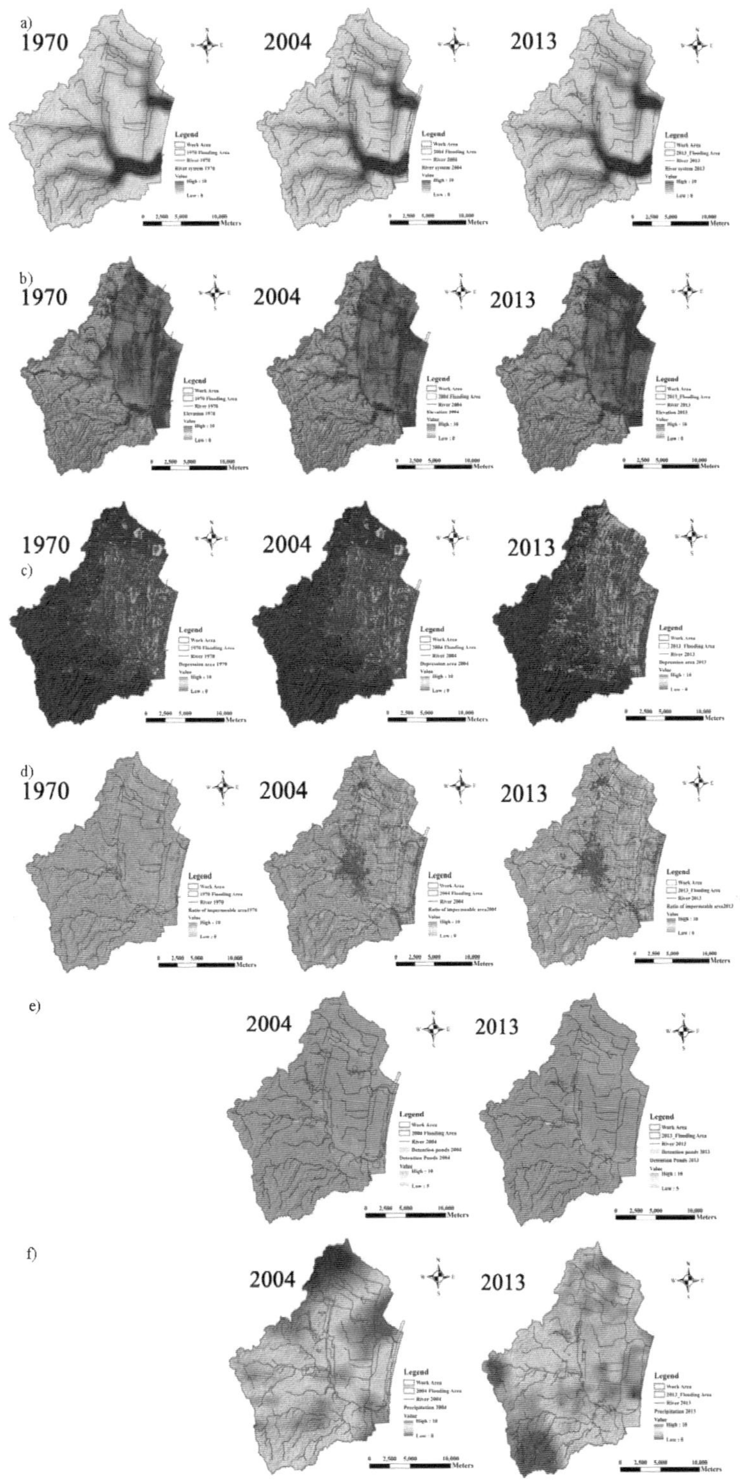

Figure 3. The images of each layer for the flood hazard assessment. The areas surrounded by pink line show the ones actually flooded. Figures in the left column are those for year 1970, the center for year 2004, and the right for year 2013. **(a)** River system layer, **(b)** elevation layer, **(c)** depression area layer, **(d)** ratio of impermeable area (Ratio of IA) layer, **(e)** detention ponds layer, and **(f)** precipitation layer.

Table 1. Assigned weights for the layers.

Layer	River system	Elevation	Depression area	Ratio of impermeable area	Detention ponds	Precipitation	Consistency ratio
Condition map	0.300	0.233	0.233	0.167	0.067	–	0.012
Assessment map	0.225	0.175	0.175	0.125	0.050	0.250	0.014

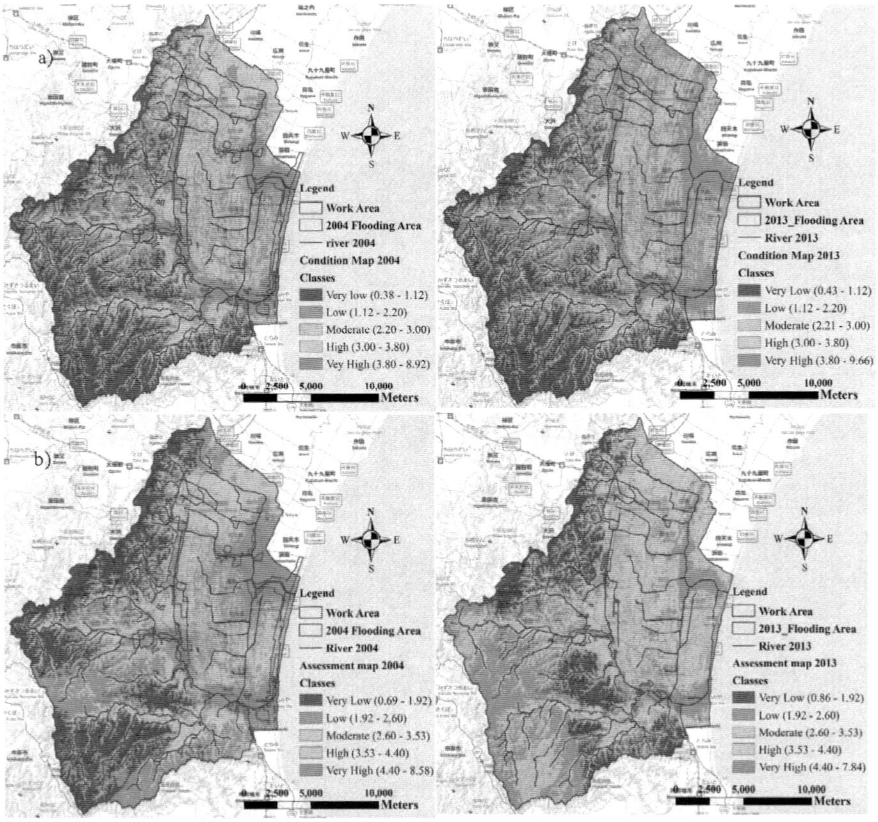

Figure 4. (a) Flood hazard condition map and **(b)** flood assessment map obtained by this study, for left (2004) and right (2013). The areas surrounded by pink line show the ones actually flooded (Chiba Prefecture, 2005, 2013).

and higher river density. The comparison between the actual flood areas reported by the local authorities (Chiba Prefecture, 2005, 2013) and the map constructed in this study showed that most of the flooded areas were located in the very high hazard potential area.

4 Discussion and summary

By comparing the results for three different periods, 1970, 2004 and 2013, the urban area enlarged significantly, the area effected by detention ponds became lager, while the change of the river system, the depression area and that of the elevation layer were small. For precipitation, it was difficult

Table 2. The areal percentage of each class for the condition and assessment maps.

Class name	Condition map		Assessment map	
	2004	2013	2004	2013
Very low	18.89 %	18.36 %	19.49 %	11.75 %
Low	20.53 %	20.44 %	20.48 %	22.21 %
Moderate	20.04 %	16.07 %	20.00 %	35.28 %
High	20.52 %	17.87 %	20.06 %	19.72 %
Very High	20.02 %	27.25 %	19.96 %	11.04 %

to compare because of differences in the resolution of the data and the rather large spatial difference of precipitation patterns. From Table 2, it was found that the percentage of very high hazard potential was increased from year 2004 to 2013 for the condition map with slight decreases of the high and moderate potential classes. However, other classes did not show big changes. Due to the large difference of spatial patterns of precipitation between year 2004 and 2013, the change in the assessment map was difficult to compare.

The areas with highest flood potential are in the southeastern, the northeastern and the central parts of the study area. The land cover change may increase the flood hazard potential. Construction of flood hazard maps with different periods described in this paper can assist decision makers and planners to evaluate the study area with respect to the temporal and spatial changes of different factors. This study can be applied as an early stage assessment of the flooding problem. A flood model will be developed to simulate more detailed change of the flood hazard influenced by land subsidence.

Acknowledgements. We would like to express our sincere gratitude to the Keiyo Natural Gas Association (Japan), the National Natural Science Foundation of China (41401539), and Science Research Foundation for the Returned Overseas Chinese Scholars (State Education Ministry of China, (2015) No. 1098) for supporting this study. Residential maps provided by ZENRIN CO., LTD are used as the CSIS Joint Research (No. 524) using spatial data provided by Center for Spatial Information Science, The University of Tokyo.

References

Chen, H. L., Ito, Y., Sawamukai, M., and Tokunaga, T.: Flood hazard assessment in the Kujukuri Plain of Chiba Prefecture, Japan, based on GIS and multicriteria decision analysis, Nat. Hazards, 78, 105–120, 2015.

Chiba Prefecture: Flooding report in Chiba Prefecture, Chiba, 12–12, 2005 (in Japanese).

Chiba Prefecture: Flooding report in Chiba Prefecture, Chiba, 18–18, 2013 (in Japanese).

Doornkamp, J. C.: Coastal flooding, global warming and environmental management, J. Environ. Manage., 52, 327–333, 1998.

Environmental and Community Affairs Department: Chiba Prefecture, Leveling results in Chiba Prefecture, reference date: February 1970–1 January 2007, 2007 (in Japanese).

Hunt, A. and Watkiss, P.: Climate change impacts and adaptations in cities: A review of the literature, Climatic Change, 104, 13–49, 2011.

IPCC, Climate change 2013: The physical science basis, Cambridge, edited by: Stocker, T. F., Qin, D., Plattner, G. K., Tignor, M. M. B., Allen, S. K., Boschung, J., Nauels, A., Xia, Y., Bex, V., Midgley, P. M., Cambridge, Cambridge University Press, 659–741, 2013.

Obanawa, H., Tokunaga, T., Rokugawa, S., Deguchi, T., and Nakamura, T.: Land subsidence at the Kujukuri Plain in Chiba Prefecture, Japan: Evaluation and monitoring environmental impacts, IAHS Publ., 339, 17–22, 2010.

Saaty, T. L.: The Analytic Hierarchy Process, Mc Graw Hill Company, New York, 1980.

Proc. IAHS, 372, 163–167, 2015
proc-iahs.net/372/163/2015/
doi:10.5194/piahs-372-163-2015

Open Access PIAHS

An analysis on the relationship between land subsidence and floods at the Kujukuri Plain in Chiba Prefecture, Japan

Y. Ito[1], H. Chen[2], M. Sawamukai[3], T. Su[1], and T. Tokunaga[1]

[1]Graduate School of Frontier Sciences, The University of Tokyo, 5-1-5 Kashiwanoha, Kashiwa City, Chiba, 277-8563, Japan
[2]School of Environmental Science and Engineering, Zhejiang Gongshang University, Hangzhou, 310018, China
[3]VisionTech Inc., Tsukuba City, Ibaraki, 305-0045, Japan

Correspondence to: Y. Ito (yu_ito@geoenv.k.u-tokyo.ac.jp)

Published: 12 November 2015

Abstract. Surface environments at the Kujukuri Plain in Chiba Prefecture, Japan, in 1970, 2004, and 2013, were analyzed and compared to discuss the possible impact of land subsidence on the occurrence of floods. The study area has been suffered from land subsidence due to ground deformation from paleo-earthquakes, tectonic activities, and human-induced subsidence by groundwater exploitation. Meteorological data, geomorphological data including DEM obtained from the airborne laser scanning (1-m spatial resolution), leveling data, and the result of our assessment map (Chen et al., 2015) were used in this study. Clear relationship between floods and land subsidence was not recognized, while geomorphological setting, urbanization, and change of precipitation pattern were found to contribute to the floods. The flood prone-area is distributed on the characteristic geomorphological setting such as floodplain and back swamp. It was revealed that the urban area has been expanded on these geomorphological setting in recent years. The frequency of hourly precipitation was also shown to be increased in the past ca. 40 years, and this could induce rapid freshet and overflow of small- and medium-sized rivers and sewerage lines. The distribution of depression areas was increased from 2004 to 2013. This change could be associated with the ground deformation after the Tohoku earthquake ($M_w = 9.0$) in 2011.

1 Introduction

Land subsidence causes several problems which include changes in elevation, groundwater salinization in coastal area, structural damage, and increase in the potential for flooding (Galloway et al., 1999; Ng et al., 2015). Land subsidence has occurred at many areas in Japan. The Kujukuri Plain, Chiba Prefecture, Japan (Fig. 1), is one of the areas experiencing land subsidence which is caused by ground deformation from paleo-earthquakes, tectonic activities, and human-induced subsidence by groundwater abstraction. The maximum value of accumulated subsidence is ca. 1 m at the Mobara City during the period from 1969 to 2014 (Chiba Prefecture, 2014). This coastal area is relatively flat with its elevation lower than 10 m. The lowland consists of three main landforms (Fig. 1) (Moriwaki, 1979): (1) beach ridge; (2) sand dune; and (3) back swamp and flood plain. Therefore, the impact of land subsidence on the surface environment has been concerned in this area; one of its effects may include the increase of the frequency of flooding. In the south Kujukuri plain area, floods have repeatedly occurred in the past. There were at least 16 large flood events in this area during the period between 1970 and 2013. The frequency of large flood events had relatively low before 1970. Figure 2 shows the flood inundation areas of the 1 July 1970, 8–11 October 2004, 16 October 2013.

In this study, we analyzed and compared the factors of the floods in 1970, 2004, and 2013 by using GIS to assess the possible impact of land subsidence on the occurrence of floods. The three periods were selected because of the following reasons; 1970 is after huge land modification, 2004 is

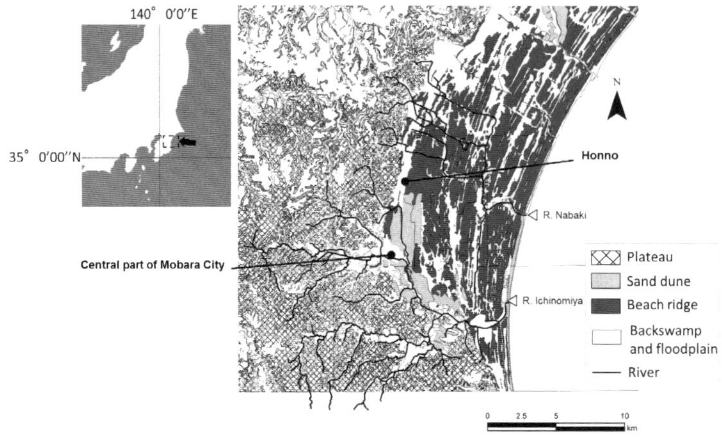

Figure 1. Geomorphological map of the Kujukuri Plain. Study area is Nabaki and Ichinomiya Rivers watershed.

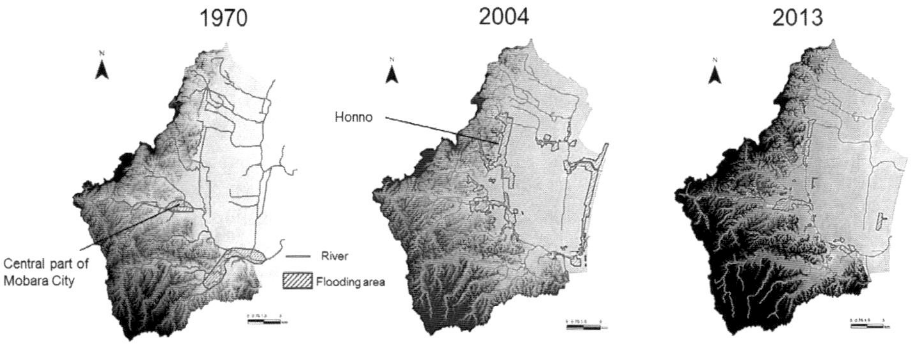

Figure 2. Flood inundation area in 1970, 2004, and 2013. Flood area was digitalized from the flood report of Chiba prefecture (1971, 2005, and 2015).

prior to the 2011 off the Pacific coast of Tohoku Earthquake ($M_W = 9.0$), and 2013 is after the earthquake.

2 Analyses of the factors related to floods

Three major factors which were related to floods were analyzed – climatic, geomorphologic, and human factors.

2.1 Climatic factor

The intensity and frequency of precipitation may affect the frequency and magnitude of floods (Fitzpatrick et al., 1999). Spatial distribution of the Radar/raingauge-Analyzed precipitation data obtained from Japan Meteorological Agency in 2004 and 2013, indicate that the spatial rainfall pattern could affect the flooding area. The flooding area in 2004 was mainly recognized in coastal area and the central part of the Mobara city (Fig. 2) due to high intensity rainfalls in the lower reach of the Nabaki River and the higher tidal level. In 2013, the upper reaches of the Ichinomiya River and the central part of the Mobara city were flooded (Fig. 2) because the rainfall was concentrated in the upper reaches of the Ichinomiya River. The lower reaches were not flooded because the tidal level was low and the rainfall was not strong. The results of our assessment maps in 2004 and 2013 (Chen et al., 2015) also suggested that the flood areas were affected by the spatial rainfall pattern.

Frequency of hourly precipitation of more than 50 mm h^{-1} is shown to be increased during the past ca. 40 years within the watersheds, and this could induce rapid freshet and overflow of small- and medium-sied rivers and sewerage lines.

2.2 Geomorphological factor

Alluvial and coastal lowlands have a potential floods risk (Umitsu, 2012). Flood inundation area is mainly recognized on floodplains, back swamp, Holocene terrace, and reclaimed land in three periods (Fig. 3 and Table 1). Coastal plain and beach ridges are also flooded.

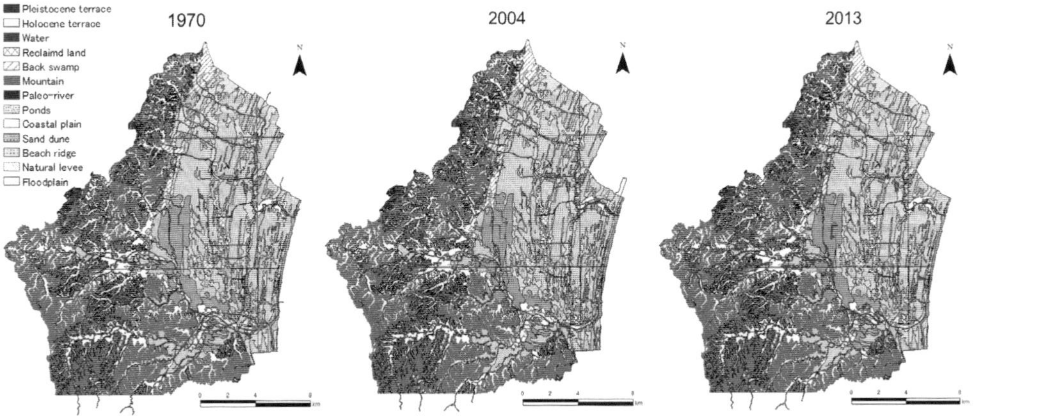

Figure 3. Geomorphological classification map in 1970, 2004, and 2013 based on land condition map (1 / 25 000) of Geospatial Information Authority of Japan. Red line indicates flood inundation area.

Table 1. Ratio of flood inundation area by geomorphological condition.

Geomorphological classification	1970	2004	2013
Holocene terrace	29.01 %	5.74 %	13.74 %
Flood plain	23.06 %	16.42 %	31.74 %
Natural levee	17.34 %	1.88 %	4.99 %
Beach ridge	6.92 %	25.94 %	5.07 %
Reclaimed land	7.57 %	9.19 %	10.22 %
Water	5.60 %	2.88 %	5.11 %
Sand dune	4.15 %	11.66 %	5.00 %
Ponds	2.40 %	0.82 %	0.52 %
Coastal plain	2.29 %	11.91 %	3.95 %
Back swamp	–	12.69 %	14.82 %
Mountain	0.94 %	0.41 %	3.27 %
Paleo-river	0.62 %	0.36 %	1.39 %
Pleistocene terrace	0.09 %	–	–
Others	–	0.11 %	0.17 %

In particular, in the central part of the Mobara City which is flood-prone area (Fig. 2), the floods occur on the flood plains surrounded by slightly highland (relative elevation: +1–2 m) such as natural levee and Holocene terraces. Honno, are the flood-prone area, is also situated on back swamp and is often flooded even with a light rainfall due to the area is surrounded by sandbars or beach ridges. Therefore, the flood-prone area may depend heavily on the characteristics of the geomorphological setting.

Local topographic depressions were extracted from the 1-m mesh resolution DEM. The results showed that the distribution of depressions was mostly unchanged from 1970 to 2004, however, was changed in 2013, particularly in the Nabaki River mouth and its surroundings (Fig. 4). This change might be the influence of the ground deformation after the $M_w = 9.0$ earthquake in 2011.

Table 2. Impermeable area ratios in each land cover.

Land cover classification	1970	2004	2013
Urban	0.20	0.30	0.30
Building	0.14	0.25	0.21
Farmland	0.12	0.18	0.11
Paddy	0.09	0.09	0.10
Sands	0.12	0.09	0.07
Grassland	0.07	0.09	0.10
Forest	0.07	0.07	0.08
Water	0.08	0	0.03
Bareland	0.15	–	–

2.3 Human factor

Urban development of the Mobara city and its surroundings was quite rapid during 1970–2013 (Fig. 5). The ratio of urban and building estate areas were 15.4 % in 2004 and 17.4 % in 2013, respectively, while 9.5 % in 1970. Farm lands and paddy areas were decreased during the period from 1970 to 2013. Impermeable area ratios in each land cover also change with land cover changing (Table 2). The area ratio of urban and building have increased from 0.20 and 0.14 in 1970 to 0.30 and 0.21–0.25 in 2004–2013, respectively.

The Urban areas have spread into lowland area such as flood plains and back swamps since 1970. This is clearly reflected in the relationship between the urbanization area and the landforms. The results of the analysis on the relationship between built-up area and the geomorphological changes (Fig. 6) indicate that the built-up area have expanded on vulnerable landforms with respect to floods, i.e., flood

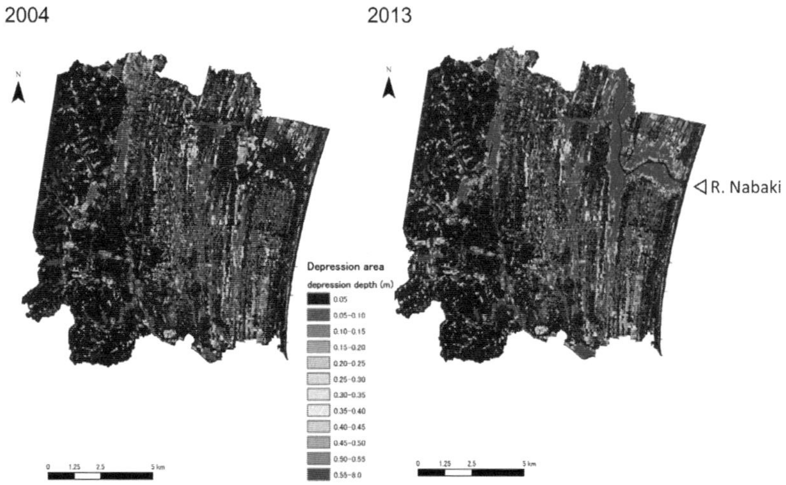

Figure 4. Distribution of depressions in 2004 and 2013. 1-m mesh digital elevation model obtained from the airborne laser scanning survey was used for the analysis.

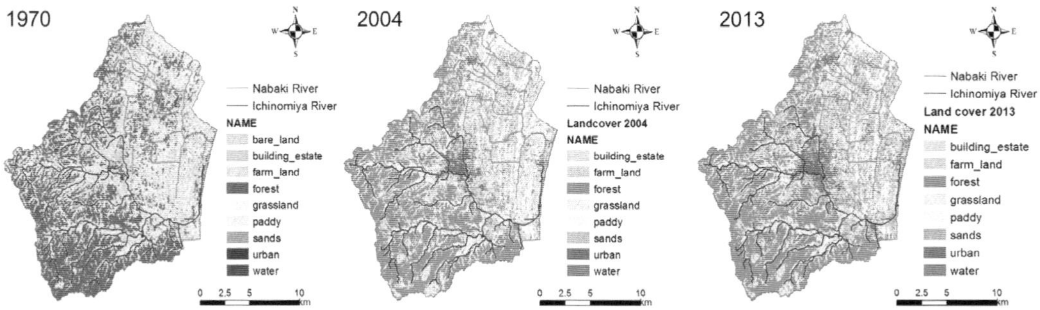

Figure 5. Land cover changes in 1970, 2004, and 2013. Land cover derived from Landsat-MSS and TM was classified by methods of maximum likelihood.

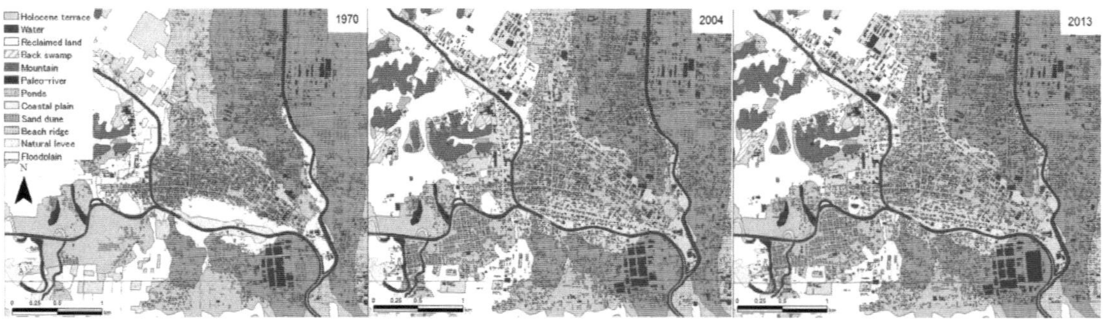

Figure 6. Expansion of the built-up area and geomorphological classification map in the central Mobara City. Red features indicate the building.

plains and paleo-channels. Majority of the flood plains in the central Mobara City was also changed to reclaimed lands since 1970 by land development. The buildings were located on the relatively less vulnerable land and ground conditions in 1970, as a result of urban development and the expansion to the flood plains and paleo-channels since 1970, the one might be changed to be prone to floods.

3 Summary

The visible change of the spatial distribution of hazard potential areas with land subsidence were not clearly recognized whereas the influence of artificial land modification and the natural geomorphological condition are considered to be the factors that explain the spatial distribution and the temporal changes of the vulnerable areas to floods. Back swamp and flood plains are surrounded by beach ridges in the center of the Mobara City and its surrounding area; its geomorphological condition suggests that water tends to accumulate in the area. Honno area could be flood even with a light rainfall because of the back swamp area surrounded by the sand bars. Urbanizations have expanded the urban area into lowlands such as flood plains since 1970, and the function to control the floods might be reduced in the area because flood plains were changed to the reclaimed land.

It is important to consider natural factors such as changes of the rainfall patterns and geomorphological factor, artificial factors such as urbanization, as well as the influence of the land subsidence all together to discuss the spatial distribution of the flood-prone area in its temporal changes.

Acknowledgements. We would like to express our sincere gratitude to the Keiyo Natural Gas Association for supporting this study. Residential maps provided by ZENRIN CO., LTD are used as the CSIS Joint Research (No. 524) using spatial data provided by Spatial Information Science, The University of Tokyo.

References

Chiba prefecture: Report of Chiba prefecture flooding, 1971 (in Japanese).

Chiba prefecture: Report of Chiba prefecture flooding, 2005 (in Japanese).

Chiba Prefecture: Report of current state of land subsidence in Chiba prefecture, 2014 (in Japanese).

Chiba Prefecture: Documents of Ichinomiya River watershed meeting, 2015 (in Japanese).

Chen, H. L., Ito, Y., Sawamukai, M., and Tokunaga, T.: Flood hazard assessment in the Kujukuri Plain of Chiba Prefecture, Japan, based on GIS and multicriteria decision analysis, Nat. Hazards, 78, 105–120, doi:10.1007/s11069-015-1699-5, 2015.

Fitzpatrick, F. A., Knox, J. C., and Whitman, H. E.: Effects of historical land-cover changes on flooding and sedimentation, North Fish Creek, Wisconsin, USGS Water-resources investigations report, 99–4083, 1–12, 1999.

Galloway, D., Jones, D. R., and Ingebritsen, S. E.: Land subsidence in the United States, U.S. Geological survey circular 1182, USA, 1999.

Moriwaki, H.: The landform evolution of the Kujukuri Coastal Plain, Central Japan, Quaternary Res., 18, 1–16, 1979 (in Japanese).

Ng, A. h.-M., Ge, L., and Li, X.: Assessments of land subsidence in the Gippsland Basin of Australia using ALOS PALSAR data, Remote Sens. Environ., 159, 86–101, 2015.

Umitsu, M.: Alluvial lowlands and floods, Geo-environment of alluvial and coastal lowlands, Kokon-Shoin Publishers, Tokyo, 47–48, 2012 (in Japanese).

Proc. IAHS, 372, 169–171, 2015
proc-iahs.net/372/169/2015/
doi:10.5194/piahs-372-169-2015

Open Access
PIAHS

Subsurface geology and aquifer system of the Nohbi Plain, central Japan

T. Makinouchi[1] and Tokai Three-Prefecture Investigation Committee on Land Subsidence[2]

[1]Department of Environmental Science and Technology, Faculty of Science and Technology, Meijo University, Nagoya, 468-8502, Japan

[2]Chubu Geographical Survey Institute, Geospatial Information Authority of Japan, Nagoya, 460-0001, Japan

Correspondence to: T. Makinouchi (takmak@meijo-u.ac.jp)

Published: 12 November 2015

Abstract. The Nohbi Plain is a typical coastal plain in Japan. The subsurface of this plain consists of alternating beds of marine clays in transgressive stages and fluvial gravels in regressive stages, and forms a groundwater basin. The aquifers are represented by the gravel beds in the regressive stages.

1 Introduction

The Nohbi Plain, a typical coastal plain in Japan, is situated in the Pacific coast side. This plain is approximately 50 km long in the N–S direction and about 35 km wide in the E-W direction. Three major rivers, the Kiso, Nagara, and Ibi Rivers flow into this plain (Fig. 1). The subsurface geology and aquifer systems of this plain is described in this paper. Details of land subsidence in this plain are reported in other papers.

2 Topography of the Nohbi Plain

Three topographic features, alluvial fan, flood plain, and delta, typical of a coastal plain, extend in this plain from upstream to downstream (Fig. 1). The eastern margin of this plain is fringed with terraces (Fig. 1).

3 Outline of the subsurface geology

Subsurface geology of the Nohbi Plain consists of the Nan'yo, Nohbi, First Gravel, Atsuta, Second Gravel, Ama, Third Gravel, Yatomi Formations, Plio-Pleistocene Tokai Group, and Miocene marine sediments, in descending order (Figs. 2, 3, and 4).

The Holecene Nan'yo Formation is divided into three members, uppermost mud, upper sand, and lower marine clay members. The lower marine clay member was deposited in the transgressive stage of the postglacial age. The Nohbi Formation, latest Pleistocene, is composed of sand and mud beds. The First Gravel Formation is composed of riverbed gravels during the regressive stage of the last glacial age. The Atsuta Formation, late Pleistocene, consists of upper sand and lower marine clay members. The lower marine clay member was deposited during the transgressive stage in the last interglacial age.

The Second Gravel Formation is composed of riverbed gravels during the regressive stage of the glacial age before the last interglacial. The Ama and Yatomi Formations consist of alternating beds of gravel, sand, and mud (Fig. 4).

The Plio-Pleistocene Tokai Group consists of alternating beds of gravel, sand, and mud with intercalations of many volcanic ash and lignite beds, and was deposited in Lake Tokai (paleolake), a freshwater sedimentary basin. The Lake Tokai sedimentary basin covered a wide region of the present Nohbi Plain and the surrounding areas (Fig. 5). Miocene marine sediments are mainly composed of sand and mud beds deposited under the paleo-Seto Inland Sea.

Among these strata, the marine clay members of the Nan'yo and Atsuta Formations have largely shrank, and about 80 % or more of the land subsidence is owing to the shrinkage of these members.

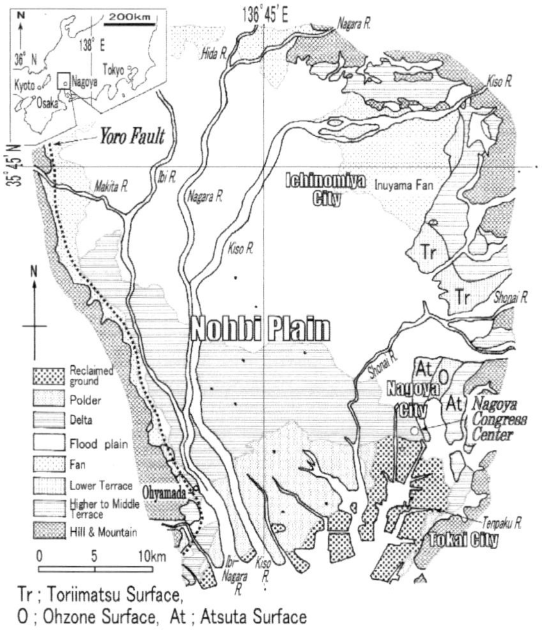

Figure 1. Topography of the Nohbi Plain (Makinouchi et al., 2013).

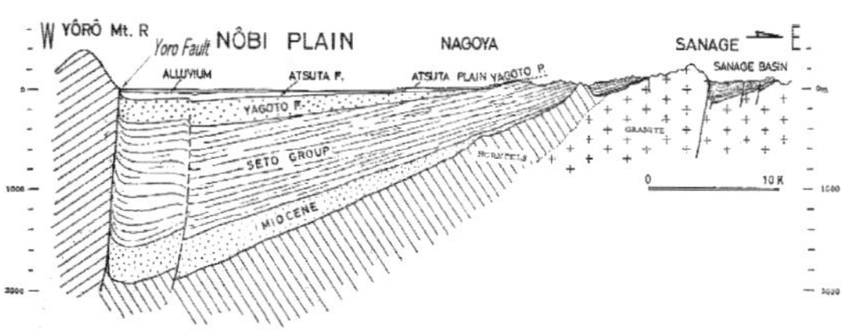

Figure 2. Quaternary block faulting and geologic profile of the Nohbi Plain (Kuwahara, 1968).

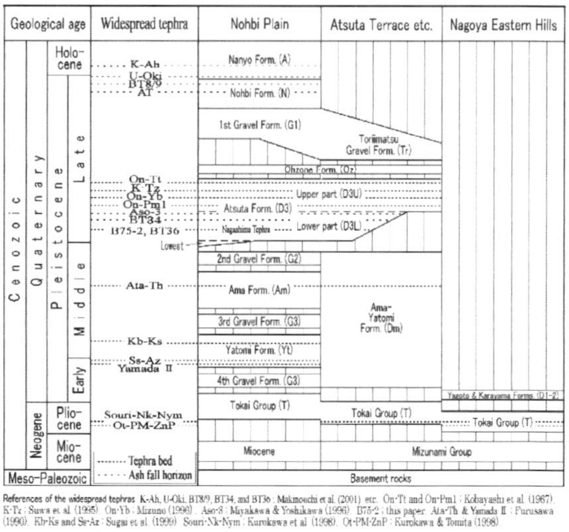

Figure 3. Subsurface stratigraphy of the Nohbi Plain and the surrounding areas (Makinouchi et al., 2013).

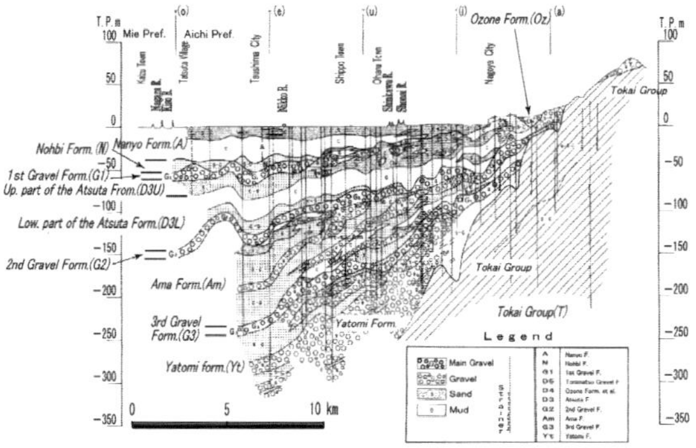

Figure 4. E-W geologic profile of the Nohbi Plain modified from Kuwahara (1985).

4 Depositional process in the Nohbi Plain

Westward tilting and glacial eustasy control the depositional process in this plain since middle Pleistocene time. The tilting forms a sedimentary basin deepened westward (Nohbi Tilting Basin, Figs. 2 and 4). The glacial eustasy influences the sedimentary facies, namely, marine facies in the transgressive stage and fluvial facies in the regressive stage.

Gradual subsidence due to tilting forms shallow depositional space, in which marine clay beds have deposited during the transgressive stages, and fluvial sand and gravel beds have deposited during the regressive stages. As the result of westward tilting and glacial eustasy, gravel, sand, and marine clay beds have alternately accumulated in this basin (Fig. 4), and many formations become thicker and deeper toward the west due to the tilting (Fig. 4).

5 Nohbi Groundwater Basin

The Nohbi Groundwater Basin consists of three overlapping groundwater sub-basins (Fig. 2). The lower sub-basin is made up of Miocene marine sediments. The Miocene system yields saline water. The middle sub-basin is composed of the Plio-Pleistocene Tokai (Seto) Group. The upper sub-basin, Nohbi Basin, was formed by a block faulting which began at middle Pleistocene time (Fig. 6) and have tilted westward (Nohbi Tilting Block).

6 Aquifer system of the Nohbi Basin

In the Nohbi Basin, upper sand member of the Nan'yo Formation yields superficial unconfined groundwater. The base

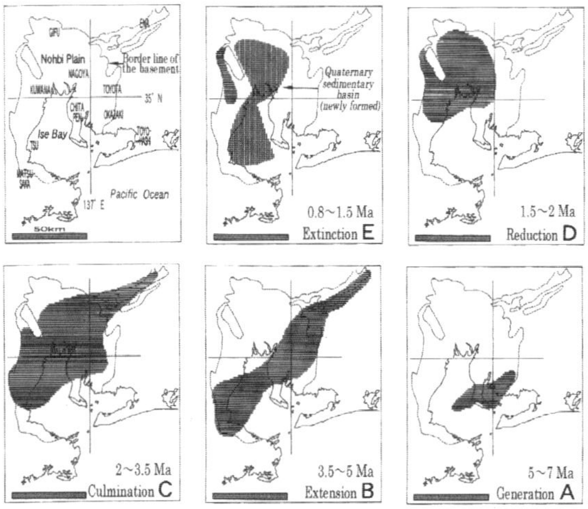

Figure 5. Change in territory (depositional area) of the Lake Tokai sedimentary basin (Makinouchi, 2001). 1 Ma: one million years ago.

depth of this member is an altitude of −30 to −10 m, and this member is 10 to 20 m thick.

Confined groundwater is abundantly supplied from fluvial gravel aquifers of the First (G1), Second (G2), and Third (G3) Gravel Formations, and also gravel beds of the Ama and Yatomi Formations. Especially, the G1 and G2 Formations are the most widely exploited ones in this basin.

The base depth of G1 Formation has an altitude of −90 m (west) to +20 m (east), and this formation is 10 to 30 m thick. The base depth of G2 Formation ranges from −200 m (west) to −10 m (east) in the altitude, and this formation is 10 to 20 m thick.

The Ama Formation and underlying sediments have many intercalations of gravel beds as aquifer units. But each aquifer unit is not easily distinguished from others, and the areal extent of a certain aquifer unit is hardly pursued. Therefore, these aquifer units are generally treated as an aquifer system in this basin.

In the middle and lower basins, the Tokai (Seto) Group and Miocene sediments yield abundant hot water.

7 Conclusions

Westward tilting and glacial eustasy control the depositional process in the Nohbi Basin. The tilting forms a sedimentary basin deepened westward. The glacial eustasy influences the sedimentary facies, namely, marine clays in the transgressive stages and fluvial gravels in the regressive stages. The aquifers are represented by the gravel beds in the regressive stages, and these aquifers yield abundant confined groundwater.

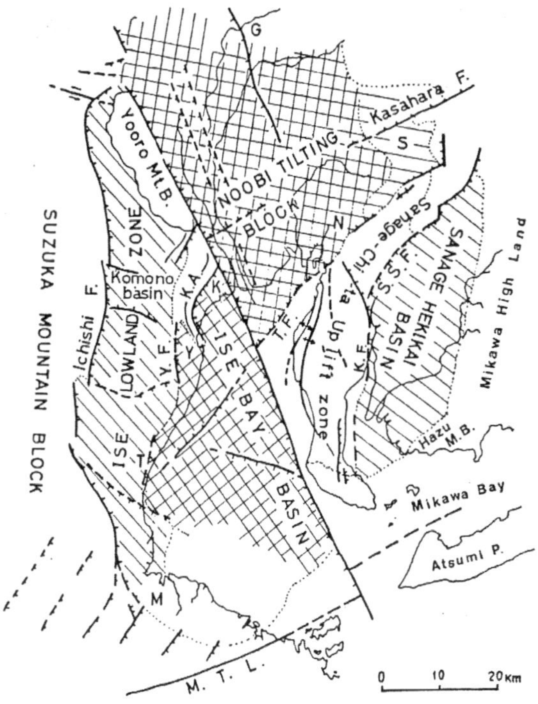

Figure 6. Quaternary block faulting in the Nohbi Plain and Ise Bay region (Kuwahara, 1969).

Acknowledgements. We express our sincere thanks to working staffs of the Tokai Three-Prefecture Investigation Committee on Land Subsidence.

References

Kuwahara, T.: The Noobi Basin and its Fault Block Movements, Quatern. Res., 7, 235–247, 1968.

Kuwahara, T.: The Ise Bay – Sedimentary Basin in the Environs of Ise Bay and Neotectonics in Southwest Japan, Symposium Kaigan Heiya (Symposium on Coastal Plain), Geological Society of Japan, Tokyo, Japan, 21–30, 1969.

Kuwahara, T.: Groundwater basin in the Nohbi Plain, Ground Subsidence and Ground Water in the Nohbi Plain, Nagoya Univ. Press, Nagoya, Japan, 35–76, 1985.

Makinouchi, T.: Stratigraphy of the Tokai Group and evolutional process of the Lake Tokai sedimentary basin, Science Report of the Toyohashi Museum of Natural History, No. 11, Toyohashi Museum of Natural History, Toyohashi, Japan, 33–39, 2001.

Makinouchi, T., Tsukamoto, M., Danhara, T., Yamashita, T., Uchizono, T., and WG of Research Committee for the Ground of Nohbi Plain: Subsurface geology in the eastern part of the Nohbi Plain, Aichi Prefecture, central Japan, J. Geol. Soc. Jpn., 119, 335–349, 2013.

Proc. IAHS, 372, 173–178, 2015
proc-iahs.net/372/173/2015/
doi:10.5194/piahs-372-173-2015

Potential of Holocene deltaic sequences for subsidence due to peat compaction

E. Stouthamer and S. van Asselen

Dept. of Physical Geography, Faculty of Geosciences, Utrecht University, Utrecht, the Netherlands

Correspondence to: E. Stouthamer (e.stouthamer@uu.nl)

Published: 12 November 2015

Abstract. Land subsidence is a major threat for the livability of deltas worldwide. Mitigation of the negative impacts of subsidence, like increasing flooding risk, requires an assessment of the potential of the deltas' subsurfaces for subsidence. This enables the prediction of current and future subsidence and optimization of sustainable management strategies. In this paper we present a method to determine the amount of compaction within different Holocene deltaic peat sequences based on a case study from the Rhine-Meuse delta, the Netherlands, showing the potential of these sequences for subsidence due to peat compaction.

1 Introduction

Land subsidence is a major threat to hundreds of millions of people living in deltas worldwide. It leads to increasing flooding risk, damage to buildings and infrastructure, intrusion of salt water, and hence, has a negative impact on the livability in deltas. Natural subsidence rates, caused by tectonics, isostasy, sediment compaction and oxidation of organic material are generally in the order of $\mathrm{mm\,yr^{-1}}$ (van Asselen et al., 2011). Human-induced subsidence rates, due to extraction of groundwater, oil and gas, drainage for land reclamation, and loading by for example roads and buildings are commonly an order of magnitude higher ($\mathrm{cm\,yr^{-1}}$; e.g. van Asselen, 2011; Erban et al., 2014). Consequently, in many populated deltas the contribution of subsidence to relative sea-level rise is much larger than eustatic (absolute) global sea-level rise, which is in the order of $\mathrm{mm\,yr^{-1}}$ (Church et al., 2013).

The potential of delta sequences for both natural and human-induced subsidence due to compaction and oxidation of peat, and hence subsidence magnitude and rate, is highly variable in time and space, mainly depending on the composition of the subsurface. Subsurface composition and characteristics are determined by the palaeogeographical development (i.e. the spatial and temporal distribution of depositional environments and hence deposits), post-depositional processes of erosion (i.e. preservation) and loading, and physical, geochemical and biological processes, including compaction and soil formation. Based on the 3-D distribution of specific sequences within a delta and their physical properties, the potential for subsidence due to peat compaction of these sequences at the spatial scale of an entire delta can be determined.

To mitigate the negative impacts of human-induced subsidence it is essential to know the potential of the subsurface for compaction and oxidation. This enables the prediction of current and future subsidence under different delta management strategies and to develop and optimize sustainable management strategies. In this paper we present a method to determine the amount of peat compaction for typical Holocene distal deltaic organo-clastic sequences based on a case study from the Rhine-Meuse delta, the Netherlands, showing the potential of these sequences for subsidence due to peat compaction.

2 Subsidence potential due to peat compaction

Holocene delta sequences are usually very heterogeneous, and composed of alternating layers consisting of mixtures of sand, silt, clay and peat. The main factors influencing the potential for subsidence due to peat compaction are (1) the thickness of organic, mainly peat and clay, layers in the subsurface, (2) the organic matter content of these layers, and (3) the effective stress, which is a function of the weight of the overburden and pore water pressures (van Asselen et al.,

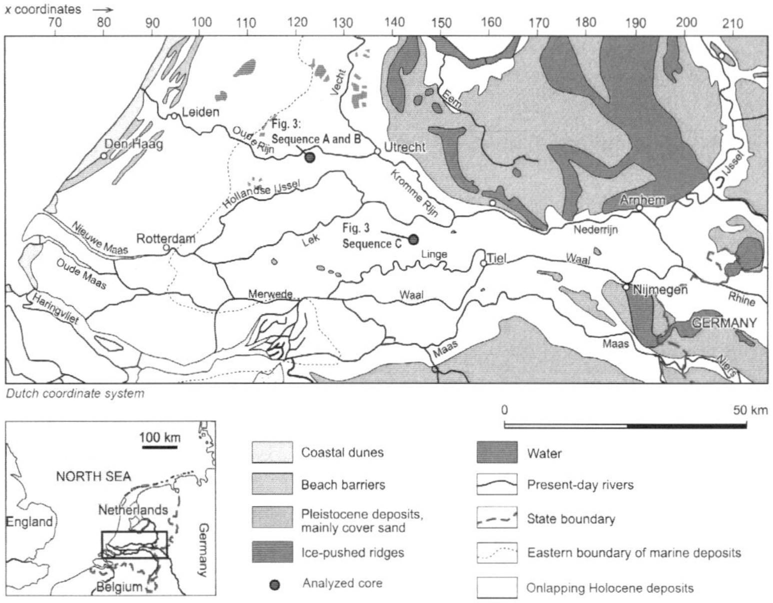

Figure 1. The Holocene Rhine-Meuse delta, the Netherlands; rivers run E–W. x coordinates according to the Dutch map coordinate system (in kilometers, after Stouthamer and Berendsen, 2001 and Stouthamer et al., 2011).

2010). In addition, the Holocene sedimentation and groundwater table history is extremely important, because this determines if organic layers have experienced past compaction and/or oxidation, and hence their present geotechnical properties. Lowering of the phreatic groundwater level, for e.g. agricultural purposes, leads to a drop in hydrostatic pressure, causing diminishing pore water pressures and consequently an increase in effective stress (= total weight − pore water pressure; Terzaghi, 1943) and compaction of the peat sequence. At the depth interval between the former, higher, and new, lower, groundwater level, oxidation of organic material occurs. This leads to a decrease in volume and collapse of the original soil structure that existed before groundwater level was lowered. Both compaction and oxidation contribute to total subsidence. Sequences that are composed of thick high-organic peat layers, especially in the upper meters of the sequence have the highest potential for compaction and oxidation and hence subsidence. Compaction rates will be much lower if a peat layer has already experienced compaction by loading. Rates of compaction due to loading are highest within decades to a few centuries after loading (van Asselen et al., 2011).

3 Study area: Rhine-Meuse delta, the Netherlands

The Rhine-Meuse delta is located in the Netherlands, in the tectonically subsiding Southern North Sea Basin. The estimated Holocene-averaged basin subsidence rate is 0.1–0.3 mm yr^{-1} (Cohen, 2005). The Holocene delta was jointly formed by the rivers Rhine and Meuse. The delta apex is located near the Dutch-German border (East). The Rhine and Meuse debouch into the North Sea (West) (Fig. 1). The distance between the modern delta apex and the Rhine-Meuse river mouth is ca. 150 km. The Holocene sequence is approximately 2 m thick near the delta apex and just over 20 m at the present river mouth and coastal barrier. The upstream part of the delta is confined by higher topography of sandy Pleistocene deposits (ice-pushed pre-Saalian glacial sandy Rhine and Meuse deposits) and relatively narrow; 15–25 km wide. Downstream the delta widens to 60 km and grades laterally into tidal-estuary and lagoonal coastal plain environments. The southern and central-northern part of the delta are bordered by higher topography formed by Late Glacial cover sand deposits. Between 1100 and 1300 AD all Rhine and Meuse distributaries in the delta were embanked, stopping flood basins from receiving sediments (van de Ven, 2003). Flooding and channel-migration activity has since remained restricted to the embanked flood plains.

The Holocene subsurface of the delta, fed with sediment by the mixed-load rivers Rhine and Meuse, is notoriously heterogeneous. Channel-belt sand bodies dissect areas of overbank deposits (dominantly clayey, but holding sand and silt too). Multiple generations of channel belts avulsed adding further complexity (Stouthamer and Berendsen, 2000). Downstream in the delta, the overbank clastics alternate with peat layers (Fig. 2). The distal Holocene coastal,

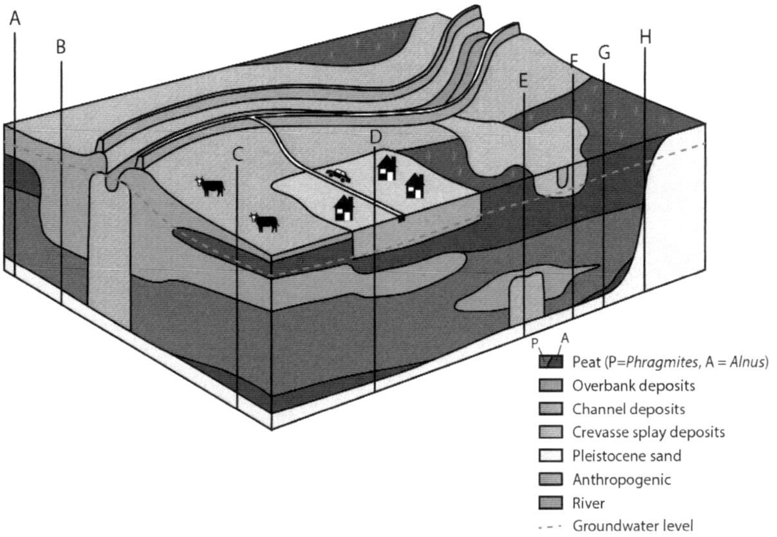

Figure 2. Schematic representation of the lithological subsurface composition of the distal (western) part of the Holocene Rhine-Meuse delta. The vertical lines (A–H) indicate different sequences with a varying potential for subsidence due to compaction and oxidation depending on the subsurface composition and characteristics.

tidal, estuarine and fluvial deposits are onlapping the Pleistocene cover sand deposits in the North and the South and braided river deposits in the central part of the delta (Fig. 1).

4 Methods: determining the potential for compaction

The Holocene sequence composition of the Rhine-Meuse delta is well-known from previous studies (e.g. Berendsen and Stouthamer, 2000; Stouthamer, 2001; Cohen, 2003; Gouw and Erkens, 2007; Gouw, 2008; Bos, 2010; Stouthamer et al., 2011). These studies allow to distinguish key clastic-organic sequences (Fig. 2), which have different potentials for subsidence due to peat compaction.

The degree of peat compaction has been determined at different locations in the organo-clastic central part of the Rhine-Meuse delta (van Asselen, 2011; Van Asselen, unpublished work; here we present three sequences A–C in Figs. 1 and 3), by comparing the dry bulk density of compacted and uncompacted peat of similar organic-matter content (Bird et al., 2004; van Asselen et al., 2009; van Asselen, 2011). The dry bulk density and organic-matter content of compacted peat in a delta sequence are measured from sediment cores at 5-cm intervals using a 1 cm × 1 cm × 5 cm sampler. The cores are extracted with a 100 cm × 6 cm wide gouge auger, and subsequently cut in half lengthways using a thin stretched wire, to sample of the inner least disturbed part. The dry bulk density of each 5 cm^3 peat sample has been determined by drying it at 105 °C and weighing it on an electronic scale with an accuracy of 0.001 g (dry bulk density = weight / 5 cm^3). The organic-matter content has been determined by loss-on-ignition (LOI = ((dried weight − ashed weight) / dried weight) × 100 %; cf. Heiri et al., 2001). The dried peat samples were heated at 550 °C for 4 h and subsequently weighted to determine the ashed weight.

Uncompacted peat samples were obtained from the Biebrza National Park (Poland) and the Rhine-Meuse delta (van Asselen, 2011). The Biebrza National Park is a wetland that has experienced minimal human disturbance, where similar peat types occur as in the Holocene Rhine-Meuse delta. The fresh, uncompacted peat has been sampled using a special device that was designed for this purpose (van Asselen and Roosendaal, 2009). The dry bulk density and LOI of uncompacted peat samples were determined following the same procedure as for compacted 5 cm^3 samples. Based on these data an equation for calculating the dry bulk density of uncompacted peat of any LOI has been constructed:

$$\rho_{\text{dry, uncompacted}} = a - ce^{-(b/\text{LOI})}, \qquad (1)$$

in which a, b and c are fitted parameters (van Asselen, 2011). The calculated dry bulk density was subsequently used to calculate the decompacted thickness of each 5 cm^3 compacted peat sample:

$$h_{\text{decomp}} = \frac{\rho_{\text{dry, comp}}}{\rho_{\text{dry, uncomp}}} \times 5. \qquad (2)$$

The percentage of compaction is expressed as the ratio between the volume reduction

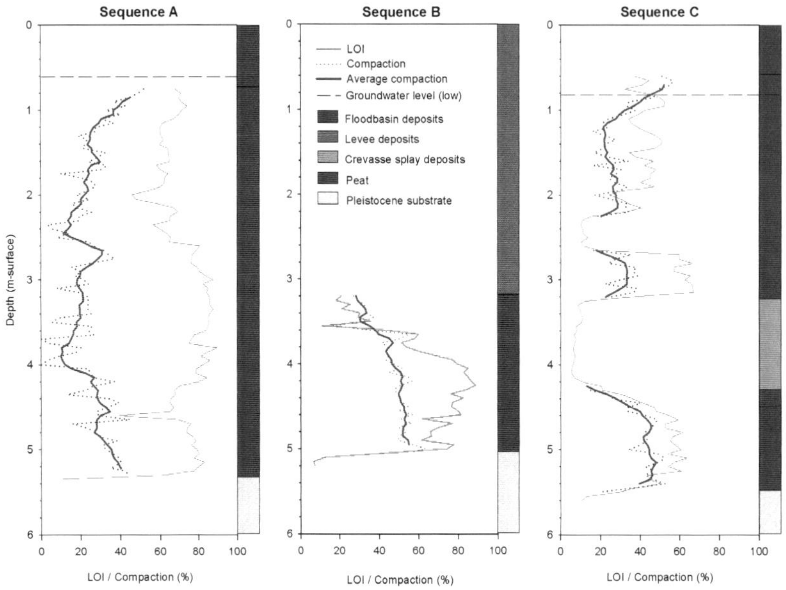

Figure 3. Lithological composition, LOI and amount of compaction of cores OR-I (sequence B), OR-II (sequence A), and CB-I (sequence C) (after van Asselen, 2011). These sequences are representative for the sequences A, B and C in Fig. 2.

$(v_{\text{red}} = (1 \times 1 \times h_{\text{decomp}}) - (1 \times 1 \times 5))$ and the calculated decompacted volume of a $5\,\text{cm}^3$ peat sample $(v_{\text{decomp}} = 1 \times 1 \times h_{\text{decomp}})$:

$$\text{compaction} = \frac{v_{\text{red}}}{v_{\text{decomp}}} \times 100\,\%. \tag{3}$$

5 Results: compaction of sequences A–C

Sequence B is located close to the river (Fig. 2) and is characterized by approximately 3 m of natural levee deposits on top of a 2-m-thick peat layer between 3–5 m below the surface that is situated on the incompressible Late-Pleistocene sand substrate (Fig. 3). Sequence A and C are located at further distance from the river (Fig. 2). Sequence A is characterized by a 60 cm thick layer of flood basin deposits on a 4.5 m thick peat layer on the Late-Pleistocene sand substrate (Fig. 3). Sequence C comprises alternating layers of peat and fluvial deposits; 50 cm of flood basin deposits at the top, 30 cm flood basin deposits at 2.20–2.40 m depth and 1 m of crevasse-splay deposits at 3.20–4.20 m depth.

Fluctuating LOI-values within the sequences can be explained by the varying clastic content; deposits with a high clastic content show low LOI-values (e.g. flood basin deposits, clayey peat intervals), deposits with a low clastic content show high LOI-values (e.g. peat). For sequences B and C, in which peat layers are considerably compacted, the amount of peat compaction is positively related to LOI (van Asselen, 2011). For example, this is observed in the compacted peat layer in sequence B, where lower compaction values are observed in the low-organic top of the peat layer.

Relatively high peat compaction values occur at the top of the peat layer in sequence A at 120–80 cm depth and at the top of the upper peat layer in sequence C at 120–70 cm depth (Fig. 3). This can be explained by artificial lowering of the phreatic groundwater level in this agricultural area, which causes compaction and oxidation that both lead to an increase in the dry bulk density in the zone above and just below groundwater level (see Sect. 2). In sequence B the peat layer is situated too deep for groundwater level lowering to have any effect. Furthermore, high values occur beneath the natural levee and crevasse-splay deposits in sequence B and C respectively, due to the overburden weight. Also, at the base of sequence A the amount of compaction increases, which may be attributed to palaeo-groundwater level fluctuations, (auto)compaction and/or fluctuations in the hydrostatic pressure (Fig. 3).

6 Discussion: assessing the potential of Holocene deltaic sequences for subsidence due to compaction

The calculated compaction values for the sequences A–C indicate the potential for future subsidence due to compaction for comparable delta settings. For example, strongly compacted peat beneath thick natural levee deposits (Fig. 2, sequence B) is less vulnerable for future subsidence due to compaction than an uncompacted thick peat sequence with-

Table 1. Relative potential of key sequences for peat compaction, based on lithological composition. For the lithological built-up of the sequences see Fig. 2.

Sequence	Compaction potential
G	High
A	
F	
E	
C	
D	
H	
B	Low

out clastic overburden (Fig. 2, sequence G). Based on the lithological sequence composition and its properties, the relative potential for future subsidence due to peat compaction can be estimated (Table 1). In addition, the depth of the groundwater level can affect peat compaction: a drop in the pore water pressure following groundwater level lowering may induce additional compaction, and oxidation, of the peat sequence (van Asselen et al. 2009, 2010). This should be taken into account in assessing the compaction potential at locations where the groundwater table has been significantly lowered.

In summary, assessing the relative potentials for peat compaction of Holocene deltaic sequences requires determining the (1) lithological composition of the sequence, (2) the geotechnical properties of the different lithologies, which is predominantly a function of organic-matter and clastic content, and post-depositional processes of loading, and physical, geochemical and biological processes, determining the present degree of compaction and geotechnical properties, and (3) the effective stress, which is a function of the weight of the overburden and pore water pressure that is influenced by respectively the built-up of the sequence and its properties and fluctuations in groundwater level or the absence thereof. Based on this information, the future potential for additional subsidence due to peat compaction can be estimated.

7 Conclusion

The potential for subsidence due to peat compaction of Holocene deltas is mainly determined by the 3-D distribution of different lithologies, and associated geotechnical properties, in the subsurface. In this study the amount of compaction within 3 sequences that are representative for distal deltaic settings is determined to assess the potential of comparable sequences in other deltas. In all sequences A–C the peat layers are compacted. Relatively high compaction values occur in peat layers within approximately 120 cm below surface, which are subjected to compaction and oxidation following groundwater level lowering. Furthermore, high amounts of compaction occur in peat layers beneath natural levee and crevasse-splay deposits, and at the base of the Holocene sequences, due to the weight of the overburden. In compacted peat layers, high-organic peat generally compacts more than low-organic peat.

References

Berendsen, H. J. A. and Stouthamer, E.: Late Weichselian and Holocene palaeogeography of the Rhine–Meuse delta, the Netherlands, Palaeogeogr. Palaeocl., 161, 311–335, 2000.

Bos, I. J.: Architecture and facies distribution of organic-clastic lake fills in the fluviodeltaic rhine-meuse system, the Netherlands, J. Sediment. Res., 80, 339–356, 2010.

Bird, M. I., Fifield, L. K., Chua, S., and Goh, B.: Calculating sediment compaction for radiocarbon dating of intertidal sediments, Radiocarbon 46, 421–435, 2004.

Church, J. A., Clark, P. U., Cazenave, A., Gregory, J. M., Jevrejeva, S., Levermann, A., Merrifield, M. A., Milne, G. A., Nerem, R. S., Nunn, P. D., Payne, A. J., Pfeffer, W. T., Stammer, D., and Unnikrishnan, A. S.: Sea Level Change, in: Climate Change 2013: The Physical Science Basis. Contribution of Working Group I to the Fifth Assessment Report of the Intergovernmental Panel on Climate Change, edited by: Stocker, T. F., Qin, D., Plattner, G.-K., Tignor, M., Allen, S. K., Boschung, J., Nauels, A., Xia, Y., Bex, V., and Midgley, P. M., Cambridge University Press, Cambridge, UK, New York, NY, USA, 2013.

Cohen, K. M.: Differential subsidence within a coastal prism: late-Glacial – Holocene tectonicsin The Rhine Meuse delta, the Netherlands, PhD thesis Utrecht University, Utrecht, the Netherlands, Geographical Studies, 316, 2003.

Cohen, K. M.: 3D Geostatistical interpolation and geological interpretation of paleo-groundwater rise in the Holocene coastal prism in the Netherlands, in: River Deltas – Concepts, models, and examples, edited by: Giosan, L. and Battacharya, J. P., SEPM Special Publication, 83, 341–364, 2005.

Erban, L. E., Gorelick, S. M., and Zebker, H.: Groundwater extraction, land subsidence, and sea-level rise in the Mekong Delta, Vietnam, Environ. Res. Lett., 9, 1–6, 2014.

Gouw, M. J. P.: Alluvial architecture of the Holocene Rhine-Meuse delta, the Netherlands, Sedimentology, 55, 1487–1516, 2008.

Gouw, M. J. P. and Erkens, G.: Architecture of the Holocene Rhine-Meuse delta (the Netherlands) – A result of changing external controls, Neth. J. Geosci., 86, 23–54, 2007.

Heiri, O., Lotter, A. F., and Lemcke, G.: Loss on ignition as a method for estimating organic and carbonate content in sediments: Reproducibility and comparability of results, J. Paleolimnol., 25, 101–110, 2001.

Stouthamer, E.: Sedimentary products of avulsions in the Holocene Rhine-Meuse Delta, the Netherlands, Sediment. Geol., 145, 73–92, 2001.

Stouthamer, E. and Berendsen, H. J. A.: Factors controlling the Holocene avulsion history of the Holocene Rhine-Meuse delta (the Netherlands), J. Sediment. Res. A, 70, 1051–1064, 2000.

Stouthamer, E., Cohen, K. M., and Gouw, M. J. P.: Avulsion and its implications for fluvial-deltaic architecture: insights from the Holocene Rhine-Meuse delta. in: From River to Rock Record: The preservation of fluvial sediments and their subsequent interpretation, edited by: Davidson, S. K., Leleu, S., and North, C. P., Society for Sedimentary Geology, 11, 215–232, 2011.

Terzaghi, K.: Theoretical Soil Mechanics, John Wiley and Sons, New York, USA, 510 pp., 1943.

van Asselen, S.: The contribution of peat compaction to total basin subsidence: Implications for the provision of accommodation space in organic-rich deltas, Basin Res., 23, 239–255, 2011.

van Asselen, S. and Roosendaal, C.: A new method for determining the bulk density of uncompacted peat from field settings, J. Sediment. Res., 79, 918–922, 2009.

van Asselen, S., Stouthamer, E., and Van Asch, Th. W. J.: Effects of peat compaction on delta evolution: A review on processes, responses, measuring and modelling, Earth-Sci. Rev., 92, 35–51, 2009.

van Asselen, S., Stouthamer, E., and Smith, N. D.: Factors controlling peat compaction in alluvial floodplains – A case study in the cold-temperate Cumberland Marshes, Canada, J. Sediment. Res., 80, 155–166, 2010.

van Asselen, S., Karssenberg, D., and Stouthamer, E.: Holocene peat compaction in fluvial lowlands: Implications to subsidence within deltas, Geophys. Res. Lett., 38, L24401, doi:10.1029/2011GL049835, 2011.

van de Ven, G. P.: Leefbaar laagland. Geschiedenis van de waterbeheersing en landaanwinning in Nederland, Utrecht: Uitgeverij Matrijs, 456 pp., 2003.

Proc. IAHS, 372, 179–182, 2015
proc-iahs.net/372/179/2015/
doi:10.5194/piahs-372-179-2015

Open Access PIAHS

Future Deltas Utrecht University research focus area: towards sustainable management of sinking deltas

E. Stouthamer and S. van Asselen

Dept. of Physical Geography, Faculty of Geosciences, Utrecht University, Utrecht, the Netherlands

Correspondence to: E. Stouthamer (e.stouthamer@uu.nl)

Published: 12 November 2015

Abstract. Deltas are increasingly under pressure from human impact and climate change. To deal with these pressures that threat future delta functioning, we need to understand interactions between physical, biological, chemical and social processes in deltas. This requires an integrated approach, in which knowledge on natural system functioning is combined with knowledge on spatial planning, land and water governance and legislative frameworks. In the research focus area *Future Deltas* of Utrecht University an interdisciplinary team from different research groups therefore works together. This allows developing integrated sustainable and resilient delta management strategies, which is urgently needed to prevent loss of vital delta services.

1 Introduction

Deltas are valuable but vulnerable areas (e.g. Syvitski, 2008; Vörösmarty et al., 2009; Giosan et al., 2014). They provide essential services to humans, such as fertile land for food production, biodiversity, drinking and irrigation water and a river network for transportation. Because of all these benefits, over 500 million people live in deltas worldwide. These people are however increasingly under pressure of human and climate change, introducing impacts of different nature: sea-level rise, land subsidence, ecosystem degradation, increasing flood risk, pollution, loss of biodiversity, conflicting land claims and declining livability. To find solutions for sustainable management of deltas under pressure, the research focus area *Future Deltas* has been established (2014–2017). Every four years, Utrecht University defines new research focus areas to stimulate interdisciplinary research, that should act as nuclei for broad (inter)national collaborations.

Future Deltas aims to understand and predict how deltas under pressure of massive urbanization, land-use and climate change, sea level rise and subsidence will change in the future and how societies can successfully deal with this. To be able to predict and anticipate to the impacts of these changes in future deltas, it is crucial to have insight and profound understanding of the functioning of both natural and societal aspects of delta systems. This enables us to answer questions like: how does river embankment influence floodplain development and coastal sedimentation? What is the impact of deep and shallow groundwater extraction due to expanding cities on subsidence and saltwater intrusion? Who is responsible for causes of negative impacts, how are mitigation strategies organized, how and at what level are decisions on delta management taken?

To address the challenges deltas world-wide face and to facilitate integrated sustainable and resilient management, an integrated approach combining mechanistic knowledge of natural system functioning, comprising interacting physical, chemical, and biological processes, with specialized knowledge on spatial planning, land and water governance and legislative frameworks is required. Such an integrated approach has been lacking to date and forms an interdisciplinary challenge that will be engaged within Future Deltas (Fig. 1).

2 Research focus

In Future Deltas we initially focus on problems in deltas related to subsidence and loss of ecosystem services. Both topics include physical, chemical, biological and governance aspects. Our ultimate aim is to generate and integrate knowledge on these two topics and to apply this in the development of coherent conceptual and numerical delta models to predict delta dynamics and inform management and governance. The integrated delta models will provide tools to de-

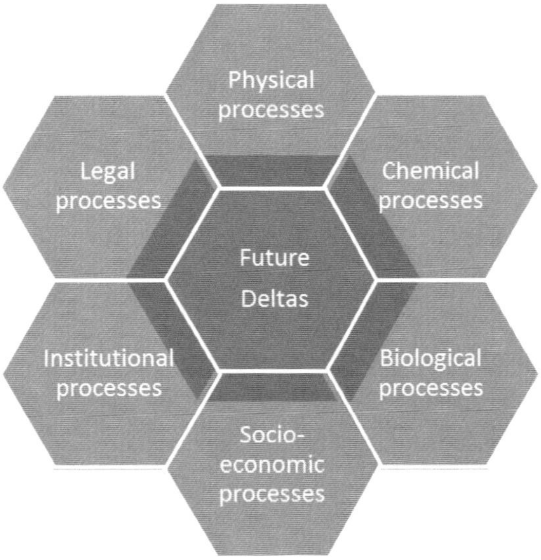

Figure 1. Mono-disciplinary research fields will be integrated in Future Deltas, generating understanding of the functioning of the delta system as a whole, including physical, chemical, biological, socio-economic, institutional and legal processes and their interactions.

sign and evaluate the impact of new, innovative strategies for delta management such as related to subsidence, wetland restoration and the recovery of ecosystem services. On the longer term, we envisage a broadening of the focus area by the inclusion of additional topics that address new emerging priorities.

2.1 Subsidence

Although land subsidence poses a far more immediate and severe problem than for instance climate-change-induced sea-level rise (i.e. the rate of subsidence is in many populated deltas worldwide an order of magnitude higher ($cm\,yr^{-1}$) than the rate of eustatic (global) sea-level rise ($mm\,yr^{-1}$; Church et al., 2013), the processes and consequences of land subsidence have yet to receive the full and integrated scientific and public awareness. Delta subsidence causes a range of problems, including flooding, salt water intrusion, fresh water scarcity, damage to buildings and infrastructure, loss of cultivated land, wetlands and biodiversity, degradation of fishing areas, and rapid shoreline retreat. Combined with a future sea-level rise, submerging of deltaic areas is projected to increase by 50 % by the end of the 21st century (Syvistki et al., 2009).

2.2 Ecosystem services

Mankind relies heavily on deltas and associated water bodies and wetlands for the delivery of essential ecosystem services, including primary production and agriculture, environmental filtering and water quality improvement, flood abatement, coastal protection and safety, nature and biodiversity conservation, and critical interactions with the atmosphere and climate through greenhouse gas reduction and carbon sequestration. It is imperative that provisioning of delta services be included in future planning and governance schemes that integrate ecosystem services into the discussion of land use rights and the governance of limited resources. The knowledge generated within this topic will facilitate adaptation decisions for nature management, biodiversity restoration, water resources management, sustainable agriculture, urban planning, and area developments yielding safer and more resilient delta systems for the future.

3 Research themes

Our research is structured into three main themes that are described in the next sections (see also Fig. 2).

3.1 Understanding drivers

The scientific challenge of this theme is to understand the complex interactions between physical, biological and chemical processes, and human interventions that affect subsidence and loss of ecosystem services in deltas. There are numerous natural and human-induced drivers of land subsidence. Natural drivers are tectonics, isostasy, and sediment compaction. Human-induced drivers include extraction of groundwater and hydrocarbons, peat oxidation due to artificial groundwater level lowering, loading by buildings and infrastructure, and sediment starvation due to upstream sediment trapping. Human forcing of subsidence is influenced by socio-economic conditions and the governance system.

Subsiding land in deltas potentially affects ecosystem functioning by degradation or loss of agricultural soils, reduced freshwater supply by intrusion of salt water, deterioration of water quality by eutrophication, loss of biodiversity and loss of coastal protection zones. Also, groundwater level lowering in peat soils causes subsidence due to peat oxidation, which leads to considerable greenhouse gas emissions thereby generating a positive feedback on climate change, as the expectations are that climate change will further accelerate peat oxidation and subsidence under prolonged periods of drought and higher temperatures.

In Future Deltas it is our aim to unravel and quantify the different contributors to subsidence in different deltas, and to understand how they relate to loss of ecosystem services under a changing climate.

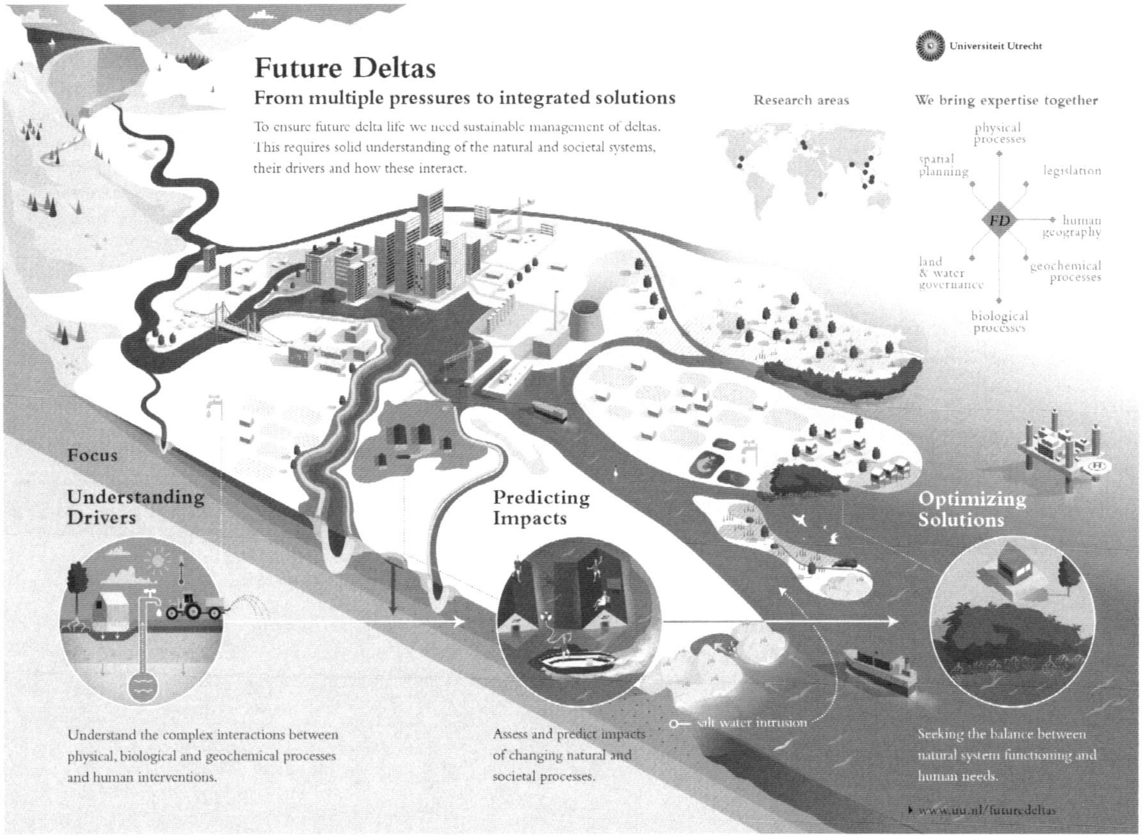

Figure 2. Future Deltas focusses on understanding drivers of change, predicting impacts and optimizing solutions in deltas.

3.2 Predicting impacts

The scientific challenge of this theme is to assess and predict impacts of subsidence and loss of ecosystem services in deltas. These impacts not only affect livability in deltas but also potentially reduce a delta's capability to adapt to and recover from external and internal forcing of change. If thresholds are crossed a delta system may collapse, thereby losing its valuable services.

Subsidence has a direct impact on dynamics of water, sediment, nutrients and vegetation, and hence on the boundary conditions of ecosystem functioning. We aim to assess and predict these changing dynamics and their impacts on water security including flood risk, food security, land use and land loss, and on socio-economic and livelihood conditions.

3.3 Optimizing management solutions

The scientific challenge of this theme is to analyse, evaluate and optimize strategies to deal with impacts of subsidence and loss of ecosystem services. Sustainable delta management strategies are needed to ensure that future delta societies benefit from vital services provided by deltas while not exhausting or destroying them. Delta managers aiming to restore or mitigate subsidence not only need fundamental knowledge concerning processes that drive changes in deltas and associated impacts, but also must take into account the socio-economic and governance system and legislative frameworks operating in the area.

We evaluate strategies to deal with impacts of subsidence and loss of ecosystem services in deltas as the basis, to improve and optimize delta management. We focus especially on regulatory and policy approaches that support or hamper change in different deltas, awareness building, financing the prevention or restoration of subsidence. We also emphasize issues related to burden sharing and diverse and fragmented public and private responsibilities such as water management, land use, nature conservation, drinking water supply, urban and rural development and agriculture to prevent and restore damages.

Dealing with uncertain risks and shifts in land use towards more sustainable and resilient deltas call for public and private involvement and responsibilities. To address these responsibilities in an appropriate and legitimate way, based on shared and mutual responsibilities on different levels and scales and taking equitable burden and benefit sharing as leading principles, it is essential to incorporate scientific knowledge concerning natural processes into policy development, environmental legislation and decision-making.

4 Further information

Future Deltas is a close cooperation between the faculties Geosciences, Science and Law, Economics and Governance of Utrecht University. Organisations that are interested in our integrated approach are cordially invited to join the Future Deltas research programme. Further information on Future Deltas can be obtained from Esther Stouthamer (e.stouthamer@uu.nl) and/or from our website www.uu.nl/futuredeltas.

Acknowledgements. We thank our Future Deltas colleagues that all have contributed to this paper: H. Middelkoop, P. Hoekstra, M. van der Vegt, G. A. Kowalchuk, M. M. Hefting, M. B. Soons, P. P. J. Driessen, F. S. J. van Laerhoven, H. F. M. W. van Rijswick, F. A. G. Groothuijse, E. B. Zoomers, and R. Fletcher.

References

Church, J. A., Clark, P. U., Cazenave, A., Gregory, J. M., Jevrejeva, S., Levermann, A., Merrifield, M. A., Milne, G. A., Nerem, R. S., Nunn, P. D., Payne, A. J., Pfeffer, W. T., Stammer, D., and Unnikrishnan, A. S.: Sea Level Change, in: Climate Change 2013: The Physical Science Basis, Contribution of Working Group I to the Fifth Assessment Report of the Intergovernmental Panel on Climate Change, edited by: Stocker, T. F., Qin, D., Plattner, G.-K., Tignor, M., Allen, S. K., Boschung,J., Nauels, A., Xia, Y., Bex, V., and Midgley, P. M., Cambridge University Press, Cambridge, UK and New York, NY, USA, 2013.

Giosan, L., Syvitski, J. P. M., Constantinescu, S., and Day, J.: Climate change: Protect the world's deltas, Nature, 516, 31–33, 2014.

Syvitski, J. P. M.: Deltas at risk, Sustain. Sci., 3, 23–32, 2008.

Syvitski, J. P. M., Kettner, A. J., Overeem, I., Hutton, E. W. H., Hannon, M. T., Brakenridge, G. R., Day, J., Vörösmarty, C., Saito, Y., Giosan, L., and Nicholls, R. J.: Sinking deltas due to human activities, Nat. Geosci., 2, 681–686, 2009.

Vörösmarty, C. J., Syvitsky, J. P. M., Day, J., de Sherbinin, A., Giosan, L., and Paola, C.: Battling to save the world's river deltas, B. Atom. Sci., 65, 31–43, 2009.

Proc. IAHS, 372, 183–187, 2015
proc-iahs.net/372/183/2015/
doi:10.5194/piahs-372-183-2015

Open Access PIAHS

Compaction parameter estimation using surface movement data in Southern Flevoland

P. A. Fokker[1], J. Gunnink[1], G. de Lange[2], O. Leeuwenburgh[1], and E. F. van der Veer[1]

[1]TNO, Utrecht, the Netherlands
[2]Deltares, Utrecht, the Netherlands

Correspondence to: P. A. Fokker (peter.fokker@tno.nl)

Published: 12 November 2015

Abstract. The Southern part of the Flevopolder has shown considerable subsidence since its reclamation in 1967. We have set up an integrated method to use subsidence data, water level data and forward models for compaction, oxidation and the resulting subsidence to estimate the driving parameters. Our procedure, an Ensemble Smoother with Multiple Data Assimilation, is very fast and gives insight into the variability of the estimated parameters and the correlations between them. We used two forward models: the Koppejan model and the Bjerrum model. In first instance, the Bjerrum model seems to perform better than the Koppejan model. This must, however, be corroborated with more elaborate parameter estimation exercises in which in particular the water level development is taken into account.

1 Introduction

The Southern part of the Flevopolder (the Netherlands) was reclaimed in 1967–1968. After lowering the groundwater level, the sediments compacted, resulting in surface subsidence. The compaction is a combination of elastic and viscoplastic compaction of clay and peat, and of peat oxidation. A good prediction of future compaction and an extrapolation to other areas naturally requires knowledge of the subsurface lithology, the compaction parameters and the phreatic water level development. The present paper describes an exercise of parameter estimation using a limited number of borehole data in the Southern Flevopolder area. An Ensemble Smoother with Multiple Data Assimilation was employed to this end.

2 Available data

The Flevopolder is a large polder in the Netherlands, which was reclaimed from the IJsselmeer in 1967 (Fig. 1). The surface has an elevation of about 3 m below sea level. The composition of the subsurface had been mapped on a number of places prior to the reclamation. It is a variable layered structure which we categorized into four generic types: clay, humic clay, peat, and sand. For the present study, 10 measurement locations were selected, at which the surface level had been monitored yearly between 1967 and 1993 by geodetic levelling.

Data about the phreatic groundwater level are scarce. One of the few measurements covering the complete period is represented in Fig. 2. It shows the fast lowering of the water level at reclamation of the polder, followed by a gradual decrease during the next 7 years. This can be explained by the gradual adjustment of the phreatic level at the measurement location to the drainage level in the ditches.

3 Forward model

For the subsidence prediction we employed a combination of compaction in the clay, the humic clay and the peat, and oxidation in the humic clay and peat layers above the phreatic level. We used models for primary and secondary settlement for stresses larger than the pre-consolidation stress: the Koppejan model and the Bjerrum model, both with an instantaneous lowering of the groundwater level to a fixed value below the original surface level (Koppejan, 1948; Bjerrum, 1967). In combination with peat oxidation, the two models then predict the following compaction strain and associated

Table 1. Assimilation results.

	Koppejan Case 1 Prior	Koppejan Case 1 Estimate	Koppejan Case 2 Prior	Koppejan Case 2 Estimate	Bjerrum Prior	Bjerrum Estimate
Compaction parameters						
C_p (clay)	30 ± 5	33 ± 4	30 ± 20	50 ± 10		
CR (clay)					0.15 ± 0.08	-0.03 ± 0.06
C_p (humic clay)	30 ± 5	46 ± 4	30 ± 20	65 ± 10		
CR (humic clay)					0.15 ± 0.08	-0.12 ± 0.05
C_p (peat)	20 ± 4	30 ± 3	25 ± 20	57 ± 12		
CR (peat)					0.15 ± 0.08	-0.06 ± 0.04
C_s (clay)	320 ± 20	330 ± 17	320 ± 80	380 ± 55		
C_α (clay)					0.030 ± 0.015	0.010 ± 0.005
C_s (humic clay)	120 ± 12	140 ± 11	120 ± 80	270 ± 55		
C_α (humic clay)					$0.030\pm.015$	$0.022\pm.009$
C_s (peat)	80 ± 8	90 ± 6	100 ± 50	180 ± 34		
C_α (peat)					0.030 ± 0.015	0.015 ± 0.008
Oxidation parameters						
υ_{ox} (humic clay)	0.010 ± 0.005	0.030 ± 0.002	0.020 ± 0.010	0.041 ± 0.003	0.010 ± 0.005	0.032 ± 0.002
υ_{ox} (peat)	0.015 ± 0.005	0.021 ± 0.004	0.025 ± 0.010	0.027 ± 0.007	0.015 ± 0.005	0.025 ± 0.003
Phreatic levels (m below surface in 1968):						
ZF 3	2.0 ± 0.5	2.35 ± 0.36	2.0 ± 0.5	2.24 ± 0.33	2.0 ± 0.5	1.99 ± 0.35
ZF 6	2.0 ± 0.5	2.10 ± 0.25	2.0 ± 0.5	1.84 ± 0.33	2.0 ± 0.5	2.18 ± 0.25
ZF 11	2.0 ± 0.5	1.98 ± 0.09	2.0 ± 0.5	1.82 ± 0.17	2.0 ± 0.5	2.30 ± 0.19
ZF 15	2.0 ± 0.5	2.27 ± 0.08	2.0 ± 0.5	2.17 ± 0.08	2.0 ± 0.5	2.71 ± 0.11
ZF 19	2.0 ± 0.5	1.90 ± 0.06	2.0 ± 0.5	1.92 ± 0.08	2.0 ± 0.5	2.27 ± 0.09
ZF 20	2.0 ± 0.5	2.51 ± 0.15	2.0 ± 0.5	2.54 ± 0.06	2.0 ± 0.5	2.98 ± 0.19
ZF 26	2.0 ± 0.5	1.96 ± 0.07	2.0 ± 0.5	1.96 ± 0.20	2.0 ± 0.5	2.37 ± 0.10
ZF 30	2.0 ± 0.5	2.26 ± 0.07	2.0 ± 0.5	2.27 ± 0.07	2.0 ± 0.5	2.52 ± 0.11
ZF 33	2.0 ± 0.5	1.92 ± 0.08	2.0 ± 0.5	1.80 ± 0.08	2.0 ± 0.5	2.14 ± 0.10
ZF 36	2.0 ± 0.5	1.40 ± 0.06	2.0 ± 0.5	1.39 ± 0.05	2.0 ± 0.5	1.45 ± 0.12

subsidence as a function of time:

$$\varepsilon_{\mathrm{v,Koppejan}} = \left[\frac{1}{C_\mathrm{p}} + \frac{1}{C_\mathrm{s}}\log\frac{t}{t_\mathrm{ref}}\right] \times \ln\frac{\sigma'}{\sigma'_0} + \left[1-\exp(-\upsilon_\mathrm{ox}t)\right]$$

$$\varepsilon_{\mathrm{v,Bjerrum}} = \mathrm{CR}\log\frac{\sigma'}{\sigma'_0} + C_\alpha\log\frac{t}{t_\mathrm{ref}} + \left[1-\exp(-\upsilon_\mathrm{ox}t)\right]$$

Here, σ' and σ'_0 are the actual and original effective vertical stresses; C_p and C_s are the primary and secondary compression coefficients above pre-consolidation pressure in the Koppejan model; υ_ox is the oxidation rate; CR and C_α are the virgin compressibility (above pre-consolidation pressure) and the creep parameter or coefficient of secondary compression in the Bjerrum model.

4 Inverse model

The present study aimed at the assessment of the uncertainty of the model parameters in order to improve the history match of the ground level displacement and the reliability of the associated predictions. We considered as uncertain parameters the primary and secondary compression coefficients, the oxidation rates and the phreatic water levels. We assumed the measured lithology at every location as fixed. Of the four lithologies identified, we assumed the sand to be incompressible. Both models were thus left with 18 uncertain parameters: primary and secondary creep parameters for clay, for humic clay, and for peat (6 parameters); oxidation rates for humic clay and for peat (2 parameters); and groundwater levels for each location (10 parameters).

We define the vector $\boldsymbol{m}$ as the collection of adjustable model parameters; the vector $\boldsymbol{d}$ as the collection of data (surface level measurements vs. time) and the functional $G(\boldsymbol{m})$, working on the model parameters, as the forward model. The inverse problem is then formulated as the task of estimating the vector $\boldsymbol{m}$ for which $G(\boldsymbol{m})$ approaches the data vector $\boldsymbol{d}$ best. With additional information present in the form of a

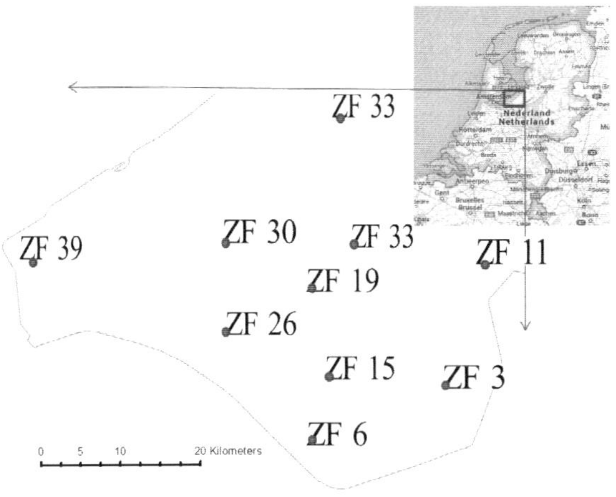

Figure 1. Map of the SE Flevopolder with the selected locations.

prior model ($\boldsymbol{m}_0$) and covariance matrices of the measurements ($\mathbf{C}_d$) and of the prior model ($\mathbf{C}_m$), the conventional least-squares solution is obtained by maximizing the objective function J given by Tarantola (2005) (or by minimizing the exponent in the expression, $-\log(J)$):

$$J = \exp\left[-\frac{1}{2}(\boldsymbol{m}-\boldsymbol{m}_0)^T\mathbf{C}_m^{-1}(\boldsymbol{m}-\boldsymbol{m}_0) - \frac{1}{2}(\boldsymbol{d}-G(\boldsymbol{m}))^T \mathbf{C}_d^{-1}(\boldsymbol{d}-G(\boldsymbol{m}))\right].$$

We used an ensemble approach in which the mean and the covariance of the model vector $\boldsymbol{m}_0$ are mapped on an ensemble of N_e vectors. Different approaches exist to obtain an estimate of the model vector. A global update of the model using all available data can be achieved in a single step; this procedure is called an Ensemble Smoother (Emerick and Reynolds, 2012) – it is, however, suboptimal for non-linear problems as ours. When both data and parameters are assimilated to newly incoming data in subsequent time steps, the procedure is called a filter. EnKF is an example of a filter, but we did not use EnKF because all data were available at all times. Emerick and Reynolds (2013) and Tavakoli et al. (2013) state that the best approach for non-linear systems such as ours is to use an Ensemble Smoother with Multiple Data Assimilation (ES-MDA). In ES-MDA, the ensemble smoother is applied iteratively multiple times. In this way, correlations between parameters that result from the smoother are retained in subsequent steps. The advantage of ES-MDA above EnKF is that the data are used as given and that the procedure is computationally less demanding.

5 Results

The results of the estimation exercise with both models are presented in Table 1. For the Koppejan model we report results that were achieved after starting from two initial distributions – the first one with closer bounds than the second.

The results show that the main adjustments for the Koppejan model are in the oxidation rates and in the phreatic water levels. These parameters have the largest impact on the surface movement rates. The adjustments of the oxidation rates are larger for the less constrained case, as should be expected. For the phreatic water levels, the adjustments are similar. There is quite a large variability in the resulting estimate of the phreatic water level among the different locations – this is related to the types of lithology present around the phreatic water level: peat oxidation has the largest influence on subsidence and boreholes with peat will result in a better constrained water level because the sensitivity to it is larger.

For the Bjerrum model, the largest effect is also related to the peat oxidation rates and the water levels, but the compaction parameters are also significantly constrained. For these parameters, the values of CR and C_α are anti-correlated – an increase in one of them can be partially compensated by a decrease in the other.

Figure 3 visualizes the data, the priors and the inversion results. While a large degree of scatter is present for the prior distribution of parameters, the smoother succeeds in constraining this uncertainty and obtaining a reasonable fit to the data. The assimilation is better when using the Bjerrum model.

6 Discussion

We wish to address some issues here that are striking from the results as presented above. The first is the relative insensitivity to the compaction parameters in the Koppejan model; the second is the appearance of unphysical negative numbers for the immediate compaction parameter CR in the Bjerrum model; the third is the high oxidation rates resulting for the humic clay and for the peat – the former even being the largest. In our opinion, these issues are related to a fourth issue: the fact that the match of the curves using estimated parameters is suboptimal in the first 6 or 7 years. We feel there is a strong relation with the fact that the parameter estimation with the present dataset was accomplished under the assumption of an immediate drop of the water level to a further constant value. This is a simplification that must be addressed in a follow-up study. Although only limited data on the water level are available, Fig. 2 shows that this assumption is not generally applicable. A first-order approximation could be to distribute a 1 m level drop to the final value over the first 7 years after reclamation. We are currently building such an extension into our models.

7 Conclusions

We estimated compaction parameters for clay, humic clay, and peat, oxidation rates for peat and for humic clay,

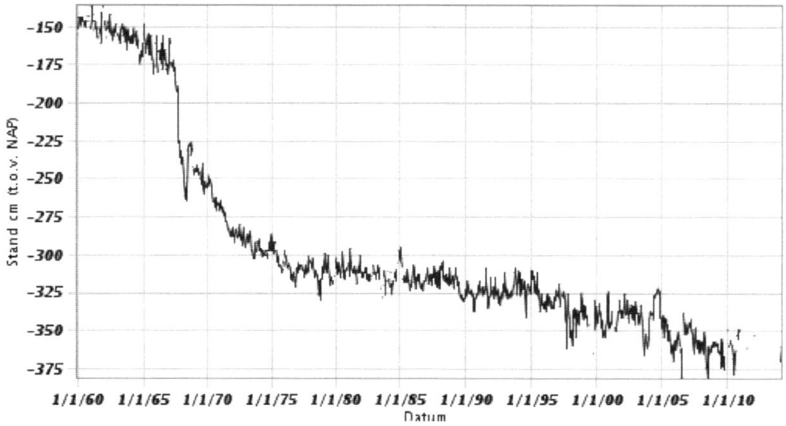

Figure 2. Hydraulic head in well B26E00030001, located on the boundary between the previously reclaimed Eastern Flevoland polder and the subsequently reclaimed Southern Flevoland polder. The reclamation of the Southern Flevoland polder in 1967 is followed by a gradual decrease of the level during approximately 7 years; after which the level remains virtually constant until 1995.

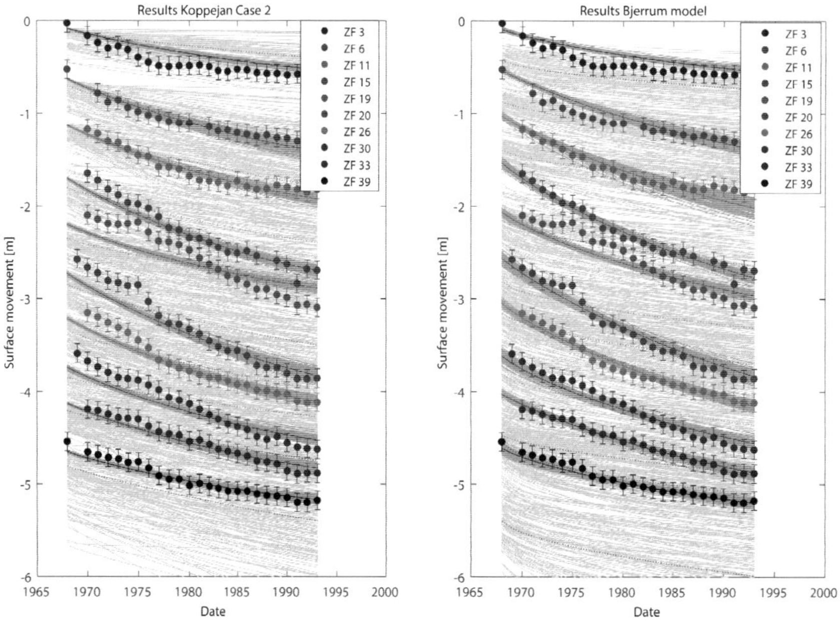

Figure 3. Surface movement data and predictions before and after data assimilation. Subsequent series are shifted downward by 0.5 m. Colored circles: data; colored dashed lines: average prior estimate; colored solid lines: average assimilated estimate. Light grey lines: prior ensemble predictions. Dark grey lines: posterior ensemble predictions. Case 2, with less constrained prior gives larger prior spread and better estimates than Case 1.

and phreatic water levels for 10 locations in the Southern Flevopolder, using surface movement parameters between the moment of reclamation in 1967 and 1993. The procedure, using an Ensemble Smoother with Multiple Data Assimilation, is very fast and gives insight into the variability of the estimated parameters and the correlations between them. In first instance, the Bjerrum model seems to perform better than the Koppejan model. This must, however, be corroborated with more elaborate parameter estimation exercises in which in particular the water level development is taken into account.

References

Bjerrum, L.: Engineering geology of Norwegian normally consolidated marine clays as related to settlements of buildings, Géotechnique, 17, 81–118, 1967.

Emerick, A. and Reynolds, A. C.: Ensemble smoother with multiple data assimilation, Comput. Geosci., 55, 3–15, 2012.

Emerick, A. and Reynolds, A. C.: Investigation of the sampling performance of ensemble-based methods with a simple reservoir model, Comput. Geosci. 17, 325–350, 2013.

Koppejan, A. W.: A Formula Combining the Terzaghi Load Compression Relationship and the Buisman Secular Time Effect, Proc. Int. Conf. Soil Mech. Found. Eng., Rotterdam, 3, 32–38, 1948.

Tarantola, A.: Inverse Problem Theory and Methods for Model Parameter Estimation, SIAM, Paris, France, 2005

Tavakoli, R., Yoon, H., Delshad, M., ElSheikh, A. H., Wheeler, M. F., and Arnold, B. W.: Comparison of ensemble filtering alrorithms and null-space Monte Carlo for parameter estimation and uncertainty quantification using CO_2 sequestration data, Water Resour. Res., 49, 8108–8127, 2013.

Proc. IAHS, 372, 189–198, 2015
proc-iahs.net/372/189/2015/
doi:10.5194/piahs-372-189-2015

Open Access
PIAHS

Sinking coastal cities

G. Erkens[1,2], T. Bucx[1], R. Dam[3], G. de Lange[1], and J. Lambert[1]

[1]Deltares Research Institute, Utrecht, the Netherlands
[2]Utrecht University, Utrecht, the Netherlands
[3]WaterLand Experts, Amsterdam, the Netherlands

Correspondence to: G. Erkens (gilles.erkens@deltares.nl)

Published: 12 November 2015

Abstract. In many coastal and delta cities land subsidence now exceeds absolute sea level rise up to a factor of ten. A major cause for severe land subsidence is excessive groundwater extraction related to rapid urbanization and population growth. Without action, parts of Jakarta, Ho Chi Minh City, Bangkok and numerous other coastal cities will sink below sea level. Land subsidence increases flood vulnerability (frequency, inundation depth and duration of floods), with floods causing major economic damage and loss of lives. In addition, differential land movement causes significant economic losses in the form of structural damage and high maintenance costs for (infra)structure. The total damage worldwide is estimated at billions of dollars annually.

As subsidence is often spatially variable and can be caused by multiple processes, an assessment of subsidence in delta cities needs to answer questions such as: what are the main causes? What is the current subsidence rate and what are future scenarios (and interaction with other major environmental issues)? Where are the vulnerable areas? What are the impacts and risks? How can adverse impacts be mitigated or compensated for? Who is involved and responsible to act?

In this study a quick-assessment of subsidence is performed on the following mega-cities: Jakarta, Ho Chi Minh City, Dhaka, New Orleans and Bangkok. Results of these case studies will be presented and compared, and a (generic) approach how to deal with subsidence in current and future subsidence-prone areas is provided.

1 Introduction

Currently, global mean absolute sea- level rise is around $3\,\mathrm{mm\,yr^{-1}}$, and projections until 2100 based on Intergovernmental Panel on Climate Change (IPCC) scenarios expect a global mean absolute sea-level rise in the range of 3–$10\,\mathrm{mm\,yr^{-1}}$ (Church and White, 2011; Slangen, 2012). However, currently observed subsidence rates in coastal megacities are in the range of 6–$100\,\mathrm{mm\,yr^{-1}}$, and projections until 2025 expect similar subsidence rates (Fig. 1).

In coastal cities around the world, land subsidence increases flood vulnerability (flood frequency, inundation depth, and duration of floods), and hence contributes to major economic damage and loss of lives. Land subsidence is additionally responsible for significant economic losses in the form of structural damage and high maintenance costs; it affects roads and transportation networks, hydraulic infrastructure, river embankments, sluice gates, flood barriers, pumping stations, sewage systems, buildings, and foundations. The total damage associated with subsidence worldwide is estimated at billions of dollars annually.

There are no indications that neither subsidence nor the resulting damage will reduce in the near future. In fact, both are likely to increase. Ongoing urbanization and population growth in delta areas, in particular in coastal mega-cities, continues to fuel economic development in subsidence-prone areas. Consequently, economic development drives both the growing demand for groundwater, thereby increasing subsidence rates, and the growth of the total value of assets at risk. These impacts are aggravated on the long term in coastal areas, by expected future climate change impacts, such as sea-level rise, increased storm surges, and changes in precipitation.

In this paper, we focus on land subsidence in the urban environment, rather than land subsidence in rural agricultural areas, where the drivers may be similar, but the impact very

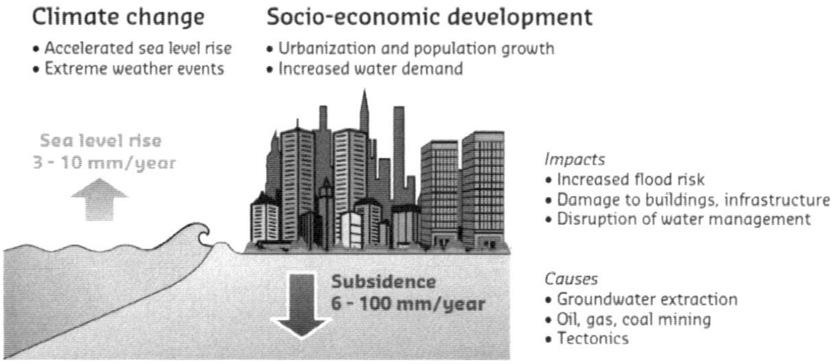

Figure 1. Drivers, processes and impacts of land subsidence in coastal cities. Land subsidence can exceed global absolute sea-level rise (SLR) with a factor 10.

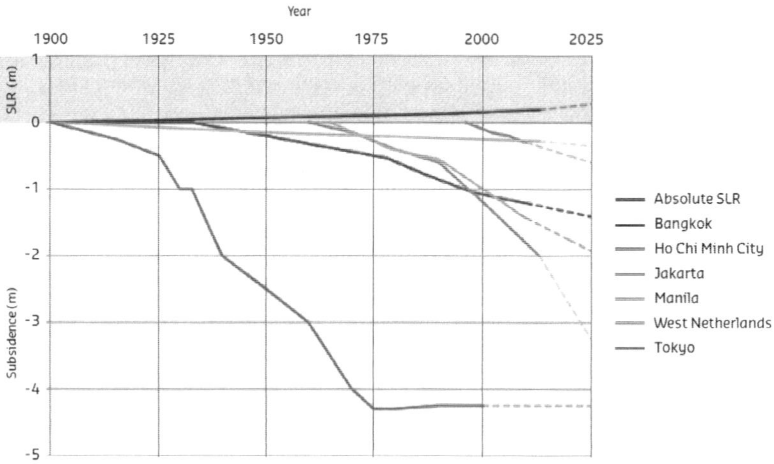

Figure 2. Subsidence history (cumulative) in a series of coastal cities around the world. Absolute sea level rise is depicted as reference. Subsidence can differ considerably within a city area, depending on groundwater levels and subsurface characteristics. Values provided here can be seen as average for the local subsidence hotspots. Some cities are currently seeing an acceleration of subsidence as a result of economic growth. Tokyo stands out as an example where subsidence has stopped after successful mitigation measures were implemented. The caption of Table 1 provides references.

Table 1. Subsidence in coastal cities. Estimated additional mean cumulative subsidence until 2025 (mm) are linear interpolations of the current rates, notwithstanding any policy changes. Sources: Bangkok: MoNRE-DGR (2012), Aobpaet et al. (2013); Ho Chi Min City: van Trung and Minh Dinh (2009); Jakarta: Bakr (2011); Manila: Eco et al. (2011); West Netherlands: van de Ven (1993); Tokyo: Kaneko and Toyota (2011).

City	Mean cumulative subsidence in period 1900–2013 (mm)	Mean current subsidencerate (mm yr^{-1})	Maximum subsidence rate (mm yr^{-1})	Estimated additional mean cumulative subsidence until 2025 (mm)
Jakarta	2000	75–100	179	1800
Ho Chi Minh City	300	up to 80	80	200
Bangkok	1250	20–30	120	190
New Orleans	1130	6	26	> 200
Tokyo	4250	≈ 0	239	0

different. Figure 2 and Table 1 show that land subsidence rates widely vary from city to city. In many cases, the underlying processes and the relative contribution of the different drivers is not well understood. Similar to the level of technical understanding, policy formulation and governmental engagement in cities is equally diverse. Whereas some cities are in an early state of research and policy development on land subsidence, others have already implemented measures mitigating subsidence and the resulting damage. The observed different stages in development mean that cities can learn from each other, thereby avoiding re-inventing the wheel. Cities that actively pursue a policy on subsidence have valuable experiences to share with cities that have just started to address their subsidence.

This is exactly the thought behind the assessment that was carried out for this research. We compared five cities regarding their state of subsidence research and policy development: Jakarta, Ho Chi Minh City, Dhaka, New Orleans and Bangkok. The assessment aimed at getting insight into the processes causing subsidence in the urban environment, obtaining a (generic) research agenda for this topic, and listing best practice cases. Results of these case studies will be presented and a (generic) approach how to cope with subsidence in current and future subsidence-prone areas is provided.

2 Results of the review

For the quick assessment we used published reports on both the technical and the policy aspects of subsidence in the focus cities. In addition, we interviewed local scientist and policy-makers to obtain their perspective. It became quickly clear that all cities tried to answer similar questions. We compiled the interview results into seven interrelated questions (Fig. 3). They include questions such as: what are the main causes for subsidence? How much is the current subsidence rate and what are future scenarios? Where are the vulnerable areas? What are the impacts and risks? How can adverse impacts be mitigated or compensated for? Who is involved and responsible to act? How to monitor the effect of the implemented measures? The interrelation between the questions is indicated with the arrows in Fig. 3. The indicated interrelation does not necessarily mean that each question needs to be answered in a specific order, but it merely indicates that each answer may be valuable input for a next question.

In this paper we follow this framework (in seven steps) and illustrate how these questions are addressed in example cities, thereby discussing both technical and policy aspects of subsidence. In this way, this framework could serve as a blue print for cities to shape their policy and research agenda regarding subsidence.

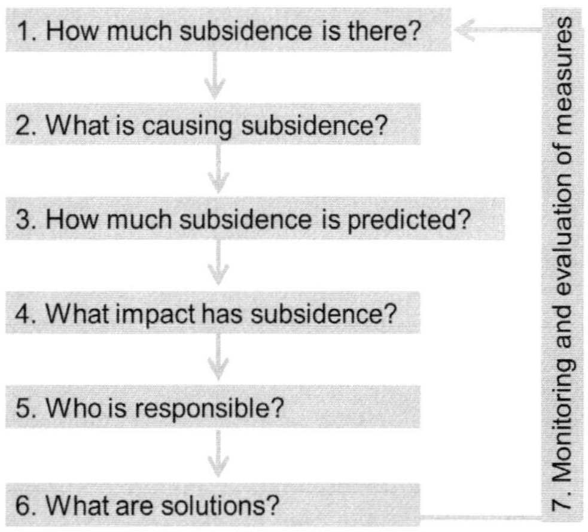

Figure 3. Seven questions that need to be addressed to pursue a successful policy to coop with subsidence. This is loosely based on the policy cycle, a popular framework to analyse policy development.

3 Measuring and monitoring

The first step towards a successful strategy for subsidence is to establish if a certain area is actually subsiding. This may not me evident from the field, particular if subsidence is non-differential and no structural damage (cracks, tilting) is observed in buildings or infrastructure. Typically, the loss of elevation, which may have been observed, is interpreted as the result of climate-driven sea level rise instead of the result of subsidence.

To determine land subsidence rates, accurate measuring techniques are required. Continuous subsidence monitoring provides the necessary insight into changes – ranging from minor to very significant changes – in the topography of the urban area. These observations are also essential to validate subsidence prediction models in a later stage.

The following geodetic observation methods are being used:

- optical levelling;
- Global Positioning System (GPS) surveys;
- Laser Imaging Detection and Ranging (LIDAR);
- Interferometric synthetic aperture radar (InSAR) satellite imagery.

Following early work with systematic optical levelling nowadays GPS surveys and remote sensing techniques (LIDAR and InSAR) are deployed with impressive results. In contrast to surveys, LIDAR and InSAR images give a spatially resolved subsidence signal. InSAR images date back to the 1990s and can now be used to establish subsidence since that time. Application of this technique in soft soil areas is for the moment limited to the build-up environment, as a result of the need for stable reflectors. Ideally, multiple observation techniques are combined, for instance absolute measurements from GPS and Optical Levelling Spatially can be combined with remotely sensed, relative displacement measurements from InSAR. In this way, spatially resolved subsidence maps with respect to a global reference frame can be produced. InSAR measurement can therefore not replace periodic and systematic ground surveys, as they remain essential for ground truthing subsidence rates derived by remote sensing and as an independent source for validating subsidence prediction models.

Systematic observation of elevation forms the base for subsidence monitoring systems. Monitoring results can be used to develop a so-called dynamic digital elevation model (dDEM). This is not a static, one-time only (preferably high resolution) recording of the local topography, but an elevation model that can be corrected and updated from time to time, and that can be used in hydraulic models for flood prediction and urban water management.

All techniques mentioned above measure land surface elevation change, but give no information on the source of the subsidence. Subsidence benchmarks or extensometers can provide in-situ information of ground movement, as they record the volume reduction across a certain stretch in the subsurface, or even of individual geological layers. Ideally the benchmarks or extensometers need to be connected to surface movement observations, for instance by using a combined extensometer and continuous GPS station (e.g. Wang et al., 2014). Monitoring total subsidence at these "supersites", where a terrestrial network of site specific measurement stations is combined with remote sensing, forms the backbone of a spatially resolved subsidence measurement system (Allison et al., 2014). To support subsidence modelling, hydraulic heads of different aquifer systems and the phreatic groundwater level need to be monitored at these super sites as well. Measurements of geotechnical parameters at the same site provide additional necessary input for model studies.

4 Unravelling the subsidence signal

Subsidence can have natural as well as anthropogenic causes. The natural causes include tectonics, loading by ice sheets, by sediments, of by the ocean/sea (isostatic adjustment), and natural sediment compaction (autocompaction). Anthropogenic causes include compression of shallow soft layers by loading (with buildings for instance), or as a result of drainage and subsequent oxidation and consolidation of organic soils and peat. Alluvial or coastal sediments consisting of alternating layers of sand, clay, and peat are specifically compressible and vulnerable for oxidation. This is related to the physical characteristics of these sediments and makes low-lying coastal and delta areas specifically prone to subsidence. In deeper layers subsidence is caused by extraction of resources such as oil, gas, coal, salt, and groundwater.

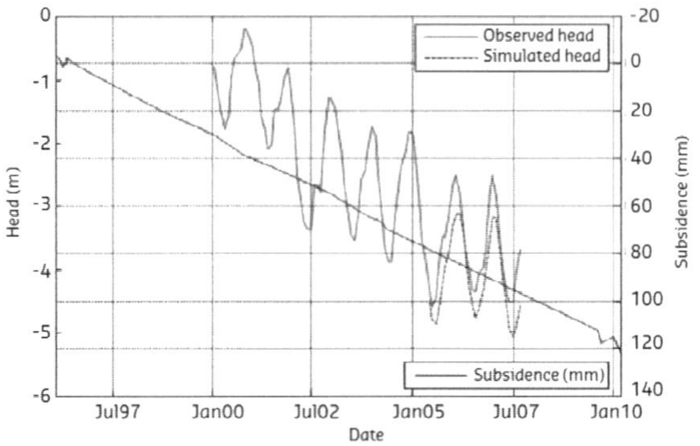

Figure 4. A distinct relation between falling hydraulic heads and subsidence in Ho Chi Minh City (Vietnam). This is indicative of an important contribution of groundwater over-exploration to subsidence, although it is not necessarily the only component contributing to the total subsidence signal.

In most of the large delta cities where land subsidence is severe (Jakarta, Ho Chi Minh City, Bangkok, Dhaka, Shanghai, and Tokyo), the main cause is extraction of groundwater. Rapidly expanding urban areas require enormous amounts of water for domestic and industrial water supply. This need often leads to over-exploitation of groundwater resources, especially when surface waters are seriously polluted (Jakarta, Dhaka). Dhaka (Bangladesh) is an example of a city that started to discover that it subsided after the flood frequency increased. In this rapidly expanding city data on subsidence and its impacts are currently largely lacking. Large-scale extractions cause groundwater levels to fall by 2–3 $m\,yr^{-1}$. At present, 87 % of the supplied water is from groundwater extraction, and it has been acknowledged that a shift to using surface water instead is necessary. However, treating the polluted surface water is much more technically complex and expensive than extracting groundwater.

Although groundwater extraction is often not the sole source of subsidence, studies in many cities have revealed a distinct relation between falling groundwater levels and subsidence, indicative of an important contribution of aquifer compaction (Fig. 4). The resulting spatial pattern of subsidence and its progress over time are strongly related to the local composition of the subsurface and the number and positions of groundwater abstraction wells.

New Orleans (USA) is a prominent example of a city where an array of processes contributes to the total subsidence of the city. The Mississippi Delta subsides as a result of natural processes, such as autocompaction, faulting, sediment loading and isostacy (e.g. Törnqvist et al., 2008; Yu et al., 2012). Within the urban area of New Orleans, there additionally is anthropogenic induced subsidence as a result of drainage of shallow soft soils (Stuurman and Erkens, 2015)

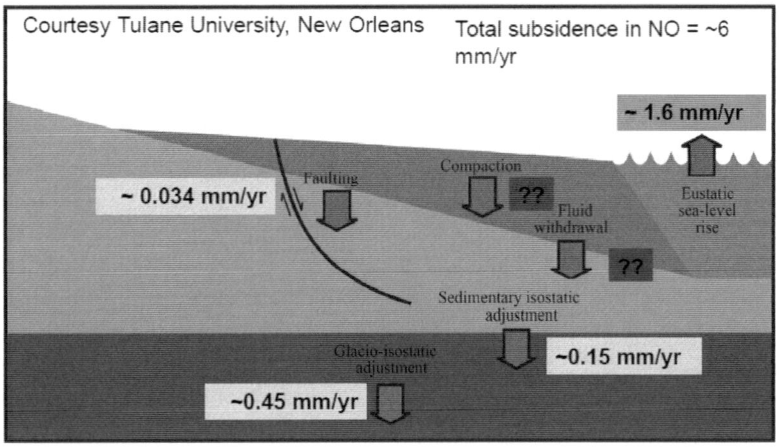

Figure 5. Subsidence components in the urban area of New Orleans. Values are derived from studies of Tulane University, New Orleans, and are indicative. The total subsidence rate is derived from InSAR measurements (Dixon et al., 2006). It shows that in the urban area natural subsidence forms the smaller portion of the total subsidence and that human induced subsidence dominates.

and extraction of deeper groundwater in confined aquifers, for industrial use mainly (Dokka, 2011). After drainage of the organic rich soils, they start to oxidize and lose volume, and this process will continue to cause subsidence as long as organic material is available in the drained subsoil.

The average measured subsidence rate in the city of New Orleans (including the urban area of Jefferson and St Bernard Parishes) is $6\,\mathrm{mm\,yr^{-1}}$ (Dixon et al., 2006). Many studies try to quantify one or more of the different components contributing to the total measured subsidence. Figure 5 shows how this may look for New Orleans, when components are quantified step by step (source: Tulane University, New Orleans). In-situ observation data may provide an independent valuable source of information to unravel the total subsidence signal, as argued in Sect. 3. Another approach to unravel the subsidence signal is inverse modelling, whereby with the use of a careful inversion scheme, the available knowledge on the geology and hydrological dynamics of a system can be quantitatively constrained with subsidence observations (e.g. Fokker et al., 2007).

From Fig. 5 also follows that in the urban area of New Orleans, human induced subsidence has a much larger contribution to the total subsidence signal than natural subsidence. This is often the case, as natural subsidence rates are mainly limited to tens of millimeters per year, to millimeters per year in exceptional cases. Human induced subsidence rates can easily reach centimeters per year, to even tens of centimeters per year (e.g. Jakarta). For policy development this is an important notion: it is worthwhile to implement measures to reduce human-induced subsidence.

5 Modelling subsidence to make predictions

In step three, once the causes for land subsidence have been established (see Sect. 4), predictions can be made to get insight in future land subsidence. Land subsidence modelling and fore-casting tools are being progressively developed that enable quantitative assessment of medium- to long- term land subsidence rates, and determination of multiple causes. Modelling tools are ideally complemented with monitoring techniques (i.e., GPS leveling, the use of InSAR -monitoring techniques), see Sect. 3.

Because land subsidence is in many places closely linked to excessive groundwater extraction, we focus in this paper on modelling of aquifer compaction. One of the most widely used computer program to simulate vertical compaction in models of regional ground-water flow is MODFLOW SUB-WT (Leake and Galloway, 2007). MODFLOW SUB-WT is developed by the US Geological Survey and uses changes in groundwater storage in subsurface layers (aquifers and aquitards) and accounts for temporal and spatial variability of geostatic and effective stresses to determine layer compaction.

In soft soils, such as unconsolidated Holocene layers of peat and clay, the classical consolidation theory by Terzaghi is unable to explain observed consolidation behaviour. These lithology form the aquitards and interbed units in confined aquifer complex systems, albeit often more consolidated, which start to compact after groundwater is extracted from the confined aquifers. Creep deformation is one of the typical processes that occur when the effective stress is increased in clay or peat soils. The creep deformation (also known as secondary strain) of soils is a secondary consolidation process that leads to a reduction in void ratio at con-

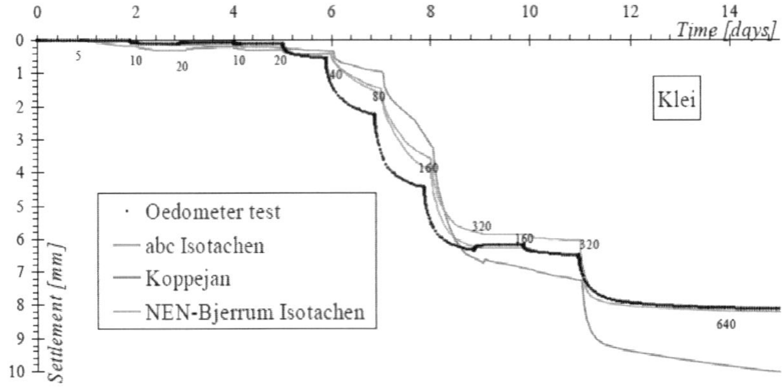

Figure 6. The performance of a series of models used to calculate settlement compared to Oedometer test results. Prediction made by classic models such as Koppejan fit the measurements less well compared to models based on the isotachs method, such as NEN-Bjerrum or abc-Isotachs.

stant effective stress, and consequently, to the development of an apparent pre-consolidation pressure (Den Haan, 1994). It is seen as visco-plastic behaviour and is considered a slow process, compared to primary or elastic consolidation. The inclusion of creep behaviour in numerical descriptions of the consolidation process has a long history, which is excellently described in Bakr (2015). An important aspect of the creep based models is that, due to secondary compression, there is a family of stress-strain curves rather than a single curve describing the relationship between stress and strain. Each of these curves, called "time lines" (i.e. isochrones), corresponds to a different duration of the applied load in a standard oedometer test. For soft soils, model predictions that make use of the isochrones method tend to match the oedometer test results best, specifically on longer time periods (Fig. 6).

Deltares Research Institute modified the US Geological Survey SUB-WT module by including isotachs (line of equal speed) based consolidation predictions. This model, MODFLOW SUB-CR (SUBsidence Creep), is used to determine medium- to long-term land subsidence trends under different scenarios of groundwater usage. In this way, the consequences of groundwater extraction for urban flood management become clear.

Because the SUB-CR model works with isotachs to calculate consolidation, it differs from the SUB-WT model in two ways:

- It predicts on the longer term more consolidation, thus subsidence, in clay and peat layers, as creep is a slow and largely irreversible component of subsidence
- Creep may continue for some time even after the hydraulic heads increased, introducing a time lag in consolidation.

As a result of these differences, aquifers with many fine-grained interbeds, creep forms a considerable part of the total amount of settlement over time and should not be neglected. An example is the subsidence predictions conducted for Jakarta, Indonesia, using isotachs-based consolidation calculations. Bakr (2015) calculates the subsidence occurring in four future groundwater management scenarios for Jakarta. The four scenarios are:

1. drawdown for all aquifers are kept zero till 2100 by maintaining piezometric levels at their values of 2010 (no change);
2. drawdown for all aquifers increase 5 m every 5 years from 2010 till 2030 (business as usual);
3. piezometric heads are recovered for all aquifers by 2015 to their values of 1995 (recovery),
4. piezometric heads are recovered for all aquifers by 2015 to the maximum level of all aquifers in 1995 (full recovery).

In Table 2, we report the predicted cumulative subsidence for Jakarta as calculated by Bakr (2015), calculated with the inclusion of creep. The results indicate that (i) if the hydraulic head declines continue with the current rate (scenario 2) parts of North Jakarta will sink an additional 3.9 m, and (ii) even if hydraulic heads remain the same (scenario 1) or are restored (scenarios 3 and 4) subsidence continues, up to 2.3 or 2.4 m in the recovery scenarios in 2100. This residual subsidence is the result of both delayed pore water pressure dissipation and visco-plastic creep compaction. This means that even if effective stresses do not change, land subsidence will continue till all layers reach hydrostatic equilibrium and creep compaction of all layers vanishes by time due to aging (Bakr, 2015).

This has important implications for policy development in the city of Jakarta. The significant predicted subsidence in the

Table 2. Cumulative subsidence (m), modelled including creep behaviour, for 4 groundwater management scenarios for Jakarta, Indonesia, by Bakr (2015). Because of the slow creep rates, subsidence continues even after hydraulic heads are restored (scenarios 3 and 4).

Year	Scenario 1 (no change)	Scenario 2 (business as usual)	Scenario 3 (recovery)	Scenario 4 (rapid recovery)
2020	1.97	2.48	1.74	1.73
2025	2.08	2.75	1.80	1.77
2030	2.18	2.92	1.85	1.81
2100	3.01	3.91	2.43	2.30

business as usual scenario justifies a subsidence mitigation policy. But the forecasted subsidence values in the recovery scenarios indicate that for the remaining residual subsidence an adaptation strategy must be developed too. Because of these far reaching implications for policy development, it is important that subsidence predictions are as accurate as possible. Although the inclusion of creep behaviour in prediction models aimed to increase accuracy of predictions, they are also sensitive for the geo(hydro)logical schematisation used in the model. This is because the fine-grained interbeds and aquitards are most sensitive to creep, and their exact distribution both in vertical and lateral direction determines the model outcome. The 3-D distribution of fine-grained deposits in the subsurface, and their geo-mechanical properties, are therefore key to reliable subsidence predictions for cities.

6 Impact and damage

With subsidence predictions for different management scenarios (step 3, see Sect. 5), damage estimates (step 4 in the framework, Fig. 3) provide additional information for policy decisions. The estimation of costs associated to subsidence is very complex. Subsidence is a "hidden threat" because in practice, costs appear on financial sheets as part of ad hoc investments or planned maintenance schemes, but are not labelled as subsidence-induced damage. Dedicated damage estimates of subsidence can help to raise awareness among policymakers and initiate policy development.

Generally, two (very different) types of damage as a result of subsidence can be recognised: (i) increased flood risk (due to increased flood frequency, floodwater depth, and duration of inundation) and more frequent rainfall-induced floods due to ineffective drainage systems, and (ii) damage to buildings, foundations, infrastructure (roads, bridges, dikes), and subsurface structures (drainage, sewerage, gas pipes, etc.). The former is mainly the result of non-differential subsidence, which is characteristic for large subsidence bowls that exist when groundwater or hydrocarbons at greater depth are extracted. Examples of cities that have increased flood risk as a result of subsidence include Jakarta, Ho-Chi-Minh and Bangkok. The second type of damage, to structures, is the result of differential subsidence. This commonly happens when fault systems are (re)activated, or when the subsidence is the result of shallow processes (loading or drainage of soft soils). Examples of cities in which structures are damaged include New Orleans, Venice (Italy) and Amsterdam (the Netherlands). Note that the construction site preparation and construction costs in soft-soil areas should be considered as subsidence-related costs, as these are mainly incurred to prevent consolidation. On the longer term, however, cumulative subsidence of soft soils may also increase flood risk as for instance happened in the Netherlands (subsidence over the last ~ 1000 years) or in New Orleans (subsidence over the last ~ 150 years). The extent of the damage is different in the two cases: increased flood risk usually applies to a larger area than structural damage that applies to single structures or parts of the network. The owner of the problem is also different: it is the local government who is investing in reducing flood risk, whereas local communities, (utility) companies or even home owners pay for the damage to (infra)structures.

Making an estimation of costs associated with subsidence is notoriously complex. Some bulk estimates are available. For instance, in China, the average total economic loss due to subsidence is estimated at around USD 1.5 billion per year, of which 80–90 % is from indirect losses. In Shanghai, over the period 2001–2010, the total loss cumulates to approximately USD 2 billion. In the Netherlands, new estimates based on subsidence modelling, try to unravel the bulk costs. For instance, it is calculated that damage to foundations (as a result of subsidence) has been more than EUR 5 billion thus far, and might reach EUR 40 billion in 2050 (although this is a theoretical maximum, Hoogvliet et al., 2012). The communities in soft soil areas in the Netherlands spend EUR 0.25 billion per year more on maintenance than the communities on supportive soils. This values consists of EUR 0.17 billion per year maintenance for roads and water networks and EUR 0.08 billion per year for sewage systems (Lambert et al., 2014). The total damage associated with subsidence worldwide is unknown, but estimated based on the aforementioned values suggest billions of dollars annually. Because of ongoing economic and urban development, the potential damage costs of subsidence will increase considerably in the future, especially in subsidence-prone areas such as flood plains.

Damage estimates form the core of cost-benefit analyses. For subsidence, cost-benefit analyses will help to systematically calculate and compare benefits and costs of a decision or government policy on the short and long term. Being a gradual process, usually mitigation measures for subsidence are costly on the short term, but cost-effective on the longer term. Cost-benefit analyses could provide this insight in a quantitative way.

7 Measures and monitoring

Once the damage caused by subsidence is quantified (Sect. 6), the responsible actors (step 5 of the framework, Fig. 3) can work out a policy on subsidence (step 6), that should be evaluated after implementation (step 7). In this section we focus on action necessary for steps 6 and 7.

There are generally two policy strategies for subsiding cities: mitigation and adaptation – analogue to the climate change policy discussions. A successful strategy, however, probably includes both. Mitigation only works for human-induced subsidence (Sect. 4). Typical mitigation measures include restrictions of groundwater extraction, artificial recharging aquifers, or raising (phreatic) water levels in areas with organic rich soils, thereby reducing oxidation of organic matter. Building with lighter materials decreases the load on soft soils, thereby decreasing consolidation and subsidence (Lambert et al., 2014).

For the human induced subsidence that cannot be mitigated, either because of technical difficulties (for instance the use of lighter building materials in high rise buildings), or because of financial reasons (i.e. the mitigation costs are too high), an adaptation strategy should be considered. This is also true for residual subsidence after a successful mitigation of subsidence (see Sect. 5) or for natural subsidence, where mitigation is not possible.

Adaptation must focus on reducing the impact of subsidence, for instance by decreasing the vulnerability of a certain asset to the negative impacts of subsidence. For increased flood risk as a result of subsidence, adaptation measures include the strengthening or heightening of embankments, building on mounds or piles, or conduct spatial planning in such a way that new constructions are only built on elevated areas. For damage to structures, adaptation strategies may include the use of flexible pipes and cables (specifically for connection points), the use of better foundations for structures, or again careful spatial planning, whereby building is limited to areas with supportive soils (for instance channel belt deposits within a delta).

Adaptation strategies are commonly applied in subsiding coastal cities, for instance most of them have network of embankments that reduces the flood risk. Cities that pursue an active policy on subsidence mitigation are less common, but successful examples do exist. In Tokyo, after taking regulations measures restricting the groundwater use were imposed in the early 1960s, the groundwater levels began to rise as a result (Fig. 7). Subsidence came to hold 10 years later as a result of the delayed response in the compacting layers (see also Sect. 5). The restrictions on groundwater use meant that a replacement water source had to be found. Dams were constructed in several river basins that were designated for water resources development. During the 1970s and 80s numerous dams were built to provide storage to avoid future water scarcity and to supply the growing cities with sufficient water. Beginning in the 1960s an additional investment in waste water treatment was initiated.

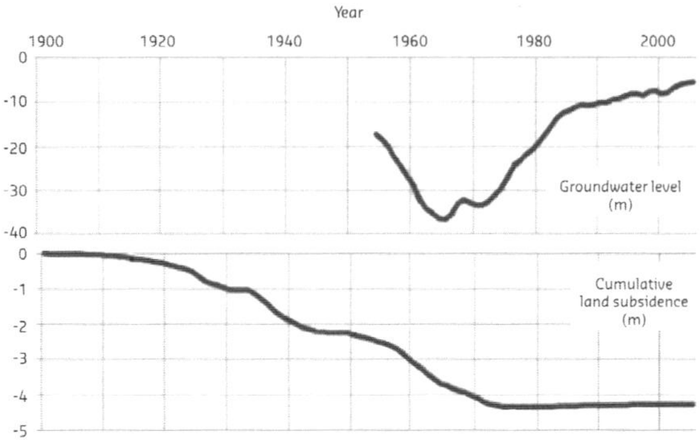

Figure 7. Land subsidence and groundwater levels in the Tokyo area (Japan), modified after Kaneko and Toyota (2011). The effect of the reduction of groundwater extraction on groundwater levels is clearly visible. Note that land subsidence completely stops 10 years after the groundwater level recovery started.

Shanghai in China is another example of a city with a successful subsidence mitigation strategy. Following the increased understanding of the close relationship between groundwater extraction and land subsidence in Shanghai (e.g. Shi et al., 2008), groundwater levels were restored with active recharge techniques. Although this approach reduced the further lowering of groundwater tables and limited subsidence, it did not completely eliminate the effects of subsidence on infrastructure, roads, and buildings. The Shanghai case shows that, with active and substantial recharge, sustainable groundwater use is achievable, without severe subsidence, provided that average yearly pumping rates are in balance with the average yearly recharge.

In Bangkok, Thailand, regulation of and restrictions on groundwater extraction have successfully reduced severe land subsidence. A specific law (the Groundwater Act) was enacted in 1977. The most affected areas were designated as Critical Zones, and the government was given more control over private and public groundwater use in these areas. Groundwater use charges were first implemented in 1985 and have gradually increased. Currently about 10 % of the total water use is supplied by groundwater extraction, and this mainly used by the industry in Bangkok. In most urban areas, subsidence is now reduced to 1 cm yr^{-1}, with local increased subsidence rates of 2 cm yr^{-1} in the aforementioned industrial sites.

Jakarta (Indonesia) and Ho-Chi-Minh City (Vietnam) are considering similar subsidence mitigation strategies. In the Greater Jakarta area, metropolitan authorities and technical agencies are advocating the reduction of groundwater extraction in vulnerable areas. The goal is to completely phase out the use of groundwater by taxing groundwater consumption. This would require developing an alternative water supply for large industrial users or relocation of large groundwater users, outside the so-called "critical zones". The number of "unregistered" users is still a problem. To some extent, spatial planning measures have been applied to avoid subsidence-prone areas, but the fast growth of informal settlements has made many of these plans obsolete. Recently, the Jakarta province government started to clear out the water management structures to reduce flood risk. In 2015, the Governor of Jakarta announced the reduction of the usage of deep groundwater in all government and public buildings, as a first step in the transition to piped water supply. The expected delayed response of subsidence to groundwater head recovery (Bakr, 2015; Sect. 5) asks for accurate subsidence prognosis. They form a vital component for any integrated flood management and coastal defence strategy (Dam, 2012).

Although land subsidence in Ho Chi Minh City has been observed since 1997, there is still – similar to Jakarta – considerable disagreement about the underlying processes and impacts. This is partly due to poor land level and groundwater extraction monitoring data (Ho Chi Minh City Flood and Inundation Management, 2013). Restrictions of groundwater extraction have been initiated, but it is too early to observe any effects.

In the Netherlands, with arguably the longest history of human induced subsidence in the world (since 1000 AD), the focus has been on adaptation strategies for more than nine centuries. In the coastal peatlands, after $\sim$ 1200 AD, adaptation measures included improving drainage (digging canals), the closing of (tidal) creeks and rivers, raising dikes and creating polders, and the improvement of foundations of buildings and infrastructure. Only in the last 50 years, with ever increasing damage to structures, mitigation measures were implemented. Nowadays, groundwater is sustained as shallow as possible in the peatlands. This means careful land use planning (less productive grassland and nature development in the most sensitive areas and considering alternative crops elsewhere) and the inlet of fresh water in polders in dry periods. Complete mitigation of subsidence is probably not possible, because that would end agriculture in a major part of western and northern Netherlands. The associated high economic losses are socially and culturally not acceptable. In the northern part of the Netherlands, gas extraction results in significant subsidence. Here policy similarly developed towards mitigation measures, albeit on a shorter time scale. Gas extraction started in the 1960's, but until about 2010 the governmental response to subsidence was limited to adaptation of the surface water management system. After 2010, subsidence was accompanied by more frequent and powerful induced seismicity (earthquakes). The resulting damage of houses and other constructions forced the government in 2014 to start additional mitigation measures in the form of a significant reduction of gas exploitation in the most critical fields. Again, full mitigation was very difficult as stopping of the gas exploitation would endanger the national energy supply and would reduce the gas revenues by several billion of euro's per year. In addition, even if the gas exploitation was completely phased out, the subsidence and earthquakes are likely to continue. Concluding, for the Netherlands full mitigation of subsidence is far more expensive than implementing adaptation measures and an adaptation strategy combined with limited mitigation is a much more feasible option.

For all measures taken to reduce subsidence or its impacts, it is important that the effectiveness of these efforts is monitored. This implies that a subsidence monitoring network (see Sect. 3) need to be installed before the measures are implemented. The monitoring data form an important contribution to any subsidence monitoring network that has been established in step 1 (Sect. 3). Preferably, the monitoring data and analytical results (of the various modelling tools) are stored in a central database.

8 Concluding remarks

- In urban areas, human induced land subsidence dominates the total subsidence signal.
- Land surface elevation measurements need to be combined with in-situ measurements in order to be able to unravel the total subsidence signal.
- There are two types of damage as a result of subsidence: increased flood risk (with non-differential subsidence) and damage to structures (with differential subsidence).
- Analogue to climate change policies, a successful policy on subsidence consists of adaptation measures (reducing the damage and vulnerability) and mitigation measures (actively reducing subsidence).
- Delayed response of aquitards and interbed compaction may introduce unwanted additional subsidence after implementing mitigation measures, which is presently unaccounted for.

Acknowledgements. This study is based on research conducted by scholars working in the cities mentioned in this paper. They shared data and insights from which this study greatly benefitted. They are all gratefully acknowledged.

References

Allison, M., Yuill, B., Törnqvist, T., Amelung, F., Dixon, T., Erkens, G., Stuurman, R. J., Milne, G., Steckler, M., Syvitski, J., and Teatini, P.: Coastal subsidence: global risks and research priorities, EOS Transactions, 2014.

Aobpaet, A., Caro Cuenca, M., Hooper, A., and Trisirisatayawong, I.: InSAR time series analysis of land subsidence in Bangkok, Thailand, Int. J. Remote Sens., 34, 8, 2013.

Bakr, M.: Land subsidence in North Jakarta – preliminary analysis results, part of Jakarta Coastal Defencse Strategy (JCDS) study, JCDS Atlas, 2011.

Bakr, M.: Influence of Groundwater Management on Land Subsidence in Deltas: A Case Study of Jakarta (Indonesia), Water Resour. Manage., 29, 1541–1555, doi:10.1007/s11269-014-0893-7, 2015.

Church, J. A. and White, N. J.: Sea-Level Rise from the Late 19th to the Early 21st Century, Surv. Geophys., 32, 585–602, 2011.

Dam, R.: Jakarta Coastal Defencse Strategy (JCDS) study, Activity Report: Land subsidence and adaptation/mitigation strategies, JCDS Bridging Phase, 2012.

Den Haan, E. J.: Vertical compression of soils, PhD thesis, Technical University of Delft, Delft, the Nehterlands, 1994.

Dixon, T. H., Amelung, F., Ferretti, A., Novali, F., Rocca, F., Dokka, R., Sella, G., and Kim, S.-W.: Subsidence and flooding in New Orleans – A subsidence map of the city offers insight into the failure of the levees during Hurricane Katrina, Nature, 441, 587–588, 2006.

Dokka, R. K.: New Orleans The role of deep processes in late 20th century subsidence of New Orleans and coastal areas of southern Louisiana and Mississippi, J. Geophys. Res., 116, 2011.

Eco, R. C., Lagmay, A. A., and Bato, M. P.: Investigating ground deformation and subsidence in northern Metro Manila, Philippines using Persistent Scatterer Interferometric Synthetic Aperture Radar (PSInSAR), American Geophysical Union, Fall Meeting, San Francisco, CA, USA, 5–9 December 2011, G23A-0822, 2011.

Fokker, P. A., Muntendam-Bos, A.-G., and Kroon, I. C.: Inverse modelling of surface subsidence to better understand the Earth's subsurface, First Break, 25, 8, 2007.

Ho Chi Minh City Flood and Inundation Management: Final Report, Volume 2: IFRM Strategy, Annex 3: Land Subsidence, Royal Haskoning-DHV and Deltares, 2013.

Hoogvliet, M., van de Ven, F., Buma, J., van Oostrom, N., Brolsma, R., Filatova, T., Verheijen, J., and Bosch, P.: Schades door watertekorten en – overschotten in stedelijk gebied – Quick scan van beschikbaarheid schadegetallen en mogelijkheden om schades te bepalen, Deltares report 1205463-000, 128 pp., 2012 (in Dutch).

Kaneko, S. and Toyota, T.: Long-Term Urbanization and Land Subsidence in Asian Megacities: An Indicators System Approach, in: Groundwater and Subsurface Environments: Human Impacts in Asian Coastal Cities, 2011.

Lambert, J. W. M., van Meerten, J. J., Woning, M. P., and Eijbersen, M. J.: Verbeterde onderhoud strategie infrastructuur in slappe bodemgebieden, Deltares report 1209950-000, 31 pp., 2014 (in Dutch).

Leake, S. A. and Galloway, D. L.: MODFLOW ground-water model, User guide to the Subsidence and Aquifer-System Compaction Package (SUB-WT) for water-table aquifers: US Geological Survey, Techniques and Methods 6–A23, 42 p., 2007.

MoNRE-DGR (Ministry of Natural Resources and Environment, Department of Groundwater Resources): The study of systematic land subsidence monitoring on critical groundwater used area project, Project report, 2012.

Shi, X., Wua, J., Yea, S., Zhangb, Y., Xuea, Y., Weic, Z., Lic, Q., and Yud., J.: Regional land subsidence simulation in Su-Xi-Chang area and Shanghai City, China, Engineering Geology, 100, 27–42, 2008.

Slangen, A. B. A.: Towards regional projections of twenty-first century sea-level change based on IPCC SRES scenarios, Clim. Dynam., 38, 5–6, 2012.

Stuurman, R. J. and Erkens, G.: New Orleans' soft soils need water and solid management, J. Hydrogeol., in prepraration, 2015.

Törnqvist, T. E., Wallace, D. J., Storms, J. E. A., Wallinga, J., van Dam, R. L., Blaauw, M., Derksen, M. S., Klerks, C. J. W., Meijneken, C., and Snijders, E. M. A.: Mississippi Delta subsidence primarily caused by compaction of Holocene strata, Nat. Geosci., 1, 173–176, 2008.

van der Ven, G. P.: Man-made lowlands, history of water management and land reclamation in the Netherlands, Uitgeverij Matrijs, Utrecht, the Netherlands, 1993.

van Trung, L. and Minh Dinh, H. T.: Monitoring Land Deformation Using Permanent Scatterer INSAR Techniques (case study: Ho Chi Minh City), 7th FIG Regional Conference, Vietnam, 2009.

Wang, G., Yu, J., Kearns, T. J., and Ortega, J.: Assessing the Accuracy of Long-Term Subsidence Derived from Borehole Extensometer Data Using GPS Observations: Case Study in Houston, Texas, J. Surv. Eng., 140, 2014.

Yu, S.-Y., Törnqvist, T. E., and Hu, P.: Quantifying Holocene lithospheric subsidence rates underneath the Mississippi Delta, Earth Planet. Sci. Lett., 331–332, 21–30, 2012.

Proc. IAHS, 372, 199–205, 2015
proc-iahs.net/372/199/2015/
doi:10.5194/piahs-372-199-2015

Regional and local land subsidence at the Venice coastland by TerraSAR-X PSI

L. Tosi[1], T. Strozzi[2], C. Da Lio[1], and P. Teatini[1,3]

[1]Institute of Marine Sciences, National Research Council, Venice, Italy
[2]GAMMA Remote Sensing, Gümligen, Switzerland
[3]Dept. of Civil, Architectural and Environmental Engineering, University of Padova, Italy

Correspondence to: L. Tosi (luigi.tosi@ismar.cnr.it)

Published: 12 November 2015

Abstract. Land subsidence occurred at the Venice coastland over the 2008–2011 period has been investigated by Persistent Scatterer Interferometry (PSI) using a stack of 90 TerraSAR-X stripmap images with a 3 m resolution and a 11-day revisiting time. The regular X-band SAR acquisitions over more than three years coupled with the very-high image resolution has significantly improved the monitoring of ground displacements at regional and local scales, e.g., the entire lagoon, especially the historical palaces, the MoSE large structures under construction at the lagoon inlets to disconnect the lagoon from the Adriatic Sea during high tides, and single small structures scattered within the lagoon environments. Our results show that subsidence is characterized by a certain variability at the regional scale with superimposed important local displacements. The movements range from a gentle uplift to subsidence rates of up to 35 mm yr^{-1}. For instance, settlements of 30–35 mm yr^{-1} have been detected at the three lagoon inlets in correspondence of the MoSE works, and local sinking bowls up to 10 mm yr^{-1} connected with the construction of new large buildings or restoration works have been measured in the Venice and Chioggia historical centers. Focusing on the city of Venice, the mean subsidence of 1.1 ± 1.0 mm yr^{-1} confirms the general stability of the historical center.

1 Introduction

Cities in lowlying coastlands and deltaic regions are the sites most susceptible to land subsidence worldwide. The influence of land subsidence on these communities and environments can be very significant, increasing vulnerability to saltwater intrusion and flooding, threatening agriculture and ecological systems. Presently, land subsidence caused mainly by extraction of groundwater is threatening Jakarta (Ng et al., 2012), Ho Chi Minh City (Erban et al., 2014), Bangkok (Phien-wej et al., 2006), Shanghai (Dong et al., 2014), i.e. rapidly expanding urban areas which require huge amounts of water for domestic and industrial water supply.

Similarly to that occurring in these urban centres in developing countries, Venice experienced anthropogenic land subsidence due to aquifer overexploitation in the past (Carbognin et al., 2004). Although the effect of groundwater pumping ended a few decades ago, a reliable and detailed knowledge of land subsidence affecting the historical center is even more important today as vulnerability has continuously increased over the years. The city of Venice and its surrounding lagoon is presently one of the sites most sensitive to land subsidence worldwide. Even a few mm yr^{-1} loss of elevation with respect to the mean sea level can threaten the city's survival and irrevocably change the natural environment of the lagoon.

This is the reason why monitoring land subsidence in the Venice area at a very high accuracy has always been important. Recent studies have highlighted a significant heterogeneity of the subsidence at the local scale, both connected to the subsoil characteristics and human interventions (Strozzi et al., 2009; Tosi et al., 2012, 2013). Thus, in addition to the millimetric accuracy, a metric spatial resolution is also required for improving the understanding of the natural and human-induced components.

In this study, we use a long-term stack of TerraSAR-X scenes processed by PSI (Persistent Scatterer Interferome-

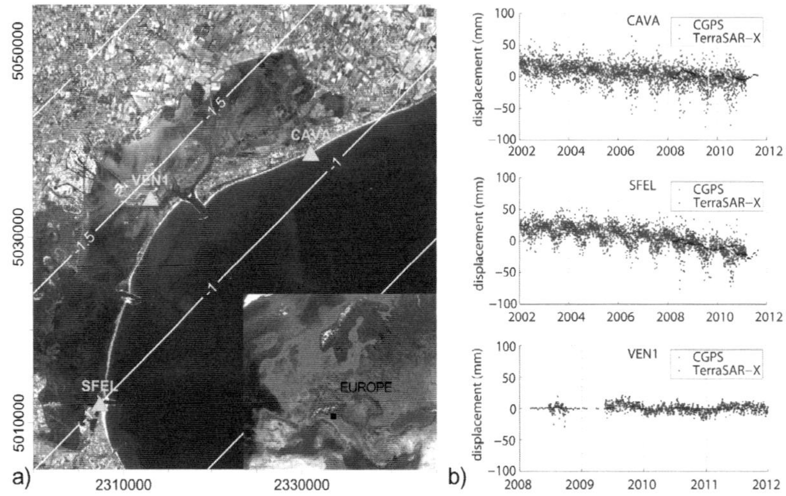

Figure 1. **(a)** Data calibration: plane model based on the Venice (VEN1), Chioggia (SFEL), and Cavallino (CAVA) CGPS displacements, used to correct the PSI outcome. **(b)** Comparison between the CGPS and the calibrated PS time series.

try) to monitor land subsidence at the Venice coastland from 2008 to 2011 and highlight the local movements of historical palaces and monuments in the Venice center, single small structures scattered within the natural lagoon environments, and the MoSE large structures under construction at the lagoon inlets.

2 Methods

The Interferometric Point Target Analysis (IPTA; Werner et al., 2003; Wegmüller et al., 2004; Teatini et al., 2005), a PSI implementation chain, is applied on a stack of 90 TerraSAR-X stripmap images for the period 2008 to 2011 with a regular 11-day revisiting time. In order to correct the so-called flattening problem, i.e. the slight phase tilt resulting by the inaccuracy in estimation of the orbital baseline due to the imperfect knowledge of the satellite positions (Teatini et al., 2012a, b), IPTA results have been calibrated using three permanent GPS stations (CGPS) well distributed in the monitored area, i.e. VEN1 at Venice, CAVA on the northern Cavallino littoral strip, and SFEL in the nearby of Chioggia at the southern tip of the lagoon (Fig. 1a). Figure 1a shows the tilting plane used to calibrate the PSI solution on the CGPS records. Even if the slope is relatively small, the corrections between 1 to 2 mm yr^{-1} are significant for the Venice coastal area. Figure 1b points out the satisfactory match between the CGPS time series and the average time series of the persistent scatterers (PS) in the surrounding of the GPS stations once calibrated the PSI solution.

IPTA provides land displacements along the line-of-sight (LOS) between the satellite and the targets. Due to the small incidence angle of the radar signal, IPTA measurements are less sensitive to the horizontal components of the displacement than to the vertical one. Nevertheless, the occurrence of differential horizontal displacements, for example because of the tectonics, would affect the SAR interferometry measurement. The network of the CGPS stations shows however that the Venice coastland is uniformly moving north-eastward (Teatini et al., 2005; Tosi et al., 2010). Therefore, the differential component of the horizontal displacement rate perpendicular to the TerraSAR-X inclination orbit is smaller than 1 mm yr^{-1}, and negligible with respect to the vertical displacement rate. In our results the difference between the LOS and the vertical components of the movement (about 15 % for TerraSAR-X, i.e. less than 1 mm yr^{-1} for the large majority of the PS) is neglected.

3 Results

3.1 Land subsidence at the regional scale

Figure 2a shows the PSI solution in terms of average displacement rates on the whole coastal area of Venice. An impressive number of 1 933 690 PS, i.e. 2275 km^{-2}, has been detected. The movement velocity ranges from a gentle uplift ($\sim$2 mm yr^{-1}) to large subsidence (up to 30 mm yr^{-1}).

Similarly to previous results obtained with ENVISAT images (e.g., Teatini et al., 2012a, b), the monitoring outcome reveals that this portion of the Northern Adriatic coastland is characterized by a certain variability in terms of land subsidence at the large (or regional) scale. A number of zones with different features in the displacement rates is distinguished:

- the north-western zone (A circle in Fig. 2a): this area is characterized by uniform stability. The average sub-

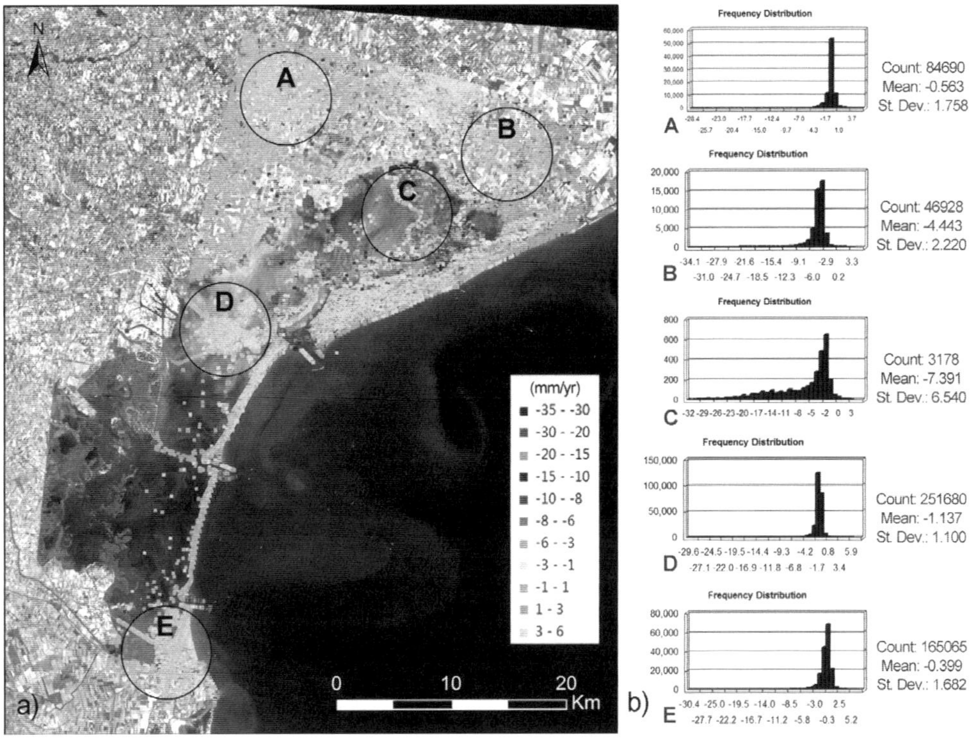

Figure 2. (**a**) Average land displacements (mm yr^{-1}) for the Venice coastland obtained by PSI on TerraSAR-X images acquired between March 2008 and August 2011. Positive values mean uplift, negative values land subsidence. A Landsat image obtained from the US Geological Survey – Earth Resources Observation and Science (EROS) Center is used as base map. (**b**) Frequency distribution of the displacement rates for the PS located in the A-E circular areas ($r = 3$ km) highlighted on the map.

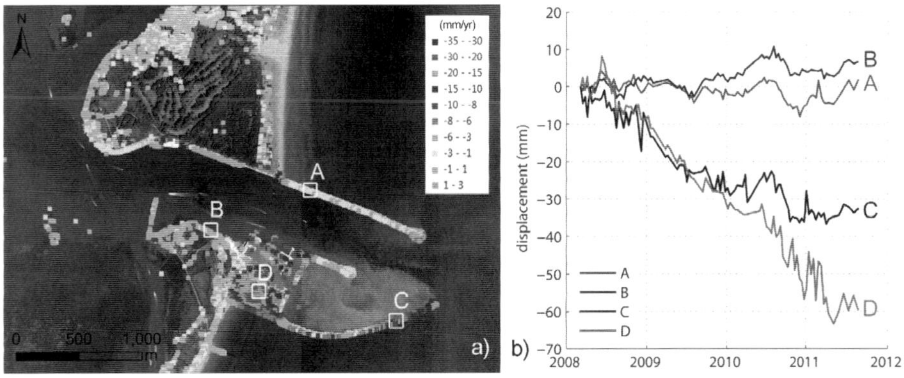

Figure 3. (**a**) Mean displacement rates between March 2008 and August 2011 from TerraSAR-X interferometry at Malamocco inlet. The image background is an aerophotograph acquired in 2007. Negative values indicate settlement, positive mean uplift. (**b**) Displacement history in the satellite line-of-sight direction for four scatterers (A, B, C, D). See location 1a in Fig. 1.

sidence η_r is less than $1\ \text{mm yr}^{-1}$ and the standard deviation σ_η is about $1\text{–}2\ \text{mm yr}^{-1}$ (Fig. 2b). Only local bowls of relatively large displacements caused by human activities (groundwater pumping, development of new urban and industrial zones) are detected;

- the north-eastern zone (B circle in Fig. 2a): a relatively uniform ($\sigma_\eta \approx 2\ \text{mm yr}^{-1}$) land subsidence on the order

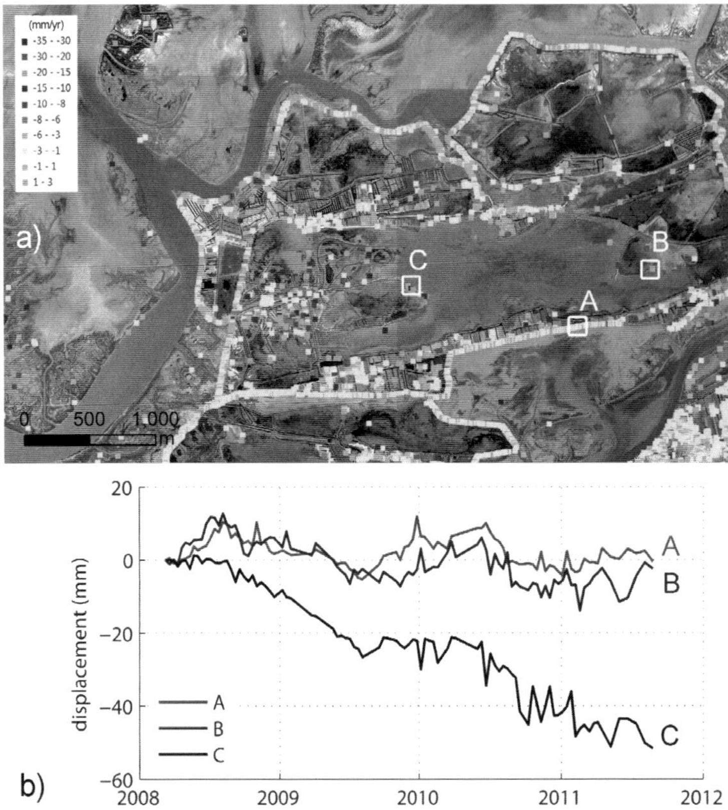

Figure 4. (a) Mean displacement rates between March 2008 and August 2011 from TerraSAR-X interferometry at Valle Paleazza, Northern Venice lagoon. The image background is an aerophotograph acquired in 2007. Negative values indicate settlement, positive mean uplift. **(b)** Displacement history in the satellite line-of-sight direction for three scatterers (A, B, C). See location 2a in Fig. 1.

of 4–6 $\mathrm{mm\,yr^{-1}}$ is threatening this part of the coastal area. Land subsidence due to aquifer exploitation occurs mainly along coastal sector; also processes related to the recent land reclamation (the zone was reclaimed between 1900 and 1930) likely contribute to the total movements;

- the northern lagoon (C circle in Fig. 2a): it is the zone characterized by the highest variability ($\sigma_\eta \approx 6\,\mathrm{mm\,yr^{-1}}$) with land displacements ranging between -1 and $-30\,\mathrm{mm\,yr^{-1}}$. The largely heterogeneous nature of the shallow Holocene deposits strongly impacts the movements;

- Venice and the central lagoon (D circle in Fig. 2a): the historical centre and the surrounding lagoon is relatively stable. On the average $\eta_r = 1\,\mathrm{mm\,yr^{-1}}$, with a very narrow v ariability of $\sigma_\eta \approx 1\,\mathrm{mm\,yr^{-1}}$. The Holocene deposits are relatively thin (about 5 m) and groundwater withdrawals have been precluded since the early 1970s;

- Chioggia and the southern coastland (E circle in Fig. 2a): moving to the south, land subsidence remains generally small. The city of Chioggia is stable (σ_η ,$< 1\,\mathrm{mm\,yr^{-1}}$), and a certain subsidence occurs mainly along the coastline in the soundings of the mouths of the main rivers (the Brenta and Adige rivers).

3.2 Ground displacements at the local scale

A significant local scale variability of the land displacements is superposed on the regional trend of land subsidence. In the following, a few cases showing the typical local movements observed along the Venice coast are presented and discussed.

3.2.1 The MoSE constructions at the lagoon inlets

Starting from the middle 1990s, the jetties at the three inlets were reinforced, and since 2005, in the framework of the MoSE works (i.e. the project of mobile barriers for the temporarily closure of the lagoon to the sea), they have been strongly reshaped and supplemented by offshore breakwa-

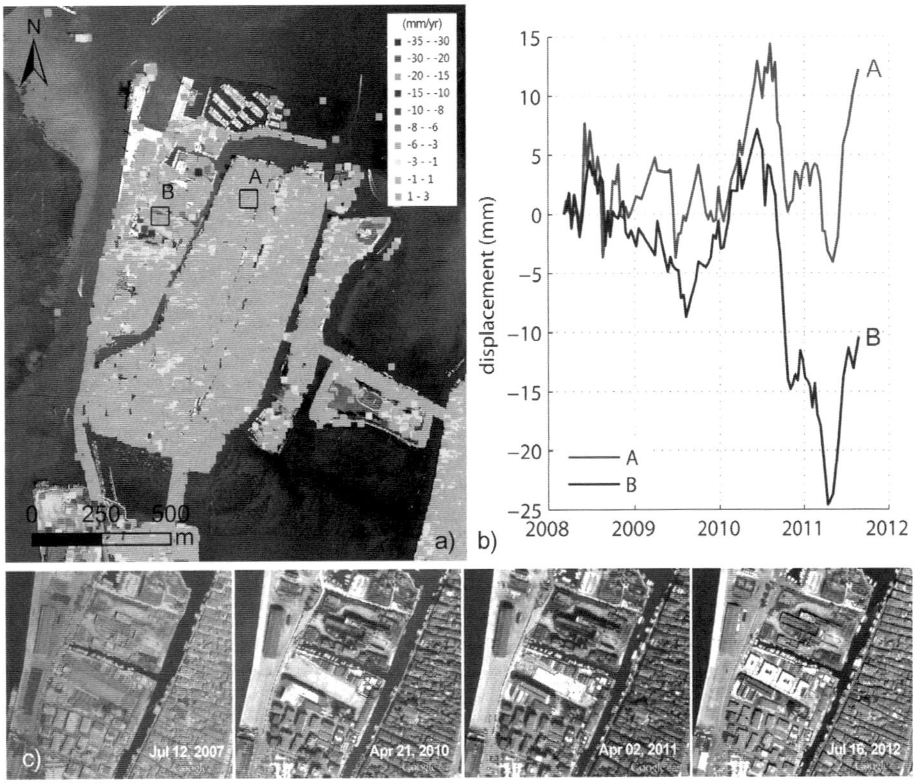

Figure 5. **(a)** Mean displacement rates between March 2008 and August 2011 from TerraSAR-X interferometry at Chioggia. The image background is an aerophotograph acquired in 2007. Negative values indicate settlement, positive mean uplift. **(b)** Displacement history in the satellite line-of-sight direction for two scatterers (A, B). **(c)** Satellite Google Earth images showing the building construction carried out at location B from 2010, cause of the large consolidation highlighted by PSI measurements in the zone from mid-2010 (Google Earth, data source: European Space Imaging, Image NASA; Digital Globe, Image). See location 3a in Fig. 1.

ters. Presently the works at the inlets are still ongoing and monitoring land displacements in these sectors, has a double aim, namely (i) the evaluation of possible effects of the works on the littoral environment, and (ii) the quantification of the consolidation of the new coastal structures. The former issue can support the public authorities in monitoring the environmental impacts of these giant works, the latter can help the engineering companies building the MoSE to check the absolute and differential displacements that can threaten the integrity and efficiency of the structures.

Figure 3a shows a large variability of the displacement velocities characterizing the different degree of the consolidation of the structures at the Malamocco inlet. Sinking rates less than 3 $\mathrm{mm\,yr^{-1}}$ characterize the parts of the jetties where older restoration works were carried out (A, B), while sinking rates up to 30 $\mathrm{mm\,yr^{-1}}$ are measured in the newer structures mainly built in the southern side (C, D). Figure 3b shows the displacement history of four scatterers selected in correspondence of old restoration works and new structures. The scatterers A and B highlight the ground stability after the old restoration in these parts of the inlets. The scatterer C, located in the new structure of the outer jetty, sinks about 20 $\mathrm{mm\,yr^{-1}}$ in the whole monitored period, the lower values detected after 2010 show that the primary consolidation has likely been completed. Instead scatterer D, in the newest portion of the lock, is still experiencing primary consolidation due to the new load, with velocity rates of up to 30 $\mathrm{mm\,yr^{-1}}$. Notice that the works at the inlet generally do not affect the land subsidence in the littoral surroundings.

3.2.2 The natural marshlands and the fish farms

Monitoring of land subsidence in natural marshlands and fish farms is a major challenge. The use of traditional methodologies (levelling, GPS) are impractical in these environment. SAR- based methodologies represent a unique approach in this context. Specifically, the X-band acquisitions, which are characterized by a higher resolution with respect to the C-band scenes, have demonstrated a high effectiveness. Figure 4 shows the impressive capability of IPTA to detect

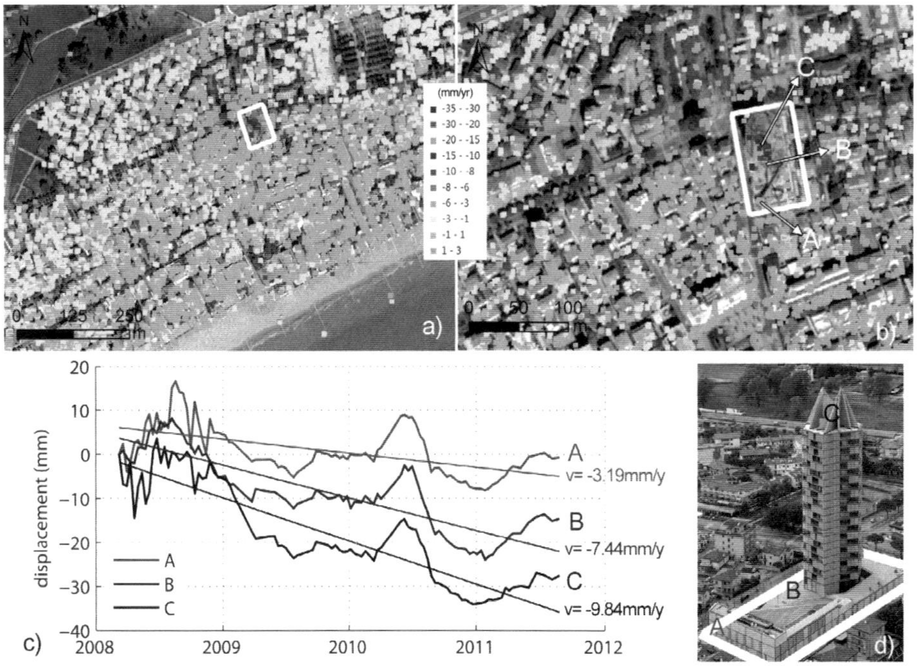

Figure 6. **(a, b)** Mean displacement rates between March 2008 and August 2011 from TerraSAR-X interferometry at Jesolo. The image background is an aerophotograph acquired in 2007. Negative values indicate settlement, positive mean uplift. **(c)** Displacement history in the satellite line-of-sight direction for three scatterers (A, B, C). **(d)** Photograph of Aquileia Tower, Jesolo, which is the cause of the largest movements. See location 4a in Fig. 1.

the displacements of thousands of radar targets in the Valle Paleazza, a fish farm in the northern Venice lagoon. The majority of the PS are located on artificial embankments bounding these lagoon zones whose water level is artificially managed, but several radar reflectors are also located within the fish farm and also on the natural marshland located to the west. The displacement rates are extremely variable, with the largest subsidence usually detected in the inner zones while the fish farm embankments, built up in the previous centuries, are more stable.

3.2.3 The lagoon and coastal urban areas

Apart from Venice, which has been the focus of several studies on land subsidence most recently (Tosi et al., 2013), other important cities and tourist villages are located in the area. Two main tourist centres are Chioggia and Jesolo at the southern and northern tip of the Venice lagoon, respectively.

The subsidence maps detected by PSI for these two cities are provided in Figs. 5 and 6, respectively. Chioggia is almost stable and Jesolo subsides at an average rate equal to 3 mm yr^{-1}. However, local bowls of large subsidence, up to 10 mm yr^{-1}, have been detected in both urban areas. Detailed investigations point out that these occurrences are usually connected with the construction of new large buildings that exert new heavy loads on the land surface, leading to consolidation of the shallow subsoil below the foundations.

4 Conclusions

PSI on a large stack of stripmap TerraSAR-X images from 2008 to 2011 has proven particularly effective for monitoring land subsidence in the Venice coastland. The interferometric processing of this dataset allows for the characterization of the land movements at both the regional and local scales. The sufficiently large area covered by the images allows the use of a few CGPS stations to calibrate the radar solution. The high resolution and acquisition frequency guarantees the possibility of accurately detecting local-scale short-time anomalies in the subsidence pattern. Our results clearly show the large variability characterizing the consolidation of the structures at the inlets; hence, the effect of the primary consolidation is clearly visible close to the construction of new buildings due to the ongoing urbanization of the coastline. Venice is still experiencing natural land subsidence with rates in the range 1.1 ± 1.0 mm yr^{-1}, mainly because of the heterogeneous nature and age of the lagoon subsoil.

TerraSAR-X acquisitions over the Venice coastland continued at a regular interval of 11 days after August 2011. We plan to update the PSI analysis using 49 new images acquired through November 2013. In the newest PSI analysis particular attention will be given to short observation periods, the occurrence of appearing and disappearing scatterers, and to the thermal dilation component of displacement (Monserrat et al., 2011) for a better characterization of the behaviour of large structures.

Acknowledgements. This work has been developed in the framework of the Action 2 (SP3-WP1) funded by the Flagship Project RITMARE – The Italian Research for the Sea – coordinated by the Italian National Research Council and funded by the Italian Ministry of Education, University and Research within the National Research Program 2011–2013. Data courtesy: (1) TerraSAR-X, Project COA0612© DLR "Assessing vertical movements of natural tidal landforms and anthropogenic structures at the Venice Lagoon inlets" and (2) CAVA and SFEL CGPS time series, Magistrato alle Acque (Venice Water Authority) through its concessionary Consorzio Venezia Nuova, and (3) VEN1 time series, Nevada Geodetic Laboratory (NGL), obtained from http://geodesy.unr.edu/.

References

Carbognin, L., Teatini, P., and Tosi, L.: Relative land subsidence in the lagoon of Venice, Italy, at the beginning of the new millennium, J. Mar. Sys., 51, 345–353, 2004.

Dong, S., Samsonov, S., Yin, H., and Ye, S.: Time-series analysis of subsidence associated with rapid urbanization in Shanghai, China measured with SBAS InSAR method, Environ. Earth Sci., 72, 677–691, 2014.

Erban, L. E., Gorelick, S. M., and Zebker, H. A.: Groundwater extraction, land subsidence, and sea-level rise in the Mekong Delta, Vietnam, Environ. Res. Lett., 9, 084010, doi:10.1088/1748-9326/9/8/084010, 2014.

Monserrat, O., Crosetto, M., Cuevas, M., and Crippa, B.: The Thermal Expansion Component of Persistent Scatterer Interferometry Observations, Geosci. Remote Sens. Lett., 8, 864–868, doi:10.1109/LGRS.2011.2119463, 2011.

Ng, A. H.-M., Ge, L., Li, X., Abidin, H. Z., Andreas, H., and K. Zhang: Mapping land subsidence in Jakarta, Indonesia using persistent scatterer interferometry (PSI) technique with ALOS PALSAR, Int. J. Appl. Earth Obs., 18, 232–242. 2012.

Phien-wej, N., Giao, P. H., and Nutalaya, P.: Land subsidence in Bangkok, Thailand, Eng. Geol., 82, 187–201, 2006.

Strozzi, T., Teatini, P., and Tosi, L.: TerraSAR-X reveals the impact of the mobile barrier works on Venice coastland stability, Remote Sens. Environ., 113, 2682–2688, 2009.

Teatini, P., Tosi, L., Strozzi, T., Carbognin, L., Wegmüller, U., and Rizzetto, F.: Mapping regional land displacements in the Venice coastland by an integrated monitoring system, Remote Sens. Environ., 98, 403–413, 2005.

Teatini, P., Tosi, L., Strozzi, T., Carbognin, L., Cecconi, G., Rosselli, R., and Libardo, S.: Resolving land subsidence within the Venice Lagoon by persistent scatterer SAR interferometry, Phys. Chem. Earth. Pt. A/B/C, 40, 72–79, 2012a.

Teatini, P., Tosi, L., and Strozzi, T.: Comment on "Recent subsidence of the Venice Lagoon from continuous GPS and interferometric synthetic aperture radar" by Y. Bock, S. Wdowinski, A. Ferretti, F. Novali, and A. Fumagalli., Geochem. Geophys. Geosys. 13, Q07008, doi:10.1029/2012GC004191, 2012b.

Tosi, L., Teatini, P., Strozzi, T., Carbognin, L., Brancolini, G., and Rizzetto, F.: Ground surface dynamics in the northern Adriatic coastland over the last two decades, Rend. Fis. Acc. Lincei, 21, S115–S129, 2010.

Tosi, L., Teatini, P., Bincoletto, L., Simonini, P., and Strozzi, T.: Integrating geotechnical and interferometric SAR measurements for secondary compressibility characterization of coastal soils, Surv. Geophys., 33, 907–926, 2012.

Tosi, L., Teatini, P., and Strozzi, T.: Natural versus anthropogenic subsidence of Venice, Nat. Sci. Reports, 3, 2710, doi:10.1038/srep02710, 2013.

Wegmüller, U., Werner, C., Strozzi, T., and Wiesmann, A.: Multitemporal interferometric point target analysis, Ser. Remote Sens., 3, 136–144, 2004.

Werner, C., Wegmüller, U., Strozzi, T., and Wiesmann, A.: Interferometric Point Target Analysis for deformation mapping. Proc. of the Geoscience and Remote Sensing Symposium (IGARSS 2003), IEEE International, 7, 4362–4364, 2003.

Proc. IAHS, 372, 207–210, 2015
proc-iahs.net/372/207/2015/
doi:10.5194/piahs-372-207-2015

Open Access
PIAHS

Prevention partition for land subsidence induced by engineering dewatering in Shanghai

J. X. Wang[1,2], **X. T. Liu**[1], **and T. L. Yang**[3]
[1]Tongji University, Shanghai, China
[2]CCCC Key Lab of Environment Protection & Safety in Foundation Engineering of Transportation, Guangzhou, China
[3]Shanghai Institute of Geological Survey, Shanghai, China
Correspondence to: J. X. Wang (wang_jianxiu@163.com)

Published: 12 November 2015

Abstract. Land subsidence in shanghai has been found for more than 70 years. In the early years, it was mainly caused by groundwater exploitation. In recent years, engineering dewatering in shallow ground (within 90 m) has become a major source for land subsidence in the rapid urbanization course. A management partition of land subsidence induced by foundation pit dewatering for the first confined aquifer was suggested.

1 Introduction

Shanghai is the economic and financial center of China. In 1939, land subsidence was first observed and its maximum value has reached 2.0 m until now. In the early years, the subsidence was caused by groundwater extraction for water supply. Later, the exploitation was restricted in order to control the land subsidence. However, during the rapid urbanization in recent years, lots of underground engineering were constructed, the engineering dewatering in shallow ground (within 90 m) has become a major source contributing to increasing land subsidence.

Xu et al. (2009) analyzed the geohazards which may occur during the construction and maintenance of infrastructures in Shanghai, and the confined water was one of the main causes. Shen and Xu (2011) used numerical simulation to predict the behavior of land subsidence in Shanghai due to pumping of groundwater. Wang et al. (2012, 2013), Zhang et al. (2013), Luo et al. (2008) and Zhou et al. (2010) used numerical simulation to analyze the land subsidence caused by foundation pit dewatering. However, most previous studies focused on a single foundation pit. The land subsidence of engineering dewatering in a city's scale is not concerned too much.

Table 1. Stratigraphic characteristics of central city in Shanghai.

Layer No.	Composition	Note
②	Clay	
③	Muddy clay	
④	Muddy clay	
$⑤_1$	Silty clay	
$⑤_2$	Sandy silt	The micro-confined aquifer
$⑤_{3-4}$	Silty clay	
⑥	Clay	
⑦	Sand	The first confined aquifer
⑧	Clay	
⑨	Sand	The second confined aquifer

2 Geological partition for strata combination

The Quaternary sediments (Table 1) of Shanghai belong to the deposit of Yangtze River and tidal current of East China Sea. The foundation pit dewatering in Shanghai mainly relates to the first confined aquifer (layer ⑦). The aquifer is formed in the Late Pleistocene and mainly consists of sand. In normal deposit environments, layer ⑥ is distributed above layer ⑦, but it is replaced by layers $⑤_2$ and $⑤_{3-4}$ in the ancient river erosion area. Layer ⑧ is distributed between layers ⑦ and ⑨, but in some area, it does not exist where the layer ⑦

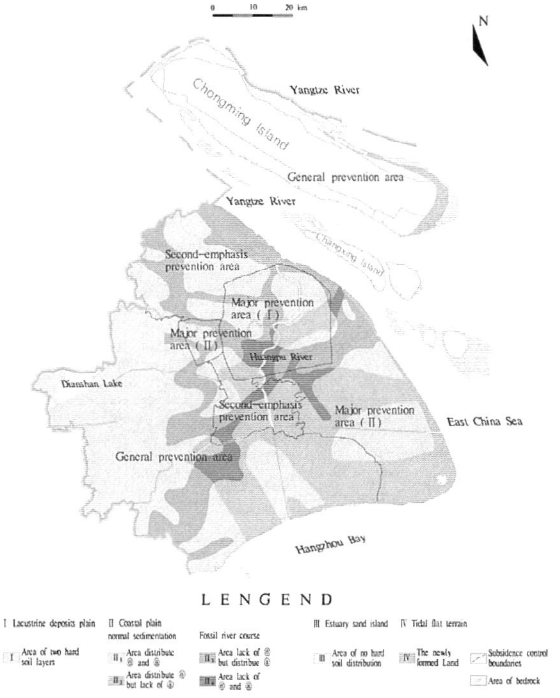

Figure 1. Geological partition for the first aquifer in Shanghai.

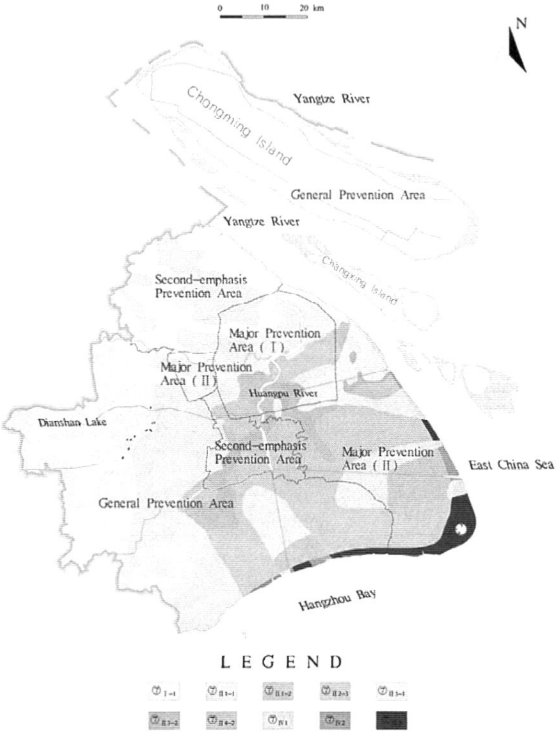

Figure 2. Prevention partition for land subsidence induced by engineering dewatering.

is connected with layer ⑨. So the first confined aquifer was divided into 7 geological partitions (Fig. 1 and Table 2).

3 Prevention partition for land subsidence

According to the urban development planning (SMCPA, 2012), the land subsidence prevention was divided into 4 partitions: emphasized prevention area I, emphasized prevention area II, second emphasized prevention area and general prevention area.

4 Prevention partition for land subsidence induced by engineering dewatering

In foundation pit dewatering, the curtain (diaphragm wall) is the most effective way to control the drawdown and land subsidence outside the pit (Wang et al., 2009). Considering the curtain's depth of present technology, the first confined aquifer was divided into 3 partitions in vertical direction. The dividing depth is 60 and 70 m separately.

A prevention partition for land subsidence induced by engineering dewatering for the first confined aquifer was suggested (Fig. 2). It includes 11 sub-partitions (Table 3). The maximum drawdown for different foundation pits was suggested for each partition (Table 4).

5 Conclusions

Engineering dewatering has become a major factor of land subsidence in Shanghai. A prevention partition of land subsidence induced by engineering dewatering for the first confined aquifer was developed and suggested.

Table 2. Geological partition for strata combination.

Deposit environment			Characteristic	Composition
Partition	Sub-partition			
I (Lacustrine deposits plain)	–		Layers ⑥ and ⑧	Silty sand
II (Coastal plain)	Normal sedimentation	II_1	Layers ⑥ and ⑧	Sand
		II_2	Layer ⑥, without Layer ⑧	
	Ancient river sedimentation	II_3	Layer ⑧, without layer ⑥	
		II_4	without layers ⑥ and ⑧	
III (Estuary sand island)	–		without layers ⑥ and ⑧	Without
IV (Tidal flat terrain)	–		Newly formed land	Sand

Table 3. Prevention partition for land subsidence induced by engineering dewatering.

Sedimentary characteristics and strata combination			Bottom depth of the confined aquifer "B" (m)	
Partition	Sub-partition	Combination	Number	Depth (m)
I	–	Layers ⑥ and ⑧	1	$30<B\leq 60$
II	II_1	Layers ⑥ and ⑧	1	$30<B\leq 60$
			2	$B>60$
	II_2	Layer ⑥, without Layer ⑧	3	⑦ and ⑨ connected
	II_3	Layer ⑧, without layer ⑥	1	$30<B\leq 60$
			2	$B>60$
	II_4	Without layers ⑥ and ⑧	3	⑦ and ⑧ connected
IV	–	Newly formed land	1	$30<B\leq 60$
			2	$B>60$
			3	⑦ and ⑧ connected

Table 4. Control indexes of drawdown for partitions for land subsidence induced by engineering dewatering.

Partition	Prevention partition for land subsidence		
	Emphasized prevention partition (m)	Second-emphasized prevention (m)	General prevention partition (m)
⑦$_{II\,1-1}$	2.0	2.0	2.0
⑦$_{II\,1-2}$			
⑦$_{II\,2-3}$	1.0	1.0	1.5
⑦$_{II\,3-1}$	1.5	2.0	2.0
⑦$_{II\,3-2}$			
⑦$_{II\,4-3}$	0.5	1.0	1.5
⑦$_{IV\,2}$	1.0	–	1.5

Acknowledgements. This work is supported by the LSMP Open Funding (KLLSMP201403, KLLSMP201404), and the research grant (No. 201311045-04) from the Special Fund for Land and Resources-scientific Research in the Public Interest of China.

References

Luo, Z. J., Zhang, Y. Y., and Wu, Y. X.: Finite element numerical simulation of three-dimensional seepage control for deep foundation pit dewatering, J. Hydrodyn., 20, 596–602, 2008.

Shen, S. L. and Xu, Y. S.: Numerical evaluation of land subsidence induced by groundwater pumping in Shanghai, Can. Geotech., 48, 1378–1392, 2011.

SMCPA: The 12th Five-Year Plan for the land subsidence of Shanghai, SMCPA China, 2012.

Wang, J. X., Hu, L. S., Wu, L. G., Tang, Y. Q., Zhu, Y. F., and Yang, P.: Hydraulic Barrier Function of the Underground Continuous Concrete Wall in the Pit of Subway Station and its Optimization, Environ. Geol., 57, 47–453, 2009.

Wang, J. X., Feng, B., Liu, Y., Wu, L. G., Zhu, Y. F., Zhang, X. S., Tang, Y. Q., and Yang, P.: Controlling subsidence caused by de-watering in a deep foundation pit, B. Eng. Geol. Environ., 71, 545–555, 2012.

Wang, J. X., Feng, B., Yu, H. P., Guo, T. P., Yang, G. Y., and Tang, J. W.: Numerical study of dewatering in a large deep foundation pit, Environ. Earth. Sci., 69, 863–872, 2013.

Xu, Y. S., Shen, S. L., and Du, Y. J.: Geological and hydrogeological environment in Shanghai with geohazards to construction and maintenance of infrastructures, Eng. Geol., 109, 241–254, 2009.

Zhang, X. S., Wang, J. X., Wong, H., Leo, C. J., Liu, Q., Tang, Y. Q., Yan, X. L., Sun, W. H., Huang, Z. Q., and Hao, X. H.: Land subsidence caused by internal soil erosion owing to pumping confined aquifer groundwater during the deep foundation construction in Shanghai, Nat. Hazards., 69, 473–489, 2013.

Zhou, N. Q., Vermeer, P. A., Lou, R. X., Tang, Y. Q., and Jiang, S. M.: Numerical simulation of deep foundation pit dewatering and optimization of controlling land subsidence, Eng. Geol., 114, 251–260, 2010.

Proc. IAHS, 372, 211–216, 2015
proc-iahs.net/372/211/2015/
doi:10.5194/piahs-372-211-2015

Land subsidence of clay deposits after the Tohoku-Pacific Ocean Earthquake

K. Yasuhara[1] **and M. Kazama**[2]

[1]Institute for Global Change Adaptation Science, Ibaraki University, 2-1-1 Bunkyo, Mito, Ibaraki, 310-8512, Japan
[2]Graduate School of Engineering, Tohoku University, 2-1-1 Katahira Aoba-ku Sendai-shi, Miyagi, 980-8577, Japan

Correspondence to: K. Yasuhara (kazuya.yasuhara.0927@vc.ibaraki.ac.jp)

Published: 12 November 2015

Abstract. Extensive infrastructure collapse resulted from the cataclysmic earthquake that struck off the eastern coast of Japan on 11 March 2011 and from its consequent gigantic tsunami, affecting not only the Tohoku region but also the Kanto region. Among the geological and geotechnical processes observed, land subsidence occurring in both coastal and inland areas and from Tohoku to Kanto is an extremely important issue that must be examined carefully. This land subsidence is classifiable into three categories: (i) land sinking along the coastal areas because of tectonic movements, (ii) settlement of sandy deposits followed by liquefaction, and (iii) long-term post-earthquake recompression settlement in soft clay caused by dissipation of excess pore pressure. This paper describes two case histories of post-earthquake settlement of clay deposits from among the three categories of ground sinking and land subsidence because such settlement has been frequently overlooked in numerous earlier earthquakes. Particularly, an attempt is made to propose a methodology for predicting such settlement and for formulating remedial or responsive measures to mitigate damage from such settlement.

1 Introduction

Land subsidence and ground sinking are as remarkable and extremely severe phenomena as the gigantic tsunami that struck Tohoku and Kanto in Japan after the 2011 Off the Pacific Coast of Tohoku Earthquake (hereinafter the "Tohoku Earthquake"). Moreover, the mechanisms of both events are extremely complex because several factors induced land subsidence and ground sinking from tectonic, geologic, and geotechnical aspects, exerting overlapping influences that persist even now, four years after the earthquake.

Roughly speaking, this land subsidence and ground sinking is classifiable into three categories: (i) land sinking along coastal areas because of tectonic movements (Imakiire and Koarai, 2012, Yasuhara et al., 2012;, (ii) settlement of sandy deposits followed by liquefaction, and (iii) long-term post-earthquake recompression settlement in soft clays caused by the dissipation of excess pore pressure. Land sinking described in category (i) induces inundation of wide coastal areas, particularly during severe storm-surges and strong typhoons. However, land subsidence in the latter two categories (ii) and (iii) occurs in inland areas, strongly influencing local residents. A difference is apparent between the mechanisms of settlement of both types in that settlement included in category (ii) is related to saturated sandy deposits, whereas settlement in category (iii) is associated with saturated clayey deposits, which cause them to take a long time to cease. Based on differences among the causes and mechanisms related to the three kinds of land subsidence described above, suitable countermeasures against subsidence should be taken, corresponding to their different characteristics.

This paper presents two case histories of post-earthquake settlement of clay deposits from among the three categories of ground sinking and land subsidence described above because normally such settlement has been overlooked in numerous historical earthquakes.

2 Post-earthquake settlement of cohesive soil deposits in Miyagi

Generally speaking, liquefaction-induced settlement described in the previous section ceases in a very short time. As described later, post-earthquake long-term settlement is caused mainly by consolidation of clayey soils situated under sand deposits simply because the hydraulic conductivity of clay layers is much lower than that of sand deposits. For that reason, a long time is necessary for the dissipation of excess pore pressure generated during earthquakes (Yasuhara et al., 2001; Yasuhara and Matsuda, 2002; Matsuda et al., 2014).

Figure 1 depicts the variations of settlement at bench marks after 1974, when the total amount of settlement was greater than 200 mm up to 2010. Settlement measuring locations are benchmarks located in the eastern part of Sendai on the Quaternary alluvial lower plain of unconsolidated layers of around 50 m. It is apparent from Fig. 1 that a sudden increase of settlement occurred after the 1978 Miyagioki Earthquake. It then progressed at a higher rate of settlement than before the earthquake. The sudden increase of settlement in 1978 resulted from the collapse of clay soil particle structures: an immediate effect of the earthquake action.

Acceleration of post-earthquake settlement as shown in Fig. 1 originated from the superimposition of land subsidence before the earthquake and consolidation settlement caused by the dissipation of excess pore pressure generated during the earthquake.

Figure 2 depicts an example of variations of settlement over time before and after the Tohoku Earthquake in 2011. The tendency presented in Fig. 2 is almost identical as that presented previously in Fig. 1.

Kazama (2014) reported that post-earthquake settlement of clay deposits are caused by collapse of clay soil particle structures immediately after the earthquake and the recovery of collapsed structures to the new state of clay structures This mechanism for post-earthquake settlement of clay soils is understood from Fig. 3, which portrays the progress of settlement in the form of void ratio, e vs. logarithmic of vertical stress, log p. To validate the proposed concept for estimating post-earthquake settlement, undrained cyclic triaxial tests followed by drainage were conducted by Kazama using undisturbed and remolded clay specimens taken from the site where settlement had taken place. Subsequently, the results from undrained cyclic triaxial tests followed by drainage under the conditions of 0.4 for cyclic stress ratio and 0.2 Hz. for cyclic frequency are presented for comparison in Fig. 3 along with results from odometer tests on the undisturbed and remolded specimens of the same clay as that used in cyclic triaxial tests.

Void ratios after termination of undrained cyclic triaxial tests followed by drainage are shown on the vertical line corresponding to the consolidation pressure with $e - \log p'$ curves in Fig. 3. Three dotted circles are shown after each

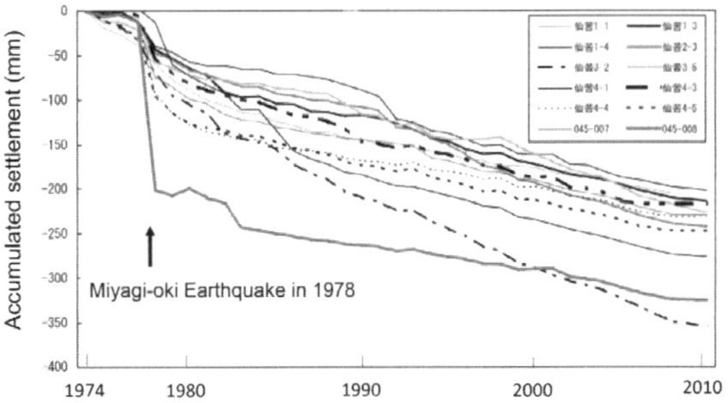

Figure 1. Examples of land subsidence variation in Sendai before and after the earthquake.

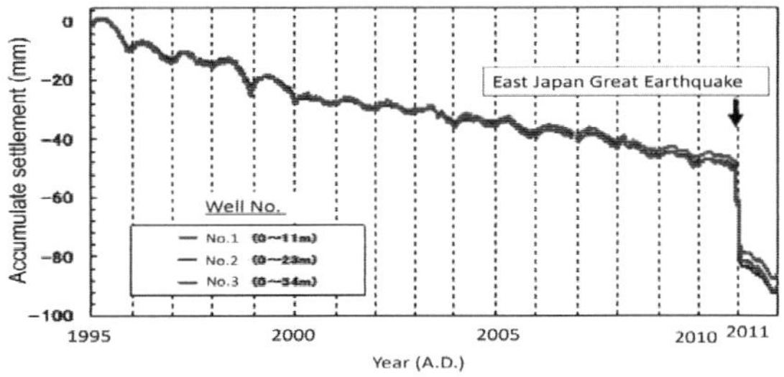

Figure 2. Variation of settlement before and after the Tohoku Earthquake (at Kamomachi of Sendai).

undrained loading followed by drainage, which simulates the circumstances under which three earthquakes struck separately after some intervals.

If initial void ratios under a current consolidation pressure p_i' for undisturbed and remolded conditions corresponds to cases before and after the earthquakes are assumed as $e_{i,\mathrm{u}}$ and $e_{\mathrm{f,r}}$, respectively, then post-earthquake settlement S_{post} is given as

$$S_{\mathrm{post}} = \frac{e_{i,\mathrm{u}} - e_{\mathrm{f,r}}}{1 + e_{i,\mathrm{u}}} H \tag{1}$$

where H stands for the clay layer height. If one assumes 2.15 and 1.85 for $e_{i,\mathrm{u}}$ and $e_{\mathrm{f,r}}$, respectively, and 10 m for the clay layer height, then one obtains 47.6 cm for post-earthquake settlement S_{post}.

Another methodology can be used to estimate the post-earthquake settlement of clay layers proposed by Yasuhara et al. (1991, 2001). It is given as

$$S_{\mathrm{post}} = 0.225 C_c\, H \log\left(\frac{1}{1 - \frac{u_{\mathrm{cy}}}{p'_c}}\right), \tag{2}$$

where C_c stands for the compression index of clay soil, H signifies the clay layer height, u_{cy} represents excess pore pressure generated by the earthquake, and p_c' denotes the consolidation or overburden pressure. When we take values

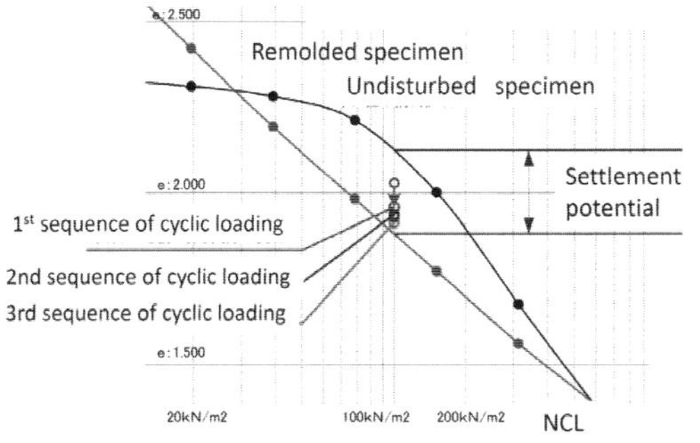

Figure 3. Void ratio vs. consolidation pressure relations for prediction of post-earthquake settlement of clay deposits (from Kazama, 2014).

from Fig. xx for C_c, H, u_{cy} and p'_c as 1.06, 5.0 m, 0.5 p'_c and 110 kN m^{-2}, then we obtain 33.9 cm for S_{post}. It is assumed here for calculation using Eq. (2) that the normalized excess pore pressure in the clay layer, u_{cy}/p'_c is 0.5. Therefore, the amount of settlement will be less than this value, 05, when $u_{cy}/p'c$ is smaller than 0.5. The amount of settlement of 33.9 cm obtained using Eq. (2) is less than the 47.6 cm obtained using Eq. (1).

3 Post-earthquake settlement of cohesive soil deposits in Ibaraki

3.1 Outlines of post-earthquake settlement of residential areas beside the River Dykes

Another case study in Ibaraki is that where clay deposits near the river dyke have undergone post-earthquake settlement causing damage to many private residences. The horizontal profile of the river dyke and the residential areas portrayed in Fig. 4 indicate a thick clay layer with xx m deposits on the sand layer with the higher ground water level (GWL). clay is characterized by the following. Highly sensitive with high liquidity index High water content with high undrained strength Post-earthquake settlement is expected to result from consolidation followed by dissipation of excess pore pressures generated by a strong earthquake. The average index properties of clay are presented in Table 1. This

Characteristic (ii) does not make sense from geotechnical perspective. It is extremely unusual although characteristic (i) does make sense. This clay has high structures of clay particles, which readily undergo deformation by disturbance, leading to radical deformation and settlement. For comparison with Kujikawa clay, Table 1 also includes index properties of Joban clay, deposited beneath the Kanda area of Joban express highway, which was constructed almost 35 years ago. Both clays are similar: both are highly plastic and become considerably soft with water contents beyond the liquid limit, although it cannot be said that the water content of either clay is very high.

Table 1. Index properties of clay beneath river dykes in Ibaraki.

Index	unit	Kujigawa clay	Jyouban clay
Void ratio e		2.17	2.40
Density of soil particle ρ_s	g/cm^3	2.67	2.68
Wet density ρ_t	g/cm^3	1.52	-
Natural water content w_i	%	81.0	93.7
Liquid limit w_L	%	73.0	70.7
Plasticity limit w_L	%	37.0	28.8
Plasticity index I_p		36.0	41.9
Liquidity index I_L		1.20	1.20
Sensitivity ratio S_t		17.0	15.0
Unconfined compression strength q_u	kN/m^2	90.9	70.5

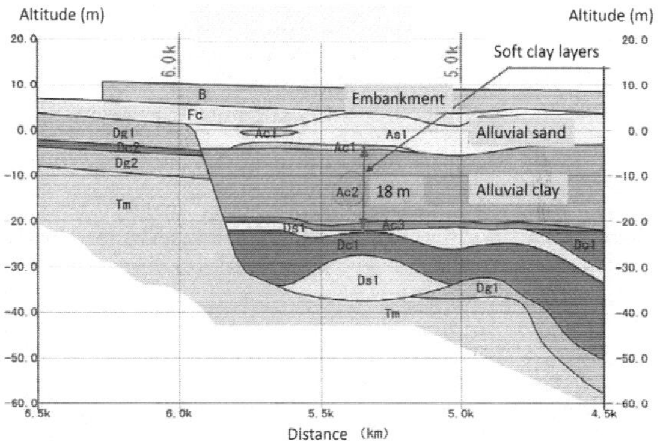

Figure 4. Geological profile of the objective site.

A set of typical observed settlement vs. time relations continuously before and after the earthquake is presented in Fig. 5. Settlement before the earthquake is caused mainly by land subsidence because of ground water abstraction. However, settlement after the earthquake is very complex but perhaps caused by combining the following:

i. land sinking by crustal movement,

ii. immediate settlement followed by liquefaction of upper sand deposits, which is not time-dependent, and

iii. time-dependent settlement of clay followed by dissipation of excess pore pressure generated by earthquakes.

Those two components are present in addition to land subsidence.

The current paper, however, addresses issue (iii) because it emphasizes an examination of post-earthquake time-dependent settlement of clay deposits, which corresponds to the latter part of the settlement vs. time curves depicted in Fig. 6. The rates of settlement before and after the earthquake are quite noticeable. The settlement has been exacer-

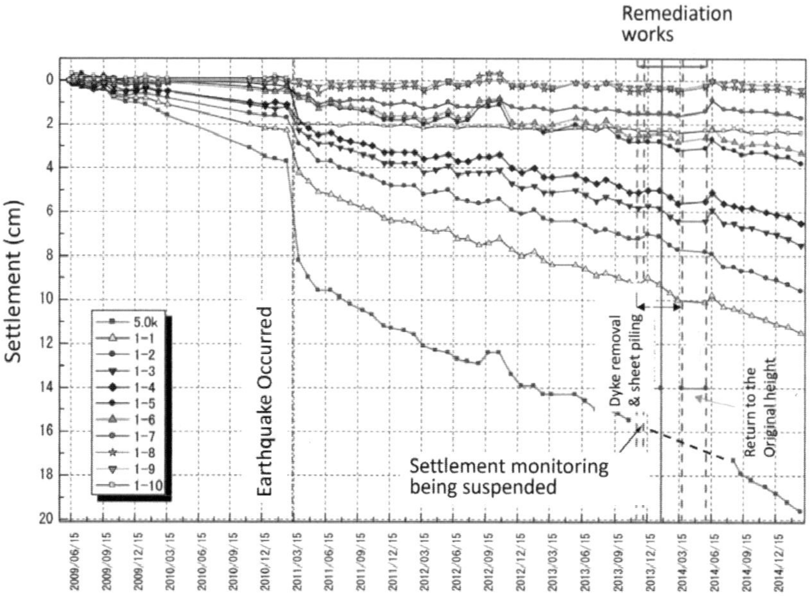

Figure 5. Settlement vs. time curves of river dykes.

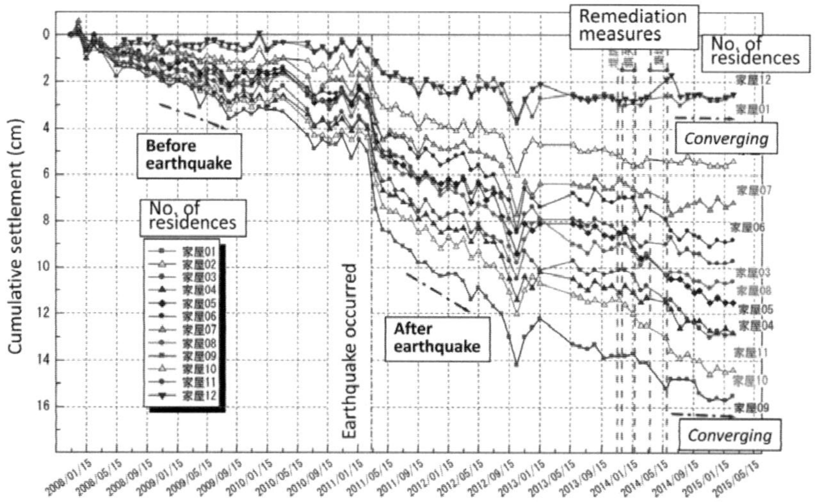

Figure 6. Settlement vs. time curves for residences.

bated by earthquakes because of the post-earthquake settlement of clay deposits caused by delayed dissipation of excess pore pressure in clay deposits after the earthquake. Generally speaking, such settlement takes a long time to cease. Therefore, it is necessary for engineers and researchers to ascertain when the rate of residual settlement after the earthquake becomes negligible for residents not to feel at risk in everyday life.

3.2 Results from settlement monitoring and numerical analysis

Since 2007, the Governmental Office in charge of river management in the objective area has been monitoring variations of settlement with time inside and outside the river dykes. Results of settlement monitoring indicate the following:

i. Settlement of residences accumulated up to 14 cm for the prior six years starting in 2007.

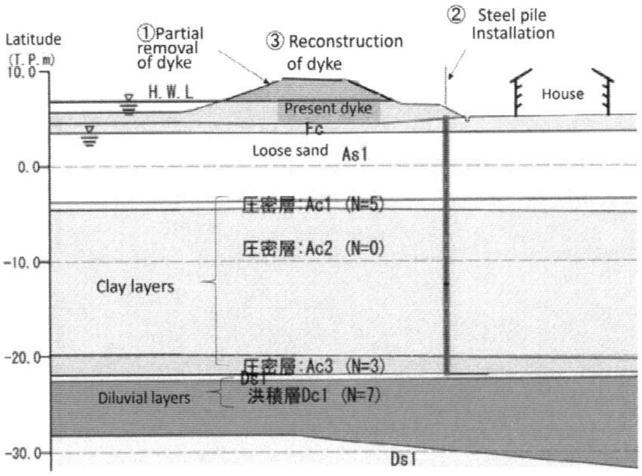

Figure 7. Remediation for reducing post-earthquake settlement of soft clay deposits.

ii. Settlement increased to around 2 cm after the earthquake in 2011.

According to elasto-visco-plastic two-dimensional numerical analysis for settlement of river dykes and residences conducted in addition to settlement monitoring:

i. The predicted total settlement of river dykes will be greater than 90 cm.

ii. Settlement must persist for more than 30 years.

iii. Inclination of residences also continues for a long time.

4 Effects of damage mitigation countermeasures

To prevent the effects of river dykes on post-earthquake settlement of residences, the following countermeasures were undertaken (see Fig. 7).

i. Remove 2.2 m of dykes to reduce dyke self-weight.

ii. Thereafter install sheet piles aside from the river dykes into the hard stratus, called A_{c3} layers, to intercept the effects of river dyke self-weight on residences.

iii. Return the dyke height to the original height with 4.1 m.

As might be readily apparent from the results depicted in Figs. 5 and 6, the following behavior was observed after carrying out the countermeasures stated above.

i. Remaining excess pore pressure on grounds both outside and inside the dykes became less than that before partial removal of dykes.

ii. Installation of sheet piles is associated with increased ground settlements beneath the dykes, but does not increase settlement of the ground outside the dykes, but rather might decrease settlement of the ground of residents.

iii. Filling up to the original height of dykes decreased settlement of the ground of residents because of the intercept of stress distribution of dykes to residential ground.

As a summary of the tendencies stated above, probably the countermeasures adopted herein have been effective to date for reducing post-earthquake settlement of residents near the river dykes. However, monitoring of settlements should be conducted because the influence of groundwater abstraction on continuous settlement of residential areas remains uncertain.

5 Conclusion

The paper presents an outline of the present situation of ground sinking, land subsidence, and settlement induced by tectonic movements on a global scale and ground movements on a local scale induced by the Tohoku Earthquake in 2011. Particularly, the paper describes a specific examination of post-earthquake settlement of clay layers observed in some locations in the Tohoku and Kanto regions. This paper describes two case histories related to this category of settlement: One in Miyagi for a methodology for predicting settlement and another in Ibaraki for effective countermeasures for settlement. Investigation of the two case histories suggests that careful attention should be devoted to the fact that such settlement is time-dependent and that monitoring of settlement should therefore continue for a long period. These phenomena show different tendencies from those of two kinds of settlement originating from tectonic movement, followed by liquefaction.

Acknowledgements. The authors are grateful for financial support from a Grant-in-Aid for Scientific Research from the Ministry of Education, Culture, Sports, Science and Technology (FY2014–FY2017, Project No. 26281055), Japan, whose representative is Makoto Tamura, Associate Professor of Ibaraki University, Japan.

References

Imakiire, T. and Koarai, M.: Wide-area land subsidence caused by "the 2011 OFF the Pacific Coast of Tohoku Earthquake", Soils Found., 52, 842–855, 2012.

Joint Editorial Committee in Tohoku Branch for the Report on the Great East Japan Earthquake Disaster 2013, Part 3 Geohazards, Ch. 8, Land sinking and inundation, 3-212–3-218, 2013 (in Japanese).

Kazama, M.: Geotechnical subjects seen to the damage of the 2011 off the Pacific Coast of Tohoku Earthquake, Proc. 11th Symp. on Ground Improvement, Journal of Material Science, Japan, 1–20, 2014 (in Japanese).

Matsuda, H., Nhan, T. T., Nakahara, D., Thien, D. Q., and Tuyen, T. H.: Post-cyclic recompression characteristics of a clay subjected to undrained uni0directional and multi-directional cyclic shears, Proc. 10th US National Conf. on Earthquake Engineering, Earthquake Eng. Research Institute, Anchorage, AK, 2014.

Ohara, S. and Matsuda, H.: Study on the settlement of saturated clay layer induced by cyclic shear, Soils Found., 28, 103–113, 1988.

Yasuhara, K. and Andersen, K. H.: Recompression of normally consolidated clay after cyclic loading, Soils Found., J. JGS, 31, 83–94, 1991.

Yasuhara, K. and Matsuda, H.: 5. Dynamic properties of cohesive soils, J. of Japan Geotechnical Society, 46, 59–64, 1998 (in Japanese).

Yasuhara, K. and Usui, T.: Potential application of geosynthetics for reconstruction following coastal land subsidence induced by the Great East Japan Earthquake, Geosynthetics Journal, Japan Chapter of IGS, 27, 69–77, 2012 (in Japanese).

Yasuhara, K., Murakami, S., Toyota, N. and Hyde, A. F. L.: Settlements in fine-grained soils under cyclic loading, Soils Found., J. of JGS, 41, 25–36, 2001.

Proc. IAHS, 372, 217–220, 2015
proc-iahs.net/372/217/2015/
doi:10.5194/piahs-372-217-2015

Open Access
PIAHS

Liquefaction-fluidization induced land subsidence: impact of the 2011 Tohoku earthquake on reclaimed land around Tokyo bay area, Japan

A. Kagawa, K. Furuno, T. Kusuda, Y. Sakai, T. Yoshida, and O. Kazaoka

Research Institute of Environmental Geology, Chiba (RIEGC), 3-5-1 Inagekaigan, Mihama, Chiba city, Japan

Correspondence to: A. Kagawa (kagawaa2016@gmail.com)

Published: 12 November 2015

Abstract. The 2011 Tohoku earthquake caused major liquefaction-induced, ground deformation of the reclaimed land surrounding Tokyo Bay. In this area, liquefaction was visibly manifest by sand boils, ejection of sandy water, land subsidence and floating underground tanks. The level measurements show a correspondence between the degree of liquefaction-fluidization and the amount of subsidence. The strata most susceptible to liquefaction are hydraulically emplaced dredged fill and artificial strata on thick uncompacted Holocene deposits. On the other hand, the phenomena of seismic isolation coursed by liquefaction had saved the single-family houses from collapse.

1 Introduction

The 2011 off the Pacific coast of Tohoku Earthquake caused heavy damage by Tsunami and geologic disaster on east Japan. Tokyo bay area is underlain by hydraulically emplaced dredged fill and artificial strata resting on thick uncompacted Holocene deposits. In this area, liquefaction was visibly manifest by sand boils, ejection of muddy groundwater, land subsidence and floating underground tanks. In partly, the banks on the edge of the reclaimed lands had broken by the outflowing boiled sand pressure. The authors have investigated liquefaction-induced phenomena in southeast part of the Kanto Groundwater Basin, and report the disaster of Urayasu City on north part of Tokyo bay area (Fig. 1).

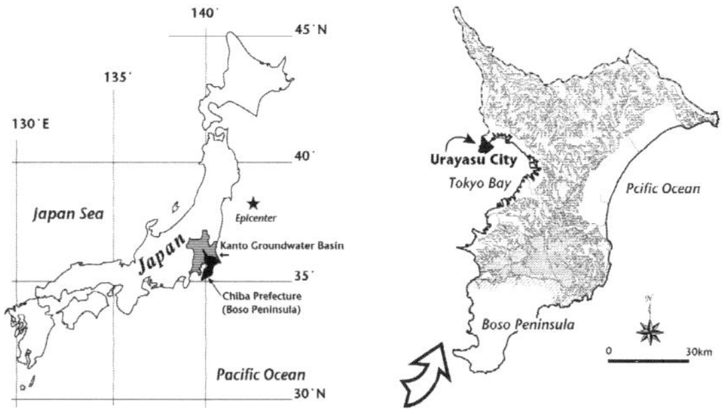

Figure 1. Location of Urayasu City and the epicenter of the 2011 Tohoku earthquake.

2 Level measurements

Urayasu city, located along the northeast coast of the Tokyo bay, is composed of Holocene deltaic strata and man-made strata (Fig. 2). The authors surveyed by levelling the elevation differences of a number of the benchmarks before (2010) and after (2011) the earthquake. The results show that about 4 cm of subsidence developed in the area with Holocene deposits, and more than 17 cm in the reclaimed land (Fig. 3). Moreover, the authors investigated the differences between land subsidence of the ground level and the remains of the pre-earthquake surface in correspondence of the piled buildings (Fig. 4). The value of subsidence is representative of the degree of liquefaction-fluidization.

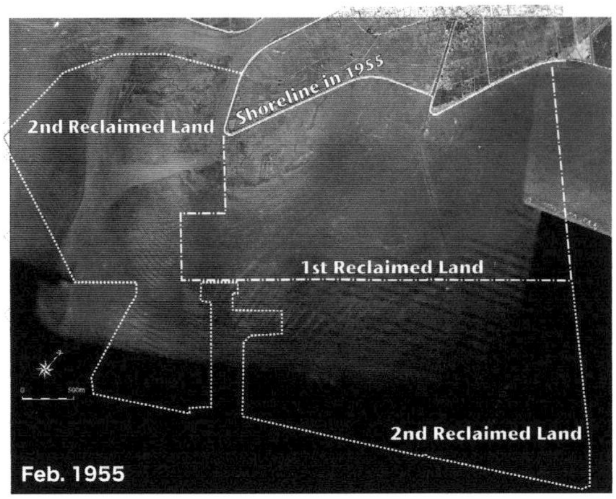

Figure 2. Original shoreline and the reclaimed land in Urayasu City (The aerial photo by Kokusai-kogyo co. ltd.).

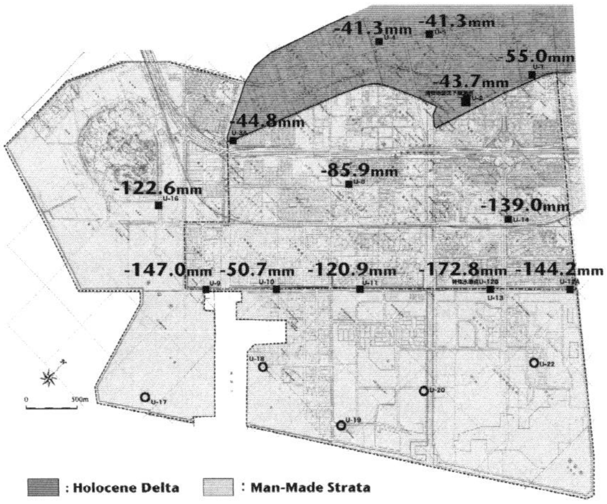

Figure 3. Elevation change of bench marks in Urayasu City (January 2011–2012).

3 Chidori block – the latest reclaimed land in Urayasu city

Chidori block, located in the southeast part of Urayasu city, is the latest hydraulically emplaced dredged fill, and has been used for the logistic warehouse and the incinerating office. This area has struck the severe liquefaction-fluidization hazards, land subsiding, sand boiling, telegraph poles tilting and broken road (Fig. 5). At the center of the Chidori block, land subsidence reached 95 cm. The historic aerial photographs show a correspondence between the dredged trench and the rates of subsidence (Fig. 6). The differences between the thickness of the man-made strata and Holocene sediments played an important role on the subsidence amount and distribution (Fig. 7).

4 Conclusions

The 2011 Tohoku earthquake caused major liquefaction and, ground deformation of the reclaimed land surrounding Tokyo Bay. In this area, liquefaction was visibly manifest by sand boils, ejection of sandy water, land subsidence and floating underground tanks. The banks on the edge of the reclaimed lands partially broke by the outflowing boiled sand pressure. Based on surface benchmarks and subsurface markers on 10-m deep piles, near-surface compaction was quantified in about 14 cm. Locally, land subsidence reached 95 cm. The level measurements show a correspondence between the degree of liquefaction-fluidization and the amount of subsidence. The strata most susceptible to liquefaction are hydraulically emplaced dredged fill and artificial strata on thick uncompacted Holocene deposits. Much reclaimed land around Tokyo Bay had been densely urbanized, thus land

Figure 4. Values of land subsidence caused by the 2011 Tohoku earthquake in Urayasu City (RIEGC, 2011).

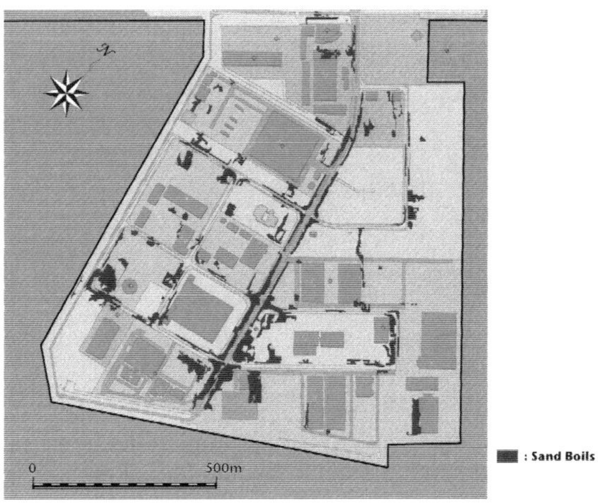

Figure 5. Distribution of sand boils in Chidori block, Urayasu City (RIEGC, 2011).

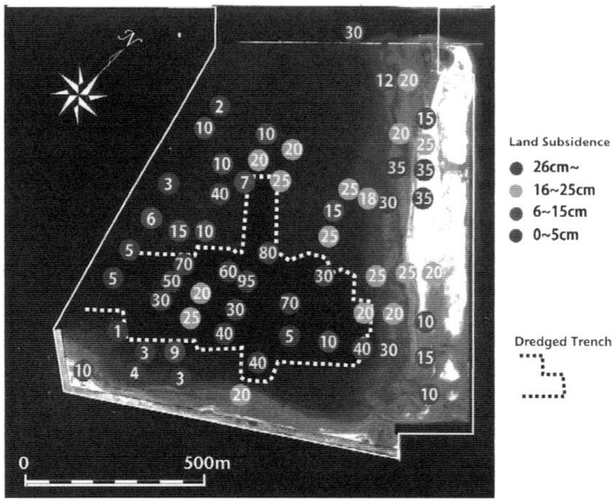

Figure 6. Subsidence values and dredged trench in Chidori block, Urayasu City (RIEGC, 2011).

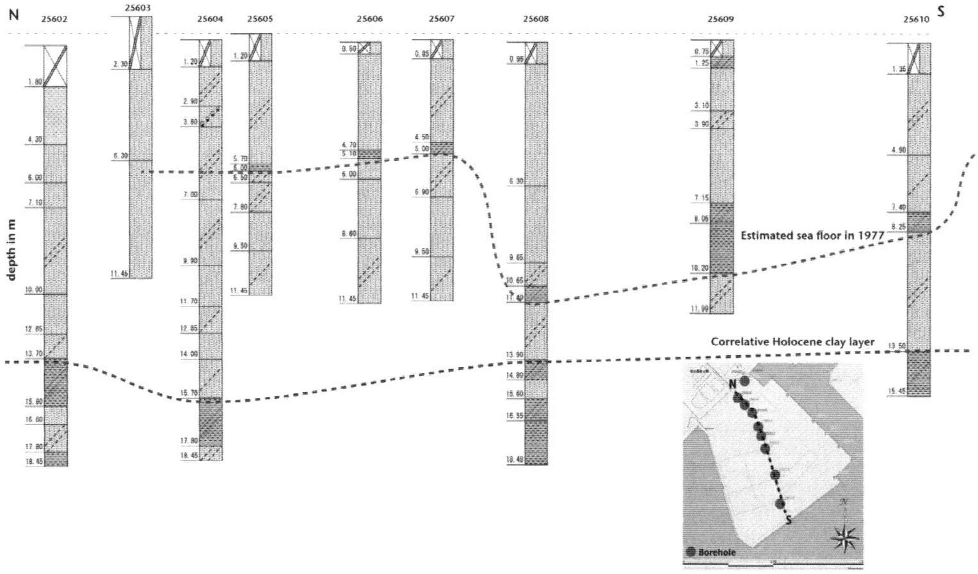

Figure 7. Geologic section in Chidori block, Urayasu City (RIEGC, 2011).

subsidence caused major damage to structures and infrastructure. On the other hand, the phenomena of seismic isolation coursed by liquefaction had saved the single-family houses from collapse (Kagawa et al., 2000). Seismically induced liquefaction may be not a rare event, and its impact on water-saturated man-made strata and reclaimed land is inherently costly.

Acknowledgements. We sincerely appreciate colleague of Research Institute of Environmental Geology, Chiba who have investigated the liquefaction-fluidization hazards with us.

References

Kagawa, A., Furuno, K., Kusuda, T., Satoh, K., Sakai,Y., Kamura, K., Kazaoka, O., Morisaki, M., and Nirei, H.: Land Subsidence in artificial islands due to liquefaction caused by the Kobe Earthquake in 1995, Japan, Proceedings of the Sixth International Symposium on Land Subsidence, 45–51, 2000.

Research Institute of Environmental Geology, Chiba, Japan: Investigation report of liquefaction-fluidization phenomena caused by The 2011 off the Pacific coast of Tohoku earthquake in Chiba prefecture, No.G-8, 2011.

Proc. IAHS, 372, 221–226, 2015
proc-iahs.net/372/221/2015/
doi:10.5194/piahs-372-221-2015

Subsidence by liquefaction-fluidization in man-made strata around Tokyo bay, Japan: from geological survey on damaged part at the 2011 off the Pacific Coast of Tohoku Earthquake

O. Kazaoka[1], S. Kameyama[2], K. Shigeno[3], Y. Suzuki[3], M. Morisaki[1], A. Kagawa[1], T. Yoshida[1], M. Kimura[1], Y. Sakai[1], T. Ogura[1], T. Kusuda[a], and K. Furuno[a]

[1]Research Institute of Environmental Geology, Chiba (RIEGC), Japan
[2]Environmental Protection division of Chiba Prefectural Government, Japan
[3]Meiji Consultante Co., Ltd, Japan
[a]Former member of Research Institute of Environmental Geology, Chiba

Correspondence to: O. Kazaoka (osamu.kazaoka@gmail.com)

Published: 12 November 2015

Abstract. Geological disaster by liquefaction-fluidization happened on southern part of the Quaternary Paleo-Kanto submarine basin at the 2011 Earthquake off the Pacific Coast of Tohoku. Liquefaction-fluidization phenomena occurred mainly in man-made strata over shaking 5+ intensity of Japan Meteorological Agency scale. Many subsided spots, 10–50 m width, 20–100 m length and less than 1 m depth, by liquefaction-fluidization distributed on reclaimed land around northern Tokyo bay. Large amount of sand and groundwater spouted out in the terrible subsided parts. But there are little subsidence and no jetted sand outside the terrible subsided part. Liquefaction-fluidization damaged part at the 1987 earthquake east off Chiba prefecture re-liquefied and fluidized in these parts at the 2011 great earthquake. The damaged area were more wide on the 2011 earthquake than the 1987 quake. Detailed classification maps of subsidence by liquefaction-fluidization on the 2011 grate earthquake were made by fieldwork in Chiba city around Tokyo bay. A mechanism of subsidence by liquefaction-fluidization in man-made strata was solved by geological survey with continuous large box cores on the ACE Liner and large relief peals of the cores at a typical subsided part.

1 Introduction

Strong shaking hit the Kanto district and the Tohoku district at the 2011 off the Pacific Coast of Tohoku Earthquake. Strong shaking over 5+ intensity of Japan Meteorological Agency scale distributed widely in southeast part of the Neogene Kanto Basin (JMA, 2011; Fig. 1). The shaking 5+ intensity of JMA scale correlate with about 8 intensity of MSK scale. Liquefaction-fluidization phenomena caused on reclaimed land by hydraulic fill in the strong shaking area at the Neogene Kanto Basin (RIEGC, 2011c; Fig. 1).

2 Distribution of liquefaction-fluidization site on Boso Peninsula

Liquefaction-fluidization sites were distributed on north part of the Boso peninsula, southeastern Neogene Kanto Basin at the earthquake (RIEGC, 2011c; Fig. 2). Many of these liquefaction- fluidization sites had been caused liquefaction-fluidization phenomena at the 1987 east off Chiba prefecture earthquake (RIEGC, 2011a; RIEGC,2011c; Fig. 3). Each liquefaction-fluidization sites at the 2011 earthquake is more wide than the liquefaction-fluidization sites at the 1987 earthquake.

Liquefaction-fluidization sites concentrate on belts with 0.5 km wide, NE-SW extending on reclaimed land along

Published by Copernicus Publications on behalf of the International Association of Hydrological Sciences.

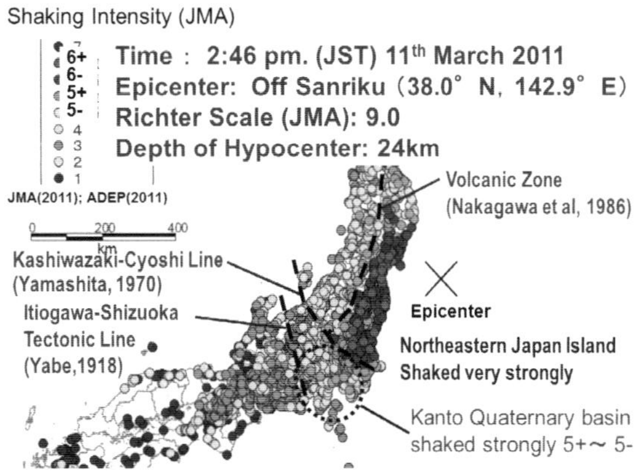

Figure 1. Distribution of JMA intensity on the 2011 Earthquake off the Pacific Coast of Tohoku and geological structure.

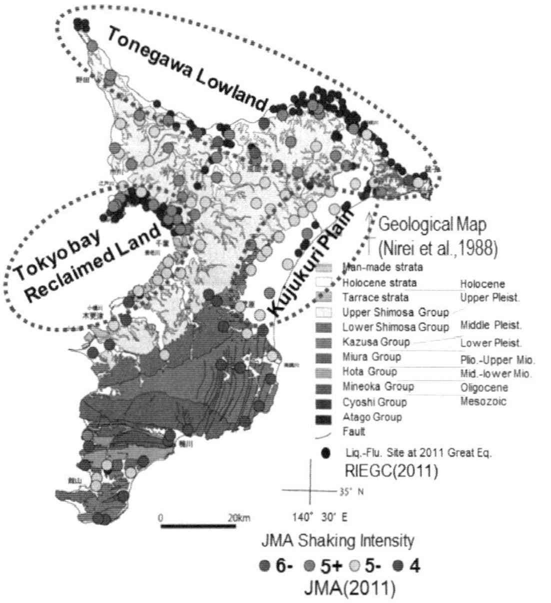

Figure 2. Distribution of liquefaction-fluidization phenomena and JMA intensity by the 2011 Earthquake.

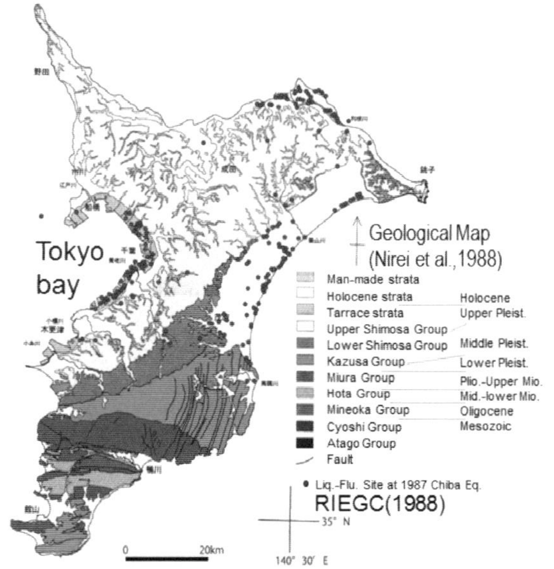

Figure 3. Distribution of liquefaction-fluidization phenomena by the 1987 east off Chiba prefecture earthquake.

Tokyo bay (RIEGC, 2011a, c; Fig. 4). These belts may be on the thick part of the Alluvium formation after the Wurm glacial stage (Kazaoka, 2011; RIEGC, 2011b; Fig. 5).

Liquefaction-fluidization phenomena caused partially with subsidence on the reclaimed land of northern Tokyo bay at the 2011 earthquake. Many subsided spots, 10–50 m width, 20–100 m length and less than 1 m depth, by liquefaction-fluidization distributed on reclaimed land around northern Tokyo bay. Large amount of sand and groundwater spouted out in the terrible subsided parts. But there are little subsidence and no jetted sand outside the terrible subsided part. Detailed classification maps of subsidence by liquefaction-fluidization on the grate earthquake were made by fieldwork in Chiba city around Tokyo bay (RIEGC, 2011c; Fig. 6).

3 Mechanism of subcidence by liquefaction-fluidization in manmade-strata

A mechanism of subsidence by liquefaction-fluidization in man-made strata was solved by geological survey with continuous large box cores on the ACE Liner (Fig. 7) and continuous boring cores in Chiba city (Fig. 6). Continuous box core samples from surface to 5–7 m depth could be taken at the each 3–5 m length from little subsided part to terrible subsided part (Fig. 8). Detailed litho-stratigraphy, sedimentary structure, dencity and hardness of strata were studied on the continuous box core samples, continuous boring core samples and large relief peels of the cores (RIEGC, 2014). These data indicate that:

1. The thickness of man-made strata is about 12 m.
2. Man-made strata is composed of Dumped Association, Upper Filling Association and Lower Filling Association (RIEGC, 2014; Figs. 9, 10). Two Filling Associations were made by sand pump method from bottom sediments in the Tokyo bay. Upper Filling Association consists of lowermost, lower, upper and uppermost bundle.
3. Litho-facies of each man-made strata is as follows.

Dumped Association: This association is composed of 1.5–2.2 m thick sandy silt to silty fine sand layers with silt-

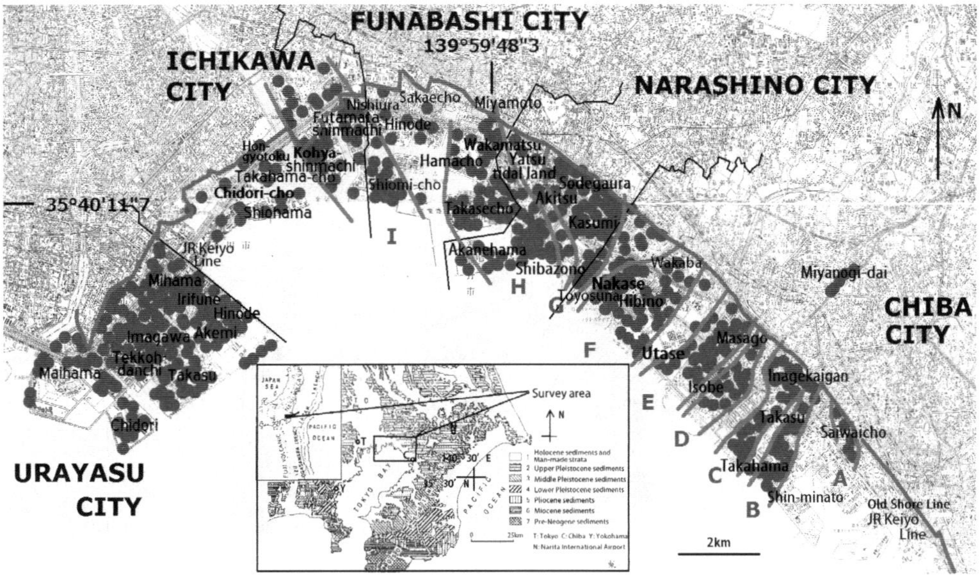

Figure 4. Distribution of liquefaction-fluidization phenomena. Blue spot means liquefaction-fluidization site (Kazaoka et al., 2011).

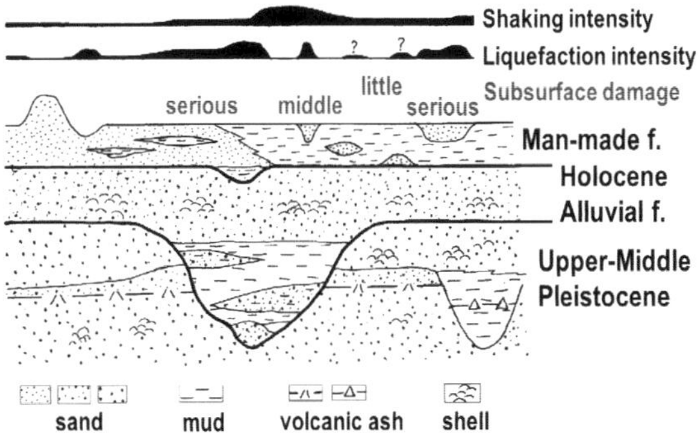

Figure 5. Idealized geological cross section on reclaimed land of Tokyo bay. Intensity of shaking and liquefaction show line wideth (Kazaoka, 2011).

stone brocks and rock gravels. Sand dike of yellowish brown sand and gray sand distribute rarely (RIEGC, 2014; Fig. 11).

Uppermost Bundle of Upper Filling Association: this bundle is composed of 0.2–0.8 m thick yellowish brown laminated fine to medium sand layers. Upper part of this bundle lost partially primary sedimentary structures and loose. The feature described above suggests that the losting structure part is liquefaction-fluidization part. The base of this bundle consists of laminated coarse-very coarse sandy shell fragment layers.

Upper Bundle of Upper Filling Association: this bundle is composed of 0.4–1.8 m thick gray laminated medium sand layers. Shell fragment layers often interbeded in this sand

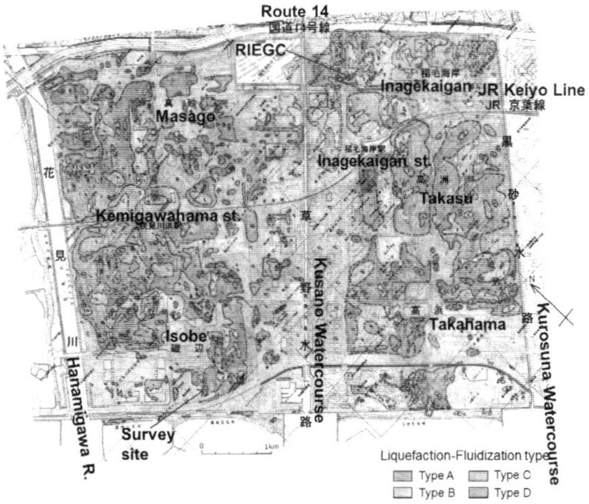

Figure 6. Classification of subsidence by liquefaction-fluidization on the 2011 earthquake in the reclaimed land in Chiba city. A Type: subsidence > 0.3 m, B type: 0.3 m < subsidence < 0.1 m, C Type: 0.1 m < subsidence < few centimeter, D Type: no subsidence.

layers. The sand layers lost widely primary sedimentary structures and very loose or relatively dence.

Lower Bundle of Upper Filling Association: this bundle is composed of 0–1.8 m thick gray silt layers.

Lowermost Bundle of Upper Filling Association: this bundle is composed of 0.7–1.8 m thick gray shelly medium sand layers. Shell fragment layers often interbedded in the shelly sand layers. In upper part of this bundle, sand layer lost

Figure 7. Continuous sampling by ACE liner and continuous man-made strata samples on this survey site.

Figure 8. Distribution of subsidence and sampling points by ACE liner. Ground surface subside a little at the sampling points No. 4, 5, 9 and 10. The surface subsided 0.4 m height at the sampling points No. 1, 2, 6 and 7. Roofs were deformed by subsidence.

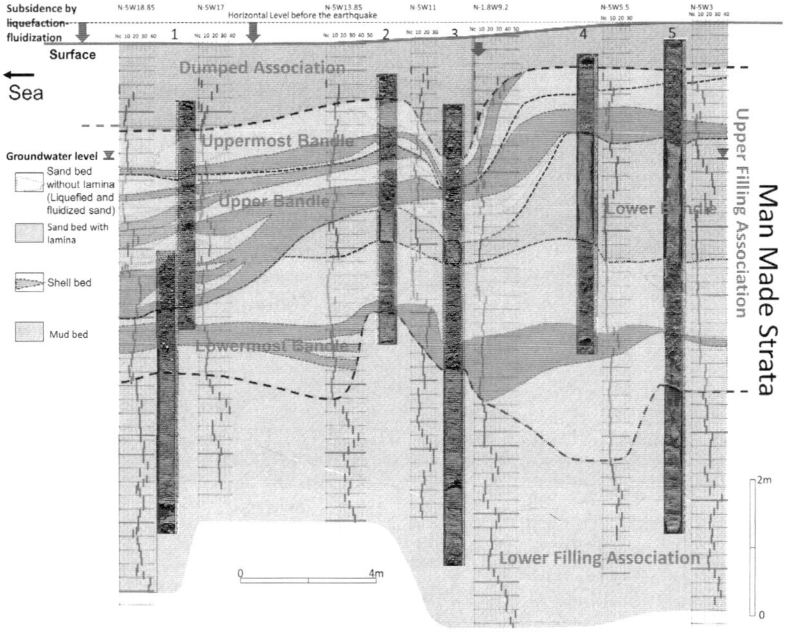

Figure 9. Geological cross section from No. 1 to No. 5 sampling points of Fig. 8.

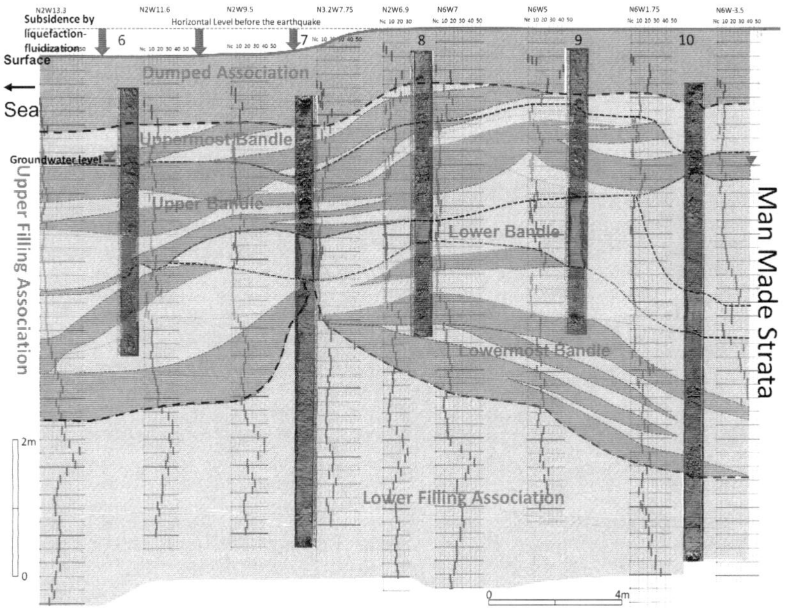

Figure 10. Geological cross section from No. 6 to No. 10 sampling points of Fig. 8.

Figure 11. Jetted yellowish gray sand along fisher on No. 4 point of Fig. 2.

widely primary sedimentary structures. The medium sand injected in the upper silt layers.

Lower Filling Association: this association is composed of 0.5–3.5 m thick yellowish gray laminated relatively dence matrix free good sorted fine-medium sand layers. The sand layers lost partially primary sedimentary structure. The liquefaction-fluidization part is cut by the Upper Filling Association.

4. Liquefaction-fluidization parts are in lowermost bundle, upper bundle and uppermost bundle of the Upper Filling Association.

5. Subsidence part is entirely identical with distribution of liquefaction-fluidization part in upper bundle of the Upper Filling Association.

Acknowledgements. We thank Yuji Takanashi and Yutaka Yazawa, former Chief of Chiba Environmental Research Center and Hiroaki Hiura, Chief of Chiba Environmental Research Center for helpful to study and survey liquefaction-fluidization.

References

Japan Meteorological Agency: The seismological bulletin of the Japan Meteorological Agency for March 2011, Japan Meteorological Agency, 321 pp., 2011.

Kazaoka, O.: Liquefaction-fluidization damage and geological structure of Man-made strata, Symposium for Disaster in man-made strata and 2011 Tohoku Earthquake, Japanese Society of Geo-pollution Science, Medical Geology and Urban Geology, 1–21, 2011. Kazoka, O., Furuno, K., Kagawa, A., Kusuda, T., Sakai, Y., Yoshida, T., Kato, A., Yamamoto, M., and Takanashi, Y.: Distribution of Geological Disaster by Liquefaction-Fluidization Phenomena at The 2011 off the Pacific coast of Tohoku Earthquake on Tokyo Bay Reclaimed Land, Jour. Geo-Pol. Sci., Med. Geol. and Urban Geol., 7, 10–21, 2011.

Kazaok, O., Kameyama, S., Morisaki, M., Shigeno, K., Suzuki Y., Kagawa A., Yoshida T., Kimura, M., Sakai, Y., and Ogura, T.: Characteristic of Man-made strata in liquefaction-fluidization site at the 2011 off the Pacific coast of Tohoku Earthquake: from geologaical survey on reclaimed land in Isobe district around Tokyo bay, Chiba city, Proceedings of the 24th Symposium on Geo-Environments and Geo-Technics, Japanese Society of Geo-pollution Science, Medical Geology and Urban Geology, 9–14, 2014.

Nirei, H., Kusuda, T., Suzuki, K., Kamura, K., Furuno, K., Hara, Y., Satoh, K., and Kazaoka, O.: The 1987 East off Chiba Prefecture Earthquake and its Hazard, Memoir of Geological Society of Japan, Geological Society of Japan, no. 35, 31–46, 1990.

RIEGC: Liquefaction-Fluidization Phenomena on Boso Peninsula at the 2011 Earthquake off the Pacific Coast of Tohoku, Central Japan: Part 1. RESEARCH REPORT G-8, Chiba Environmental Research Center, 1-1–1-8, 2011a.

RIEGC: Liquefaction-Fluidization Phenomena on Boso Peninsula at the 2011 Earthquake off the Pacific Coast of Tohoku, Central Japan: Part III, RESEARCH REPORT G-8, Chiba Environmental Research Center, 3-1–3-26, 2011b.

RIEGC: Liquefaction-Fluidization Phenomena on Boso Peninsula at the 2011 Earthquake off the Pacific Coast of Tohoku, Central Japan: Part IV. RESEARCH REPORT G-8, Chiba Environmental Research Center, 4-1–4-69, 2011c.

RIEGC: Liquefaction-Fluidization Phenomena on Boso Peninsula at the 2011 Earthquake off the Pacific Coast of Tohoku, Central Japan: Part V. RESEARCH REPORT G-8, Chiba Environmental Research Center, 5-1–5-8, 2012.

RIEGC: Liquefaction-Fluidization Phenomena on Boso Peninsula at the 2011 Earthquake off the Pacific Coast of Tohoku, Central Japan: Part VI., Chiba Environmental Research Center, 4P (http://www.pref.chiba.lg.jp/wit/chishitsu/ekijoukahoukoku/documents/6-2.pdf), 2014.

Proc. IAHS, 372, 227–229, 2015
proc-iahs.net/372/227/2015/
doi:10.5194/piahs-372-227-2015

Visualization of three dimensional earth fissures in geological structure

L. Zhu[1,2], J. Yu[3], Y. Liu[1,2], H. Gong[1,2], Y. Chen[4], and B. Chen[1,2]

[1]College of Resources Environment and Tourism, Capital Normal University, Beijing, China
[2]Laboratory Cultivation Base of environment process and digital simulation, Beijing, China
[3]Key Laboratory of Earth Fissures Geological Disaster, Ministry of Land and Resources, Jiangsu, China
[4]Land and Water Laboratory, Canberra, Australia

Correspondence to: L. Zhu (hi-zhulin@163.com)

Published: 12 November 2015

Abstract. This paper proposes a new method for visualizing the earth fissures of geological structure in three dimensional (3-D) domains on the basis of the seismic data and features information of earth fissures. The seismic data were interpreted for obtaining the stratagraphic data with various lithological information and the depth of the earth fissures. The spatial distribution of the ground fissures including the dip, strike and width were digitalized on an ArcGIS platform. Firstly, the 3-D geological structure was rebuilt using the Generalized Tri-Prism (GTP) method which is a real solid method for displaying geological structures. The GTP method can reflect the inner material of the strata and can simulate complicated geological structures such as faults and stratagraphic pinch outs. The upper and lower surfaces of each stratum consist of Triangle Irregular Networks (TIN). The inner solid between the two surfaces are a series of triangular prisms. Secondly, since the width of the ground fissure gradually decreases with depth, multiple edge lines of the earth fissures on the bottom stratum surface are deduced on the basis of the fissure characteristics. Then, the model of the earth fissures consisting of a series of triangular pyramids can be constructed using these points and the edge lines. A cutting operation was carried out on the 3-D geological structure using this ground fissures model. If the surfaces of the ground fissures model intersects with the GTPs in the geological structure model, new GTPs were generated within the local regions. During this process, the topological relations between TIN, triangular prism and lines were reconstructed so that the visualization of ground fissures in the geological structure model is realized. This method can facilitate the mechanism for studying fissures and avoid the gaps between the fissure solid and the geological structure to accurately reflect their 3-D characteristics.

1 Introduction

Earth fissures are induced by natural or man-made factors such as volcano, ice, water, and exploration of the subsurface resources (water, oil and gas). Inland areas, especially in many alluvial basins, earth fissures are mainly caused by tensional forces related to differential land subsidence, which is mainly related to long-term groundwater pumping in excess of aquifer recharge such as in Arizona and Utah. An earth fissure generally exhibits characteristics including orientation, discrete rupture, local range and long extension, although horizontal displacements across earth fissures are relatively small. In China, the frequency and magnitude of earth fissures is much greater and has caused associated hazards. For example, the earth fissures have damaged infrastructures including roads, gas lines and buildings, which has presented a danger to the public. Many experts have worked on measurement methods, mechanism analysis, risk assessment, modeling and forecasting, as well as mitigation methods of earth fissures. Visualization of fissures is a perquisite to these studies.

Three-dimensional (3-D) visualization in hydrogeology has been frequently used for inspecting field data and for understanding model results (Voss, 1999). Compared to numer-

Published by Copernicus Publications on behalf of the International Association of Hydrological Sciences.

ical values, 3-D visualization is sometimes more effective for researchers to learn the distribution of geologic phenomena in 3-D domains. Generally there are two methods including the surface and solid methods for displaying the geological structure. Surface methods such as multiple DEM (digital elevation model) are popular and easier to be realized, while, the model from this method can not reflect the inner materials or attributs in geologic structures. Solid methods can overcome this shortcoming and models from this method are realized in three dimensions. Generalized Tri-Prism (GTP) was proposed for the layered strata (Wu, 2004), which can be used to simulate complicated geological structures. There are few studies on the visualization of earth fissures. Earth fissures are generally shown by photos or schematic drawings. Zhu et al. (2007) used the polygon of earth fissure to rebuild the earth surface model in the Suzhou-Wuxi-Changzhou area (short for Su-Xi-Chang). Li et al. (2010) expressed ground fissures' spatial continuous curved surfaces by using the linear-distributed ground fissures data and geological model with a surface method of Triangle Irregular Networks (TIN) in Fenwei Plain. However, geologic models in these studies are quasi-3-D.

OpenGL is a 3-D graphics software library developed by SGI Inc. (Silicon Graphics). OpenGL is composed of hundreds of commands and functions for creating real-time 3-D graphics, which can be used to generate 3-D object with high fidelity, and can realize the interactive operations with the simulated objections. To enhance the fidelity, lighting conditions including ambient illumination and diffuse light are defined to create a 3-D scene.

This paper aims to investigate a 3-D visualization method to exhibit the distribution of earth fissures. Yingguoan in Su-Xi-Chang was taken as an example area. Visual C++ was used as a developed platform and OpenGL was employed to create real-time 3-D graphics for the visualization of a geologic model with earth fissures.

2 Description of case study area

There are numerous earth fissures in the Su-Xi-Chang area, Jiangsu Province, China, which lies in the alluvial plain of the Yangtze River Delta. The earth fissures that developed in 1991 in Yinguoan are typical and ere induced by steep bedrock surfaces and differential subsidence. There is a fracture belt extending 2000 m and 50–150 mm width along the NE direction, which shows obvious tensile characteristics (Yu et al., 2004, 2013). A 3-D seismic exploration, covering a survey area of 1.7 km^2, was carried out to detect the earth fissures in this area. These seismic data were interpreted to extract the detailed information on stratagraphy and the 3-D shape of earth fissures, such as width, depth and length on the basis of the seismic velocity and logging.

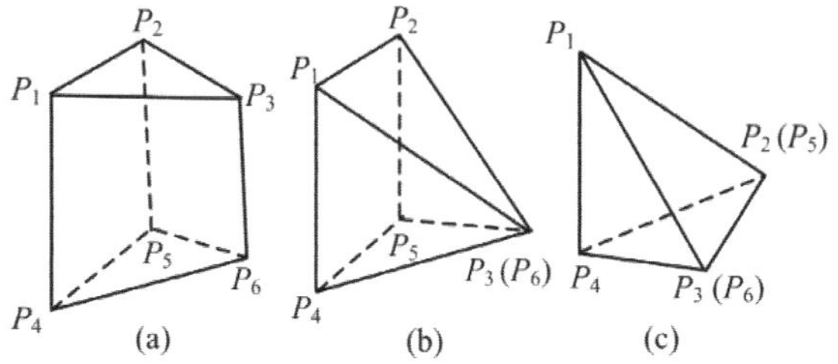

Figure 1. Different GTP elements. **(a)** The up surface and the down surface of GTP are not necessarily parallel. **(b)** The edge of P3 and P6 converges to one point. **(c)** The edge of P2 and P5 also converges to another point.

3 Methodology

Detailed information of Aquifer II was interpreted from 3-D seismic data. Scattered points of six layers were obtained. Each point has a lithologic description and coordinates (X, Y, Z). Information of earth fissures includes the outlines of the fissures on the ground, extension and dip. The earth fissures are assumed to be a "V" shaped.

The detailed process of visualizing the earth fissures using the GTP method is as follows.

Firstly, a geologic model without fissures was constructed. The up surface and the down surface of GTP are not necessarily parallel (Fig. 1a) and there are other special shapes. For example, the edge of P_3 and P_6 converges to one point (Fig. 1b) and P_2 and P_5 also to another point (Fig. 1c), which can simulate complicated geological structures such as faults and stratagraphic pinch outs. There are six points in each GTP stored in the database, which is easy for query. For example, P_3 and P_6 (Fig. 1b, c) are stored separately but with the same coordinates. The up and down surfaces of each GTP, defined by Triangle Irregular Networks such as $P_1P_2P_3$ and $P_4P_5P_6$, consist of the surface of each stratum. The inner solid between two surfaces are of series of triangular prisms. These triangular prisms can further be split.

Secondly, an earth fissure model was built. An earth fissure is a polygon on the ground and the width decreases linearly with depth. The vertices of the polygon are assumed to be symmetrical. The earth fissure volume was constructed by a series of triangular pyramids.

Thirdly, a geological model with earth fissures was developed. The geologic model and the earth fissures model were combined using Boolean operators. Three types of cutting were involved in the process. They are whole cutting, special cutting and false cutting (Li et al., 2008). No cutting will be carried out if six points of GTP are at the same side of the cutting plane. The cutting type is determined by the intersection points of the cutting plane and the GTP. Then the origin GTP is re-build and the topology relationship of GTP is renewed.

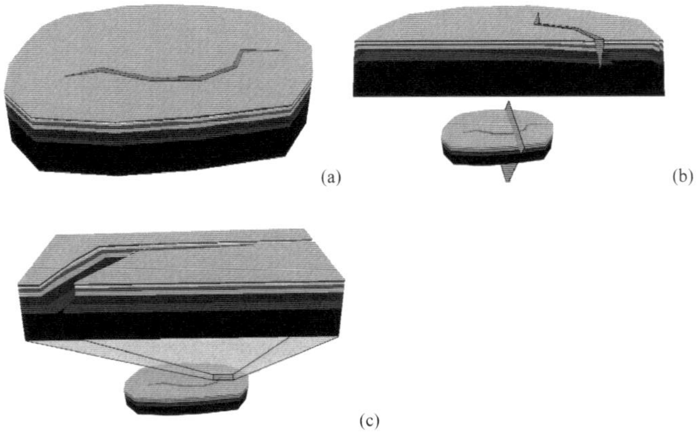

Figure 2. Geologic structure with earth fissure model. **(a)** The geological body with earth fissures model. **(b)** The spatial relations between the strata and the fissures. **(c)** The combination of bottom part of **(b)** and a close-up of the model.

4 Results and conclusions

GTP represents the element for visualization. The top and bottom triangular surfaces, and the lateral quadrilateral are drawn by using the OpenGL functions of glBegin(GL_TRIANGLES) and glBegin(GL_QUADS). The modeling results can be displayed as the "skeleton" pattern by using wireframe in OpenGL. At the solid mode, each surface of GTP with different materials in the geological model can be specified with different colors. The earth fissures are plotted as a series of triangles in grey. The geological body with earth fissures model (Fig. 2a) can be clipped with any line, which is convenient for researchers to learn the spatial relations between the strata and the fissures (Fig. 2b). The model can also be zoomed in and out (Fig. 2c).

This paper presents a process to visualize the geological model with earth fissures based on the GTP method. The geological model can be subdivided by a fissure model to derive the geological structure with each earth fissure. We can set attributes such as permeability or storage to each GTP element, which is in a real 3-D domain. This method can simulate complicated geological phenomena. In this study, the earth fissures are assumed to be "V" shaped, while, the real distribution of fissures is more complex.

Acknowledgements. This study is funded by National Natural Science (No. 41201420, 41130744) and Key Laboratory of Earth Fissures Geological Disaster, Ministry of Land and Resources, Geological Survey of Jiangsu Province, Nanjing.

References

Li, B., Zhang, J., Xu, Y., Lv, Y., Zhang, L., Wang, W., Yan, F., Xu, M., Xing, C., and Wang, S.: Study on modeling and 3-D-visualization of ground fissures' spatial curved surfaces, Image and Signal Processing (CISP), 6, 2838–2844, 2010.

Li, C.-C., Wang, B.-S., and Xue, H.-Z.: Coal geological body three-dimensional modeling and cutting based on GTP, Journal of China Coal Society, 33, 1268–1271, 2008 (in Chinese).

Wu, L.: Topological relations embodied in a generalized tri-prism (GTP) model for a 3-D geosciences modeling system, Comput. Geosci., 30, 405–418, 2004.

Yu, J., Wang, X. M., Su, X. S., and Yu, Q.: The mechanism analysis on ground fissure disaster formation in Suzhou-Wuxi-Changzhou area, Journal of Jilin University, Earth Science Edition, 34, 236–241, 2004 (in Chinese).

Yu, J., Su, X., Zhu, L., Duan, F.-Z., Pan, Y., Gao, L., and Wu, S.-L.: Research on 3-D visualized strata model virtual reality system of land subsidence in Suzhou-Wuxi-Changzhou area, 8th International Symposium on Land Subsidence, 17–22 October 2010, Queretaro, Mexico, 339, 108–113, 2010

Yu, J., Zhu, J. Q., Gong, X. L., and Yang, Y.: The potential mechanisms of Yingguoan earth fissure using 3D seismic exploration data, 20th International Congress on Modelling and Simulation, 1–6 December 2013, Adelaide, Australia, 2751–2757, 2013.

Voss, C.: Editor's Message: Scientific visualization in hydrogeology, Hydrogeol. J., 7, 153–154, 1999.

Zhu, L., Su, X., Lin, X., Gong, H., and Yu, J.: Application of VR-GIS in studying ground fracture in the western region of Wuxi area, Journal of Jilin University, Earth Science Edition, 2, 317–321, 2009 (in Chinese).

Proc. IAHS, 372, 231–234, 2015
proc-iahs.net/372/231/2015/
doi:10.5194/piahs-372-231-2015

Study of the deformation mechanism of the Gaoliying ground fissure

G. Cheng[1], H. Wang[1], Y. Luo[2], and H. Guo[1]

[1]China Institute of Geo-environment Monitoring, Beijing 100081, China
[2]Beijing Institute of Geo-environment Monitoring, Beijing 100195, China

Correspondence to: G. Cheng (chenggm@mail.cigem.gov.cn)

Published: 12 November 2015

Abstract. The Gaoliying ground fissure in Beijing has caused building cracking and road damage, and has seriously influenced city construction. Based on investigations and trenching, the influences of the fault and the variation of groundwater levels on the formation mechanism of the Gaoliying ground fissure were investigated by using FLAC3D. The results indicated that (1) the surface location of Gaoliying fissure is controlled by the underlying normal fault activity, and over pumping further exacerbates development of the ground fissure; (2) when the groundwater level declines, obvious differential settlement occurs at both sides of the ground fissure, in which greater settlement occurs in the vicinity of the hanging wall, the greater the distance from the hanging wall, the smaller the ground subsidence, however smaller ground subsidence occurs in the vicinity of the footwall, the greater the distance from the footwall, the greater the ground subsidence; (3) the vertical velocity of the ground fissure triggered by the fault activity and groundwater decline ranges from 15.5 to 18.3 mm a^{-1}, which is basically in line with the monitoring data. The fault activity contributes about 28–39 %, and the groundwater contributes about 61–72 % to the deformation of the ground fissure, respectively.

1 Introduction

Ground fissures are common in many parts of the world, and have been particularly well documented in the United States and in China (Holzer, 1980; Zellmer et al., 1985; Peng, 2012). Many fissures appear similar to those caused by groundwater withdrawal and related ground subsidence (Lister and Secrest, 1985; Lee et al., 1996). In China, the earliest ground fissures were detected at the end of 1950s (Peng et al., 2013). Now ground fissures have been a main geological hazard developed in north China (Wang, 2000; Wang et al., 2001), especially in Gaoliying district in Beijing, which has led to great losses to local economies and has created a large potential danger to human living (Jia and Guo, 2007).

Jia (2010) and Jia et al. (2011) suggests that the Gaoliying ground fissure is controlled by the Gaoliying fault, and its formation is the result of fault creep deformation and the differential settlement caused by groundwater recession. However, there are still some critical problems not yet solved, for example, How does the activity of underlying faults generate and aggravate the formation and the activity of the ground fissures? What important roles do the tectonic activity and the over-pumping play in the formation of the ground fissures? The influence of fault activities and the groundwater level changes on the Gaoliying fissure based on the measured data was simulated using FLAC3D in this paper. The results provide insight as to the specific contribution of fault activities and groundwater on ground fissure formation, and infers the mechanism for deformation of the fissure.

2 Outline of Gaoliying ground fissure

The Gaoliying fissure in Beijing was first discovered in the 1990s, and the fissure trend is approximately from northeast 45° to northeast 60°, which is consistent with the underlying Huangzhuang–Gaoliying fault. It extends north-eastward from the Xiwang Village to the North sixth ring road (Fig. 1). The fissure has a length of about 6000 m, with a width ranging from a few millimeters to hundreds of millimeters, in which the maximum width is 200 mm. The obvious ground

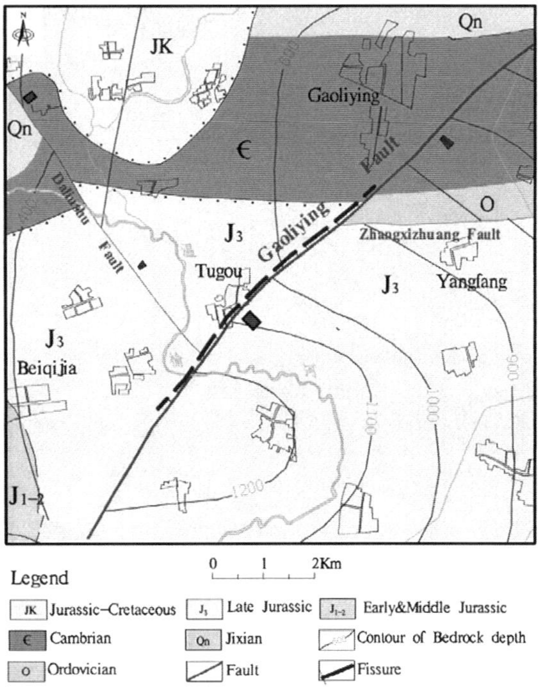

Figure 1. The plan distribution map of Gaoliying fissure and fault.

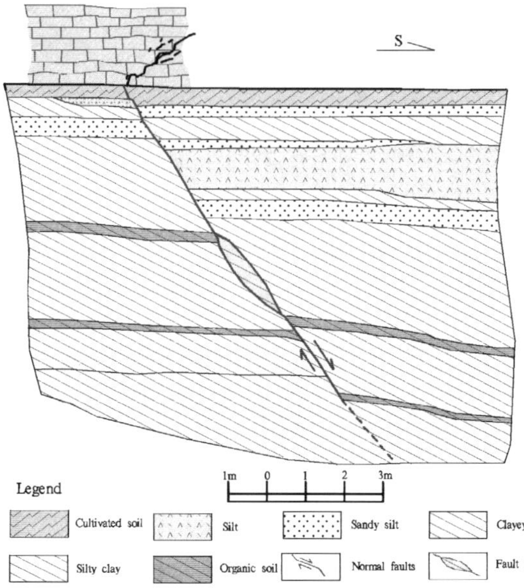

Figure 2. The profile of Gaoliying ground fissure according to trench data.

dislocation occurred at both sides of the fissure, in which the northwest is higher than the southeast, and has caused buildings cracking and has destroyed roads. The data from a ground fissure monitoring station shows that the maximum vertical displacement rate of Gaoliying fissure is $18\,\mathrm{mm\,a^{-1}}$.

3 Geological background

The Gaoliying fault is an important active fault in the Beijing plain area. Its trend is NNE with a length of 132 km. The fault is the main control fault of the west boundary of Beijing depression in the Tertiary, and Shunyi and Huairou secondary depressions were formed during the Quaternary period. The fault is a normal fault with dip-slip characteristics.

According to the characters of the Quaternary kinematics analysis, the Gaoliying fissure is divided into three parts: the Miaocheng–Gaoliying section (North section, 40 km), the Huangzhuang section (Middle section, 52 km), and the Laishui section (South section, 40 km). The Gaoliying fissure is within the Huangzhuang section. Shallow seismic exploration reveals that the fall of the Quaternary system on both sides of the fault near Gaoliying fissure ranges from 140 to 280 m. The seismic exploration at Tugou village indicates that the thickness of the Quaternary sediments at the footwall and hanging wall is respectively 460–500 and 660–740 m, in which the thickness difference on both sides of the fault is 200–240 m. The Gaoliying fault location corresponds to the Gaoliying ground fissure. An exploratory trench disclosed that the dislocation distance of the fault at the surface layer is much as 3.56 m (Fig. 2). Boring data revealed that the dislocation distance of fault in the shallow strata is smaller, whereas the dislocation distance of fault in the deep strata is larger. The dislocation distance of fault at the depth of about 50 m is approximately 8 m.

The study area is located on the Wenyu Alluvial Fan plain area, its surface is covered with Quaternary strata. The deposit is relatively simple, and is dominated by fluvial sediments. Table 1 shows the results of geophysical exploration data on both sides of the fault and bedrock bench mark data acquired by the ground settlement monitoring station on both sides of ground fissures.

4 Numerical model

4.1 Conceptual model

The numerical model is 2000 m in X direction, 1000 m in Y direction and 20 m in Z direction. The model contains 11 440 elements and 11 886 grid points (Fig. 3). Horizontal translation is confined at the lateral boundary and vertical translation in the footwall is confined, whereas vertical translation in the hanging wall is free at the bottom boundary in the model.

Table 1. The simplified strata used in the numerical model.

Stratum name	Silty clay	Silt	Sand	Silty clay	Sandy and silty	Silty clay	Sandy and silty	Bedrock
Thickness of hanging wall (m)	25	35	70	70	150	250	250	150
Thickness of footwall (m)	20	30	50	50	50	100	150	550
Separation of the hanging wall and footwall top (m)	0	5	10	30	50	150	300	400

Table 2. The physical and mechanical parameters of each stratum in the model.

Material	Density ($kg\,m^{-3}$)	Elastic modulus (GPa)	Poisson's ratio	Cohesion (kPa)	Strength of extension (kPa)	Internal friction angle (°)	Dilatancy angle (°)
Silty cla	1750	0.05	0.35	50	20	20	10
Silt	1850	0.06	0.33	40	10	22	10
Sand	1950	0.1	0.3	10	5	25	10
Silty clay	1850	0.2	0.33	50	20	22	10
Sandy and silty	1950	0.5	0.3	30	10	20	10
Silty clay	1950	0.5	0.3	60	30	24	10
Sandy and silty	2000	0.8	0.3	40	20	21	10
Bedrock	2500	10	0.25	10 000	3000	40	10
Fracture surface	1850	0.1	0.2	2	0.2	0	0

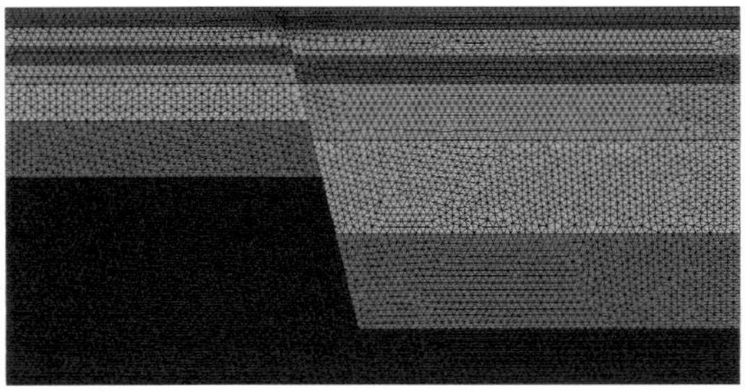

Figure 3. Numeric simulation structure of Gaoliying fissure.

4.2 Material parameters

In order to simplify the calculation, the Mohr–Coulomb constitutive relation is used for the analysis, and parameter characteristics of different materials are shown in Table 2. In addition, the fracture surface is treated as a weak interface, and its material parameters are separately set.

4.3 Modelling schemes

Based on the bedrock bench mark data from Tianzhu monitoring station and the Baxianzhuang monitoring station on both sides of the Gaoliying fissure, the annual differential settlements are 8, 6.5, 4.8, and 6.7 mm respectively between 2009 and 2012, which indicates that bedrock on both sides of the fault has been in a state of activity, in which the hanging wall declines relative to the footwall. The long-term groundwater observation hole near the Gaoliying fissure shows that the groundwater level is about 20 m, and the rate of decline of the water level can reach $2\,m\,a^{-1}$.

Therefore, three kinds of modeling schemes are considered in this simulation, i.e., the relative decline rate of bedrock between the hanging wall and the footwall is $4.8\,mm\,a^{-1}$ (Scheme 1) and $8\,mm\,a^{-1}$ (Scheme 2), and the groundwater decline rate is $2\,m\,a^{-1}$ (Scheme 3).

4.4 Simulation results

Figures 4 and 5, show that the surface location of the ground fissure is controlled by the underlying normal fault, and it is in agreement with the fault. When the activity velocity of the bedrock inside the fault is $4.8\,mm\,a^{-1}$, the maximum differential settlement of the ground fissure is $4.4\,mm\,a^{-1}$; when the activity velocity of the bedrock inside the fault is $8\,mm\,a^{-1}$, the maximum differential settlement of the ground fissure is $7.2\,mm\,a^{-1}$ (Fig. 6).

From Fig. 6, it can be seen that when the groundwater level declines, subsequent differential settlement occurs on both sides of the ground fissure, in which greater ground subsidence occurs in the vicinity of the hanging wall. The greater the distance from the hanging wall the smaller the ground subsidence, however small ground subsidence also occurs in the vicinity of the footwall where the greater the distance from the footwall, the greater the ground subsidence, and the maximum differential settlement of the ground fissure is as much as 11.1 mm.

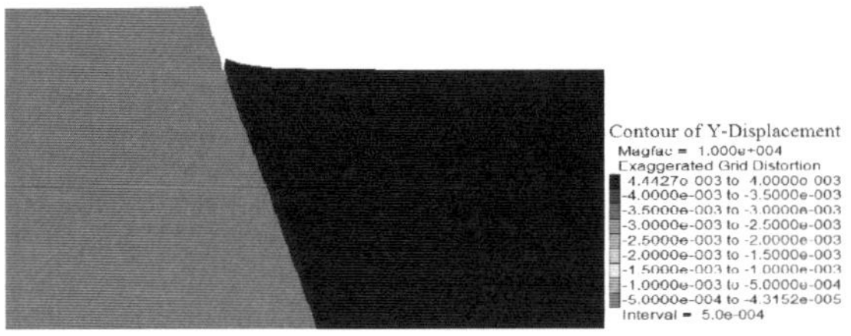

Figure 4. The annual settlement in scheme 1 (scaled up to 10 000 times of the original.

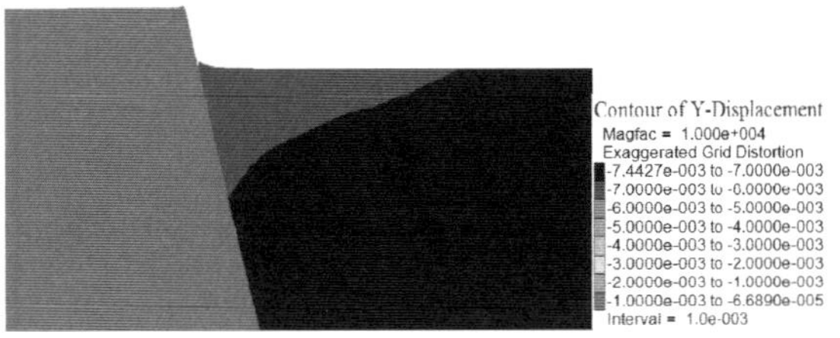

Figure 5. The annual settlement in scheme 2 (scaled up to 10 000 times of the original).

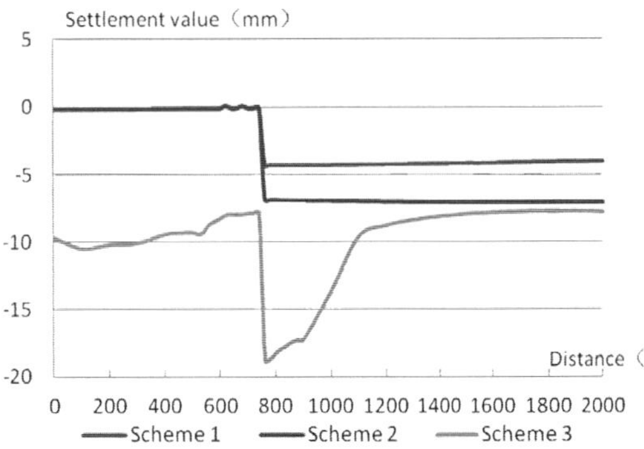

Figure 6. The annual ground settlement at both sides of the ground fissure under different schemes.

The vertical velocity of the ground fissure triggered by the fault activity and over pumping (groundwater level decline) can reach to 15.5–18.3 $mm\,a^{-1}$, which is basically in line with the monitoring data. In which the fault activity contributes about 28–39 %, and the groundwater contributes about 61–72 % to deformation of the ground fissure.

5 Conclusions

1. The surface location of the Gaoliying fissure is controlled by the underlying normal fault activity. Differential settlement occurs at both sides of the fault due to the fault activity, which induces the formation of ground fissure, and over pumping further exacerbates development of the ground fissure.
2. When the groundwater level declines, differential settlement occurs on both sides of the ground fissure, and greater settlement occurs in the vicinity of the hanging wall. The greater the distance from the hanging wall, the smaller the ground subsidence, however small ground subsidence also occurs in the vicinity of the footwall and the greater the distance from the footwall, the greater the ground subsidence.
3. The vertical velocity of the ground fissure triggered by the fault activity and groundwater decline ranges from 15.5 to 18.3 $mm\,a^{-1}$, which is basically in line with the monitoring data. The fault activity contributes about 28–39 %, and the over pumping contributes about 61–72 % to the deformation of the ground fissure, respectively.

Acknowledgements. This study was supported by the National Science Foundation of China (No. 41172266) and Geological Investigation Project of China (No. 12120115044901).

References

Holzer, T. L.: Faulting caused by groundwater level declines, San Joaquin Valley, California, Water Resour. Res., 16, 1065–1070, 1980.

Jia, S.: Annual report of land subsidence monitoring in Beijing, Beijing Institute of Hydrogeology and Engineering Geology, 2010 (in Chinese).

Jia, S. and Guo, M.: The Relation between Huangzhung-Gaoliying fault and fissure by Gaoliying Trench, City Geology, 4, 24–28, 2007 (in Chinese).

Jia, S., Wang, H., and Ye, C.: Investigation and survey methods appropriate for ground fissures in Beijing, Journal of Engineering Geology, 19 (suppl.), 104–111, 2011.

Lee, C., Zhang, J., and Zhang, Y.: Evolution and origin of the ground fissures in Xian, China, Eng. Geol., 43, 45–55, 1996.

Lister, A. L. and Secrest, D. C.: Giant desiccation cracks and differential surface subsidence, Red Lake Playa, Mohave County, Arizona, Bulletin of the Association of Engineering Geologists, 22, 299–314, 1985.

Peng, J.: The Geo-hazard of Xi'an Ground Fissures. Science Press, Beijing, 2012 (in Chinese).

Peng, J., Chen, L., Huang, Q., Men, Y., Fan, W., and Yan, J.: Physical simulation of ground fissures triggered by underground fault activity, Eng. Geol., 155, 19–30, 2013.

Wang, J., Wang, C., and Liu, K.: Progress in ground fissures and its hazard research, Advances in Earth Sciences, 16, 303–313, 2001 (in Chinese).

Wang, J. M.: Theory of Ground Fissures Hazards and Its Application, Science and Technology Press of Shaanxi, Xi'an, 2000 (in Chinese).

Zellmer, J. T., Roquemore, G. R., and Blackerby, B. A.: Modern tectonic cracking near Garlock Fault, California, Bull. Geol. Soc. Am., 96, 1037–1042, 1985.

Proc. IAHS, 372, 235–238, 2015
proc-iahs.net/372/235/2015/
doi:10.5194/piahs-372-235-2015

Discussion on the origin of surface failures in the Valley of Aguascalientes, México

M. Hernández-Marín[1], N. González-Cervantes[1], J. Pacheco-Martínez[2], and D. H. Frías-Guzmán[3]

[1]Departamento de Geotecnia e Hidráulica, Universidad Autónoma de Aguascalientes, Aguascalientes, México
[2]Departamento de Construcción y Estructuras, Universidad Autónoma de Aguascalientes, Aguascalientes, México
[3]Doctorado en Ciencias de los Ámbitos Antrópicos, Universidad Autónoma de Aguascalientes, Aguascalientes, México

Correspondence to: M. Hernández-Marín (mhernandez@correo.uaa.mx)

Published: 12 November 2015

Abstract. Surface failures have been observed in the valley of Aguascalientes since the early 1980's. Although, groundwater pumping began in the early 1950's but became intensive until the late 1970's, when many of the surface failures appeared. For this reason, surface deformation (surface failures and land subsidence) has been associated to groundwater withdrawal. Recent observations, however, suggest that some of these surface discontinuities are the result of natural geologic stresses rather than those associated groundwater decline, at least in its origin. Geologically, this valley is a tectonic graben flanked by two north-to-south trending normal faults, and seismicity of low intensity has been detected into the valley evidencing that the study area is seismically active. In this work, we present and discuss evidence showing tectonic discontinuities on the surface or at a shallow depth. Evidence presented in this investigation brings up uncertainty about the purely pumping-induced origin of fissuring in the valley, and allows addressing some key questions, for instance, are the surface failures caused by the combined stresses of pumping and tectonism? Is the current tectonism sufficient to cause some surface or deep failures?

1 Introduction

Mechanisms to origin surface discontinuities (faults or fissures) in urbanized valleys are variable. In the valley of Aguascalientes, three possible causes for the occurrence of ground failures (fissures and faults) were suggested: groundwater pumping, tectonic stresses and creep-type displacement (Aranda-Gómez, 1989). Most of the surface discontinuities were first observed by the early 1980's, when damages to several constructions were documented (Aranda-Gómez and Aranda-Gómez, 1985; UNAM-SOP, 1994; Lermo et al., 1996). Due to the close relation in time between of massive occurrences of surface discontinuities and the starting of the heavy groundwater pumping, it has been accepted that groundwater depletion is the main cause for the occurrence of these discontinuities. However, recent observations suggest ages of several decades, centuries or even thousands of years of formation. Here we present and discuss some of these observations. Figure 1 shows an outlook of the valley of Aguascalientes, main urban areas, and surface discontinuities.

2 Observations on the Oriente Fault

The named Oriente Fault (oriente means eastern in Spanish), flanks the Aguascalientes valley on the east, as observed in Fig. 1. In some branches, this fault is observed as a fissure and in others as a fault. In few locations is not shown at the surface (hidden by recent sediments?), but in others is observed as parallel alignments of two or three discontinuities. It is one of the most active in the valley, as evidence of that, the paved crossing roads over the trace failure trace are continuously repaired. It is also one of the most affected surface discontinuities by weathering because in some points,

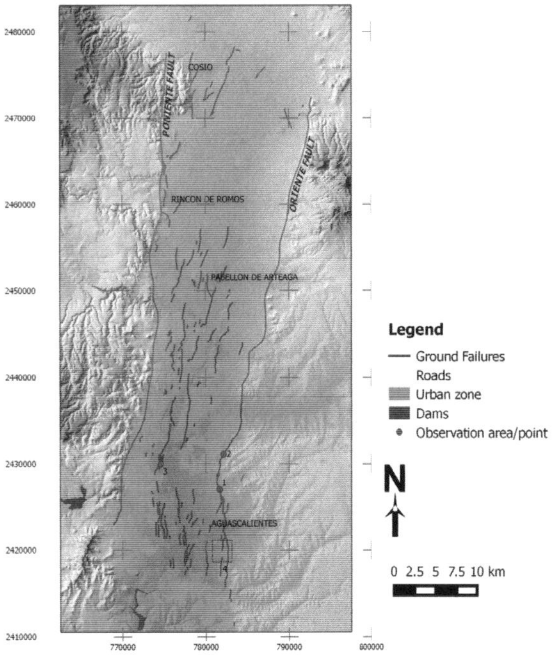

Figure 1. Main features of the Aguascalientes basin. Areas and points of observation in the map are referred according to figures: point 1 to Fig. 2, area 4 to Fig. 3, point 2 to Fig. 4, and point 3 to Fig. 5.

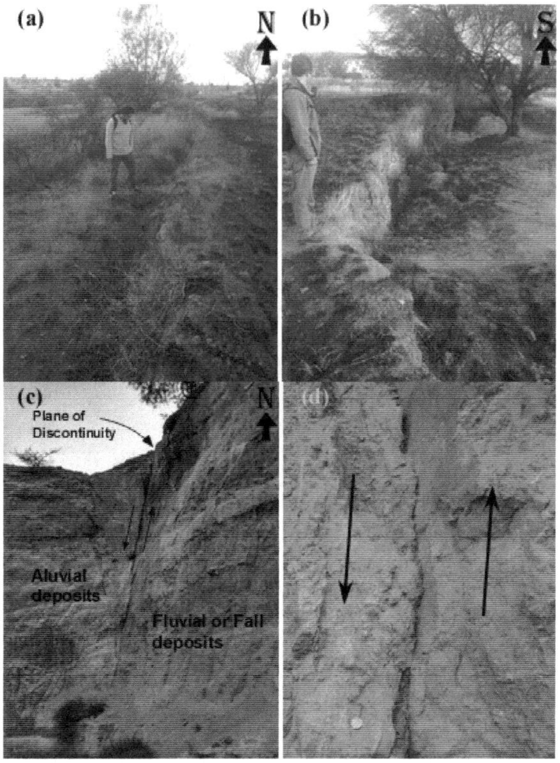

Figure 2. Fotographs depict some characterisitics of the Oriente fault inside the city of Aguascalientes. Difference between the base and the surface in the quarry in panels **(c)** and **(d)** is close to 8 m.

the opening is of more than 1 m and the observable depth can reach 2 m. Photographs in Fig. 2a and b show the geometric characteristics of this fault (point 1 in Fig. 1). Another important point of discussion can be observed in the quarry of Fig. 2c and d. There, figures show an approximate quarry depth of 8 m, and as observed, alluvial sediments on the left are not shown on the right, this mean that there is a fault offset higher than 8 m. The fact that the fault plane cuts the entire upper sedimentary sequence indicates continuous activity of the fault in recent years. Also, the relative large fault offset may indicate that this activity started many years ago, prior groundwater pumping.

3 Comparison of aerial photographs

As presented in Pacheco-Martínez et al. (2013), the comparison of images containing a fault in two different years suggests that the origin of some surface discontinuities is natural, prior the pumping-induced subsidence process. In Fig. 3, the photograph of the left is from the early 1970's, before the occurrence of heavy pumping, however, the alignment of bushes clearly suggests an incipient discontinuity at the surface. It probably was well formed at depth by then. The discussion on this observation lies on the activity of this discontinuity at this particular point: since this fault flanks the valley and contributed to the origin of the valley the graben, then it is easy to think that this fault is tectonic in origin, However, the discussion here is the fact that tectonic activity could continue by the early 1970's, therefore this fault might be geologically defined as active.

4 Formation of refilled cracks

The observed refilled cracks with carbonates in the Oriente fault reinforce the idea of quaternary tectonic activity stressing on the valley, which generated faults and cracks currently reactivated due to induced stresses by groundwater decline. As discussed in Pacheco-Martínez et al. (2013), a representative 1 cm-width crack in a region with climate similar to that in the valley of Aguascalientes, can last between 100 and 500 years per linear meter to be refilled with caliche, a time sufficiently long before the initiation of heavy pumping. As observed in the figure all cracks are closely parallel between them and their orientation quite similar to that of the main discontinuity at regional scale. Point 2 in Fig. 4 shows the location of observation for these refilled cracks.

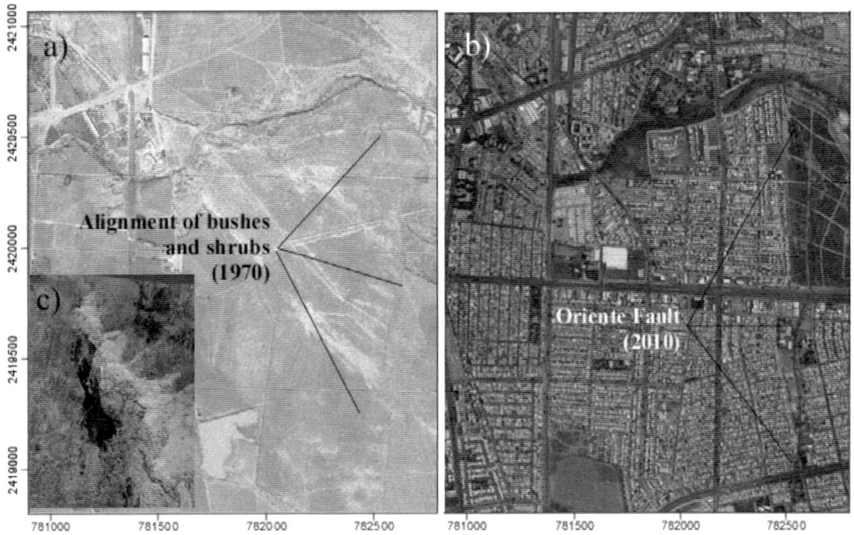

Figure 3. Comparison of two zones before and after intensive pumping (modified from Pacheco-Martínez et al., 2013). The particular location of this area is indicated in Fig. 1 with number 4.

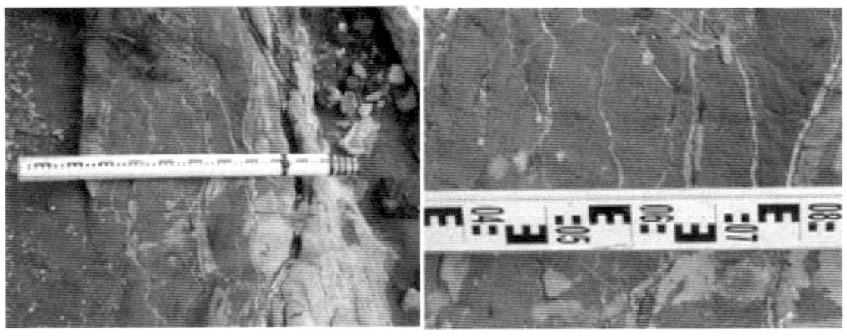

Figure 4. Filled cracks inside the zone of influence of the Oriente Fault. Refill consisted in groundwater caliche with a maximum thickness of 1 cm.

5 Fault offset on upper sedimentary layers

In order to elucidate on the existence of buried cracks potentially becoming in surface discontinuities, it is a common practice in the Aguascalientes Valley to excavate trenches. This activity is usually performed to complement the geophysical studies and allows either confirming or dismissing buried ground discontinuities. When a buried soil crack is confirmed, they can observe their path, describe the mechanical damage of the adjacent geologic material, and define the potential influence zone on both sides of the discontinuity. Figure 5 shows a fault cutting the upper sediments into a trench. An offset of more than 1.5 m was observed. This fault however is not the Oriente fault. This observation indicates that intra-basin faults may be also older than the pumping-induced subsidence process. This fault may represent an old fault reactivated by groundwater pumping.

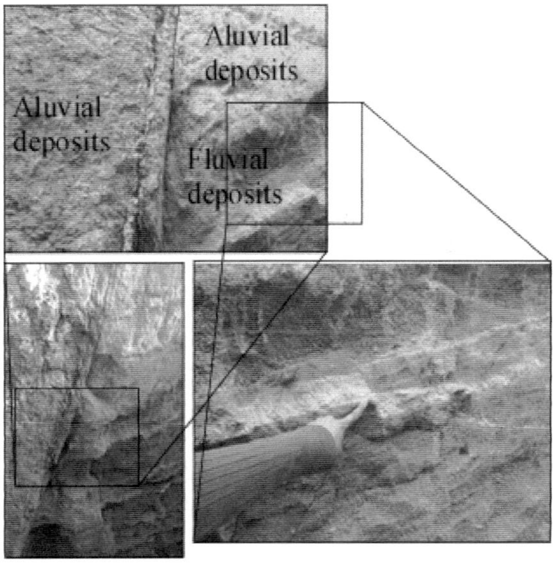

Figure 5. Trench excavated to physically observe a fault at depth. This is point 4 in Fig. 2. Observations indicate that hanging wall seems to be on the west of the fault, not to the east, which is toward the center of the valley.

6 Future work

Future work regarding surface deformation in the valley of Aguascalientes includes the geophysical and geohydrological characterization of more zones affected by surface discontinuities, with particular emphasis in the oriente fault. Also, more observations on excavations with discontinuities are expected in order to understand more the origin of these structures in the valley. The new code of construction in the valley requires the physical observation of the upper (< 2 m depth) sediments in order to detect or dismiss potential discontinuities. These excavations will help to observe more details in this type of structures.

7 Summary and conclusions

Some observations regarding the origin of surface discontinuities in the valley of Aguascalientes are presented and discussed. Most of observations are on the Oriente fault, which flanks the semi-graben of Aguascalientes in its eastern portion. This fault presents recent activity and appears to be active in the last decades or even centuries. Intensive pumping started in the late 1970's; then, in few years many surface discontinuities were observed. However, observation presented in this work suggests that some of these discontinuities were originated long time before. There is doubt about how many of these discontinuities were originated long before intensive pumping, but there is certain that most of them appeared on the surface simultaneously to intensive pumping and continue developing at these days. The evidence presented in this work, suggests that some of the current active surface discontinuities were originated by stresses other than pumping (tectonic?).

Acknowledgements. Diego H. Frías-Guzmán is thankful to CONACYT for financial support.

References

Aranda-Gómez, J. J.: Geología preliminar del graben de Aguascalientes, Revista Mexicana de Ciencias Geológicas, 8, 22–32, 1989.

Aranda-Gómez, J. M. and Aranda-Gómezomez, J. J.: Análisis del agrietamiento en la ciudad de Aguascalientes, Informe anual preparado para el Departamento de Geotecnia del Centro Tecnológico, Autonomous University of Aguascalientes, Aguascalientes, México, unpublished results, 1985.

Lermo, J., Nieto-Obregon, J., and Zermeño-De-Leon, M. E.: Faults and fractures in the Aguascalientes Valley. Preliminary microzonification. Proceedings of the Eleventh World Conference on Earthquake Engineering. Acapulco, Mexico, Elsevier Science Ltd. Ed., Paper No. 1651, 8 pp., 1996.

Pacheco-Martínez, J., Hernandez-Marín, M., Burbey, T. J., González-Cervantes, N., Ortíz-Lozano, J. A., Zermeño-De-Leon, M. E., and Solís-Pinto, A.: Land subsidence and ground failure associated to groundwater exploitation in the Aguascalientes Valley, Mexico, Eng. Geol., 164, 172–186, doi:10.1016/j.enggeo.2013.06.015, 2013.

UNAM-SOP: Monitoreo sísmico y análisis estructural de las grietas en la ciudad de Aguascalientes, Informe técnico preparado por el Instituto de Ingeniería de la UNAM para la Secretaria de Obras Publicas del estado de Aguascalientes, Aguascalientes, México, Internal project No. 142/93, unpublished results, 1994.

Proc. IAHS, 372, 239–242, 2015
proc-iahs.net/372/239/2015/
doi:10.5194/piahs-372-239-2015

Research of features related to land subsidence and ground fissure disasters in the Beijing Plain

Y. Yang[1,2], Y. Luo[2], M. Liu[2], R. Wang[2], and H. Wang[3]
[1]School of Earth and Space Sciences, Peking University, Beijing, China
[2]Beijing Institute of Hydrogeology and Engineering Geology, Beijing, China
[3]China Institute of Geo-Environment Monitoring, Beijing, China
Correspondence to: Y. Yang (yangyanhunan1982@126.com)

Published: 12 November 2015

Abstract. This study overviews the development history, current situation, impact, and hazards of land subsidence and ground fissure disasters in the Beijing Plain (BP) and focuses on the disaster distribution and features of disaster-causing mechanisms. Currently, the BP is still in a rapid developmental stage of land subsidence. The development and distribution of land subsidence are affected by various factors including the thickness of Quaternary compressible clay, groundwater overexploitation, and the rapid development of urban construction. The causes of ground fissures in the BP are complex and diverse, with evidence of structure fissures, non-structure fissures, and mixed genesis fissures. Investigations of the Gaoliying ground fissure have shown that this fissure has evidence of fracture activity, with vertical deformation that is more significant than horizontal deformation. Furthermore, this ground fissure has characteristics of inter-annual periodicity and annual jumping. The land subsidence and ground fissures are all under structural control and impact each other, and more severe hazards may be induced under the superimposition of these two types of disasters. Effective measures and suggestions for disaster prevention and control are recommended on the basis of this study.

1 Introduction

Land subsidence is one of the most pervasive developmental geological disasters in the Beijing Plain (BP). The subsidence area has accounted for approximately 2/3 of the entire BP up to 2014, with a maximum accumulative subsidence of over 1.6 m and a maximum subsidence rate of more than 150 mm year^{-1} (Yang et al., 2010). The BP is still in the rapid developmental stage of land subsidence. The land subsidence monitoring in the BP started in the 1980s, and has been greatly improved over the past decade. Several monitoring methods have been utilized including InSAR, GPS, extensometers, and high-precision level measurements. Furthermore, the monitoring network has covered almost the entire plain.

The distribution and development of ground fissures in the BP are apparently affected by geological structures, and most of them are caused by multiple geneses. In recent years, exploratory trench excavations, automatic monitoring, and level measurements of ground deformation have been conducted to obtain more detailed information of these fissures. Furthermore, a special monitoring station has been constructed in the Shunyi district.

The land subsidence and ground fissures can potentially damage buildings and other constructions, reduce the value of land use, threaten the safety of track traffic, decrease the ground levels as well as reduce the capability of urban flood control and drainage, resulting in extensive economic and social losses (Yang et al., 2013).

2 Characteristics of land subsidence

2.1 Distribution

The land subsidence in the BP mainly develops in the lower and middle parts of the diluvial–alluvial plain. The connecting parts among the diluvial–alluvial fans of the Wenyu River, Chaobai River, and Yongding River have become one of the largest subsidence regions in the BP, resulting in

Published by Copernicus Publications on behalf of the International Association of Hydrological Sciences.

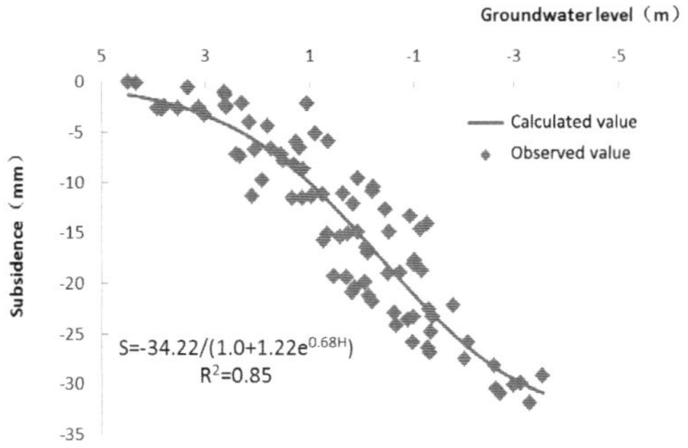

Figure 1. The relationship between groundwater levels and land subsidence in the stratum 66–94 m below the ground surface.

the greatest subsidence area and the maximum subsidence amount and subsidence rate.

The most serious subsidence areas in the BP coincide well with the distribution of the thickness of Quaternary compressible clay. The Quaternary depositional environment in the BP is complex, with sediments displaying alternating sequences of sand and clay layers and complex structures (Jia et al., 2011). The land subsidence develops significantly in areas where the total thickness of compressible layers is greater than 150 m.

2.2 The development of land subsidence is controlled by variations of groundwater levels

The overexploitation of groundwater is the most important factor that contributes to the emergence and development of land subsidence in the BP. Monitoring data obtained from measurements of land subsidence and groundwater levels over several years indicate similar decreasing trends (Fig. 1). In regions where a continuous decrease of groundwater levels is observed, the land subsidence develops rapidly and the accumulated subsidence increases continuously. However, in regions where the groundwater levels increase, the subsidence rate becomes slower and the ground may experience minor rebounding in some local areas.

2.3 Effects of city construction on land subsidence

With the rapid development of urban construction and continuous growth of the urban population, water demand has increased significantly; consequently, groundwater depression cones and subsidence bowls have occurred. Therefore, urban construction has led to the formation of ground settlement and resulted in the rapid development of land subsidence.

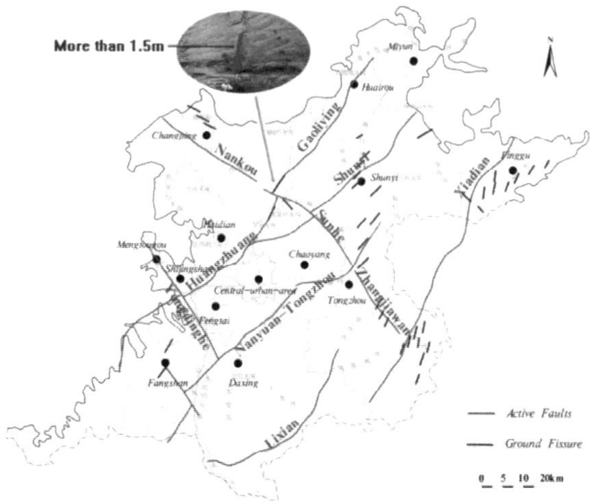

Figure 2. The distribution of ground fissures in the BP.

3 Characteristics of ground fissures

3.1 Distribution

Beijing is located at the junction between the Yanshan fold and thrust belt (mountain area) and the Huabei fault sag (plain area), and has experienced multi-stage tectonic movements such as the Indosinian, Yanshan, and Himalayan events. Historically, several strong earthquakes have occurred in the area, creating ground fissures in different regions such as Shunyi, Pinggu, Changping, and Tongzhou.

There are two main groups of active faults in the BP: the NE-trending fault belt comprising the Huangzhuang–Gaoliying fault, Shunyi fault, Xiadian fault, and the NW-trending Nankou–Sunhe fault. The ground fissures mainly occur along active faults (Fig. 2).

3.2 Causes of ground fissures

Ground fissures are divided into structure fissures, non-structure fissures, and mixed genesis fissures according to the origin type. All three types of ground fissures can be found in the BP, and the mixed genesis fissures are the most common type. The mixed-genesis fissures are mainly controlled by endogeneic forces in the Earth and can be intensified by human activities. The Shunyi ground fissure belt, Gaoliying ground fissure, Tugou ground fissure belt, and Yangfang–Miaojuan ground fissure belt belong to the class of mixed genesis fissures. Structure fissures are caused by endogenic geological processes. Their formation, size, and level of destruction have direct relationships with earthquakes, fault activities, and the regional tectonic stress field. For example, after the Tangshan earthquake in 1976, some ground fissures occurred in the BP, which damaged buildings as well as houses and caused panic among the population. Non-structure fissures

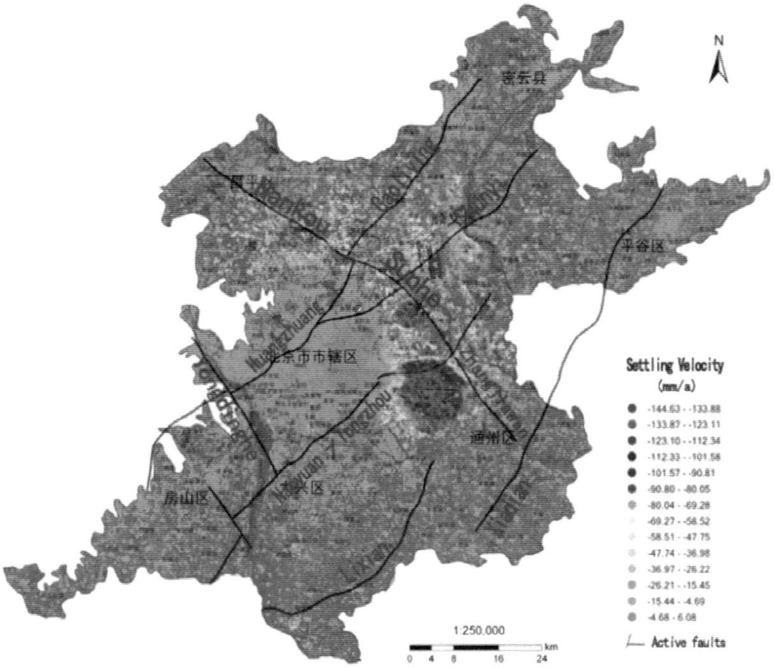

Figure 3. Distribution of subsidence rates and active faults in the BP.

are caused by exogenic geological processes, which have different origins and types. In the BP, the non-structure fissures include collapsing ground fissures, special soil ground fissures, and soil–liquefaction ground fissures.

3.3 Activity characteristics of ground fissures

The Gaoliying ground fissure, for example, is under the joint influence of both creeping slip of the Huangzhuang–Gaoliying fracture and differential settlement of the ground surface. Exploratory trenches show that this fracture has crossed over the entire Quaternary strata. The fracture plane is exposed at the ground surface and has the same location as the ground fissure. The monitoring results of differential settlement on both sides of the fault correspond well with the monitoring results of regional land subsidence.

The surface displacements of the Gaoliying fissure were obtained from the KLA-1 type automatic telemetering system. The results indicate that the ratio of the horizontal torsional displacement, horizontal extension, and vertical differential settlement show obvious three-dimensional activity characteristics (Wang et al., 2013). Some engineering characteristics have also been observed such as similar disaster impacts, limited disaster-causing width, and progressive disaster processes (Liu et al., 2014).

4 The relationship between land subsidence and ground fissures

4.1 All controlled by geological structures

As shown in Fig. 2, most ground fissures are distributed along active fractures, indicating that they are greatly controlled by geological structures.

The land subsidence in the BP is also controlled by geological structures. Figure 3 shows the distribution of the subsidence rate from InSAR data and the active faults in the BP. The northwestern subsidence area is mostly controlled by the Nankou–Sunhe fracture and Huangzhuang–Gaoliying fracture, where the development trend of the subsidence is essentially parallel to the trend of the fractures. The northeastern subsidence area is mostly controlled by the Shunyi fracture. The eastern East–Balizhuang–Dajiaoting subsidence area is controlled by the Shunyi fracture and the Nanyuan–Tongzhou fracture. The main influence from tectonic movements causes the tectonic units to decline entirely and slowly under the regional stress field. Furthermore, the relative rise and decline of the hanging walls or the footwalls have some impact on the difference in the subsidence rate between the two sides of the fractures.

The distribution of the average subsidence rates in two cross-sections at the Huangzhuang–Gaoliying fracture and Nanyuan–Tongzhou fracture are shown in Fig. 4. The defor-

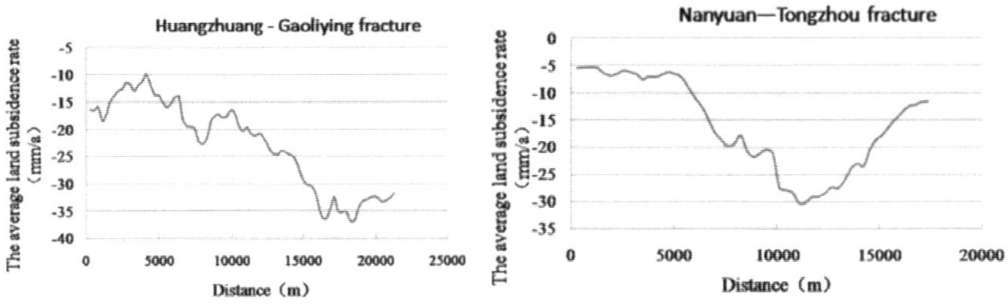

Figure 4. Deformation gradient map in two cross-sections at the Huangzhuang–Gaoliying fracture and Nanyuan–Tongzhou fracture.

mation gradient of the ground surface is relatively large along the two cross-sections, with maximum differences of nearly 28 and 25 mm year^{-1}, respectively.

4.2 Interplay between land subsidence and ground fissures

Mixed genesis fissures are the most common type in the BP. In the developmental areas of ground fissures, the lithology is commonly different on both sides of the fissure. When the influence of surface load and soil self-weight impacts and changes, the original groundwater flow field, obvious differential settlement may occur due to the different stratigraphic textures and saturated conditions on either side of the fissure. In contrast, ground fissures may be induced as the differential settlement increases to a certain degree. The damage of ground fissures will increase greatly with the joint function of the differential settlement and active faults, which will intensify the damage to buildings. Therefore, the developmental areas of ground fissures coincide well with the serious areas of land subsidence; i.e., the ground fissures are very active in areas with obvious differential settlement.

5 Summary and conclusions

Land subsidence and ground fissures are widely distributed in the BP. They are affected by several factors, likely controlled by geological structures, and have impacts on each other. In the near future, land subsidence and ground fissure disasters will rapidly develop, threatening urban construction and safety.

The common characteristics of land subsidence and ground fissures are slow and irreversible. Thus, prevention is a priority and it should be efficiently combined with control procedures. When new cities and towns are planned and major linear infrastructure and buildings are built, it is necessary to avoid the areas of severe land subsidence and ground fissures as much as possible. The design for these projects must be strict to reduce the influence of these hazards in the new development areas.

Acknowledgements. Thanks to my institute and all colleagues, for their efforts to finish this report results. Thank Ye Shu-jun, Gong Shi-liang and Zhang Jin-jiang, who have been encouraging.

References

Jia, S., Wang, H., Ye, C., and Liu, M.: Investigation and survey methods appropriate for ground fissure in Beijing, Journal of Engineering Geology, 19, 104–111, 2011.

Liu, M., Jia, S., Chen, Z., and Guo, G.: Study of the activity and impact of the Gaoliying ground fissure on the Beijing plain, Shanghai Land & Resources, 35, 53–57, 2014.

Wang, H., Yang, Y., Liu, M., Jia, S., and Tian, F.: Analysis of the activity of the Gaoliying ground fissure (Beijing) based on automatic monitoring, Shanghai Land & Resources, 34, 64–67, 2013.

Yang, Y., Jia, S., and Wang, H.: The status and development of land subsidence in Beijing plain, Shanghai Geology, 4, 23–28, 2010.

Yang, Y., Jia, S., Wang, H., and Zhou, Y.: Analysis on impact of land subsidence on planned new cities in Beijing, City Planning Review, 11, 67–71, 2013.

Proc. IAHS, 372, 243–248, 2015
proc-iahs.net/372/243/2015/
doi:10.5194/piahs-372-243-2015

Open Access
PIAHS

Subsidence monitoring with geotechnical instruments in the Mexicali Valley, Baja California, Mexico

E. Glowacka[1], O. Sarychikhina[1], V. H. Márquez Ramírez[2], B. Robles[3], F. A. Nava[1], F. Farfán[1], and M. A. García Arthur[1]

[1]Centro de Investigacion Cientifica y Educacion Superior de Ensenada, Ensenada, Mexico
[2]UNAM Campus, Centro de Geociencias, Juriquilla, Querétaro, Mexico
[3]Instituto Mexicano de Tecnología de Agua, Jiutepec, Morelos, Mexico

Correspondence to: E. Glowacka (glowacka@cicese.mx)

Published: 12 November 2015

Abstract. The Mexicali Valley (northwestern Mexico), situated in the southern part of the San Andreas fault system, is an area with high tectonic deformation, recent volcanism, and active seismicity. Since 1973, fluid extraction, from the 1500–3000 m depth range, at the Cerro Prieto Geothermal Field (CPGF), has influenced deformation in the Mexicali Valley area, accelerating the subsidence and causing slip along the traces of tectonic faults that limit the subsidence area. Detailed field mapping done since 1989 (González et al., 1998; Glowacka et al., 2005; Suárez-Vidal et al., 2008) in the vicinity of the CPGF shows that many subsidence induced fractures, fissures, collapse features, small grabens, and fresh scarps are related to the known tectonic faults. Subsidence and fault rupture are causing damage to infrastructure, such as roads, railroad tracks, irrigation channels, and agricultural fields.

Since 1996, geotechnical instruments installed by CICESE (Centro de Investigación Cientifica y de Educación Superior de Ensenada, B.C.) have operated in the Mexicali Valley, for continuous recording of deformation phenomena. Instruments are installed over or very close to the affected faults. To date, the network includes four crackmeters and eight tiltmeters; all instruments have sampling intervals in the 1 to 20 min range.

Instrumental records typically show continuous creep, episodic slip events related mainly to the subsidence process, and coseismic slip discontinuities (Glowacka et al., 1999, 2005, 2010; Sarychikhina et al., 2015).

The area has also been monitored by levelling surveys every few years and, since the 1990's by studies based on DInSAR data (Carnec and Fabriol, 1999; Hansen, 2001; Sarychikhina et al., 2011).

In this work we use data from levelling, DInSAR, and geotechnical instruments records to compare the subsidence caused by anthropogenic activity and/or seismicity with slip recorded by geotechnical instruments, in an attempt to obtain more information about the process of fault slip associated with subsidence.

1 Introduction

The Mexicali Valley is located, in the southern part of the San Andrés fault system, within the southern part of the Salton Trough, on the border between the North America and Pacific tectonic plates. The Valley is characterized by recent volcanic hydrothermal processes, active tectonics, and high seismicity. Moreover large earthquakes concentrate along the major, Imperial and Cerro Prieto faults, while scattered seismicity, (mainly swarms), and deformation are observed in the Pull-apart Cerro Prieto Basin (Lomnitz et al., 1970; Nava and Glowacka, 1994; Suárez-Vidal et al., 2008). Fluid extraction began in the Cerro Prieto Geothermal Field (CPGF) in 1973, and brine injection therein began in 1989; these processes have been influencing deformation, stress, and seismicity of the area (Majer and McEvilly, 1981; Glowacka and Nava, 1996; Fabriol and Munguía, 1997; Glowacka et al., 1999, 2005; Trugman et al., 2014).

The subsidence area is limited to the area between the Imperial, Saltillo, Cerro Prieto and Morelia faults (Fig. 1). This zone, also known as Cerro Prieto basin, is larger than

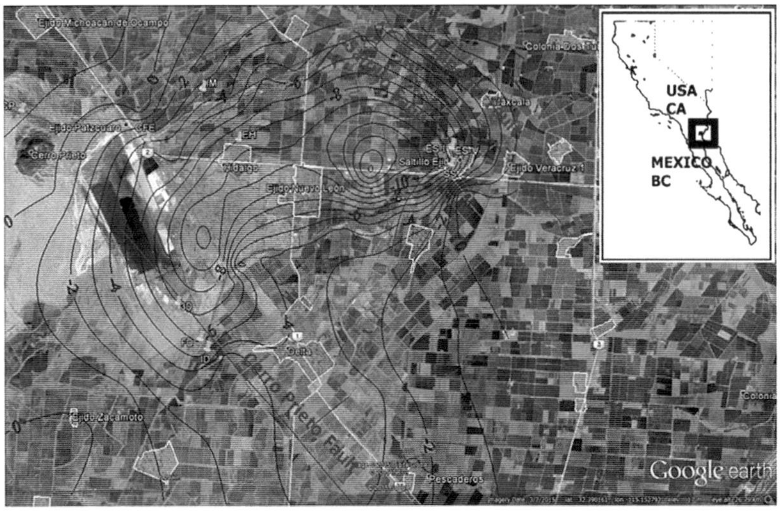

Figure 1. Instruments (yellow pins) and the subsidence rate blue isolines (cm yr^{-1}) for 1997–2006 (modified from Glowacka et al., 2011). Red lines indicate tectonic faults. The inset shows the geographical location of the study area.

the extraction zone, suggesting the existence of a recharging region. Since the sunken area is limited by tectonic faults, Glowacka et al. (1999, 2005, 2010a), suggest that these faults constitute a boundary of the subsiding region, due to differential compaction and/or due to poor permeability in the direction perpendicular to the faults that acts as a groundwater barrier.

The CPGF has 720 MW production capacity, or 1/3 of electricity production in Baja California; however, the anthropogenic subsidence caused by deep extraction of fluids in the CPGF causes damage to roads, railways and irrigation systems, increasing the natural hazards in this area that is quite vulnerable because of its tectonic situation.

The history of subsidence at the CPGF area has been well documented. Geodetic studies in the Mexicali Valley began in the 1960's. Leveling surveys have been done in the area of the CPGF and the Mexicali Valley since 1977 (Velasco, 1963; Lira and Arellano, 1997; Glowacka et al., 1999, 2012). Subsidence in the CPGF has also been measured via DInSAR (Differential Synthetic Aperture Radar Interferometry) by Carnec and Fabriol (1999) and Hanssen (2001) using ERS1/2 images acquired during 1993–1997 and 1995–1997, respectively, and interpreted as an anthropogenic effect of fluid extraction.

Sarychikhina et al. (2011) applied the DInSAR technique using C-band ENVISAR ASAR data acquired between 2003 and 2006, to determine the extent and amount of land subsidence in the Mexicali Valley near the CPGF. Recently Sarychikhina et al. (2015) applied DInSAR technique, together with leveling results and geotechnical instruments data and modeling, to estimate seismic and aseismic deformation in Mexicali Valley for the period 2006–2009. The current subsidence rate, evaluated from DInSAR data is of the order of 12 cm yr^{-1} in the production area and 18 cm yr^{-1} for the recharging zone (Sarychikhina et al., 2015; Sarychikhina and Glowacka, 2015a).

2 Geotechnical instruments monitoring

To study the spatial and temporal distribution of crustal deformation in the Mexicali Valley, CICESE installed a network of geotechnical instruments, starting in 1996. Since then, the REDECVAM (Red de Deformaciones de la Corteza en el Valle de Mexicali) network has included four creepmeters (wide range extensometer) and nine tiltmeters installed over different periods in different places; all instruments have sampling intervals in the 1 to 20 min range (Nava and Glowacka, 1999; Glowacka et al., 2002, 2010b, c; Sarychikhina et al., 2015). Creepmeters (Geokon model 4420) were installed mainly on vertical planes perpendicular to faults, in order to record vertical displacement. Biaxial tiltmeters (Applied Geomechanics, models 711, 712, and 722) were installed in shallow vaults or wells, close to faults.

Data from a creepmeter and a tiltmeter installed on the Saltillo fault (Fig. 1) show fault vertical displacement rate at ~ 5.3 cm yr^{-1} until 2003 and ~ 7.3 cm yr^{-1} since then. The distance-time relationship between changes in extraction at the CPGF and the displacement rate change found at the Saltillo fault suggests that the fault is probably affected by extraction through diffusive transmission of pore pressure changes, with a characteristic hydraulic diffusivity (Glowacka et al., 2010a).

Vertical displacement at the Saltillo fault consists of continuous creep and episodic slip events, sometimes concen-

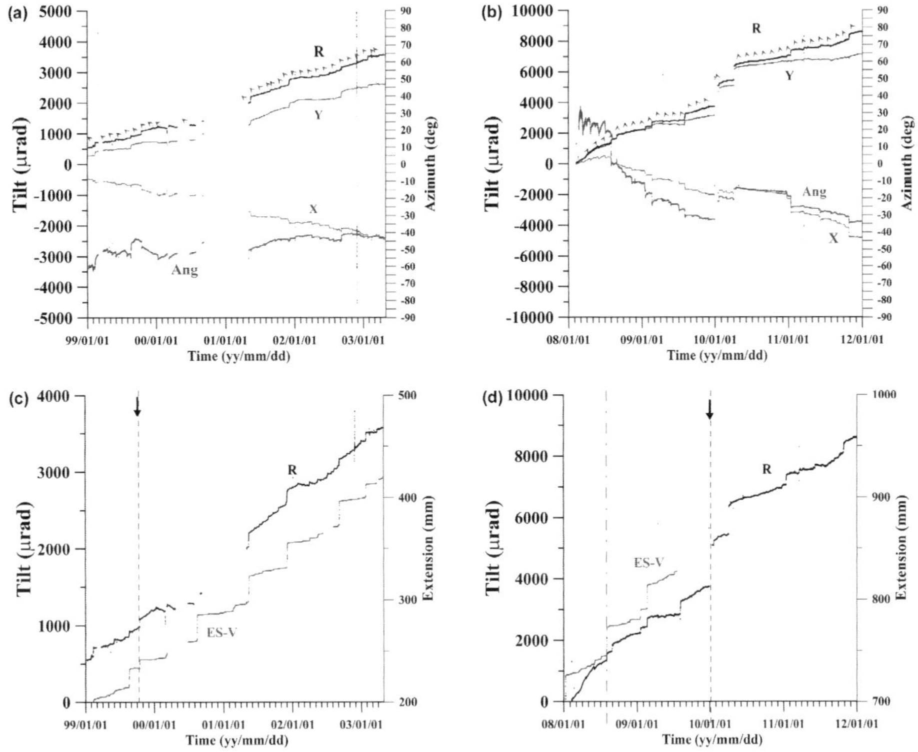

Figure 2. Tilt observations. X and Y are the East-West and North–South components of tilt, respectively, and R is the resultant tilt magnitude in the azimuth direction, Ang, for 1999–2003 **(a)** and 2003–2012 **(b)**. X, Y and R are referred to the left axes, while azimuth Ang is referred to the right axes. Azimuth is also shown by the direction in which brown arrowheads point, with North straight up. Extension ES-V and magnitude R for 1999–2003 **(c)** and 2008–2012 **(d)**. The vertical arrow indicates the $M = 7.2$ hector Mine earthquake in **(c)**, and $M = 5.8$ 2009 earthquake in **(d)**. The dotted-line indicates slip event in August 2008.

trated in suites (Nava and Glowacka, 1999). Episodic slips have 1 to 3 cm magnitudes and 1–3 days duration, separated by months of monotone creep, and release about 50 % of total slip. The episodic fault slip in the Saltillo fault appears mainly as slip-predictable, normal, aseismic slip (e.g. Glowacka et al., 2001, 2010a). Using reservoir model, geological structure, and subsidence data, Glowacka et al. (2010a), speculatively proposed the depth range of slip events to be from 1 to 2.5 km.

Some of the slip events are triggered by distant earthquakes. This includes, for example, Hector Mine Earthquake (of magnitude 7.2 and at a distance 260 km away 15 on 1999), as described by Glowacka et al. (2002), and, probably, by the Canal de las Ballenas earthquake (of magnitude 6.9 and a distance $\sim$400 km away 2009), as described by Glowacka et al. (2015). These studies suggest that distant earthquakes can influence fault slip related to the subsidence. Both, sporadic and triggered slip events confirm slip predictable fault behavior.

Detailed analysis of tiltmeter and creepmeter data done by Sarychikhina et al. (2015), identified which part of subsidence observed by DInSAR in the Mexicali valley during 2006–2009 was caused by local seismicity. Glowacka et al. (2015) estimated that during the 2006–2009 period of relatively high seismicity, the anthropogenic subsidence was of the order of 80 % of the total subsidence, which is close to estimates done by Glowacka et al. (2005), Camacho Ibarra (2006), and Sarychikhina et al. (2015).

The goal of this paper is to find if there are differences, in the signal characteristics observed on the creepmeter and/or tiltmeter records, between continuous creep, sporadic slip, triggered slip and coseismic slip recorded on the Saltillo fault.

3 Results

In the following we will analyze records from ES-V creepmeter and ES-I biaxial tiltmeter installed very close to each other on the Saltillo fault. A biaxial tiltmeter (ES-I) was installed in 1998, very close to the creepmeter and recorded ground tilt until 2003 (Fig. 2a), and after a small reconstruction, again since 2008 (Fig. 2b). E–W (X) and N–S (Y) inclination recorded on ES-I tiltmeter are presented on Fig. 2a (1999–2003) and Fig. 2b (2008–2012).

For the tilt we calculate the resultant magnitude, R, and the azimuth, Ang, as

$$R = \left(X^2 + Y^2\right)^{1/2}, \quad (1)$$

$$\text{Ang} = \arctan(\Delta X / \Delta Y), \quad (2)$$

where X and Y are the measured tilts in the N–S and E-W directions, respectively; $\Delta X = X \cos\theta + Y \sin\theta$, and $\Delta Y = Y \cos\theta - X \sin\theta$, and θ is the correction for the azimuth orientation of the Y axis from North. $\theta = -10°$ for ES-I.

The tilt magnitude, R, and the ES-V vertical extension recorded by the creepmeter, are shown in Fig. 2c (1999–2003) and Fig. 2d (2008–2012).

Figure 3 shows R and Ang for the August 2008 slip episode (Fig. 3a), the slip event triggered by the $M = 7.2$ Hector Mine earthquake (Fig. 3b) and by the local $M = 5.8$, 30 December 2009, earthquake (Fig. 3c). While episodic slip events and triggered slip events show very similar behavior of R and Ang, the slip related to the 5.8 local earthquake has different behavior.

4 Discussion

Figure 2c and d show R and creepmeter extension ES-V observed between 1999–2003 and 2008–2012, respectively. The similarity between R and ES-V seen on both Fig. 2 suggests that tilt increases when vertical displacement increases on the fault, as observed during continuous creep and slip events.

There is not such an evident tendency for azimuthal behavior (Fig. 2a and b); the azimuth oscillated during 1999–2000, slightly increased during 2001–2003, and diminished with time during 2008–2012, except during December 2009–April 2010, which will be discussed later.

However, a significant azimuth change can be seen between Fig. 2a and b. Tilt tends to be more north-oriented for the 2008–2012 period. This phenomenon can be related to the subsidence amplitude increase in the recharge area observed for 1993–2009 by Sarychikhina and Glowacka (2015b) (this book).

Three kinds of deformation are shown in Fig. 3. The typical slip event shown on Fig. 3a is characterized by an R increase, related to the extension (subsidence) increase and azimuth decrease, with about one day duration. Triggered by the $M = 7.2$ HME, the slip event shown on Fig. 3b is characterized by abrupt R increase, extension (subsidence) increase, and azimuth decrease. This is followed by the slow deformation with the same orientation, similar to the regular slip event, but with smaller magnitude, expected since the slip predictable character of deformation suggested by Glowacka et al. (2002). The deformation shown on Fig. 3c was caused by a local $M = 5.8$ earthquake and about 1 h of deformation was recorded after the earthquake before the instrument went out of range. Within the time precision of the tiltmeter (one sample every 4 min) the immediate R increase followed by postseismic R increase can be observed. However, the azimuth shows immediate increase and postseismic decrease.

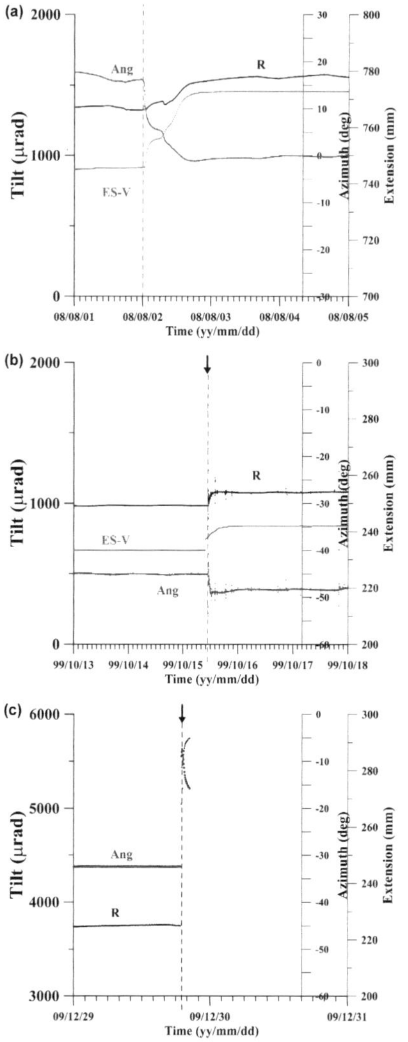

Figure 3. Tilt observations vs. time for episodic slip **(a)**, triggered slip **(b)** and coseismic slip **(c)**. R is the resultant tilt magnitude in the azimuth direction Ang. R is referred to the left axes, while azimuth Ang is referred to the right axes. Extension ES-V is shown for the period when the creepmeter was working. The vertical arrow indicates the $M = 7.2$ 1999 hector Mine earthquake **(b)** and the $M = 5.8$, 2009 earthquake **(c)**.

Although creepmeters were not working during the $M = 5.8$, 2009 earthquake, Sarychikhina et al. (2015), using data from InSAR and leveling, showed there was subsidence (about 20 cm) and slip associated with this earthquake. They suggested that the observed deformation could be caused by triggered aseismic slip on the Saltillo fault or by earthquake-triggered soft sediments subsidence.

5 Conclusions

Our results show that the tilt magnitude is proportional to the extension recorded by a creemeter, which may eventually allow to substitute results when one instrument is missing.

Since magnitude and azimuth behaviors are very similar for episodic and triggered slip we suggest that they occur along the same fault segment and the same depth.

However, the different tilt azimuthal behavior for local earthquakes suggests that the slip on the fault has a different behavior, possibly including a horizontal component or/and that the slip was deeper than for episodic slip. More dense instrumentation with a higher frequency of measurements could be useful to confirm this suggestion.

Acknowledgements. This research was sponsored in part by CONACYT, project 105907 and CICESE internal funds.

References

Camacho Ibarra, E.: Análisis de la deformación vertical del terreno en la región de confluencia del sistema de fallas Cerro Prieto – Imperial en el periodo 1962–2001, MSc Thesis, Centro de Investigación Científica y Educación Superior de Ensenada, Ensenada, B.C., Mexico, 135 pp., 2006

Carnec, C. and Fabriol, H.: Monitoring and Modeling Land Subsidence at the Cerro Prieto Geothermal Field, Baja California,Mexico, Using SAR Interferometry, Geophys. Res. Lett., 26, 1211–1214, 1999.

Fabriol, H. and Munguia, L.: Seismic Activity At The Cerro Prieto Geothermal Area (Mexico) From August 1994 To December 1995, And Relationship With Tectonics And Fluid Exploitation, Geophys. Res. Lett., 24, 1807–1810, 1997.

Glowacka, E. and Nava, F. A.: Major earthquake in Mexicali valley, Mexico, and Fluid Extraction at Cerro Prieto Geothermal Field, Bull. Seismol. Soc. Am., 86, 93–105, 1996.

Glowacka, E., González, J., and Fabriol, H.: Recent Vertical deformation in Mexicali Valley and its Relationship with Tectonics, Seismicity and Fluid Operation in the Cerro Prieto Geothermal Field, Pure Appl. Geophys., 156, 591–614, 1999.

Glowacka, E., González, J. J., Nava, F. A., Farfán, F., and Díaz de Cossío, G.: Monitoring Surface Deformations in the Mexicali Valley, B.C., Mexico, Proceedings of Tenth International Symposium on Deformation Measurements, Orange, California, USA, 175–183, 2001.

Glowacka, E., Nava, F. A., Diaz de Cossio, G., Wong, V., and Farfan, F.: Fault slip, Seismicity and Deformation in the Mexicali Valley (B.C., Mexico) after the 1999 Hector Mine Earthquake ($M = 7.1$), Bull. Seismol. Soc. Am., 92, 1290–1299, 2002.

Glowacka, E., Sarychikhina, O., and Nava, F. A.: Subsidence and stress change in the Cerro Prieto Geothermal Field, B.C., Mexico, Pure Appl. Geophys., 162, 2095–2110, 2005.

Glowacka, E., Sarychikhina, O., Suárez, F., Nava, F. A., and Mellors, R.: Anthropogenic subsidence in the Mexicali Valley, B.C., Mexico, and the slip on Saltillo Fault, Environ. Earth Sci., 59, 1515–1523, 2010a.

Glowacka, E., Sarychikhina O., Suárez, F., Nava, F. A., Farfan, F., Batani, G., and Garcia-Arthur, M. A.: Anthropogenic subsidence in the Mexicali Valley, B.C., Mexico, caused by the fluid extraction in the Cerro Prieto geothermal Field and the role of faults, Proceedings, World Geothermal Congress, Bali, Indonesia, 25–29, 2010b.

Glowacka, E., Sarychikhina, O., Nava, F. A., Suarez, F., Ramirez, J., Guzman, M., Robles, B., and Farfan, F.: Guillermo Diaz De Cossio Batani, Continuous monitoring techniques of fault displacement caused by geothermal fluid extraction in the Cerro Prieto Geothermal Field (Baja California, Mexico), in: Land Subsidence, Associated Hazards and the Role of Natural Resources Development, edited by: Correón-Freyre, D., Cerca, M., and Galowey, D., Land Subsidence, Associated Hazards and the Role of Natural Resources Development, Proceedings of EISOLS 2010, 17–22 October 2010, Querétaro, Mexico, IAHS Publ., 339, 326–332, 2010c.

Glowacka, E., Sarychikhina, O., Vazquez-Gonzalez, R., Nava, F. A., García-Hernández, A., Pérez, A., Aguado, C., López-Hernández, M., Farfan, F., Diaz De Cossio, G., and Orozco, L.: Catálogo de eventos de slip registrados con los instrumentos geotérmicos y piezómetros en la cuenca de Cerro Prieto, Reunión Anual Unión Geofísica Mexicana, 31, p. 121, 2011.

Glowacka, E., Sarychikhina, O., Robles, B., Suarez Vidal, F., Ramírez Hernández, J., and Nava, F. A.: Estudio geológico para definir la línea de hundimiento cero y monitorear la subsidencia de los módulos 10, 11 y 12 en el Valle de Mexicali, en el distrito de riego 014, Reporte Final, Rio Colorado, B.C., 2012.

Glowacka, E., Sarychikhina, O., Márquez Ramírez, V. H., Nava, F. A., Farfán, F., and García Arthur, M. A.: Deformation Around the Cerro Prieto Geothermal Field Recorded by the Geotechnical Instruments Network REDECVAM During 1996–2009, WGC 2015, Proceedings CD, Melbourne, Australia, 2015.

González, J., Glowacka, E., Suárez, F., Quiñones, J. G., Guzmán, M., Castro, J. M., Rivera, F., and Félix, M. G.: Movimiento reciente de la Falla Imperial, Mexicali, B. C. Ciencia para todos Divulgare, Universidad Autónoma de Baja California, 6, 4–15, 1998.

Hanssen, R. F.: Radar Interferometry: Data Interpretation and Error Analysis, Kluwer Academic Publishers, Dordrecht, the Netherlands, 328 pp., 2001.

Lira, H. and Arellano, J. F.: Resultados de la nivelación de precisión realizada en 1997, en el campo geotérmico Cerro Prieto, Informe Técnico RE 07/97, Comisión Federal de Electricidad, Residencia de Estudios, México, 28 pp., 1997.

Lomnitz, C., Mooser, F., Allen, C. R., Brune, J. N., and Thatcher, W.: Seismicity and tectonics of the northern Golf of California region, Mexico, preliminary results, Geof. Int., 10, 37–48, 1970.

Majer, E. L. and Mcevilly, T. V.: Seismological Studies at the Cerro Prieto Geothermal Field, 1978–1982, Proc. Fourth Symp. on the Cerro Prieto Geothermal Field, Baja California, Mexico, Comisión Federal de Electricidad, Guadalajara, Mexico, 145–151, 1981.

Nava, F. A. and Glowacka, E.: Automatic identification of seismic swarms and other spatio-temporal clustering from catalogs, Comput. Geosci., 20, 797–820, 1994.

Nava, F. A. and Glowacka, E.: Fault slip triggering, healing, and viscoelastic afterworking in sediments in the Mexicali_Imperial Valley, Pure Appl. Geophys., 156, 615–629, 1999.

Sarychikhina, O. and Glowacka, E.: Application of DInSAR Stacking Method for Monitoring of Surface Deformation Due to Geothermal Fluids Extraction in the Cerro Prieto Geothermal Field, Baja California, Mexico, Proceedings, World Geothermal Congress, Melbourne, Australia, 2015a.

Sarychikhina, O. and Glowacka, E.: Spacio-temporal evaluation of aseismic grond deformation in the Mexicali Valley (Baja California, Mexico) from 1993 to 2010, using differential SAR interferometry, NISOLS, Nagoya, Japan, 2015b.

Sarychikhina, O., Glowacka, E., Mellors, R., and Suárez-Vidal, F.: Land subsidence in the Cerro Prieto Geothermal Field, Baja California, Mexico, from 1994 to 2005. An integrated analysis of DInSAR, leveling and geological data, J. Volcanol. Geoth. Res., 204, 76–90, 2011.

Sarychikhina, O., Glowacka, E., Robles, B., Nava, F. A., and Guzmán, M.: Estimation of Seismic and Aseismic Deformation in Mexicali Valley, Baja California, Mexico, in the 2006–2009 Period, Using Precise Leveling, DInSAR, Geotechnical Instruments Data, and Modeling, Pure Appl. Geophys., 172, 3139–3162, 2015.

Suárez-Vidal, F., Mendoza-Borunda, R., Naffarrete-Zamarripa, L. M., Ramírez, J., and Glowacka, E.: Shape and dimensions of the Cerro Prieto pull-apart basin, Mexicali, Baja California, Mexico, based on the regional seismic record and surface structures, Int. Geol. Rev., 50, 636–649, 2008.

Trugman, D. T., Borsa, A. A., and Sandwell, D. T.: Did stresses from the Cerro Prieto Geothermal Field influence the El Mayor-Cucapah rupture sequence?, Geophys. Res. Lett., 41, 8767–8774, 2014.

Velasco, J.: Levantamiento Gravimétrico, Zona Geotérmica de Mexicali, Baja California. Technical Report, Consejo de Recursos Naturales no Renovables, 55 pp., 1963.

Proc. IAHS, 372, 249–253, 2015
proc-iahs.net/372/249/2015/
doi:10.5194/piahs-372-249-2015

Open Access
PIAHS

Characterization of earth fissures in South Jiangsu, China

S. Ye[1], Y. Wang[1], J. Wu[1], P. Teatini[2], J. Yu[3], X. Gong[3], and G. Wang[3]

[1]School of Earth Sciences and Engineering, Nanjing University, Nanjing, China
[2]Department of Civil, Environmental and Architectural Engineering, University of Padova, Padova, Italy
[3]Key Laboratory of Earth Fissures Geological Disaster, Ministry of Land and Resources, (Geological Survey of Jiangsu Province), Nanjing, China

Correspondence to: S. Ye (sjye@nju.edu.cn)

Published: 12 November 2015

Abstract. The Suzhou-Wuxi-Changzhou (known as "Su-Xi-Chang") area, located in the southern part of Jiangsu Province, China, experienced serious land subsidence caused by overly exploitation of groundwater. The largest cumulative land subsidence has reached 3 m. With the rapid progress of land subsidence since the late 1980s, more than 20 earth fissures developed in Su-Xi-Chang area, although no pre-existing faults have been detected in the surroundings. The mechanisms of earth fissure generation associated with excessive groundwater pumping are: (i) differential land subsidence, (ii) differences in the thickness of the aquifer system, and (iii) bedrock ridges and cliffs at relatively shallow depths. In this study, the Guangming Village Earth Fissures in Wuxi area are selected as a case study to discuss in details the mechanisms of fissure generation. Aquifer exploitation resulted in a drop of groundwater head at a rate of 5–6 m yr^{-1} in the 1990s, with a cumulative drawdown of 40 m. The first earth fissure at Guangming Village was observed in 1998. The earth fissures, which developed in a zone characterized by a cumulative land subsidence of approximately 800 mm, are located at the flank of a main subsidence bowl with differential subsidence ranging from 0 to 1600 mm in 2001. The maximum differential subsidence rate amounts to 5 mm yr^{-1} between the two sides of the fissures. The fissure openings range from 30 to 80 mm, with a cumulative length of 1000 m. Depth of bed rock changes from 60 to 140 m across the earth fissure. The causes of earth fissure generation at Guangming Village includes a decrease in groundwater levels, differences in the thickness of aquifer system, shallow depths of bedrock ridges and cliffs, and subsequent differential land subsidence.

1 Introduction

Excessive exploitation of groundwater in Suzhou-Wuxi-Changzhou (known as "Su-Xi-Chang") area, located in the southern part of Jiangsu Province, China, started in the 1970s. It has resulted in serious land subsidence and earth fissure hazard that caused significant damage and heavy economic losses. The first earth fissure case in the Su-Xi-Chang area was reported in 1989 at Henglin town of Changzhou City. Twenty earth fissures occurred in the Su-Xi-Chang area between 1990 and 2000. The rapid development of earth fissures from 1990 to 2000 destroyed many buildings and the local governments decided to stop groundwater pumping in 2000 to alleviate land subsidence and earth fissures. Currently, a total of 25 earth fissures have been found in Su-Xi-Chang area, mainly located in the Wujin County in Changzhou and Xishan County in Wuxi (Fig. 1) (Liu et al., 2004). According to geological surveys, the orientation of the earth fissure in this area are mainly NE (NNE), EW and SN direction (Fig. 1), showing no obvious unified direction in the region. All the fissures are in a dense banded distribution and their orientation are in perfect accordance with the underlying bedrock. The width of the earth fissure zone generally ranges from 30 to 100 m, and the length varies between 200 and 600 m, and several fissures can reach lengths of 1000 m (Yu et al., 2004).

A significant amount of research has been done to study the occurrence of earth fissures in this area. Liu et al. (2004)

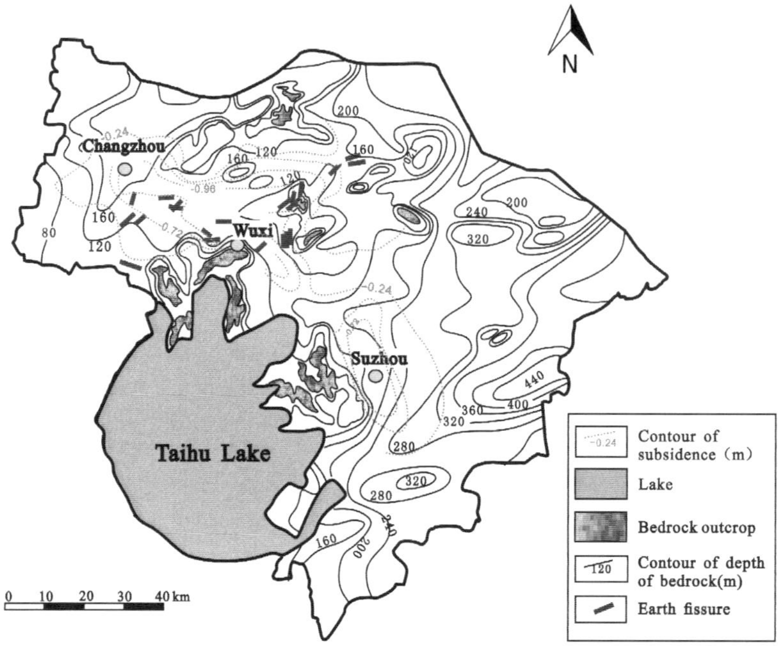

Figure 1. The distribution of earth fissures in Su-Xi-Chang area. The bedrock outcrops and depth (m from the mean sea level) and the cumulative land displacement from 1980 to 2000 are mapped (after Liu et al., 2004).

Figure 2. The ground rupture **(a)** and damaged house **(b)** at Guangming Village (red line in **(a)** shows the direction of the earth fissure).

identified the factors related to the development of earth fissure including bedrock hills, buried terraces, Karst collapse, structural differences of the aquifer system, and over-exploitation of groundwater. Wang et al. (2009, 2010) reported earth fissures triggered by excessive groundwater withdrawal and the variability of bedrock depth in Hetang town, Jiangyin City. Yu et al. (2006) analyzed the mechanisms leading to occurrence of some typical earth fissures using three-dimensional seismic exploration techniques. Based on the inversion of the 3-D seismic data, the sedimentary structure and morphology of the bedrock surface were obtained. Yu et al. (2004) reported that earth fissures in the Su-Xi-Chang area develop under the combined effects of the visco-elasto-plastic deformation of some sedimentary layers, buried bedrock surface morphology, thickness differences of some highly compressible sedimentary layers, and excessive exploitation of groundwater.

In Su-Xi-Chang area, the second confined aquifer was the main over-exploited unit. Groundwater pumping has formed a large, wide ranging cone of land subsidence because of long term over-exploitation of groundwater. The groundwater levels in the drawdown cone declined to -80 m in 1990s, and the largest cumulative subsidence was close to 2000 mm (Zhang et al., 2008). The groundwater levels recovered gradually from the previous lowest value to about -75 m in 1994 and -58 m in 2003, since the implementation of severe restrictions of groundwater exploitation by the Jiangsu government in 1995 (Zhang et al., 2008). The Jiangsu government finally decided to prohibit all groundwater exploitation in the Su-Xi-Chang area from 2000. As a result, land subsidence became controlled in some areas, and the rate of land subsidence decreased from 110 mm yr^{-1} in 1994 to 10–0 mm yr^{-1} in 2006 (Wang et al., 2009). The movement rates for these fissures slowed gradually, and the number of earth fissure are no longer increasing except for a few active earth fissures. The most active fissures in Su-Xi-Chang area are located in the Guangming Village, Xishan District of Wuxi City, with the maximum subsidence rate of about 5 mm yr^{-1}.

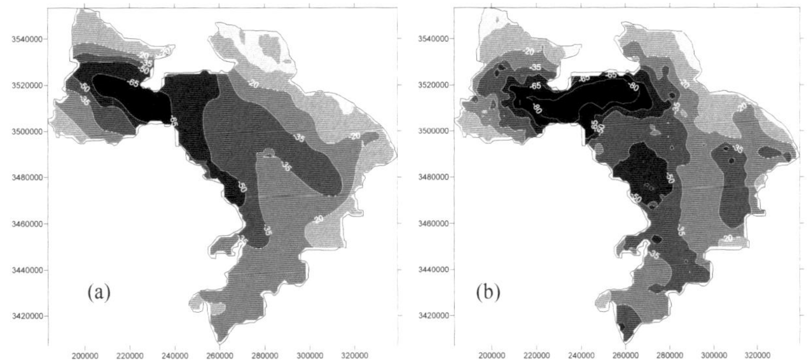

Figure 3. Groundwater level (m above mean sea level) in the second confined aquifer in 1995 **(a)** and 2001 **(b)**. The red line represents the earth fissure at Guangming Village.

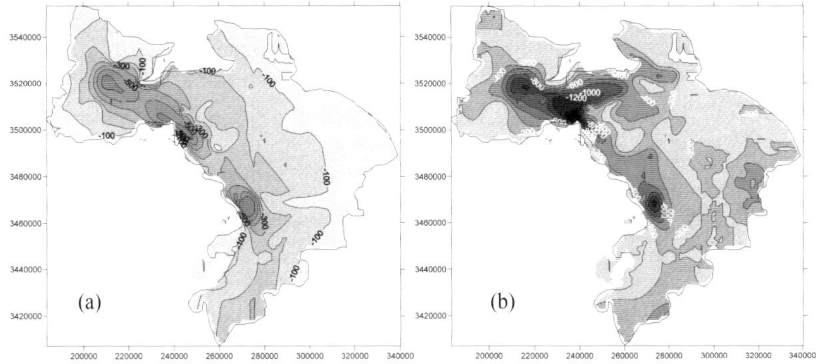

Figure 4. Cumulative land subsidence (mm) from 1980 to 1995 **(a)** and 2001 **(b)**. The red line represents the earth fissure at Guangming Village.

2 Earth fissures at Guangming Village

The first earth fissure at Guangming Village was discovered in 1998. Earth fissuring advanced significantly since 2000, with tilting farmlands and ground fractures at the early stages of development. Fissuring increased drastically since 2007 with more than 400 newly damaged houses (Fig. 2). The orientation of the earth fissures at Guangming Village are mostly in NE direction which is along the ridge tip.

Regional and local groundwater levels

The second confined aquifer is the main withdrawn aquifer in the Su-Xi-Chang area. Groundwater was overly exploited in the 1990s, and the total pumped groundwater reached 380 million m^3 in 1995. The groundwater levels in the second confined aquifer dropped from -35 to -75 m, with values between -60 and -55 m at Guangming Village (Fig. 3a). The total pumping rate decreased from 1996 to 2000 to 250 million m^3 in 2000. The groundwater exploitation was prohibited after 2001 in the Su-Xi-Chang area. However, the groundwater levels in some areas still continued to drop. For example, the groundwater levels in the second confined aquifer dropped to -95 m in the center of the drawdown cone and the groundwater level at Guangming Village ranged from -75 to -70 m (Fig. 3b).

Regional and local cumulative land subsidence

The lowering of the groundwater levels resulted in a severe land subsidence. The cumulative land subsidence in 1995 reached to 700 mm at the center of the subsidence bowl surrounding Guangming Village, and between 400–500 mm at Guangming Village (red line in Fig. 4a). The cumulative land subsidence in 2001 peaked to 1600 mm at the center of subsidence bowl located about 20 km to the east of the Guangming Village, and to 700–800 mm at Guangming Village (Fig. 4b). From the regional cumulative land subsidence maps, one can see that earth fissures in Guangming Village are in the middle of the subsiding area.

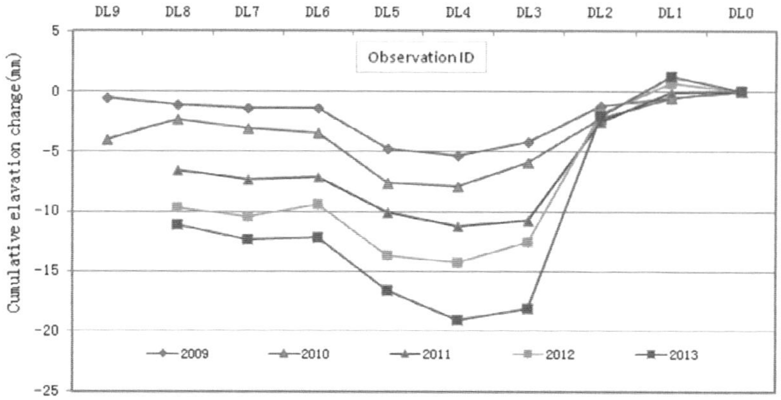

Figure 5. Cumulative elevation changes from 2009 to 2013 at the ten leveling benchmarks established across the fissure at Guangming Village. DL0 is used as reference point. The distance between DL0 and DL9 is approximately 80 m.

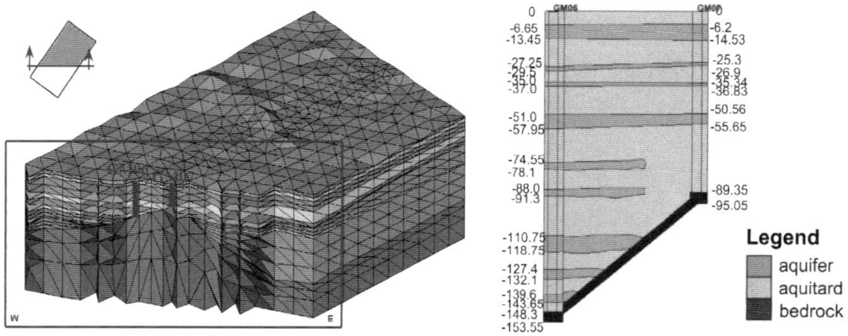

Figure 6. Perspective view of the Finite-Element model developed by Franceschini et al. (2015) to simulate the earth fissure generation at Guangming Village (**a**). The geological setting is highlighted in various colours: rock in grey, aquifers in yellow, and aquitards in blue). Hydrogeological profile between boreholes GM06 and GM07 (**b**). The numbers provide the buried depths of the top of each layer.

Local differential land subsidence

The farmland is obviously tilting at Guangming Village, which signifies the occurrence of differential land subsidence. Ten leveling points were established in 2007 along a trail that crosses one of the earth fissures. Because no deep stable benchmark was set at Guangming Village, the leveling benchmark DL0 is used as a reference point. The other 9 levelling points show a significant different elevation compared to DL0 (Fig. 5) over the monitoring interval from 2009 to 2013. Benchmark DL4 always shows the highest land subsidence values, and the average differential land subsidence is $5\,\mathrm{mm\,yr^{-1}}$.

Geological setting with buried rock ridge

The thickness of the aquifer system in the Su-Xi-Chang area is not uniform because of buried rock ridges whose depth varies significantly in some local areas. Due to this peculiar geological setting, the thickness of the aquifer system clearly differs across the rock ridges, with thicker sediments further away from ridge (Fig. 6a). For example, the thickness of the aquifer system at borehole GM07 is 95 m and that at GM06, only 95 m apart, is 153 m (Fig. 6b). The hydrostratigraphic section in Fig. 6b also shows that there are more aquifers at GM06 than at GM07.

3 Conclusions

Earth fissures are serious problems in Su-Xi-Chang area. They are distributed along the buried rock ridge primarily in the main NE direction. Excessive groundwater exploitation triggers earth fissuring. Earth fissuring at Guangming Village typically demonstrates the causes and mechanism of earth fissure generation and development, which includes groundwater level decrease, buried rock ridges with differing depths, the heterogeneous aquifer system, and subsequent differential land subsidence. A numerical model to simulate the generation and development of earth fissures at Guangming Village is being developed (Franceschini et al., 2015).

Acknowledgements. Funding supported by the Special Project for the Public Scientific Research of the Ministry of Land and Resources of China No. 201411096, KLEFGD No. 201403, NSFC No. 41272259 and NSF of Jiangsu Province No. BK2012730 is appreciated. Pietro Teatini was partially supported by the University of Padova, Italy, within the 2014 International Cooperation Programme. The international collaboration is supported by UNESCO IGCP Project No. 641 (M3EF3 – Deformation and fissuring caused by exploitation of subsurface fluids).

References

Franceschini, A., Teatini, P., Janna, C., Ferronato, M., Gambolati, G., Ye, S., and Carreon-Freyre, D., Modelling ground rupture due groundwater withdrawal: applications to test cases in China and Mexico, Proceeding of the Ninth International Symposium on Land Subsidence, Nagoya, November 2015.

Liu, C., Yuan, X. J., and Zhu, J. Q.: Earth fissures in Su-Xi-Chang area, China University of Geosciences Press, 2004 (in Chinese).

Wang, G. Y., You, G., Shi, B., Yu, J., Li, H. Y., and Zong, K. H. : Earth fissures triggered by groundwater withdrawal and coupled by geological structures in Jiangsu Province, China, Environ. Geol., 57, 1047–1054, 2009.

Wang, G. Y., You, G., Shi, B., Qiu, Z. L., Li, H. Y., and Tuck, M.: Earth fissures in Jiangsu Province, China and geological investigation of Hetang earth fissure, Environ. Earth Sci., 60, 35–43, 2010.

Yu, J., Wang, X., Su, X., and Yu, Q.: The mechanism analysis on ground failure disaster formation in Suzhou- Wuxi – Changzhou area, Journal of Jilin University (Earth Science Edition), 34, 236–241, 2004 (in Chinese).

Yu, J., Chen, G, Wang, X. M., He, H. S., Xu, X. L., and Wu, J. H.: Formation mechanisms analysis on ground failure based on 3D seismic exploration data in Yinguo'an area, Hydrogeology & Engineering Geology, 34, 117–123, 2006 (in Chinese).

Zhang, Y., Xue, Y. Q, Wu, J. C., Yu, J., Wei, Z. X., and Li, Q. F.: Land subsidence and earth fissures due to groundwater withdrawal in the Southern Yangtse Delta, China, Environ. Geol., 55, 751–762, 2008.

Proc. IAHS, 372, 255–260, 2015
proc-iahs.net/372/255/2015/
doi:10.5194/piahs-372-255-2015

Open Access
PIAHS

Satellite geodesy tools for ground subsidence and associated shallow faulting hazard assessment in central Mexico

E. Cabral-Cano[1], D. Solano-Rojas[2], T. Oliver-Cabrera[2], S. Wdowinski[2], E. Chaussard[3], L. Salazar-Tlaczani[1], F. Cigna[4], C. DeMets[5], and J. Pacheco-Martínez[6]

[1]Departamento de Geomagnetismo y Exploración, Instituto de Geofísica, Universidad Nacional Autónoma de México, México DF, 04510, México
[2]Rosenstiel School of Marine and Atmospheric Science, University of Miami, Miami, FL 33149, USA
[3]Dept. of Earth and Planetary Science, University of California Berkeley, Berkeley, USA
[4]British Geological Survey, Earth Hazards & Observatories, Earth and Planetary Observation and Monitoring, Nottingham, UK
[5]Department of Geoscience, University of Wisconsin-Madison, Madison, Wisconsin 53706, USA
[6]Departamento de Construcción y Estructuras, Universidad Autónoma de Aguascalientes, Aguascalientes, México

Correspondence to: E. Cabral-Cano (ecabral@geofisica.unam.mx)

Published: 12 November 2015

1 Motivation

While subsidence has affected Mexico City for over a century, other cities in central Mexico have been subjected to ground subsidence since the 1980's, as a result of their large urban expansion, population increase and aggressive groundwater extraction rates. The continuous ground subsidence results in severe damage to urban infrastructure and civil structures. Unfortunately the damage cost assessment and vulnerability are difficult to evaluate, because of the variable geographic extent and the time-continuous nature of the process, which have different characteristics than other phenomena characterized by localized, short duration events such as earthquakes or floodings with better media coverage.

A recent SBAS survey based on ALOS-1 PALSAR data along central Mexico (Chaussard et al., 2014) indicates that this process is rapidly becoming a widespread phenomenon that potentially affects a population over 25 million inhabitants. InSAR-derived information and its cartographic products from geodetic imaging will be key to develop a better groundwater extraction strategy in areas that undergo extreme subsidence rates, propose better mitigation strategies and ultimately improve water resource management.

In Mexico, subsidence induced damage to houses or other urban infrastructure is not eligible for federal emergency relief funds, because it is not considered a naturally occurring process but rather an anthropogenically induced one. Furthermore, this phenomenon is localized and usually managed only at the local city or county-level administrations, while most other large impact natural hazards, as earthquakes, volcano eruptions, or flooding events, are usually managed by federal agencies. This work is aimed towards a better recognition of subsidence as a major hazard in Mexico, assess the number of inhabitants affected and the spatial extent of the subsiding areas affected by this process. The space-based subsidence imaging can be used to improve water resources management (e.g. Hernández-Espriú et al., 2014) in an area that comprises a number of rapidly growing urban areas and assess its associated surface faulting geological hazard.

We envision that this work will serve as a test case where water-usage policy makers can have access to custom cartographic products suitable for data fusion and data mining from third party regional water management agencies databases and include satellite geodesy imaging techniques as a routine monitoring technique for aquifer assessment and its sustainable management. Within this context, space geodetic techniques, most notably InSAR complemented by GPS are quite suitable for this task. InSAR provides an unsurpassed synoptic view of the earth's dynamic surface; for

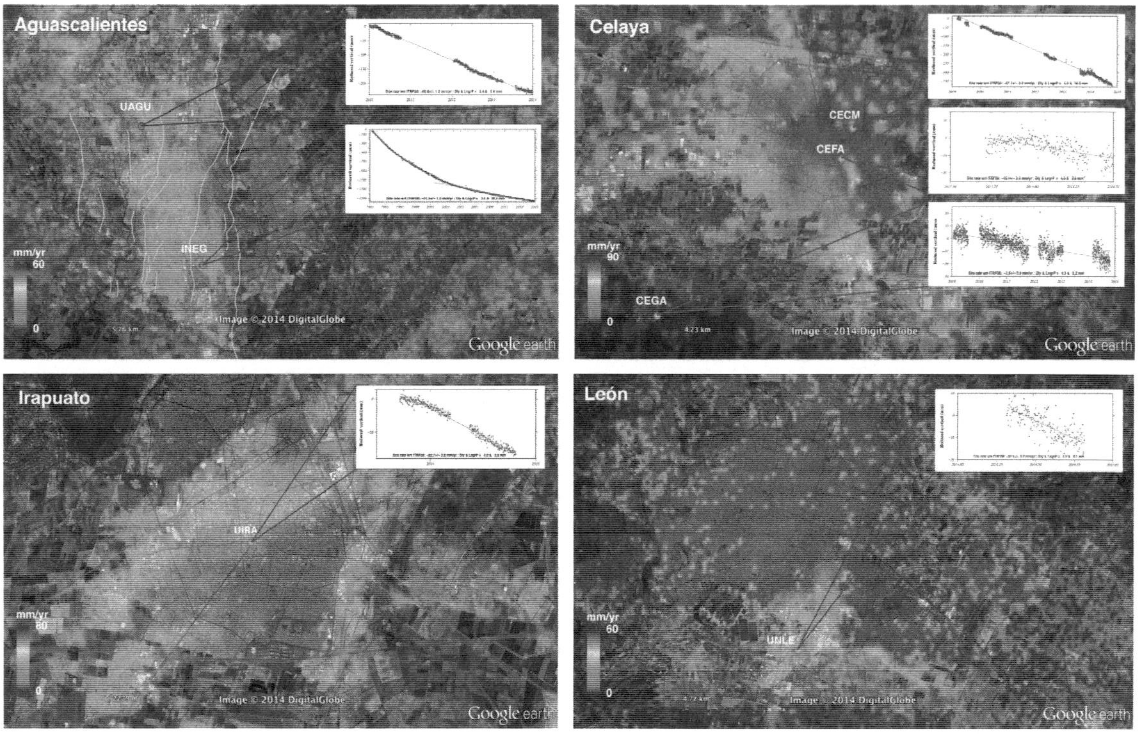

Figure 1. Ground subsidence velocity maps (modified from Chaussard et al., 2014) and cGPS vertical component time series for the cities of Aguascalientes, Celaya, Irapuato, and León.

example, the most recent InSAR analysis in central Mexico (Chaussard et al., 2014) shows that land subsidence can be identified in 17 cities. However, different satellite sensors and sometimes widely spaced data availability make it difficult to derive long-term time series, or measure rapid changes or nonlinear variations of subsidence velocities. To alleviate this situation, higher temporal resolution subsidence observations of associated fault motion has been pursued using continuously operating GPS stations. As a complement to the InSAR observations, we have developed a GPS network that covers 6 urban centers to better characterize the temporal variability of ground subsidence, using several processing schemes that include both real-time solutions using RTNet as well as daily solutions with IGS precise orbits using Gipsy-Oasis.

Our current efforts are aimed to evaluate the applicability of high spatial resolution, space-based subsidence imaging based on operational SAR sensors (namely TSX and ALOS-2) to improve water resources exploitation and better management of these resources within a region that comprises a number of rapidly growing urban areas. In this work we are presenting several cases of rapidly subsiding cities to their respective municipal and state level administrations in order to set the stage for a future routine use of remote sensed data for the long term management of hydrologic resources

Another important objective is to provide an independent observation tool for estimating structural damage in the urban areas to be studied. The accumulated subsidence in Mexico City has resulted in continuous damage to buildings and infrastructure, which is hard to monitor. In our previous PSI studies we were able to refine methodologies to define potential areas of subsidence associated faulting based on the InSAR horizontal gradient analysis (Cabral-Cano et al., 2010a). We could also detect different subsidence behaviors occurring in different building types due to either suboptimal foundation calculations or evaluate longer term results of subsidence mitigation civil works in colonial-time historic buildings (Osmanoglu et al., 2011). Recent SBAS analysis throughout the region (Fig. 1) indicates substantial subsidence magnitudes either in their central downtown areas or nearby suburban developments.

2 Significance

Study of the subsidence phenomena, its consequences and mitigation efforts in Mexico have been carried out by different entities and are thus detached from each other. Academic institutions have undergone the study of regional subsidence and the aquifer vulnerability, while municipal wa-

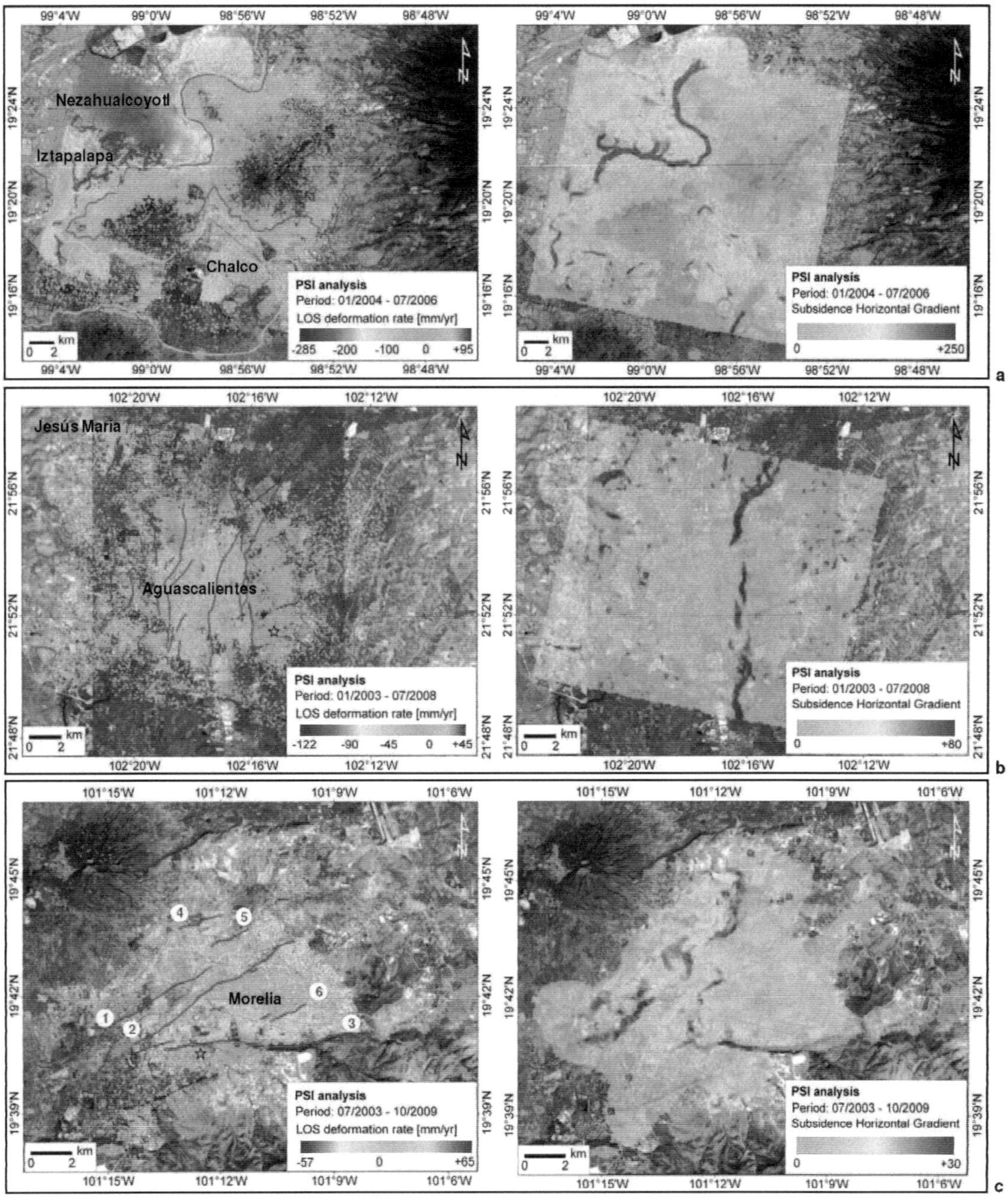

Figure 2. LOS deformation rates (left panels) and subsidence horizontal gradients (right panels) in the urban areas of Chalco **(a)**, Aguascalientes **(b)** and Morelia **(c)** in Central Mexico, using ENVISAT 2004–2006, 2003–2008 and 2003–2009 data, respectively (after Cigna et al., 2011). Red lines show mapped faults. The geotechnical zonation of Chalco **(a)** shows the boundaries of lacustrine (light blue lines) and transitional (blue lines) geotechnical units. InSAR derived ground subsidence velocity maps (left panels) were processed using a horizontal subsidence gradient analysis approach that provides an excellent tool to discriminate areas with high potential of surface faulting (right panels). Notice the close correspondence between mapped faults on the corresponding left panels and the high gradient areas shown on the left panels. Further data merging with population, hydrology and meteorology data sets allow the creation of risk maps using a simple risk matrix (see Fig. 3).

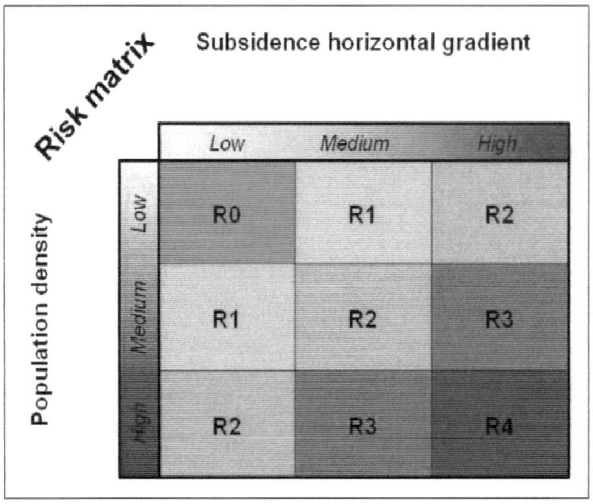

Figure 3. Risk matrix used to produce the hazard zonation map shown on Fig. 4, relating subsidence gradient and population density.

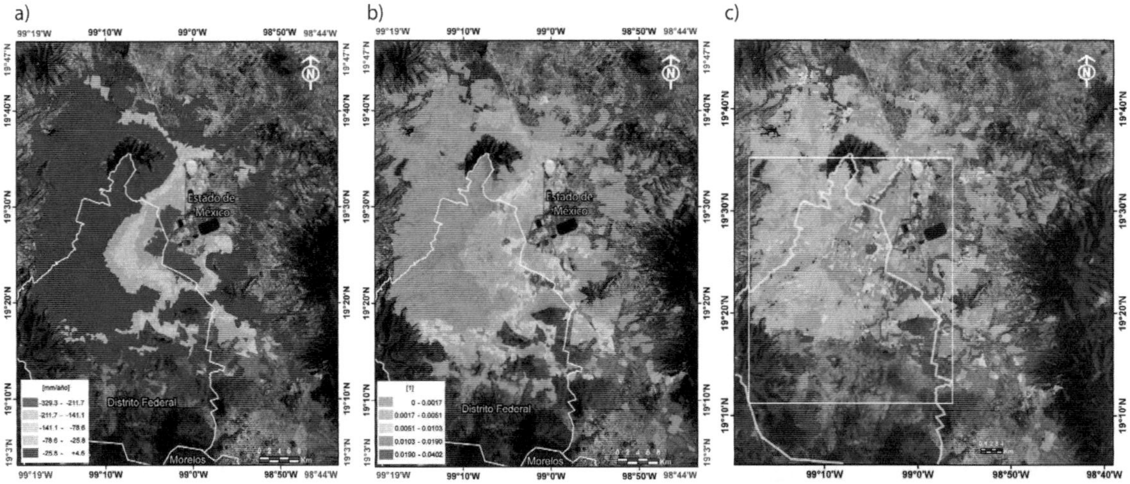

Hazard Zones	Area [km2]	Area [%]	Population	Population [%]
R0	536.1	33.2	1300114	7.7
R1	648.9	40.2	9213257	54.7
R2	77.8	4.8	509066	3.0
R3	269.8	16.7	4585714	27.2
R4	81.2	5.0	[illegible]	7.3

Figure 4. Example of hazard zonation for surface faulting associated to ground subsidence in the Mexico City Metropolitan Area. Left panel: ground subsidence magnitude. Central panel: subsidence horizontal gradient zonation. Right panel: hazard zonation. In this example, 21.7 % (351.0 km^2) of the city's urban area and 34.5 % of its population (4 709 305 inhabitants) are under a high to very high risk of being affected by faulting associated with subsidence (right panel).

ter management agencies are concerned with providing water for the continuously growing. However this scenario implies that municipal water agencies have long ignored that extreme aquifer exploitation and poor management conveys a heavy financial impact due to damage to housing, and urban infrastructure in ground subsidence affected areas. Moreover, mitigation efforts focus their efforts on the short-term engineering solutions but tend to ignore the driving force.

This work is aimed to: (a) showcase a methodological approach to use both imaging remote sensing and GPS data to provide a detailed panorama of the spatial and time variability of subsidence, and use products such as the horizontal subsidence gradient from InSAR derived using several cities in central Mexico as an example and provide an initial assessment of hazard due to surface faulting associated with ground subsidence. (b) Construct the observational capability to generate geodetic-hydrological datasets that will serve to better model future scenarios and foster a sustainable use of water resources. (c) Define vulnerable areas of subsidence related faulting, create hazard maps and help refine municipal land use codes for the studied metropolitans in such a way that water management agencies, have high resolution cartographic products that will serve for the development of a long term, sustainable strategy and mitigation effort plans that will eventually restore aquifers and stop ground subsidence.

3 Approach

Our approach for hazard assessment of shallow faulting associated with subsidence (Fig. 2) is based on ground subsidence derived from InSAR vertical displacement data following Cabral-Cano et al. (2011). These data were further analyzed using a horizontal subsidence gradient analysis approach (Cabral-Cano et al., 2010a; Fig. 2 right panels).

Our previous studies in Mexico City (Cabral-Cano et al., 2008; Osmanoglu et al., 2011) and Morelia (Cigna et al., 2012) highlighted the fact that spatial gradients of land subsidence, rather than magnitude, are the primary factors to be analyzed for subsidence hazard assessment (Cabral-Cano et al., 2010a, b). Differential ground displacement induces shear stresses in urban structures and consequent development of fractures and damages. This parameter can therefore be used to qualitatively assess subsidence hazard and to perform, together with the analysis of the elements at risk and their vulnerability, a qualitative mapping of subsidence risk that should be necessarily taken into account for any future urban development plans and land use codes As shown on Fig. 2, This provides an excellent tool to discriminate areas with high potential of surface faulting. We then combine these hazard indicators with the local population information from census data, through a risk assessment matrix. To compute the subsidence gradients, surface displacement data points from the Persistent scatterer interferometric analysis is gridded into a regular mesh, after which the non-dimensional horizontal subsidence gradient is then computed for each one of the grid cells (Cabral-Cano et al., 2011; Cigna et al., 2011; see Fig. 2). Hazard and elements at risk can be then combined through a 2-D risk matrix (Fig. 3), assuming that the coexistence of high differential subsidence rates and high density of population leads to higher levels of risk. The combination of these products with socioeconomic and population data sets allow the creation of risk maps such as the one shown on Fig. 4. The resulting cartographic products can be integrated into a geographical information system suitable for distribution to local, state and federal as a geocoded, high- resolution product suitable for detailed spatial analysis and better hazard assessment.

These ground subsidence risk maps (e.g. Fig. 4) represent valuable decision elements to help local authorities design

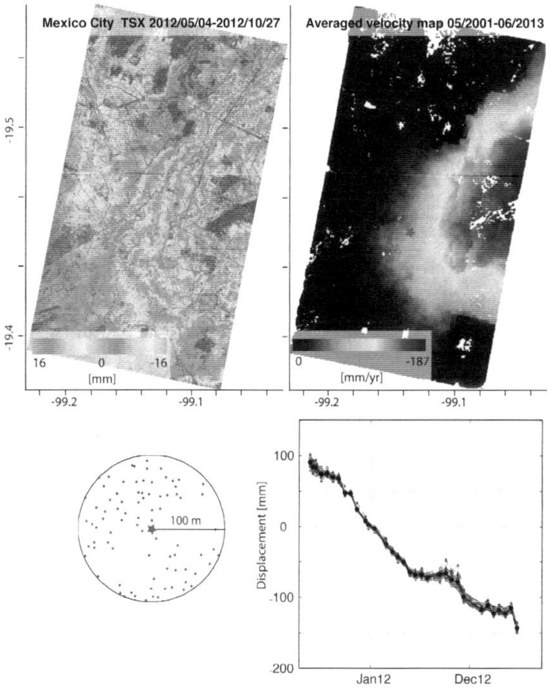

Figure 5. Top left panel: wrapped interferogram showing displacement in Mexico City from 4 May to 27 October 2012. Top right panel: averaged velocity map from PS analysis in the study area from mid 2011 to mid 2013. The star shows the location where time series were extracted. Bottom left panel: PS density in a radius of 100 m around the point selected. Bottom right: time series of the PS showed in the left panel respect to a master SAR scene from 6 February 2012.

the best risk mitigation and management strategies for cities under intense subsidence. On the other hand, the analysis of the subsidence gradient map can also be successfully used as an indicator of hazard and to support the planning of future development and land use zoning in urban areas. This approach allows the definition of zones where new development and urban expansion should be discouraged because of their highest levels of differential subsidence.

While InSAR based information provides an unsurpassed synoptic analysis, in some cases different satellite sensor and sometimes widely spaced data availability make it difficult to derive long-term time series, or measure rapid changes or nonlinear variations of subsidence velocities. In order to accommodate for these deficiencies, the higher temporal resolution subsidence observations has been pursued using continuously operating GPS stations. For this purpose, we have developed a continuous GPS network that covers 6 urban centers (e.g. Fig. 1) to detect short duration variations using different processing schemes that include both real-time solutions using RTNet as well as daily solution using Gipsy-

Oasis. These stations also serve as validation and calibration sites for InSAR analysis. Data from these stations is open and freely available at the TLALOCNet data archive (http://tlalocnet.udg.mx).

4 Continuous ground subsidence monitoring

The continuous nature of the subsidence process necessary requires frequent updates on velocity and timely detection of any ground subsidence acceleration. New SAR sensors availability facilitate this task by providing new opportunities for ground subsidence velocity map updates. Our work in progress includes the updated subsidence analysis using TerraSAR-X and ALOS-2 data. For the Stripmap mode scenes from Mexico City obtained by the TerraSAR-X satellite (Fig. 5), we used the Stanford Method for Persistent Scatterers (StaMPS) to extract ground deformation displacements from time series of 35 Synthetic Aperture Radar acquisitions from mid 2011 to mid 2013. Our results include displacement maps with a ground resolution close to 3 m, and derived Persistent Scatterers time series. In Fig. 5 we observe very localized deformation patterns in some areas of the city; this allows us to analyze the vulnerability of some structures to differential deformation as well as ground failure. Our field observations give us a reference of the importance of identifying areas with increased vulnerability to suffer deformation and stress due to differential ground subsidence.

5 Conclusions

Ground subsidence hazard analysis in central Mexico urban areas show that this process affects several cities and a substantial percentage of their inhabitants. However due to the rapid urban expansion and the corresponding continuous overexploitation of their underlying aquifers, the process seems unlikely to be be reversed in the short term. This situation requires constant subsidence monitoring. The new generation of SAR satellite sensors and free data policies have the potential for providing with the necessary data for a continuous subsidence monitoring effort.

Acknowledgements. S. Wdowinski and E. Cabral-Cano acknowledge support through NASA-ROSES grant NNX12AQ08G. E. Cabral-Cano is supported by UNAM-PAPIIT projects IN104213-2, and IN109315-3. This material is partly based on data provided by the UNAVCO Facility with support from the National Science Foundation (NSF) under award EAR-1338091.

References

Cabral-Cano, E., Dixon, T. H., Miralles-Wilhelm, F., Sánchez-Zamora, O., Díaz-Molina, O., and Carande, R. E.: Space Geodetic Imaging of Rapid Ground Subsidence in México City, B. Am. Geol.l Soc., 120, 1556–1566, doi:10.1130/B26001.1, 2008.

Cabral-Cano, E., Osmanoglu, B., Dixon, T., Wdowinski, S., DeMets, C., Cigna, F., and Díaz-Molina, O.: Subsidence and fault hazard maps using PSI and permanent GPS networks in central Mexico, Int. Assoc. Hydrol. Sci., 339, 255–259, 2010a.

Cabral-Cano, E., Arciniega-Ceballos, A., Díaz-Molina, O., Cigna, F., Osmanoglu, B., Dixon, T., Demets, C., Vergara-Huerta, F., Garduño-Monroy, V. H., Ávila-Olivera, J. A., and Hernández-Quintero, E.: Is there a tectonic component on the subsidence process in Morelia, Mexico?, Int. Assoc. Hydrol. Sci., 339, 164–169, 2010b.

Cabral-Cano, E., Diaz Molina, O., and Delgado-Granados, H.: Subsidencia y Mapas de Peligro por Fallamiento Superficial: Un Ejemplo en el área Nororiental de la Zona Metropolitana de la Ciudad de México, Bol. Sociedad Geol. Mex., 63, 53–60, 2011.

Cigna, F., Cabral-Cano, E., Osmanoğlu, B., Dixon, T. H., and Wdowinski S.: Detecting subsidence-induced faulting in Mexican urban areas by means of Persistent Scatterer Interferometry and subsidence horizontal gradient mapping. In: IEEE International, Geoscience & Remote Sensing Symposium (IGARSS) 2011, Vancouver, Canada, 24-029 July 2011, 2125–2128, doi:10.1109/IGARSS.2011.6049585, 2011.

Cigna, F., Osmanoğlu, B., Cabral-Cano, E., Dixon, T. H., Ávila-Olivera, J. A., Garduño-Monroy, V. H., DeMets, C., and Wdowinski, S.: Monitoring land subsidence and its induced geological hazard with Synthetic Aperture Radar Interferometry: a case study in Morelia, Mexico, Remote Sens. Environ., 117, 146–161, doi:10.1016/j.rse.2011.09.005, 2012.

Chaussard, E., Wdowinski, S., Cabral-Cano, E., and Amelung, F.: Land subsidence in central Mexico detected by ALOS InSAR time-series, Remote Sens. Environ., 140, 94–106, doi:10.1016/j.rse.2013.08.038, 2014.

Hernández-Espriú, A., Reyna-Gutiérrez, J. A., Sánchez-León, E., Cabral-Cano, E., Carrera-Hernández, J., Martínez-Santos, P., Macías-Medrano, S., Falorni, G., and Colombo, D.: DRASTIC-Sg Model, a new extension to the standard DRASTIC approach for mapping groundwater vulnerability in urban aquifers subject to differential land subsidence. Application to Mexico City, Hydrogeol. J., 22, 1469–1485, doi:10.1007/s10040-014-1130-4, 2014.

Osmanoglu, B., Dixon, T. H., Wdowinski, S., Cabral-Cano, E., and Jiang, Y.: Mexico City Subsidence Observed with Persistent Scatterer InSAR, Int. J. Appl. Earth Obs., 13, 1–12, doi:10.1016/j.jag.2010.05.009, 2011.

Proc. IAHS, 372, 261–265, 2015
proc-iahs.net/372/261/2015/
doi:10.5194/piahs-372-261-2015

Open Access
PIAHS

Mass movement processes triggered by land subsidence in Iztapalapa, the eastern part of Mexico City

M. González-Hernández[1], **D. Carreón-Freyre**[2], **R. Gutierrez-Calderon**[1], **M. Cerca**[2], **and W. Flores-Garcia**[1]

[1]Centro de Evaluación de Riesgos Geológicos CERG, Iztapalapa, Mexico City, Mexico
[2]Centro de Geociencias de la UNAM, Juriquilla, Querétaro, Mexico

Correspondence to: D. Carreón-Freyre (freyre@geociencias.unam.mx)

Published: 12 November 2015

Abstract. Geological and structural conditions in the Basin of Mexico coupled with natural and anthropogenic factors, such as groundwater exploitation, provokes land subsidence and differential deformation. The study area is located in to the north of Iztapalapa, a municipality within Mexico City, in a site called "El Eden" with irregular topography. Where volcanic sequences overlie the lacustrine deposits of clays and silts and show displacements by the action of gravity. The displacement zone was delimited at the top of the slope by the formation of circular tensile fractures with stair -shaped geometries. At the base of the slope, compressive processes damaged housing, sidewalks and inclined light poles and trees. A NW-SE system of fractures was identified in which displacement velocities vary from a few millimeters to several centimeters per year. Which affects urban facilities. In this work a conceptual model of deformation is presented that integrates the geological and mechanical factors leading to landslide and land subsidence. A geophysical survey leads to evidence of how land subsidence processes increase the sliding slope.

1 Introduction

Mexico City (MC) is located within the Basin of Mexico, an endorheic basin whose morphology is the result of the interaction of volcanic and sedimentary processes. Nowadays the highest records of subsidence and ground fracturing are located at the northeast side of the city. More than 30 % of the total surface has been damaged by processes related to the subsoil deformation within the Iztapalapa area. From a geomorphologic point of view the study area is comprised of three units (Fig. 1): (1) the highland, formed by volcanic deposits, (2) the slope zone, consisting of alluvial sequences and reworked volcanic deposits and (3) the lacustrine plain, formed by silty clayey sequences interbedded with fluvial and volcanic materials. In the zones 2 and 3, the geological conditions coupled with groundwater extraction, seismicity, static and dynamic loads, facilitate the development of land subsidence and ground fracturing, causing significant damage to urban infrastructure.

In Iztapalapa subsidence ranges from 10 to 40 cm $year^{-1}$ (López-Quiroz, 2009) and three systems of fracturing affect the 10 % of its total surface (Carreón-Freyre, 2010). The study area called "El Eden", of about 0.15 km^2, is located at the-northeast sector of Iztapalapa became highly damaged by localized deformation and fracturing. The subsoil is composed of sequences of volcanic materials from the Santa Catarina volcanic range, interbedded with pyroclastics, fluvial and lacustrine deposits (that correspond to the former Texcoco Lake). The topography is irregular and the volcanic sequences that overlie the lacustrine deposits of clays and silts displaced by the action of gravity. The movement zone was identified at the top of the slope by the formation of circular tensile fractures with stair shaped geometries.

At the base of the slope, compressive processes damaged housing, sidewalks and inclined light poles and trees. In this work a conceptual model of deformation is presented that integrates the geological and mechanical factors leading to landslide and land subsidence. The average depth of the contact zone and areas of fracturing and compression and a NW-SE system of fractures are identified.

Published by Copernicus Publications on behalf of the International Association of Hydrological Sciences.

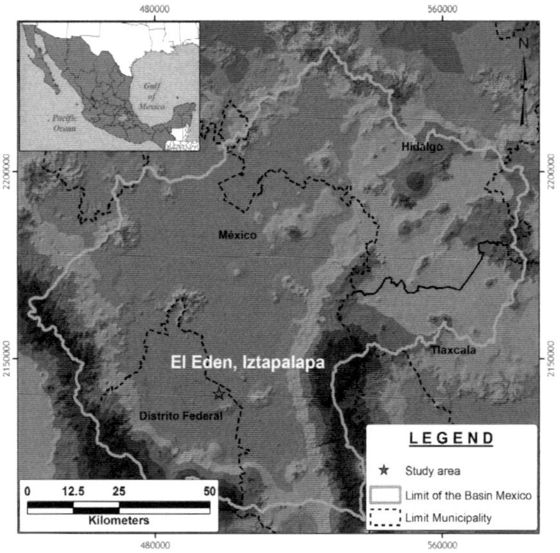

Figure 1. Location of the Mexico basin (in green) in the central part of Mexico. Whithin the Basin of Mexico is located the Mexico City (blackk dotted line, also called Distrito Federal) and the Iztapalapa municipality (red line). The location of "El Eden" Study Site is shown with a magenta star.

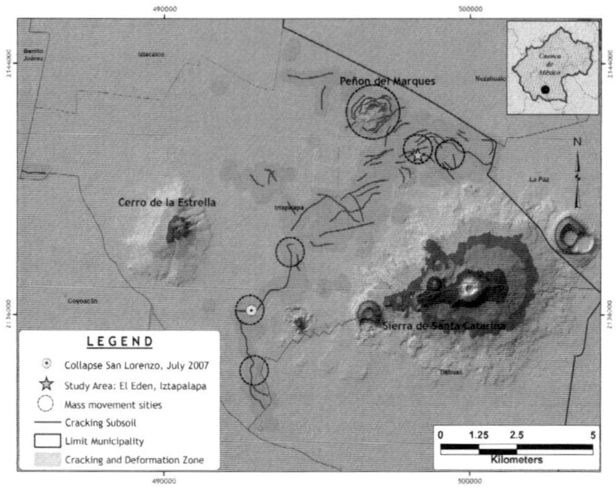

Figure 2. Detailed mapping of fracturing in Iztapalapa (red lines). The *Santa Catarina* Range is located to the south-eastern sector of Iztapalapa and the *Peñon del Marques* volcano is located to the north-east. Black circles correspond to the six indentified mass movements.

2 Geological setting

The morphological features of Mexico City are due to the interaction of faulting systems and volcanic activity from the Oligioceno to Miocene (Mooser at al., 1974). The subsoil is mainly composed of sedimentary rocks of the Upper Cretaceous, Tertiary and Quaternary volcanic and recent fluvio-lacustrine sequences. In the Iztapalapa zone higher structures correspond to volcanic buildings composed by andesites, basalts and pyroclastic rocks of the Pliocene-Pleistocene *Santa Catarina* Range, *Cerro de la Estrella* and *Peñón del Marques*. Along these structures the former Texcoco Lake covered the volcanic plain that was filled by lava flows, andesitic-basaltic pyroclastics (tuffs and ashes) as well as alluvial and lacustrine deposits (sand, silt and clays). From the mapping of subsidence and ground fracturing three main facture systems were identified in Iztapalapa: (a) the NE-SW system, aligned with the *Santa Catarina* Range (b) the WNW-ESE system aligned with the edge of *Texcoco* Lake and, (c) the system of tensile fractures surrounding the *Peñón del Marqués* (Fig. 2).

3 Ground fracturing in slope areas of Iztapalapa

Fractures mapped in Iztapalapa are concentrated around the border of the Santa Catarina Range and the Peñón del Marques volcano. These systems exhibit preferential directions of fracturing NE-SW and NW-SE, changes in the direction suggest that they are associated with the shoreline of the former Texcoco Lake and follow contacts between lake sediments (clays and silts) and buried lavas or pyroclastics (flow, surges and fall deposits) (Fig. 2). After mapping-in slope borders of the Santa Catarina range, six mass movement areas were identified. These areas are showing gravitational displacements that can be associated with the paleomorphology of the volcanic structures formed at the same period as Texcoco Lake, and that may be covered by volcanic-sedimentary deposits (Fig. 2).

In these areas slopes vary from 3 to 10° and the volcanic deposits are often interbedded with fluvio-lacustrine sequences (Fig. 3). The lithological contacts become sliding planes that determine a lateral spreading and fracturing of the surface materials associated with vertical and/or horizontal displacements. The contrasting physical and mechanical properties of lacustrine sequences (low shear strength and high compressibility) and pyroclasts materials (non-plastic, porous and non-cohesive), the irregular topography and coupled with groundwater exploitation facilitates the development of land subsidence. This phenomenon causes gradual changes in the thicknesses of the lacustrine sequences in turn modifying the original slope, this condition trigger a gravity process of "mass movement" with orders of magnitude ranging from some mm to many cm per year. These "masses" displace overlying irregular surfaces that are lithological contacts and generates shear fissures between both geological materials. The morphological features of this phenomenon consist of tensile fractures in the upper zone, with variable length and irregular geometry, often concave, with significant vertical displacements (0.3 to 1 m in average). The stress fractures have variable length, opening and direction. While

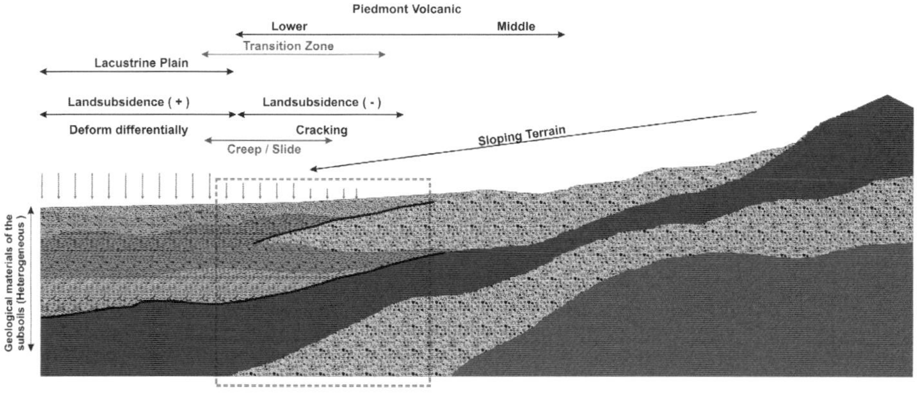

Figure 3. Schematic geological section of the piedmont zones in the *Santa Catarina* Range. Phenomena of land subsidence in the lacustrine plain (left) associated with mass movements processes (right) over a detachment surface are shown.

in the front area a compression process has been mapped by damage on sidewalks and walls, as well as distortion structural elements of houses. The depth of the lithological contact acting as a detachment surface depends on the fabric and morphology of the deposit (Fig. 3).

4 Field mapping at "El Eden" study site

From the six sites with identified mass movement processes, the site with the highest reported damage is called "El Eden", located at the northeast sector of Iztapalapa (Fig. 4). The subsoil sequence is composed by silty clays and volcanic deposits, lava and fall, identified through information collected from previous geotechnical boreholes and outcrops nearby the study area. The zones identified within "El Eden" mass movement are shown in the map of Fig. 4. The orange area, of about 38 000 m^2, corresponds to a tensile zone with NW-SE stepped fractures spaced every 45 m. The yellow area, of about 35 000 m^2, indicates a high deformed compressive zone. The levelling survey in the study site shows an irregular topography that can be associated with the paleo-morphology of a buried volcanic structure, part of the same Santa Catarina Range. Altitude variations ranges from 0.5 to 8 m and slopes obtained by the interpolation of levelling points vary from 0 to 12°.

A geophysical survey was performed to characterize the subsurface structure. The 2-D MASW method was applied for obtaining S wave patterns, with the analysis variation of velocities of shear waves in depth, it was possible to identify the depth of lithological contacts and the thicknesses of the prospected layers. A 40 m length profile with an estimated penetration depth of 15 m is shown in Fig. 5. A sequence of 3 layers was characterized: (1) A high plasticity lacustrine clay, at a depth from 0 to 2 m, with an S wave velocity less than 180 $m\,s^{-1}$; (2) Black sands interbedded with silt and pyroclastic fall deposits, at depths from 2 to 6 m depth, with

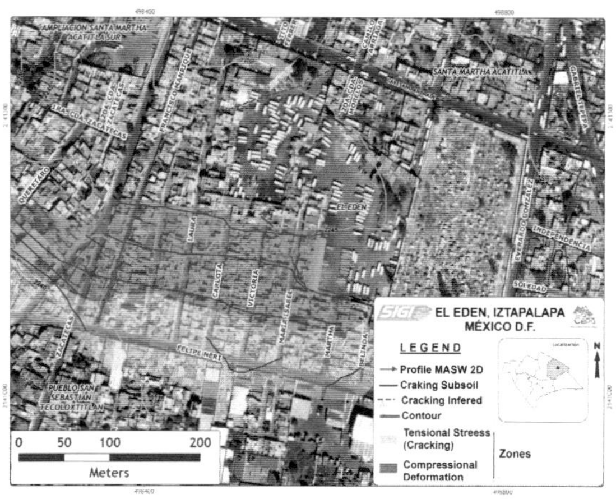

Figure 4. Location of "El Eden" Study Site in Iztapalapa. Orange colour indicate the tension area and the yellow colour the compressive area. Fractures are shown as red lines.

a s wave velocity varying than 180 to 300 $m\,s^{-1}$; and, (3) fractured lava and pyroclastic flows with S wave velocities greater than 300 $m\,s^{-1}$. The detachment surface of the mass movement was located a depth 3 m.

5 Conclusions

Land Subsidence in the Iztapalapa area of Mexico City is associated with mass movement processes in the slope-area of the two main volcanic structures, the *Santa Catarina* Range and the *Peñon del Marques* volcano. Slope changes due to land subsidence in the lacustrine plains causes instability in areas where the subsoil is composed of high plasticity clays from Texcoco Lake, along with interbedded sandy and silty ashes and fractured lava flows.

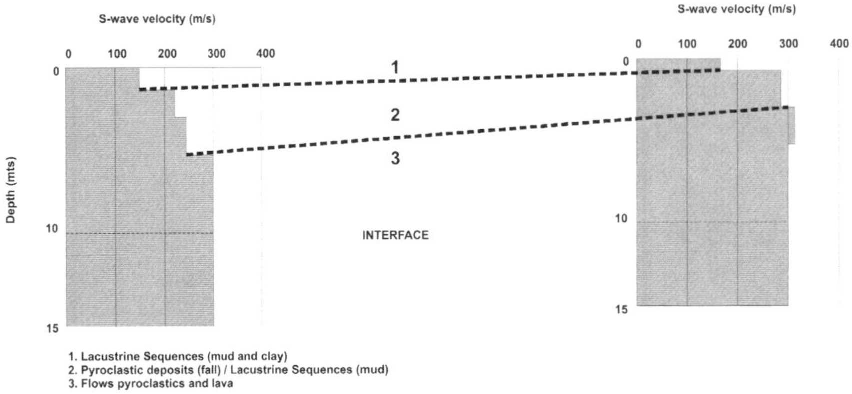

Figure 5. 1-D S-Wave velocities model from the 2-D MASW Method showing a 15 m depth of investigation. Three layer according to the shear velocities were identified (1) lacustrine clays, (2) sand, volcanic ashes; (3) volcanic flows and fractured lavas.

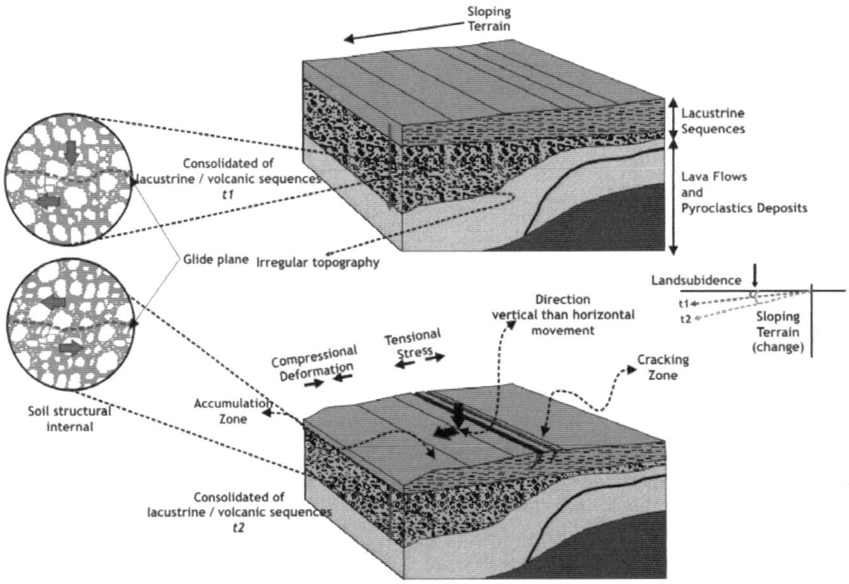

Figure 6. Conceptual model of the mass movement processes associated to land subsidence. **(a)** Initial conditions before consolidation of lacustrine sediments; **(b)** Mass movement showing the tension and compression zones in the upper and lower part of the slope respectively. Black arrows indicate the movement direction.

A conceptual model of the mass movement in the "El Eden" study site is presented (Fig. 6). A sliding slope was delimited into two strips, the upper zone and the front area. The upper zone is under tensional strees because of lateral spreading with stepped fracturing systems and parallel geometries. The front zone is under compressional stress causing significant damage to urban structures. The clayey layers behave as detachment surfaces that could be identified by the 2-D MASW method because of its low S wave velocities.

The mass movements associated with land subsidence processes are of greatest importance in Iztapalapa because the damage caused to housing and urban facilities (roads and pipes), where the demaged area covers approximately 74 000 m^2.

Acknowledgements. The authors acknowledge the support of the DGAPA-UNAM research project No. IN114714 and, the field support of the staff from *Centro de Evaluación del Riesgo Geológico* (CERG): Cesar M. Tovar Medina, Enrique Luna Sánchez, Felix A. Centeno-Salas and Said R. Zacarías Ramírez.

References

Carreón-Freyre, D.: Land subsidence processes and associated ground fracturing in Central Mexico, in: Land Subsidence, Associated Hazards and the Role of Natural Resources Development (Proceedings of EISOLS 2010, Querétaro, México), edited by: Carreón-Freyre, D., Cerca, M., and Galloway, D. L., Red Book Series Publication 339, IAHS Press, CEH Wallingford, UK, ISBN: 978-1-907161-12-4, ISSN: 0144-7815, 149–157, 2010.

López-Quiroz, P., Marie-Pierre Doin, M.-O., Tupin, F., Briole, P., and Nicolas, J.-M.: Time series analysis of Mexico City subsidence constrained by radar interferometry, J. Appl. Geophys., 69, 1–15, doi:10.1016/j.jappgeo.2009.02.006, 2009.

Mooser, F., Nairn, A. E. M., and Negendank, J. F. W.: Paleomagnetic investigations of the Tertiary and Quaternary igneous rocks: VIII A paleomagnetic and petrologic study of volcanics of the Valley of Mexico, Geologischen Rundschau, 63, 451–483, 1974.

Proc. IAHS, 372, 267–272, 2015
proc-iahs.net/372/267/2015/
doi:10.5194/piahs-372-267-2015

Open Access
PIAHS

Application of high resolution geophysical prospecting to assess the risk related to subsurface deformationin Mexico City

F. A. Centeno-Salas[1,3], D. Carreón-Freyre[2], W. A. Flores-García[3], R. I. Gutiérrez-Calderón[3], and E. Luna-Sánchez[1]

[1]Posgrado en Ciencias de la Tierra, Universidad Nacional Autónoma de México (UNAM), Mexico City, Mexico
[2]Laboratorio de Mecánica Geosistemas (LAMG), Centro de Geociencias, UNAM, Querétaro, Mexico
[3]Centro de Evaluación de Riesgo Geológico (CERG), Delegación Iztapalapa del Distrito Federal, Mexico

Correspondence to: D. Carreón-Freyre (freyre@geociencias.unam.mx)

Published: 12 November 2015

Abstract. In the eastern sector of Mexico City the sub soil consists of high contrasting sequences (lacustrine and volcanic inter bedded deposits) that favor the development of erratic fracturing in the surface causing damage to the urban infrastructure. The high-resolution geophysical prospecting are useful tools for the assessment of ground deformation and fracturing associated with land subsidence phenomena.

The GPR method allowed to evaluate the fracture propagation and deformation of vulcano-sedimentary sequences at different depths, the main electrical parameters are directly related with the gravimetric and volumetric water content and therefore with the plasticity of the near surface prospected sequences. The active seismology prospection consisted in a combination of Seismic Refraction (SR) and Multichannel Analysis of Surface Waves (MASW) for the estimation of the velocity of the mechanical compressive (P) and the shear (S) waves. The integration of both methods allowed to estimate the geomechanical parameters characterizing the studied sequence, the Poisson Ratio and the volumetric compressibility.

The obtained mechanical parameters were correlated with laboratory measured parameters such as plasticity index, density, shear strength and compressibility and, GPR and seismic profiles were correlated with the mapped fracture systems in the study area. Once calibrated, the profiles allowed to identify the lithological contact between lacustrine and volcanic sequences, their variations of thicknesses in depth and to assess the deformation area in the surface. An accurate determination of the geometry of fracturing was of the most importance for the assessment of the geological risk in the study area.

1 Introduction

The ground-fracturing phenomenon related with land subsidence has increased recently in Mexico City due to natural and anthropogenic processes. One of the areas with the highest differential subsidence is the lacustrine plain of Iztapalapa (Carreón-Freyre, 2011). The high-resolution geophysical prospecting, such as Ground Penetrating Radar (GPR) and active seismology, are useful tools for the assessment of ground deformation. In the eastern sector of Mexico City the sub soil consists of high contrasting sequences (lacustrine and volcanic inter bedded deposits) that favor the development of erratic fracturing in the surface causing damage to the urban infrastructure.

The GPR method is based on the generation of short electromagnetic pulses generated by a transmitter that penetrate in the subsoil, and are reflected according with the properties of optics laws, toward the receiving antenna. This method allowed to evaluate the fracture propagation and deformation of vulcano-sedimentary sequences at different depths (Daniels et al., 1998). By the use of the Common Mid Point (CMP) arrangement was possible to measure the propagation velocity of electromagnetic waves through the contrasting layers and allowed to build high-resolution geological

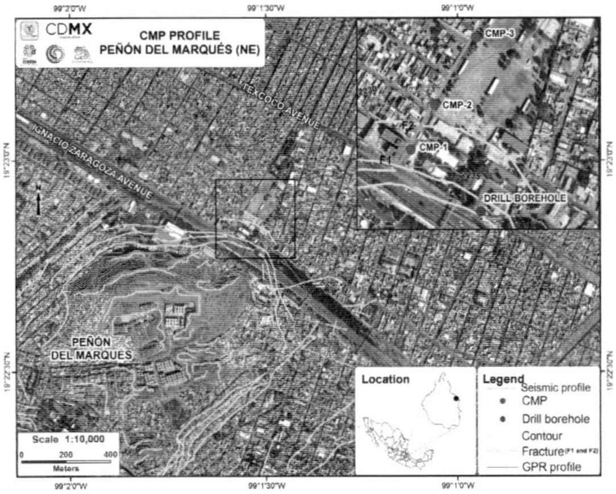

Figure 1. Location of the study area, east of Mexico City with spatial distribution of ground-fracturing and geophysical study.

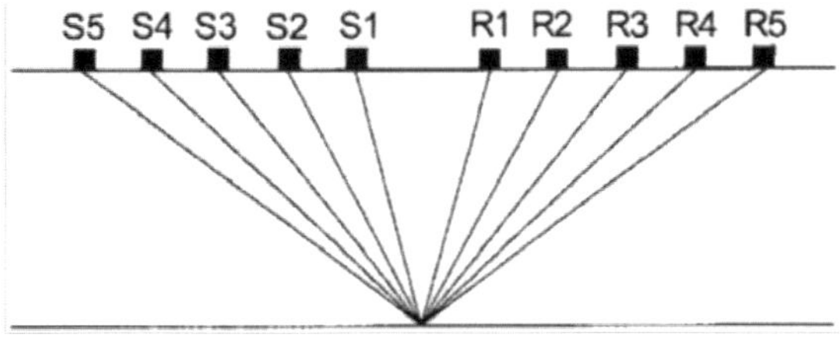

Figure 2. Common Mid Point (CMP) acquisition technique, where "S" denotes the transmitter location and "R" denotes the receiver location (Huisman, 2003).

sections (Grandjean et al., 2000). The energy is reflected to the surface by the contrasting electrical properties of each lithological sequence; the main electrical parameters are dielectric constant, ε, the magnetic permeability, μ, and the electrical conductivity, σ, which are directly related with the gravimetric and volumetric water content and therefore with the plasticity.

The MASW seismic method allows the record of the surface waves generated by a mechanical source in the surface (hammer) by the analysis of the dispersion curve of the fundamental mode, and eventually the higher modes. This curve can be inverted in a vertical section of shear strength because the surface waves are directly related by a factor of 0.97 with shear wave moreover, these take usually 70 % of the total seismic energy. The active seismology prospection consisted in a combination of Seismic Refraction (SR) and Multichannel Analysis of Surface Waves (MASW) for the estimation of the velocity of the mechanical compressive (P-) and the shear (S-) waves. The variation in depth of the shear velocity of seismic waves is directly related with some mechanical parameters of the prospected sequence such as: bulk modulus, Young's modulus, shear modulus and Poisson's ratio, used for the estimation of their deformation and potential fracturing.

2 High-resolution geophysical studies: GPR and MASW

The study area is located in the Basin of Mexico, east of the city, within a transition zone between volcanic and lacustrine materials (Fig. 1). The volcanic deposits come from a volcanic cone called "Peñón del Marques" composed by pyroclastic rocks and lava. Lake sediments are mainly clays, with large amounts of water content retained in their structure and with high plasticity values. The geophysical profiles were acquired at the base of the volcanic structure, over a high deformed and fractured area. GPR and seismic profiles are parallel, separated by 50 m. In the same area a geotechnical borehole was drilled for the characterization of the lithological sequence. In Fig. 1 the location of the study area is shown.

2.1 GPR prospecting

GPR fundamentals are based in the electromagnetic induction, where time is a decisive variable in conjunction with the spatial coordinates x, y, z. The induction field is of the same frequency as the transmitter current and varies in time within the prospected media, following electromagnetic laws, it creates secondary currents emerging from the prospected media to be measured by a receiver (Nabighian, 1988).

This is a high resolution method because it is capable to get 64 samples per meter, allowing to identify areas of deformation and fracture by reflection profiles. The Common Mid Point (CMP) technique was used to estimate the propagation velocity of the electromagnetic waves in different media and allowed the estimation of other physical properties. The CMP technique consists on separating the transmitter and receiver at different constant distances, as shown in Fig. 2. Three CMP profiles were acquired to characterize the spatial variation of the geological materials by the analysis of their velocity. Figure 3 shows the field data, processing, and interpretation of velocities from CMP. A GPR profile of 200 m length (radargram) was acquired complement the punctual measurements. The radargram shown in Fig. 4 was acquired with a 100 MHz antenna with 32 scans m^{-1}, each scan contains 1024 samples. The horizontal resolution is about 3 cm and the vertical resolution is less than 1 cm. This resolution allowed the characterization of the lithological variations related with the deformation and fracturing of the studied sequence through the change in the waveform.

The radargram shows three main layers: (a) a layer of volcanic sand 2 m width; (b) a lacustrine layer corresponding to clayey materials of variable thickness; and (c) a pyroclastic deposit is the subspace corresponding to volcanic sand until 10 m depth. The radargram shows a distortion of the se-

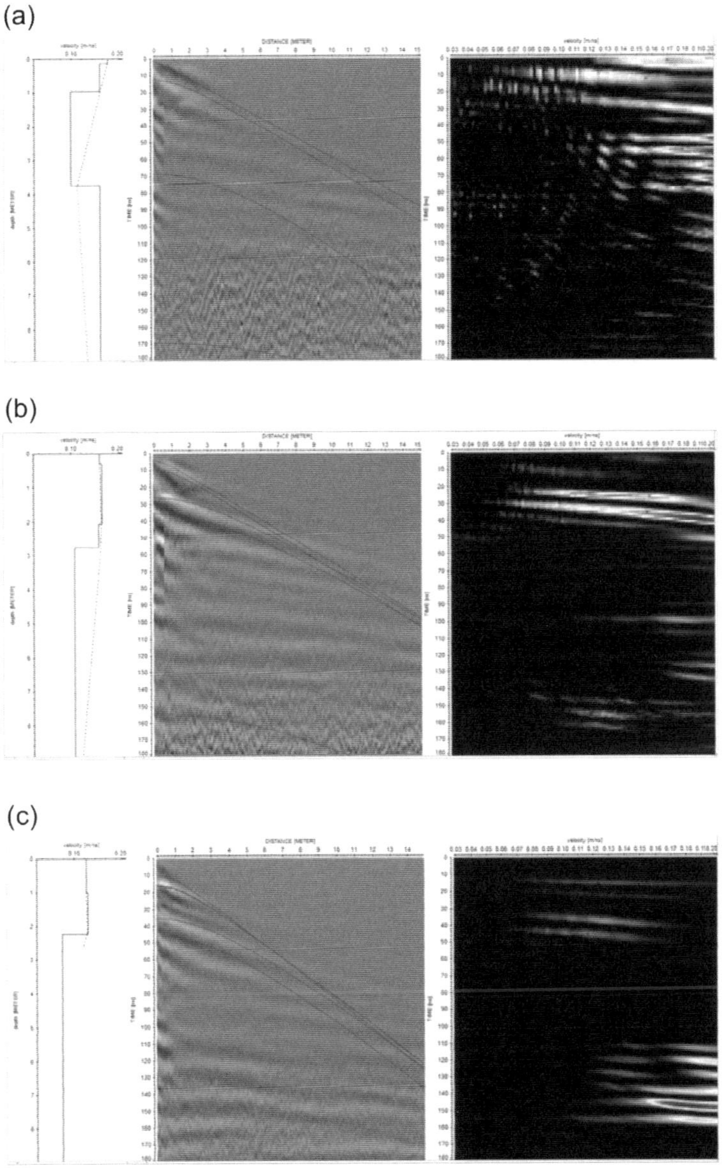

Figure 3. Data and results of the three CMP, on the left side the velocity, the field data on central side, and the processing on right side. **(a)** First CMP; **(b)** second CMP; **(c)** third CMP.

quence because of the high plasticity of the clayey materials, and fracturing recorded in the lower sandy layer.

2.2 MASW and Refraction Seismic (SR) profiles

The MASW method deals with surface waves in the low frequencies (1–30 Hz) and uses a shallow range of investigation (approximately 20 m depth). These superficial waves are related with S-wave by a factor of 0.97. Shear velocity is directly linked to the material stiffness and is one of the most critical mechanical parameters. The shear-wave velocity (V_S) is the best indicator for the detection of fracture zones because of the dramatic decrease of velocities.

The SR method involves the analysis of the travel times of first wave arrivals through the prospected layer to surface. The interpretation of the seismic data involves resolving the number of layers of the sequence, according to the velocity

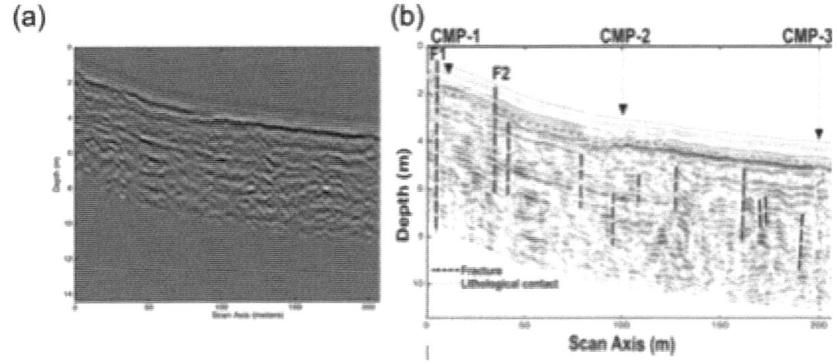

Figure 4. Reflection Profile of GPR. **(a)** Profile without geological interpretation; **(b)** profile with geological interpretation and CMP data correlation.

Table 1. Mechanical properties inferred from wave velocities of body.

Lithologhy	P-wave velocity ($m\,s^{-1}$)	S-wave velocity ($m\,s^{-1}$)	Bulk module ($kg\,cm^{-2}$)	Shear module ($kg\,cm^{-2}$)	Young module ($kg\,cm^{-2}$)	Poisson module	Water content (%)
Volcanic sequence (L_1)	500	270	65 000–90 000	20 000–34 000	180 000–260 000	0.04–0.16	11–47
Volcanic sequence (L_2)	280–400	90–180	15 000–30 000	15 000–30 000	100–600 000	0.48	226–258
Volcanic sequence (L_3)	480–552	300	65 00–90 000	65 000–90 000	180 000–260 000	0.04–0.16	14–50

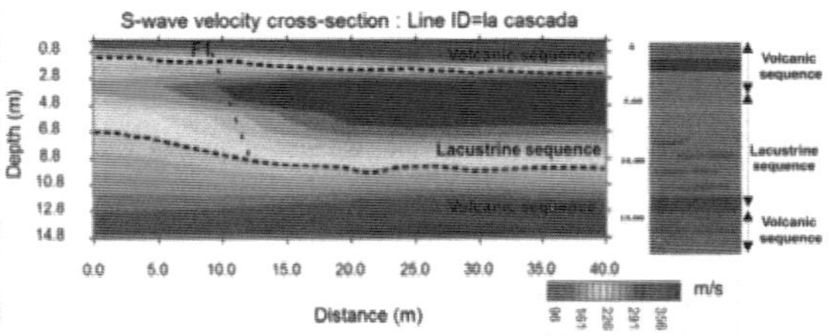

Figure 5. S-wave model, acquired from MASW method and a stratigraphic correlation with the borehole.

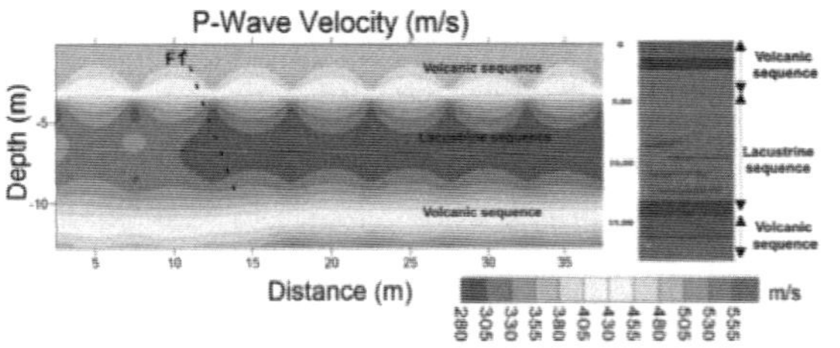

Figure 6. P-wave model, acquired from the determination of the first arrivals method and a stratigraphic correlation with the borehole. SR Method.

variation of each layer, and the travel-time from a given refractor to the ground surface.

MASW and SR profiles were acquired using 16 geophones of 12.5 Hz acquisition frequency, spaced every 2.5 m with a total length of 40 m. The S-wave velocity is determined by analyzing the Rayleigh wave and, P-wave velocity by calculating the first arrivals. Figures 5 and 6 show the S-wave model and P-wave model respectively. In Table 1 the relationship of P-wave and S-wave velocity is presented.

The rate of the two body waves (P and S) determine the spatial distribution of the different mechanical parameters. Equation (1) describe the compression waves and Eq. (2) the shear wave.

$$V_p = \sqrt{\frac{\lambda + 2\mu}{\rho}} \quad (1)$$

$$V_s = \sqrt{\frac{\mu}{\rho}}, \quad (2)$$

with $\lambda = \kappa - \frac{2}{3}\mu$.

The parameter λ is known as the first Lamé constant, which has no physical sense. The V_p and V_s parameters are the P-wave and S-wave velocities, μ is the modulus of rigidity (or shear), κ is the bulk modulus (or incompressible) and ρ is the density. These mechanical parameters result from the theoretical basis of the dynamic elasticity that studies the behavior of a solid upon the application of a force, both compression and shear, in one or in all three spatial dimen-

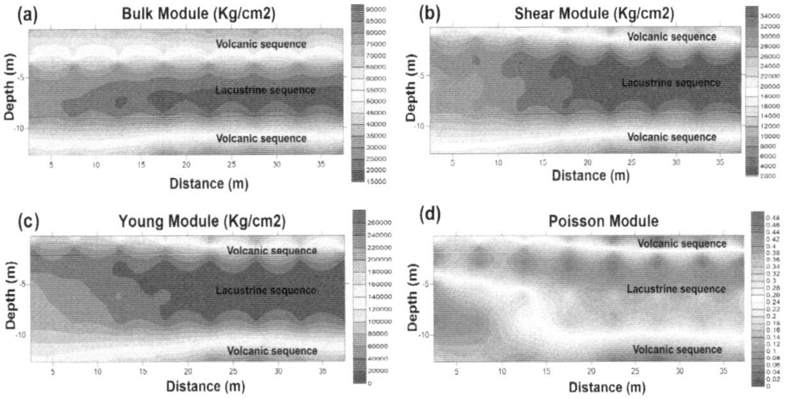

Figure 7. Distribution of mechanical parameters obtained from modeling of the P- and S-wave (SR Method): **(a)** spatial model of Bulk module; **(b)** spatial model of shear modulus; **(c)** spatial model of Young's modulus; **(d)** model spatial of Poisson constant.

Table 2. Velocities obtained from the CMP data.

Lithologhy	CMP-1 Depth velocity		CMP-2 Depth velocity		CMP-2 Depth velocity	
	(m)	($m\,ns^{-1}$)	(m)	($m\,ns^{-1}$)	(m)	($m\,ns^{-1}$)
Volcanic sequence (L_1)	0–1	0.16	0–2.8	0.15	0–2.3	0.13
Volcanic sequence (L_2)	1–3.8	0.085	2.8	0.09	2.3	0.05
Volcanic sequence (L_3)	3.8	0.16	–	–	–	–

sions (Velis, 2007). Seismic velocities are directly related with these mechanical properties for a prospected media.

The bulk modulus, rigidity, and Young modulus, can be determined either from static or dynamic experiments, involving the travelling of seismic waves through the soil. The bulk modulus of a soil measures the resistance to uniform compression. It is defined as the ratio of the infinitesimal pressure increase to the resulting decrease of the volume. The shear modulus is the coefficient of elasticity for a shearing force. It is defined as the ratio of shear stress to the unit displacement per sample length. Mathematically is the tangential force exerted and the related strain change. The elastic constant, called the Poisson's ratio (σ) measure the compressibility of a material perpendicular to the applied stress, or the ratio of vertical and horizontal strains. This can be expressed in terms of properties measured in the field, as velocities of P-waves and S-waves and in terms of μ and λ, as shown below (Aki, 2002).

$$\sigma = \frac{\lambda}{2(\lambda + \mu)} \tag{3}$$

with

$$\mu = V_s^2 \rho; \quad \lambda = V_p^2 \rho - 2\mu = V_p^2 \rho - 2V_s^2 \rho$$

$$\sigma = \frac{V_p^2 \rho - 2V_s^2 \rho}{2(V_p^2 \rho - 2V_s^2 \rho + V_s^2 \rho)} \tag{4}$$

$$= \frac{V_p^2 - 2V_s^2}{2(V_p^2 - 2V_s^2)}$$

$$= \frac{(V_p/V_s)^2 - 2}{2(V_p/V_s)^2 - 2}.$$

If V_s is small then Poisson's ratio equals 0.5, indicating either a fluid, because shear waves do not pass through fluids, or a material that maintains constant volume regardless of stress, also known as an ideal incompressible material.

The Young modulus, E, describes the elastic properties of a solid undergoing tension or compression in only one direction. Young's modulus is a measure of the ability of a material to withstand changes in length when under lengthwise tension or compression (see Eq. 5).

$$E = 3\kappa(1 - 2\sigma) \tag{5}$$

If seismic data are available, the spatial distribution of mechanical parameters of the prospected sequence can be modeled to create pseudo-geologic sections as shown in Fig. 7. The distribution of the obtained mechanical parameters (κ, μ, σ and E) was correlated with the log of the studied sequence and with physical measurements from borehole samples.

The Poisson constant distribution, varying from 0.38 to 048, and S-wave velocities, varying from 90 to 180 $m\,s^{-1}$, are correlated with the lacustrine layer having the larger water content, approximately 200 %, in the sequence. The low values of E modulus, κ modulus and μ modulus show that the lacustrine material is highly deformable (i.e. low bearing capacity). The volcanic sand layer shows higher values of κ, μ and E, less deformable, having lower water content values, 30 % in average.

The distribution of the mechanical parameters (Fig. 7 and Table 2) also allowed identifying the transition zone by the increase the thickness of the lacustrine sequence which contact correspond to the location of fracture F_1.

3 Conclusions

The results of applying the GPR, MASW and SR methods simultaneously show the persistence of the clayey lacustrine material with variable thickness encased by volcanic materials. This material was characterized in the laboratory a present gravimetric large water contents varying from 200 to 300 %.

The contrasting mechanical properties between the two types of material may cause slippage over lithological contacts, causing the propagation of deformation and fracturing of the sequence, as shown in the GPR profile of Fig. 4. This can also be correlated with the distribution of the S-wave velocities presented in Fig. 5, where the lower values (about 100 $m\,s^{-1}$) corresponds to the location of fracture F_1.

The application of seismology and electromagnetic high resolution methods allow to characterize deformation and fracturing in a subsidence area. GPR studies allowed the monitoring of the deformation zone and fracturing and to correlate lithological variations through CMP profiles. Seismology shows to be very useful to quantify the mechanical behavior of the studied sequences.

The physical and mechanical parameters of the high deformability of the lacustrine sequence showing large water content and low shear strength. Fracturing showed to be strongly related with the lithological contact between materials having contrasting mechanical parameters in heterogeneous subsidence areas such as the east zone of Mexico City.

Acknowledgements. The authors thank his colleagues at the Centre for Evaluation of Geological Risk and the Center of Geosciences for the field support and technical suggestions for improving this article. The authors acknowledge the support of the DGAPA-UNAM research project No. IN114714.

References

Aki, K. and Richards, P.: Quantitative seismology: theory and methods: W. H. Freeman and Co. University Science Book, 2nd Edn., 67–288, 2002.

Carreón-Freyre, D.: Identificación y Caracterización de los Diferentes Tipos de Fracturas que Afectan el Subsuelo de la Delegación Iztapalapa en el Distrito Federal, Informe Técnico, Ingeniería Geologica, AI México, 3–9, 2011.

Daniels, D., Gunton, D., and Scott H.: Introduction to subsurface radar, IEEE Proceedings, 135, 278–320, 1998.

Grandjean, G., Gourry, J., and Bitri, A.: Evaluation of GPR techniques for civil-engineering applications: study on a test site, J. App. Geophys., 45, 141–156, doi:10.1016/S0926-9851(00)00021-5, 2000.

Huisman, J., Hubbard, S., Redman, J., and Anna, P.: Measuring Soil Water Content with Ground Penetrating Radar: A Review, Vadose Zone J., 2, 476–491, 2003.

Nabighian, N.: Electromagnetic Methods in Applied Geophysics: Theory, Society of Exploration Geophysics, Tulsa, OK, USA, 1, 131–141, 1998.

Velis, D.: Improved AVO fluid detection and lithology discrimination using Lame petrophysical parameters "$\lambda\rho$", "$\mu\rho$", and "λ/μ fluid stack" from P and S inversions, 67tn Annual International Meeting, SEG, Expanded Abstracts, 183–186, 1997.

Proc. IAHS, 372, 273–279, 2015
proc-iahs.net/372/273/2015/
doi:10.5194/piahs-372-273-2015

Open Access
PIAHS

Analysis of the variation of the compressibility index (Cc) of volcanic clays and its application to estimate subsidence in lacustrine areas

D. Carreón-Freyre[1], M. González-Hernández[2], D. Martinez-Alfaro[3], S. Solís-Valdéz[1], M. Vega-González[1], M. Cerca[1], B. Millán-Malo[4], R. Gutiérrez-Calderón[2], and F. Centeno-Salas[3]

[1]Centro de Geociencias de la UNAM Juriquilla. Queretaro, Mexico
[2]Centro de Evaluación de Riesgos Geológicos CERG, Iztapalapa, Mexico City, Mexico
[3]Posgrado en Ciencias de la Tierra, UNAM Juriquilla, Queretaro, Mexico
[4]Centro de Física Aplicada y Tecnología Avanzada, UNAM Juriquilla, Queretaro, Mexico

Correspondence to: D. Carreón-Freyre (freyre@geociencias.unam.mx)

Published: 12 November 2015

Abstract. An analysis of the deformation conditions of lacustrine materials deposited at three sites in the volcanic valley of the Mexico City is presented. Currently geotechnical studies assume that compressibility of granular materials decreases in depth due to the lithostatic load. That means that the deeper the sample the more rigid is supposed to be, this assumption should be demonstrated by a decreased Compression Index (Cc) in depth. Studies indicate that Mexico City clays exhibit brittle behaviour, and have high water content, low shear strength and variable Cc values. Furthermore, groundwater withdrawal below the city causes a differential decrease in pore pressure, which is related to the physical properties of granular materials (hydraulic conductivity, grain size distribution) and conditions of formation. Our results show that Cc for fine grain materials (lacustrine) can be vertically variable, particularly when soils and sediments are the product of different volcanic materials. Lateral and vertical variations in the distribution of the fluvio-lacustrine materials, especially in basins with recent volcanic activity, may be assessed by Cc index variations. These variations can also be related to differential deformation, nucleation and propagation of fractures and need to be considered when modelling land subsidence.

1 Introduction

Mexico City is the best known example of land subsidence in Mexico (Fig. 1). The city is located within the geological province of the Transmexican Volcanic Belt (Ferrari et al., 2012), over a desiccated lacustrine plain bounded by faults and volcanic edifices. The near-surface lithology in the Mexico basin is highly heterogeneous and includes fluvio-lacustrine sediments with particle sizes varying from gravel, sand, and silts to clays, with interbedded layers of pyroclastic rocks and lava flows (Carreón-Freyre, 2010, Carreón-Freyre et al., 2011). Analyses of groundwater withdrawal and of the associated land subsidence have been documented systematically in Mexico City since the 1940s (Carrillo, 1947). As a consequence of descending water level, the Basin of Mexico has been divided in several clay bearing fluvio-lacustrine sub-basins, which include Mexico City, Tlahuac-Chalco, Texcoco, Xochimilco, and Zumpango basins.

Although subsidence in Mexico City is thought to be related to high compressibility rates of the clayey sequences when the pore pressures decrease (Mesri et al., 1976), there is a lack of systematic studies to explain the compositional and mechanical behavior variations in clayey sequences and their role on the estimation of vertical deformation. The Compression Index (Cc; Lambe and Whitman, 1969) is obtained from the void ratio vs log of normal effective stress plot in consolidation tests (ASTM, 1994), and was developed for the estimation of settlements for fine-grained deposits. The evaluation of vertical and lateral variations in the Compression Index Cc may facilitate the assessment of the potential of land subsidence in clay-bearing sequences. Several authors (see Haigh et al., 2013; and references therein) found that Cc

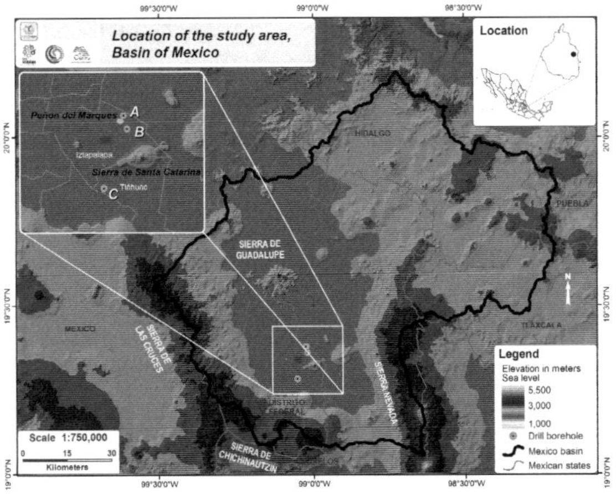

Figure 1. Location of the study areas in the Basin of Mexico (black polygon), Iztapalapa and Tlahuac. The geotechnical borehole locations are shown by blue circles: (A) La Cascada, (B) Lucio Blanco, (C) Tlahuac. Lacustrine planes are shown in green and volcanic structures in yellow and red. Grey lines indicate administrative boundaries.

can be correlated with water content, Atterberg Limits, and void ratio variations (e). The presented analysis of the Cc was made on silt and clayey deposits because of the strong interdependence of water content and compressibility with mineralogy, mechanical behaviour, and structure.

Three silty and clayey sequences, located in Iztapalapa and Tlahuac localities (Fig. 1), are compared to identify the conditions of formation and evolution that led to their physical and mechanical properties. The systematic characterization of these sequences included compressibility, grain size distribution, mineralogy of clays (analyzed by X-ray diffraction), water content, and other physical parameters.

2 Lacustrine deposits in the Iztapalapa and Tlahuac-Chalco sub-basins

The near surface stratigraphy of Iztapalapa and Tlahuac areas comprise fluvial and lacustrine sequences (silt and clay bearing) interbedded with tephra and lava (Zeevaert, 1953). The main volcanic structures are the Sierra de Santa Catarina volcanic complex, that separates the Iztapalapa and Thalhuac sub-basins and, the Peñón del Marques located at the north of the Iztapalapa plain. Weathering and hydrothermal alteration of basaltic rocks and volcanic ashes generate different types of clays (Hillier, 1995; Velde, 1995; Wada, 1987). The influence of mineralogy composition of clays on the mechanical behavior of lacustrine materials in Mexico City has been widely studied (Diaz-Rodriguez et al., 1998; Peralta y Fabi, 1989). The physical characteristics and composition of the lacustrine deposits have been documented from geotechnical and geophysical surveys (Carreón-Freyre and Cerca 2006; Hernández-Marin et al., 2005; Carreón-Freyre et al., 2003). Clays are metastable materials, their forming processes are sequential and the existence of each type is restricted to specific temperature and time spans. The environmental conditions of a lake are a consequence of its salinity, pH, water depth, temperature, and water residence time (with restricted drainage). These conditions varied dramatically in very short periods because of the intense local volcanic activity (mafic or silicic in composition, effusive and/or pyroclastic rocks) in the sub-basins of the Basin of Mexico and their interaction with the lake deposits. The water level of the former lakes also played an important role to differentiate lacustrine conditions in adjacent sub-basins, such as Iztapalapa and Tlahuac.

A common clay mineral assumed to predominate in the sediments of the Basin of Mexico is montmorillonite, from the smectite family (Diaz Rodriguez, et al., 1998; Peralta y Fabi, 1989), that is characterized by: (1) 2 : 1 crystalline layered structure (2 tetrahedral sheets to 1 octahedral sheet) (Velde, 1995); (2) high values of plasticity and compressibility index (Hernandez-Marin and Carreón-Freyre, 2002); (3) large amounts of water retained in structure (generally more than 100 %); and, (4) high solids density ($2.5\,g\,cm^{-3}$). Smectite formation is favored in lake areas with restricted drainage and low saline-alkaline conditions (Hillier, 1995). In such environments, iron (Fe) oxidizes and accumulates. When the lake level increases, iron reduces and becomes soluble forming green clay sequences (Carreón-Freyre et al., 2006). Mafic minerals (e.g., olivine and pyroxene) from volcanic rocks may form chlorite and talc. Moreover, alteration of pumicite-rich volcanic ashes produces allophane and imogolite, clay minerals of low crystalline order having a gel-like structure, low vertical compressibility, and susceptibility to shear deformation (Wesley, 2001). These materials have a characteristic amorphous spectrum (with no specific clay reflexions) of X-Ray Difraction (XRD) analysis (Carreón-Freyre et al., 2006). If the environmental conditions favor dehydration, these materials become gibbsite, halloysite, and talc (Velde, 1995). Heterogeneity of mineral content (mica, pyroxene, and plagioclase) suggests a close similarity with the volcanic source and less chemical equilibrium. The close relation between mineralogical conditions of the clayey materials, their ability to retain water, and their compressibility has been doumented by Haigh et al. (2013), Robinson (1999) and Warren and Rudolph (1997).

3 Characterization of the studied sequences

The three study sites were selected after detailed mapping of the fluvio-lacustrine sediments and volcanic rocks. Geological mapping was complemented with a topographic survey of the sites with evidence of ground surface deformation, where vertical displacements varied from 50 cm to 3 m. The different sites have approximately the same elevation. Samples

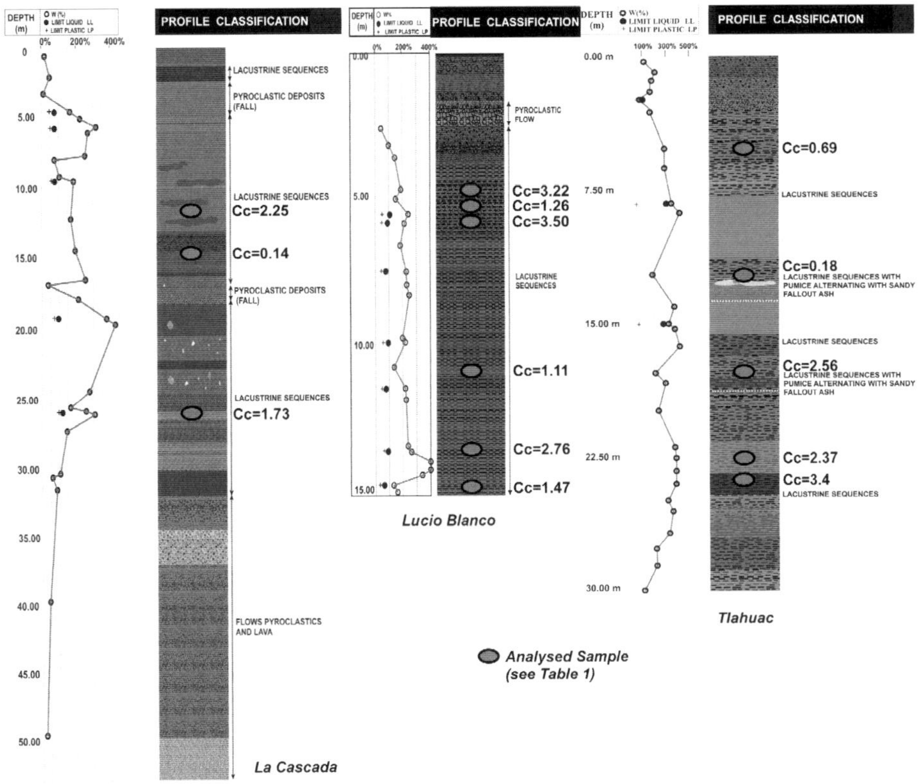

Figure 2. Lithological correlation of three fluvio-lacustrine and volcanic sequences in the Iztapalpa (La Cascada and Lucio Blanco) and Tlahuac sub-basins, obtained from geotechnical boreholes. Green ellipses indicate the location in depth of non-altered samples used for physical and mechanical characterization (see Fig. 1 for location).

were collected through three geotechnical boreholes named La Cascada (50 m depth), Lucio Blanco (15 m depth), and Tlahuac (30 m depth) (Figs. 1 and 2).

3.1 Physical properties and clay mineralogy

Physical properties of the collected samples were determined in the Laboratorio de Edafología and the Laboratorio de Mecánica de Geosistemas (LAMG), of the Centro de Geociencias, UNAM. These determinations included: gravimetric water content (W_{grav} %), grain-size distribution by sieving and fine-grain distribution using the Bouyoucos densitometer, organic matter content (%), and solid density (S_s) by the pycnometer method. Particles with diameters smaller than 2 microns were segregated from samples and dried at room temperature. They were crushed in an agate mortar and sieved with a 200 mesh. Carbonates were removed by washing with buffer solution of sodium acetate-acetic acid (Buhrke et al., 1998). Subsequently the samples were washed with deionized water and mild sonic vibration was applied. Particle dispersion was propitiated by the addition of sodium hexametaphosphate. Samples for XRD analysis were prepared according to the standard methodology proposed by Buhrke et al. (1998); and were analyzed at the Laboratorio de Difracción de Rayos X, Centro de Fisica Aplicada y Tecnologia, UNAM. The microstructure of some samples were also analyzed using a Scanning Electron Microscope (SEM) and an Transmission Electron Microscope (TEM) in the Laboratorio de Fluidos Corticales, Centro de Geociencias.

3.2 Mechanical properties: compressibility

More than 30 borehole samples were tested by the standard incremental consolidation method in order to determine their compressibility (Cc) and overconsolidation ratio (OCR, overconsolidated having values > 1), but only 14 representative samples are discussed in this work (Table 1 and Figs. 2 and 3). Samples were classified according to the Unified Soil Classification System (USCS). Liquid limit (LL), plastic limit (PL), and plasticity Index (PI) were determined. Each type of soil has a certain water retention capacity and, therefore, its plasticity index may reflect their mineralogical condition. The pneumatic consolidometer used for the compressibility tests (ASTM, 1998) included measurements of

Table 1. Physical and mechanical characterization of the studied lacustrine sequences. S_s, solid density; W_{grav}, gravimetric water content; GSD, Grain Size Distribution: C, Clay, M, Silt, S, Sand; e_0 and e_f, initial and final void ratio, OCR, Overconsolidation Ratio, Cc, Compressibility Index, USCS, Unified Soil Classification System: MH, High Plasticity Silt, SM, Silty Sand.

Depth m	S_s g cm^{-3}	W_{grav} %	LL-PI/IP %	GSD C/M/S	$e_0 - e_f$	Δe	OCR	Cc	USCS
Tláhuac									
6.0	2.20	106	74–59/15	14/60/26	2.5–2.1	0.4	13	0.69	MH
12.15	2.21	187	76–57/19	19/43/38	3.5–3.3	0.2	2.8	0.18	MH
17.05	2.27	228	123–90/33	11/32/57	5.3–2.4	2.9	1	2.56	MH
22.6	2.23	260	95–67/28	15/62/23	5.2–2.5	2.7	1	3.13	MH
22.9	2.05	260	136–90/46	15/62/23	7.6–3.2	4.4	1	3.40	MH
La Cascada									
12.65	2.43	203	87–61/26	16/34/50	4.7–2.2	2.5	1.8	2.25	MH
14.65	2.53	45	–	7/15/78	0.8–0.7	0.1	3.7	0.14	SM
26.84	2.26	130	103–74/29	16/38/46	3.0–1.4	1.6	10	1.73	MH
L. Blanco									
4.7	2.39	190	–	12/29/59	5.1–2.8	2.3	12.5	3.22	SM
5.6	2.26	234	103–74/29	24/29/47	0.8–0.0	0.8	5.5	1.26	MH
5.85	2.21	222	96–72/24	46/31/23	4.9–1.8	3.1	6.5	3.5	MH
11.5	2.23	213	75–59/16	19/36/45	3.2–1.8	1.4	20.5	1.1	MH
14.1	2.23	302	101–67/34	18/37/45	5.9–2.2	3.7	1.4	2.76	MH
14.8	2.36	120	59–47/12	6/41/53	2.9–1.7	1.2	3.6	1.47	SM

variations of axial pressure, back pressure, pore pressure and displacement with electronic sensors (Hernandez-Marin and Carreón-Freyre, 2002). The description of its operation and data acquisition software is detailed in Flores et al. (2010). Compression Index (Cc) and Overconsolidation Ratio (OCR) were obtained from graphs of void ratio vs log effective stress.

Results obtained for representative samples (14 out of more than 30) of the three studied sequences are presented in Table 1. Mostly, sediments are high plasticity silts (MH) with some intercalations of uniform sand and gravel deposits of black colour and ash layers (sandy silts) that correspond to pyroclastic deposits. Water content varies from 0 to 100 % for pyroclastic deposits and from 200 to 300 % for lacustrine deposits (Fig. 2). Variations of colour from red to olive green indicate the oxidation or reduction of iron (Fe) present in clays. Likely, local variations of pH, temperature, and salinity produced different clayey materials within the same depositional environment. In the La Cascada samples green colours of the clayey sediments denote a reduction condition indicating a water cover. The Lucio Blanco site is located in a channel that connected with the Texcoco sub-basin from the north within the lacustrine area of Iztapalapa (Fig. 1). Variations of the water level in the channel might have varied conditions from dry to humid, and the presence of Fe and magnesium (Mg) from basaltic rocks propitiated the crystallization of smectite clays, such as montmorillonite. In the Tlahuac site, the sequence is composed mainly by pumice and ash (silicic pyroclastic deposits) that rapidly generated fine grain materials, rich in silica (SiO_2), and with a low level of crystallization. The initial void ratio (e_0) for each layer indicates the potential compressibility and is directly related to the grain size and mineralogy.

Three samples from different lacustrine layers at 12, 14, and 27 m depth in La Cascada borehole are overconsolidated. The highest Cc value and the higher water content correspond to the 12 m sample and the plasticity index (IP) is similar to that of 27 m sample which has the lower Cc value (Table 1, Fig. 3b). The sequence is interpreted to be formed by clayey well crystallized lacustrine sediments.

In the Lucio Blanco borehole, six samples were tested from 4.7, 5.6, 5.85, 11.5, 14.1, and 14.8 m depth. The highest Cc value corresponds to the 5.85 depth but does not present the highest water content or IP. The lowest Cc value corresponds to a sandy material that is a pyroclastic flow deposit (Fig. 3c and d). Pyroclastic deposits in general behave as non-cohesive potentially collapsible material. The sequence was interpreted as a wetland area within a lake channel, the red colour (oxidation) of the lacustrine deposits indicated large variations of water level.

In the Tláhuac borehole, five samples were tested from 6, 12, 17, 22.6, and 22.9 m depth. The highest Cc value corresponds to the deepest layer and the two lowest to the shallower layers. Also in this case the most compressible layer does not present the highest water content, neither the highest IP (Fig. 3a). The sequences is interpreted to be mainly

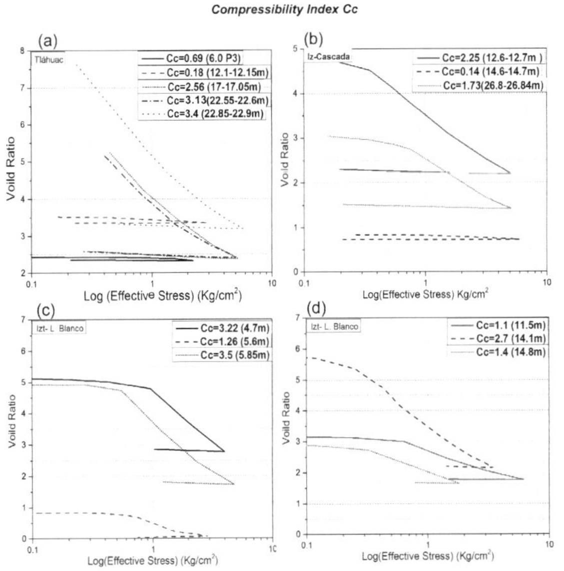

Figure 3. Compressibility plots of samples from the three borehole samples. **(a)** Five normally consolidated samples from 6 to 22.9 m depth in the Tlahuac site, only the material in the surface is over consolidated and have the smaller Cc value and the deeper samples have the higher Cc values; **(b)** three over consolidated samples from 12.6 to 26.8 m depth in La Cascada site, the lowest Cc value is placed at 14 m depth and the highest at 12 m depth; (**c** and **d**) six over consolidated samples from 4.7 to 14.8 m depth in the Lucio Blanco site, the lower Cc value corresponds to a 11.5 m depth sample and the higher at 5.85 m depth. Note also high variations in Cc for the same depth (e.g. at 12 m depth: 0.18, 2.25, and 1.1) for the different sequences.

formed by amorphous materials, formed by the alteration of ashes, that may explain the apparently contradictory values.

3.3 Mineralogy

For better understanding of the Cc variations we carried out XRD analysis in order to identify mineralogy; SEM and TEM imaging to characterize structure. Figure 4 shows some of the results for the Lucio Blanco and Tlahuac sites. The XRD records of the sample from 14 m depth in the Lucio Blanco site, show heterogeneous mineralogical composition of the entire sample: mica, illite, anorthite, silica, talc, and possibly nontronite. The clay fraction sample (after washing salt) indicates the presence of montmorillonite and cristobalite (silica) (Fig. 4a). Anorthite, a calcium plagioclase, and mica are minerals formed during volcanic eruptions that have not been completely altered. Cristobalite may have a volcanic origin or be part of the diatomaceous frustule (e.g., Marsal and Mazari, 1959). The montmorillonite suggests saturated conditions during its formation. The presence of talc indicates alteration of the basic volcanic minerals at high salinity and temperatures below 100 °C (Velde, 1995).

The XRD record for the Tlahuac samples indicates the presence of vermiculite which also is related to the alteration of volcanic minerals (an example from 22 m depth is presented in Fig. 4a). Amorphous material and low-crystallized materials cannot be identified by XRD but SEM and TEM can identify such materials. The SEM image presented in Fig. 4b corresponds to a silty clay from 6 m depth (Fig. 4b1), and to a low crystallized material from 22 m depth (Fig. 4b1 and b2). The Fig. 4b3 was obtained using an TEM and shows the lack of structure. This material is similar to allophane and is a product of the alteration of acid volcanic ashes.

4 Discussion

The results of the physical characterization presented in this paper indicate that the conditions of formation and the evolution of lacustrine sequences control mechanical behaviour, and ultimately differential subsidence in the Basin of Mexico, and particularly in Mexico City. The silty-clay fraction has a highly heterogeneous composition including partially altered volcanic ash, illite, smectite, and allophane.

These materials are usually assumed to be highly compressible and with low resistance to shear. However, complex consolidation of these materials is also influenced by the displacement of free water, adsorbed water and/or intermolecular water, which holding force is dependent on the mineralogical characteristics of the solid phase. In such lacustrine materials having a metastable structure and mineralogy the mechanical behavior cannot be represented only by one plasticity or compressibility index. Lacustrine clay heterogeneity due to changes in depositional environment should be considered for a proper assessment of the compressibility index variations. This parameter is the most important because it is widely used for the estimation of land subsidence and, hence of the fracturing susceptibility of sediments.

A common assumption is that Cc decreases with depth due to lithostatic load; however, the analysis of the data presented in this work suggest a different mechanical behaviour for the lacustrine sediments. The mixture of silt and sand composed of silica, talc, and plagioclase with smectite clays (montmorillonite) presents variable consolidation conditions. Sandy lenses are less compressible, although the silt and clay content is higher. For the case of Tlahuac, it should be noted that the sand corresponds to pyroclastic materials composed of pumice that may degrade and change structure in hundreds or tens of years, and this explains the presence of low crystallized materials in the whole sequence. These materials show a lower plasticity than the smectites, higher contraction limits and a brittle behavior. The lateral and vertical compressibility variations are one of the causes of the substantial differential deformation observed in the study area. The high water

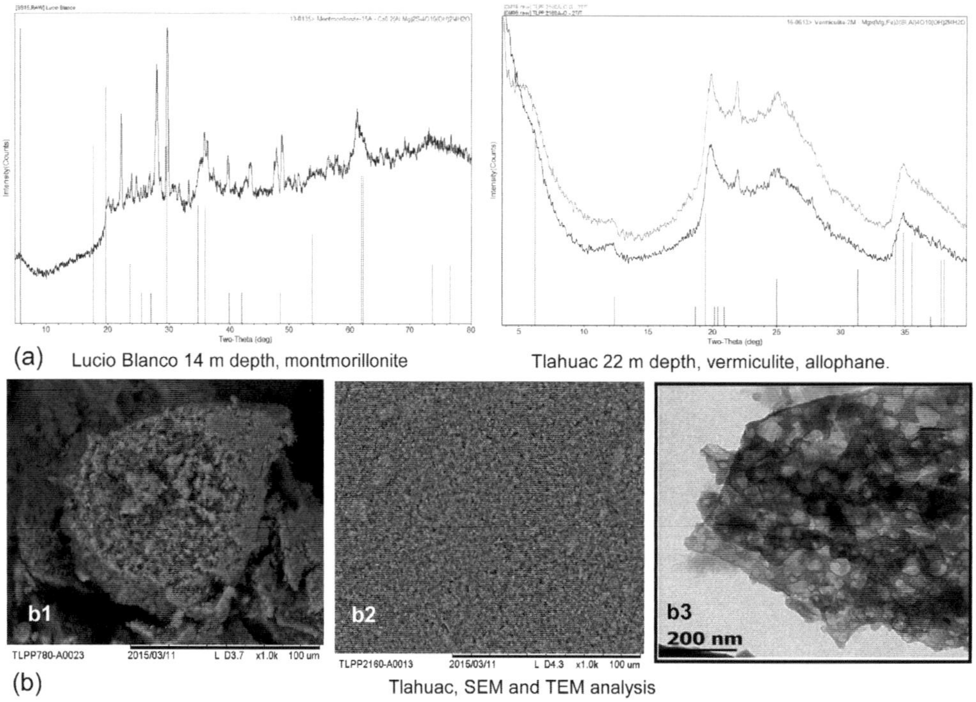

Figure 4. Characterization of the mineralogy and structure of clayey materials. **(a)** XRD analysis of clays from the Lucio Blanco site (14 m depth) showing the montmorillonite signature and from Tlahuac (22 m depth) showing the signature of vermiculite and low crystallized material; **(b)** SEM images from Tláhuac deposits: **(b1)**, is a silty material at 6 m depth; **(b2)**; is a low crystallized material at 22 m depth and, the **(b3)** image, taken using an TEM, shows the lack of structure in material similar to allophane.

content values suggest that these materials have low shear strength and fracture easily. Once fractured, the space created is filled with material coming from the surface creating a tensional state of stress that propitiates the propagation of fracturing to the surface.

Furthermore, the studied silt-clay sequences are commonly interbedded with fluvial and pyroclastic granular materials and/or volcanic rocks, which increase the mechanical heterogeneity of sequences and favours differential deformation. Fracturing of the subsurface in the Basin of Mexico, (e.g. Iztapalapa and Tlahuac) might be provoked by several factors, such as groundwater withdrawal and regional faulting; nevertheless, the gradual deformation related to differential consolidation between the different types of fluvio-lacustrine materials in volcanic areas also play an important role and should be better studied.

Acknowledgements. The authors acknowledge the work of characterization of samples in the laboratory of Mr. Ricardo Carrisoza, laboratory technician in the Laboratorio de Mecánica de Geosistemas (LAMG) at the Centro de Geociencias. The field support of Mr. Cesar Tovar CERG-Iztapalapa is also appreciated. The authors also thank support from the UNAM DGAPA Project PAPIIT No. 114174.

References

American Society for Testing Materials (ASTM): D2435-96 Standard method for one-dimensional consolidation properties of soils, ASTM Standards, USA, 5 pp., 1998.

Buhrke, V. E., Jenkis, R., and Smith, D. K.: A practical guide for the Preparation of specimens for X-ray Fluorescene and X-Ray diffraction Analysis, Wiley-VCH, New York, 333 pp., ISBN: 0-471-19458-1, 1998.

Carreón-Freyre, D.: Land subsidence processes and associated ground fracturing in Central Mexico, in: Land Subsidence, Associated Hazards and the Role of Natural Resources Development (Proceedings of EISOLS 2010, Querétaro, México), edited by: Carreón-Freyre, D., Cerca, M., and Galloway, D. L., Red Book Series Publication 339, IAHS Press, CEH Wallingford, UK, ISBN: 978-1-907161-12-4, ISSN: 0144-7815, 149–157, 2010.

Carreón-Freyre, D. and Cerca, M.: Delineating the near-surface geometry of the fracture system affecting the valley of Queretaro, Mexico: Correlation of GPR signatures and physical properties of sediments, Near Surface Geophysics, EAGE (European Assoc. of Geoscientists and Engineers), 4, 49–55, 2006.

Carreón-Freyre, D. C., Cerca, M., and Hernández-Marín, M.: Correlation of near-surface stratigraphy and physical properties of clayey sediments from Chalco Basin, Mexico, using Ground Penetrating Radar, J. Appl. Geophys., 53, 121–136, 2003.

Carreón-Freyre, D. C., Hidalgo-Moreno, C., and Hernández-Marín, M.: Mecanismos de fracturamiento de depósitos arcillosos en zonas urbanas. Caso de deformación diferencial en Chalco, Estado de México, Boletín de la Sociedad Geológica Mexicana, Número Especial de Geología Urbana, Tomo LVII, 2, 237–250, 2006.

Carreón-Freyre, D., González-Hernández, M., Cerca M., and Gutiérrez-Calderón, R. I.: Caracterización geomecánica de los suelos de Iztapalapa, México, para evaluar el fracturamiento causado por deformación diferencial, Proceedings of the 14th Pan-American Conference on Soil Mechanics and Geotechnical Engineering, 2–6 October 2011, Toronto, Canada, 2011.

Carrillo, N.: Influence of artesial wells in the sinking of Mexico City, en Volumen Nabor Carrillo "El hundimiento de la Ciudad de México y el Proyecto Texcoco", Comisión Impulsora y Coordinadora de la Investigación Científica Anuario, 47, 7–14, 1947.

Díaz-Rodríguez, A., Lozano-Santacruz, R., Dávila-Alcocer, V. M., Vallejo, E., and Girón, P.: Physical, chemical, and mineralogical properties of México City sediments: a geotechnical perspective, Can. Geotech. J., 35, 600–610, 1998.

Ferrari, L., Orozco-Esquivel, T., Manea, V., and Marina, M.: The dynamic history of the Trans-Mexican Volcanic Belt and the Mexico subduction zone, Tectonophysics, 522–523, 122–149, doi:10.1016/j.tecto.2011.09.018, 2012.

Flores, C. O., Gómez, R. E., Hernández, M. S., and Carreón Freyre, D.: Automatización de un consolidómetro neumático, Memorias de la XXV Reunión Nacional de Mecánica de Suelos e Ingeniería Geotécnica, Mexico, ISBN: 978-968-5350-23-5, 8–16, 2010.

Haigh, S. K., Vardanega, P. J., and Bolton, D.: The plastic limit of clays, Géotechnique, 63, 435–440, doi:10.1680/geot.11.P.123, 2013.

Hernández-Marín, M., Carreón-Freyre., D., and Cerca Martínez, M.: Mechanical and physical properties of the montmorillonitic and allophanic clays in the near-surface sediments of Chalco Valley, Mexico: Analysis of contributing factors to land subsidence, Proceedings of the 7th International Symposium on Land Subsidence SISOLS 2005, 23–28 October 2005, Shanghai, China, ISBN: 7-5323-8209-5, Vol. I, 276–285, 2005.

Hillier, S.: Erosion, sedimentation and sedimentary origin of clays, in: Composition and mineralogy of Clay Minerals, edited by: Velde, B., Springer-Verlag, Berlin, Heidelberg, New York, 162–214, ISBN: 3-540-58012-3, 1995.

Lambe, W. T. and Whitman, R. V.: Soil Mechanics, John Wiley and Sons Inc., NY, 553 pp., 1969.

Marsal, R. J. and Mazari, M.: El subsuelo de la Ciudad de México, Instituto de Ingeniería, U.N.A.M. I, II, Mexico, 505 pp., 1959.

Mesri, G., Rokhse, A., and Bonor, B. F.: Composition and compressibility of typical samples of Mexico City clay, Geotechnique, 25, 527–554, 1976.

Peralta y Fabi, R.: Sobre el origen de algunas propiedades mecánicas de la formación arcillosa superior del Valle de México, Simposio sobre Tópicos Geológicos de la Cuenca del Valle de México, Mexico, 282 pp., 1989.

Robinson, R. G.: Consolidation analysis with pore pressure measurements, Géotechnique, 49, 127–132, 1999.

Velde, B.: Origin and Mineralogy of Clays, Clays and the environment, Springer-Verlag, Berlin, Heidelberg, New York, 334 pp., ISBN: 3-540-58012-3, 1995.

Wada, K.: Minerals formed and mineral formation from volcanic ash by weathering, Chem. Geol., 60, 17–28, 1987.

Warren, C. J. and Rudolph, D. L.: Clay minerals in basin of Mexico lacustrine sediments and their influence on ion mobility in groundwater, J. Contam. Hydrol., 27, 177–198, 1997.

Wesley, L. D.: Consolidation behaviour of allophane clays, Géotechnique, 51, 901–904, 2001.

Zeevaert, L.: Estratigrafía y problemas de ingeniería en los depósitos de arcilla lacustre de la Ciudad de México, Memoria del Congreso Científico Mexicano, 5, 58–70, 1953.

Proc. IAHS, 372, 281–284, 2015
proc-iahs.net/372/281/2015/
doi:10.5194/piahs-372-281-2015

Open Access
PIAHS

The effect of the depth and groundwater on the formation of sinkholes or ground subsidence associated with abandoned room and pillar lignite mines under static and dynamic conditions

Ö. Aydan[1] **and T. Ito**[2]

[1]Department of Civil Engineering and Architecture, University of the Ryukyus, Nishihara, Japan
[2]Toyota National College of Technology, Toyota, Aichi, Japan

Correspondence to: Ö. Aydan (aydan@tec.u-ryukyu.ac.jp)

Published: 12 November 2015

Abstract. It is well known that some sinkholes or subsidence take place from time to time in the areas where abandoned room and pillar type mines exist. The author has been involved with the stability of abandoned mines beneath urbanized residential areas in Tokai region and there is a great concern about the stability of these abandoned mines during large earthquakes as well as in the long term. The 2003 Miyagi Hokubu and 2011 Great East Japan earthquakes caused great damage to abandoned mines and resulted in many collapses. The author presents the effect of the depth and groundwater on the formation of sinkholes or ground subsidence associated with abandoned room and pillar lignite mines under static and dynamic conditions and discusses the implications on the areas above abandoned lignite mines in this paper.

1 Introduction

Some caving, collapses and subsidence take place from time to time in the areas where abandoned room and pillar type mines exist. If the area is urbanized, it presents a great concern to the municipalities as well as to the people living in those areas. Furthermore, if the seismicity of areas is high, the risk of collapses due to earthquakes may amplify those concerns. The 2003 Miyagi Hokubu and the 2011 Great East Japan earthquakes clearly indicated that such areas may experience roof collapses and subsidence as well as the ejection of underground waters from the mines (Aydan and Kawamoto, 2004; Aydan and Tano, 2012). The author has been involved with the stability problems of some room and pillar abandoned mines in Tokai Region of Japan and he has been investigating their performance and responses in the long term as well as during earthquakes.

The backfilling of the abandoned lignite mines of room and pillar type is considered and implemented as a countermeasure against the formation of sinkholes or surface subsidence. The major issue is the depth of the abandoned lignite mines for the implementation of backfilling. When abandoned mines are deep, the surface effects of failed abandoned mines may be quite small due to bulking of the failed overburden rock. It is also claimed that if the mines are submerged in groundwater, the degradation of surrounding rock becomes negligible and there is no necessity of backfilling. Although such a claim may be true under static conditions, the model tests and observations show that the formation of sinkholes or surface ground subsidence may occur in the areas of abandoned room and pillar lignite mines due to earthquakes (Aydan and Kawamoto, 2004; Aydan and Tano, 2012).

The author is concerned with the depth and ground water on the formation of sinkholes or ground subsidence in the areas of abandoned lignite mines under static and dynamic conditions in this paper. The possible causes of sinkhole formation or ground subsidence under static and dynamic conditions in the areas of abandoned lignite mines have been summarized. Some case history data from Japan and have been compiled and presented as empirical guidelines for the risk assessment of abandoned lignite mines with respect to

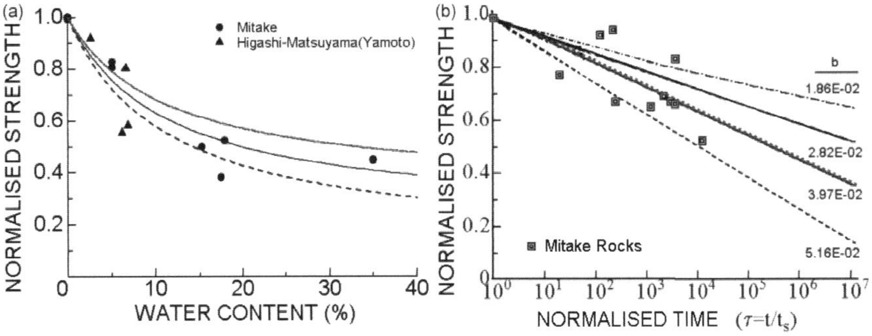

Figure 1. Degradation of strength of rocks of abandoned mines in Mitake with water content and long-term creep loading (b is empirical coefficient in the empirical long-term strength relation). **(a)** Effect of water content. **(b)** Long-term strength.

Figure 2. Large sinkholes occurred recently over the abandoned lignite mine in Mitake town.

sinkholes or surface ground subsidence. Furthermore, some suggestions on the quality of backfilling are presented and discussed.

2 Degradation characteristics of surrounding rocks

The surrounding rocks and lignite itself in abandoned lignite mines are sedimentary rocks and very prone to degradation due to cyclic variations of water content during absorption and desorption. Such variations cause flaking of surrounding rocks from roof and pillars of lignite seam and results in the reduction of the support area of pillars and thinning of roof layers. Furthermore, the increase of water content results in the reduction of the compressive strength and elastic modulus of soft ocks surrounding (Fig. 1a).

The pillars are subjected to sustained load due to overburden so that the creep phenomenon of surrounding rocks (Fig. 1b) and lignite seam is another cause of the formation of sinkholes or surface subsidence. When the lignite mines are abandoned, they may become wholly or partially submerged in groundwater. The authors experimentally showed that the long-term strength of lignite and surrounding soft rocks drastically reduce upon submergence in groundwater.

3 Case history data on room and pillar mines

New constructions over the abandoned mines and earthquakes from time to time cause additional static and dynamic forces on the rock masses of pillars and roofs. As a result of these actions, partial collapses of roof layers, the cross-sectional decrease of supporting pillars and even the failure of pillars occurs eventhough their effects may not be observed on the ground surface (Aydan and Tano, 2012). There are some occurrences of sinkholes and large ground subsidence from time to time in Mitake town (Fig. 2).

The authors compiled various data on the formation of sinkholes and subsidence over abandoned mines reported in literature. Sinkholes are generally caused by the degradation of roof layers under different environmental conditions, groundwater flow and vibration effects. On the other hand, subsidence involves failure of rock pillars, and overburden depth and degradation of rock materials are critical parameters. Figure 3a shows the case history data plotted in the space of overburden (H) and settlement (S) normalized by the seam extraction height (h). The majoriy of sinkholes are observed when the overburden is less than 10–15 times the seam extraction height. This implies that the overburden would be 20–30 m for a seam extraction height of 2 m. However, subsidence may occur at any depth, provided that pillars fail and may initiate after longer time following the termination of mining. There are some examples of delayed failures after about 40–130 years.

Aydan and Tano (2012) reported some examples of the damage to abandoned mines in Tokai and Tohoku regions by several earthquakes with a moment magnitude greater than 6. Figure 3b shows the relation between the moment mag-

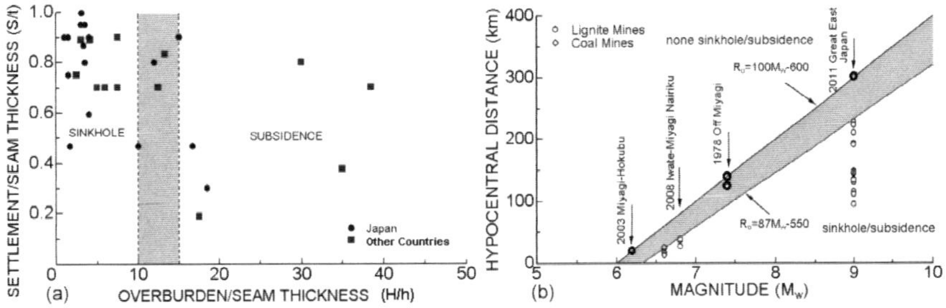

Figure 3. (**a**) Normalized seam height versus normalized settlement; (**b**) moment magnitude of earthquakes versus the hypocentral distance of sinkhole/subsidence localities.

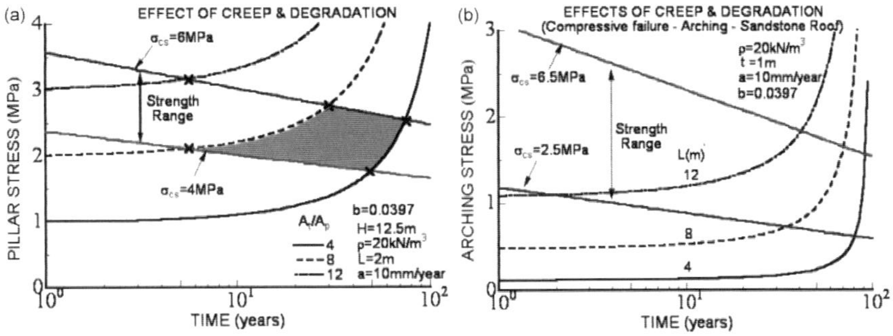

Figure 4. Effects of degradation and creep loading on failure time of pillars and roofs (σ_{CS}: short-term uniaxial compression strength; ρ: unit weight of surrounding rock; A_t: supported area; A_p; pillar height; L: room width; a: empirical coefficient of degradation function; b: empirical coefficient of long-term strength function, see Aydan et al., 2006 for details). (**a**) Pillar. (**b**) Roof consisting sandstone.

nitude of the earthquake and hypocentral distances of sinkholes/subsidence localities. It is of great interest that an envelope defined by linear relations can be inferred between damaged and undamaged regions of the moment magnitude versus hypocentral distance reation. Figure 3 may serve as a guideline to assess the risk of sinkhole or large subsidence due to abandoned mines and quarries exploited using the room and pillar method as a function of earthquake magnitude. However, there is still a great necessity to compile more data to confirm the validity of the proposed relation. Besides the possibility of caving of abandoned mines during earthquakes due to either pillar failure and/or roof failure, the groundwater in mines may present additional effects on the susceptibility pf submerged abandoned mines. These effects may be observed as sloshing, which may further weaken the surrounding rockmass and cause additional collapses.

4 Considerations on sinkhole formation and ground subsidence

4.1 Static condition

The pillar failure would fundamentally induce ground subsidence while the failure of roof may induce either sinkholes or subsidence depending upon the depth of lignite seam. The tributary area method is simple yet effective method to assess the overall stability of the pillars. Aydan et al. (2006) proposed a procedure for evaluating the stability of pillars subjected to the effect of degradation of lignite under the cyclic wetting and drying and the creep loading. Figure 4a shows both effects of long-term strength and degradation rate. Compared with the failure time of pillar under the condition of creep loading only, the failure time of pillars become shorter. The shaded region (Fig. 4a) corresponds to the estimated conditions of the abandoned mine that the authors have been investigating in Mitake area (Aydan et al., 2006). For the lower bound strength range, the stability problems of pillars would occur after 5.6–50 years and after 30–75 years for the upper bound strength range.

The stability of roof layers can be checked against various forms of failure. Although roof layers may crack due to bending stresses, their ultimate failure would be governed by arching phenomenon (although shear failure may be conceived, it is not generally likely to occur) within the roof layers. If the fiber stress exceeds the compressive strength at the crest of the arch within the roof layer, the roof layer will collapse. The effect of degradation can be imposed on the thickness of roof layer while the effect of long-term strength can be imposed on the compressive strength of roof layers.

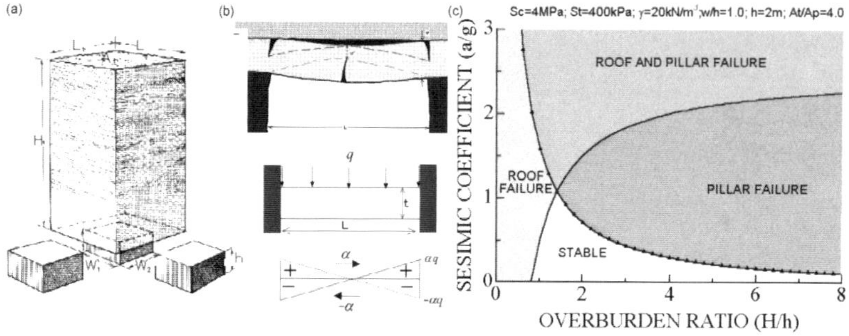

Figure 5. Mathematical model and example of computation for the stability of pillars and roof layers under dynamic conditions. See Aydan et al. (2006) for details. **(a)** Pillar. **(b)** Roof. **(c)** Example of computation.

Figure 4b shows some computational results for the effect of creep loading and degradation for sandstone roof.

Groundwater in abandoned mines can be considered to be preventing the degradation of pillars and roof layers. Furthermore, the effective vertical stress acting on the pillar would decrease due to the fluid pressure support provided by groundwater and pillar stress would be given as

$$\sigma_p{}' = \frac{1}{1-e}(\gamma_r H - \gamma_w H_w), \quad (1)$$

where γ_r, γ_w, H, H_w and e are unit weight of rock, water, overburden, water depth and excavation ratio, respectively. If abandoned mine is fully submerged and water height is equivalent to overburden height, the effective stress acting on the pillar would be decreased by about 43 %. However, as noted from Fig. 1a, the strength of the surrounding lignite layer decreases about 60–75 % under fully saturated conditions. In other words, groundwater would further decrease the overall stability of abandoned lignite mines.

4.2 Dynamic condition

Aydan et al. (2006) proposed a practical method for evaluating the stability of pillars and roof of abandoned room and pillar lignite mines using the seismic coefficient approach. Figure 5a and b show the assumed models for the dynamic stability of pillars and roofs. In this approach, the pillar was assumed to fail under the maximum compressive stress. The load condition under horizontal shaking is assumed to consist of gravitational load inducing the bending stresses and linearly varying axial stress along the roof axis from tension to compression due to horizontal shaking. On the basis of this assumption, the seismic coefficient at the time of roof layer failure by bending is obtained. Figure 5 shows computed diagrams for the relation between overburden ratio and seismic coefficient for various failure modes for the chosen parameters shown in the same figure. These results indicate that the shallow mines are prone to roof failure while the deeper mines are prone to pillar failure for actual strong ground motions during earthquakes.

5 Conclusions

The authors considered the effect of the depth and groundwater on the formation of sinkholes and ground subsidence associated with abandoned lignite mines on under static and dynamic conditions and presented some case history data under both static and dynamic conditions. It is shown that sinkhole formation or subsidence of abandoned room and pillar mines, which may take some time, can definitely occur. While sinkhole formation is generally limited to depths 10–15 times the extracted seam height, subsidence can occur over abandoned mines of any depth provided that the pillars fail under both static and dynamic conditions. While the groundwater decreases vertical stress acting on the pillars under fully submerged conditions, the overall stability of the abandoned room and pillar mines would be decreased due to the reduction of strength of the lignite seam and surrounding rocks under saturated condition.

Acknowledgements. The authors sincerely acknowledge the advices on this study by Emeritus Professor Toshikazu Kawamoto of Nagoya University.

References

Aydan, Ö. and Kawamoto, T.: The damage to abandoned lignite mines caused by the 2003 Miyagi-Hokubu earthquake and some considerations on its causes, 3rd Asian Rock Mechanics Symposium, 30 November–2 December 2004, Kyoto, 525–530, 2004.

Aydan, Ö. and Tano, H.: The damage to abandoned mines and quarries by the Great East Japan Earthquake on March 11, 2011, in: Proceedings of the International Symposium on Engineering Lessons Learned from the 2011 Great East Japan Earthquake, 1–4 March 2012, Tokyo, Japan, 981–992, 2012.

Aydan, Ö., Daido, M., Ito, T., Tano, H., and Kawamoto, T.: Instability of abandoned lignite mines and the assessment of their stability in long term and during earthquakes, 4th Asian Rock Mechanics Symposium, Singapore, Paper No. A0355 (on CD), 2006.

Proc. IAHS, 372, 285–290, 2015
proc-iahs.net/372/285/2015/
doi:10.5194/piahs-372-285-2015

Physical experiments of land subsidence within a maar crater: insights for porosity variations and fracture localization

M. Cerca[1], **L. Rocha**[2], **D. Carreón-Freyre**[1], **and J. Aranda**[1]

[1]Centro de Geociencias, Universidad Nacional Autónoma de México, Campus Juriquilla, Querétaro, Qro., 76230, México

[2]Posgrado en Ciencias de la Tierra, Centro de Geociencias, Universidad Nacional Autónoma de México, Campus Juriquilla, Querétaro, Qro., 76230, México

Correspondence to: M. Cerca (mcerca@geociencias.unam.mx)

Published: 12 November 2015

Abstract. We present the results of a series of physical models aiming to reproduce rapid subsidence (at least 25 m in 30 years) observed in the sediments of a maar crater caused by extraction of groundwater in the interconnected adjacent aquifer. The model considered plausible variations in the geometry of the crater basement and the measured rate of groundwater extraction (1 m per year in the time interval from 2005 to 2011) in 15 wells located around the structure. The experiments were built within a rigid plastic bowl in which the sediments and rocks of the maar sequence were modeled using different materials: (a) plasticine for the rigid country rock, (b) gravel for the fractured country rock forming the diatreme fill and, (c) water saturated hollow glass microbeads for the lacustrine sedimentary fill of the crater. Water table was maintained initially at the surface of the sediments and then was allowed to flow through a hole made at the base of the rigid bowl. Water extraction provoked a sequence of gentle deformation, fracturing, and faulting of the surface in all the experiments. Vertical as well as lateral displacements were observed in the surface of the experiments. We discuss the results of 2 representative models. The model results reproduced the main geometry of the ring faults affecting the crater sediments and helps to explain the diversity of structures observed in relation with the diatreme geometry. The surface of the models was monitored continuously with an optical interferometric technique called structured light projection. Images collected at nearly constant time intervals were analyzed using the ZEBRA software and the obtained interferometric pairs permitted to analyze the full field subsidence in the model (submilimetric vertical displacements). The experiments were conducted at a continuous flow rate extraction and show a also a linear subsidence rate. Comparison among the results of the physical models and the fault system associated to subsidence in the maar show that fault geometry in the sedimentary sequence imitates closely the geometry of the volcanic basement.

1 Introduction

Rincón de Parangueo is a Quaternary volcanic maar, a crater formed by a phreatomagmatic eruption, located in the northern portion of the Michoacán-Guanajuato volcanic field in the central sector of the Trans Mexican Volcanic Belt (Fig. 1, location of the maar: 20°25.839′ N, −101°14.917′ W). Volcanic maars are craters located above a feeder magmatic dike that poses into contact magma with groundwater causing a gaseous explosion and resulting in a rock filled fracture with a conic shape, called diatrema (Lorenz, 1986). Although the literature on maars is extensive the studies of post-eruption subsidence in the lacustrine sequences filling the craters are rare (e.g., Suhr et al., 2006; Pirrung et al., 2008). Rincón de Parangueo was one of four maar craters nearby the locality of Valle de Santiago village that used to have a perennial lake and nowadays is partially occupied by a brine lake. During the 1980's progressive desiccation began as a consequence of water overdraft from the regional aquifer that used to feed the lake (Escolero-Fuentes and Alcocer-Duran, 2004;

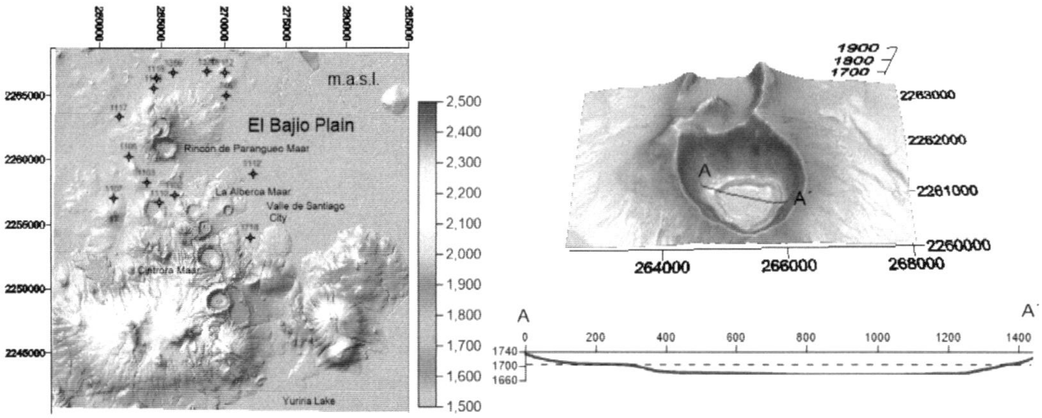

Figure 1. Digital elevation model obtained by LIDAR data from INEGI (Institute of Informatics and Geography; Mexico) showing the volcanic cluster in the area of Rincón de Parangueo maar. Perspective view of the crater and the annular fault system affecting the bottom. Topographic profile crossing the crater showing the initial lake surface (dashed line) and the present day leveling (red line).

Aranda-Gómez et al., 2013). Detailed geologic mapping in the sediments of the lake bed shows clear evidence of rapid active subsidence contemporaneous to desiccation (Aranda-Gómez et al., 2010, 2013; Aranda-Gómez and Carrasco-Nuñez, 2014; Cerca et al., 2014). Deformation is characterized by a main ring-shaped normal fault accompanied by gravitational sliding and gliding of mud blocks, and folding at the foot of the slides. Sinking of the maar crater is associated with the formation of vertical fractures and lateral mass movements to the depocenter, resulting in the formation of tensile fractures. Likewise, normal faulting exerts a major influence in the location of today depocenter.

Sinking at the bottom of the crater produced a ring-fault system which is exposed near the former coast of the lake, on a pair of parallel 15 m high topographic scarps. Attitudes of laminations in the mud exposed in the scarp are consistent with down sagging of the sediment caused by sinking of the lake's depocenter. Tilt angles of the laminations vary between 10 and 35° towards the basin. Normal faults exposed at the topographic scarp produced in some areas deformation domains that are consistent with the presence of listric detachment structures, which produced rollover anticlines with axes located near the former coast of the lake. Lake sediments are composed by a sequence presenting rhythmic layering with changes of colors between beige and dark brown; beige layers are composed carbonate micritic mud with crystals of 2–5 µm (Kienel et al., 2009). The dark brown layers are composed by clay, silt, and organic material. Also interbedded can be observed more prominent layers with black coloration constituted mainly silt and fine sand. In cores, the lower part of the observed sequence is composed of deformed layers affected by fractures with a small displacements. Mechanically, the whole sequence is characterized by low compressible silty sediments, and the gravimetric water content increases gradually with depth up to more than 40 %.

Water withdrawal in the aquifer system adjacent to the volcanic field has been pointed out as the cause for deformation and desiccation of the lake since the 1980's decade (Escolero-Fuentes and Alcocer-Duran, 2004; Aranda-Gómez et al., 2013; and references therein). Fragmentation of the country rock due to diatreme formation is likely to be contained within the maar–diatreme and less likely to propagate down to deeper levels into the feeder dike. Nevertheless, alignment and overlap of the numerous craters at the study area indicates the presence in the subsurface of multiple parallel and orthogonal feeding dykes. Coalescence of the shallow structures suggested by vent distribution might be the cause of a high hydraulic connectivity below the volcanic edifices with the regional aquifer. Faults and fractures within the country rock enhance the hydraulic conductivity and might serve as drainage lines into the regional aquifer system. Thus, the maar may be divided into two significant hydrogeological units, a deforming lacustrine sedimentary unit, acting as an aquitard and, the high fractured diatreme hydraulically connected with the regional aquifer system. Nowadays groundwater extraction and evaporation exceed the inflow of groundwater into the crater. Therefore, the bottom of the crater will either be dry except for runoff from precipitation or develop into a brine lake where accumulation of salts will take place in the depocenter. The drawdown of the groundwater table will continue indefinitely because the region depends on groundwater for human consumption.

The present day elevation at the bottom of the crater (1868 m a.s.l.) is lower than the adjacent Bajio plain (1710–1730 m a.s.l). The variations in the piezometric levels were recorded by the National Water Commission (CNA) in selected wells around the Parangueo maar for the time period

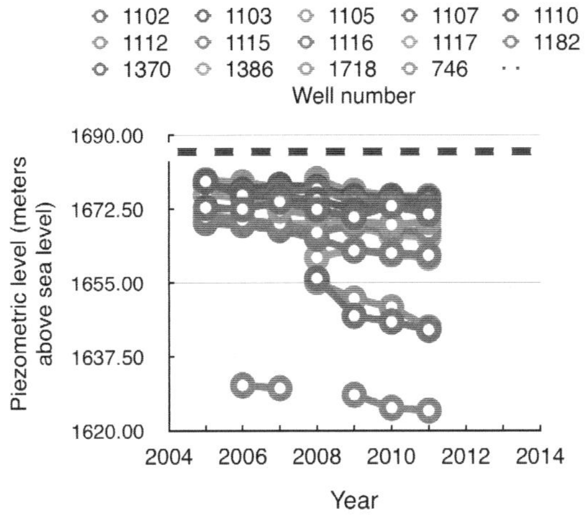

Figure 2. Piezometric level evolution for the time interval between 2005 and 2011 for selected extraction wells measured around the Rincón de Parangueo maar. Location of the wells is indicated in Fig. 1.

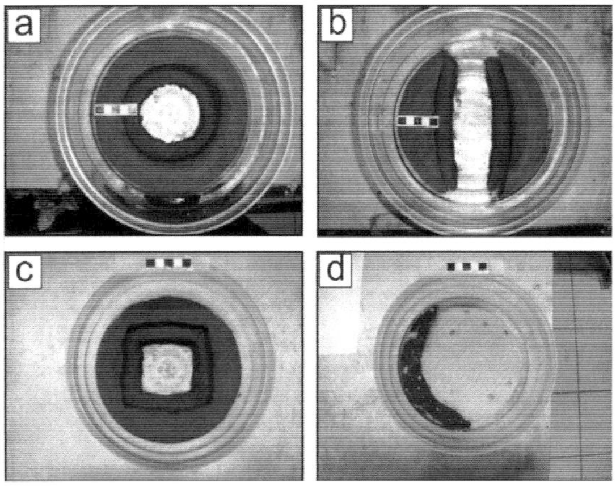

Figure 3. Photographs of the experimental setup showing the plasticine molded to produce variations in the geometry of the rigid country rock. The bowl was then filled to the level of the plasticine with gravel and covered with hollow glass microbeads to obtain a plain surface. **(a)** Ring faults circular, **(b)** parallel faults simulating a magmatic dike shape, **(c)** ring faults square, **(d)** asymmetric faults set up.

between 2005 and 2011. The observations were made twice in a year at both dry and raining seasons, we present here the measurements made during the dry season (Fig. 2). All piezometers record a similar depleting evolution of 1 m per year, but the cluster of wells 1370, 1182 and 746, located northeast of the Rincón de Parangueo maar record a deeper piezometric level. In summary, the piezometric data suggest an significant hydraulic gradient towards the northeast and the data can be interpolated to a lineal decrease in the water table for the last 30 years. In this work we aim at reproducing the subsidence observed in the Rincón de Parangueo maar by constructing a series of scaled physical models.

2 Experimental subsidence bowl

Experiments were conducted using a plastic bowl simulating the overall half-sphere shape of the maar crater. Plasticine was used to mold and vary systematically the different geometries of volcanic basement (rigid country rock) assumed to be below the sedimentary sequence. Variations in the geometry included ring circular faults, ring square faults, parallel faults, and asymmetric basement (Fig. 3). The internal part of the bowl was filled with gravel to simulate the highly fractured country rock in the volcanic conduct. Posteriorly, the bowl was filled up to an horizontal level with a fine grain size and well sorted, slightly cohesive granular media (hollow glass microbeads) completely saturated in water, simulating the initial conditions of the lake maar in the 1980's when the phreatic level reached the surface. In nature, porosity changes during subsidence are related to effective stresses. The water head decrease in the maar was of about 30 m in the last 35 years. For scaling we have measured porosity of samples and assumed the initial porosity. Experiments are a simplification of nature and assume that: (1) the lacustrine material is relatively homogeneous, (2) the voids of the lacustrine sediments are filled with water below the piezometric surface and, (3) changes in the effective stress produced by the drop of piezometric level are the main cause for the subsidence in the crater maar. The water table is also assumed to be parallel to the surface of the lacustrine sediments at the beginning of the process in nature and in the model

$$\Delta\sigma^*_{\text{ef}} = \frac{\Delta\sigma_{\text{ef}}\text{model}}{\Delta\sigma_{\text{ef}}\text{nature}}, \tag{1}$$

where σ_{ef} is the change in the effective stress and σ^*_{ef} is the relation among the variations in the effective stresses in model and nature, in this case the dimensionless scaling ratio is of the order of 5.33×10^{-4}. If the total stress is constant through time the differences in the effective stresses are related to the pore pressure drop, which in turn is proportional to the change in the groundwater level.

The experimental time for all models was about 20 min with a flow rate maintained constant at $1.57 \times 10^{-7}\,\text{m}^3\,\text{s}^{-1}$ that was induced by opening of a 2 mm in diameter tube at the base of the bowl. Subsidence in the physical models was monitored at constant time intervals of 15 s using an optical system that allows obtaining subtle changes (submilimetric) in the experimental relief using the ZEBRA software (based on the work by Barrientos et al., 2008). A structured light (black and white fringes) pattern was projected on the surface of the model and the digital images recorded by a CCD sensor. Comparison between pairs of images (deformed state

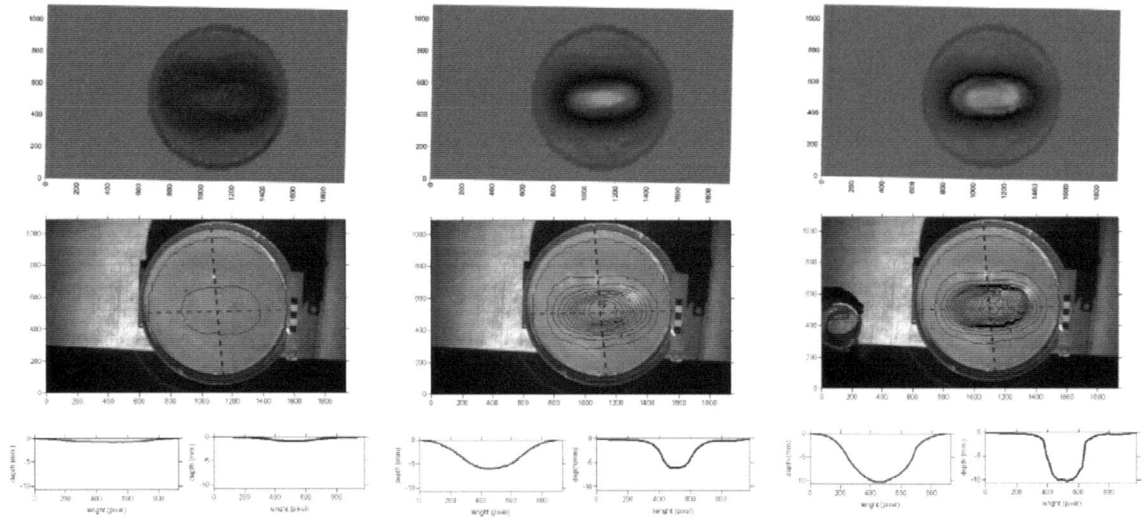

Figure 4. Example of the model results at the start, middle and end of the experiment. The images show the digital elevation model obtained with the ZEBRA software, a photograph of the model with overlapped contours and the topographic profile obtained in two orthogonal directions.

and reference image) were analyzed by the ZEBRA software using optical interferometric methods.

3 Results

All of the experimental runs showed a comparable evolution: (a) formation of an initial gentle depression without fracture formation in the central part of the bowl at the beginning of the water level decrease; (b) nucleation of small and isolated fractures after 5 min of the experiment run; (c) fracture coalescence to form a continuous ring fault with a geometry similar to the basement; (d) propagation of the fractures toward the external and/or internal part of the model. Differences in the model results depended mainly on the geometry of the diatreme.

Here, we discuss the results of two representative experiments. The first experiment discussed was of the type b in Fig. 3, parallel faults representing a diatreme with the shape of a dike. A central elliptical depression formed in the first 5 min, after 6 min of flow the parallel faults formed at the surface of the model. At minute 9 the parallel faults linked to form the ring fault with an elliptical shape. After 13 min the model ended its deformation; nevertheless some images were recorded after 30 min to corroborate that there was no further deformation. Figure 4 represents the evolution of the experiment in the digital elevation model, photographs with contour levels, and two topographic profiles obtained from the DEM's.

Subtle changes in the subsidence rate and geometry can be analyzed in the topographic profiles that were constructed using 25 images processed against a reference image of the initial experimental conditions. Figure 5 shows differences in the evolution of subsidence in the profile parallel to the basement faults (RED) and the profile orthogonal to the parallel faults (BLUE). Localization of deformation occurs in the BLUE profile at an earlier time (500 s) and differences in compaction are attributable to the presence of the rigid basement and thinner sedimentary deposits. The inset of Fig. 5 (RED) shows the subsidence model obtained for the center of the model that follows a nearly lineal behavior.

The second experiment was designed to explore experimentally the influence of an asymmetric basement and the influence of tilted surfaces that could act as detachment planes during subsidence. The setup considered the structural differences observed in the crater (Aranda-Gómez and Carrasco-Nuñez, 2014; Cerca et al., 2014), mainly the formation of rollover folds along listric fault surfaces and the formation of shortening domes. This deformational features suggest an important lateral spreading of the lacustrine sequences during subsidence. Another experimental difference was that the second sedimentary layer was thicker than in other model runs. The overall evolution of the model was similar to the first experiment, but in this case the lateral sliding of the material formed extensional fractures in a zone of the ring fault and small folds toward the depocenter, implying that lateral displacements are important to accommodate deformation. Figure 6 shows a photographic sequence of the second experimental setup, the ellipse indicates the zone of folding and the nucleation of the fractures; as well as the initial depression formed.

Obtaining an accurate digital elevation model of the experiments allowed to perform further analysis of the data.

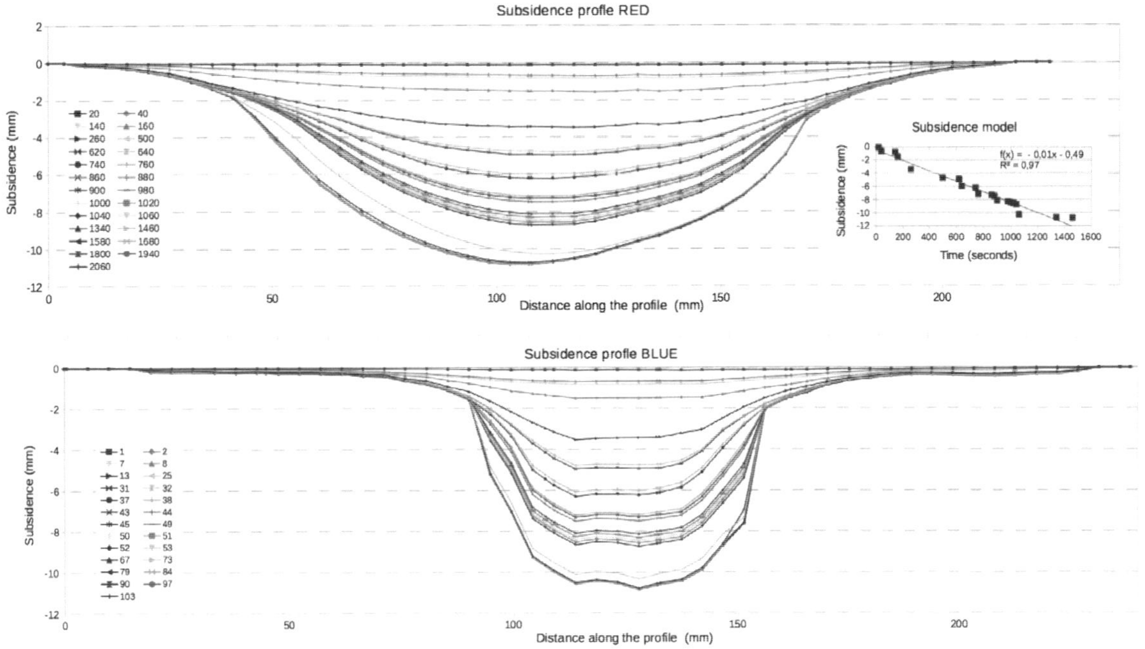

Figure 5. Topographic profiles obtained in the model in the direction: **(a)** parallel to the basement faults (RED) and, **(b)** the profile orthogonal to the parallel faults (BLUE).

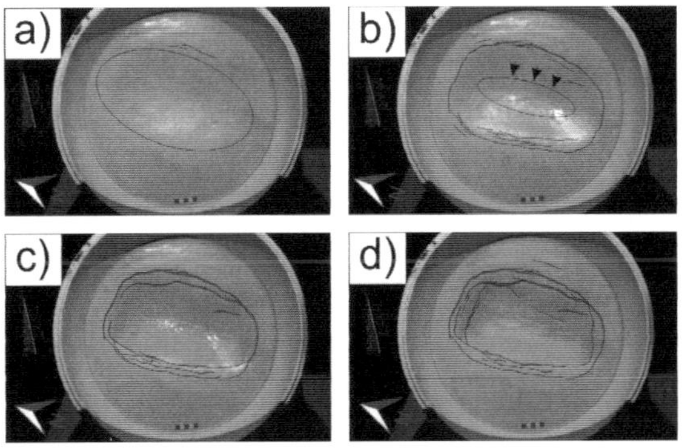

Figure 6. Photographic sequence of the second experiment showing the stages of deformation observed during desiccation of the model crater: **(a)** formation of a depression, **(b)** formation of fractures at the border of the depression and the displacement of material toward the depocenter, **(c)** growing and coalescence of faults, **(d)** desiccation of the model and formation of shortening structures in the depocenter.

For instance, the development of slopes after desiccation of the crater is important for the formation of extensional structures. In Fig. 7 we show the analysis of the subsidence and slope development of the images c and d of Fig. 6. The opening of the fractures and the formation of small folding along the axis of the elliptical depression suggest lateral spreading of the sediments in the late stages of deformation.

The comparison of model results with the deformation observed in the Rincón de Parangueo maar suggest that faults formed above the geometric borders of the rigid country rock. The ring fault pattern of faults in the mar suggest that the diatreme has a non uniform circular geometry. The presence of at least two major fault scarps in nature may indicate that the diatreme has two fault steps in the country rock.

4 Conclusion

Subsidence in the experiments formed patterns directly related to the geometry of the rigid country rock modeled below the sedimentary cover. Thus varying systematically the geometry of the diatreme and analyzing the results give important insights on the evolution of deformational features associated with land subsidence. Fractures and faults formed nearly above the border of the rigid basement, its position depending on the inclination of the border. The subsidence that can be generated in a crater is directly proportional to the thickness of the sedimentary lacustrine sequence. Groundwater extraction in the regional aquifer may trigger the deformation and needs to be regulated for a sustainable management of the crater lake. Finally, scaled physical experiments provided an important tool for understanding of subsidence mechanisms, for testing hypothesis of the causes of defor-

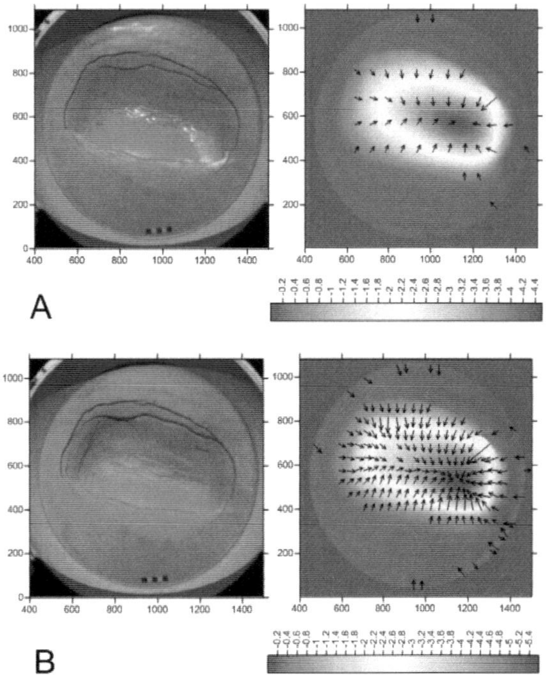

Figure 7. Interferometry analysis of selected photographs from Fig. 6 (c – growing of faults and d – dessication). **(a)** DEM of the image at the moment when the water level reached the bottom of the crater, color scale represent the subsidence in centimeters and the arrows indicate the magnitud and direction of the slope, the in-plane scale is in pixels. **(b)** DEM of the image at a later desiccation time and deformation (color scale and arrows as above), note that the development of faulting has an effect in slope.

mation of the terrestrial surface, and for the development of monitoring techniques.

Acknowledgements. Ricardo Carrizosa of the Laboratorio de Mecánica de Geosistemas (LAMG) is thanked for his help with the experimental setup.

References

Aranda-Gómez, J. J. and Carrasco-Nuñez, G.: The Valle de Santiago maars, México: the record of magma-water fluctuations during the formation of a basaltic maar (La Alberca) and active post-desiccation subsidence at the bottom of a maar lake (Rincón de Parangueo): Intra-conference field trip guidebook, 5th International Maar Conference, 17–22 November, Centro de Geociencias, Universidad Nacional Autónoma de México, Querétaro, México, 30 pp., 2014.

Aranda-Gómez, J. J., Levresse, G., Pacheco-Martínez, J., Ramos-Leal, J. A., Carrasco-Núñez, G., Chacón-Baca, E., González-Naranjo, G., Chávez-Cabello, G., Vega-González, M., Origel-Gutiérrez, G., and Noyola-Medrano, C.: Active subsidence at the bottom of a recently desiccated crater-lake and its environmental impact: Rincón de Parangueo, Guanajuato, México: Field trip guidebook, Eighth International Symposium on Land Subsidence, Universidad Nacional Autónoma de México, Centro de Geociencias, Querétaro, Qro., México, 1–48, 2010.

Aranda-Gómez, J. J., Levresse, G., Pacheco-Martínez, J., Ramos-Leal, J. A., Carrasco-Nuñez, G., Chacón-Baca, E., González-Naranjo, G. A., Chávez-Cabello, G., Vega-González, M., Origel, G., and Noyola-Medrano, C.: Active sinking at the bottom of the Rincón de Parangueo Maar (Guanajuato, México) and its probable relation with subsidence faults at Salamanca and Celaya, Boletín de la Sociedad Geológica Mexicana, 65, 169–188, 2013.

Barrientos, B., Cerca, M., García-Márquez, J., and Hernández-Bernal, C.: Three-dimensional displacement fields measured in a deforming granular-media surface by conbined frige projection and speckle photography, J. Opt. A-Pure Appl. Op., 10, 10 pp., 2008.

Cerca, M., Carreón-Freyre, D., Aranda-Gómez, J. J., and Rocha-Treviño, L.: GPR profiles for characterizing subsidence deformation in lake sediments within a mar crater, 15th International Conference on Ground Penetrating Radar GPR2014, Brussels, Belgium, 30 June–4 July, 4 pp., 2014.

Escolero-Fuentes, O. A. and Alcocer-Durand, J.: Desecación de los lagos cráter del Valle de Santiago, Guanajuato, in: El agua en México vista desde la Academia, edited by: Jiménez, B., Marín, L., Morán, D., Escolero, O., and Alcocer, J., Academia Mexicana de Ciencias, México, D.F., 99–116, 2004.

Kienel, U., Wulf Bowen, S., Byrne, R., Park, J., Böhnel, H., Dulski, P., Luhr, J. F., Siebert, L., Haug, G. H., and Negendank, J. F. W.: First lacustrine varve chronologies from Mexico: impact of droughts, ENSO and human activity since AD 1840 as recorded in maar sediments from Valle de Santiago, J. Paleolimnol., 42, 587–609, doi:10.1007/s10933-009-9307-x, 2009.

Lorenz, V.: On the growth of maars and diatremes and its relevance to the formation of tuff rings, B. Volcanol., 48, 265–274, 1986.

Pirrung, M., Büchel, G., Lorenz, V., and Treutler, H.-C.: Post-eruptive development of teh Ukinrek East Maar since its eruption in 1977 A.D. in the periglacial area of south-west Alaska, Sedimentology, 55, 305–334, 2008.

Suhr, P., Goth, K., Lorenz, V., and Suhr, S.: Long lasting subsidence and deformation in and above maar-diatreme volcanoes – a never ending story, Z. Dtsch. Ges. Geowiss., 157, 491–511, 2006.

Proc. IAHS, 372, 291–296, 2015
proc-iahs.net/372/291/2015/
doi:10.5194/piahs-372-291-2015

Open Access
PIAHS

Continuous monitoring of an earth fissure in Chino, California, USA – a management tool

M. C. Carpenter

formerly at: U.S. Geological Survey, Tucson, Arizona, USA

Correspondence to: M. C. Carpenter (mccarp@dakotacom.net)

Published: 12 November 2015

Abstract. Continuous measurements of deformation have been made in Chino, California across an earth fissure and nearby unfissured soil since 2011 in two buried, horizontal, 150 mm pipes, 51 m long, which are connected by sealed boxes enclosing vertical posts at mostly 6 m intervals. Horizontal displacements and normal strain are measured in one line using nine end-to-end quartz tubes that are attached to posts and span fissured or unfissured soil. The free ends of the tubes are supported by slings and move relative to the attachment post of the next quartz tube. Linear variable differential transformer (LVDT) sensors measure the relative movements. Five biaxial tilt sensors were also attached to selected posts in that line. Relative vertical movement was measured at nine locations along the line in the second pipe using low-level differential pressure sensors. The second pipe is half full of water giving a free water surface along its length. Data are recorded on a Campbell CR10 using multiplexers.

The quartz-tube horizontal extensometers have exhibited more than 3 mm of predominantly elastic opening and closing in response to about 32 m of seasonal drawdown and recovery, respectively, in an observation well 0.8 km to the south. The nearest production well is 1.6 km to the west. The horizontal strain was 5.9×10^{-5} or 30 % of the lowest estimate of strain-at-failure for alluvium. Maximum relative vertical movement was 4.8 mm. Maximum tilt in the fissure zone was 0.09 arcdeg while tilt at a separate sensor 100 m to the east was 0.86 arcdeg, indicating a wider zone of deformation than is spanned by the instrumentation. High correlation of horizontal displacements during drawdown, and especially recovery, with change in effective stress supports differential compaction as the mechanism for earth-fissure movement.

The continuous measurements of horizontal strain coupled with water-level fluctuations and vertical borehole extensometry can provide a real-time adaptive management tool for restricting pumping if strain approaches the lower limit of strain-at-failure or a stress-strain curve deviates from the previous mostly elastic regimen.

1 Introduction

A zone of earth fissures opened as early as 1974 on the California Institute for Men prison in Chino, California. Renewed fissuring occurred in 1991 on prison grounds and northward for 1.6 km in a zone as much as 150 m wide (Stewart et al., 1998). In the 1990s, the fissure split a house and drained a liquid manure pond at a dairy, possibly directly contaminating the shallow aquifer. The house was condemned and destroyed. As a result of management of groundwater pumping by the Chino Basin Watermaster, fissuring has been minimal since the 1990s and limited to asphalt parking lots and vacant properties. The structural setting of the Chino earth fissure is a north-south, strike-slip, fault-zone, groundwater barrier that bounds a subsidence bowl to the west and concentrates differential subsidence over a limited horizontal distance (Wildermuth Environmental, 2007).

2 Instrumentation

In October 2011, instruments were installed in a west to east line at 5500 Daniels St., Chino, California, in a vacant lot, which had been the location of the dairy and destroyed house (Fig. 1). The instruments provide a continuous record of horizontal displacements, differential vertical movement, and tilt across the earth fissure and nearby unfissured soil. Two buried, horizontal, 150 mm pipes, 51 m long, are connected by sealed boxes enclosing vertical, 3-m, grounding rod posts driven into the ground at mostly 6 m intervals. Data are recorded on a Campbell CR10 datalogger using multiplexers.

Horizontal displacements are measured in one line using nine end-to-end quartz tubes that are attached to posts and span fissured or unfissured soil. Tubes are 15 mm OD, 13 mm ID. Trans-Tek ±6 mm, 0242-0000 linear variable differential transformer (LVDT) sensors measure the relative displacements. The free ends of the tubes are supported by beaded-chain slings and move relative to the attachment post of the next quartz tube. A tube is attached to a post by means of an aluminum strip that extends behind the end of its attachment post, and the LVDT is attached to a nonmagnetic stainless steel strip that extends beyond the end of its post. The length of aluminum times its coefficient of expansion ($2.4 \times 10^{-5}\,°C^{-1}$) plus the length of stainless steel times its coefficient of expansion ($1.2 \times 10^{-5}\,°C^{-1}$) equals the length of quartz times its coefficient of expansion ($5.7 \times 10^{-7}\,°C^{-1}$). This compensating design reduces temperature effects by as much as an order of magnitude.

Relative vertical movement was measured at nine locations along the line in the second pipe using Honeywell Micro Switch 24PCEFA6D ±34 hPa (±0.5 psi) low-level differential pressure sensors. Differential silicon-strain-gage pressure sensors exhibit a voltage shift in response to a pressure change common to both ports. For low-level sensors used for high precision, a secondary common-mode calibration must be done over the range of barometric-pressure change. An adjustment can be applied or sensors can be selected with identical shifts. The second pipe is half full of water giving a free water surface along its length.

Five Applied Geomechanics (now Jewell Instruments) 902-TH biaxial tilt sensors were attached to selected posts in the quartz-tube line. Temperature sensors incorporated in the tilt sensors provided the temperature measurements for the instrument enclosures. Biaxial tilt sensors provide the easiest information for the extent to which deformation is restricted to a vertical plane orthogonal to the fissure.

Figure 1. Instrument layout at 5500 Daniels horizontal extensometer site, Chino, California. Q1–Q8 are quart-tube horizontal extensometers. LL1-LL10 are liquid-level sensors. E04-N65f are biaxial tilt sensors.

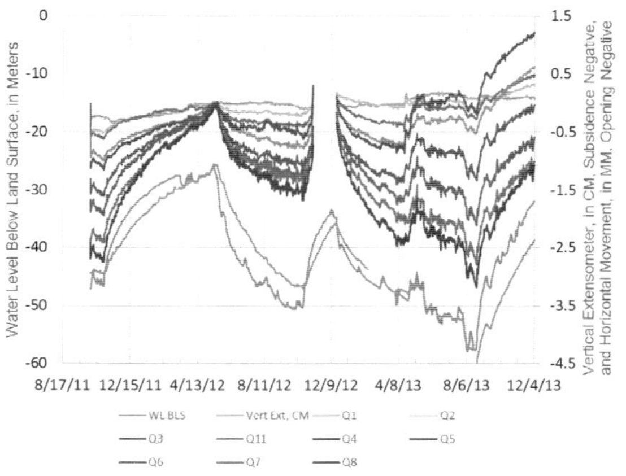

Figure 2. Quartz-tube horizontal movement compared with water-level fluctuation in an observation well and deep-extensometer compaction 0.8 km to the south.

3 Results

From west to east the quartz-tube spans are designated Q1–Q3, Q11, and Q4–Q8 (Fig. 2). Q11 spans the historical fissure zone. Curves represent cumulative displacement along the line. For example, Q8 is the sum of all displacements Q1–Q8. The quartz-tube horizontal extensometers have exhibited more than 3 mm of predominantly elastic opening and closing in response to about 32 m of seasonal drawdown and recovery, respectively, in an observation well 0.8 km to the south. The nearest production well is 1.6 km to the west, screened in a deep confined aquifer at depths of 91–140 m and 152–298 m. The horizontal strain was 5.9×10^{-5} or 30 % of the lowest estimate of strain-at-failure for alluvium (Jachens and Holzer, 1982).

Coefficients of determination for the pairs water-level decline and deep extensometer, water-level decline and horizontal displacements (Q8), and deep extensometer and horizontal displacements (Q8) range from 0.89 to 0.99 (Table 1). The lower correlation between horizontal displacements and both water-level decline and the deep extensometer during drawdown 2013 reflects closing of and major deviation of span Q4 during that period.

The deviation of span Q4 began during drawdown 2012 when that span closed while spans to the east of Q5 opened uniformly. That deviation indicates concentration of fissure opening across span Q5. A much larger closing across Q4 occurred during drawdown beginning in April 2013. That closing, little movement across spans Q1 and Q2, and fairly

Table 1. Coefficients of determination for periods of recovery and drawdown for water-level fluctuations, deep extensometer compaction, and horizontal movements.

	Water-level decline and deep extensometer	Water-level decline and horizontal movements	Deep extensometer and horizontal movements
Recovery winter 2011–2012	0.93	0.95	0.99
Drawdown 2012	0.95	0.97	0.93
Drawdown 2013	0.95	0.89	0.89
Recovery winter 2013	0.98	0.97	0.98

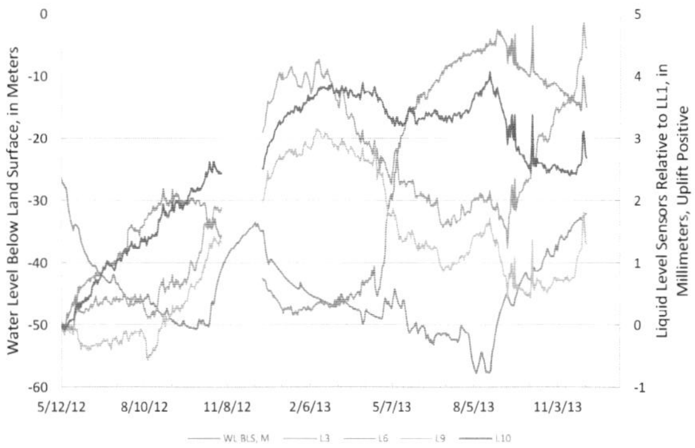

Figure 3. Vertical movement of liquid-level sensors with respect to LL1 compared with water-level fluctuation in observation well 0.8 km to the south. Sensor locations are in Fig. 1.

uniform opening and closing across spans Q3, Q11, and Q5–Q8 indicates that most fissure movement was concentrated across span Q5, at least 6 m east of the historic fissure zone, Q11.

The liquid-level, vertical-movement instrumentation is a work in progress. Sensors were built using the wire bundle of the cable as the vent tube. This technique had worked exceptionally well in the past, but excessive noise existed in the liquid-level data, presumably because of tight bends in the cable restricting pressure equalization in the reference side of the sensors. Venting the pipe to the atmosphere, piercing the cable, and inserting a syringe needle with a compliant diaphragm (sandwich bag) improved five of the sensors, LL1, LL3, LL6, LL9, and LL10. Movement of the other sensors is with respect to LL1 (Fig. 3).

The relation between liquid-level sensors and water-level fluctuation is not clearly defined. Maximum relative vertical movement was 4.7 mm at LL3. During drawdown 2012 all sensors exhibited uplift with respect to LL1, and LL3 reversed during late drawdown. During drawdown 2013, LL6, LL9, and LL10 exhibited similar reversing movement during early drawdown followed by decline during late drawdown, and LL3 experienced uplift delayed by more than 4 months. In contrast with horizontal strain, vertical movement in the liquid level sensors is delayed with respect to water-level fluctuations. Sensors on opposite sides of the fissure generally exhibit opposite movement.

Tilt sensors exhibited response to water-level fluctuation that was delayed by about 1 month (Fig. 4). E04 and N05f on the west side tracked water level fluctuation, while E06, E07f, and N65f exhibited opposite movement. Only tilts at E04 and E07f are in the vertical plane orthogonal to the fissure. Tilts at N05f, E06 and N65f were 75, 40, and 75 arcdeg off axis, respectively.

Maximum tilt in the fissure zone was 0.09 arcdeg while tilt at a separate sensor 100 m to the east was 0.86 arcdeg, indicating a wider zone of deformation than is spanned by the instrumentation (Fig. 5). Tilt at Station 23 tracks water-level fluctuation with a 1-month delay similar to the sensors near the fissure zone.

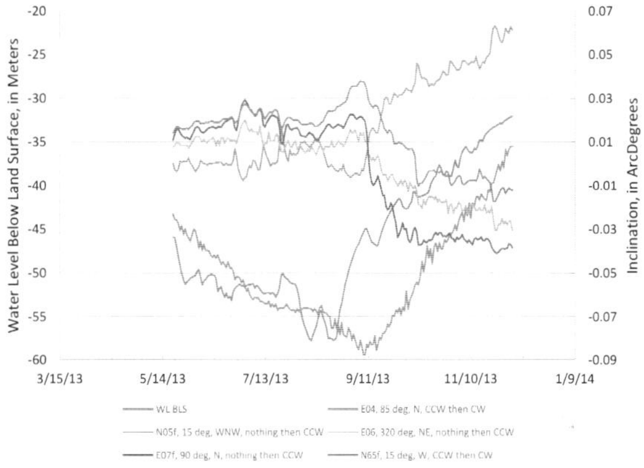

Figure 4. Biaxial tilt compared with water-level fluctuation in observation well 0.8 km to the south. Sensor locations are in Fig. 1. The legend gives the azimuth of the vertical plane of tilt, viewing direction, and sense of rotation for two segments of each graph. CW is clockwise; CCW is counterclockwise. Response to water-level fluctuation is delayed about one month and sensors on opposite sides of the fissure zone respond with the opposite sense. Only E04 and E07f exhibited tilt in the vertical plane orthogonal to the fissure.

4 Fissure mechanisms

Among several mechanisms proposed for earth fissuring, differential subsidence caused by differential aquifer compaction at depth remains the most substantiated (Fig. 6). Lateral change in thickness of the compressible aquifer-aquitard system, lateral change in compressibility within the system, or abrupt lateral change in water-level decline can cause fissuring. The conventional model treats the sediments overlying the zone of differential compaction as a bending plate. In that model, maximum horizontal strain is at the surface (Jachens and Holzer, 1982). Field evidence at many fissures indicates that fissures can have greater horizontal strain at

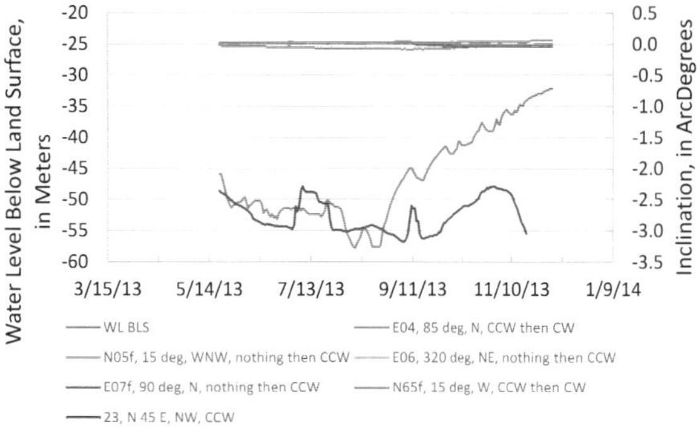

Figure 5. Biaxial tilt compared with water-level fluctuation in nearby observation well. Sensor locations are in Fig. 1. Station 23 is 100 m east of the fissure. The legend gives the azimuth of the vertical plane of tilt, viewing direction, and sense of rotation for two segments of each graph. CW is clockwise; CCW is counterclockwise. Deformation extends at least 100 m east of the fissure zone, and tilt at Station 23 is approximately ten times greater than tilt in the fissure zone.

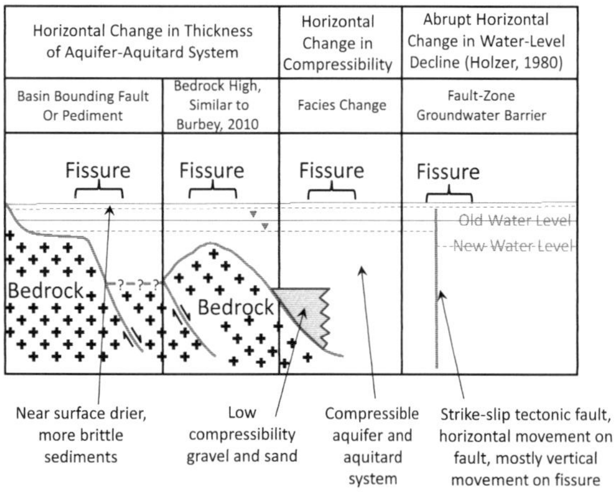

Figure 6. Schematic of differential compaction settings for earth fissures.

depth than at the surface (Carpenter, 1999). An earlier model of upward propagation of stress and strain in an elastic body (Sanford, 1959) solves this dilemma. In addition to the elastic model, Sanford performed physical experiments that demonstrate the patterns of deformation that can develop at the land surface in response to differential offset at depth (Fig. 7). High correlation of horizontal strain with change in effective stress supports differential compaction as the mechanism for earth-fissure movement.

Horizontal seepage stress is another mechanism proposed for earth fissuring (Helm, 1994). In Carpenter (1993), water-level fluctuations in pairs of piezometers on opposite sides of a fissure compared with horizontal strain across the fissure

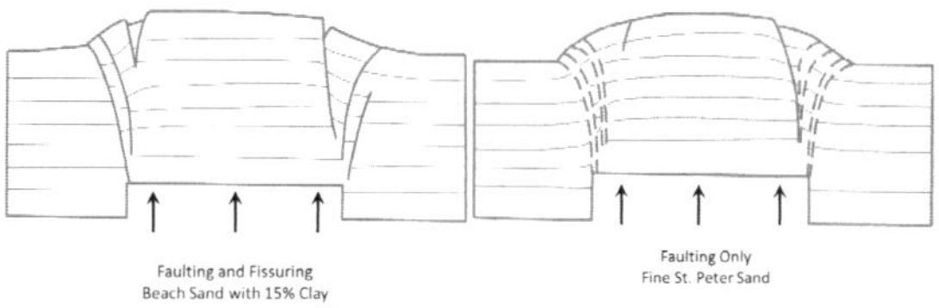

Figure 7. Results of sandbox experiments demonstrating fissuring and faulting consistent with shear and tensile failure of an elastic body exhibiting differential offset at depth (modified from Sanford, 1959 by Carpenter, 1993).

enabled a test of seepage stresses as a fissure mechanism. The coefficients of determination between water-level fluctuation in the piezometers (change in effective stress) and horizontal strain ranged from 0.83 to 0.85, while the coefficients of determination for horizontal seepage stresses and horizontal strain ranged from 0.03 to 0.17. In this study, coefficients of determination between water-level fluctuations and horizontal strain range from 0.89 to 0.97 (Table 1). During water-level recovery, fissures close (Fig. 2). Unless proponents of seepage stresses have a mechanism of water flowing back uphill during recovery dragging aquifer matrix, the mechanism of seepage stresses should be dismissed.

5 Discussion

The precise measurements of this study are incorporated in a larger monitoring network of semi-annual levelling and EDM with monument spacing of more than 400 m. With error bars of ±3 mm horizontal, the seasonal elastic deformation of about 30 % of strain-at-failure would be invisible in that network. This difficulty illustrates the need for closer monument spacing and greater precision in zones of known or suspected fissuring. Monument spacing of 12 m would enable delimitation of a zone to be instrumented, and infilling of monuments with 2 m spacing would enable spacing of quartz-tube monuments.

At present, the horizontal quartz tubes are the best established technique for continuous monitoring (Davis et al., 1969; Wolff, 1970). Displacement resolution can be finer than 0.1 µm. Biaxial tilt sensors (Haneberg and Friesen, 1995) are the easiest to install but, as point measurements, may be the least definitive technique. The liquid-level technique adapted after Riley (1970) remains unverified by precise levelling but has great promise.

Continuous measurements of horizontal displacements using quartz tubes presents a unique method for real-time adaptive management to prevent fissure opening or reopening. Municipal groundwater pumping can be restricted if horizontal strain approaches the lower limit of strain-at-failure or, coupled with water-level fluctuations and vertical borehole extensometry, stress-strain curves deviate from the previous mostly elastic regimen. Fissures have caused catastrophic damage to buildings and infrastructure. A municipal water agency that actively produces groundwater from fine-grained aquifer units runs the risk of causing such damage. This monitoring tool can be used to keep their operations safely below renewed fissuring.

The clarity of response of the horizontal-displacement data demonstrates the superiority of resolution of quartz tubes over EDM surveys and the superiority of continuous measurements over periodic surveys. The seasonal elastic deformation, which constitutes a significant percentage of strain at failure would be lost in the error bars of repeated surveys.

6 Recommendations

Wherever both horizontal and vertical deformation have been made at earth fissures, they have been of comparable magnitude (Carpenter, 1993, 1999). Fissures termed surface faults by Holzer (1980) that have significant vertical offset can pose particular hazards. Fissure opening of as much as 25 mm has been documented in a few hours during major frontal storms (Carpenter, 1999). A sudden vertical offset of 50–75 mm in a concrete highway during a storm with limited visibility could cause a multi-vehicle accident resulting in several deaths. Repeated precise levelling and EDM surveys where known or suspected fissures cross highways would indicate locations where GPS receivers could be placed along the highway on opposite sides of the fissure. Pairs of receivers could be monitored automatically for a threshold change that would trigger a warning sign or highway closing.

Fissures have been known to close over many years. That is, monuments on opposite sides have converged in repeated surveys in addition to infilling of the void space by collapse and deposition. In many cases the closure has been accompanied by water-level recovery (Carpenter, 1999). This presents the intriguing possibility of fissure mitigation by "wait and see" or determination of a rate of water-level decline that could be sustained without causing renewed fissuring. T. Holzer's old USGS project and the old USGS Subsidence Research Project EDM and levelling measurements at numerous fissures in the Western United States could be remeasured using original instruments and state of the art instruments, located by precise GPS, and compared with long-term water-level fluctuations. Application of a Bingham-substance model with appropriate plastic, elastic, and viscous material properties could lead to a greater understanding of long-term fissure behaviour (LSCE et al., 2014).

Fissure systems are varied and complex. Two sets of published records of continuous measurements of fissure movement have been made to date (Carpenter, 1993, and this study). As droughts lengthen and basins become more depleted, fissuring will increase. Several more sets of continuous measurements would add immensely to the understanding of fissures and increase the possibility of finding mitigating measures. In particular, comparison of fissures associated with fault-zone groundwater barriers, fissures associated with known subsurface structure, and fissures without known subsurface structure would be desirable.

A study would begin with repeated EDM and levelling over monuments with 40 m spacing on a line 500 m long (LSCE et al., 2014). Deformation within that line would determine the locations of more closely spaced surveying monuments that would be used for selection of quartz-tube posts and liquid-level stations with 6-m spacing. Diameter of the quartz tubes should be increased to 18 mm OD, 15 mm ID for improved rigidity. Permanent instrument and target monuments of 100 mm pipe crash posts for precise EDM would be installed beside the quartz-tube posts. Semiannual repeated precise EDM (Leica TC2003 or TDM5000) and levelling (Leica DNA03) would be compared with continuous quartz-tube horizontal displacements, liquid-level vertical movements, and tilt. Precise surveying would extend over greater lengths than the instruments providing continuous record. Surveying and continuous measurements would provide cross checks at the limit of resolution of the surveying instruments.

Acknowledgements. Funding for this work is provided by Wildermuth Environmental, Incorporated under the auspices of the Chino Basin Watermaster. Richard Wilson and Eric Lindberg provided field assistance and reviews of this paper. George Pardo also provided field assistance.

References

Burbey, T. J.: Mechanisms for earth fissure formation in heavily pumped basins, Land Subsidence, Associated Hazards and the Role of Natural Resources Development,Proceedings of EISOLS 2010, 17–22 October 2010, Queretaro, Mexico, IAHS Publ. 339, 3–8, 2010.

Carpenter, M. C.: Earth-fissure movements associated with fluctuations in ground-water levels near the Picacho Mountains, south-central Arizona, 1980-84, U.S. Geological Survey Professional Paper 497-H, 49 pp., available at: http://pubs.usgs.gov/pp/0497h/report.pdf, last access: 15 October 2015, 1993.

Carpenter, M. C., South-central Arizona, in: Land Subsidence in the United States, edited by: Galloway, D., Jones, D., and Ingebritsen, S., U.S. Geological Survey Circular 1182, 65–78, available at: http://pubs.usgs.gov/circ/circ1182/pdf/09Arizona.pdf, last access: 15 October 2015, 1999.

Davis, S. N., Peterson, F. L., and Halderman, A. D.: Measurement of small surface displacements induced by fluid flow, Water Resour. Res., 5, 129–138, 1969.

Haneberg, W. C. and Friesen, R. L.: Tilts, strains, and ground-water levels near an earth fissure in the Mimbres Basin, New Mexico, Geol. Soc. Am. Bull., 107, 316–326, 1995.

Helm, D. C.: Hydraulic forces that play a role in generating fissures at depth, Bulletin of the Association of Engineering Geologists, 31, 293–304, 1994.

Holzer, T. H.: Faulting caused by groundwater level declines, San Joaquin Valley, California, Water Resour. Res., 16, 1065–1070, 1980.

Jachens, R. C. and Holzer, T. L.: Differential compaction mechanism for earth fissures near Casa Grande, Arizona, Geol. Soc. Am. Bull., 93, 998–1012, 1982.

LSCE (Luhdorff & Scalmanini Consulting Engineers), Borchers, J. W., and Carpenter, M. C.: Land subsidence from groundwater use in California, California Water Foundation, 151 pp., available at: http://californiawaterfoundation.org/uploads/1397858208-SUBSIDENCEFULLREPORT_FINAL.pdf, last access: 15 October 2015, 2014.

Riley, F. S.: Land-surface tilting near Wheeler Ridge, Southern San Joaquin Valley, California, U.S. Geological Survey Professional Paper 497-G, 29 pp., 1970.

Sanford, A. R.: Analytical and experimental study of simple geologic structures: Geol. Soc. Am. Bull., 70, 19–52, 1959.

Stewart, C. A., Colby, N. D., Kent, R. T., Eagan, J. A., and Hall, N. T.: Earth fissuring, ground-water flow, and water quality in the Chino basin, California, in: Dr Joseph F. Poland symposium, Land subsidence case studies and current research: proceedings of the Dr Joseph F Poland symposium on land subsidence; 1995, Sacramento, CA, edited by: Borchers, J. W., Association of Engineering Geologists Special Publication 8, 195–205, 1998.

Wildermuth Environmental: MZ-1 Subsidence Management Plan, variously paginated, Chino Basin Watermaster, Rancho Cucamonga, California, USA, available at: http://ayala.wildermuthenvironmental.com:8888/AyalaPark/documents/20071017_MZ1_Plan.pdf, last access: 15 October 2015, 2007.

Wolff, R. G.: Relationship between horizontal strain near a well and reverse water level fluctuation, Water Resour. Res., 6, 1721–1728, 1970.

Proc. IAHS, 372, 297–303, 2015
proc-iahs.net/372/297/2015/
doi:10.5194/piahs-372-297-2015

Introduction to GPS geodetic infrastructure for land subsidence monitoring in Houston, Texas, USA

G. Wang, J. Welch, T. J. Kearns, L. Yang, and J. Serna Jr.

Department of Earth and Atmospheric Sciences, University of Houston, Houston, Texas, 77204, USA

Correspondence to: G. Wang (gwang@uh.edu)

Published: 12 November 2015

Abstract. Houston, Texas is one of the places that first employed high-accuracy GPS technology for land subsidence monitoring beginning in the late 1980s. Currently, there are over 170 permanent GPS stations located in the Houston metropolitan area. This article summarizes the current GPS geodetic infrastructure in the Houston metropolitan area, which is comprised of three components: a dense GPS network with 170 permanent stations, a stable Houston reference frame (SHRF14), and sophisticated software packages for post positioning processing. Average land subsidence and groundwater-level altitude changes during the past 10 years (2005–2014) also are presented in this paper.

1 Introduction

Land subsidence in the Houston, Texas area has been occurring for almost a century. Accumulated subsidence of over 3 m during the past century has been observed in a large area of southeast Harris County, including Houston downtown, the cities of Pasadena, Baytown, Texas City, and Galveston (Kasmarek et al., 2009). The Houston area is also suffering from faulting problems. More than 150 historically active faults have been identified in the Houston area (Shah and Lanning-Rush, 2005). Ground deformations associated with subsidence and faulting cause moderate to severe damage to hundreds of residential, commercial, and industrial structures in the Houston area each year. As a result, the foundation repair industry has been thriving in this area since the 1990s. Subsidence is of particular concern in low-lying downtown, the Houston ship channel, and the Galveston coastal areas because of the increased risk of flooding from storm surges (Coplin and Galloway, 1999). To prevent land subsidence, the Texas Legislature created the Harris-Galveston Subsidence District (HGSD) in 1975, the Fort Bend Subsidence District (FBSD) in 1989, the Lone Star Groundwater Conservation District (LSGCD) in 2001, and the Brazoria County Groundwater Conservation District (BCGCD) in 2003. One of the primary obligations of these districts is to regulate and reduce groundwater withdrawal for the purpose of minimizing ground subsidence, which contributes to infrastructure damage and flooding.

In the early 1990s, the HGSD established a surveying network of 20 permanent GPS monuments for the purpose of subsidence monitoring. These GPS sites are also called Port-A-Measure (PAM) stations (Zilkoski et al., 2003). The PAM network has been continuously expanded and the number of total stations has reached 80 as of 2014. On average, HGSD continuously collected GPS data at each PAM site for one week every month prior to 2005. It now collects data for one week every other month on average. In addition to the PAM GPS network, there are over 40 Continuously Operating Reference Stations (CORS) in the Houston area (Wang and Soler, 2013). These CORS are operated by a joint effort of the National Geodetic Survey (NGS) at the National Oceanic and Atmospheric Administration (NOAA), the Texas Department of Transportation, the City of Houston, and other local agencies. Recently, the University of Houston established a real-time GPS network (HoustonNET) with 50 permanent real-time stations. The primary purpose of the HoustonNET is to provide a platform for studying multiple urban natural hazards, including subsidence, faulting, salt dome uplift, flooding, and to improve hurricane intensity forecasts in the Gulf Coast region. As of 2014, there are approximately 170 permanent GPS stations in the Houston metropolitan area (Fig. 1). Raw GPS data (RINEX files) from

Published by Copernicus Publications on behalf of the International Association of Hydrological Sciences.

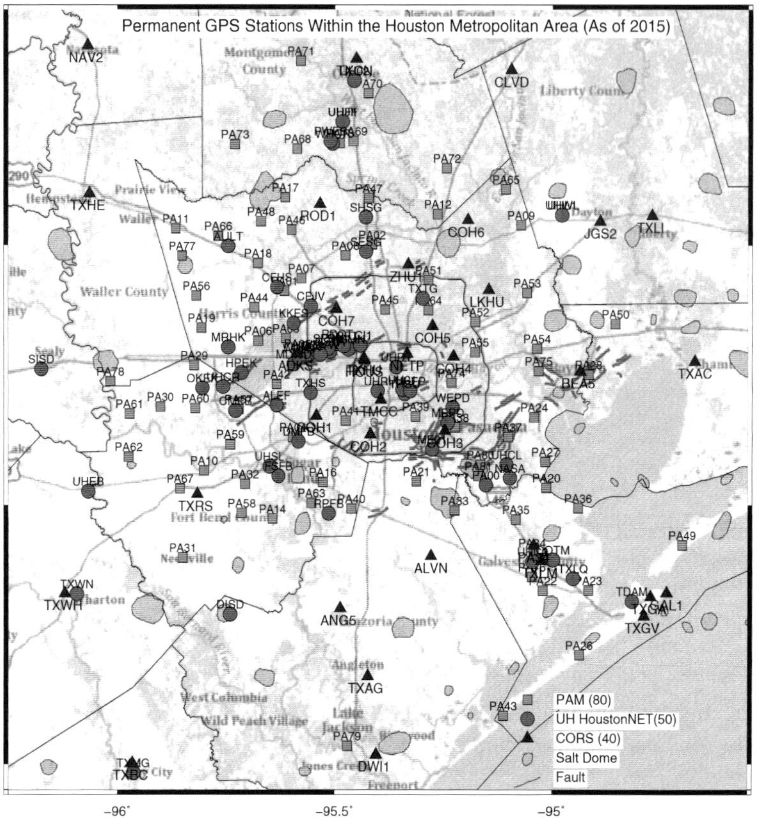

Figure 1. Map showing the locations of current permanent GPS stations within the Houston metropolitan area. The locations of salt domes and fault lines are sourced from American Association of Petroleum Geologists (2011) and from USGS (Shah and Lanning-Rush, 2005), respectively.

these stations are available to the public through HGSD and data archiving facilities at UNAVCO (https://www.unavco.org/data) and NGS (http://geodesy.noaa.gov/CORS).

2 Stable Houston Reference Frame of 2014 (SHRF14)

Surveying-level GPS units record satellite signals and do not directly provide high-precision positions. Additional information, such as precise satellite orbits and clock information, the modeling of atmospheric condition (dry and wet temperature and pressure) and a suitable mapping function, and complex calculations are required for obtaining precise positions. Several sophisticated GPS data post-processing software packages are available to the research community. Some such packages include GAMIT/GLOBK, GIPSY/OASIS, and Bernese. This study applied the precise point positioning (PPP) method employed by the GIPSY/OASIS (V6.3) software package (Bertiger et al., 2010) for calculating daily positions. The precision of PPP solutions has dramatically increased over the last decade. This is primarily attributed to highly precise satellite orbit and clock data provided by the International Global Navigation Satellite System (GNSS) Service (IGS, http://www.igs.org), and new algorithms used to resolve phase ambiguity with a single receiver. The PPP method has attracted broad interests in ground deformation monitoring because of its operational simplicity and high accuracy (e.g., Wang 2013; Wang et al., 2014; Wang and Soler, 2014; Yu et al., 2014). Our recent GPS study indicated that the repeatability (RMS) of the daily ambiguity-resolved PPP solutions achieved 2–3 mm in the horizontal components and 6–8 mm in the vertical component within the Houston metropolitan area (Wang et al., 2013).

A complex part of GPS is that it initially provides position coordinates within a global reference frame, such as the International GNSS Service reference frame of 2008 (IGS08). In general, a global reference frame is realized with an approach of minimizing the overall movements of a large number of selected frame stations distributed worldwide. As a result, the majority of sites keep moving within a global reference frame. Several continental-scale reference frames fixed on the North American plate have been developed by the geodetic community, such as the North American Datum of 1983 (NAD83) (Snay and Soler, 2000; Soler and Snay, 2004; Pearson and Snay, 2013), the Stable North American Reference Frame (SNARF) (Blewitt, 2008), and the North American Reference Frame of 2012 (NA12) (Blewitt et al., 2013). The average horizontal velocities in the Houston metropolitan area are approximately 15 mm year^{-1} toward the southwest within IGS08 and 2 mm year^{-1} toward the northeast within NAD83 (Wang et al., 2013). There are no stable sites that have velocities of 0 mm year^{-1} within the Houston metropolitan area with respect to the NAD83 and IGS08 reference frames.

The need for a consistent and stable reference frame has become critically important in the Houston metropolitan area because of the broad extent of subsidence and faulting activities since the 1980s. It is difficult to identify long-term stable benchmarks (reference sites) inside of the subsidence area. In order to integrate long-term GPS data collected by different agencies in different areas with different equipment into a uniform reference frame, a Stable Houston Reference Frame was established in 2012 (Wang et al., 2013). The local reference frame was realized by a 14-parameter simultaneous Helmert transformation from the global reference frame IGS08. This study updated the previous local reference frame by improving the geometrical configuration of reference stations and by including two additional years (2013 and 2014) of continuous data. All reference stations are located outside the Houston metropolitan area and they have a positional data set of 5 to 10 years. Figure 2 depicts the locations of 10 reference stations for the updated local reference frame, the stable Houston reference frame of 2014 (SHRF14). The positional coordinates of a site within the SHRF14 can be calculated through the following equations:

$$
\begin{aligned}
x(t)_{\mathrm{SHRF}} &= T_x(t) + [1 + s(t)] \cdot x(t)_{\mathrm{IGS08}} \\
&\quad + R_z(t) \cdot y(t)_{\mathrm{IGS08}} - R_y(t) \cdot z(t)_{\mathrm{IGS08}} \\
y(t)_{\mathrm{SHRF}} &= T_y(t) - R_z(t) \cdot x(t)_{\mathrm{IGS08}} + [1 + s(t)] \cdot y(t)_{\mathrm{IGS08}} \\
&\quad + R_x(t) \cdot z(t)_{\mathrm{IGS08}} \\
z(t)_{\mathrm{SHRF}} &= T_z(t) + R_y(t) \cdot x(t)_{\mathrm{IGS08}} - R_x(t) \cdot y(t)_{\mathrm{IGS08}} \\
&\quad + [1 + s(t)] \cdot z(t)_{\mathrm{IGS08}}
\end{aligned}
$$

Here, $T_x(t)$, $T_y(t)$, $T_z(t)$, $R_x(t)$, $R_y(t)$, $R_z(t)$, and $s(t)$ are the seven Helmert reference frame transformation parameters at epoch t. $T_x(t)$, $T_y(t)$, and $T_z(t)$ are translations along x, y, and z axes; $R_x(t)$, $R_y(t)$, and $R_z(t)$ are counterclockwise rotations about these three axes; $s(t)$ is a differential scale change between IGS08 and SHRF. These seven parameters at a specific epoch (t). They can be obtained from the following equations:

$$
\begin{aligned}
T_x(t) &= T_x(t_0) + T'_x \cdot (t - t_0) \\
T_y(t) &= T_y(t_0) + T'_y \cdot (t - t_0) \\
T_z(t) &= T_z(t_0) + T'_z \cdot (t - t_0) \\
R_x(t) &= R_x(t_0) + R'_x \cdot (t - t_0) \\
R_y(t) &= R_y(t_0) + R'_y \cdot (t - t_0) \\
R_z(t) &= R_z(t_0) + R'_z \cdot (t - t_0) \\
s(t) &= s(t_0) + s' \cdot (t - t_0)
\end{aligned}
$$

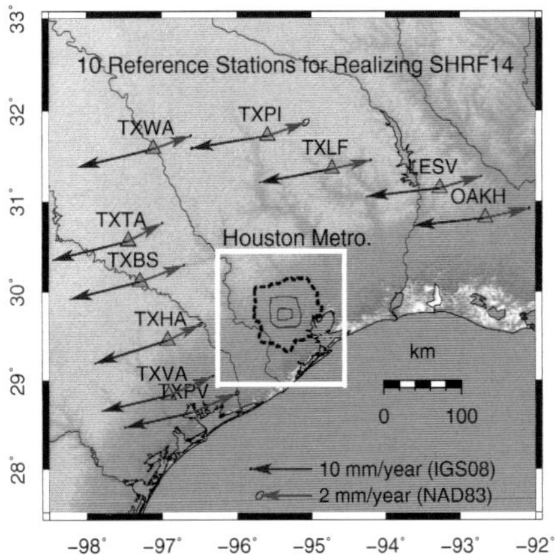

Figure 2. Map showing the locations of 10 reference stations used to realize the Stable Houston Reference Frame of 2014 (SHRF14). The vectors represent the horizontal velocity vectors with respect to IGS08(blue) and NAD83(red). The white box represents the Houston metropolitan area.

Here, t_0 denotes a specific epoch that was used to align the two reference frames. It is set as 2013.0 in this study. $T_x(t_0)$, $T_y(t_0)$, $T_z(t_0)$, $R_x(t_0)$, $R_y(t_0)$, $R_z(t_0)$, and $s(t_0)$ are the seven Helmert-parameters at epoch t_0, which equal zeros since a site has the same coordinates at this epoch with respect the both reference frames. T'_x, T'_y, T'_z, R'_x, R'_y, R'_z, and s' are the first-time derivatives (rates) of these seven parameters, which are listed in Table 1.

3 Current land subsidence

Figure 3 depicts the land subsidence contours in the Houston metropolitan are for the time period from 2005 to 2014. Continuous observations (>3 years) from 100 permanent GPS stations and 11 extensometer sites were used in calculating the contours. The color areas represent the groundwater regulation zones enforced by the local groundwater management agencies. The HGSD regulates the groundwater use in the Harris and Galveston Counties, which are divided into three regulation areas: Areas I, II, and III. The FBSD regulates groundwater use in Fort Bend County, which is divided into three areas: northern (Area A), southern (Area B), and the Richmond-Rosenberg subarea (Area RR). There is no considerable land subsidence in Brazoria County. The subsidence in the HGSD's Areas I and II has ceased

Table 1. Fourteen Parameters for Helmert Reference Frame Transformation from IGS08 to SHRF14.

Parameter	Unit	IGS08 to SHRF14
		$t_0 = 2013.0$
$T_x(t_0)$	m	0.0
$T_y(t_0)$	m	0.0
$T_z(t_0)$	m	0.0
$R_x(t_0)$	radian	0.0
$R_y(t_0)$	radian	0.0
$R_z(t_0)$	radian	0.0
$s(t_0)$	unitless	0.0
dT_x	m year^{-1}	−1.490584E-02
dT_y	m year^{-1}	−1.328112E-02
dT_z	m year^{-1}	−2.326649E-02
dR_x*	radian year^{-1}	4.177140E-09
dR_y*	radian year^{-1}	−5.198959E-09
dR_z*	radian year^{-1}	−1.699222E-09
ds	1 year^{-1}	0.0

* Counterclockwise rotations of axes (x, y, z) are positive.

(< 3 mm year^{-1}) except for in a localized subsidence area located in the La Marque area (5–10 mm year^{-1}). Groundwater pumping is strictly regulated in Areas I and II. Only 10 % of the total water demand has been permitted to be withdrawn from groundwater in Area I and only 20 % has been permitted to be withdrawn in Area II since 1999. Currently, two subsidence bowls have formed in The Woodlands-Spring area and the Katy area. The Woodlands-Spring subsidence bowl is located in both the northern portion of HGSD's Area III and the southern portion of LSGCD regulation area (the Montgomery County). Subsidence of up to 25 mm year^{-1} is occurring in the center of the subsidence bowl. There was no groundwater regulation in the HGSD's Area III until 2010. Current HGSD's regulation plan requires that no more than 20 % of the total water demand can be withdrawn from the ground in Area III. Montgomery County is the fifth fastest growing county in Texas. Water for residents and businesses use was almost exclusively from groundwater before 2014. Subsidence of up to 20 mm year^{-1} is occurring within the subsidence bowl in the Katy area, which is located in the FBSD's Area A and southern Waller County. No groundwater regulation plans have been implemented in Waller County. The Katy area is growing rapidly, but it is not scheduled to utilize surface water resources in any of the existing groundwater reduction plans. Groundwater was the sole water source for residents and businesses as of 2014.

4 Groundwater-level altitudes

The Chicot and Evangeline aquifers are two major fresh water aquifers underlying the Houston metropolitan area and consequently the most of the subsidence has occurred as a direct result of groundwater withdrawal from these two

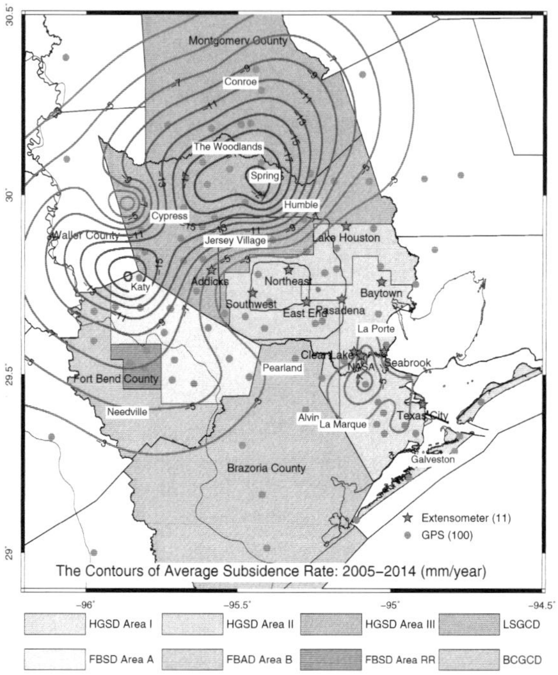

Figure 3. Average subsidence rate occurring from 2005 to 2014 in the Houston metropolitan area, Texas. The colored areas indicate the current groundwater conservation zones regulated by the Harris-Galveston Subsidence District (HGSD), the Fort Bend Subsidence District (FBSD), the Lone Star Groundwater Conservation District (LSGCD), and the Brazoria County Groundwater Conservation District (BCGCD).

aquifers (e.g., Coplin and Galloway, 1999; Kasmarek et al., 2014; Kearns et al., 2015). The two aquifers are hydraulically connected. Long-term groundwater-level changes that occur in one aquifer can affect the groundwater-level in the other one. As a result, the areas where water levels have declined or risen are approximately spatially coincident for the Chicot and the Evangeline aquifers. Figure 4 depicts the groundwater-level altitudes in the Chicot and Evangeline aquifers at the end of 2014 and the changes of groundwater-level altitudes during the past 10 years (2005–2014). Water-level measurements from 175 wells screened in the Chicot aquifer and 330 wells screened in the Evangeline aquifer were used to construct the contour maps. All groundwater data was obtained from the USGS National Water Information System (NWIS) (http://groundwaterwatch.usgs.gov).

Recent investigations indicate that the local preconsolidation heads of the groundwater within the Chicot and Evangeline aquifers in the Houston metropolitan area are approximately 35 m below the land surface (−35 m) (Kearns et al., 2015). The preconsolidation head is estimated as the approximate local groundwater level prior to excessive groundwater pumping that began in the 1940s. Figure 4 indicates that both

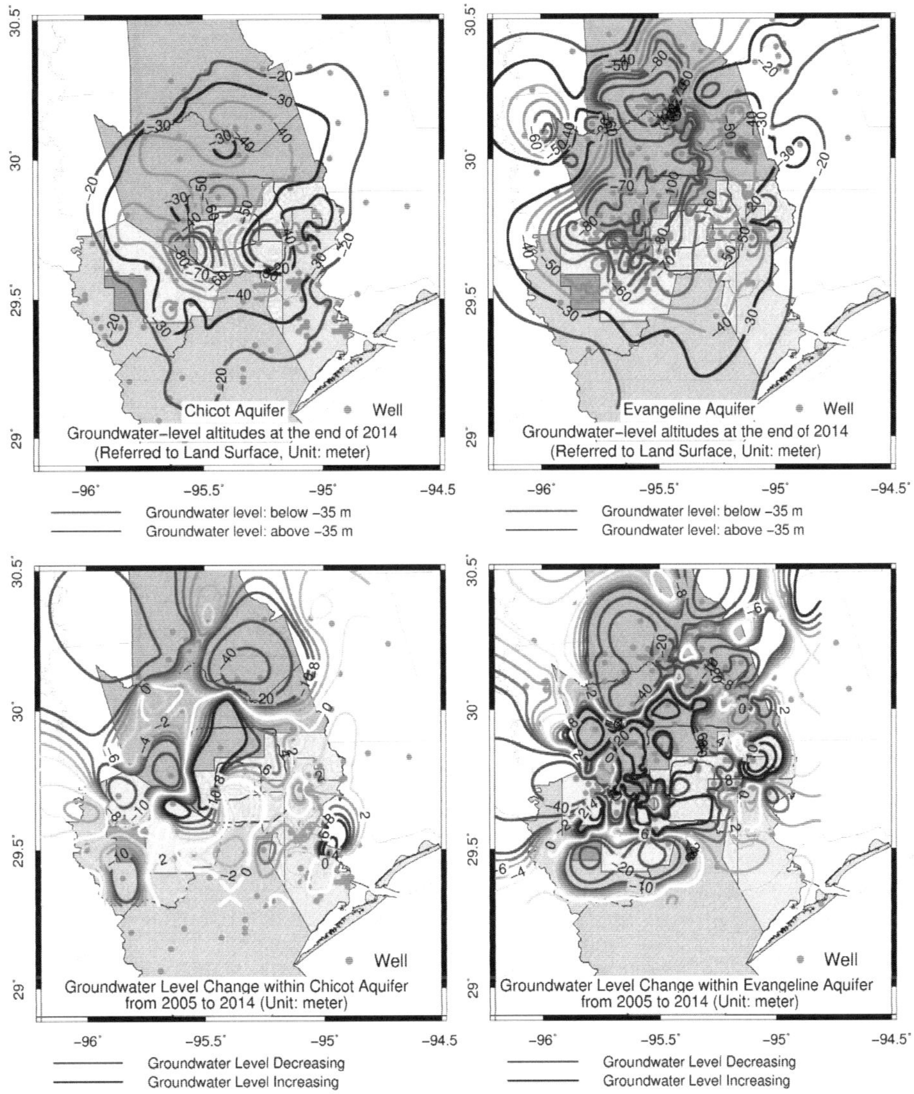

Figure 4. Contour maps showing groundwater-level altitudes within the Chicot and Evangeline aquifers at the end of 2014 and the changes of groundwater-level altitudes from 2005 to 2014. The colored areas represent the groundwater regulation zones as shown in Fig. 3.

the Chicot and Evangeline water-level altitudes are above the preconsolidation head (-35 m) in HGSD's Area I, the majority of FBSD's Area B, and in Brazoria County as of 2014. The Chicot groundwater-level altitudes are below the preconsolidation head in HGSD's Area III, the western HGSD's Area II, and southern Montgomery County. The Evangeline groundwater level is below the preconsolidation head within a much larger area. The deepest Evangeline groundwater level is approximately 100 m below the land surface in The Woodland-Spring area and the Katy area. The Evangeline and Chicot groundwater levels have been rising in HGSD's Areas II and III during the past 10 years. The average groundwater level is rising at a rate of 1 to 3 m year^{-1} in the Evangeline aquifer. The rising rate of the Chicot groundwater level is slower. The groundwater-level altitudes in the FBSD's Area B and Montgomery County declined at an approximate rate of 2 to 4 m year^{-1} over the past 10 years.

A comparison of current land subsidence, groundwater level altitudes, and their changes indicates a close correlation between groundwater-level altitude and aquifer compaction (subsidence). There is no considerable subsidence in the areas where the Chicot and Evangeline groundwater-level altitudes are close to or above −35 m. In the areas where the groundwater-level altitudes are below −35 m and the groundwater levels are still declining, the rate of subsidence is rapid (> 10 mm $year^{-1}$). In the areas where the groundwater-level altitudes are below −35 m, but the groundwater levels are rising, the rate of subsidence is moderate (< 10 mm $year^{-1}$). It appears that subsidence will continue to occur as long as the groundwater levels stay below about −35 m.

5 Discussion and conclusions

The Houston metropolitan area represents one of the largest subsidence areas in the USA (Galloway et al., 1999). The current dense GPS and groundwater monitoring networks in this area have provided direct measurements used for the purpose of studying the correlation between land subsidence and groundwater withdrawal and for the scientific management of groundwater and land resources. The groundwater regulations enforced by HGSD have successfully halted subsidence in areas that were once heavily subsiding (HGSD'S Area I and II) and have slowed the rate of subsidence in recently developed suburban area (HGSD's Area III). Groundwater levels have been stabilized in HGSD's Area I (above −35 m) and have been rising in HGSD's Area II and the majority of Area III. Based on the success if the subsidence districts in mitigating the occurrence, it is evident that groundwater resources and subsidence are manageable.

Accurate monitoring of subsidence over long period of time and of large area is vital to provide calibration data for subsidence modeling, prediction, and management purposes. A precise local geodetic infrastructure can be defined with three fundamental components: a dense GPS network, a stable local reference frame, and sophisticated software packages for data processing. This article updated the local reference frame with an improved geometrical configuration of reference stations and two additional years of data. SHRF14 provides a consistent and accurate reference frame for studying local ground deformations over space and time and serves the broad research and surveying communities. The local reference frame will be incrementally improved and synchronized with the updates of the IGS reference frame.

This study demonstrated an approach of using publicly available GPS data to conduct millimeter-accuracy land subsidence studies in an urban environment. Numerous CORS networks have been installed in urban areas all over the world during the past two decades for the purpose of conducting accurate land surveying and for scientific applications. It is hoped that this study will promote the applications of GNSS techniques in subsidence monitoring and contribute to the reduction of subsidence hazards in other subsidence-prone areas.

Acknowledgements. This study was supported by an NSF CAREER award EAR-1229278 and an NSF MRI award EAR-1242383. The authors acknowledge USGS and HGSD for providing groundwater, extensometer, and GPS data to the public.

References

American Association of Petroleum Geologists: Salt tectonism of the US gulf coast basin (CD-ROM), Geographic Information System-Upstream Digital Reference Information Library (GIS-UDRIL), available at: http://www.datapages.com/associated-websites/documents/gisudrilbrochurejan2013 (last access: 5 October 2015), 2011.

Bertiger, W., Desai, S., Haines, B., Harvey, N., Moore, A., Owen S., and Weiss, J.: Single receiver phase ambiguity resolution with GPS data, J. Geod., 84, 327–337, 2010.

Blewitt, G.: Overview of the SNARF Working Group, its activities, and accomplishments, Report of the Ninth SNARF Workshop, available at: https://www.unavco.org/projects/past-projects/snarf/snarf.html (last access: 5 October 2015), 2008.

Blewitt, G., Kreemer, C., Hammond, W. C., and Goldfarb, J. M.: Terrestrial reference frame NA12 for crustal deformation studies in North America, J. Geodyn., 72, 11–24, 2013.

Coplin, L. and Galloway, D.: Land subsidence in the United States. Houston Galveston, Texas: managing coastal subsidence, in: Galloway, edited by: D., Jones, D. R., and Ingebritsen, S. E., Land subsidence in the United States: US Geological Survey Circular 1182, 35–48, 1999.

Galloway, D., Jones, D., and Ingebritsen, S.: Land subsidence in the United States, USGS Circular 1182, US Geological Survey, Reston, VA, 35–48, 1999.

Kasmarek, M., Gabrysch, R., and Johnson, M.: Estimated land-surface subsidence in Harris County, Texas, 1915–1917 to 2001, USGS Sci. Invest. Map 3097, 2 sheets, US Geological Survey, Reston, available at: http://pubs.usgs.gov/sim/3097/ (last access: 5 October 2015), 2009.

Kasmarek, M., Johnson, M., and Ramage, J.: Water-level altitudes 2013 and water-level changes in the Chicot, Evangeline, and Jasper aquifers and compaction 1973–2012 in the Chicot and Evangeline aquifers, Houston-Galveston region, Texas, USGS Sci. Invest. Map 3263, US Geological Survey, Reston, VA, 1–30, 2014.

Kearns, T. J., Wang, G., Bao, Y., Jiang, J., and Lee, D.: Current Land Subsidence and Groundwater Level Changes in the Houston Metropolitan Area, Texas (2005–2012), J. Surv. Eng., 05015002, 1–16, 2015.

Pearson, P. and Snay, R.: Introducing HTDP 3.1 to transform coordinates across time and spatial reference frames, GPS Solut., 17, 1–17, 2013.

Shah, S. D. and Lanning-Rush, J.: Principal faults in the Houston, Texas, Metropolitan Area, USGS, Scientific Investigations Map 2874, available at: http://pubs.usgs.gov/sim/2005/2874/ (last access: 5 October 2015), 2005.

Snay, R. A. and Soler, T.: Modern Terrestrial Reference Systems. Part 2: The evolution of the NAD 83, Professional Surveyor, 20, 16–18, 2000.

Soler, T. and Snay, R. A.: Transforming positions and velocities between the International Terrestrial Reference Frame of 2000 and North American Datum of 1983, J. Surv. Eng., 130, 49–55, 2004.

Wang, G.: Millimeter-accuracy GPS landslide monitoring using precise point positioning with single receiver phase ambiguity resolution: a case study in Puerto Rico, J. Geod. Sci., 3, 22–31, 2013.

Wang, G. and Soler, T.: Using OPUS for measuring vertical displacements in Houston, TX, J. Surv. Eng., 139, 126–134, 2013.

Wang, G. and Soler, T.: Measuring Land Subsidence Using GPS: Ellipsoid Height vs. Orthometric Height, J. Surv. Eng., 141, 1–12, 2014.

Wang, G., Yu, J., Ortega, J., Saenz, G., Burrough, T., and Neill, R.: A stable reference frame for ground deformation study in the Houston metropolitan area, Texas, J. Geod. Sci., 3, 188–202, 2013.

Wang, G., Yu, J., Kearns, T. J., and Ortega, J.: Assessing the accuracy of long-term subsidence derived from borehole extensometer data using GPS observations: case study in Houston, Texas, J. Surv. Eng., 140, 1–7, 2014.

Yu, J., Wang, G., Kearns, T. J., and Yang, L.: Is there deep-seated subsidence in the Houston-Galveston area?, Int. J. Geophys., 2014, 942834, 1–11, doi:10.1155/2014/942834, 2014.

Zilkoski, D. B., Hall, L., Mitchell, G., Kammula, V., Singh, A., Chrismer, W., and Neighbors, R.: The Harris-Galveston coastal subsidence district/national geodetic survey automated global positioning system subsidence monitoring project, Proc., US Geological Survey Subsidence Interest Group Conf., Galveston, Texas, 27–29 November 2001, USGS OpenFile Report, 03–308, 13–28, 2003.

Proc. IAHS, 372, 305–309, 2015
proc-iahs.net/372/305/2015/
doi:10.5194/piahs-372-305-2015

Open Access
PIAHS

Different scale land subsidence and ground fissure monitoring with multiple InSAR techniques over Fenwei basin, China

C. Zhao[1,2], Q. Zhang[1,2], C. Yang[1,2], J. Zhang[1,2], W. Zhu[1,2], F. Qu[1], and Y. Liu[1]

[1]College of Geology Engineering and Geomatics, Chang'an University, Xian Shaanxi, China
[2]Key Laboratory of Western China's Mineral Resources and Geological Engineering, Ministry of Education, No. 126 Yanta Road, Xian, Shaanxi, China

Correspondence to: C. Zhao (zhaochaoying@163.com)

Published: 12 November 2015

Abstract. Fenwei basin, China, composed by several sub-basins, has been suffering severe geo-hazards in last 60 years, including large scale land subsidence and small scale ground fissure, which caused serious infrastructure damages and property losses. In this paper, we apply different InSAR techniques with different SAR data to monitor these hazards. Firstly, combined small baseline subset (SBAS) InSAR method and persistent scatterers (PS) InSAR method is used to multi-track Envisat ASAR data to retrieve the large scale land subsidence covering entire Fenwei basin, from which different land subsidence magnitudes are analyzed of different sub-basins. Secondly, PS-InSAR method is used to monitor the small scale ground fissure deformation in Yuncheng basin, where different spatial deformation gradient can be clearly discovered. Lastly, different track SAR data are contributed to retrieve two-dimensional deformation in both land subsidence and ground fissure region, Xi'an, China, which can be benefitial to explain the occurrence of ground fissure and the correlation between land subsidence and ground fissure.

1 Introduction

Fenwei basin, composed by Fenhe basin in Shanxi province and Weihe basin in Shaanxi province, Northwest China, is located in the complex intersection of Ordos block, North China block and South China block. S-shaped Fenwei basin with 1200 km in length and 30 to 60 km in width, can be segmented into Datong, Xinzhou, Taiyuan, Linfen, Yuncheng, and Weihe six sub-basins from north to south. Approximate North-North-East faults control the hazard activies of each sub-basins.

Due to the frequent fault activities and complex geotectonics in this region, severe land subsidence and ground fissures have been occurred for around 60 years. Statistics show around 170 locations and 400 ground fissures have been detected in 56 counties within Fenwei basin. Besides, Fenwei basin is one of the three largest and severest land subsience regions in China. Consquently, great econmic losses and infrastracture damages have been caused in this basin. So different InSAR techniques are employed to monitor the large scale land subsidence and small scale ground fissure deformation in Fenwei basin.

2 Methodologies

2.1 PS-InSAR technique

PS-InSAR successfully overcomes the problems of temporal and spatial decorrelation and heterogeneous atmospheric delay, which are main shortage in the conventional differential InSAR processing.

In this paper, we apply StaMPS-InSAR software for SAR data processing to achieve deformation results. The key steps are summarized as follows. Readers can refer to the papers as Hooper et al. (2004, 2007), Hooper (2006) and StaMPS software manual if they are interested in the details.

Suppose N interferograms are generated from $N+1$ SAR images acquired at different epochs on the same track and same study area. The master image is selected by considering both temporal, spatial baseline and the influence of the Doppler central frequency (Hooper et al., 2007; Zebker and Villaseno, 1992). Usually, the master image minimizes the sum decorrelation of all the interferograms. The external digital elevation model (DEM), such as Shuttle radar topography mission (SRTM) DEM is used to remove the effects of topography. From the coregistered SAR SLC images and DEM subtracted interferograms, PS points below amplitude dispersion and phase stability thresholds are determined.

The following steps are only focused on these PS pixels. For a generic i-interferogram, the wrapped differential interferometric phase of a PS pixel x can be described as:

$$\psi_{\text{int},x,i} = W\left\{\varphi_{\text{def},x,i} + \varphi_{\text{atm},x,i} + \Delta\varphi_{\text{orb},x,i} + \Delta\varphi_{\theta,x,i} + \varphi_{\text{n},x,i}\right\} \quad (1)$$

where, $W\{\cdot\}$ indicates modulo 2π radian operation, $\varphi_{\text{def},x,i}$ is the phase change due to surface movement in line-of-sight direction, which is the mainly useful part for deformation analysis. While the other parts in Eq. (1) are nuisance terms, that is, $\Delta\varphi_{\theta,x,i}$ is the residual phase due to look angle error, $\varphi_{\text{atm},x,i}$ is the phase due to the difference in atmospheric delay between passes, $\Delta\varphi_{\text{orb},x,i}$ is the residual phase due to satellite orbit inaccuracies, $\varphi_{\text{n},x,i}$ is a noise term due to variability in scattering, thermal noise, co-registration errors and uncertainty in the position of the phase center in azimuth. Three dimensional phase unwrapping is applied to get more reliable unwrapping results (Hooper and Zebker, 2007). These nuisance terms can be well isolated by using low-pass filter in spatial domain and high-pass filter in time domain. Moreover, low-pass filter in time domain within a given Gaussian width can also recover the nonlinear deformation phase. Then annual deformation rate and time-series results can be trivially constructed.

2.2 Combined PS and SBAS-InSAR technique

Standard SB methods (Berardino et al., 2002; Lanari et al., 2004) work with interferograms that are first multilooked and then individually phase unwrapped. While in order to maitain highest possible resolution, Hooper (2008) refered to slowly decorrelating filtered phase (SDFP) pixels as the targets of SBAS methods. That is pixels whose phase when filtered decorrelates little over short time intervals are SDFP. The SBAS interferograms are formed by recombination of the resampled SLC images meeting both spatial and temporal baseline thresholds. First filtering in azimuth to exclude non-overlapping Doppler spectrum and in range to reduce the effects of geometric decorrelation. The geometric phase to be subtracted for each small baseline interferogram is calculated as the difference between the relevant single-master interferograms for both the flat-Earth and topographic phase.

SDFP pixels are identified among the candidate pixels in the same way that PS pixels are, using the algorithm of Hooper et al. (2007). Then SDFP and PS are combined to maximize the reliability of the unwrapped phase. The unwrapped phase of the SB interferograms is inverted to derive a time series of phase change for each pixel in a least square fashion because of only one SBAS dataset is considered.

2.3 Two dimensional deformation with ascending and descending data

Because both InSAR ascending and descending data are from near-polar orbits, which makes it not sensitive to the deformation field in the north-south direction (Wright et al., 2004; Samsonov et al., 2006; Jung et al., 2011). Thus the deformation components in the east-west and vertical directions are estimated by fusing the descending and ascending InSAR Line-of-sight displacements. The measurement equations can be written as:

$$\begin{bmatrix} v_\text{e} \\ v_\text{u} \end{bmatrix} = \mathbf{U}\cdot\begin{bmatrix} d_\text{e} \\ d_\text{u} \end{bmatrix} - R_\text{los} \quad (2)$$

where $\boldsymbol{d} = (d_\text{e}, d_\text{u})^T$ is the 2-dimensional deformation vector (east, up) in a local reference frame; $\mathbf{U}$ is a matrix consisting of LOS vectors $(\sin\theta\cdot\cos\phi, \cos\theta)$, where θ is the radar incidence angle and φ is the satellite heading angle; R_los represents the LOS measurement, and (v_e, v_u) is the observation residual. By minimizing the observation residual with least square norm, the deformation vector $\boldsymbol{d} = -(U^T\cdot U)^{-1}\cdot U^T\cdot R_\text{los}$ is obtained.

3 Data

In total 74 archived C-band Envisat ASAR data acquired from 2006 to 2010 from seven descending tracks and 22 L-band ascending ALOS PALSAR data covering part of Weihe basin are employed to monitor the large scale land subsidence and small scale ground fissure deformation (Fig. 1).

4 Results

4.1 Large scale land subsidence deformation

4.1.1 Fenwei basin

The combined PS and SBAS technique is applied to calculate the entire Fenwei basin land subsidence with 67 000 km^2. The annual land subsidence rate map in entire Fenwei basin is shown in Fig. 2, from which the land subsidence area is up to 42 000 km^2. The maximum subsidence rates in different sub-basins range from 2 cm yr^{-1} to over 10 cm yr^{-1}, while the area with subsidence rate larger than 2 cm yr^{-1} is around 2000 km^2. Besides, high speed railway connecting Datong in the north of Shanxi province and Xian, capital of Shaanxi province is designed inevitablely to pass through land subsidence regions. As shown in Fig. 3 the maximum land subsidence rate presented in Qixian, Taiyuan basin was up to

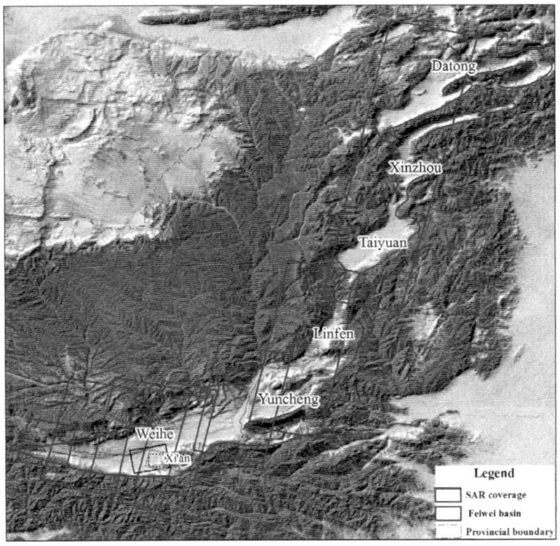

Figure 1. SAR coverage sketch map of Fenwei basin.

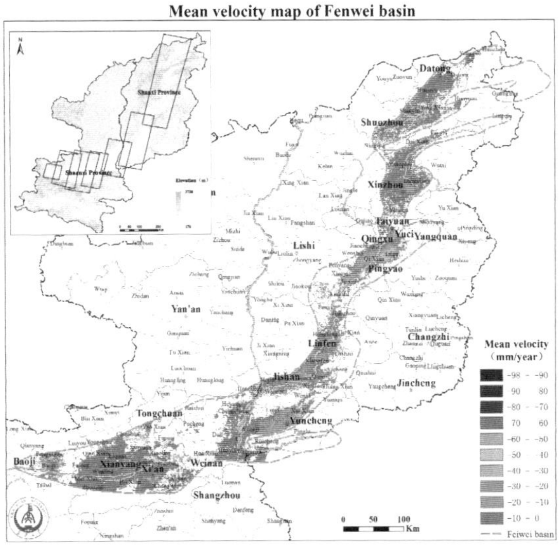

Figure 2. Annual land subsidence rate map in entire Fenwei basin.

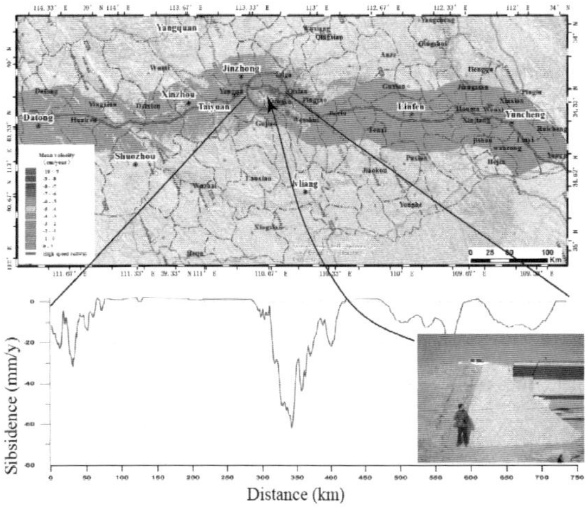

Figure 3. Deformation map along Datong-Yuncheng high speed railway (HSR). Insets are land subsidence profile in Qixian county and field photo of joint of bridge and foundation of HSR under construction.

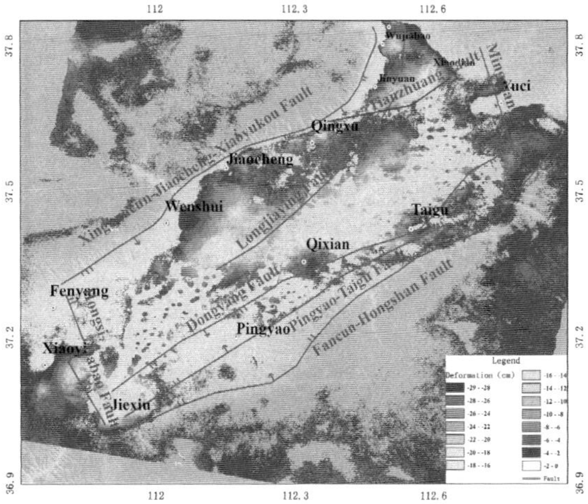

Figure 4. Land subsidence map over Taiyuan basin.

$6\,\mathrm{cm\,yr^{-1}}$, accordingly, the design had altered from ballastless track on bridge to ballast track on fundation (see inset photo of Fig. 3).

4.1.2 Taiyuan basin

Taiyuan basin is one of the severest deformation regions in Fenwei basin, closely check in Fig. 4, we can see that the land subsidence in Taiyuan basin is controlled by the normal faults surrounding the basin. Seven counties, including Taiyuan, Qingxu, Jiaocheng, Qixian, Taigu, Wenshui and Jiexiu, are sufferring from obvious land subsidence. Moreover, some counties have also been affected by the ground fissures deformation.

4.2 Small scale ground fissure deformation

4.2.1 Yuncheng basin

Xia county in Yuncheng basin has been suffering ground fissures hazard for several years. Figure 5 is the annual deformation map calculated from Envisat data during 2008 to 2010. It is clearly to see the normal fault Zhongtiaoshan controls the deformation region. And different deformation mag-

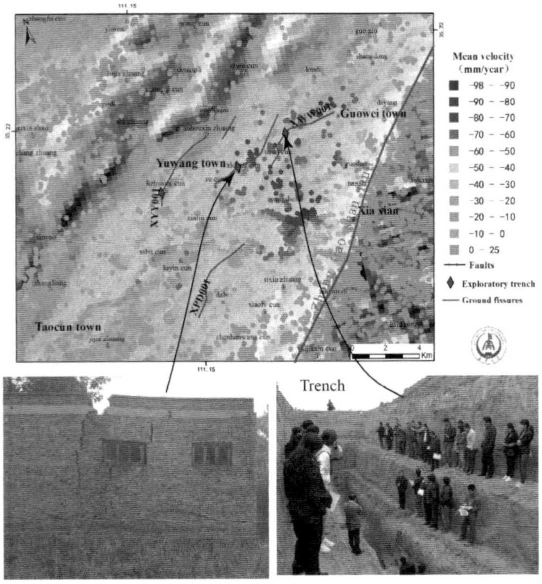

Figure 5. Ground fissure deformation map in Xia County, Yuncheng basin.

nitudes can be monitored in the two sides of each ground fissures. Inset photos show the trench of one fissure, which emerged in 2000, and had around 20 cm surface dislocation and 10 cm opening width (Liu, 2014). The maximum land subsidence occurred close to the Zhongtiaoshan normal fault with maximum subsidence rate around 10 cm yr^{-1}.

4.2.2 Xian city

Up-to-date, in total 14 ground fissures have been occurred in Xian city, which affect the areas of around 250 km^2. Therefore special measures of the constructions in this region must be taken to mitigate the influence of ground fissures movements.

The 2-D deformation components are shown in Fig. 6. The vertical displacements (Fig. 6b) range between −110 and 20 mm, which are much larger than the east-west motions (Fig. 6a). This confirms that the vertical motion dominates the deformation field in Xian. Areas with larger east-west displacements correspond to regions of larger vertical displacements where active ground fissures are present (Fig. 6). Horizontal displacements cannot be ignored in land subsidence measurement when the deformation is large, particularly when ground fissures are active (Qu et al., 2014).

5 Conclusions

Multiple InSAR techniques, including combined SBAS and PS-InSAR, two-dimensional inversion from ascending and descending are employed to large scale land subsidence and

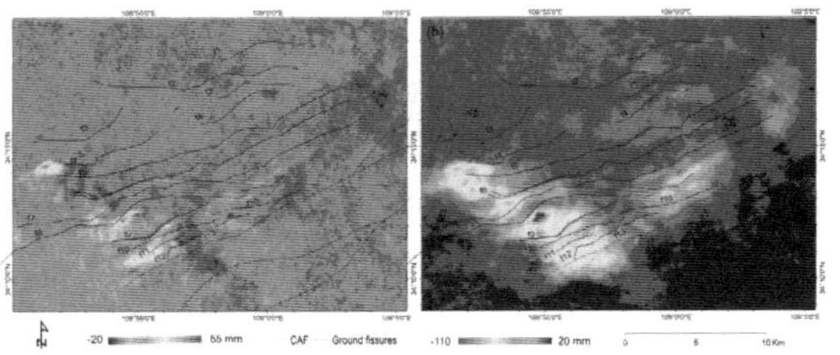

Figure 6. 2-D deformation map in 2009 by using ascending ALOS and descending Envisat ASAR data, whose coverages are marked on Fig. 1 in pink dash line: **(a)** the east-west component of the deformation field; **(b)** the vertical component of the deformation field.

small scale ground fissure deformation calculation. High accuracy InSAR results can not only show different magnitude land subsidence in different sub-basins, but also the different deformation gradient and direction crossing localized ground fissures. The results have been and will be refer to the engineering design and construction in order to mitigate the damage and losses. While, high resolution SAR data will be considered to monitor the up-to-date deformation and fine deformation features regarding small scale fissures.

Acknowledgements. This research is funded by projects of the Natural Science Foundation of China (NSFC) (No. 41372375, 41274005 and 41202245), and the project of the Ministry of Land & Resources, China (No. 1212011220186). Three arc-seconds SRTM DEM are freely downloaded from the website http://rmw.recordist.com/index.html. StamPS package is freely downloaded from the website http://homepages.see.leeds.ac.uk/~earahoo/stamps/index.html. The archived Envisat ASAR data are provided by ESA Category 1 project.

References

Berardino, F., Fornaro, G., Lanari, R., and Sansosti, E.: A new algorithm for surface deformation monitoring based on small baseline differential SAR interferometry, IEEE T. Geosci. Remote, 40, 2375–2383, 2002.

Hooper, A.: Persistent Scaterrer Radar Interferometry for Crustal Deformation Studies and Modeling of Volcanic Deformation, PhD thesis, Stanford University, Palo Alto, 2006.

Hooper, A.: A multi-temporal InSAR method incorporating both persistent scatterer and small baseline approaches, Geophys. Res. Lett., 35, L16302, doi:10.1029/2008GL034654, 2008.

Hooper, A. and Zebker, H.: Phase unwrapping in three dimensions with application to InSAR time series, J. Opt. Soc. Am. A, 24, 2737–2747, 2007.

Hooper, A., Zebker, H., Segall, P., and Kampes, B.: A new method for measuring deformation on volcanoes and other natural terrains using InSAR persistent scatterers, Geophys. Res. Lett., 31, 611–615, 2004.

Hooper, A., Segall, P., and Zebker, H.: Persistent scatterer InSAR for crustal deformation analysis, with application to

Volcán Alcedo, Galapagos, J. Gephys. Res., 112, B07407, doi:10.1029/2006JB004763, 2007.

Jung, H. S., Lu, Z., Won, J., Poland, M., and Miklius, A.: Mapping three-dimensional surface deformation by combining multiple aperture interferometry and conventional interferometry: application to the June 2007 eruption of Kīlauea Volcano, Hawaii, IEEE Geosci. Remote Sens. Lett., 8, 34–38, 2011.

Lanari, R., Mora, O., Mununta, M., Mallorqui, J., Berardino, P., and Sansonsti, E.: A Small Baseline Approach for Investigating Deformation on Full resolution Differential SAR Interferograms, IEEE T. Geosci. Remote, 42, 1377–1386, 2004.

Liu, Y.: The Study on the Mass Mechanism of Ground Fissures in Piedmont Fault on the Plate–Case of Yuncheng Basin Xiaxian Ground Fissures, master thesis, Chang'an University, Xi'an, 2014.

Qu, F., Zhang, Q., Lu, Z., Zhao, C., Yang, C., and Zhang, J.: Land subsidence and ground fissures in Xi'an, China 2005–2012 revealed by multi-band InSAR time-series analysis, Remote Sens. Environ., 155, 366–376, 2014.

Samsonov, S. and Tiampo, K.: Analytical optimization of DInSAR and GPS dataset for derivation of three-dimensional surface motion, IEEE Geosci. Remote Sens. Lett., 3, 107–111, 2006.

Wright, T. J., Parsons, B. E., and Lu, Z.: Toward mapping surface deformation in three dimensions using InSAR, Geophys. Res. Lett., 31, L01607, doi:10.1029/2003GL018827, 2004.

Zebker, H. A. and Villaseno, J.: Decorrelation in interferometric radar echoes, IEEE T. Geosci. Remote, 30, 950–959, 1992.

Proc. IAHS, 372, 311–314, 2015
proc-iahs.net/372/311/2015/
doi:10.5194/piahs-372-311-2015

Exploitation of the full potential of PSI data for subsidence monitoring

M. Crosetto[1], N. Devanthéry[1], M. Cuevas-González[1], O. Monserrat[1], and B. Crippa[2]
[1]Centre Tecnològic de les Telecomunicacions de Catalunya (CTTC), Castelldefels, Barcelona, Spain
[2]Department of Earth Sciences, University of Milan, Milan, Italy
Correspondence to: M. Crosetto (mcrosetto@cttc.cat)

Published: 12 November 2015

Abstract. Persistent Scatterer Interferometry (PSI) is a remote sensing technique used to measure and monitor land deformation from a stack of interferometric SAR images. The main products that can be derived using the PSI technique are the deformation maps and the time series of deformation. In this paper, an approach to apply the PSI technique to a stack of Sentinel-1 images is described. Moreover, the problems encountered during the processing are detailed and an explanation of how they were dealt with is provided. Finally, Sentinel-1 deformation maps and time series obtained over the metropolitan area of Mexico DF are shown.

1 Introduction

Satellite-based radar interferometry represents a powerful tool to detect and monitor ground deformation. For instance, successful land deformation estimates have demonstrated the potential of this technique for a wide range of applications related to landslides (Colesanti et al., 2003; Hilley et al., 2004; Bovenga et al., 2012), land subsidence caused by groundwater pumping (Tomás et al., 2005; Stramondo et al., 2007; Bell et al., 2008; Heleno et al., 2011) and mining (Colesanti et al., 2005; Jung et al., 2007), or urban monitoring (Crosetto et al., 2008; Vallone et al., 2008; Cigna et al., 2012).

This paper is focused on the application of Persistent Scatterer Interferometry (PSI; Ferretti et al., 2000, 2001) for land subsidence detection and measurement. Differential Interferometric Synthetic Aperture Radar (DInSAR) and PSI techniques have undergone a major development in the last two decades (Crosetto et al., 2015), mainly related to the progress accomplished through the exploitation of C-band data from ERS-1/2, Envisat and Radarsat. The data acquired by these satellites cover long periods of time, which is a key aspect to guarantee a long-term deformation monitoring. The advent of very high resolution X-band data in 2007 led to a major step forward in the PSI techniques (Crosetto et al., 2015), including the capability to generate a dense sampling of Persistent Scatterers (PSs), a higher sensitivity to small displacements and a remarkable quality improvement of the time series with respect to the C-band. Furthermore, a new significant improvement is expected to occur as a result of the data acquired by the C-band sensor onboard the Sentinel-1 satellite, launched on 3 April 2014. Sentinel-1 will improve the data acquisition performance with respect to previous sensors, especially for what concerns the area coverage, revisiting time and data throughput, considerably increasing the deformation monitoring potential of the PSI technique.

This article describes the first PSI processing experience of the authors with Sentinel-1 data by defining the processing strategy and illustrating the first results obtained over the metropolitan area of Mexico DF, i.e. the accumulated deformation map and the time series of deformation.

2 Data processing

The key steps of the procedure used to process the Sentinel-1 imagery are described in this section. The first stage of the processing is the generation of the interferograms. The key step of this stage is image coregistration, which in the case of Terrain Observation with Progressive Scan (TOPS) images (Prats-Iraola et al., 2012; Meta et al., 2010) requires a high accuracy. When this accuracy is not achieved, large linear terms appear in the interferograms. In this study, an approach that works burst-wise and does not require precise orbits is used, which causes an azimuth linear term in each interferogram. This linear term has been estimated image-wise and removed from the interferograms. To clean the linear term from the interferograms, a phase unwrapping using the Minimum Cost Flow method (Costantini, 1998; Costantini et al., 1999) was performed. This step has been executed over the interferograms multilooked 2 in azimuth and 10 in range in order to reduce the noise. Then, a direct integration, which consists in retrieving the image phases with respect to a given image, is done over the unwrapped interferograms. Finally, the linear planes have been estimated from the phase images and subsequently removed from the original interferograms.

Two key steps of the PSIG (PSI chain of the Geomatics Division of CTTC) procedure (Crosetto et al., 2011; Devanthéry et al., 2014) have been applied to derive the accumulated deformation map and the deformation time series: the so-called Cousin PSs (CPSs) selection and the 2 + 1-D phase unwrapping. These two steps are briefly explained below:

- CPS selection: the aim of the CPS selection is to derive a set of PSs characterized by a moderate spatial phase variation to ensure a correct phase unwrapping and guarantee a uniform CPS coverage. This selection works at full resolution and is suitable to cover wide areas. It consists of a candidate CPSs selection step based on an iterative phase similitude criterion and a phase unwrapping consistency check. The procedure consists in an iterative search propagation criterion that starts with the selection of one or more seed PSs located on the ground, with no motion or thermal expansion and characterized by small noise. The candidate CPSs are those PS candidates that satisfy the condition:

  $$\left|\Delta\varphi_{i,\mathrm{SEED}}^{\mathrm{unw}}\right|_{90\,\%} < \mathrm{Thr}$$

  where $|\Delta\varphi_{i,\mathrm{SEED}}^{\mathrm{unw}}|_{90\,\%}$ is the 90th percentile of the unwrapped phase differences between the seed and the ith PS candidate over the minimum independent interferogram configuration network, and Thr is the phase difference threshold, which is defined as a function of the distance from the seed to account for the atmospheric effects.

- 2 + 1-D phase unwrapping: this step is performed over the selected CPSs to derive the deformation measurements and some quality index tools of the estimations. It consists in a spatial 2-D phase unwrapping performed on each interferogram using the Minimum Cost Flow method (Costantini, 1998; Costantini et al., 1999), and a 1-D phase unwrapping performed pixel wise over the interferogram network. It uses an iterative Least Squares (LS) procedure (Baarda, 1968; Förstner, 1986; Björck, 1996), which exploits the integer nature of the unwrapping errors. It is based on the estimation of the following system of equations:

  $$\Delta\varphi_{\mathrm{MS}} = \widetilde{\varphi}_{\mathrm{S}} - \widetilde{\varphi}_{\mathrm{M}}$$
  $$\widetilde{\varphi}_0 = 0$$

  where $\Delta\varphi_{\mathrm{MS}}$ are the unwrapped interferometric phases (the observations), "S" and "M" are the slave and master images (the unknowns).

  The system is solved by means of an iterative LS driven by two key parameters: the residuals $\hat{\varphi}_{\mathrm{res}}$, which are the difference between the observations and estimated phases, and the redundancy of the network. If there is sufficient system redundancy, typically more than 10 interferograms per image, the procedure detects and corrects the unwrapping errors.

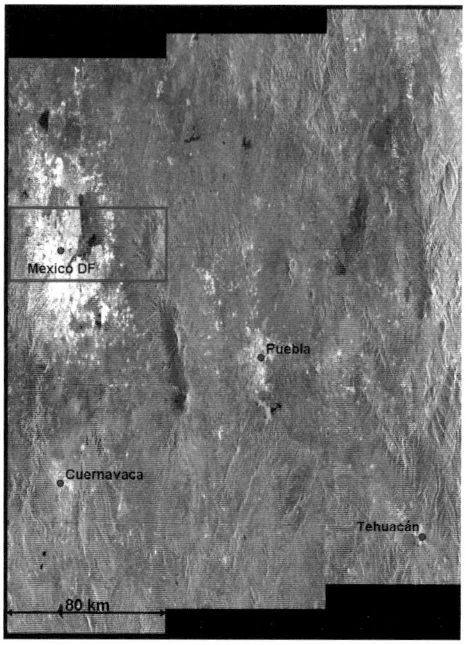

Figure 1. SAR amplitude of the Mexico dataset. Scene acquired by the Sentinel-1 satellite covering approximately 250 by 180 km. The red square indicates the area processed using the PSI technique.

In this study, ten Interferometric Wide swath (IW) Single Look Complex (SLC) Sentinel-1 descending images spanning the period between October 2014 and January 2015 were used to derive the deformation map and time series.

3 Results

The results of the burst-wise procedure for the processing of Sentinel-1 data are illustrated in this section. A IW SLC Sentinel-1 scene covers approximately 250 × 180 km and is composed of three swaths containing nine burst each. The

Figure 2. Three examples of the sample density of Cousin Persistent Scatterers (CPSs) achieved in the urban area of the processed burst. Each example shows an orthoimage and the SAR amplitude image of the same area with the CPSs processed with the PSI technique.

results of the PSI processing of one burst (Fig. 1), which encompass a significant portion of Mexico DF, are shown in this paper. The spatial resolution of these images is approximately 3 by 20 m in range and azimuth, respectively. The area processed cover approximately 80 by 25 km and a density of 575 CPS km^{-2} has been achieved (Fig. 2), with a total of 720 882 CPSs processed. Sentinel-1 SAR imagery offers wide-area coverage associated to a relatively high spatial resolution, therefore allowing to obtain a comprehensive outlook of the deformation phenomena occurring in wide areas while, at the same time, maintaining the capability to measure individual features such as buildings or infrastructures. It is also important to note that the PSI technique is opportunistic, which means that it is only able to estimate deformation over the available CPSs, i.e. those points where PSI phases maintain good quality over time to get reliable deformation estimates.

Figure 3 shows a deformation map of the burst under study derived from 10 Sentinel-1 images and a total of 45 interferograms. Positive values (blue) indicate displacements towards the SAR, while the negative ones (red) denote displacements away from the SAR sensor. Note that these values refer to the SAR Line-of-Sight (LOS). The map shows a large area affected by subsidence (red) with displacement values of up to 9 cm occurred during the four months of observations.

The time series of deformation (TS) is one of the most interesting and advanced products derived with the PSI technique. TSs display the deformation history of each CPS over the observed period. Several examples of TS of deformation

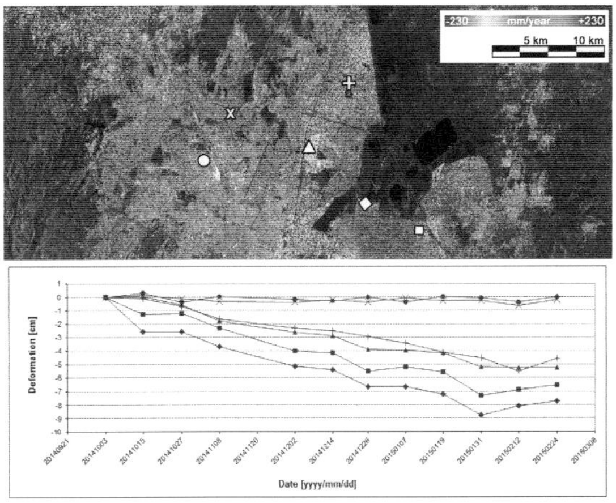

Figure 3. Deformation map (top panel) of the processed burst encompassing a large area of Mexico DF superimposed over a SAR amplitude image, and six examples of time series of deformation (bottom panel) from the same area.

are shown in Fig. 3. These TSs show different behavior; some of them display a stable behaviour, which is related to the green colour in the deformation map, while others show subsidence values of up to 9 cm which are associated to red in the deformation map.

4 Conclusions

The study presented here is the first attempt of the authors to process Sentinel-1 imagery using the PSI technique. A brief description of the processing chain is provided followed by an analysis of the results achieved, i.e. the deformation map and time series of deformation.

The PSI processing chain described in this work was carried out without performing any additional steps at the image processing level. However, the chain employed in this study is reliable to run burst-wise in cases where the coherence is relatively high.

Acknowledgements. This work has been partially funded by the Spanish Ministry of Economy and Competitiveness through the project MIDES "Monitorización y análisis sistemáticos de deformaciones del terreno, estructuras e infraestructuras sobre areas muy extensas" (Ref. CGL2013-43000-P).

References

Baarda, W.: A Testing Procedure for Use in Geodetic Networks, Rijkscommissie voor Geodesie, Delft, the Netherlands, 1968.

Bell, J. W., Amelung, F., Ferretti, A., Bianchi, M., and Novali, F.: Permanent scatterer InSAR reveals seasonal and longterm aquifer-system response to groundwater pumping

and artificial recharge, Water Resour. Res., 44, W02407, doi:10.1029/2007WR006152, 2008.
Björck, Å.: Numerical Methods for Least Square Problems, Siam, Philadelphia, PA, USA, 1996.
Bovenga, F., Wasowski, J., Nitti, D. O., Nutricato, R., and Chiaradia, M. T.: Using COSMO/SkyMed X-band and ENVISAT Cband SAR interferometry for landslides analysis, Remote Sens. Environ., 119, 272–285, 2012.
Cigna, F., Osmanoglu, B., Cabral-Cano, E., Dixon, T. H., Ávila-Olivera, J. A., Garduño-Monroy, V. H., DeMets, C., and Wdowinski, S.: Monitoring land subsidence and its induced geological hazard with Synthetic Aperture Radar Interferometry: A case study in Morelia, Mexico, Remote Sens. Environ., 117, 146–161, 2012.
Colesanti, C., Ferretti, A., Prati, C., and Rocca, F.: Monitoring landslides and tectonic motions with the Permanent Scatterers Technique, Eng. Geol., 68, 3–14, 2003.
Colesanti, C., Le Mouélic, S., Bennani, M., Raucoules, D., Carnec, C., and Ferretti, A.: Detection of mining related ground instabilities using the permanent scatterers technique – a case study in the east of France, Int. J. Remote Sens., 26, 201–207, 2005.
Costantini, M.: A novel phase unwrapping method based on network programming, IEEE T. Geosci. Remote., 36, 813–821, 1998.
Costantini, M., Farina, A., and Zirilli, F.: A fast phase unwrapping algorithm for SAR interferometry, IEEE T. Geosci. Remote, 37, 452–460, 1999.
Crosetto, M., Biescas, E., Duro, J., Closa, J., and Arnaud, A.: Generation of Advanced ERS and Envisat Interferometric SAR Products Using the Stable Point Network Technique, Photogramm. Eng. Rem. S., 74, 443–450, 2008.
Crosetto, M., Monserrat, O., Cuevas, M., and Crippa, B.: Spaceborne Differential SAR Interferometry: Data Analysis Tools for Deformation Measurement, Remote Sensing, 3, 305–318, 2011.
Crosetto, M., Monserrat, O., Cuevas-González, M., and Devanthéry, N.: Persistent Scatterer Interferometry: a review, ISPRS J. Photogram. Remote Sens., submitted, 2015.
Devanthéry, N., Crosetto, M., Monserrat, O., Cuevas-González, M., and Crippa, B.: An approach to Persistent Scatterer Interferometry, Remote Sensing, 6, 6662–6679, 2014.
Ferretti, A., Prati, C., and Rocca, F.: Nonlinear subsidence rate estimation using permanent scatterers in differential SAR interferometry, IEEE T. Geosci. Remote, 47, 2202–2212, 2000.
Ferretti, A., Prati, C., and Rocca, F.: Permanent scatterers in SAR interferometry, IEEE T. Geosci. Remote, 39, 8–20, 2001.
Förstner, W.: Reliability, gross error detection and self-calibration, ISPRS Commission III Tutorial on Statistical Concepts for Quality Control, ISPRS Int. Arch. Photogramm., 26, 1–34, 1986.
Heleno, S. I. N., Oliveira, L. G. S., Henriques, M. J., Falcão, A. P., Lima, J. N. P., Cooksley, G., Ferretti, A., Fonseca, A. M., Lobo-Ferreira, J. P., and Fonseca, J. F. B. D.: Persistent Scatterers Interferometry detects and measures ground subsidente in Lisbon, Remote Sens. Environ., 115, 2152–2167, 2011.
Hilley, G. E., Bürgmann, R., Ferretti, A., Novali, F., and Rocca, F.: Dynamics of slow-moving landslides from Permanent Scatterer análisis, Science, 304, 1952–1955, 2004.
Jung, H., Kim, S., Jung, H., Min, K., and Won, J.: Satellite observation of coal mining subsidence by permanent scatterer analysis, Eng. Geol., 92, 1–13, 2007.
Meta, A., Mittermayer, J., Prats, P., Scheiber, R., and Steinbrecher, U.: TOPS imaging with TerraSAR-X: Mode design and performance analysis, IEEE T. Geosci. Remote, 48, 759–769, 2010.
Prats-Iraola, P., Scheiber, R., Marotti, L., Wollstadt, S., and Reigber, A.: TOPS interferometry with TerraSAR-X, IEEE T. Geosci. Remote, 50, 3179–3188, 2012.
Stramondo, S., Saroli, M., Tolomei, C., Moro, M., Doumaz, F., Pesci, A., Loddo, F., Baldi, P., and Boschi, E.: Surface movements in Bologna (Po Plain – Italy) detected by multitemporal DInSAR, Remote Sens. Environ., 110, 304–316, 2007.
Tomás, R., Márquez, Y., Lopez-Sanchez, J. M., Delgado, J., Blanco, P., Mallorquí, J. J., Martínez, M., Herrera, M., and Mulas, J.: Mapping ground subsidence induced by aquifer overexploitation using advanced Differential SAR interferometry: Vega Media of the Segura river (SE Spain) case study, Remote Sens. Environ., 98, 269–283, 2005.
Vallone, P., Giammarinaro, M. S., Crosetto, M., Agudo, M., and Biescas, E.: Ground motion phenomena in Caltanissetta (Italy) investigated by InSAR and geological data integration, Eng. Geol., 98, 144–155, 2008.

Proc. IAHS, 372, 315–321, 2015
proc-iahs.net/372/315/2015/
doi:10.5194/piahs-372-315-2015

Open Access PIAHS

Evolution of the techniques for subsidence monitoring at regional scale: the case of Emilia-Romagna region (Italy)

G. Bitelli[1], F. Bonsignore[2], I. Pellegrino[3], and L. Vittuari[1]

[1]University of Bologna, DICAM – Dept of Civil, Chemical, Environmental and Materials Engineering, Bologna, Italy
[2]ARPA-Emilia-Romagna – Regional Agency for Environmental Prevention in Emilia-Romagna, Bologna, Italy
[3]Regione Emilia-Romagna, Direzione Generale Ambiente e Difesa del Suolo e della Costa, Bologna, Italy

Correspondence to: G. Bitelli (gabriele.bitelli@unibo.it)

Published: 12 November 2015

Abstract. The recent decades have seen a significant evolution of the methodologies and techniques for the monitoring of subsidence on a regional scale: from the traditional levelling technique to GNSS and finally to SAR interferometry. The case study of Emilia-Romagna, Italy, is a prime example of this evolution.

As known, the Emilia-Romagna plain is subject to a phenomenon of subsidence with a natural and an anthropogenic component, both of varying amounts depending on the area. The first contributes a few mm/year; the second, particularly evident in the last 60 years, is mainly correlated to excessive withdrawal of fluids from underground and reaches higher values (in the past, subsidence rates of several cm per year were observed in the Po delta and near Bologna).

The geodetic monitoring of subsidence started in the 1950s by different entities, establishing and measuring levelling networks of varying size and with various characteristics, mainly located where the phenomenon was most clearly manifest. These local initiatives were not able to provide a consistent understanding of the phenomenon throughout the entire Emilia-Romagna plain.

The first regional-scale monitoring of the Emilia-Romagna plain was initiated in 1999, with a large levelling network (about 3000 km) and a coupled network of 60 GNSS points. In subsequent years, the monitoring approach has mainly focused on the most modern remote sensing techniques integrated with each other, with the adoption of the method DInSAR calibrated to a GNSS Continuous Operating Reference Stations (CORS) database. The application of DInSAR methods resulted in subsidence maps with a greater level of detail.

The paper analyzes the methodology choices made during 1999–2012, through three successive campaigns that adopted and integrated the different techniques.

1 Introduction

Monitoring of subsidence at regional level was done in the last decades with different approaches and methodologies, following the development of the technology in the land surveying disciplines.

From the scientific literature regarding subsidence measurement, and more specifically from the papers presented at the latest editions of the International Symposia on Land Subsidence, the evolution is evident from the three main techniques adopted in this process: geodetic levelling, GNSS and SAR interferometry.

Whereas geodetical infrastructures are available, the integration of the three techniques can constitute the best approach to the problem, when datum definition and transformations permit all the datasets can be aligned into a single reference frame. Large levelling networks are still very widespread and utilize a well-established method, especially useful when historical data for a large number of benchmarks are available and can be updated; the accuracy level is lower than 1 mm per km. GNSS networks (Hofmann-Wellenhof et al., 2008), mainly implemented today by permanent stations, can provide point information with a lower spatial density and good accuracy, depending on different surveying strategies and the time span of the acquisition. For example, using current technologies of measurements and of data analysis, GNSS permanent stations are able to provide measurements of velocity components with an accuracy of about $1\,\mathrm{mm\,year^{-1}}$ in the vertical component, after at least 3 years of continuous acquisition (Kontny and Bogusz, 2012).

SAR interferometry, using the Persistent Scatterers approach (Ferretti et al., 2001; Hanssen, 2001; Ketelaar, 2009), can provide high density point information about the rate of the phenomenon with precision on the order of 1–2 mm per year.

The Emilia-Romagna region, Italy, is a prime example of this evolution. The Emilia-Romagna plain is subject to the phenomenon of subsidence with a natural and an anthropogenic component, both of varying amounts depending on the areas. The first contributes a few mm/year; the second, particularly evident in the last 60 years, is mainly correlated to excessive withdrawal of fluids from underground and reaches higher values.

Since the 1950s different agencies have designed and managed subsidence monitoring networks, measured by geodetic levelling, in the areas in which the phenomenon had become particularly evident (Caputo et al., 1970; Pieri and Russo, 1984). These local monitoring networks highlighted maximum subsidence rates of more than $25\,\mathrm{cm\,year^{-1}}$ during 1951–1962 in the Po delta, more than $11\,\mathrm{cm\,year^{-1}}$ during 1974–1981 near Bologna and more than $4\,\mathrm{cm\,year^{-1}}$ during 1972–1977 in the Ravenna area (Bissoli et al., 2010). More recently, rates of subsidence have significantly decreased but still remain important in some places.

This paper describes the experience carried out using geodetic techniques to monitor the phenomenon and the results of the three regional campaigns completed since 1999 by ARPA-Emilia-Romagna (Regional Agency for Environmental Prevention in Emilia-Romagna Region) on behalf of the Emilia-Romagna Region and in collaboration with the DICAM Department of Bologna University.

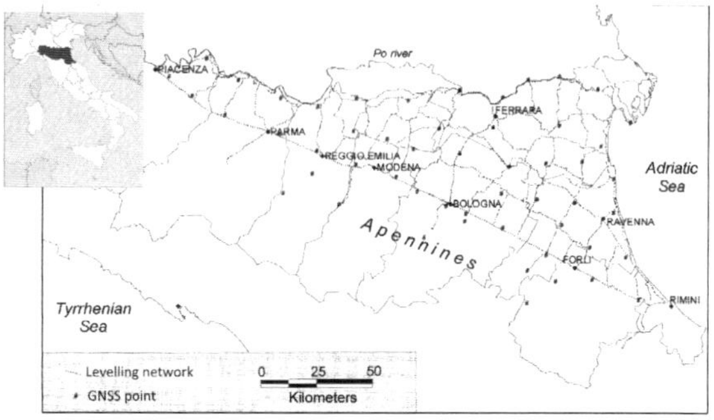

Figure 1. Levelling network and GNSS points for the 1999 campaign.

2 The first campaign (1999 and 2002)

The first campaign designed and instituted the geodetic infrastructure composed by two networks (Fig. 1): a high precision levelling network composed of over 2300 benchmarks, and a GNSS network consisting of about 60 points to be surveyed by long static sessions (Benedetti et al., 2000; Bitelli et al., 2000).

The levelling network was conceived with the aim to connect, where possible, a number of existing networks, dealing with a very heterogeneous situation: each authority or agency used different specifications and, therefore, different precision for their networks, different observation periods and different reference benchmarks. The 1999 levelling survey, accomplished in 75 days, was one of the first uses of large networks entirely surveyed by digital levels; the a-posteriori mean square error was less than 1 mm per km.

The scope of GNSS surveys was two-fold: to improve the gravimetric geoid model in the area and to establish quick and inexpensive control of some points in order to decide, for example, the measurement repetition frequencies of some levelling lines.

Both the levelling and the GNSS networks were measured for the first time in 1999; the measurement of the GNSS network was repeated in 2002 (Bonsignore et al., 2003).

A first attempt to evaluate the subsidence from this first survey was done by comparing some benchmarks heights with those from previous surveys, after some homogenization procedures in the vertical datum. The results largely showed the main trends of the phenomenon.

A specialized relational database was developed as an archiving and querying system, coupled with a WebGIS implementation, to improve the management of the huge amount of data coming from historical and 1999 and 2002 records.

3 The second campaign (2005)

In 2005, the knowledge of the subsidence phenomenon was updated through the integration of two techniques: high precision geodetic levelling of a subnet of the regional network (more than 1000 benchmarks, about 50 % of the 1999 network; Fig. 1), and radar interferometric analysis, adopting the PSInSAR™ technique (Ferretti et al., 2001) in the standard PS analysis low resolution method. The study covered the whole plain territory of the region, about $11\,000\,\mathrm{km^2}$. The interferometric analysis focused, in particular, on two distinct periods: the first time interval 1992–2000 refers to the processing of data from the European Space Agency's ERS1 and ERS2 satellites, while the second one, for 2002–2006, refers

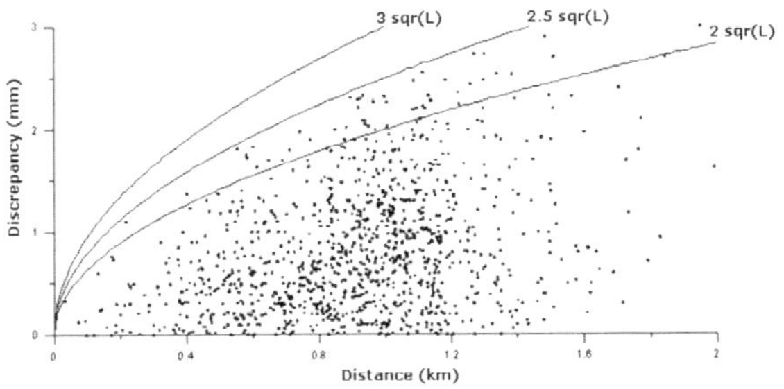

Figure 2. Levelling campaign 2005: discrepancies vs. distance for the forward and backward runs of the section height differences.

to the processing of data from Envisat (ESA) and Radarsat (Canadian Space Agency) satellites.

The levelling results and the analysis of the discrepancies between forward and backward runs of the benchmark height differences showed the good quality of the campaign (Fig. 2), with the indication that it was possible to reduce the threshold level on the discrepancy to the value of $2 \times \sqrt{L}$.

A critical aspect for the comparison of results from the levelling campaign and from the interferometry was the choice of a reference point to be considered fixed in subsequent measurements or processing: the origin was fixed at a benchmark situated near Sasso Marconi (Bologna), in the Apennines foothills, located in a position almost barycentric (in a longitudinal sense) with respect to the network and considered stable in previous large surveys.

During the analysis of the results (Bissoli et al., 2010), some tests were performed – within a GIS – to have an objective evaluation of the intrinsic reliability of the radar results and of the comparison with results of the levelling campaigns for similar periods (levelling 1999–2005 vs. PSInSAR™ 2002–2006). The interferometric technique provided measurements of vertical relative motion characterized by a precision of the same order of those obtainable using levelling. The distribution of measurement points, in the case of urban environments, however was considerably higher than the number of benchmarks usually measured per square kilometer using levelling techniques. Standard deviation and coherence values derived from the interferometric processing for the PS points were determined for the three types of radar sensors used. As expected, the estimated velocities for the period 2002–2006 showed higher standard deviation values and lower coherence than during 1992–2000, because of the reduced number of SAR images available.

The results associated with PS points were converted to a grid model by geostatistical interpolation, in order to provide a more expressive representation of the subsidence velocity for the entire Emilia-Romagna plain (Fig. 3).

4 The third campaign (2011)

The third and last campaign, which will be described in more detail, was accomplished by integrating SAR interferometry and GNSS data collected at permanent stations (Bitelli et al., 2014). SAR analysis was performed by TRE – Tele-Rilevamento Europa through the SqueeSAR™ technology, patented algorithm PSInSAR™ second generation (Ferretti et al., 2011), using a RADARSAT-1 ascending dataset, acquired between 2006 and 2011. The analysis led to the identification of more than 2 000 000 measurement points, with an average point density of 200 MP km^{-2}. The full resolution data was sub-sampled on a 100×100 m grid using the radar coherence as selection parameter in order to obtain around 320 000 PS (Permanent Scatterers) and DS (Distributed Scatterers) points, more than double the previous campaign. Based on previous knowledge of the area, supported by GNSS measurements, it was assumed that horizontal motion was negligible, and the velocity of movement derived by SAR had been reprojected from the Line Of Sight to the vertical.

4.1 SqueeSAR™ dataset assessment and outliers rejection

In order to filter the outliers present in the SqueeSAR dataset which degrade the results, some procedures were implemented to identify points characterized by "abnormal" velocities compared to their surroundings, indicative of different phenomena not related to the subsidence of a regional nature, object of this study.

These "abnormal" points may arise due to extremely localized movements related to sagging of individual structures, settling of recent constructions, or in some cases, particularly in agricultural areas, they could be attributable to changes in soil moisture that involve phase shifts in the SAR signal, erroneously identified as movements.

Identifying "abnormal" points is an important process, which must be conducted with objective methods based on statistics. Given the large amount of data to be processed, the main part of the analysis must be automated, but ultimate identification of outliers should be supported by specific assessments guided by a visual inspection.

4.2 Definition and calibration of velocity datum

In planning this campaign, a first attempt was to overcome the previous situation for the definition and calibration of the velocity datum. The study of subsidence, in a similar way as any other geodetic phenomenon that is described with respect to a geographically defined reference frame, requires the definition of a geometrical and velocity datum to allow the comparison between observations collected at different times (epochs). This datum can be defined according to geometrical criteria or external measurements (i.e. GNSS time

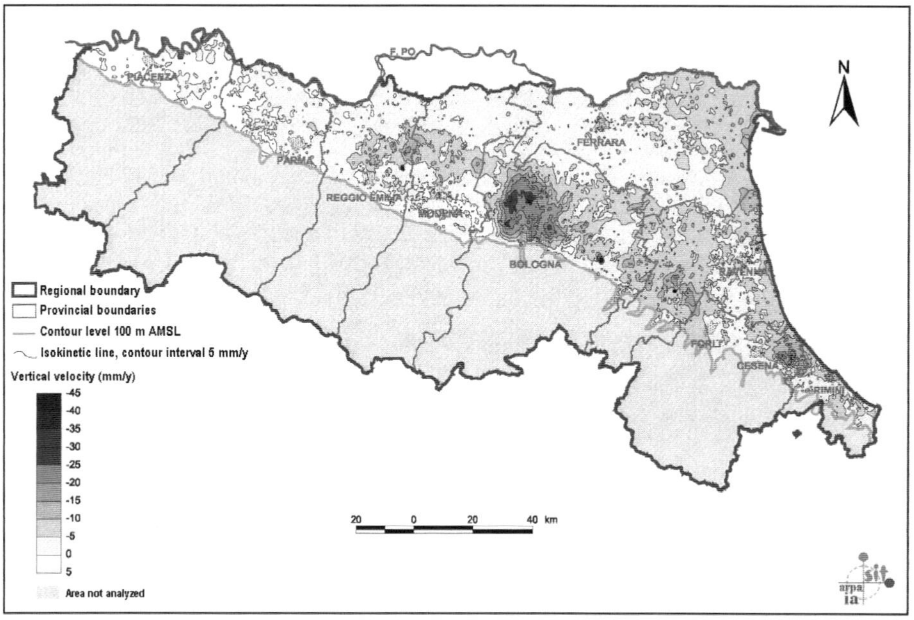

Figure 3. Subsidence map (mm year^{-1}) for the period 2002–2006.

series), or according to more qualitative criteria related to geological setting.

For the levelling networks, the vertical datum was defined, for example, through the a priori knowledge of the elevation of at least one reference point, taking into account its vertical velocity if it is not 0 mm year^{-1}. As previously mentioned, a levelling network was designed to monitor the regional subsidence, which was referenced to a benchmark located in the Apennines foothills close to Bologna. The stability of this benchamrk was verified through the comparison of vertical velocity obtained through the geodetic observation collected at the co-located GNSS-VLBI Observatory of Medicina (Bo), with respect to the estimation of the vertical velocity obtained from the repeated geodetic levelling between the reference benchmark and the GNSS-VLBI Observatory.

In the third campaign, no geodetic levelling measurement were done, and the datum was defined, as much as possible, in an "absolute" way for the PSInSAR™ analysis through a network of 17 GNSS permanent stations (Fig. 4), located in the study area and belonging to three different geodetic networks:

- three stations belonging to global networks IGS-EUREF (Global and European Service);
- four stations belonging to the "Rete Nazionale Integrata Gps" (RING) of the INGV Institute, materialized mainly for studies of Geodynamics;
- ten stations belonging to the network of permanent GNSS stations established by Foger – Foundation of professional land surveyors of Emilia-Romagna, for technical purposes and, in particular, for real time positioning.

GNSS-derived positioning time series were processed from acquisitions made every 30 s during 2007–2011. The daily solutions of each station were computed with respect to ITRF2008. For each component (North, East, Up) of the time series, an average linear velocity (mm year^{-1}) was estimated.

Six stations served initially to establish the datum of the interferometric analysis, and the other 11 were used for validation through the comparison between the rate of vertical movement provided by GNSS and rates derived from the interferometric measurement points (MP) distributed close to each GNSS station. These points were individuated by a GIS procedure, because the locations were not generally coincident.

The selection of the MP for the comparison was made for a small area analyzed at full resolution using the SqueeSAR™ method in the neighbourhood of the GNSS stations (Fig. 5). The velocities of the selected points (subjected to a preliminary visual screening) were then averaged, giving each point a weight based on its coherence. The resulting values of the differences with GNSS measurements for the six calibration stations were then calculated. Since InSAR results can be affected by low frequency artefacts due to uncompensated atmospheric component or orbital inaccuracy, a first order surface was estimated and removed from the differences between InSAR and GNSS velocities. The cal-

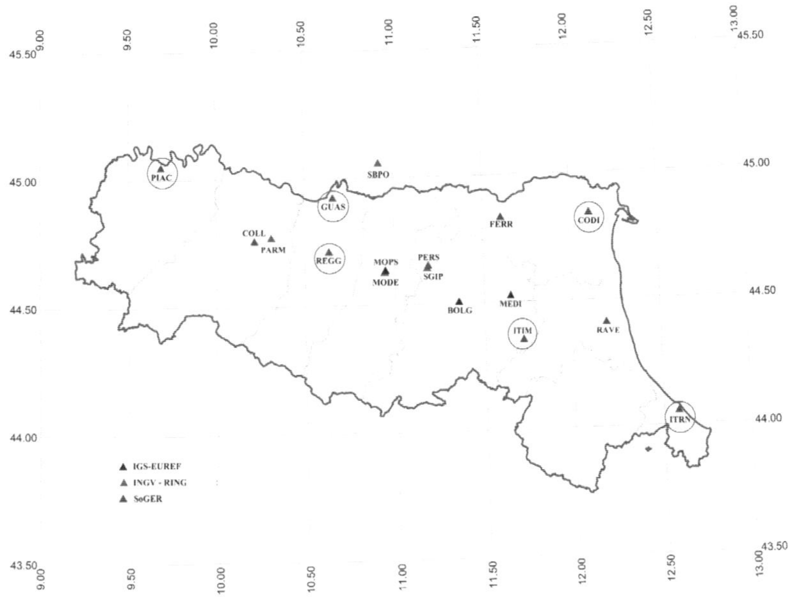

Figure 4. Positions of the GNSS permanent stations used. The symbols with a circle indicate the stations used as calibration to establish the datum of the interferometric analysis.

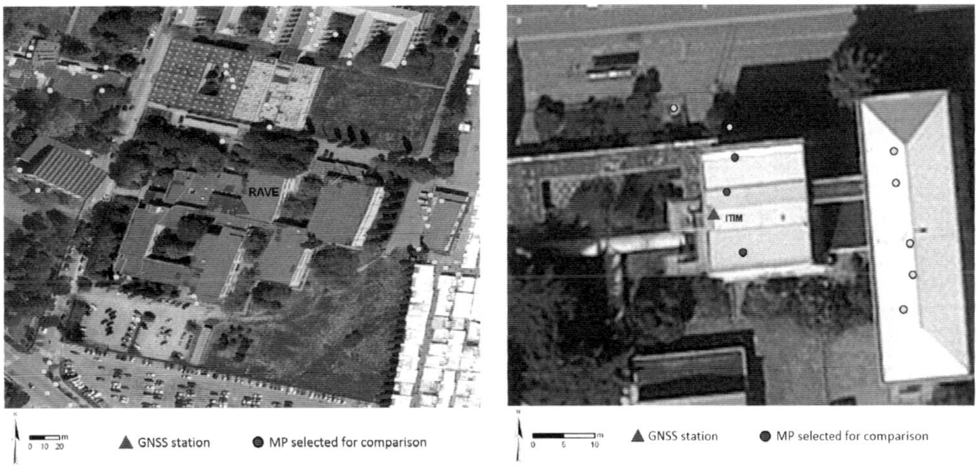

Figure 5. Distribution of measurement points identified by the analysis SqueeSAR™ at full resolution in the neighborhood of two GNSS stations; selected points are in red. The positions of GNSS antennas are marked with red triangles.

ibrated vertical velocity was then validated with the remaining 11 GNSS stations, adopting the same procedure for the identification of GNSS and MP match. The validation highlighted the very good agreement (Table 1) between the calibrated Squee-SAR™ data and the GNSS network data (average residual $= -0.36 \pm 1.39$ mm year^{-1}). Some exception could be related to different causes: GNSS problems, low number or radar quality of the selected MP, relative position between MP and GNSS antenna, estimation inaccuracy, presence of horizontal motion component, spatial and/or temporal motion variations in the monitored period. Considering all the factors, an overall uncertainty of the whole analysis is expected, as a precaution, to be within the value of about ± 2 mm year^{-1}.

Table 1. Comparison of the residual in the velocities (mm year^{-1}) between the values obtained from the GNSS permanent stations and those derived from the analysis of the closest radar measurement points.

Station	GNSS displ. rate [mm year^{-1}]	# MP	MP displ. rate [mm year^{-1}]	Residual [mm year^{-1}]
CODI	-1.88 ± 0.17	8	-2.18 ± 1.12	-0.30
GUAS	-0.09 ± 0.72	8	0.74 ± 0.59	0.83
ITIM	-0.58 ± 0.32	3	1.01 ± 0.92	1.59
ITRN	-1.69 ± 0.31	5	-2.41 ± 0.75	-0.72
PIAC	1.03 ± 0.66	13	0.9 ± 1.08	-0.12
REGG	-0.83 ± 0.29	7	-2.10 ± 1.13	-1.27
BOLG	-2.34 ± 1.83	7	-4.25 ± 1.17	-1.91
MEDI	-1.28 ± 0.21	5	-1.82 ± 0.93	-0.54
MOPS	-1.69 ± 1.72	9	-2.19 ± 0.70	-0.50
MODE	-3.58 ± 1.95	7	-3.15 ± 0.75	0.42
PARM	2.95 ± 0.44	19	0.65 ± 0.78	-2.29
SGIP	-4.96 ± 0.25	18	-5.74 ± 0.73	-0.78
COLL	0.48 ± 0.50	4	1.25 ± 0.57	0.77
FERR	0.92 ± 0.24	9	-2.54 ± 0.83	-3.46
PERS	-8.08 ± 0.36	15	-7.45 ± 0.79	1.34
RAVE	-4.78 ± 0.37	15	-3.57 ± 0.77	1.12

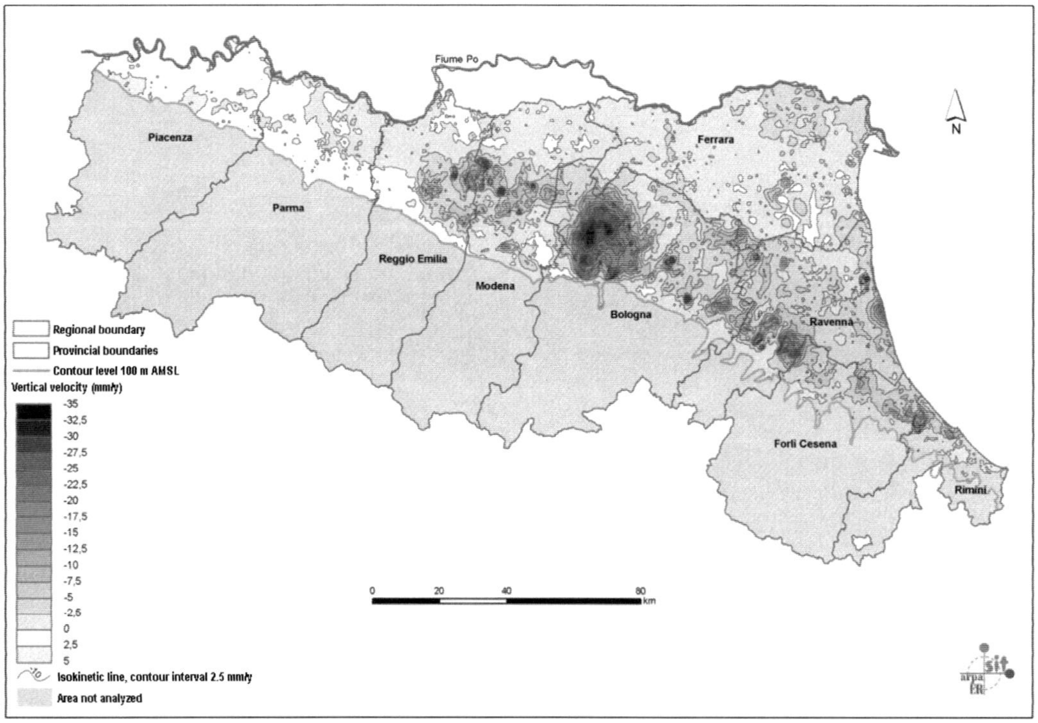

Figure 6. Subsidence map (mm year^{-1}) for the period 2006–2011.

4.3 Subsidence map production

From the final data set of 315 371 points, geostatistical interpolation methods were applied to compute a dense regular grid (100 m × 100 m) of ground vertical movements for the Emilia-Romagna plain for 2006–2011 and to subsequently generate a contour-based thematic map by isokinetic isolines with a contour interval of 2.5 mm $year^{-1}$ (Fig. 6). The results showed in different areas a reduction of the subsidence phenomenon.

5 Conclusions

A short description of the subsidence monitoring for the plain area of the Emilia-Romagna Region (Italy) was presented, describing the methodologies used in the 1999–2012 period. During three successive studies, different approaches were adopted integrating the main available geodetic techniques. Some aspects of the whole experience are described, highlighting the main findings in terms of better spatial density, accuracy, definition and calibration of a velocity datum.

The next survey is scheduled in 2016, following primarily the method adopted in the 2011 campaign.

Acknowledgements. The authors gratefully acknowledge the review process for the helpful and constructive comments and suggestions.

References

Benedetti, G., Draghetti, T., Bitelli, G., Unguendoli, M., Bonsignore, F., and Zavatti, A.: Land subsidence in the Emilia-Romagna region, Northern Italy, in: Proc. 6th Int. Symp. on Land Subsidence (SISOL), edited by: Carbognin, L., Gambolati, G., and Jonhnson, A. I., Vol. I, 61–76, Padova, 2000.

Bissoli, R., Bitelli, G., Bonsignore, F., Rapino, A., and Vittuari, L.: Land subsidence in Emilia-Romagna Region, northern Italy: recent results, in: Land Subsidence, Associated Hazards and the Role of Natural Resources Development (Proc. EISOLS, Queretaro, Mexico, October 2010), IAHS Publ. 339, 307–311, Wallingford, 2010.

Bitelli, G., Bonsignore, F., and Unguendoli, M.: Levelling and GPS networks for ground subsidence monitoring in the Southern Po Valley, J. Geodyn., 30, 355–369, 2000.

Bitelli, G., Bonsignore, F., Del Conte, S., Novali, F., Pellegrino, I., and Vittuari, L.: Integrated use of Advanced InSAR and GPS data for subsidence monitoring, in: Engineering Geology for Society and Territory, edited by: Lollino, G., Manconi, A., Guzzetti, F., Culshaw, M., Bobrowsky, P., and Luino, F., 5, III, 147–150, Springer, 2014.

Bonsignore, F., Draghetti, T., Fontana, D., Rapino, A., and Unguendoli, M.: Emilia-Romagna Subsidence Monitoring Network. GPS network: measurements 2002, in: Proc. 4th European Congress on Regional Geoscientific Cartography and Information Systems, Bologna, 17–20 June, 284–286, 2003.

Caputo, M., Pieri, I., and Unguendoli, M.: Geometric investigation of the subsidence in the Po delta, Boll. Geodesia Teorica e Applicata, 13, 187–207, 1970.

Ferretti, A., Prati, C., and Rocca, F.: Permanent scatterers in SAR interferometry. IEEE Trans Geosci Remote Sens, 39, 8–20, 2001.

Ferretti, A., Fumagalli, A., Novali, F., Prati, C., Rocca, F., and Rucci, A.: New algorithm for processing interferometric datastacks: SqueeSAR, IEEE T. Geosci. Remote., 49, 3460–3470, 2011.

Hanssen, R. F.: Radar interferometry: data interpretation and error analysis, Kluwer Academic Publishers, Dordrecht, 2001.

Hofmann-Wellenhof, B., Lichtenegger, H., and Wasle, E.: GNSS-Global Navigation Satellite Systems – GPS, GLONASS, Galileo & more, Springer-Verlag Wien, 516 pp., 2008.

Ketelaar, V. B. H.: Satellite Radar Interferometry: Subsidence Monitoring Techniques, Springer Netherlands, 243 pp., 2009.

Kontny, B. and Bogusz, J.: Models of vertical movements of the Earth crust surface in the area of Poland derived from Leveling and GNSS Data, Acta Geodyn. Geomater., 9, 331–337, 2012.

Pieri, L. and Russo, P.: The survey of soil vertical movements in the region of Bologna, in: Proc. Third Int. Symp. on Land Subsidence (Venezia, March 1984), 1984.

Proc. IAHS, 372, 323–330, 2015
proc-iahs.net/372/323/2015/
doi:10.5194/piahs-372-323-2015

Open Access
PIAHS

Subsidence monitoring system for offshore applications: technology scouting and feasibility studies

R. Miandro[1], C. Dacome[1], A. Mosconi[2], and G. Roncari[2]

[1]ENI SpA – Geodynamics department, via del Marchesato, 13, 48100 Marina di Ravenna, Ravenna, Italy
[2]ENI SpA – Geodynamics department via F. Maritano, 26, 20097 San Donato Milanese, Milan, Italy

Correspondence to: R. Miandro (roberto.miandro@eni.com)

Published: 12 November 2015

Abstract. Because of concern about possible impacts of hydrocarbon production activities on coastal-area environments and infrastructures, new hydrocarbon offshore development projects in Italy must submit a monitoring plan to Italian authorities to measure and analyse real-time subsidence evolution. The general geological context, where the main offshore Adriatic fields are located, is represented by young unconsolidated terrigenous sediments. In such geological environments, sea floor subsidence, caused by hydrocarbon extraction, is quite probable. Though many tools are available for subsidence monitoring onshore, few are available for offshore monitoring. To fill the gap ENI (Ente Nazionale Idrocarburi) started a research program, principally in collaboration with three companies, to generate a monitoring system tool to measure seafloor subsidence. The tool, according to ENI design technical-specification, would be a robust long pipeline or cable, with a variable or constant outside diameter (less than or equal to 100 mm) and interval spaced measuring points. The design specifications for the first prototype were: to detect 1 mm altitude variation, to work up to 100 m water depth and investigation length of 3 km. Advanced feasibility studies have been carried out with: Fugro Geoservices B.V. (Netherlands), D'Appolonia (Italy), Agisco (Italy).

Five design (using three fundamental measurements concepts and five measurement tools) were explored: cable shape changes measured by cable strain using fiber optics (Fugro); cable inclination measured using tiltmeters (D'Appolonia) and measured using fiber optics (Fugro); and internal cable altitude-dependent pressure changes measured using fiber optics (Fugro) and measured using pressure transducers at discrete intervals along the hydraulic system (Agisco). Each design tool was analysed and a rank ordering of preferences was performed. The third method (measurement of pressure changes), with the solution proposed by Agisco, was deemed most feasible. Agisco is building the first prototype of the tool to be installed in an offshore field in the next few years.

This paper describes design of instruments from the three companies to satisfy the design specification.

1 Introduction

Ente Nazionale Idrocarburi Upstream & Technical Services, as other companies involved in hydrocarbon exploration and production in Italy, especially for the offshore fields located in the central-northern Adriatic Sea, must submit a monitoring plan to Italian authorities to measure and analyse real-time subsidence evolution before any oil or gas production. The monitoring plan must address foreseeable impacts, direct or indirect, of hydrocarbon production on coastal areas and in the adjacent inlands.

The general geological context, where the main offshore gas field are located, is represented by young unconsolidated terrigenous sediments, partially still forming and most of the surrounding coastal lands have general heights below medium sea level, up to −5 m. In such geological environments, the subsidence, potentially caused by hydrocarbon production, has to be managed. If part of the produced subsidence cone or "bowl" extends to the coast, it can cause geologic and environmental hazard, costly third party claims and damage to the company's public image.

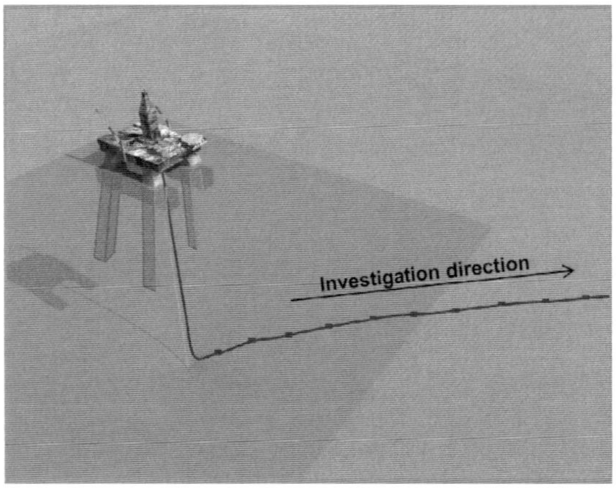

Figure 1. Sketch of cable-based subsidence monitoring system.

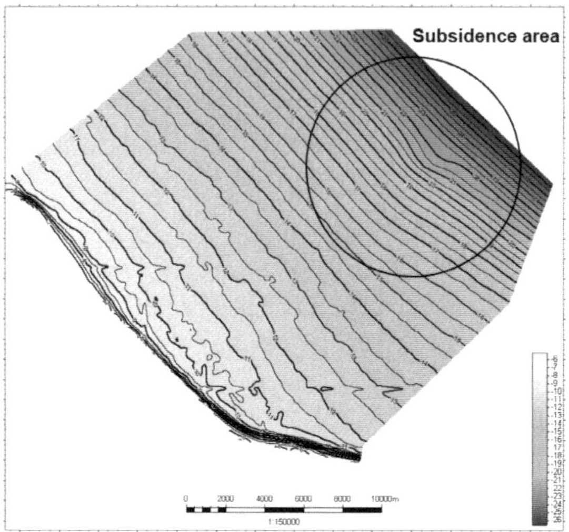

Figure 2. Subsidence case.

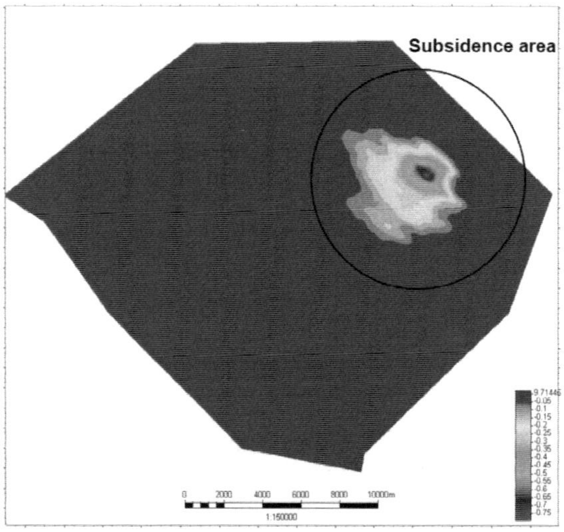

Figure 3. Isolated subsidence "bowl".

Though many tool are available for subsidence monitoring onshore, few are available for offshore monitoring. Therefore, in 2010 ENI, aiming to assure stakeholders that hydrocarbons in the offshore fields could be produced with no impacts on the coastal areas and on the region's subsidence trend, initiated a research program with three companies, to develop a monitoring system tool to measure seafloor subsidence.

The tool, according to ENI technical-specifications, would be a robust long pipeline or cable (cable based subsidence monitoring system – Fig. 1), with a variable or constant outside diameter (OD, less than or equal to 100 mm) with regularly spaced measuring points.

The design device would detect and measure seafloor movements along its deployment direction and provide temporal seafloor altimetrical variations (1 mm), measured at discrete points along the cable.

To let, each of the involved companies in the feasibilities tasks, better understand the size and extension of the subsidence that was expected to be monitored, they were provided with ancillary information from a subsidence case, registered and monitored in the Italian offshore (Figs. 2, 3). The subsidence case, monitored through a multibeam echosounder (MBES) survey, showed an anomaly in the bathymetry at water depths of about 20 and 25 m (Fig. 2). The isolated subsidence bowl (Fig. 3), extends nearly 3×5 km, with a maximum depression of nearly 80 cm. The GPS station, placed on the production platform, showed a subsidence velocity of about 35 mm each year. Geo-mechanical modelling results for this case, indicate a subsidence rate of about 15 mm a^{-1} approximately 2 km distant from the bowl centre. The companies were requested to monitor subsidence as a function of time and location, with a vertical accuracy of 1 mm in real time.

Three companies were asked to perform advanced tool-development feasibilities studies: Fugro Geoservices B.V. (The Netherlands); D'Appolonia (Genova, Italy); Agisco (Liscate (MI), Italy). In addition, to the expected subsidence case above, the companies were provided with the following technical specifications and constraints:

- Cable working condition up to 100 m water depth (ca. 1 MPa).
- Cable OD preferably not over 100 mm.
- Maximum tool length: 3 km. Modular cable system, extendable in length, was preferable.

- Maximum distance among discrete measurement points between 100 and 200 m.
- Measurement of altitude (z)-changes vs time (dz/dt). Absolute cable z position not required.
- Only cable vertical movements, downward or upward, envisaged.
- The cable had to be placed in a trench about 1 m below the sea floor and subsequently covered with soil material. The soil material fixed the cable position integral to the soil.
- Possible to reference one cable end to a point with known movement (CGPS on platform).
- Data acquisition unit and electronics to be placed on board the production platform.
- Monitoring system lay-out geometry comprises segments aligned with the most critical direction of the subsidence bowl, from maximum toward minimum with a fixed offshore platform resting on legs.
- Monitoring for a minimum period of 10 years after installation.

2 Fugro proposed tool

Fugro conducted a feasibility study, in collaboration with the Netherlands Organisation for Applied Scientific Research (TNO), of a cable-based subsidence monitoring system equipped with fiber optic sensors.

The choice for fiber optic sensors was based on high accuracy, ease of data multiplexing and most importantly, the absence of electronics in a Fiber Optic Monitoring Cable (FOMC) (Measures, 2001).

After a preliminary inventory of requirements, three fiber optic based measurements concepts different concepts were generated:

1. Cable shape determination through strain measurement (Fig. 4).
   - I. The shape of a cable and the deformation of the shape can be determined using fiber optic sensors that measure the changes in strain ε at a certain distance r_f from the neutral line of the cable, the strain at a specific location x and time t, $\varepsilon(x,t)$, is sampled at spatial interval Δx. In this technique subsidence of the cable along its length at any time, $s(x,t)$, can be reconstructed from the measured deformation of the cable.
   - II. Several fiber optic methods exist to measure the local strain ε in the cable:
     - i. Fiber Bragg Gratings (FBG).

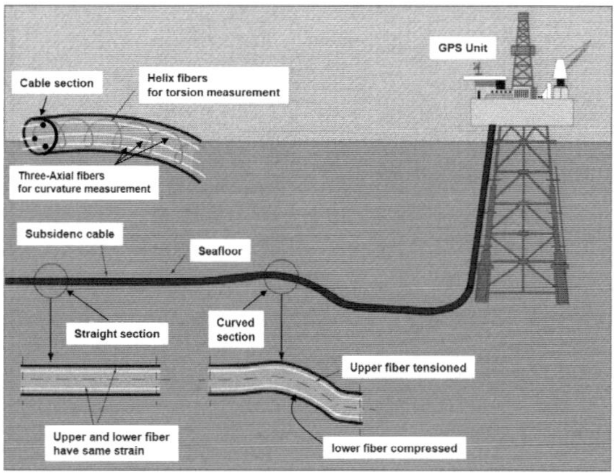

Figure 4. Schematic overview of subsidence measurement using strain measurement.

     - ii. Distributed strain sensing using Raman or Brillouin scattering (known as DTS).
     - iii. Fiber interferometry.
   - III. For this measurement concept (1), exact knowledge of the cable orientation is essential. This can be achieved by using:
     - i. A cylindrical cable that could torque during the production process and/or when the cable is placed in the trench on the sea-bottom followed by a very accurate determination of this torque, or
     - ii. An oblate cable of which the orientation can be kept constant during production and during placement.
2. Cable inclination measurement (Fig. 5)
   - I. When part of the cable moves in a vertical direction the inclination of the part just before and just after the moving part of the cable will change.
   - II. Different options exist to measure this change of inclination through fiber optic sensors.
3. Pressure measurement (Fig. 6)
   - I. When part of the cable moves in a vertical direction the pressure in this part of the cable will change relative to the pressure in other parts of the cable.
   - II. Different options exist to measure these pressure changes via fiber optic sensors.

For each measurements concept, feasibility studies determined that in theory, the required accuracy could be achieved by each of these concepts, however the steps required to develop a prototype system for each concept differed. Subsequently an inventory of the failure risks was made, followed

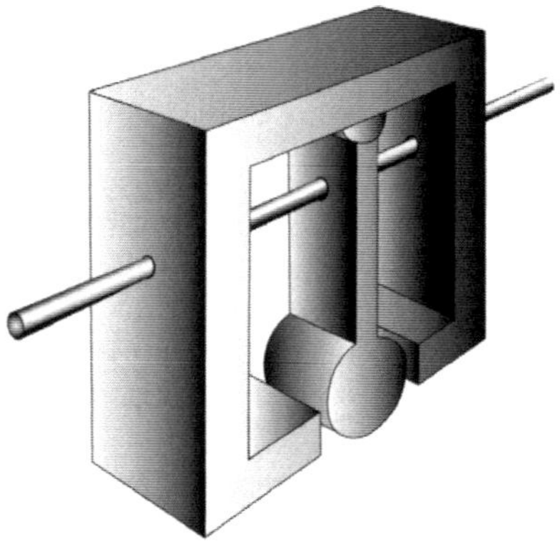

Figure 5. Schematic overview of an inclinometer based on a pendulum p. The pendulum can only rotate over an angle α along an axis perpendicular on the drawing. The rotation α of this pendulum is measured with optical fibers.

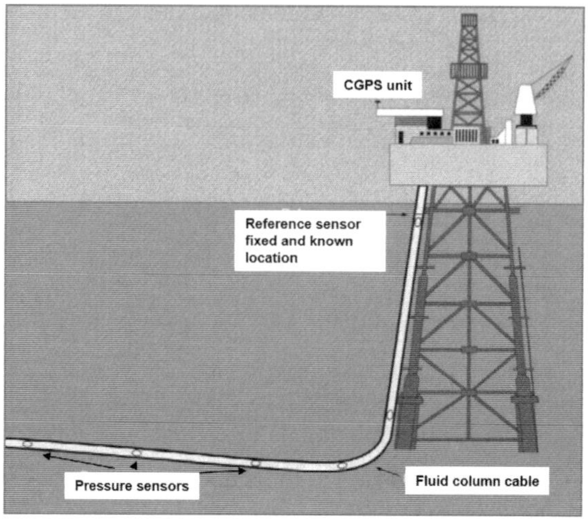

Figure 6. Schematic overview of subsidence measurement via pressure measurement in a liquid filled cable. The cable is open at one end and is filled with a liquid. The small ellipses in the cable denote the pressure measurement positions. The vertical position of the platform is continuously monitored with GPS equipment.

by a preliminary evaluation of each risk. Some risks were considered acceptable, for others, additional development studies were deemed necessary to avoid unacceptable failures. The required development process, determined through technology readiness level (TRL) analysis, was greatest for strain measurements and least for pressure measurements.

In addition to the TRL analysis above, in order to find a trade off among the different measurement concepts, the concepts were compared on a number of different aspects, like, for instance: measurement accuracy; development (risks, time, costs); capex (sensor and cable production, cable deploy, electronic); opex (maintence, calibration, etc.); modularity, etc. (only aspects that were expected to be different for the different concepts were compared). This comparison resulted in favour of pressure measurement in a liquid filled tube as a preferred concept, followed by inclination and strain. The final conclusion was to proceed with a test phase, building a demonstration system, based on fiber optic pressure measurements in a liquid filled cable.

3 D'Appolonia proposed tool

D'Appolonia conducted its feasibility study for a system to monitor seafloor subsidence performing the following tasks:

- Review of available methods.
- Technology review.
- Analysis of scientific literature and patents.
- Description of theoretical cases for system definition and methods selection.
- Assessment of the following selected two methods:

  a. Measurement of the seafloor inclination by means of an array of tiltmeters, and integration of the tilt over distance to obtain the vertical displacement.

  b. Measurement of the differential pressure along a pipe, decoupled from the external pressures, provided with high-precision pressure gauges at suitable spacing.

The first method has been used onshore and offshore (Anderson et al., 1997; Fabian and Villinger, 2007), the second method is in a development stage for offshore applications (Mes, 1989; Stenvold et al., 2006). The first method was considered the most promising and was further developed. The feasibility study was therefore continued focusing on the measuring of the altimetrical sea floor variation through tiltmeter arrays.

The further following tasks were performed:

- Evaluation of subsidence at an offshore pilot site, and confirmation of the suitability of a tiltmeter-based system to measure offshore subsidence.
- Definition of tiltmeter system requirements.
- Selection of an "off-the-shelf" tiltmeter with respect to the specified requirements, and verification of its performance in the laboratory.
- Design and modification of the tiltmeter for this application, and design of a complete system with all the needed components.

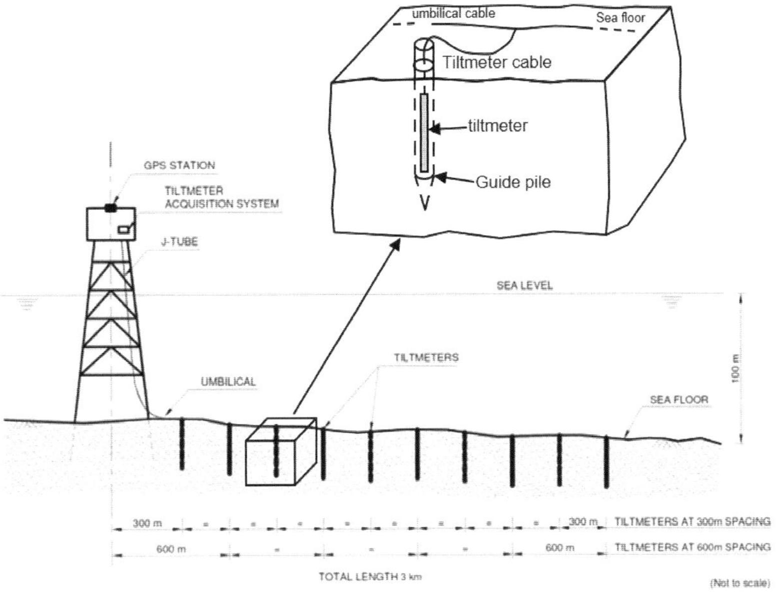

Figure 7. D'Appolonia system layout.

A generalized layout of a complete system it is shown in Fig. 7. It included the following parts:

- GPS station for the measurement of vertical displacement of the platform (Setan and Othman, 2006). GPS is already present on all ENI's platforms.
- J-tube for the umbilical.
- Array of tiltmeters fixed to the seafloor at a predetermined direction and spacing.
- Umbilical cable, including electrical wires and strong rope, connecting the tiltmeters to the data acquisition system installed on the platform.

Design of system component like: tiltmeter, guide pile (for tiltmeters installation), tiltmeter cable, umbilical cable were also provided.

Finally the requirements for offshore installation were as well investigated. In particular this part evaluated the:

- Optimal direction to deploy tiltmeter array.
- Platform characteristics.
- Necessary pre-installation bathymetric surveys of the site.
- Protective measure for the system.
- Installation procedures for the system.

At the end of the study D'Appolonia stated to be ready for the production of working system.

4 Agisco proposed tool

Agisco is a long-standing experienced ENI's contractor, that cooperated with ENI since the beginning of the 90's, providing all the customized equipments for the soil shallow compaction monitoring of the onshore Adriatic coast. Agisco, is a qualified company in design, installation and implementation of geotechnical measurement instrumentation for monitoring soil, subsoil and civilian works.

Agisco did not produce a feasibility study but proposed enhancing an existing proprietary tool, already successfully used to monitor possible horizontal bending of dams, harbours and airport runways. The measuring principle of the proposed tool was based on the measurement of the pressure in a column of a liquid acting on several measuring points. The device used pressure sensors to measure the relative vertical displacement of these points with respect to a reference point. The displacement between two sensors was obtained from the proportional relation between the pressure difference and displacement. The theoretical scheme of the devices, included: a reference tank, a hydraulic circuit and one or more measuring points. The main problem with the existing tool is that it is open to the atmosphere, condition difficult or even impossible to achieve underwater. This problem could be overcome by using a "closed" circuit, isolated from the atmosphere.

Therefore, Agisco proposed to redesign the existing tool. The new design had to take into account all the related problems of underwater deployment and also consider a different tool shape and sturdier structure, allowing a safe, secure, failure-free tool installation at the specific water depth (100 m). In fact, feasibility of off-shore tool installation, is significant at least from the economical point of view.

The proposed "submarine hydraulic profile gauge" using the pressure measurement principle and a "closed" hydraulic circuit was selected for further developmental evaluation. The design was quite challenging and time consuming, requiring a number of tests to verify the suitability of alternative solutions for the closed hydraulic circuit. The best solution was the use of a "compensator" (Fig. 8) which is the real core and innovation of Agisco's proposed system, which is covered by patents in the European Union and in the United States of America. The concept of the compensator is that it has to allow for volume changes of the liquid when no real changes in the altitude of the hydraulic circuit and when altitude-dependent pressure changes occur. Source of altitude-independent volume changes could be thermal liquid expansion, shocks or hose squeezing. Moreover "shock-wave" could be induced into the circuit during installation and/or maintenance operations. Pressure changes induced by volume changes or shock-waves may be very high and damage the pressure transducers, therefore they had to be efficiently limited.

The design of the compensator was developed in parallel with the research on two other key issues which characterize a hydraulic device: (1) the hydraulic fluid selection and (2) the thermal compensation.

The research led to the final configuration which enables use one of a single compensator for each circuit.

The liquid to be used in hydraulic profilers had to meet with some critical specifications:

- Stable properties during the measuring or monitoring time period (10 years at least).
- Air-free (no gas bubbles) during the device life time. Gas can modify tool response time and change the circuit pressure.
- Known density and temperature-dependent density relation with high accuracy. Small changes in density cause significant variation in measured pressure and therefore in system accuracy.
- No microbial growth to avoid gas production and change in density.

In order to select the best liquid, a comprehensive research effort has considered a large selection of possible liquid and blends. A number of different mixtures have been tested and their physical characteristics have been investigated and quantified.

Final attention has been focusing on water-glycol mixtures, which have low freezing point and good stability. Laboratory test results, performed on different blends, are being used to derive temperature-dependent density relationship, one of the most important and critical parameters.

Figure 8. Agisco compensator.

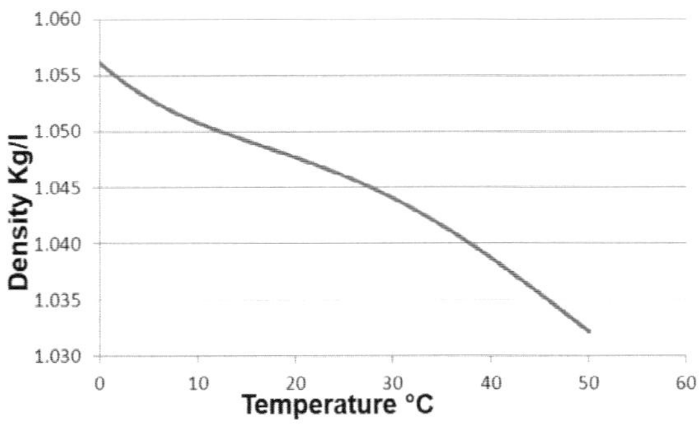

Figure 9. Liquid density vs. temperature relationship.

The results showed that liquid density changes nonlinearly with temperature (Fig. 9), the relationship has been derived starting from the reference equation by F. J. Millero and A. Poisson (Millero and Poisson, 1981). This is one of the key results, which have been used to "compensate" pressure measurement along the circuits. For off-shore application it is not so critical unless (a) the circuit is crossing water layers at different temperature or (b) part of the circuit is onshore or outside (the reference) and part is offshore (the measuring points); or (c) the offshore circuit is in shallow water, where significant temperature changes occur owing to daily and seasonal climatic conditions.

The design of the compensator needed to take into account the thermal effects and it has been studied using a specific array of thermometers embedded in the hydraulic circuits enabling measurement of the temperature distribution. Thus, the temperature profile along the whole circuit could be determined and the density of the liquid calculated to process the pressure signals. After study and selection of the two key items – liquid and temperature and compensation method, other important aspects have been studied and solutions adopted about the following system components:

- Pressure transducers.
- Hose.
- Enclosure.
- Signals.
- Cables.
- Data Acquisition Unit.
- Processing Software.

Among these, the most important item has been the pressure transducer selection. Hydraulic profilers use both absolute and differential pressure transducers. It was decided to use absolute pressure transducer to avoid the need for an equalized reference-pressure and considering that the circuits are closed.

The main specifications used as reference were:

- Robustness, because deployment had to be managed by cranes and operated by offshore devices.
- High stress tolerance to survive installation shocks and accommodate large differences in relative elevation.
- High accuracy enabling 1–10 mm overall accuracy of the device.
- Low temperature sensitivity to reduce thermal effects.
- Negligible volume change to minimize the effect on the circuit volume and to avoid liquid movement.
- High zero and high scale stability for long term monitoring.
- Enhanced output signal stability.
- High reliability to cope with long term monitoring.
- Compatibility with the hydraulic liquid.

Various pressure transducers that purportedly met the specifications were tested. The selected transducers were the piezoresistive type with a 100–500 kPa measuring range and digital output with a 20 bit analog to digital (A/D) converter. A digital output was specified for the following main reasons:

- Possibility to use a MOD-BUS transmission protocol on a single cable.
- Possibility to set the temperature compensation parameters directly inside the transducer to obtain compensated signals.

Figure 10. Configuration of profile prototype.

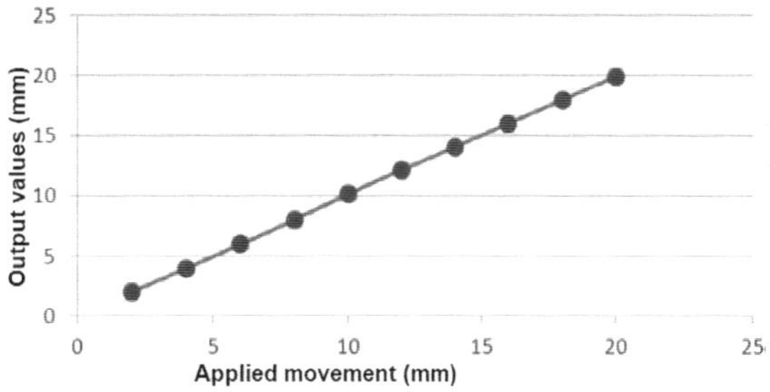

Figure 11. Calibration Sheet for measuring point.

- Possibility to use a standard 485 serial output, whose low bit rate allows long distance transmission, up to 3 km.
- Easy dynamic data acquisition.

Finally Agisco was able to produce a first prototype of the Horizontal Hydraulic Profile Gauge. The current configuration of the profiler is shown in Fig. 10 where the measuring points, the circuit and the compensator are presented.

The main specifications of the prototype tool are:

- Measuring range: up to 100 m water depth.
- Resolution: ±0.05 % F.S.
- Number of measuring points: up to 128.
- Length: standard: 500+ m; extended: 4 km.
- Degree of protection: Ingress Protection rating of IP69 K–1 MPa.
- Certifications: European Conformity (CE).

Typical results from laboratory calibration (performed by using calibrated shims) is shown in Fig. 11.

5 Conclusions

The offshore subsidence, being underwater, is usually less impacting on local coastal environments and communities and consequently less studied and with fewer tools available, for monitoring. The present work allowed ENI to explore offshore subsidence monitoring. Researching the available technology allowed an in-depth analysis of the technological opportunities in all technical fields relevant to monitoring of seafloor subsidence. The research included assessing patents/industrial property rights and the scientific and technical literature.

But, above all, feasibility studies allowed establishment of the basis to build a working tool to be used for offshore applications in subsidence monitoring. Different monitoring challenges and approaches were identified based on different measurement concepts and devices, fiber optics for measuring strain, tiltmeters for measuring tilt and a closed hydraulic system with pressure transducers for measuring differential pressures. The first two concepts were promising and challenging, but the solution proposed by Agisco for measuring differential pressure in a closed hydraulic system was deemed a more sufficiently proven technology that required minor modification to an existing proven design in order to be adapted for offshore use. Agisco is producing a working prototype tool during 2015 and ENI plans to deploy an offshore installation of the tool within the next few years.

References

Anderson, G., Constable, S., Staudigel, H., and Wyatt, F. K.: A Seafloor Long-Baseline Tiltmeter, J. Geophys. Res., 102, 20269–20285, 1997.

Fabian, M. and Villinger, H.: The Bremen ocean bottom tiltmeter (OBT) – a technical article on a new instrument to monitor deep sea floor deformation and seismicity level, Mar. Geophys. Res., 28, 13–26, doi:10.1007/s11001-006-9011-4, 2007.

Measures, R. M.: Structural Monitoring with Fiber Optic Technology, Academic Press Inc, San Diego, California, USA, Chapters 5, 6, 7, 716 pp., 2001.

Mes, M. J.: Accuracy of Offshore Subsidence Measurements with Seabed Pressure Gauges", 21st Annual Offshore Technology Conference, 1–4 May 1989, Houston, Texas, USA, Paper No. 6063, 213–222, 1989.

Millero, F. J. and Poisson, A.: International one-atmosphere equation of state of seawater, Deep-Sea Res. Pt. I, 28, 625–629, 1981.

Stenvold, T., Eiken, O., Zumberge, M. A., Sasagawa, G. S., and Nooner, S. L.: High-Precision Relative Depth and Subsidence Mapping from Seafloor Water-Pressure Measurements, Journal of the Society of Petroleum Engineers, Paper No. SPE 97752, 380–389, 2006.

Proc. IAHS, 372, 331–334, 2015
proc-iahs.net/372/331/2015/
doi:10.5194/piahs-372-331-2015

InSAR data for monitoring land subsidence: time to think big

A. Ferretti, D. Colombo, A. Fumagalli, F. Novali, and A. Rucci

TRE – Tele-Rilevamento Europa, Milano, Italy

Correspondence to: A. Ferretti (alessandro.ferretti@treuropa.com)

Published: 12 November 2015

Abstract. Satellite interferometric synthetic aperture radar (InSAR) data have proven effective and valuable in the analysis of urban subsidence phenomena based on multi-temporal radar images. Results obtained by processing data acquired by different radar sensors, have shown the potential of InSAR and highlighted the key points for an operational use of this technology, namely: (1) regular acquisition over large areas of interferometric data stacks; (2) use of advanced processing algorithms, capable of estimating and removing atmospheric disturbances; (3) access to significant processing power for a regular update of the information over large areas. In this paper, we show how the operational potential of InSAR has been realized thanks to the recent advances in InSAR processing algorithms, the advent of cloud computing and the launch of new satellite platforms, specifically designed for InSAR analyses (e.g. Sentinel-1a operated by the ESA and ALOS2 operated by JAXA). The processing of thousands of SAR scenes to cover an entire nation has been performed successfully in Italy in a project financed by the Italian Ministry of the Environment. The challenge for the future is to pass from the historical analysis of SAR scenes already acquired in digital archives to a near real-time monitoring program where up to date deformation data are routinely provided to final users and decision makers.

1 Introduction

Results from satellite InSAR make it a unique instrument for ground displacement analysis, which contributes to the portfolio of technologies of traditional monitoring techniques and can be used in many diverse applications (Ferretti, 2014). InSAR data can be used for fault characterization and calibration of geo-mechanical models in the oil and gas sector, for monitoring landslides, volcanoes and seismic faults, areas prone to sinkholes and subsidence, terrain compaction phenomena induced by tunneling works, and even for monitoring the stability of individual buildings.

Although the first results date back to the 1980's (Gabriel et al., 1989), InSAR is often seen as a "new technology" for surface deformation monitoring and the full potential of InSAR is largely under-recognized, even within the scientific community. The reasons behind this slow uptake are numerous and complex (Ferretti, 2014). In the following sections, it will be shown how most of the challenges related to an operational use of InSAR data (e.g. monitoring urban subsidence) have now been overcome and why this technology can now become a standard tool for studying subsidence and other geological processes.

2 The space segment

Knowledge dissemination of any space technology is strongly related to the availability of a reliable and robust space segment, providing proper and adequate infrastructure and data sources. For InSAR, despite the results from SEASAT (1978), the Sentinel-1A European Space Agency (ESA) SAR satellite that entered its operational phase in late 2014 is the first (civilian) sensor specifically designed for surface deformation monitoring over large areas. So far, all SAR sensors have been multi-purpose, where repeat-pass InSAR was just one of a long list of possible applications, from iceberg monitoring and oil-spill detection, to soil moisture estimation and biomass mapping.

The first InSAR results from satellite sensors were a somewhat unexpected, although extremely welcome, outcome (Gabriel et al., 1989). However, after obtaining proof of concept with SEASAT and the success of the ERS missions operated by ESA, InSAR technology still did not take-off as fast as expected. The space segment aspects of the satellite missions did not aid its evolution either. The specifications of satellites being developed in the new millennium were not in the direction required for systematic and repeatable InSAR observations. Radar interferometry needs a simple acquisition scheme, with radar sensors acquiring data regularly over a particular area using the very same acquisition mode and geometry, to allow for comparison of range values over time. SAR technicians and aerospace industries, on the contrary, were focused on a different goal: on offering many sophisticated imaging options, in terms of polarization of the electromagnetic signal wave, image resolution, incidence angle of the radar beam, etc., and not on simple, single-mode radar sensors. Even considering the more recent satellites, the requirement to task the satellite to acquire an image, while selecting the correct acquisition mode (from a long list of options), to create a homogeneous multi-temporal data-set over an area does not facilitate the diffusion of InSAR as a standard geodetic tool (Ferretti, 2014).

Despite the difficulties faced in the last two decades, the launch of the satellite platform Sentinel-1A can be considered a milestone on the path towards an operational use of InSAR. The Sentinel-1 mission, operated by the ESA, has been specifically designed for large-scale InSAR analyses (Torres et al., 2012). Sentinel-1A and the twin sensor Sentinel 1-B (to be launched in 2016) are two C-band, polar-orbiting, satellite systems for the continuation of SAR operational applications, but with significant enhancements, compared to previous ESA missions, in terms of revisitation, coverage, timeliness and reliability of service. The Interferometric Wide Swath acquisition mode can image a 250 km swath, more than twice that of the ESA ERS mission, with a repeat cycle of 12 days, compared to the 35 days of ERS. Once both sensors will be operational the effective repeat cycle for InSAR analyses will be 6 days (Salvi et al., 2012). Moreover, radar data will be delivered within an hour of acquisition, a big improvement over existing SAR systems.

Apart from the Sentinel-1 mission, it is important to point out that the number of satellite data sources is increasing steadily. Today, SAR data for operational InSAR projects can be provided by RADARSAT-2, Sentinel-1A, RISAT-1, ALOS-2, TerraSAR-X, KOMPSAT-5, Tandem-X and the COSMO-SkyMed constellation (4 SAR satellites). Although not operating with the same central frequency and the same acquisition modes, all these sensors can be used successfully for monitoring surface deformation phenomena. This scenario allows users to design redundant acquisition schemes, where the possible failure of a satellite sensor may not critically compromise the monitoring project. A more "InSAR-friendly" and robust space segment will be a key factor for a large scale application of InSAR in the near future.

3 New algorithms and a case history

Even within the geodetic community, the slow uptake of SAR interferometry has been interpreted as evidence of the fact that InSAR is still not a mature technology. InSAR analyses are thought to provide interesting but *qualitative* data, great subjects to write scientific papers, but with a limited impact on real-life applications.

In our opinion, this interpretation is incorrect and overlooks an important point: compared to what was possible in the 1990's, InSAR algorithms have significantly improved, allowing a much better estimation of the signal of interest, constantly reducing error and computational times. The development of Persistent Scatterer Interferometry (Ferretti et al., 2001, 2011; Kampes, 2006) has made it possible to extract millimeter-accuracy displacement data over millions of measurement points automatically identified by a statistical analysis operated over long temporal series of radar images acquired over the same area of interest. The impact of atmospheric disturbances on InSAR data can be strongly reduced. Validation of Persistent Scatterer (PS) data have been reported in dozens of scientific papers (see Ferretti, 2014 and references therein) and large-scale applications have been proven feasible and effective (Costantini et al., 2015).

A good example of the potential related to satellite radar data was a large-scale InSAR project financed by the Italian Ministry of the Environment. In 2007, the Italian Government began supporting the largest ever InSAR analysis to be funded by a national government. The project was designed to map unstable areas over the entire Italian territory, using PS data. In January 2008, the Italian Ministry of the Environment awarded a contract to the industrial consortium, composed of TRE, e-geos and Compulab (Costantini et al., 2015). The main project objective was to expand and update the National Cartographic Portal (PCN) with displacement data for 1992–2010. The project involved processing the entire ESA ERS and Envisat SAR image archives over Italy ($>$ 15 000 SAR images covering $>$ 330 000 km^2), which called for a dedicated geospatial database to manage millions of PS measurements, thousands of SAR images and hundreds of interferometric data-stacks (Fig. 1). All radar acquisitions have been processed successfully and geologists and geophysicists in central and local administrations can now use PS data to get qualitative and quantitative information about a range of phenomena, affecting known and new at-risk areas, including:

- slow-moving landslides that could trigger fast-moving slides,
- volcanism and related dynamics,
- subsidence and compaction,
- seismic and aseismic fault movement,
- other known and previously unidentified phenomena.

Figure 1. Large-scale mapping of unstable areas over the entire Italian territory. More than 8000 ESA ERS satellite images were processed. Colored dots correspond to measurement points. Red areas are affected by subsidence. The total area covered is 330 000 km^2. (Radar data: ERS, June 1992–June 2000. Background image: Google Earth™.)

Moreover, displacement data from medium-resolution ESA SAR data are now being updated using high-resolution COSMO-SkyMed imagery, allowing the identification of a much higher density of measurement points and a larger temporal frequency of acquisitions.

This case history could serve as an example for other nations to exploit the Sentinel radar images (or any other data sources) to create a national database of terrain movements. In general, there is growing interest in Europe across a wide range of topics in which satellite InSAR data could play a role, including: subsidence and its role in relative sea-level rise along coastal areas, induced seismicity (due to oil and gas extraction and fluid injection), shale gas exploitation, sinkhole detection and pipeline monitoring. Assuredly, the regular, long-term acquisition of InSAR data will prompt new areas of scientific exploration and stimulate new commercial applications and market sectors.

The experience related to this successful project has shown that InSAR is a mature technology and how the new processing algorithms can pave the way to new applications. Depending on the processing technique and the resolution of the SAR sensor, InSAR can be used for monitoring displacement of individual structures, but also to provide information at regional and at larger scales. The ability to detect ground displacements that have occurred over hundreds or thousands of square kilometres and even at a national scale can be extremely useful in characterizing and addressing areas prone to risk. InSAR data can allow Civil Protection authorities to analyze and evaluate different scenarios and to plan specific actions based on homogeneous and reliable measurements. Last, but not least, InSAR data can be updated regularly, as more SAR images are acquired by satellite sensors, and can provide new ways to design early warning systems covering entire nations. Also InSAR can identify specific areas where *in situ* sensors and continuous monitoring tools should be installed. In our opinion, rather than replacing conventional technologies, in the future satellite data will work more in *synergy* with airborne sensors and ground-based instruments.

4 Processing capabilities and future trends

Data processing and storage capabilities are keeping pace with the growing number of information sources. The power of cloud computing is already impacting on remote sensing projects. Customers are becoming impatient with the delayed availability of processed Earth observation data and are increasingly demanding near real-time results from monitoring systems. This will definitely become possible thanks to distributed computing. Surely, displacement data will become available soon after any new satellite acquisition.

Another recent trend in InSAR monitoring projects is the use of different SAR data sources (e.g. Sentinel-1 and COSMO-SkyMed) to increase the number of acquisitions per year over an area of interest. Although no useful interferogram can be generated using two radar sensors having different frequencies and orbits, nevertheless, their displacement measurements can be combined, whenever the incidence angle of their radar beams do not differ significantly from one another (a tolerance of a few degrees is usually acceptable). It is then possible to sample the signal of interest (i.e. the displacement field) with a temporal frequency much higher than that of the repeat orbit for individual sensors: the concept of a virtual constellation, where different satellite data sources are used in synergy to create a more effective and robust Earth observation system. For instance, one could envision a monitoring scheme with two radar sensors: one operating at L-band (e.g. ALOS-2; 23 cm wavelength) and one at C-band (e.g. Sentinel-1; 6 cm wavelength) where InSAR results of the longer wavelength sensor are used to constrain phase unwrapping procedures on interferograms generated using the smaller wavelength sensor (more sensitive to any target displacement and therefore generating interferometric fringes having a higher spatial frequency).

In general, the use of real and virtual satellite constellations will significantly increase the frequency of observation of an area of interest and will make any monitoring project more robust by mitigating consequences of satellite failure. In fact, a very unwelcome event during an InSAR monitoring project is the lack of satellite acquisitions due to satellite maneuvers (orbit correction), conflicts with higher priority requests (for dual-use systems), and temporary (or permanent) failure of the radar sensor or the downlink system. The larger the number of data sources, the more robust the acquisition scheme and the higher likelihood of good data coverage for the entire project duration.

5 Conclusions

InSAR is a multi-scale tool: depending on the processing technique and the resolution of the SAR sensor, it can be used for monitoring individual structures, but also to provide information at regional and at larger scales. Having the ability to detect ground displacements that have occurred over hundreds or thousands of square kilometres and even at a national scale can be extremely useful in characterising and addressing areas prone to risk, such as highlighting areas affected by subsidence phenomena. InSAR data can allow decision makers to analyse and evaluate different scenarios and to plan specific actions based on homogeneous and reliable measurements. InSAR data can be updated regularly, as more SAR images are acquired by satellite sensors, and can provide new ways to design early warning systems covering entire nations, where satellite information can highlight areas where in situ sensors and continuous monitoring tools should be installed. Rather than replacing conventional technologies, InSAR data will work more in synergy with airborne sensors and ground-based instruments.

Acknowledgements. The authors thank all TRE colleagues for continuing support, Fabio Rocca and Claudio Prati of Politecnico di Milano for many stimulating discussions on the applications of InSAR data, as well as the two reviewers who improved significantly the clarity of this paper.

References

Costantini, M., Minati, F., Ciminelli, M. G., Ferretti, A., and Costabile, S.: Nationwide Ground Deformation Monitoring by Persistent Scatterer Interferometry, IEEE Geoscience and Remote Sensing Society, the International Geoscience and Remote Sensing Symposium 2015 (IGARSS 2015), 25–31 July 2015, Milano, Italy, 2015.

Ferretti, A.: Satellite InSAR Data: Reservoir Monitoring from Space, EAGE Publications bv, 2014.

Ferretti, A., Prati, C., and Rocca, F.: Permanent Scatterers in SAR Interferometry, IEEE T. Geosci. Remote, 39, 8–20, 2001.

Ferretti, A., Fumagalli, A., Novali, F., Prati, C., Rocca, F., and Rucci, A.: A new algorithm for processing interferometric datastacks: SqueeSAR, IEEE T. Geosci. Remote, 49, 3460–3470, 2011.

Gabriel, A. K., Goldstein, R. M., and Zebker, H. A.: Mapping small elevation changes over large areas: differential radar interferometry, J. Geophys. Res., 94, 9183–9191, 1989.

Kampes, B. M.: Radar Interferometry – Persistent Scatterer Technique, Springer, 43–69, 2006.

Salvi, S., Stramondo, S., Funning, G. J., Ferretti, A., Sarti, F., and Mouratidis, A.: The Sentinel-1 mission for the improvement of the scientific understanding and the operational monitoring of the seismic cycle, Remote Sens. Environ., 120, 164–174, 2012.

Torres, R., Snoeij, P., Geudtner, D., Bibby, D., Davidson, M., Attema, E., Potin, P., Rommen, B., Floury, N., Brown, M., Navas Traver, I., Deghaye, P., Duesmann, B., Rosich, B., Miranda, N., Bruno, C., L'Abbate, M., Croci, R., Pietropaolo, A., Huchler, M., and Rostan, F.: GMES Sentinel-1 mission, Remote Sens. Environ., 120, 9–24, 2012.

Proc. IAHS, 372, 335–341, 2015
proc-iahs.net/372/335/2015/
doi:10.5194/piahs-372-335-2015

Spatio-temporal evolution of aseismic ground deformation in the Mexicali Valley (Baja California, Mexico) from 1993 to 2010, using differential SAR interferometry

O. Sarychikhina and E. Glowacka

Center for Scientific Research and Higher Education of Ensenada (CICESE), Ensenada, Mexico

Correspondence to: O. Sarychikhina (osarytch@cicese.mx)

Published: 12 November 2015

Abstract. Ground deformation in Mexicali Valley, Baja California, Mexico, the southern part of the Mexicali-Imperial valley, is influenced by active tectonics and human activity, mainly that of geothermal fluid extraction in the Cerro Prieto Geothermal Field. Significant ground deformation, mainly subsidence ($\sim 18\,\mathrm{cm\,yr^{-1}}$), and related ground fissures cause severe damage to local infrastructure.

The technique of Differential Synthetic Aperture Radar Interferometry (DInSAR) has been demonstrated to be a very effective remote sensing tool for accurately measuring the spatial and temporal evolution of ground displacements over broad areas. In present study ERS-1/2 SAR and ENVISAT ASAR images acquired between 1993 and 2010 were used to perform a historical analysis of aseismic ground deformation in Mexicali Valley, in an attempt to evaluate its spatio-temporal evolution and improve the understanding of its dynamic. For this purpose, the conventional 2-pass DInSAR was used to generate interferograms which were used in stacking procedure to produce maps of annual aseismic ground deformation rates for different periods. Differential interferograms that included strong co-seismic deformation signals were not included in the stacking and analysis.

The changes in the ground deformation pattern and rate were identified. The main changes occur between 2000 and 2005 and include increasing deformation rate in the recharge zone and decreasing deformation rate in the western part of the CPGF production zone. We suggested that these changes are mainly caused by production development in the Cerro Prieto Geothermal Field.

1 Introduction

Any process involving large-scale extraction of groundwater, including geothermal energy development, can be accompanied by ground deformation. Ground deformation related to geothermal energy production has been observed and studied in many geothermal fields all over the world (e.g. Mossop and Segall, 1997; Carnec and Fabriol, 1999; Fialko and Simons, 2000; Hole et al., 2007; Allis et al., 2009; Keiding et al., 2009). Ground deformation associated with geothermal energy production can be reduced or controled through monitoring combined with aquifer management (Lowe, 2012).

Differential synthetic aperture radar interferometry (DInSAR) is a space-based geodetic technique widely applied for large-scale monitoring (spatially continuous) of ground deformation at a low cost (Massonnet and Rabaute, 1993; Zebker et al., 1994; Massonnet and Feigl, 1998), and particularly for subsidence regardless of its cause (Massonnet et al., 1997; Fielding et al., 1998; Avallone et al., 1999; Carnec and Fabriol, 1999; Wright and Stow, 1999; Carnec and Delacourt, 2000). Compared with ground-based geodetic techniques, such as precise leveling and GPS surveys, which provide accurate and precise information but only at a number of measuring points in a deforming surface, DInSAR allows detecting, measuring and monitoring ground deformation over large areas ($\sim 10^4\,\mathrm{km^2}$) with fine spatial resolution ($\sim 10^2\,\mathrm{m^2}$) and high accuracy (~ 1 cm) (Gabriel et al., 1989; Bürgmann et al., 2000; Hanssen, 2001). The systematic coverage and frequent observations (e.g. approximately monthly repeat-cycle for ERS and ASAR) provided by SAR sensors allows the extraction of information on the spatial and temporal evolution of ground deformation at more frequent intervals than traditional techniques (Massonnet and Feigl, 1998; Baer et al., 2002; Tomás et al., 2005; Herrera et al., 2007). The availability since 1992 of a large archive of satellite acquisitions allows an analysis of historical deformation.

In the present study ERS-1/2 SAR and ENVISAT ASAR images acquired between 1993 and 2010 were used to perform an analysis of historical ground deformation caused by geothermal energy development in the Cerro Prieto Geothermal Field (CPGF) in Mexicali Valley, in an attempt to evaluate its spatio-temporal evolution and improve the understanding of its dynamic.

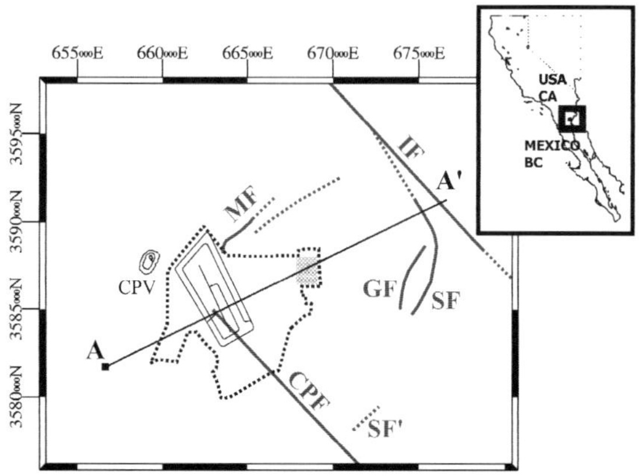

Figure 1. Plan-view map of the study area. The black thick dotted line frames the limits of the Cerro Prieto Geothermal Field. The borders of the evaporation pond and the Cerro Prieto volcano (CPV) are also shown (thin gray lines). A striped orange rectangle indicates the extraction area of CPIV which started the operation since 2000. Solid red lines are well-known surface traces of tectonic faults. CPF = Cerro Prieto Fault, GF = Guerrero Fault, IF = Imperial Fault, MF = Morelia Fault, SF = Saltillo fault. Dotted red lines are proposed surface fault traces based on mapped fissure zones from González et al. (1998), Glowacka et al. (2006, 2010a, c), Lira (2006) and Suárez-Vidal et al. (2007, 2008). SF' is continuation of Saltillo fault as proposed by Suárez-Vidal et al. (2008). Black square marks the location of the reference, considered stable, region. Black thick line corresponds to the profile A-A' illustrated in Fig. 3. Modified from Glowacka et al. (2010c).

2 Study area

Ground deformation in Mexicali Valley, Baja California, Mexico, the southern part of the Mexicali-Imperial Valley, is influenced by active tectonics and human activity. Mexicali Valley is located in a complex tectonic environment, at the tectonic boundary between the Pacific and North American plates. This area is characterized by high tectonic seismicity, heat flow and surface deformation, related to the tectonic regime of the area. Besides the tectonic deformation, extraction of fluids in the CPGF which is the oldest and largest Mexican geothermal field in operation and the second largest in the world produces deformation of large magnitude (Glowacka et al., 1999). Significant ground deformation (mainly subsidence) and related ground fissures cause severe damage to local infrastructure such as roads, irrigation canals and other facilities. A detailed plan-view map of the study area is presented in Fig. 1.

Several monitoring techniques have been employed in the study area including repeat ground-based geodetic surveys, geotechnical instruments measurements and DInSAR. Ground-based geodetic surveys (precise leveling and GPS) in Mexicali Valley began in the 1960s (Velasco, 1963) and have been carried out up to the present, with varying frequency, precision, coverage, and density (García, 1978; Grannell et al., 1979; de la Peña, 1981; Wyman, 1983; Lira and Arellano, 1997; Glowacka et al., 1999, 2012). Since 1996, the geotechnical instruments network has operated in Mexicali Valley (REDECVAM) for continuous recording of deformation phenomena (Nava and Glowacka, 1999; Glowacka et al., 2002, 2010a, b; Sarychikhina et al., 2015).

Analysis of leveling data and CPGF extraction history (Glowacka et al., 1999), and modelling of the tectonic and anthropogenic components of subsidence (Glowacka et al., 2005) suggested that the current deformation rate is mainly related to the fluid extraction, while the field observations (Gonzalez et al., 1998; Glowacka et al., 1999, 2010a, c; Suárez-Vidal et al., 2008) and geotechnical instruments data (Glowacka et al., 1999, 2010a, c) reveal that the geometry of the subsiding area is controlled by tectonic faults.

Ground deformation in the Mexicali Valley has also been measured and monitored using the DInSAR technique (Carnec and Fabriol, 1999; Hanssen, 2001; Sarychikhina et al., 2011, 2015). Carnec and Fabriol (1999) and Hanssen (2001) used ERS1/2 images acquired during 1993–97 and 1995–97, respectively. Significant local subsidence has been detected. Due to the high degree of temporal and spatial decorrelation, reliable data was obtained only over the mainly desert area of CPGF. Although agricultural activity in the area limited the investigation, interferometric monitoring revealed that the ground deformation is associated with the geothermal fluid withdrawal and was in agreement with the leveling data. More recent studies by Sarychikhina et al. (2011, 2014, 2015) and Sarychikhina and Glowacka (2015) used Envisat ASAR images acquired during 2003–10 to evaluate the spatial pattern and rate of land subsidence. The Differential Interferograms Stacking (DIS) procedure was applied to estimate the annual ground deformation rate for different periods. The results were compared to 1994–97 and 1997–2006 leveling data, revealing significant changes in ground deformation pattern and rate.

Here we extend these studies by using aditional SAR images (ERS 1/2 SAR) and the DIS procedure to obtain ground deformation rate for different periods from 1993 to 2009.

3 Methodology for deformation rate estimation

The conventional 2-pass DInSAR method using ERS 1/2 SAR and ENVISAT ASAR images was used to produce interferograms. The DIS procedure was used to estimate ground deformation rates in the study area for different time periods. Here, the DIS procedure, which consisted in averaging of multiple differential interferograms, has been applied in order to overcome limitations of conventional DInSAR, such as low coherence over long temporal separations and phase distortions introduced by atmospheric effects.

A total of 27 ERS 1/2 SAR and 52 ENVIAST ASAR images were acquired from the European Space Agency (ESA), as part of ESA CAT-1 project (ID – C1P3508). ERS 1/2 SAR images were from descending pass (track 84, frame 2961) and covered the period 1993–2000. ENVISAT ASAR images were from ascending (track 306, frame 639) and descending (track 84, frame 2961) passes and covered the period 2003–09.

Data processing was performed using the *GAMMA* commercial software, developed by Wegmüller and Werner (1997) by first performing two-pass interferometry and unwrapping of differential interferograms, and then the stacking of multiple differential interferograms and geocoding. The selection of interferograms for stacking was based on their quality in terms of high coherence and low noise level.

Unfortunately, it was impossible to form the annual stacks of coherent short temporal baseline (35–140 days) differential interferograms with reasonable coherence level from ERS 1/2 dataset because of important gaps in data acquisition and geometrical decorrelation of some short time period differential interferograms. For this reason, the pairs from ERS 1/2 were grouped into two time periods (stacks): 1993–97 and 1998–2000.

Because here we attempt to present results of only aseismic deformation the differential interferograms that presented strong co-seismic deformation signals were not included in the stacking and analysis. It was impossible to generate an aseismic LOS displacement rate map for 2008 due to a high level of moderate sized seismicity during this year.

4 Historical deformation analysis

Figure 2 illustrates the maps of annual LOS displacement rate for different time periods obtained using the DIS method. All maps of LOS displacement rate show a roughly NE–SW oriented elliptical-shaped pattern of deformation with two bowls exhibiting high deformation rates. The similarity between maps of LOS displacement rate from ascending and descending tracks (for 2007), along with the known vertical displacement from ground-based measurements including leveling surveys, suggest that the observed LOS displacements may be interpreted as reflecting mostly vertical surface displacement. However, Fig. 3, which presents the annual LOS displacement rates along the profile (A-A' in Fig. 1) across the subsidence zone, shows important differences (~ 2 cm of magnitude) between ascending and descending datasets in the central part of the CPGF. This indicates the influence of a horizontal displacement component, probably related to horizontal displacement at and beyond the margin of the subsidence bowl, as proposed by Segall (1989).

The negative rate values in Fig. 2 indicate an increase in the LOS range that corresponds to ground subsidence. The subsiding area boundaries appear to correlate with faults, as can be seen in Fig. 2 where they are superimposed onto the maps of LOS displacement rate. The first region of subsidence, with larger negative-valued LOS displacement rate for all analysed periods (Fig. 3), is located in the area between the eastern limits of the CPGF and the Saltillo fault, which was proposed as a recharge zone in previous studies (Glowacka et al., 1999, 2005; Sarychikhina, 2003). The maximum LOS displacement rate in this region increased from about -12.5 to $-17\,\text{cm yr}^{-1}$ during 1993–97, and 2005–07 and 2009, respectively. The largest negative-valued LOS displacement rate ($-17\,\text{cm yr}^{-1}$) corresponds to about $18\,\text{cm yr}^{-1}$ of ground subsidence. The area of this subsidence bowl is also evidently increasing over time. This increasing LOS displacement (subsidence) rate and areal extend of subsidence is probably related to geothermal fluid production causing continued fluid-pressure declines in the area of extraction in the CPGF that extend to the recharge zone. The 1994–97 leveling survey data zone (Glowacka et al., 1999, 2005; Sarychikhina et al., 2011) projected to the LOS direction shows a maximum LOS displacement rate of about $-8\,\text{cm yr}^{-1}$ in the recharge which is not in accord with our results. This discordance is probably due to limited coverage of the leveling network.

The second region of high subsidence rate is located in the CPGF production zone. Because the ascending and descending pass datasets present important differences in this zone, we compared the data only from the same pass.

For the descending pass data, the maximum LOS displacement rate occurring in the eastern part of the CPGF production zone is about $-10\,\text{cm yr}^{-1}$ for all compared periods (1993–97, 1998–2000, 2005–07). In the western and central parts of the CPGF production zone the LOS displacement rate is lower in 2005–07 than in 1993–97 and 1998–2000. It could be related to decreased production in the western part of CPGF (CP I). The LOS displacement rates for individual years during 2005–07 period are similar.

For ascending pass data, we found important differences in LOS displacement rate and pattern between the 2004 and the 2007, 2009 periods. However, due to the fact that the 2004

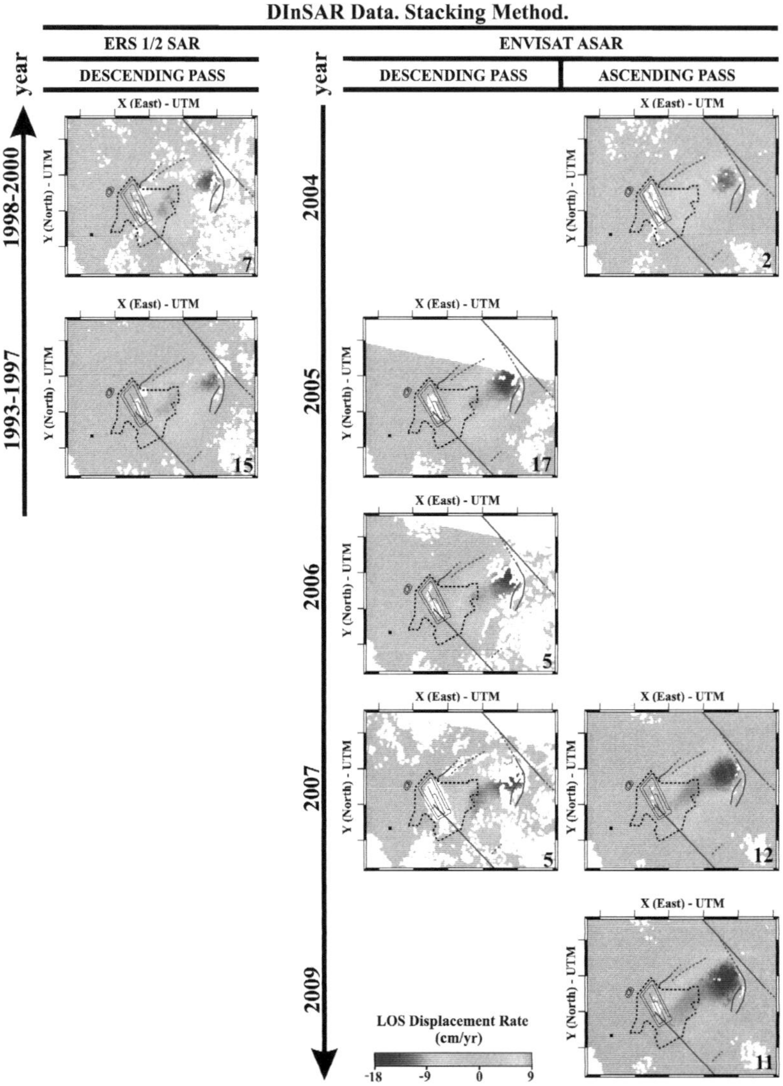

Figure 2. Maps of LOS displacement rate (cm yr^{-1}) showing the time evolution of the ground deformation field in and around the Cerro Prieto Geothermal Field in Mexicali Valley. Maps were obtained using the multiple Differential Interferograms Stacking (DIS) procedure. The numbers in the right lower corner indicate the number of differential interferograms used in stack. Negative displacement rates indicate an increase in the LOS range (ground subsidence). Black square marks the location of the reference, considered stable, region. Tectonic faults (solid and dotted red lines), CPGF limits (black thick dotted line), evaporation pond and CPV (thin gray lines) are superimposed onto the maps. The maps are in UTM coordinates, the marks are every 5 km.

LOS displacement rate was estimated using only 2 differential interferograms, we think that this difference is due to possible errors (e.g. atmospheric artifacts) in the 2004 dataset. There is no significant difference between the 2007 and 2009 LOS displacement rates.

5 Conclusions

A comparison of the annual aseismic LOS displacement rates for 1993–97, 1998–2000, 2004–07, and 2009 periods obtained using multiple Differential Interferograms Stacking (DIS) method reveals the important changes in spatial pattern and rate of ground deformation in the study area. The main changes occur between 2000 and 2005 and in-

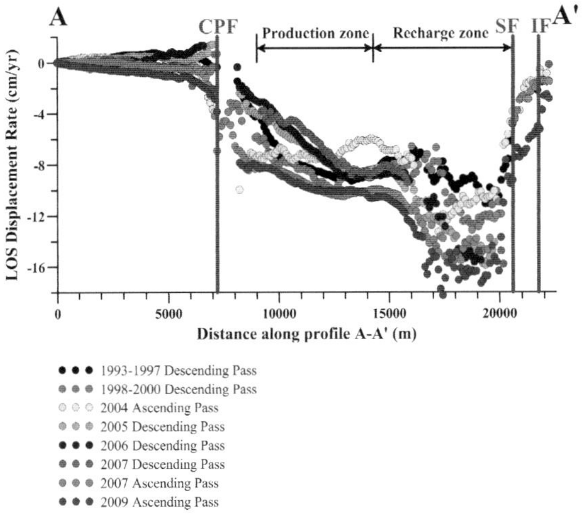

Figure 3. Comparison between LOS displacement rates derived from multiple Differential Interferograms Stacking (DIS) procedure for different periods along the profile A-A' (location shown in Fig. 1). Solid red lines indicate location of tectonic faults which cross the profile. Faults notation as in Fig. 1.

clude increasing deformation rate in the recharge zone and decreasing deformation rate in the western part of the CPGF production zone. The observed changes may be caused by changes in production in the CPGF, an increased production in the eastern part of the field due to the newest power plant (CP IV), which started operating in 2000, and decreased production in the western part of CPGF (CP I). Since 2000 there was not considerable changes in the total CPGF production, so based on the results of this study, we can state that it took a maximum 5 years for ground deformation to stabilize after the perturbation caused by changes in the CPGF production in 2000.

During 1993–2000 and 2005–09 periods, the spatial pattern and rate of deformation could be considered constant (or without significant changes).

Acknowledgements. Data from the European Space Agency's ENVISAT satellite was used to generate the interferometric data. The data were obtained as part of ESA Cat-1 Project (ID – C1P3508).

This research was sponsored in part by CONACYT, project 105907 and CICESE internal funds.

References

Allis, R. G., Bromley, C. J., and Currie, S.: Update on subsidence at the Wairakei–Tauhara Geothermal System, Geothermics, 38, 169–180, 2009.

Avallone, A., Briole, P., Delacourt, C., Zollo, A., and Beauducel, F.: Subsidence at Campi Flegrei (Italy) detected by SAR interferometry, Geophys. Res. Lett., 26, 2303–2306, 1999.

Baer, G., Schattner, U., Wachs,D., Sandwell, D., Wdowinski, S., and Frydman, S.: The lowest place on Earth is subsiding – An InSAR (interferometric synthetic aperture radar) perspective, GSA Bulletin, 114, 12–23, 2002.

Bürgmann, R., Rosen, P. A., and Fielding, E. J.: Synthetic aperture radar interferometry to measure Earth's surface topography and its deformation, Ann. Rev. Earth Planet. Sci., 28, 169–209, 2000.

Carnec, C. and Delacourt, C.: Three years of mining subsidence monitored by SAR interferometry, near Gardanne, France, J. Appl. Geophys., 43, 43–54, 2000.

Carnec, C. and Fabriol, H.: Monitoring and modeling land subsidence at the Cerro Prieto geothermal field, Baja California, Mexico, using SAR interferometry, Geophys. Res. Lett., 26, 1211–1214, 1999.

de la Peña, L. A.: Results from the first order leveling surveys carried out in the Mexicali Valley and at the Cerro Prieto field, Baja California, Proceedings, in: Proceedings of the 3rd Symposium on the Cerro Prieto Geothermal Field, Baja California, Mexico, 24–26 March 1981, San Francisco, California, USA, 281–291, 1981.

Fialko, Y. and Simons, M.: Deformation and seismicity in the Coso geothermal area, Inyo County, California: Observations and modeling using satellite radar interferometry, J. Geophys. Res., 105, 21781–21794, 2000.

Fielding, E. J., Blom, R. G., and Goldstein, R. M.: Rapid subsidence over oil fields measured by SAR interferometry, Geophys. Res. Lett., 27, 3215–3218, 1998.

Gabriel, A. G., Goldstein, R. M., and Zebker, H. A.: Mapping small elevation changes over large areas: Differential radar interferometry, J. Geophys. Res., 94, 9183–9191, 1989.

García, J. R.: Estudios de nivelación de primer orden en Cerro Prieto, in: Proceedings of the 1st Symposium on the Cerro Prieto Geothermal Field, Baja California, México, 20–22 September 1978, San Diego, California, USA, 148–150, 1978.

Glowacka, E., González, J., and Fabriol, H.: Recent vertical deformation in Mexicali Valley and its relationship with tectonics, seismicity, and the exploitation of the Cerro Prieto geothermal field, Mexico, Pure Appl. Geophys., 156, 591–614, 1999.

Glowacka, E., Nava, F. A., de Cossio, G. D., Wong, V., and Farfan, F.: Fault slip, seismicity, and deformation in Mexicali Valley, Baja California, Mexico, after the M 7.1 1999 Hector Mine earthquake, Bull. Seismol. Soc. Am., 92, 1290–1299, 2002.

Glowacka, E., Sarychikhina, O., and Nava, F. A.: Subsidence and stress change in the Cerro Prieto Geothermal Field, B.C., Mexico, Pure Appl. Geophys., 162, 2095–2110, 2005.

Glowacka, E., Sarychikhina, O., Suárez, F., Mendoza, R., and Nava, F. A.: Estudio geológico para definir la zona de hundimiento con el fin de relocalización del canal Nuevo Delta en el Valle de Mexicali, Informe Técnico, CICESE, México, 2006.

Glowacka, E., Sarychikhina, O., Nava, F. A., Suarez, F., Ramírez, J., Guzman, M., Robles, B., Farfan, F., and De Cossio Batani, G. D.:

Continuous monitoring techniques of fault displacement caused by geothermal fluid extraction in the Cerro Prieto Geothermal Field (Baja California, Mexico), in: Land Subsidence, Associated Hazards and the Role of Natural Resources, edited by: Correón-Freyre D., Cerca, M., and Galowey, D., IAHS Publ., 339, 326–332, 2010a.

Glowacka, E., Sarychikhina, O., Suárez, F., Nava, F. A., and Mellors, R.: Anthropogenic subsidence in Mexicali Valley, Baja California, Mexico, and slip on the Saltillo fault, Environ. Earth Sci., 59, 1515–1524, 2010b.

Glowacka, E., Sarychikhina, O., Suárez, F., Nava, F. A., Farfan, F., De Cossio Batani, G. D., and Garcia Arthur, M. A.: Anthropogenic subsidence in the Mexicali Valley, B.C., Mexico, caused by the fluid extraction in the Cerro Prieto geothermal Field and the role of faults, in: Proceedings of the World Geothermal Congress (WGC) 2010, 25–29 April 2010, Bali, Indonesia, 2010c.

Glowacka, E., Sarychikhina, O., Robles, B., Suárez, F., Ramírez, J., and Nava, F. A.: Estudio geológico para definir la línea de hundimiento cero y monitorear la subsidencia de los módulos 10, 11 y 12 en el Valle de Mexicali, en el distrito de riego 014, Rio Colorado, B.C., Reporte Técnico Final, Convenio con CONAGUA, CICESE, Mexico, 2012.

González, J., Glowacka, E., Suárez, F., Quiñones, J. G., Guzmán, M., Castro, J. M., Rivera, F., and Félix, M. G.: Movimiento reciente de la Falla Imperial, Mexicali, B. C, Ciencia para todos Divulgare, Universidad Autónoma de Baja California, 6, 4–15, 1998.

Grannell, R. B., Tarnman, D. W., Clover, R. C., Leggewie, R. M., Aronstam, P. S., Kroll, R. C., and Eppink, J.: Precision gravity studies at Cerro Prieto, Proceedings, Second Symposium on the Cerro Prieto, in: Proceedings of the 2nd Symposium on the Cerro Prieto Geothermal Field, Baja California, Mexico, 17–19 October 1979, Mexicali, Baja California, Mexico, 329–331, 1979.

Hanssen, R. F.: Radar Interferometry: Data Interpretation and Error Analysis, Kluwer Academic Publishers, Dordrecht, the Netherlands, 2001.

Herrera, G., Tomás, R., Lopez-Sanchez, J. M., Delgado, J., Mallorqui, J. J., Duque, S., and Mulas, J.: Advanced DInSAR analysis on mining areas: La Union case study (Murcia, SE Spain), Eng. Geol., 90, 148–159, 2007.

Hole, J., Bromley, C., Stevens, N., and Wadge, G.: Subsidence in the geothermal fields of the Taupo Volcanic Zone, New Zealand from 1996 to 2005 measured by InSAR, J. Volcanol. Geoth. Res., 166, 125–146, 2007.

Keiding, M., Hooper, A., Árnadóttir, T., Jónsson, S., and Decriem, J.: Natural and man-made deformation around geothermal fields on the Reykjanes peninsula, SW Iceland, in: Proceedings, of the Workshop Fringe 2009, 30 November–4 December 2009, ESA, Frascati, Italy, 4 pp., 2009.

Lira, H.: Características del sismo del 23 de Mayo de 2006, Informe RE-023/2006, Comisión Federal de Electricidad, Residencia de Estudios, México, 2006.

Lira, H. and Arellano, J. F.: Resultados de la nivelación de precisión realizada en 1997, en el campo geotérmico Cerro Prieto, Informe Técnico RE 07/97, Comisión Federal de Electricidad, Residencia de Estudios, México, 1997.

Lowe, M.: Subsidence in sedimentary basins due to groundwater withdrawal for geothermal energy development, Utah Geological Survey, Utah Department of Natural Resources, Open File Report 601, 13 pp., available at: http://files.geology.utah.gov/online/ofr/ofr-601.pdf (last access: 12 October 2015), 2012.

Massonnet, D. and Feigl, K. L.: Radar interferometry and its application to changes in the Earth's surface, Rev. Geophys., 36, 441–500, 1998.

Massonnet, D. and Rabaute, T.: Radar interferometry: limits and potential, IEEE T. Geosci. Remote, 31, 455–464, 1993.

Massonnet, D., Holzer, T., and Vadon, H.: Land subsidence caused by the East Mesa Geothermal field, California, observed using SAR interferometry, Geophys. Res. Lett., 24, 901–904, 1997.

Mossop, A. and Segall, P.: Subsidence at the Geysers geothermal field, N. California from a comparison of GPS and leveling surveys, Geophys. Res. Lett., 24, 1839–1842, 1997.

Nava, F. A. and Glowacka, E.: Fault slip triggering, healing, and viscoelastic after working in sediments in the Mexicali-Imperial Valley, Pure Appl. Geophys., 156, 615–629, 1999.

Sarychikhina, O.: Modelación de subsidencia en el campo geotérmico Cerro Prieto, MS thesis, CICESE, Mexico, 101 pp., 2003.

Sarychikhina, O. and Glowacka, E.: Application of DInSAR Stacking Method for Monitoring of Surface Deformation Due to Geothermal Fluids Extraction in the Cerro Prieto Geothermal Field, Baja California, Mexico, in: Proceedings of the World Geothermal Congress (WGC) 2015, 19–25 April 2015, Melbourne, Australia, 2015.

Sarychikhina, O., Glowacka, E., Mellors, R., and Suárez-Vidal, F.: Land subsidence in the Cerro Prieto Geothermal Field, Baja California, Mexico, from 1994 to 2005. An integrated analysis of DInSAR, leveling and geological data, J. Volcanol. Geoth. Res., 204, 76–90, 2011.

Sarychikhina, O., Glowacka, E., Robles, B., and Mojarro, J.: Spatio-temporal evolution of anthropogenic deformation around Cerro Prieto geothermal field in the Mexicali Valley, B.C., Mexico, between 1993 and 2009 from DInSAR and leveling, EGU General Assembly 2014, EGU2014-14709, 27 April–2 May 2014, Vienna, Austria, 2014.

Sarychikhina, O., Glowacka, E., Robles, B., Nava, F. A., and Guzmán, M.: Estimation of seismic and aseismic deformation in Mexicali Valley, Baja California, Mexico, in the 2006-2009 period, using precise leveling, DInSAR, geotechnical instruments data, and modeling, Pure Appl. Geophys., doi:10.1007/s00024-015-1067-0, in press, 2015.

Segall, P.: Earthquakes triggered by fluid extraction, Geology, 17, 942–946, 1989.

Suárez-Vidal, F., Munguía-Orozco, L., González-Escobar, M., González-García, J., and Glowacka, E.: Surface rupture of the Morelia fault near the Cerro Prieto Geothermal Field, Mexicali, Baja California, Mexico, during the Mw 5.4 earthquake of 24 May 2006, Seismol. Res. Lett., 78, 394–399, 2007.

Suárez-Vidal, F., Mendoza-Borunda, R., Nafarrete-Zamarripa, L., Rámirez, J., and Glowacka, E.: Shape and dimensions of the Cerro Prieto pull-apart basin, Mexicali, Baja California, México, based on the regional seismic record and surface structures, Int. Geol. Rev., 50, 636–649, 2008.

Tomás, R., Márquez, Y., Lopez-Sanches, J. M., Delgado, J., Blanco, P., Mallorquí, J. J., Martínez, M., Herrera, G., and Mulas, J.: Mapping ground subsidence induced by aquifer overexploitation using advanced Differential SAR Interferometry: Vega Media of the Segura River (SE Spain) case study, Remote Sens. Environ., 98, 269–283, 2005.

Velasco, J.: Levantamiento Gravimétrico, Zona Geotérmica de Mexicali, Baja California, Internal report, Consejo de Recursos Naturales no Renovables, Mexico, 1963.

Wegmüller, U. and Werner, C. L.: Gamma SAR processor and interferometry software, in: Proceedings of the 3rd ERS Scientific Symposium, 14–21 March 1997, Florence, Italy, 1686–1692, 1997.

Wright, P. A. and Stow, R. J.: Detecting mining subsidence from space, Int. J. Remote Sens., 20, 1183–1188, 1999.

Wyman, R. M.: Potential Modeling of Gravity and Leveling Data over Cerro Prieto Geothermal Field, MS thesis, California State University, Long Beach, CA, USA, 1983.

Zebker, H. A., Rosen, P. A., Goldstein, R. M., Gabriel, A., and Werner, C. L.: On the derivation of coseismic displacement fields using differential radar interferometry: The Landers earthquake, J. Geophys. Res., 99, 19617–19634, 1994.

Proc. IAHS, 372, 343–346, 2015
proc-iahs.net/372/343/2015/
doi:10.5194/piahs-372-343-2015

Open Access
PIAHS

Long-range ground deformation monitoring by InSAR analysis

S. Rokugawa and T. Nakamura

The University of Tokyo, Tokyo, Japan

Correspondence to: T. Nakamura (t-nakamura@tmi.t.u-tokyo.ac.jp)

Published: 12 November 2015

Abstract. InSAR (Interferometric Synthetic Aperture Radar) analysis is an effective technique to map 3-dimensional surface deformation with high spatial resolution. The aim of this study was to evaluate the capability of InSAR analysis when applied to ground monitoring of an environmental disaster. We performed a time series InSAR analysis using ENVISAT/ASAR and ALOS/PALSAR data and commercial software to investigate subsidence around the Kanto District of Japan. We also investigated techniques for efficient early detection of landslides in Kyushu using time series analysis that incorporated synthetic aperture radar (SAR) images. ENVISAT/ASAR data acquired from 2003–2010 and ALOS/PALSAR data acquired from 2006–2011 were used to detect poorly expressed geomorphological deformation by conducting time series analyses of periodically acquired SAR data. In addition, to remove noise caused by geographical feature stripes or phase retardation, we applied median filtering, histogram extraction processing, and clarification of the displacement with a Laplacian filter. The main functions of the InSAR time series analysis are the calculation of phase differences between two images and the inversion with smoothness constraint for the estimation of deformation along the line of sight. The results enabled us to establish criteria for the selection of suitable InSAR data pairs, and provided the final error estimation of the derived surface deformation. The results of the analysis in the Kanto District suggested that localized areas of uplift and subsidence have occurred at irregular intervals in this area. Furthermore, the method offers the possibility of early warning of environmental disasters such as landslide and abrupt subsidence. Our results confirm the effectiveness of InSAR analysis for the monitoring of ground deformation over wide areas via the detection of localized subsidence and landslides.

1 Introduction

Recently, large-scale typhoons and unusual weather have increased local heavy rain, producing conditions that are prone to large-scale landslides and causing extensive damage, because of the steep topography of many mountainous regions in Japan. Therefore, it is essential to develop techniques to detect such landslides before their occurrence to plan countermeasure. In the present study, we demonstrate the use of InSAR technique in detecting early signs of landslides and discuss about the production of related hazard maps. Firstly, we applied the method of measuring long-range ground deformation by combining InSAR and time-series analysis (Ferretti et al., 2000, 2001; Berardino et al., 2002; Deguchi et al., 2009) in the Kanto Plain, Japan, using 54 ENVISAT/ASAR images. The Kanto Plain started to subside in the 1910s, with the main causes of land subsidence related to the intake of ground-water for the use in industries, agriculture, waterworks, and other fields. In the Kujukuri area, the exploitation of soluble natural gas also counts. A subsidence of approximately 2 cm per year was detected. Then, in Kyushu area, Mt. Wanitsuka and Tsukahara dam in Miyazaki prefecture, and Imari area of Saga prefecture, interference fringes were found at the uncollapsed areas.

2 Method

InSAR is the most appropriate technique to investigate ground movements over large areas. This monitoring technique can offer many advantages, particularly in the detection of the precursory phenomena of landslide disasters. We conducted an investigation into landslide disasters occurring between 2008 and 2009 in the region around the Wanitsuka Mountains. We used a SAR PALSAR stack of 10 scenes acquired from December 2006 to April 2011. In the Imari area, northern Kyushu, we used 38 SAR interference images acquired between May 2006 and January 2011. Near the Tsukahara dam in the Mimikawa upstream region, Miyazaki prefecture, we used 22 scenes acquired in descending mode between February 2007 and April 2011 and 13 ascending scenes acquired from October 2007 and December 2010. Then, we quantitatively evaluated the reliability of the InSAR data and conducted time series analysis to clarify variations in local upheaval or settlement in order to assess ground deformation.

All the areas addressed by this study can be considered prone to landslide disaster. However, the InSAR picture that was obtained from these areas was unclear, making it an ideal location to test our newly developed detection technique. Therefore, we extracted information for a landslide by applying filters to remove noise and clarify ground movement. As a result, any place other than the point where landslide happened, I was able to detect the point considered that an initial change was caused. During this process, we corrected for the measurement error associated with InSAR and assessed the effectiveness of the technique. However, it proved difficult to remove all noise. A composite filter incorporating three individual filtering techniques has been used.

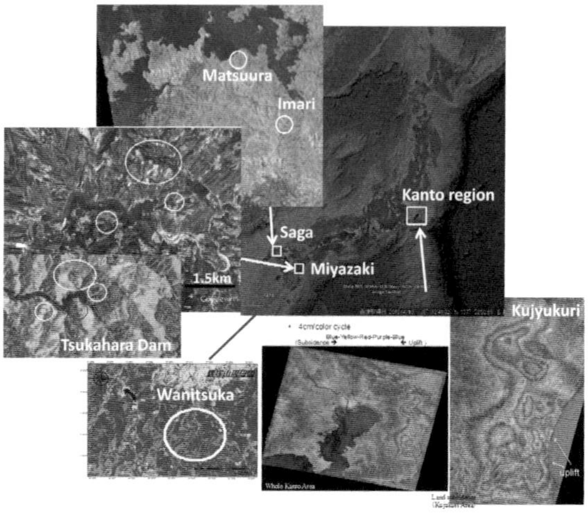

Figure 1. Map of study area and results of InSAR analysis.

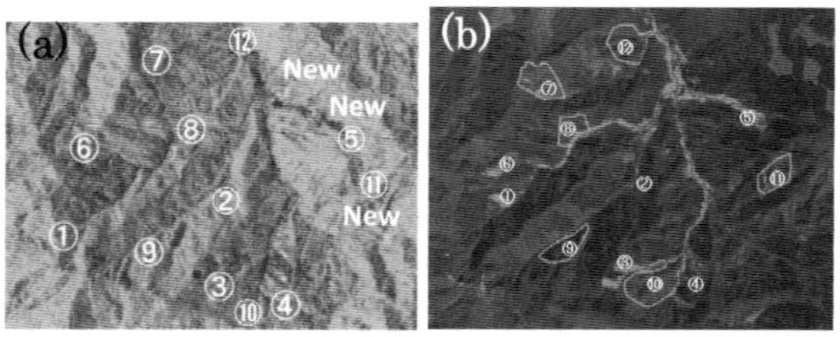

Figure 2. **(a)** The interferogram around the Wanitsuka Mountains in the Miyazaki Prefecture (ALOS/PALSAR data acquired from 2006 to 2011). **(b)** Collapsed area (red) and uncollapsed area (yellow) was interpreted based on aerial photographs acquired from 1946 to 2005.

2.1 Median filtering

We calculated a median using a 3×3 area comprising a target pixel and its neighbouring pixels; the value is obtained as object pixel value. This filter was shown to be acceptable for the removal of spot-formed noise in InSAR images.

2.2 Histogram extraction processing

We normalised the InSAR displacement data and removed all small displacements from the SAR image. I removed 80 % of low order by the ground displacement magnitude detection of the cm unit and, in the case of mm unit, removed low order 60 %.

2.3 Laplacian filter

We used a 51×51 operator derived from a four-dimensional operator that radicalised the surface displacement using a standard second derivative filter to produce a weighted-mean picture element area.

3 Result and discussion

Maps with the location of the investigated sites and the results of InSAR analysis are shown in Fig. 1 (Geological Survey of Japan, 2010). Figure 2 illustrates the interferogram for the region around the Wanitsuka Mountains and an interpretation of the collapsed area based on aerial photographs, which confirms the changes in the ground movement suggested by previous field investigation. On the basis of the InSAR results, we believe that fluctuations in the ground surface continued after collapse. Collapse induced by the typhoon of 2005 was not detected in area 7–12, although evidence suggests that subsidence continued in these areas following the typhoon. A subsidence of approximately 9 cm was found in area 4 over a period of two years and eight months; this represents the greatest change in ground elevation detected in the region, although subsidence of 4–6 cm was found in other areas. These results suggest that the technique is suitable for the assessment of such ground movements. However, problems remain in terms of the interpretation of the consequence of the research technique by the application to more areas and evaluation of the application limit. Therefore, the technique has been verified through further analysis. Figures 3 and 4 illustrate the results obtained for an InSAR analysis of the Imari area using various fil-

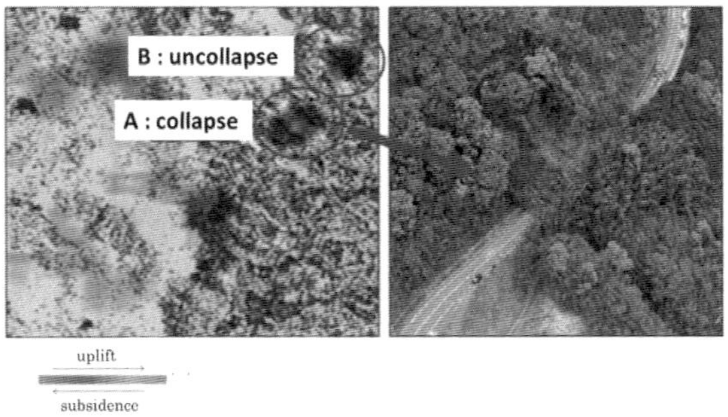

Figure 3. Imari area collapsed point (A) and uncollapsed point (B) were extracted based on InSAR analysis (left image). Actual spot of the collapse (right photo).

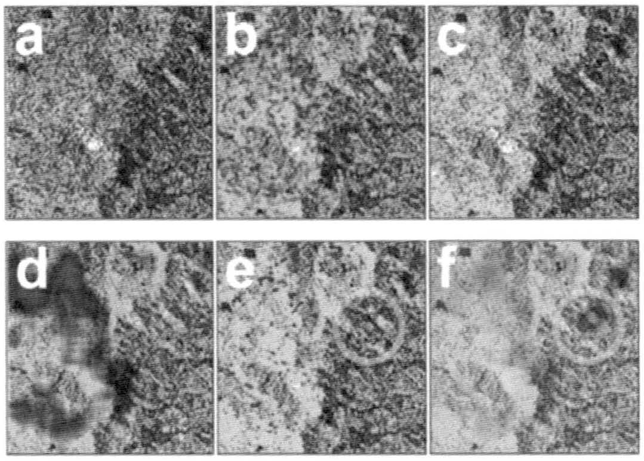

Figure 4. Imari area: effects of filter application to InSAR outcome for the Imari area. Circles indicate areas of landsliding. **(a)** InSAR result, **(b)** median filtering, **(c)** histogram extraction (upper 20 % extract of the normalised displacement), **(d)** Laplacian filter, **(e)** median filtering and histogram extraction, **(f)** median filtering and histogram extraction and Laplacian filter.

Figure 5. Collapse to the south of Tsukahara dam was extracted from ascending PALSAR image. (Base map is from Google earth)

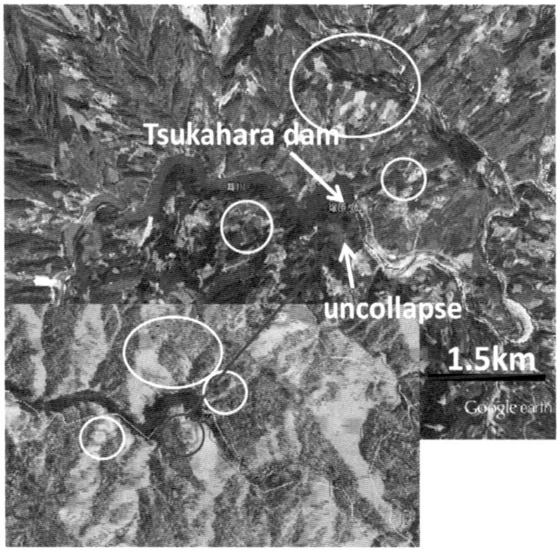

Figure 6. Uncollapsed zone to the south of Tsukahara dam was extracted from descending PALSAR image. (Base map is from Google earth)

ter processes. Landslide inception is difficult to be detected based on the outcome provided in Fig. 4a and c, because big noise has not been removed from these images. Conversely, the representation of surface displacements is improved in Fig. 4b, from which big noise (but not small noise) has been removed. Figure 4d illustrates the results obtained by emphasizing displacement and noise; surface displacements can be clearly detected in this image, indicating that the use of a Laplacian filtering is effective for noise removal. Figure 4e illustrates the results achieved by applying both median filtering and histogram extraction and suggests that noise reduction is useful in the acquisition of displacement magnitude. The location of the landslide is clearly indicated in Fig. 4f, for which all filtering techniques were applied. The results presented here indicate that InSAR is effective in detecting microdisplacements that occur during the early stages of landsliding and that detection of these displacements can be improved by application of filtering processes, at least in the Imari area. However, a point considered to have initial fluctuation other than the point where landslide happened has been detected as highlighted in Figs. 3b and 4f. Finally, different displacements were detected depending on the acquisition more of microwave radiation near the Tsukahara dam in the Mimikawa upstream region (Figs. 5 and 6). Collapse at the south of Tsukahara dam was extracted from ascending PALSAR image, as shown in Fig. 5. On the other hand, an uncollapsed area at the south of Tsukahara dam was extracted from descending PALSAR image, as shown in Fig. 6. In spite of an uncollapsed point, strong interference fringes are seen and it seems that it has a high possibility that collapse will take place in the near future.

4 Conclusions

The result of InSAR analysis shows that the position of slope collapse can be detected with high precision. Moreover, detection of the parts where the slope collapse is expected to occur in the near future is also possible. The results here presented suggest that the proposed technique can offer information useful to build up a first ground movement system for landslide detection, and then in deciding where monitoring components such as GPS equipment should be placed. We test the effectiveness of the measures described here by performing in-situ surveys in the critical zones pointed out by InSAR. The results confirm that our techniques are effective in detecting landslide-related ground displacement over wide areas. Furthermore, we have demonstrated that it is possible to construct hazard maps for large areas based on the detected ground movements. Therefore, InSAR processing may make a considerable contribution to disaster prevention.

Acknowledgements. The authors would like to thank ERSDAC for providing PALSAR L-band data.

References

Berardino, P., Fornaro, G., Lanari, R., and Sansosti, E.: A New Algorithm for Surface Deformation Monitoring Based on Small Baseline Differential SAR Interferograms, IEEE Trans. Geosci. Remote, 40, 2375–2383, 2002.

Deguchi, T., Rokugawa, S., and Matsushima: Long-term ground deformation measurement by time series analysis for SAR interferometry, Journal of the Remote Sensing Society of Japan, 29, 418–428, 2009.

Ferretti, A., Prati, C., and Rocca, F.: Nonlinear subsidence rate estimation using permanent scatters in differential SAR interferometry, IEEE T. Geosci. Remote, 38, 2202–2212, 2000.

Ferretti, A., Prati, C., and Rocca, F.: Permanent scatterers in SAR Interferometry, IEEE Trans. Geosci. Remote, 39, 8–20, 2001.

Geological Survey of Japan: AIST, Seamless Digital Geological Map of Japan (1:200,000), 2010.

Proc. IAHS, 372, 347–349, 2015
proc-iahs.net/372/347/2015/
doi:10.5194/piahs-372-347-2015

Mapping and characterization of land subsidence in Beijing Plain caused by groundwater pumping using the Small Baseline Subset (SBAS) InSAR technique

M. L.Gao [1,2,a], H. L.Gong [1,2], B. B.Chen [1,2], C. F. Zhou[1,2], K. S. Liu[1,2], and M. Shi[1,2]

[1]College of Resource Environment and Tourism, Capital Normal University, Beijing 100048, China
[2]State Key Laboratory Incubation Base of Urban Environmental Processes and Digital Simulation, Capital Normal University, Beijing 100048, China
[a]now at: Haidian, Beijing, China

Correspondence to: M. L. Gao (b-19890320@163.com)

Published: 12 November 2015

Abstract. InSAR time series analysis is widely used for detection and monitoring of slow surface deformation. In this paper, 15 TerraSAR-X radar images acquired in stripmap mode between 2012 and 2013 are processed for land subsidence monitoring with the Small Baseline Subset (SBAS) approach in Beijing Plain in China. Mapping results produced by SBAS show that the subsidence rates in the area of Beijing Plain range from −97.5 (subsidence) and to $+23.8\,\mathrm{mm\,yr^{-1}}$ (uplift), relative to a presumably stable benchmark. The mapping result also reveals that there are the five subsidence centers formed by surface deformation spreading north to south east of the downtown. An uneven subsidence patten was detected near the Beijing Capital International Airpor, which may be related to loading of buildings and the aircraft.

1 Introduction

Land subsidence has become a hazard and could lead to serious problems especially in urban areas, related to city flooding (Ng et al., 2012), structural damages to buildings and other civilian infrastructures (Chaussard et al., 2013), including local subsurface water systems (Zhang et al., 2014). Consequently, subsidence monitoring in the urban area is necessary for safety, security, economic and planning reasons.

Subsidence in Beijing was first detected in the 1960s and has progressed in several areas for decades. The subsidence is mainly caused by intense groundwater extraction according to previous studies (Chen et al., 2014; Hu et al., 2014). Ground surface deformation can be observed through the InSAR techniques, which greatly facilitate the monitoring of the land subsidence in metropolitan areas (Kagawa and Furuno, 2010; Aobpaet et al., 2013). InSAR time-series methods such as PS-InSAR and Small Baseline methods have been applied to study subsidence in Beijing (Zhang et al., 2011; Gu et al., 2014). Chen et al. (2011) concluded that seasonal and inter-annual variation of groundwater drawdowns have resulted in an uneven spatial and temporal distribution of subsidence, with a maximum subsidence rate of about $-41.08\,\mathrm{mm\,yr^{-1}}$.

This paper presents ground deformation results based on SBAS processing of 15 TerraSAR-X images covering the eastern area of the Beijing Plain, China. We processed the SLC (single look complex) images using the Stanford Method for Persistent Scatterers (StaMPS) processing software to determine the line-of-sight (LOS) displacements (native valued LOS displacements represent an increase in range and subsidence in this paper) and subsidence rates. We present the mapped subsidence rates and describe the characteristics of the subsidence in the study area.

2 Study area and data description

The study area is the Beijing Plain area (centered at 39°58′37″ N, 116°42′11″ E) in the city of Beijing with as area of about $1900\,\mathrm{km^2}$ (Fig. 1). In this area, the elevation ranges from 20 to 40 m, and there are many buildings and other civilian infrastructure.

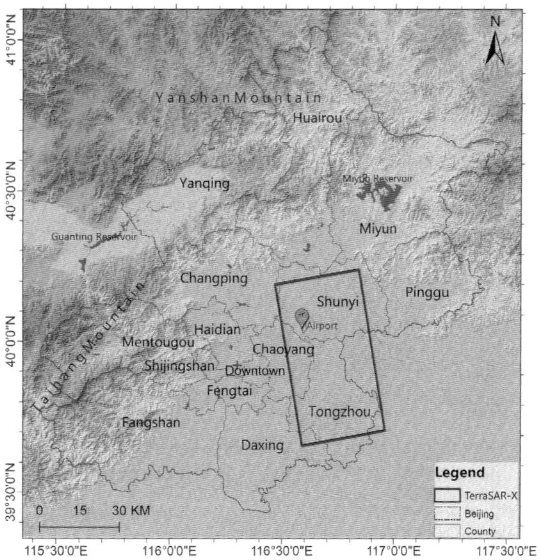

Figure 1. Map of the study area, Beijing Plain, China. The red box in the figure represents the coverage of the TerraSAR-X stripmap which covers an area of more than 1900 km^2. The blue polygons are water bodies. The red cross symbol represents the location of the referenced M1 benchmark.

The 15 SLC ascending orbit, stripmode images have an azimuth resolution of 3.3 m, and range resolution of 2.0 m. The images were acquired over Beijing during the period of January 2012 and November 2013. The ASTER GDEM V2 product with 1 arcsec geographical resolution (30 m) was used as the external DEM in the InSAR process.

3 Time-series InSAR analysis

The SBAS analysis relies on a subset of interferogram pairs that were created with small temporal (over short time intervals) and geometrical baselines to limit decorrelation noise. The noise is further reduced by applying range and azimuth filters, and spatial multi-looking. In this paper, we performed the SBAS approach on the dataset using the StaMPS/MTI toolkit. The candidate pixels are identified in the same way as PS pixels (Hooper et al., 2007), based on the spatial correlation of the phase. A subset of 30 interferogram pairs were formed temporal and geometrical baselines less than 200 days and 500 m, respectively, with an coherence threshold of 0.7. Phase unwrapping used a statistical cost approach.

4 Results and discussion

More than 220 000 sufficiently coherent pixels were identified, processed from more than 2 600 000 candidate pixels. The mean LOS deformation rates (unit: $mm\,yr^{-1}$) were derived from the unwrapped time-series phases using linear least squares. The results are relative to the benchmark M1 (as shown in Fig. 1). M1 is a levelling benchmark situated in a relatively stable area.

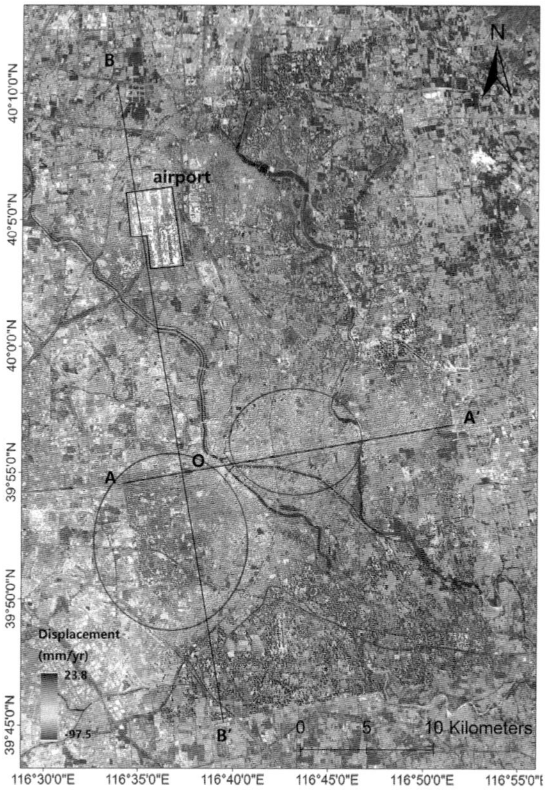

Figure 2. Mean LOS subsidence rates ($mm\,yr^{-1}$) spatial distribution. Red and blue areas represent subsidence and uplift relative to the referenced point M1, respectively. Profile A–A′, which is perpendicular to profile B–B′, goes through the two subsidence areas (marked as red circles), and intersects the profile B–B′ at O. The black box represents the location of Beijing Capital International Airport.

Figure 2 shows the mean LOS subsidence rate ($mm\,yr^{-1}$) estimations. For coherent pixel from January 2012 to November 2013, the deformation rates range from −97.5 to +23.8 $mm\,yr^{-1}$, relative to the benchmark M1. However, because no uplift is expected in this area, if we consider that M1 could be subsiding, then uplift areas shown in blue on Fig. 2 could be subsiding slower than M1. In this case, the maximum subsidence rate in eastern Beijing could be about 121.3 $mm\,yr^{-1}$.

As can be seen from the Fig. 2, fewer coherent pixels are evident in the non-urbanized rural areas and croplands in the east. Two main subsidence areas and several other subsidence features of smaller extent have formed since the

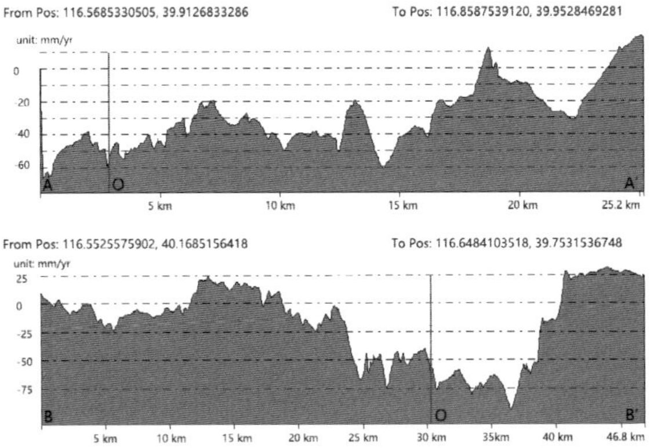

Figure 3. Subsidence rates along profile A–A′ and profile B–B′ shown in Fig. 2. The profiles intersect at the point O. (Note: negative values indicate subsidence.)

1990s. Figures 2 and 3 show that profile A–A′ intersects the two main subsidence areas, and a smaller subsidence features, while profile B–B′ intersects the southern subsidence area and two smaller subsidence features, including one at the airport. The settlements along the profile A–A′ have extended and connected to one huge subsidence area. The subsidence in this area has been attributed to the exploitation of groundwater, as the previous studies concluded. However, we also detected subsidence near the Beijing Capital International Airport with ranging from -26.2 to $13.6\,\mathrm{mm\,yr^{-1}}$. The T1 and T2 terminals constructed in the early 1990s appear to be subsiding slightly with respect to the surrounding areas, while the T3 terminal built in 2008 is uplifting relative to M1. We consider that subsidence at the airport is probably due to dynamic loading of aircraft takeoffs and landings and the static loading of the large buildings.

5 Conclusions

The SBAS technique was used with 15 TerraSAR-X SLC images to detect and characterize subsidence on the eastern Beijing Plain. Deformation measured during the period 2012∼2013, show subsidence in two main areas and several other areas of smaller extent. The maximum subsidence rate was $-97.5\,\mathrm{mm\,yr^{-1}}$. Uplift was also measured with a maximum rate of $+23.8\,\mathrm{mm\,yr^{-1}}$. We believe that the subsidence is mainly caused by the exploitation of groundwater. Uneven subsidence was detected near the of the Beijing Capital International Airport location, which may be caused by surface loading of buildings and the arcraft, apart from the groundwater depression cones. Further work is planned to evaluate subsidence near the airport, and other ground-based geodetic surveys are planned to validate these results for the Beijing Plain.

Acknowledgements. The provision of the Doris and StaMPS for data processing by TUDelft is gratefully acknowledged.

References

Aobpaet, A., Cuenca, M. C., Hooper, A., and Trisirisatayawong, I.: InSAR time-series analysis of land subsidence in Bangkok, Thailand, Int. J. Remote Sens., 34, 2969–2982, doi:10.1080/01431161.2012.756596,2013.

Chaussard, E., Amelung, F., Abidin, H., and Hong, H.: Sinking cities in Indonesia: ALOS PALSAR detects rapid subsidence due to groundwater and gas extraction, Remote Sens. Environ., 128, 150–161, doi:10.1016/j.rse.2012.10.015, 2013.

Chen, B. B., Gong, H. L., Li, X. J., Lei, K. C., Zhang, Y. Q., Li, J. W., Gu, Z. Q., and Dang, Y. A.: Spatial-temporal characteristics of land subsidence corresponding to dynamic groundwater funnel in Beijing Municipality, China, Chinese Geogr. Sci., 21, 753–64, doi:10.1007/s11769-011-0509-6, 2011.

Chen, B. B., Gong, H. L., Li, X. J., Lei, K. C., Ke, Y. H., Duan, G. Y., and Zhou, C. F.: Spatial correlation between land subsidence and urbanization in Beijing, China, Nat. Hazards, 75, 2637–2652, doi:10.1007/s11069-014-1451-6, 2014.

Gu, Z. Q., Gong, H. L., Zhang, Y. Q., Lu, X. H., Wang, S., Wang, R., and Liu, H. H.: Research on Monitoring Land Subsidence in Beijing Plain Area Using PS-InSAR Technology. Spectrosc. Spect. Anal., 34, 1898–1902, 2014.

Hooper, A., Segall, P., and Zebker, H.: Persistent scatterer interferometric synthetic aperture radar for crustal deformation analysis, with application to Volcán Alcedo, Galápagos, J. Geophys. Res., 112, B07407, doi:10.1029/2006JB004763, 2007.

Hu, B., Wang, H. S., Sun, Y. L., Hou, J. G., and Liang, J.: Long-Term Land Subsidence Monitoring of Beijing (China) Using the Small Baseline Subset (SBAS) Technique, Remote Sens., 6, 3648–3661, doi:10.3390/rs6053648, 2014.

Kagawa, A. and Furuno, K.: Land subsidence monitoring system in the southeast part of Kanto groundwater basin, Japan, in: Land Subsidence, Associated Hazards And the Role Of Natural Resources Development, Proceedings of EISOLS 2010, Querétaro, Mexico, 17–22 October 2010, IAHS Publ. 339, 339–344, 2010.

Ng, A. H.-M., Ge, L. L., Li, X. J., Hasanuddin, Z., Abidin, Andreas, H., and Zhang, K.: Mapping land subsidence in Jakarta, Indonesia using persistent scatterer interferometry (PSI) technique with ALOS PALSAR, Int. J. Appl. Earth Obs., 18, 232–242, doi:10.1016/j.jag.2012.01.018, 2012.

Zhang, X. D., Ge, D. Q., Ma, W. Y., Zhang, L., Wang, Y., and Guo, X. F.: Study the Land Subsidence Along Jinghu Highway (Beijing-Hebei) Using Ps-Insar Technique, in: Geoscience and Remote Sensing Symposium (IGARSS), 2011 IEEE International, 24–29 July 2011, Vancouver, BC, IEEE, 1608–1611, doi:10.1109/IGARSS.2011.6049538, 2011.

Zhang, Y. Q., Gong, H. L., Gu, Z. Q., Wang, R., Li, X. J., and Zhao, W. J.: Characterization of land subsidence induced by groundwater withdrawals in the plain of Beijing city, China, Hydrogeol. J., 22, 397–409, 2014.

Proc. IAHS, 372, 351–356, 2015
proc-iahs.net/372/351/2015/
doi:10.5194/piahs-372-351-2015

Open Access PIAHS

Estimate of a spatially variable reservoir compressibility by assimilation of ground surface displacement data

C. Zoccarato[1], D. Baù[2], F. Bottazzi[3], M. Ferronato[1], G. Gambolati[1], S. Mantica[3], and P. Teatini[1]

[1]Department of Civil, Architectural and Environmental Engineering, University of Padova, Padova, Italy
[2]Department of Civil & Structural Engineering, The University of Sheffield, Sheffield, UK
[3]Development, Operations & Technology, eni S. p. A., San Donato Milanese, Italy

Correspondence to: C. Zoccarato (claudia.zoccarato@dicea.unipd.it)

Published: 12 November 2015

Abstract. Fluid extraction from producing hydrocarbon reservoirs can cause anthropogenic land subsidence. In this work, a 3-D finite-element (FE) geomechanical model is used to predict the land surface displacements above a gas field where displacement observations are available. An ensemble-based data assimilation (DA) algorithm is implemented that incorporates these observations into the response of the FE geomechanical model, thus reducing the uncertainty on the geomechanical parameters of the sedimentary basin embedding the reservoir. The calibration focuses on the uniaxial vertical compressibility c_M, which is often the geomechanical parameter to which the model response is most sensitive. The partition of the reservoir into blocks delimited by faults motivates the assumption of a heterogeneous spatial distribution of c_M within the reservoir. A preliminary synthetic test case is here used to evaluate the effectiveness of the DA algorithm in reducing the parameter uncertainty associated with a heterogeneous c_M distribution. A significant improvement in matching the observed data is obtained with respect to the case in which a homogeneous c_M is hypothesized. These preliminary results are quite encouraging and call for the application of the procedure to real gas fields.

1 Introduction

Fluid extraction from aquifer systems and hydrocarbon reservoirs are among the most frequent causes of anthropogenic land subsidence. In the framework of a sustainable development of energy resources, the availability of numerical models able to reproduce the monitoring data and to predict the future development of the land settlement is nowadays of paramount importance. In this study, an ensemble-based DA method is used to infer the geomechanical parameters characterizing the rock formation of a deep gas reservoir, thus reducing the prior uncertainties of the geomechanical model response. The DA framework essentially requires three main ingredients: (i) a model to simulate the physical process of interest, (ii) a set of observation data and (iii) a suitable algorithm to incorporate these data into the model response. In this work, a 3-D finite-element (FE) geomechanical model is used to predict the land surface displacements above a gas field where displacement observations are available. The calibration focuses on the uniaxial vertical compressibility c_M, which mostly influences the occurrence of land subsidence. Partitioning the reservoir into blocks by faults provides a basis for assuming a homogeneous c_M within each block, and a heterogeneous c_M from one block to another. The effectiveness of the DA algorithm to reduce the parameter uncertainty associated with the block-to-block c_M heterogeneity is evaluated for a synthetic test case. Significant improvements are obtained with respect to the assumption of a homogeneous c_M field throughout the model domain. In this work, an Ensemble Smoother (ES) is the DA algorithm used to estimate the compressibility by inversion of land surface displacement data. The ES is deemed adequate for this purpose because its implementation (i) avoids the simulation restart necessary to apply other more common data assimilation techniques, such as the Ensemble Kalman Filter (EnKF); (ii) significantly reduces the overall computational cost required by geomechanical model and the inversion procedure. In this paper, the description of the geomechanical model and its implementation are presented is Sect. 2, whereas Sect. 3 reviews the basic concepts of the ES approach. In Sects. 4 and 5, the generation of the prior uncertain parameters and the synthetic case results are presented and discussed with both a homogeneous and a heterogeneous c_M.

2 Geomechanical model

2.1 Model description

The subsurface deformation results from the pore pressure change in space and time due to fluid injection into, or extraction from, deep reservoirs. In order to simulate the deformation up to the ground surface, we need to solve both the governing flow and the structural partial differential equations (PDEs). Geomechanical simulations are performed using a finite-element (FE) poro-elasto-plastic numerical model (e.g. Gambolati et al., 2001). In this study, an isotropic stress-strain constitutive law is used with the vertical uniaxial compressibility c_M that depends on the stress state according to the hypo-plastic hysteretic model developed by Baù et al. (2002) and Ferronato et al. (2013). The uncertain compressibility c_M is calibrated by introducing a spatially variable multiplicative factor f_{CM}, which allows scaling c_M values in the regions where fluid pressure changes occur. Poisson's coefficient ν is assumed to be known and equal to 0.3.

2.2 Model setup

A one-way coupling approach is followed with the flow model first run and the outcome subsequently used as input data for the geomechanical model. The geomechanical FE grid comprises 320 901 nodes and 1 824 768 tetrahedral elements. The model domain covers an area of 52 km × 49 km and extends to about 5 km depth. Zero-displacements conditions are prescribed on the lateral and bottom boundaries. The top of the domain is a traction-free boundary. The simulations span one year, during which the reservoir experiences a fluid extraction. The pressure data are synthetic. Figure 1 shows the reservoir domain and the grid used for the geomechanical simulations.

3 Ensemble smoother

The ES algorithm consists of Monte Carlo stochastic simulations based extension to nonlinear problems of the classic Kalman Filter (KF) (Kalman, 1960). The ES algorithm follows a two-step forecast-update process. The forecast step involves simulating an ensemble of model states **X** based upon the solution of the geomechanical FE model **Φ**, which depends upon uncertain parameters **P** and forcing terms **q** (e.g. pore-pressure):

$$\mathbf{X} = \mathbf{\Phi}(\mathbf{P}, \mathbf{q}). \tag{1}$$

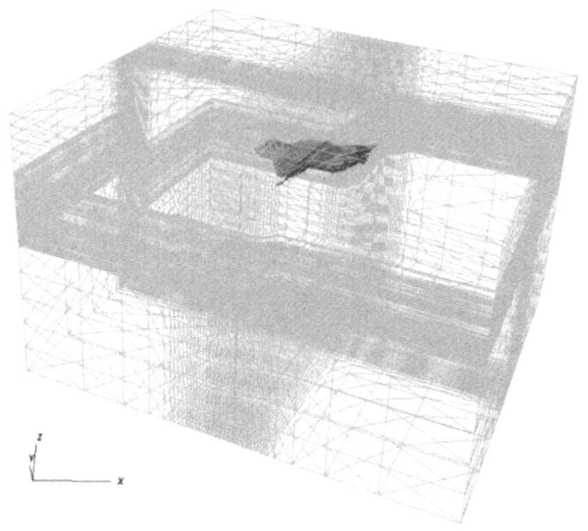

Figure 1. Axonometric view of the 3-D FE grid of the geomechanical model. The reservoir is embedded within the grid with different colors distinguishing the producing layers. The vertical exaggeration is 5.

In these simulations, each model state **X** is represented by land surface displacements from a subsidence distribution map. Each realization of the ensemble is run forward in time using random sets of the uncertain geomechanical parameters **P**, thus creating an ensemble $\mathbf{X}^f$ of model states **X**. The model results at any given location in the simulated domain are spread over a range of values, representing the uncertainty in the surface displacement prediction. In the update step, the set of measurements **z** collected to-date, i.e. point measurements of land displacement at a number of locations, is perturbed to account for measurement errors and assimilated into the forecast system state $\mathbf{X}^f$ to produce the updated state ensemble:

$$\mathbf{X}^u = \mathbf{X}^f + \mathbf{K} \cdot (\mathbf{d} - \mathbf{H} \cdot \mathbf{X}^f). \tag{2}$$

Matrix **d** includes the perturbed measurements, and the matrix **H** contains binary constants (0 or 1) that map model results at measurement locations. The matrix $\mathbf{d} - \mathbf{H} \cdot \mathbf{X}^f$ incorporates the residual at these locations between the measured and the predicted data. The matrix **K** is called the "Kalman Gain" matrix (Kalman, 1960), and has the following structure:

$$\mathbf{K} = \mathbf{C}^f \mathbf{H}^T \cdot \left(\mathbf{H} \mathbf{C}^f \mathbf{H}^T + \mathbf{R} \right)^{-1} \tag{3}$$

where $\mathbf{C}^f$ is the forecast error covariance matrix associated with the model forecast $\mathbf{X}^f$, and **R** is the measurement error covariance matrix associated with the perturbed measurements **d**. The matrix **K** plays the dual role of: (a) spreading information from measurement locations to adjacent areas,

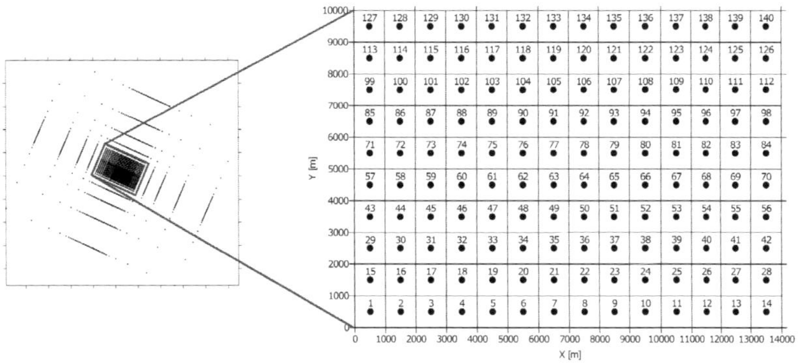

Figure 2. 2-D view of the geomechanical FE grid of Fig. 1. The enlargement view within the red rectangle refers to the area affected by the pore pressure change.

allowing for the measurements to correct the predicted values throughout the model domain; and (b) acting as a weight that scales the correction terms according to model and measurement errors. As $\mathbf{R}$ approaches zero, which means low-error measurements, the influence of $\mathbf{K}$ increases and the residual is weighted more heavily, so that the model forecast approaches the measurements. In contrast, as $\mathbf{C}^{\mathrm{f}}$ approaches zero, which indicates a relative agreement among model realizations, the influence of $\mathbf{K}$ decreases, and the residual is weighted less heavily, so that the model forecast receives little or no correction from measurements.

Within the ES algorithm, any variable incorporated into the system state matrix $\mathbf{X}^{\mathrm{f}}$ can be corrected by assimilating measurement data if a spatial correlation exists between these variables and the data (van Leeuwen, 2001). In Eq. (1), since the geomechanical parameters P dictate the behavior of the model response $\boldsymbol{\Phi}$, all uncertain parameters used in the forecast step can be included into the system state matrix $\mathbf{X}^{\mathrm{f}}$, and conditioned by land surface movements in Eq. (3). This conditioning may provide updates for the geomechanical parameters $\mathbf{P}$ that should approach those of the rock formation. Compared to other techniques for characterizing subsurface systems, the ES algorithm is quite attractive because of its low computational burden and the ability to run entirely independent of the simulation model (Bailey and Baù, 2010a, b).

4 Prior distribution of uncertain parameters

In the present study, c_{M} is assumed to be the only uncertain geomechanical parameter. Because the ES relies on a Monte Carlo approach, a prior probability distribution function (PDF) is needed to sample the prior ensemble of the multiplicative calibration factor f_{CM}. In this section, the generation of the prior PDFs in two test cases is described: (1) f_{CM} is uniform within the reservoir, and (2) f_{CM} is spatially distributed. In the latter, the heterogeneity on f_{CM} occurs only in the area shown in the enlargement of Fig. 2 where the pressure has changed due to fluid extraction.

4.1 Homogeneous compressibility (test case 1)

The calibration factor f_{CM} is assumed to be spatially uniform within the area of Fig. 2. The values of the prior f_{CM} ensemble are randomly sampled between 1 and 10 from a uniform PDF:

$$f_{\mathrm{CM}} \in U[1, 10]. \tag{4}$$

The selected range is based on the outcome of a sensitivity analysis (not shown here) carried out to investigate the possible interval of the f_{CM} variation. Figure 3 shows the spatial distribution of the mean and the standard deviation (σ_z) of the vertical displacements (u_z) from the forecast ensemble obtained by performing 100 Monte Carlo geomechanical simulations using the prior f_{CM} ensemble.

4.2 Heterogeneous compressibility (test case 2)

In test case 2, the f_{CM} is spatially distributed within the same area used in test case 1. This area is $(14 \times 10)\,\mathrm{km}^2$ wide and subdivided into 140 square cells. Each cell is characterized by a different f_{CM}. A categorical indicator algorithm, that creates random realizations of a heterogeneous f_{CM} field according to a given covariance model, is used. The f_{CM} values are drawn from a discrete uniform distribution with the prescribed ten categories ranging from 1 to 10. Each category has an equal unconditional probability (1/10). To account instead for the spatial statistical dependence between f_{CM} values on different cells, a stationary correlation model is introduced. According to this model, the probability of observing a f_{CM} category in the discrete domain [1, 10] at any given cell is conditional to the presence of the same category at surrounding cells. The correlation between two grid cells is based on an exponential isotropic function law depending

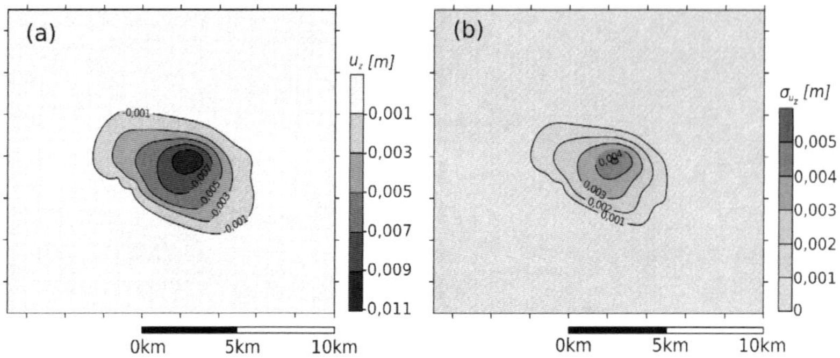

Figure 3. Test case 1: **(a)** mean and **(b)** standard deviation (σ_z) from the *forecast* ensemble of the vertical displacement (u_z).

on the distance between these cells and a prescribed correlation length, λ. The λ value has a direct influence on the degree of heterogeneity assigned to the spatial distribution of f_{CM}. Preliminary simulations have suggested the choice of $\lambda = 4000$ m. Figure 4 shows one of the 100 realizations of the generated prior ensemble of the f_{CM} field. Obviously, the mean over the ensemble in each grid block is equal to 5. As for test 1, the mean and the standard deviation of the u_z forecast ensemble are shown in Fig. 5.

5 Synthetic land subsidence data

Ground-surface displacements data are used to infer the model state and the geomechanical parameters. The observations are collected from the land subsidence map shown in Fig. 6, obtained from a geomechanical reference simulation with a prescribed and "known" compressibility distribution. The f_{CM} field is assigned on the basis of a plausible reservoir partition derived from the presence of faults and thrusts (Fig. 6). The synthetic data locations are uniformly distributed over the reservoir (Fig. 6) in the area with σ_z greater than 0.001 (see Figs. 3b and 5b). This is to account for the reservoir behavior where pressure variation occurs causing land surface deformation.

6 Results and discussion

6.1 Update of f_{CM}

In test case 1, a significant reduction of the uncertainty associated with the prior f_{CM} distribution is evident by comparing the prior and the posterior cumulative distribution functions (CDFs) plotted in Fig. 7. Defining the ensemble spread as the average absolute difference between the ensemble members and the ensemble mean, the Average Ensemble Spread (AES) for n_{MC} Monte Carlo realizations is calculated as:

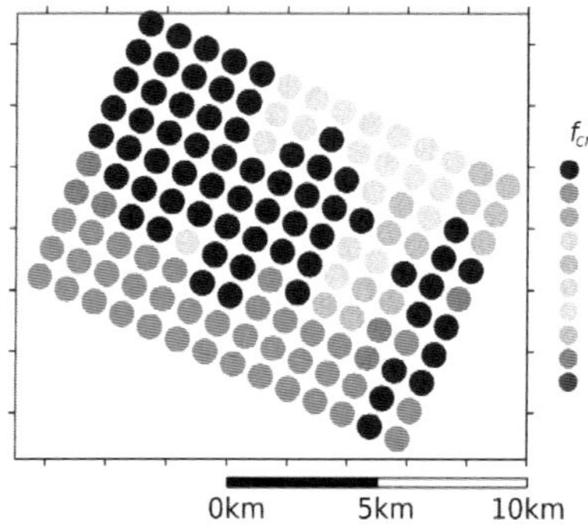

Figure 4. Test case 2: one realization out of the 2-D f_{CM} field ensemble. The compressibility varies in the area corresponding to the enlargement of Fig. 2.

$$\text{AES} = \frac{1}{n_{MC}} \sum_{i=1}^{n_{MC}} |x_i - \overline{x}| \tag{5}$$

where x_i and $\overline{x}$ are the f_{CM} value of the ith ensemble member and the ensemble mean, respectively. The prior AES, i.e. before the assimilation of data, equals 2.41 and reduces by about 93 % after assimilation. Therefore, the posterior CDF is highly constrained compared to the prior CDF with the most probable estimate value for f_{CM} equal to 1.89, corresponding to the mean, or expected value, of the updated ensemble.

In test case 2, the ES performance is evaluated by calculating the AES index over each grid block of Fig. 2. The prior AES index ranges between 2.0 and 2.9 over the domain. After assimilation, the spread of the updated ensemble significantly reduces over the area where data points are collected, while higher AES values are found in the surrounding area (see Fig. 8b). Although data are collected only over a limited portion of the domain, a sensitivity analysis (not shown here) reveals that collecting data over the entire domain does not improve the assimilation outcome. Indeed, these observa-

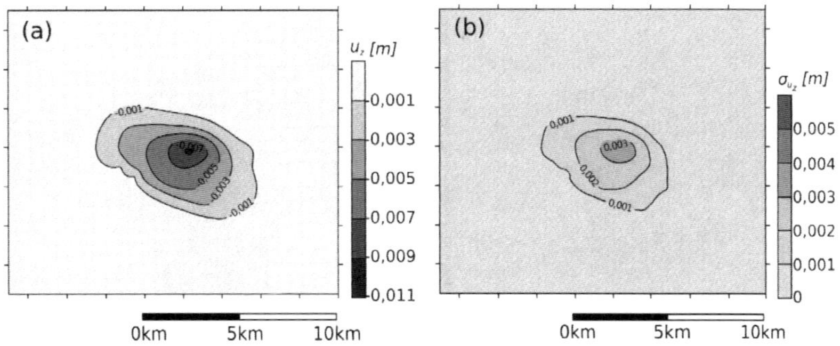

Figure 5. Test case 2: **(a)** mean and **(b)** standard deviation (σ_z) from the *forecast* ensemble of the vertical displacement (u_z).

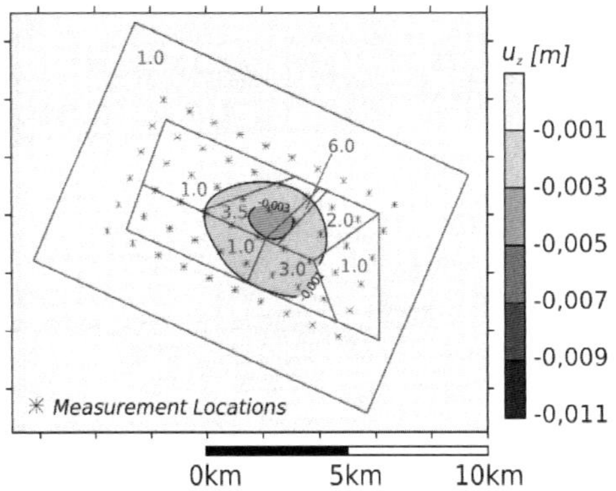

Figure 6. Spatial distribution of the synthetic land subsidence data (u_z). Assimilation data are collected at the 60 measurement locations displayed in the map. The red lines represent the trace of the faults subdividing the reservoir into blocks with the prescribed reference f_{CM} values.

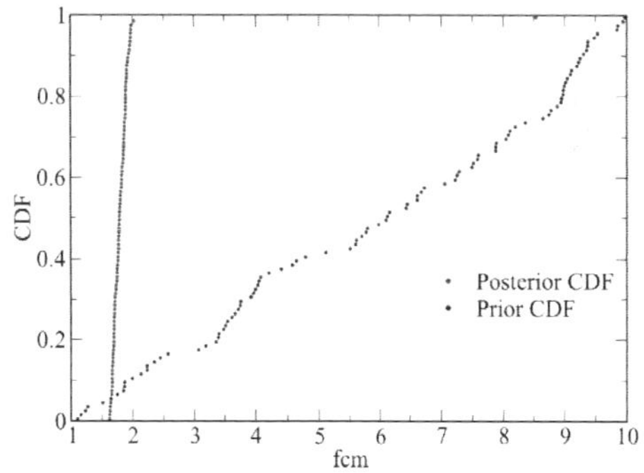

Figure 7. Test case 1: prior and Posterior CDFs from the updated f_{CM} ensemble.

tions cannot yield enough information to infer the model parameters because the deep reservoir experiences small pressure variation with a negligible influence on the land surface deformation of outer regions. The average AES over all grid blocks reduces by about 33 % after assimilation. Despite a lower relative reduction of AES compared to test 1, the inverse problem outcome is much better verified in terms of ground surface displacements (Sect. 6.2). Figure 8a depicts the spatial distribution of the mean from the updated f_{CM} field.

6.2 Forecast of ground surface displacement

The quality of the parameter estimation is validated by executing the posterior geomechanical simulations for both uniform and spatially variable c_{M}. The f_{CM} model input is the mean of the updated f_{CM} ensemble from the outcome of test cases 1 and 2. The results of this simulation are compared with the synthetic land surface data of Fig. 6 using the Normalized Root Mean Square Error (NRMSE), which represents the standard deviation of the differences between the simulated values and the observations, calculated as

$$\mathrm{NRMSE} = \frac{\sqrt{\sum_{i=1}^{N}\left(u_{zi,\mathrm{sim}} - u_{zi,\mathrm{obs}}\right)^2}}{\left(u_{z\mathrm{obs,max}} - u_{z\mathrm{obs,min}}\right)} \tag{6}$$

where u_{zi}, sim and u_{zi}, obs are the simulated and observed land ground displacement at the ith assimilation location, respectively; $u_{z\mathrm{obs,max}}$ and $u_{z\mathrm{obs}}$, min are the maximum and minimum observation values, respectively, and N is the total number of assimilation locations, i.e. 60 data points. Test case 1 and 2 gives NRMSE values equal to 15 and 3 %, respectively. Hence, the predicted vertical displacements provides lower data mismatch in test case 2 and the assumption of a heterogeneous compressibility field allows for a better fit of the synthetic observations (Fig. 9). The fitting of the model predictions to the reference data is very accurate with the heterogeneous c_{M}, while a relevant underestimation is found in the homogeneous test case for the largest subsidence values, i.e. u_z greater than 1 mm.

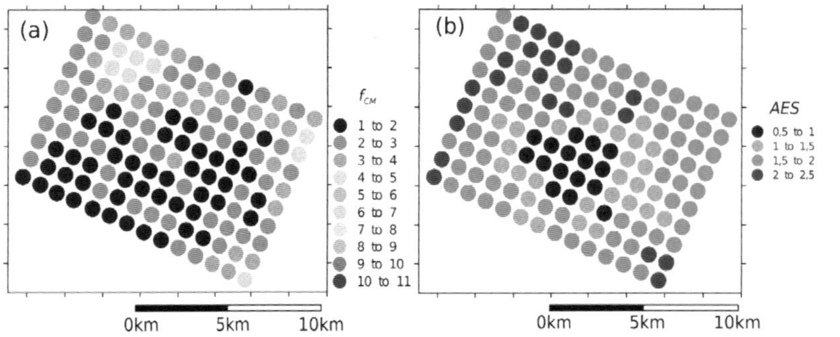

Figure 8. Test case 2: spatial distribution of **(a)** the mean and **(b)** the performance index AES from the updated f_{CM} ensemble after data assimilation.

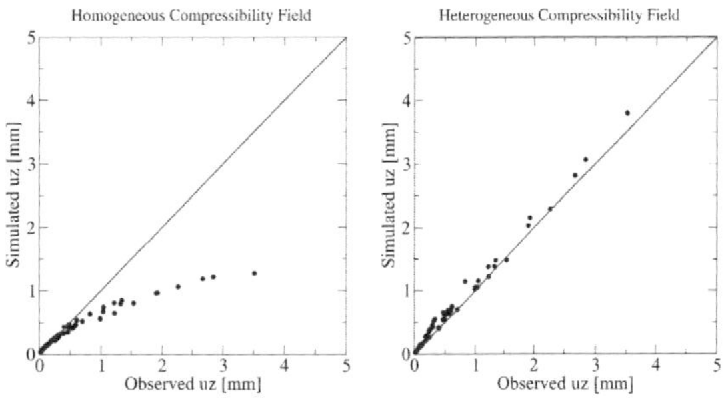

Figure 9. Simulated vs. observed values of u_z in test case 1 (homogeneous c_{M}) and test case 2 (heterogeneous c_{M}).

7 Conclusions

In this study, an ensemble-based DA approach, i.e. the ES, is used to infer the compressibility of the geomechanical model of a producing hydrocarbon reservoir by assimilating ground surface displacements. The methodology is applied herein to investigate two different conceptual models assuming (i) a homogeneous c_{M} within the reservoir and (ii) a heterogeneous reservoir compressibility. The latter assumption is made to account for the reservoir compartmentalization due to the presence of a complex fault system. Thus, c_{M} spatially varies within the reservoir and calls for the calibration of a heterogeneous c_{M} field. A test case is used to assess the validity of the proposed methodology on a reservoir with a synthetic pressure variation. The assimilation of synthetic ground-surface displacements, i.e. simulated data obtained with a priori known c_{M} distribution, provides satisfactory results. In particular, the heterogeneous c_{M} field gives a much better fitting than the homogeneous c_{M}. Hence, a more satisfactory application to a real case is expected by a DA approach with a spatially variable c_{M}.

Acknowledgements. This research was supported by eni S. p. A. within the DAG (DA in Geomechanics) project.

References

Bailey, R. T. and Baù, D.: Ensemble Smoother assimilation of hydraulic head and return flow data to estimate hydraulic conductivity distribution, Water Resour. Res., 46, W12543, doi:10.1029/2010WR009147, 2010a.

Bailey, R. T. and Baù, D.: Estimating spatially-variable first-order rate constants in groundwater reactive transport systems, J. Contam. Hydrol., 122, 104–121, doi:10.1016/j.jconhyd.2010.11.008, 2010b.

Baù D., Ferronato, M., Gambolati, G., and Teatini, P.: Basin scale compressibility of the Northen Adriatic by the radioactive marker technique, Geotechnique, 52, 605–616, 2002.

Ferronato, M., Castelletto, N., Gambolati, G., Janna, C., and Teatini, P.: II cycle compressibility from satellite measurements, Geotechnique, 63, 479–486, 2013.

Gambolati, G., Ferronato, M., Teatini, P., Deidda, R., and Lecca, G.: Finite element analysis of land subsidence above depleted reservoirs with the pore pressure gradient and the total stress formulations, Int. J. Num. Anal. Meth. Geomech., 25, 307–327, 2001.

Kalman, R. E.: A new approach to linear filtering and prediction problems, Trans. ASME , 82, 35–45, 1960.

van Leeuwen, P. J.: An ensemble smoother with error estimates, Mon. Weather Rev., 129, 709–728, 2001.

Proc. IAHS, 372, 357–360, 2015
proc-iahs.net/372/357/2015/
doi:10.5194/piahs-372-357-2015

Open Access
PIAHS

PSI-based methodology to land subsidence mechanism recognition

R. Bonì[1], **C. Meisina**[1], **C. Perotti**[1], **and F. Fenaroli**[2]

[1]Department of Earth and Environmental Sciences, University of Pavia, Via Ferrata 1, 27100 Pavia, Italy
[2]Studio Tecnico Fenaroli Fabio V. Mandalossa 3, 25055 Pisogne (BS), Italy

Correspondence to: R. Bonì (roberta.boni01@unipv.it)

Published: 12 November 2015

Abstract. A methodology based on Persistent Scatterer Interferometry (PSI) is proposed in order to disentangle the contribution of different processes that act at different spatio-temporal scales in land subsidence (i.e. vadose zone processes as swelling/shrinkage of clay soils, soil consolidation and fluid extraction). The methodology was applied in different Italian geological contexts characterized by natural and anthropic processes (i.e. a Prealpine valley and the Po Plain in northern Italy).

1 Introduction

Persistent Scatterer Interferometry (PSI) technique has significantly upgraded our ability to detect and quantify the land subsidence over wide areas. However the PSI technique shows two main limitations: (1) the PSI measurements represent the cumulative ground movements (natural plus anthropic and superficial plus deep displacements); therefore, the overlapping of several causes may make difficult the interpretation of ground deformation processes; (2) most of the geological processes are non-linear with seasonal and multi-year components, which are difficult to identify only through the use of the average rate of displacements V_{LOS} (velocity along the Line of Sight). This work presents a methodology to analyse the ground deformations in order to disentangle the contribution of different processes that act at different spatio-temporal scales (i.e. swelling/shrinkage of clay soils, soil consolidation and fluid extraction), by the integration of PSI data and ground-based measurements.

2 Study areas

Two different geological contexts located in northern Italy and characterized by natural and anthropic processes were analysed: a Prealpine valley and a sector of the Po Plain (Figs. 1 and 2).

The Valley bottom of the Oglio River (50 km^2) is located upstream the Iseo Lake and it is constituted by alluvial deposits (silt-sand, clay and sand-gravel layers) on Southalpine basement. This area is susceptible to subsidence processes, being, since the 1980, a high-urbanized area over unconsolidated alluvial materials.

The plain sector of the Oltrepo Pavese corresponds to the Southern and Central sector of the Po River Plain, with an extension of about 443 km^2. The Quaternary cover consists in the succession of impermeable sediments (silt and clay) and more permeable horizons (gravels and sandy). The area is particularly vulnerable to swelling/shrinkage of the clayey soils, subsidence due to ground water pumping and to natural consolidation of soft soils.

3 PSI datasets

PSI data elaborated by Tele-Rilevamento Europa (T.R.E.) using images acquired by ERS 1-2 (1992–2000) and RADARSAT (2003–2010) satellites were used. The datasets of the Valley bottom of the Oglio River were processed with PSInSAR™ technique and the Po Plain datasets with the SqueeSAR™ technique. The PSInSAR™ technique allows to obtain high resolution displacement measurements from individual radar reflectors or Persistent Scatterers (PS) like anthropic structure or rocks, the SqueeSAR™ technique per-

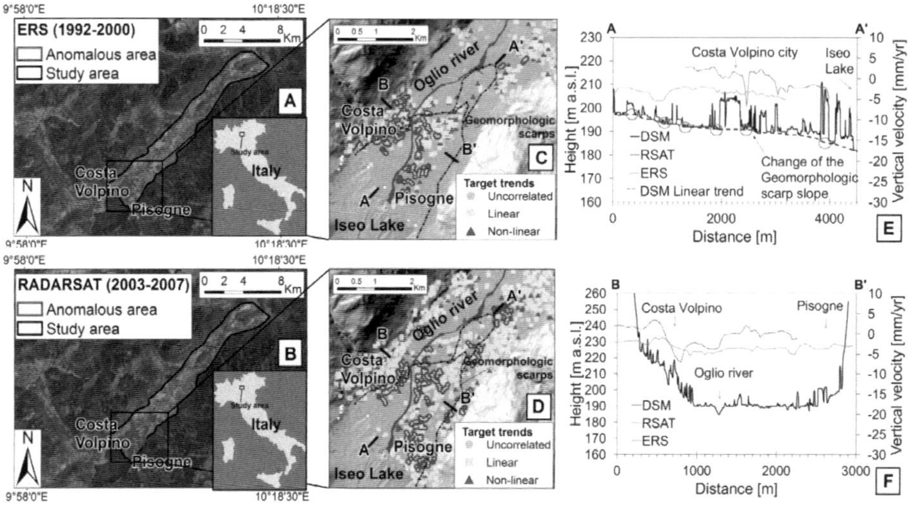

Figure 1. Anomalous areas detected by ERS descending **(a)** and RADARSAT descending **(b)**. Target trends of ERS **(c)** and RADARSAT **(d)** dataset, and Digital surface model (DSM) of the Valley bottom of the Oglio river. Cross sections along the A-A' **(e)** and B-B' **(f)** sections of the ERS and RADARSAT (RSAT) datasets.

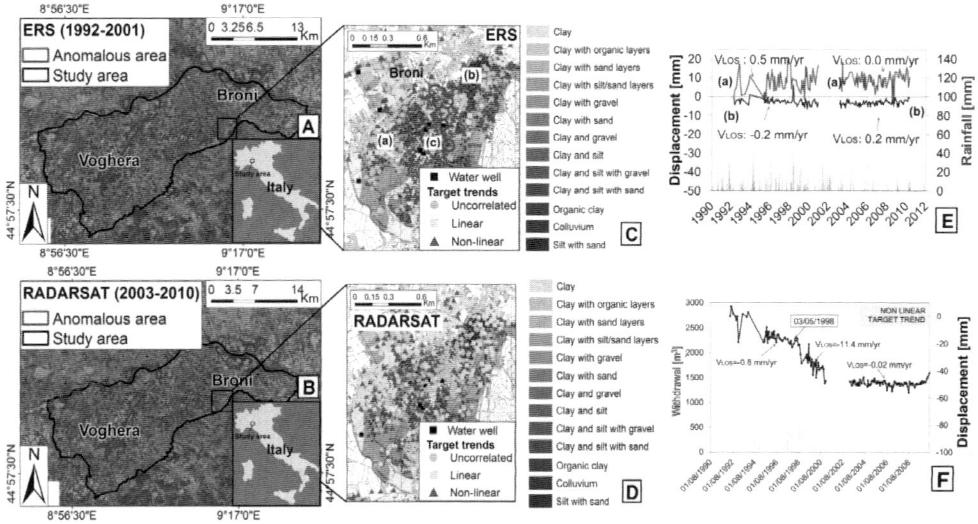

Figure 2. Anomalous areas detected by ERS descending **(a)** and RADARSAT ascending **(b)**. Target trends of ERS **(c)** and RADARSAT **(d)** dataset, and lithological map of Broni town. TS of ERS and RADARSAT datasets at the location a, b compared with daily rainfall **(e)**. Comparison of TS at location c and the groundwater exploitation **(f)**.

mits to obtain measures also from distributed scatterers (DS) like sparse vegetated areas (Ferretti et al., 2011).

4 Methodology

The procedure is divided into three steps: (1) post-processing elaborations (2) characterization of ground motion areas (3) mechanism recognition.

Deformation accuracy assessment was carried out during the first step; the stable velocity range, the vertical and horizontal components of the motion were identified. The quality of Times Series was improved and possible errors (anomalous estimates not related to real ground motion) were corrected adopting the methodology of Notti et al. (2015). Thus, in a second step, the identification of the ground motion areas was performed using spatial clustering based of the unstable V_{LOS} (anomalous areas). Trends (i.e. uncorrelated, linear, non-linear) of deformation time series (TS) were recognized by the application of the Automated Classification of Persistent Scatterers Interferometry time-series (Berti et al., 2013). Selecting the unstable linear time series, the frequency distribution of the LOS velocity was analysed and normal populations were separated through the expectation–maximization algorithm for mixtures of univariate normal function (Benaglia et al., 2009) in order to distinguish linear processes triggered by different causes (i.e. natural and anthropic).

At the third step, the PSI data were compared to predisposing (i.e. soft soil thickness geological and geotechnical model) and triggering factors (i.e. piezometric level variations etc.). Moreover, seasonal component of motion were analyzed at detailed scale; considering detrended deformation time series.

4.1 Post-processing elaborations

The velocity threshold was assessed considering the twice of the standard deviation of the PS/DS with the higher coherence (>0.8–0.9) and a common stability threshold was set to $\pm 2\,\mathrm{mm\,yr^{-1}}$ for the PSI datasets over the Prealpine valley and $\pm 1.5\,\mathrm{mm\,yr^{-1}}$ for the plain area of Oltrepo Pavese. The movements are mainly vertical in both the study areas. Only in the Po Plain datasets, anomalous displacements were detected on the 9 March 1997 and 16 July 2000 (ERS dataset), and on 10 December 2008 scene (RADARSAT dataset). The last RADARSAT (RSAT) anomalous values might be related to snowfall occurred on the day of the Synthetic Aperture Radar (SAR) acquisition.

4.2 Characterization of the subsidence areas

Spatial clustering analysis based on the unstable V_{LOS} allows to detect in the Prealpine valley 17 and 31 anomalous areas (around 1–1.5 $\mathrm{km^2}$) in the time span 1992–2000 and 2003–2010, respectively (Fig. 1a and b). In the Po Plain, 117 and 122 anomalous areas (around 7 $\mathrm{km^2}$) were identified in the time span 1992–2001 and 2003–2010 (Fig. 2a and b).

Most of unstable TS of the Prealpine valley have linear trends (Fig. 1c and d). The V_{LOS} ranges from -2 to $-8\,\mathrm{mm\,yr^{-1}}$ in the period 1992–2000 and from -2 to $-10\,\mathrm{mm\,yr^{-1}}$ in the period 2003–2010.

In the Po Plain the 51 % (ERS) and 50 % (RSAT) of unstable TS show non-linear trends within the range from -6.6 to $-1.5\,\mathrm{mm\,yr^{-1}}$ in the time interval 1992–2001, and from -7 to $-1.5\,\mathrm{mm\,yr^{-1}}$ in the period 2003–2010. The unstable linear TS show values within the range from -8 to $-1.5\,\mathrm{mm\,yr^{-1}}$ (ERS) and from -7 to $-1.5\,\mathrm{mm\,yr^{-1}}$ (RSAT) in the monitored periods (Fig. 2c and d).

Local ground motions triggered by anthropic processes (i.e. external load due to new buildings with V_{LOS} between -3 and $-13\,\mathrm{mm\,yr^{-1}}$) were recognized. In addition, the spatial distribution of the nonlinear TS in the Po Plain displays break dates in the ERS dataset (after 3 May 1998) and in the RSAT dataset (31 August 2009) related to wet and dry periods.

4.3 Mechanism recognition

The spatio-temporal analysis at large scale allowed to identify main ground motion areas located at Costa Volpino and Pisogne towns in the first test site and Voghera and Broni towns at the second one, where we carried out detailed analysis.

Different natural control mechanisms (i.e. water level fluctuation of the Iseo Lake, thickness of unconsolidated alluvial sediments) were analyzed to understand the displacements of the valley bottom of the Oglio River. The analysis highlights the geomorphologic control of the subsidence pattern; the stronger subsidence rate occurs in correspondence to a slope change (Fig. 1e and f). Subsidence due to dissolution of evaporitic rocks will also be investigated.

The comparison of the TS with the geological model of the Po Plain shows that the time series characterized by stable velocity detains seasonal component of motion (seasonal amplitude is about 10 mm) where clay deposits are found (Fig. 2e). Subsidence areas due to groundwater exploitation were identified in the 1992–2001 period, (maximum $V_{LOS} = -5\,\mathrm{mm\,yr^{-1}}$); after 2003 (RSAT dataset) the decrease of groundwater pumping results in a decreasing trend of motion (Fig. 2f).

5 Conclusions

A new methodology based on PSI data was developed and validated in various geological contexts. The procedure allowed to identify multiyear (groundwater pumping, construction of new buildings, consolidation of soft soils, geomorphological processes) and seasonal (volume change of clay soils, groundwater oscillations) ground movements and proved to be an efficient tool for distinguish different anthropic and natural processes. The methodology can be used for land use planning and civil protection purposes.

Acknowledgements. The PSI data used in this work were provided by Regione Lombardia in the framework of the project "Interpretazione geologica e idrogeologica dei dati radar derivanti dal Piano Straordinario di Telerilevamento nazionale (PST) e da dati Radarsat con particolare riferimento alle province di Lecco e Pavia" financially supported by Direzione Generale Protezione Civile Prevenzione e Polizia Locale dell'Istituto regionale di ricerca della Lombardia (IREA).

References

Benaglia, T., Chauveau, D., Hunter, D. R., and Young, D.: "mixtools": an R package for analyzing finite mixture models, J. Statist. Software, 32, 1–29, 2009.

Berti, M., Corsini, A., Franceschini, S., and Iannacone, J. P.: Automated classification of Persistent Scatterers Interferometry time series, Nat. Hazards Earth Syst. Sci., 13, 1945–1958, doi:10.5194/nhess-13-1945-2013, 2013.

Ferretti, A., Fumagalli, A., Novali, F., Prati, C., Rocca, F., and Rucci, A.: A new algorithm for processing interferometric data-stacks: SqueeSAR. Geoscience and Remote Sensing, IEEE Transactions on, 49, 3460–3470, 2011.

Notti, D., Calò, F., Cigna, F., Manunta, M., Herrera, G., Berti, M., Meisina, C., Tapete, D., and Zucca, F.: A User-Oriented Methodology for DInSAR Time Series Analysis and Interpretation: Landslides and Subsidence Case Studies, Pure Appl. Geophys., 1–25, doi:10.1007/s00024-015-1071-4, 2015.

Proc. IAHS, 372, 361–366, 2015
proc-iahs.net/372/361/2015/
doi:10.5194/piahs-372-361-2015

Open Access
PIAHS

Methods for monitoring land subsidence and earth fissures in the Western USA

K. C. Fergason, M. L. Rucker, and B. B. Panda

Amec Foster Wheeler, Phoenix, Arizona, USA

Correspondence to: K. C. Fergason (ken.fergason@amecfw.com)

Published: 12 November 2015

Abstract. Depletion of groundwater resources in many deep alluvial basin aquifers in the Western USA is causing land subsidence, as it does in many regions worldwide. Land subsidence can severely and adversely impact infrastructure by changing the ground elevation, ground slope (grade) and through the development of ground cracks known as earth fissures that can erode into large gullies. Earth fissures have the potential to compromise the foundations of dams, levees, and other infrastructure and cause failure.

Subsequent to an evaluation of the overall subsidence experienced in the vicinity of subsidence-impacted infrastructure, a detailed investigation to search for earth fissures, and design and/or mitigation of potentially effected infrastructure, a focused monitoring system should be designed and implemented. Its purpose is to provide data, and ultimately knowledge, to reduce the potential adverse impacts of land subsidence and earth fissure development to the pertinent infrastructure. This risk reduction is realized by quantifying the rate and distribution of ground deformation, and to detect ground rupture if it occurs, in the vicinity of the infrastructure.

The authors have successfully designed and implemented monitoring systems capable of quantifying rates and distributions of ground subsidence and detection of ground rupture at multiple locations throughout the Western USA for several types of infrastructure including dams, levees, channels, basins, roadways, and mining facilities. Effective subsidence and earth fissure monitoring requires understanding and quantification of historic subsidence, estimation of potential future subsidence, delineation of the risk for earth fissures that could impact infrastructure, and motivation and resources to continue monitoring through time. A successful monitoring system provides the means to measure ground deformation, grade changes, displacement, and anticipate and assess the potential for earth fissuring. Employing multiple methods, a monitoring strategy utilizes an integrated approach, including both regional and local measurements.

Various methods implemented include conventional practices and proven, instrumented in-ground sensing systems. The conventional techniques include repeat optical levelling and global positioning system (GPS) surveys, ground reconnaissance, photo-geological analysis, groundwater monitoring, and tape-extensometers. Advanced techniques include the processing and interpretation of differential interferograms of repeat-pass, satellite-based synthetic aperture radar data (InSAR), borehole tiltmeters, microseismic arrays, excavation of monitoring trenches, and time-domain reflectometry (TDR).

1 Introduction

Depletion of groundwater resources in many deep alluvial basin aquifers in the Western USA is causing land subsidence, as it does in many regions worldwide. Land subsidence can severely and adversely impact infrastructure by changing the ground elevation, ground slope (grade) and through the development of ground cracks known as earth fissures that can erode into large gullies. Earth fissures have the potential to compromise the foundations of dams, levees, and other infrastructure and cause failure.

The authors have a combined experience of over 50 years for characterizing, predicting, monitoring and remediation at project sites where ground subsidence and/or earth fissuring are impacting infrastructure (Rucker and Keaton,

1996, 1998; Keaton et al., 1998; Weeks et al., 2004; Weeks and Panda, 2004; Rucker and Fergason, 2004; Rucker and Holmquist, 2006; Rucker et al., 2008, 2013; Panda et al., 2013; Fergason et al., 2013, 2014). This includes a power generating plant in southeast Arizona where annual subsidence and earth fissure monitoring has been performed by the authors for over 20 years. The authors perform regular land subsidence and/or earth fissure related monitoring at over ten additional sites that include dams, levees, flood retarding structures, flood control basins and channels, a liquefied petroleum gas storage facility, large gold and copper mines, and at a brine cavern at risk for catastrophic collapse. The authors are currently involved in developing land subsidence and/or earth fissure monitoring for several additional projects, including the California High-Speed Rail network.

2 Monitoring methods

Various methods implemented include conventional practices and proven, instrumented in-ground sensing systems. The conventional techniques include repeat optical levelling and global positioning system GPS surveys, ground reconnaissance, photo-geological analysis, groundwater monitoring, and tape-extensometers. Advanced techniques include the processing and interpretation of InSAR, borehole tiltmeters, microseismic arrays, excavation of monitoring trenches, and TDR. The sections below summarize land subsidence and earth fissure monitoring methods that have been deployed in the Western USA.

2.1 Survey

The use of survey techniques provides a means of forming a network for various instrumentation systems, with the approach being an effective method to monitor ground deformation. Due to the often gradual and small relative movements to be measured for ground subsidence, verification of accuracy is a critical aspect of this portion of the measurement program. Multiple survey techniques can be utilized: (1) conventional optical levelling techniques, (2) real-time kinematics (RTK), and (3) static GPS observations. When constructing survey monuments for the purpose of land subsidence monitoring, proper foundations must be achieved that prevent near-surface ground deformation associated with hydrocompaction of near-surface soils, near-surface settlement of soils and/or fill material, freeze-thaw cycles, and other seasonal near-surface ground deformation.

2.1.1 Optical levelling

Conventional optical levelling remains one of the most reliable and most economic methods for obtaining highly accurate elevation data at the scale that most land subsidence occurs. Relational elevation differences between dedicated survey monuments should be determined through the use of differential digital level loops, or conventional level techniques. When plausible, these measurements should be taken in relation to a stable bedrock benchmark, however it is often sufficient to observe relative differences without a direct tie to a stable point. Elevations measured by levelling should be expected to meet an accuracy of at least ± 3 mm and monument spacing needs to be planned to achieve that level of accuracy.

2.1.2 RTK GPS

RTK GPS methods are typically the quickest and most economical survey techniques available. However, RTK GPS survey techniques do not have the level of vertical accuracy as other survey techniques. Elevations measured by RTK should be expected to meet an accuracy of at least ± 21 mm at 95 % confidence ratio and monument spacing needs to be planned to achieve that level of accuracy.

2.1.3 Static GPS

Static GPS methods have the capability of obtaining horizontal and vertical positions with relatively high accuracy and can be utilized to measure both horizontal and vertical deformation due to land subsidence and earth fissuring. Maximum care in setup on monuments and measuring height of the instruments, should be observed. When practical, a permanent or dedicated mount for a GPS antenna may be used to reduce potential setup error. A minimum accuracy of ± 5 mm horizontal and ± 20 mm vertical should be achieved and monument spacing needs to be planned to achieve that level of accuracy. Achieving such accuracy using static GPS may require that each point, including bedrock or offsite reference, be occupied for 1.5 h or more with a minimum of 5 visible satellites.

2.2 Photo-geologic analysis

Traditional low-sun-angle (LSA) aerial photography lineament analysis for the purpose of searching for earth fissures was first described in detail by Beckwith et al. (1991) and can be utilized as a regular monitoring tool to assess changes in a subsiding landscape. These features include elongated earth fissure gullies, alignments of potholes and other small depressions, lineations in the vegetative cover, deflation features, and subtle linear ground features caused by shading. A lineament analysis alone is not sufficient, and the lineament analysis must be followed by a geological reconnaissance of photo-lineaments and a terrestrial search for earth fissures.

As technology has improved, traditional LSA aerial photography has been largely replaced by high-resolution digital aerial imagery. The authors have previously assessed the use of both imagery types and determined that digital HRD imagery provides a comparable level of analysis to that provided by LSA photography (AMEC, 2013). The use of digital imagery has added benefits in that it photo-geologic lineament analyses can be performed at a lower cost and in a shorter timeframe.

The primary limitations of photo-geological lineament analysis for earth fissure searches and for use as a monitoring tool derive from the coverage and quality of the remote sensing data used for the analysis and the condition of the terrain analysed. Poor-quality remote sensing data and/or data that do not completely cover the area of interest limit the ability to conduct the analysis. Additionally, even if high-quality remote sensing data are available for the area of interest, limited temporal coverage can also limit the effectiveness of the analysis.

The condition of the analysed terrain can also significantly impact the effectiveness of the analysis. Agriculture, urbanization and other large-scale human development can obscure the ground and severely limit the effectiveness of the analysis. Other human disturbances such as off-road vehicles and historical roads can have significant impact on the terrain. Animal activity from wildlife and domestic stock such as cattle can also create significant ground disturbance. Additionally, annual and perennial vegetation can obscure lineaments. Soft, cohesionless shallow soil profiles have the ability to prevent the expression of earth fissures at the surface. Additionally, in areas where sediment is deposited by wind and/or water (such as upstream of an embankment), signs of earth fissuring may be obscured. If an earth fissure does not have a surface expression, it is unlikely that a photo-geological lineament analysis would identify it.

2.3 Geologic mapping/reconnaissance

Regular visual ground inspection is important part of a monitoring program. The inspections should be performed by an experienced person walking the project area looking for cracks, potholes or other features which may indicate earth fissuring. Visual inspections should be performed as close in time as practicable to the other field measurements included in the monitoring program. Inspections should also be performed after major storm runoff events. Locations and descriptions of cracks, potholes and other erosional features should be documented with sketches, maps and photographs as appropriate, including locations, dimensions and orientations. The appropriate level of background work should be completed before commencing field work. This includes review of InSAR and the completion of the photo-geological analysis. Additionally, an extensive data review of all relevant reports and publications, groundwater data, survey data and remote sensing data should be performed prior to the field work.

Special care should be taken in areas where giant desiccation cracking occurs. Differentiating between giant desiccation cracks and earth fissures requires experience and additional investigation such as seismic refraction profiling and trench excavation with detailed logging.

The primary limitations to the method involve impacting the ability to observe. Annual and perennial vegetation cover can limit the ability to see signs of ground cracking. Disturbance by human activity and/or livestock and wildlife can also limit the amount of observation possible. When the ground is high in moisture content, subtle signs of ground strain such as small ground cracks may not be visible.

Other limitations and uncertainties of the method are rooted in the competency of the investigation itself. The investigators must correctly interpret the background information, InSAR and other imagery, and photo-geological results to identify the areas in need of investigation. It is critical that personnel performing the investigation have previous experience in geological reconnaissance of photo-lineaments and terrestrial search for earth fissures due to the often subtle nature of important features.

2.4 Groundwater monitoring

The collection of regional groundwater levels from available sources is an important component of monitoring for land subsidence and earth fissures. Understanding the groundwater trends allows for greater understanding of local and regional ground deformation trends and can aid in predicting where future deformation may occur.

2.5 Tape extensometers

Tape extensometers can be utilized to measure horizontal displacement between two or more points with a measurement precision of ± 0.01 mm. Tape extensometer arrays utilized as part of a monitoring system can quantify the rate of deformation at a known earth fissure or can be utilized to monitor for the potential development of a new earth fissure in an area that is considered to have a high likelihood for future fissure formation. Quadrilateral arrays can be utilized to understand deformation in greater detail and determine if shear deformation is occurring.

The tape extensometer should be operated in accordance with the manufacturer's instructions, including proper tensioning and recording of ambient temperature. Grass, weeds and other possible obstructions to the tape should be cut or removed before performing readings. The author's experience with tape extensometers has shown that measurements taken during a windy day are not as accurate as those taken on calm days.

Greater horizontal movements are anticipated in a transverse direction across fissures or the trend of nearby photo-lineaments, and negligible movements are anticipated normal to the linear array. Thus, inherent checks of tape extensometer calibration and operation should be performed on the readings parallel to fissures as part of each measurement cycle.

2.5.1 Rod extensometers

In addition to tape extensometer, experimental rod extensometers have been utilized by the authors with mixed success to monitor earth fissure at one location in Central Arizona (AMEC, 2010). Two rod extensometer arrays, each straddling a previously identified earth fissure, were installed. Each rod extensometer array consists of a 72 cm diameter instrument vault that houses a reference table embedded in concrete which is founded approximately 3 m below existing grade within stable Pleistocene-age soils. Three 46 cm diameter companion end-anchor encasements are also founded on Pleistocene-age soils. The spacing between each instrument vault and its three companion end anchors is approximately 12, 18 and 30 m, with the fissure positioned between the 12 and 18 m anchors. The intent of positioning the fissure between the 12 and 18 m anchors was to provide an opportunity to segment the monitoring to assess local displacement within close proximity of the fissure and at some distance away. The third anchor at 30 m added a redundancy to the monitoring system and allowed for measurement of displacements over a greater distance away from the fissure. Horizontal measurements of ground displacement at previously identified earth fissure locations should be performed by measuring changes in rod position using a depth micrometer and the reference table measurement apparatus.

2.6 InSAR

The use of repeat-pass InSAR to characterize the distribution and rate of regional ground subsidence is of profound significance in monitoring and managing the risks associated with ground subsidence and earth fissuring. The principal of interferometry is described by Gabriel and Goldstein (1988) and there are numerous examples of applying the InSAR methodology to characterizing land subsidence where Hoffmann et al. (2001) and Panda et al. (2013) provide good examples. Interferometry has the ability to detect and quantify minute changes in the elevation of terrain by comparing phase variances of satellite-based, side-looking radar data between orbits of a similar trajectory. InSAR is used to detect the rate and distribution of subsidence occurring in an area over the time periods covered by InSAR imagery. Several InSAR-based methodologies exist and may be useful to a given site depending on site conditions and available data. These methodologies include traditional 2-pass InSAR, stacked 2-pass InSAR interpretations, and various persistent scatter (PS) InSAR methods and algorithms. In the arid western regions of the USA, traditional 2-pass InSAR is often the most effective method.

InSAR data can be analysed by direct observation of interferograms, cross-sectional presentation of vertical change, and interactive three-dimensional presentation with contouring. These analyses can be used in a variety of ways, such as to support lineament analysis and reconnaissance, to compare with survey data, and to be used as a calibration tool for subsidence modelling. InSAR data analysis is particularly useful for assessing changes in the location or rate of regional ground deformation and to help refine characterization of bedrock/alluvial interfaces and general lithologic variations in the deep alluvium (Rucker et al., 2008; Weeks and Panda, 2004). Results from InSAR will depend on the site's terrain, ground and vegetation conditions, and sources of potential ground disturbance due to cultural interference such as agriculture or construction site grading.

2.7 Borehole tiltmeters

Borehole tiltmeters (BTMs) can be utilized to monitor ground deformation, particular when large infrastructure is involved and when large-scale failure is a concern. The authors are involved in a project that utilizes borehole tiltmeters as a major monitoring component at the location of a brine cavern with a potential to catastrophically fail and form a large sinkhole (AMEC, 2014). BTMs measure the angle of movement from the horizontal position in microradians (µrad) and deliver the data collected to the onsite monitoring computer that is part of integrated early warning system.

2.8 Microseismic array

The deployment of a microseismic array can be a useful monitoring tool in some subsiding environments. The authors involved in a project that utilizes a microseismic array as a monitoring component at the location of a brine cavern with the potential to catastrophically fail and form a large sinkhole (AMEC, 2014). The microseismic array consists of the deployment of highly sensitive geophones that collect the acoustical signals generated by microseismic activity, or ground vibration, in and around the brine cavern. Each borehole microseismic array consists of four geophones or sensors, including three uniaxial geophones and one triaxial geophone. The system is designed to monitor seismic activity and local fracturing of rockmass at the site and the system's data processing routines are fully automated. Microseismic detection is based on an array of geophones instrumented at key locations. P waves and s waves are processed, and the system can automatically alert when wave amplitudes exceed set thresholds. The system transmits information from MS event signals as a voltage to a seismic recorder, which synchronizes the data with a GPS component and converts the data to a seismic signal. Processing software assigns a location and magnitude to each locatable microseismic event.

2.9 Monitoring trenches/platforms

At locations where regular inspections occur to identify signs of potential earth fissure activity, specifically designed monitoring trenches or platforms can be designed to facilitate the observation process. The intent of such a system is to provide visible access to a surface where ground cracking associated with earth fissures would be more easily observable if it were to develop. This may be in the form of permanent trenches that are excavated to a more cemented soil horizon or in the form of a large cleared area of ground. The authors annually monitor a site that includes two large monitoring trenches for this purpose. Care needs to be taken in regards to drainage of the trenches, growth of vegetation and other disturbances.

2.10 Time-domain reflectometry

TDR utilizes uses a pulsed electromagnetic signal along a coupled coaxial cable to detect reflected changes resulting from deformation. The authors have deployed a TDR system at two sites with the goal of utilizing it as an early-warning and monitoring tool for discrete ground deformation related to ground subsidence and earth fissuring (AMEC, 2006, 2010). Both travel time and signal strength are measured. Travel time is used to determine position, with signal strength being an indication of the severity of the deformation. Signal strength is an approximate measure of deformation, therefore the TDR instrumentation should be viewed as a means to detect but not fully quantify deformation. The TDR system was designed such that should sufficient ground deformation occur to create a crack in the cemented soils, that crack would transmit itself through the controlled low-strength material (CLSM) that the cable is encased in and be detected by a signal sent through the TDR cable. While the TDR system has proven effective in detecting cracks that form in the CLSM backfill material, it has proven to be more sensitive than anticipated. Cracks detected have been the result of shrinkage cracking within the CLSM material and not the cemented soils. The system was abandoned after several years due to the sensitivity to shrinkage cracks within the backfill material.

2.11 Other methods

Other methods that are often utilized to monitor land subsidence include borehole extensometers and permanent GPS stations that are widely utilized by groups such as the United States Geological Survey (LSCE et al., 2014; Galloway et al., 1999; Poland and Lofgren, 1984). These methods often provide highly accurate point-source data, however the cost of systems is often prohibitive for most applications.

Monitoring for specific types of infrastructure can allow for other methodologies to be applied to land subsidence and earth fissuring monitoring. These could include techniques utilized to monitor pipelines, sonar-based methodologies, LiDAR, techniques such as on-train instrumentation for rail infrastructure, and others. Additionally, it anticipated that many monitoring methods will become drone-based as the drone technology continues to develop.

3 Conclusions

The authors have successfully designed and implemented monitoring systems capable of quantifying rates and distributions of ground subsidence and detection of ground rupture at multiple locations throughout the Western USA for several types of infrastructure including dams, levees, channels, basins, roadways, and mining facilities. Effective subsidence and earth fissure monitoring requires understanding and quantification of historic subsidence, estimation of potential future subsidence, delineation of the risk for earth fissures that could impact infrastructure, and motivation and resources to continue monitoring through time. A successful monitoring system provides the means to measure ground deformation, grade changes, displacement, and anticipate and assess the potential for earth fissuring. Employing multiple methods, a monitoring strategy utilizes an integrated approach, including both regional and local measurements.

Various methods implemented include conventional practices and proven, instrumented in-ground sensing systems. The conventional techniques include repeat optical levelling and GPS surveys, ground reconnaissance, photo-geological analysis, groundwater monitoring, and tape-extensometers. Advanced techniques include the processing and interpretation of differential interferograms of repeat-pass, InSAR, borehole tiltmeters, microseismic arrays, excavation of monitoring trenches, and TDR.

References

AMEC (AMEC Earth & Environmental): TDR Trench Investigation and cable Installation Report, McMicken Dam Post Design Completion, Prepared for the Flood Control District of Maricopa County, Arizona, Contract FCD 2004C068, Work Assignment No. 1, 11 April 2006.

AMEC (AMEC Earth & Environmental, Inc.): McMicken Dam Instrumentation & Monitoring Plan (Revision 2), Prepared for the Flood Control District of Maricopa County, Arizona, Contract FCD 2008C016, Work Assignment No. 8. AMEC Job No. 09-117-01038, 18 June 2010.

AMEC (AMEC Earth & Environmental, Inc.): Subsidence and Earth Fissure Risk Zoning Report, Final Design Level Geohazard Investigation, McMicken Dam, Prepared for the Flood Control District of Maricopa County, Arizona, Contract FCD 2011C005, Work Assignment No. 4, 28 February 2013.

AMEC (AMEC Earth & Environmental, Inc.): Final Feasibility Study, I&W Brine Cavern, Carlsbad, New Mexico, Prepared for the Oil Conservation Division, New Mexico Energy, Minerals, and Natural Resources Department, available at: ftp://164.64.106.6/Public/OCD/Documents/I&WBrineCavernProject/I&WBrineCavernProject-FeasibilityStudyFinal/I&WFeasibilityStudy_Final_081114.pdf, August 2014.

Beckwith, G. H., Slemmons, D. B., and Weeks, R. E.: Use of Low-Sun Angle Photography for Identification of Subsidence Induced Earth Fissures in Land Subsidence, in: Land Subsidence, edited by: Johnson, A. I., Proceedings of the Fourth International Symposium on Land Subsidence, 12–17 May 1991, Arvada, CO, Oxfordshire, UK: International Association of Hydrological Sciences Press, IAHS Publication No. 200, 261–269, 1991.

Fergason, K. C., Rucker, M. L., and Greenslade, M. D.: Investigative Procedures for Assessing Earth Fissure Risk for Dams and Levees, United States Society on Dams, 33rd Annual Meeting and Conference, 11–15 February 2013, Phoenix, AZ, 2013.

Fergason, K. C., Rucker, M. L., Panda, B., and Greenslade, M.: Investigative Procedures for Assessing Subsidence and Earth Fissure Risk for Dams and Levees, in: Engineering Geology for Society and Territory, Vol. 6, edited by: Lollino, G., Giordan, D., Thuro, K. Carranza-Torres, C., Wu, F., Marinos, P., and Delgado, C., International Association for Engineering Geology and the Environment, 2014.

Gabriel, A. K. and Goldstein, R. M.: Crossed Orbit Interferometry: Theory and Experimental Results From SIRB, Int. J. Remote Sens., 9, 857–872, 1988.

Galloway, D. L. Jones, D. R., and Ingebritsen, S. E.: Land subsidence in the United States, U.S. Geological Survey Circular 1182, 177 pp., 1999.

Hoffmann, J., Galloway D. L, Zebker, H. A., and Amelug, F.: Seasonal subsidence and rebound in Las Vegas Valley, Nevada, observed by synthetic aperture radar interferometry, Water Resour. Res., 37, 1551–1566, 2001.

Keaton, J. R., Rucker, M. L., and Cheng, S. S.: Geomechanical Analysis of an Earth Fissure Induced by Ground-Water Withdrawal for Design of a Proposed Ash and Sludge Impoundment, Southeastern Arizona, in: Land Subsidence Case Studies and Current Research: Proceedings of the Dr. Joseph F. Poland Symposium on Land Subsidence, edited by: Borchers, J. W., Association of Engineering Geologists, Star Publishing Company, Belmont, California, Special Publication No. 8, 217–226, 1998.

LSCE (Luhdorff & ScalManini Consulting Engineers), Borchers, J. W., and Carpenter, M.: Land Subsidence from Groundwater Use in California, Full Report of Findings, California Water Foundation, April 2014.

Panda, B. B., Rucker, M. L., and Fergason, K. C.: InSAR as a Subsidence Characterization Tool for Flood Control Dam Studies, United States Society on Dams, 33rd Annual Meeting and Conference, 11–15 February 2013, Phoenix, AZ, 2013.

Poland, J. F. and Lofgren, B. E.: Case History No. 9.13. San Joachin Valley, California, U.S.A., in Guidebook to studies of land subsidence due to ground-water withdrawal, available at: http://wwwrcamnl.wr.usgs.gov/rgws/Unesco/PDF-Chapters/Chapter9-13.pdf, last access: 31 August 2012, 1984.

Rucker, M. L. and Fergason, K. C.: Role of Practical Geophysics to Investigate and Mitigate a Distressed Flood Control Dam, for Association of State Dam Safety Officials Annual Conference, 25–30 September 2004, Pointe South Mountain Resort, Phoenix, Arizona, 2004.

Rucker, M. L. and Holmquist, O. C.: Surface Seismic Methods for Locating and Tracing Earth Fissures and other Significant Discontinuities in Cemented Unsaturated Soils and Earthen Structures, 4th International Conference on Unsaturated Soils, 2–6 April 2006, Carefree, Arizona, 2006.

Rucker, M. L. and Keaton, J .R.: Three-Dimensional Characterization with Limited Data: An Example from Playa-Lake Basin, Southeast Arizona, in: Transportation Research Record No. 1526, Soils, Geology, and Foundations, Engineering Technologies in Geotechnical Engineering, Transportation Research Board, National Research Council, National Academy press, Washington, D.C., 1996.

Rucker, M. L. and Keaton, J .R.: Tracing an Earth Fissure Using Seismic-Refraction Methods with Physical Verification, in: Land Subsidence Case Studies and Current Research: Proceedings of the Dr. Joseph F. Poland Symposium on Land Subsidence, edited by: Borchers, J. W.,Association of Engineering Geologists, Star Publishing Company, Belmont, California, Special Publication No. 8, 207–216, 1998.

Rucker, M. L., Greenslade, M. D., Weeks, R. E., Fergason, K. C., and Panda, B. B.: Geophysical and Remote Sensing Characterization to Mitigate McMicken Dam, in: GeoCongress 2008: Geosustainability and Geohazard Mitigation (GSP 178), edited by: Reddy, K. R., Khire, M. V., and Alshawabkeh, A. N., American Society of Civil Engineers, 207–214, 2008.

Rucker, M. L., Fergason, K. C., Greenslade, M. D., and Hansen, L. A.: Characterization of Subsidence Impacting flood Control Dams and Levees, United States Society on Dams, 33rd Annual Meeting and Conference, 11–15 February 2013, Phoenix, AZ, 2013.

Weeks, R. E. and Panda, B.: Defining Subsidence-Induced Earth Fissure Risk at McMicken Dam, for Association of State Dam Safety Officials Annual Conference, 25–30 September 2004, Pointe South Mountain Resort, Phoenix, Arizona, 2004.

Weeks, R. E., Panda, B., and Fergason, K. C.: Application of Synthetic Aperture Radar Interferometry (InSAR) in Defining Mine-Related Ground Deformation and Subsidence Hazards, for 5th Biennial Workshop Mines – Interstate Technical Group on Abandoned Underground Mines (ITGAUM) – Abandoned Underground Mine Workshop, 21–23 April 2004, Tucson, Arizona, U.S. Department of Transportation, Federal Highway Administration and Arizona Department of Transportation, 2004.

Proc. IAHS, 372, 367–373, 2015
proc-iahs.net/372/367/2015/
doi:10.5194/piahs-372-367-2015

Open Access
PIAHS

Compaction and subsidence of the Groningen gas field in the Netherlands

K. van Thienen-Visser[1], J. P. Pruiksma[1,a], and J. N. Breunese[1]

[1]TNO – Geological Survey of the Netherlands, Utrecht, the Netherlands
[a]now at: TNO, Structural Reliability, Delft, the Netherlands

Correspondence to: K. van Thienen-Visser (k.vanthienen@tno.nl)

Published: 12 November 2015

Abstract. The Groningen gas field in the Netherlands is Europe's largest gas field. It has been produced since 1963 and production is expected to continue until 2080. The pressure decline in the field causes compaction in the reservoir which is observed as subsidence at the surface. Measured subsidence is characterized by a delay at the start of production. As linear compaction models cannot explain this behavior, alternative compaction models (e.g. Rate Type Compaction Model and Time Decay model) have been investigated that may explain the measured subsidence. Although the compaction models considered in this study give a good match to this delay, their forecasts are significantly different. Future measurements of subsidence in this area will indicate which type of compaction model is preferred. This will lead to better forecasts of subsidence in future. The pattern of over- and underestimation of the subsidence is similar for the compaction models investigated and tested. The pattern can be explained by differences in modeled porosity and aquifer activity illustrating the improvement of subsurface knowledge on the reservoir using subsidence measurements.

1 Introduction

The Groningen gas field, situated in the north of the Netherlands, is the largest European gas field and has been in production since 1963. Pressures have declined from about 350 bar to, on average, 100 bar in 2012. Pressure reduction in the reservoir causes compaction which is observed at the surface as subsidence. Predictions of surface subsidence at the end of field life, in 2080, reach 60 cm. Due to the proximity to sea level surface subsidence is an important issue and is therefore closely monitored. The surface subsidence measurements show a delay of subsidence at the start of production. Conventional linear elastic compaction models cannot explain this behaviour. Other compaction models, such as Rate Type Compaction models (de Waal, 1986) and Time Decay model (Mossop, 2012) can describe this behaviour. We have applied these compaction models to the Groningen reservoir in order to estimate and predict subsidence. The resulting pattern of fit between the modelled and measured subsidence provides more insight into the reservoir properties.

2 Laboratory compaction experiments

In the laboratory compaction is measured on core samples of reservoir rock. The rocks are subjected to different loading rates and deformation is measured. The amount of compaction depends on the rate of applied stress on the sample. A faster rate gives a stiffer response of the sample (less compaction). For a constant rate, ongoing compaction (creep) is observed. The result of a typical laboratory compaction experiment is shown in Fig. 1. Initially, the sample is loaded at a rate of 2300 bar h^{-1}, this is subsequently lowered to 62, 6.2 and 0.62 bar h^{-1}. The stress- strain path belonging to the imposed loading rate is followed until the loading rate is changed. After a change in loading rate, a clear direct strain response is visible. Following this direct response a more gradual response is seen until the stress-strain path corresponding to the new loading rate is reached.

The dependence of compaction on the rate of applied stress offers a possible explanation for the observed subsidence delay above depleting reservoirs and ongoing subsidence for fields after abandonment.

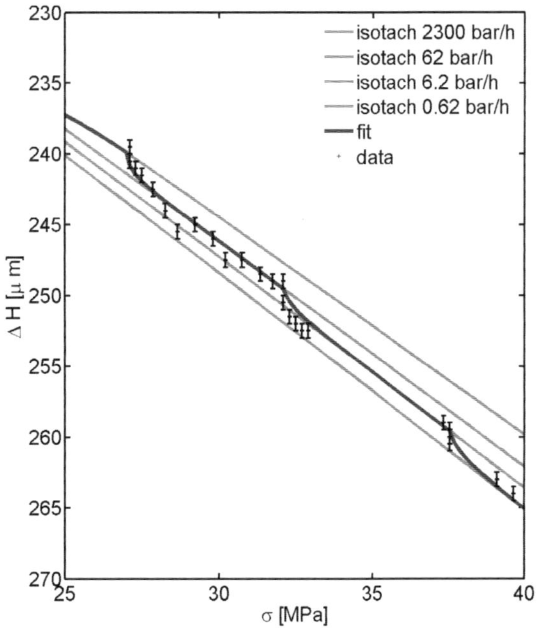

Figure 1. A compaction experiment (black) with the fit to the compaction experiment (red) with the RTiCM compaction model (de Waal et al., 2015). The lines of constant stress-strain rate are called isotachs (green) and are lines of constant loading rate.

2.1 RTiCM model

The Rate Type Compaction Model (RTCM) has been proposed by de Waal (1986) to explain rate type dependent behavior for sandstones. The original RTCM model is not able to simulate the continuous transition from one loading rate to another. Also the transition from a constant loading rate to a creep phase cannot be simulated. Therefore in Pruiksma et al. (2015), the RTCM model has been rewritten in isotach formulation, now called RTiCM. The isotach concept was developed in geotechnics (den Haan, 1996). Isotachs are lines of constant loading rate in the stress-strain diagram. Every loading rate corresponds to a unique isotach which is reached eventually if a sample is kept subjected to this loading rate. The isotach concept unifies compaction behavior for changes in loading rate with the compaction behavior for creep phases (de Waal, 2015).

The RTiCM model can be summarized in three equations:

$$\varepsilon_{\mathrm{d}} = c_{m,\mathrm{d}}(\sigma - \sigma_{\mathrm{ref}}), \tag{1}$$

$$\dot{\varepsilon}_{\mathrm{s}} = \left(\frac{\varepsilon - \varepsilon_0}{\sigma} - c_{m,\mathrm{d}}\right)\dot{\sigma}_{\mathrm{ref}}\left(\frac{\varepsilon - \varepsilon_0}{\sigma c_{m,\mathrm{ref}}}\right)^{-1/b}, \tag{2}$$

$$\varepsilon = \varepsilon_{\mathrm{d}} + \varepsilon_{\mathrm{s}}. \tag{3}$$

Here σ is the vertical effective stress in the sample, and ε is the vertical strain. This strain is composed of a vertical direct strain ε_{d}, which is linear elastic and a vertical creep strain ε_{s}. The model has three input parameters ($c_{m,\mathrm{d}}, c_{m,\mathrm{ref}}, b$) and two state parameters ($\sigma_{\mathrm{ref}}, \dot{\sigma}_{\mathrm{ref}}$). The constant ε_0 is defined by

$$\varepsilon_0 = -c_{m,\mathrm{ref}}\sigma_{\mathrm{ref}} \tag{4}$$

Equation (1) describes the direct response, while Eqs. (2) and (3) represent the rate dependent behavior.

Fit to laboratory experiment

The vertical effective stress as a function of time $\sigma(t)$ is known from the load on the sample. The vertical strain as a function of time $\varepsilon(t)$ can then be computed as the output of the model. An explicit Euler forward scheme is used for the solution (for details see Pruiksma et al., 2015). Figure 1 shows the application of the RTiCM model (de Waal et al., 2015) to the measurements in the laboratory experiment (mentioned in Sect. 2), with parameters $b = 0.010$, $c_{m,\mathrm{d}} = 0.26 \times 10^{-5}\,\mathrm{bar}^{-1}$ and $c_{m,\mathrm{ref}} = 0.46 \times 10^{-5}\,\mathrm{bar}^{-1}$.

2.2 Time Decay Model

The Time-Decay model (Mossop et al., 2012) equation reads:

$$\varepsilon(t) = c_m \Delta P(t) \cdot_t \frac{e^{-t/\tau}}{\tau}. \tag{5}$$

with ε the strain, c_m the compaction coefficient, $\Delta P(t)$ the pressure depletion as a function of time and τ a decay constant. Changes in pressure depletion lead to changes in strain with a certain delay time. Contrary to the RTiCM, the Time Decay lacks a direct response and is not based on laboratory experiments.

3 Application to field case: Groningen

Both the RTiCM and the Time Decay model have been applied to model the compaction and, from there calculate the subsidence of the Groningen gas field. Compaction is determined by the volume of gas bearing rock, the pressure depletion and the compaction coefficient which in turn are dependent on the rock type. For the Rotliegend sandstone which forms the reservoir rock in Groningen, laboratory experiments have shown a polynomial dependency on porosity (van Thienen-Visser et al., 2015). Both compaction models have been constrained using subsidence measurements over the whole of the Groningen field.

3.1 Subsidence data

The subsidence data consists of optical levelling and InSAR measurements. The optical levelling campaigns started in 1964 and were then limited to the central and southern part of the Groningen field. In subsequent campaigns in 1972 and 1987 the density and coverage of the optical levelling network was improved to cover the entire field. From 1987 to

Figure 2. Thickness (m) of the Zechstein salt layer. The difference between calculated and modeled subsidence (m) is indicated at the benchmark locations (label: subsidence diff). A red color indicates that the measured subsidence is larger than the modeled subsidence (from: TNO, 2013).

2008 repeat measurements were performed every few years. The number of measurements for each optical levelling campaign varies from 181 to 1168. The most complete campaigns are the more recent ones (2003 and 2008). InSAR uses persistent scatterers, typically buildings, to measure subsidence rates. From 1993 both InSAR as well as optical levelling has been used to model subsidence. The number of persistent scatterers varies between 0 and 1134. The available InSAR data were recorded between 1995 and 1999, and between 2004 and 2010 (TNO, 2013 and www.nlog.nl).

3.2 Reservoir description

The Groningen gas field is a Rotliegend sandstone reservoir at a depth between 2600 and 3200 m. The thickness of the reservoir gradually ranges from ~ 100 m in the south-southeast to ~ 300 m in the north–northwest. As the thickness changes, the reservoir lithology changes from mainly sandstone to more claystone. From east to west the thickness is relatively uniform (Mijnlieff and Geluk, 2011). The seal is formed by Zechstein salt covering the whole of the field and strongly varying in thickness over the field (Fig. 2). A static geological reservoir model (NAM, 2013) was used for detailed information on the porosity, reservoir thickness and depth. Porosity in the wells was determined from petrophysical analyses of well logs. The porosity of the reservoir varies from 0.12 to 0.22 leading to a compaction coefficient range between 5×10^{-6} to $2 \times 10^{-5}\,\text{bar}^{-1}$. A reservoir dynamic model (NAM, 2013) was used to match the historical gas flow and pressures in the reservoir. For both the Time Decay and the RTiCM model compaction was modeled using the dynamic reservoir model.

3.3 Calculation of subsidence

3.3.1 Approach

The above mentioned compaction models have been used to calculate subsidence at the benchmark locations at the surface. The translation from compaction to subsidence is based on a 1-D model (van Opstal, 1974). In this method, the non-compacting area around the reservoir (side-, over- and underburden) is assumed to be homogeneous and elastic. In the case of Groningen, the overburden is highly non-homogeneous as illustrated in the thickness of the Zechstein (Fig. 2). Additionally a rigid basement at a depth of 5 km is assumed, which determines the shape of the subsidence bowl (van Opstal, 1974) at the edges of the reservoir. The difference between measured and modeled subsidence is used to fit the resulting unknown parameters: τ for Time Decay and $\frac{c_{m,\mathrm{d}}}{c_{m,\mathrm{ref}}}$, $\dot{\sigma}_{\mathrm{ref}}$ for the RTiCM model, as well as a multiplier (α_{c_m}) on the porosity compaction coefficient relation for both compaction models. Additionally, the RTiCM model uses the laboratory experiments for the b value, and to define a range of $\frac{c_{m,\mathrm{d}}}{c_{m,\mathrm{ref}}}$ values used in the Red Flag procedure (Nepveu et al., 2010).

To fit the unknown parameters an ensemble of model realizations is generated that spans a large range for the parameters (α_{c_m} between 0.4 and 1.0, τ between 0 and 12 years, $\frac{c_{m,\mathrm{d}}}{c_{m,\mathrm{ref}}}$ between 0.4 and 0.8 and $\dot{\sigma}_{\mathrm{ref}}$ between 10^{-10} and $10^{-3}\,\text{bar}\,\text{yr}^{-1}$). The differences between the modeled and measured subsidence are input in a Bayesian "Red Flag" procedure (Nepveu et al., 2010). The Red Flag procedure gives the highest probability and the lowest χ^2 value to the parameter combination that fits the measured subsidence best. χ^2 is defined as

$$\chi^2 = \frac{1}{N}\sum_{i=1}^{N}\frac{(\text{measured}_i - \text{modeled}_i)^2}{\text{sig}_i^2},$$

where sig_i is the standard deviation of the ith subsidence measurement. The results of the Red Flag procedure are shown for the top three models for the RTiCM model in Table 1 and for the Time Decay model in Table 2. The χ^2 value is smaller for the RTiCM model and the parameter set is more distinct for the RTiCM model, which is indicated by the high probability (99.9 %) of the first parameter set and low probabilities (order $\times 10^{-4}$) for the second and third parameter set.

Table 1. Red Flag output for the RTiCM model.

Probability (%)	χ^2 (–)	$\frac{c_{m,\mathrm{a}}}{c_{m,\mathrm{ref}}}$ (–)	$\dot{\sigma}'_{\mathrm{ref}}$ (bar yr^{-1})	α_{c_m} (–)
99.9	0.751	0.44	3.2×10^{-4}	0.57
7.7×10^{-4}	0.753	0.49	3.2×10^{-4}	0.57
4.6×10^{-4}	0.755	0.44	1×10^{-4}	0.61

Table 2. Red Flag output for the Time Decay model.

Probability (%)	χ^2 (–)	τ (year)	α_{c_m} (–)
76.8	0.9341	4.85	0.53
21.5	0.9342	4.97	0.53
1.2	0.9345	4.734	0.53

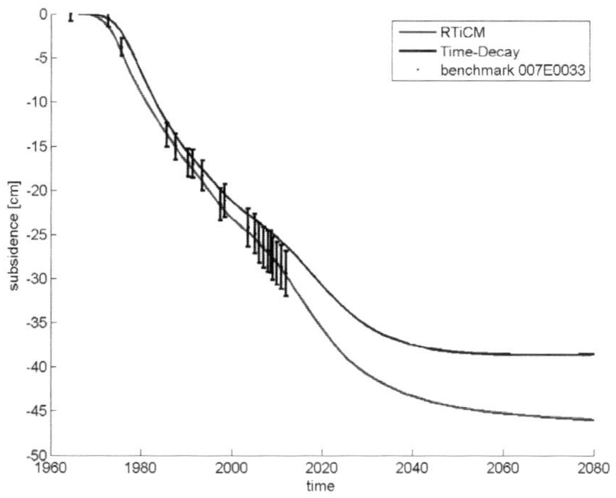

Figure 3. Different compaction models (RTiCM, linear isotach and Time Decay) and their fit to a levelling benchmark in the center of the field (from: TNO, 2013).

3.3.2 Results

Figure 3 shows the fit to one benchmark in the center of the Groningen field for both the RTiCM compaction model and the Time Decay model.

Figure 4 shows the compaction in the reservoir in the end of 2011 calculated with the RTiCM model. Superimposed on the compaction are the optical levelling benchmarks and the fit of the modeled subsidence to the measured subsidence.

Both compaction models fit the data, within one standard deviation. A further distinction between these models will only be visible, if significant changes in production rate will occur. Here it is noted that early 2014 the production in the central part of the field has been strongly reduced in view of mitigating the induced seismicity; the effect on compaction of that measure is still to be analyzed, when sufficient data become available.

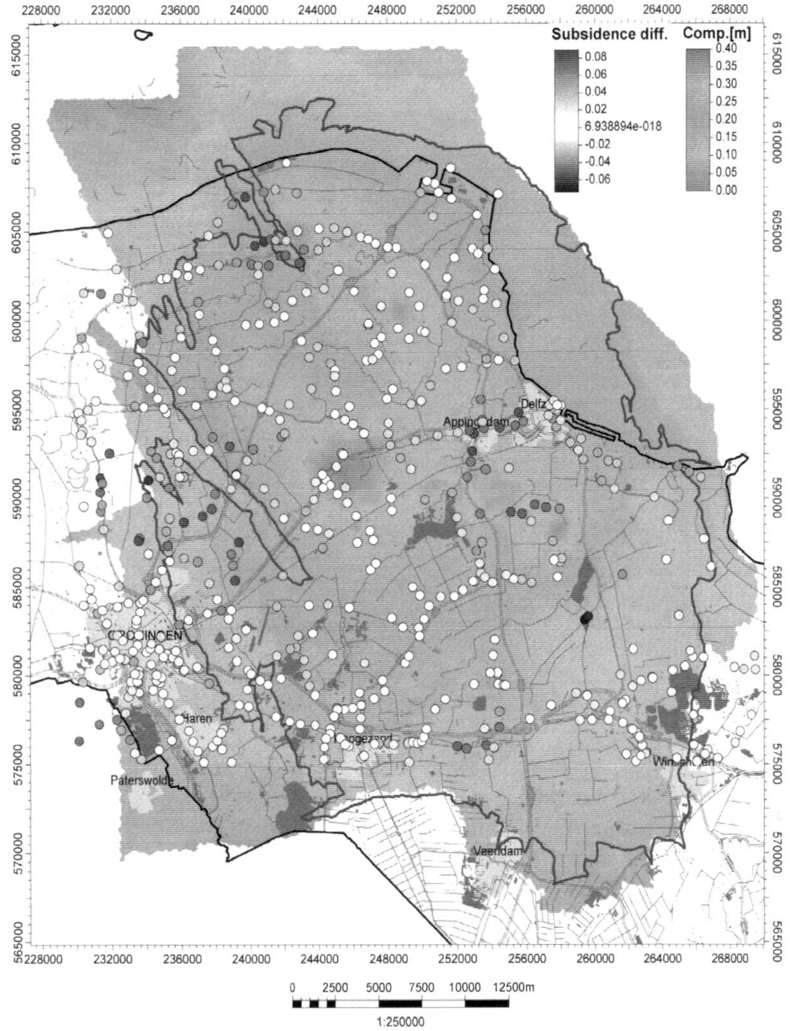

Figure 4. Compaction in the Groningen reservoir at January 2012 calculated with the RTiCM model (from TNO, 2014). The difference between calculated and modeled subsidence is indicated at the benchmark locations (label: Subsidence diff). A red color indicates that the measured subsidence is larger than the modeled subsidence.

Due to the creep part of the RTiCM model, this model will lead to larger subsidence values at the end of field life (expected in 2080).

3.4 Discussion

The compaction models (RTiCM and Time Decay) both fit the delay character of the observed subsidence in the first 10 years after the start of the gas production (Fig. 3). They underpredict the maximum subsidence in the center of the subsidence bowl by 2–3 cm for the RTiCM model and 5–6 cm for the Time Decay model at the end of 2011. Spatially both compaction models show the same pattern of overestimation and underestimation (Fig. 4). An overestimation of the subsidence occurs in the eastern part and in the northwestern part of the field. An underestimation exists in the southwestern part of the field. The differences between the compaction models are in the amplitude of maximum compaction (RTiCM larger than Time Decay) and the shape of the subsidence bowl at the edges of the field. The RTiCM model predicts a slightly steeper subsidence bowl than the Time Decay model.

As is clear from Fig. 4, relatively large misfits (up to 8 cm) occur over the field. In the van Opstal (1974) method a depth of a rigid basement is assumed, which governs the shape of

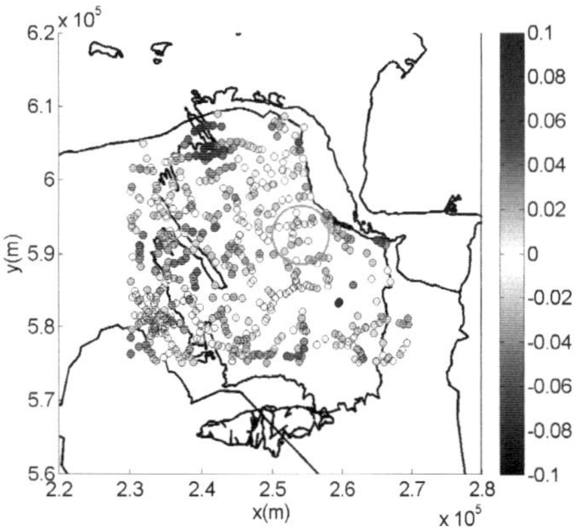

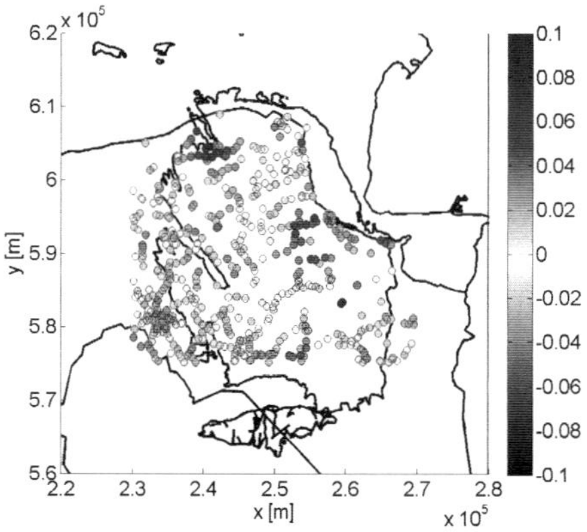

Figure 5. **(a)** Pattern of misfit (m) in the eastern part of the field, with an adaptation for the porosity of 0.85 applied in the green circle with a radius of 4 km. **(b)** Pattern of misfit (m) using a more active aquifer in the southwestern part of the field.

the subsidence bowl. A variation of rigid basement would most likely follow the depth of the reservoir which varies from south (2600 m) to north (2900 m) and therefore cannot lead to the observed mismatch patterns. Another assumption in the van Opstal (1974) method is a mechanically homogenous over- and underburden. The Zechstein salt layer which forms the seal of the gas reservoir varies greatly over small distances, which makes the overburden not homogeneous (Fig. 2). The thickness map of the Zechstein, however, does not correlate with the pattern of misfit. Also the presence of shallow soft soils such as peat and clay does not correspond to the misfit pattern (Wassing and Dost, 2012), indicating that shallow causes cannot explain the discrepancies. We conclude that the misfit pattern has to be related to the reservoir itself. Compaction in the reservoir depends on the volume of gas bearing rock, the compaction coefficient and the pressure depletion. As the compaction coefficient has a polynomial dependency on porosity a slight change in porosity will induce a large change in compaction. The pattern of the porosity (TNO, 2013) shows a high in the same area as the overestimation in the eastern part of the field. Figure 5, left, shows the pattern of misfit if a 0.85 reduction factor is applied on the porosity in a circle with a radius of 4 km in the area of high porosity. The fit to the measured subsidence is now within 2 cm, which is within two standard deviations and thus acceptable. For the other misfit areas no clear correlation with porosity is found. In both the northwestern as well as the southwestern areas an aquifer adds to the uncertainty of the reservoir model and thus to uncertainties of the pressures in those parts of the field. In the southwestern part of the field a less active aquifer (i.e. more depletion) would cause a larger modeled subsidence. In Fig. 5, right, the compaction modeled with a reservoir model consisting of a less active aquifer is shown. The misfit in the area improves considerably (van Thienen-Visser et al., 2014). The anomaly to the northwest is more difficult to explain: a more active aquifer leads to higher pressures and thus less modeled subsidence. The aquifer in the northwest is already modeled as a large active aquifer and an even larger aquifer would not fit the pressures and the water measurements in the northwest of the field. This area of the field is characterized by a couple of faults with large throws. The porosity as well as the other reservoir parameters are modelled continuously over these faults, which is not realistic due to the large throws of these faults. As a hypothesis, the mismatch in the northwestern area could possibly be explained by the presence of gas-blow-free water level, inhibiting the aquifer activity.

4 Conclusions

The delay character of subsidence measured over the Groningen field can be fitted using a Rate Type Compaction Model. The RTiCM model gives the smallest differences between measured and modeled subsidence. The pattern of over- and underestimation of the subsidence is similar for both compaction models investigated and tested here. The pattern can be explained by differences in modeled porosity and aquifer activity illustrating the improvement of subsurface knowledge on the reservoir using subsidence measurements.

Acknowledgements. We thank our colleagues of TNO, Geological Survey of the Netherlands, for their contribution to this work (especially Jitse Pruiksma, Lies Peters, Bart van Kempen, Marloes Kortekaas, Jenny Hettelaar and Maryke den Dulk).

References

den Haan, E.: A compression model for non-brittle soft clays and peat, Geotechnique, 46, 1–16, 1996.

de Waal, J.: On the rate type compaction behaviour of sandstone reservoir rock, PhD thesis, Delft University of Technology, Delft, the Netherlands, available at: http://repository.tudelft.nl/view/ir/uuid:b805782b-2eb4-4f72-98f4-f727c4ea9df0 (last access: 19 October 2015), 1986.

de Waal, J., van Thienen-Visser, K., and Pruiksma, J. P.: Rate type isotach compaction of consolidates sandstone, 49th US Rock mechanics/Geomechanics symposium (ARMA), San Franscisco, USA, 28 June–1 July, American Rock Mechanics Association (ARMA), 15-436, 2015.

Mijnlieff, H. F. and Geluk, M.: Paleotopography-governed sediment distribution: a new predictive model for the Permian Upper Rotliegend in the Dutch sector of the Southern Permian Basin, SEPM Special Publication: The Permian Rotliegend of the Netherlands 01/2011, 98, 47–160, doi:10.2110/pec.11.98.0147 2011.

Mossop, A.: An explanation for anomalous time dependent subsidence, 46th US Rock Mechanics/Geomechanics Symposium, Chicago, Illinois, USA, 24–27 June 2012, 12-518, 2012.

NAM 2013: Wijziging winningsplan Groningen 2013, inclusief technische bijlage Groningen winningsplan 2013, Versie 29 november 2013, available at: http://www.rijksoverheid.nl/documenten-en-publicaties/rapporten/2014/01/17/winningsplan-nam.html (last access: 19 October 2015), 2013.

Nepveu, M., Kroon, I. C., and Fokker, P. A.: Hoisting a Red Flag: An Early Warning Sytem for Exceeding Subsidence Limits, Math. Geosci., 42, 187–198, doi:10.1007/s11004-009-9252-2, 2010.

Pruiksma, J., Breunese, J. N., van Thienen-Visser, K., and de Waal, J. A.: Isotach formulation of the Rate Type Compaction Model for sandstone, Int. J. Rock Mech., 78, 127–132, 2015.

TNO: Toetsing van de bodemdalingsprognoses en seismische hazard ten gevolge van gaswinning van het Groningen veld, TNO report 2013 R11953, available at: http://www.nlog.nl/resources/Aardbevingen%20Groningen/TNO_rapport_Groningen_15-01-2014_gelakt_pre-scan.pdf (last access: 19 October 2015), 2013.

van Opstal, G.: The effect of base rock rigidity on subsidence due to compaction, Proceedings of the Third Congress of the International Society of Rock Mechanics, Denver, Colorado, USA, 1–7 September, 1974.

van Thienen-Visser, K., Nepveu, M., van Kempen, B., Kortekaas, M., Hettelaar, J., Peters, L., van Gessel, S., and Breunese, J.: Recent developments of the Groningen field in 2014 and, specifically, the southwest periphery of the field, TNO report 2014 R11703, available at: http://www.nlog.nl/resources/Aardbevingen%20Groningen/final%20TNO%20report%20EKL.pdf (last access: 19 October 2015), 2014.

Van Thienen-Visser, K., Muntendam-Bos, A. G., and Breunese, J. N.: Subsidence due to gas production in the Wadden Sea: How to ensure no harm will be done to nature, American Rock Mechanics Association (ARMA), 15-098, 2015

Wassing, B. B. T. and Dost., B.: Seismisch hazard van geïnduceerde aardbevingen; integratie van deelstudies, TNO/KNMI report, TNO 2012 R11139, 2012.

Proc. IAHS, 372, 375–378, 2015
proc-iahs.net/372/375/2015/
doi:10.5194/piahs-372-375-2015

Inversion of double-difference measurements from optical levelling for the Groningen gas field

P. A. Fokker and K. Van Thienen-Visser

TNO, Utrecht, the Netherlands

Correspondence to: P. A. Fokker (peter.fokker@tno.nl)

Published: 12 November 2015

Abstract. Hydrocarbon extraction lead to compaction of the gas reservoir which is visible as subsidence on the surface. Subsidence measurements can therefore be used to better estimate reservoir parameters. Total subsidence is derived from the result of the measurement of height differences between optical benchmarks. The procedure from optical height difference measurements to absolute subsidence is an inversion, and the result is often used as an input for consequent inversions on the reservoir. We have used the difference measurements directly to invert for compaction of the Groningen gas reservoir in the Netherlands. We have used a linear inversion exercise to update an already existing reservoir compaction model of the field. This procedure yielded areas of increased and decreased levels of compaction compared to the existing compaction model in agreement with observed discrepancies in porosity and aquifer activity.

1 Introduction

The Groningen gas field is a giant onshore field that has caused substantial subsidence since the start of its production in 1963. This subsidence has periodically been established by measuring the difference in height of stable benchmarks, using optical levelling. Pressures in the field have been closely monitored for reservoir management. History matching of the reservoir model on the observed pressures has resulted in a reasonably accurate pressure distribution development over the field.

There are a number of parameters in the relationship between the reservoir pressure and the subsidence which are more or less uncertain. The first one is the compaction coefficient, being dependent on the rock type and the porosity. There is also some uncertainty in the pressure estimates in some regions of the field, particularly in the connected aquifers, where pressure measurements are not available.

In the present paper we use the raw leveling difference measurements in conjunction with the prior knowledge about the Groningen gas reservoir in order to constrain the uncertainties. We employ an inverse algorithm to this end, but, instead of using interpreted heights, we use the originally measured height differences. In an earlier paper we reported the benefits this approach (Fokker and Van Thienen-Visser, 2015).

2 Available data

The Groningen gas field has been in production since 1963. It is located onshore in the Northeast of the Netherlands. Extensive geological, geophysical and reservoir engineering data have been used to history-match the reservoir characteristics like geometry, porosity and permeability. We had access to the simulated pressure field at yearly dates from 1 January 1964 to 1 January 2017. The delta pressures were multiplied by the height and the estimated compaction coefficient for each grid cell, based on lithology, pressure depletion and porosity. For each x–y location these numbers were accumulated over the reservoir layers in order to yield a prior estimate for the compaction grid at 9070 x–y locations for 54 times (Van Thienen-Visser et al., 2015). We remapped the provided compaction values to locations on a regular $400 \times 400\,\mathrm{m}^2$ grid for later manipulation. A map of the input compaction grid and the outline of the Groningen gas field in 2012 is provided in Fig. 1.

In the present study we focused on the use of data acquired through optical levelling. Usually, investigators use

Published by Copernicus Publications on behalf of the International Association of Hydrological Sciences.

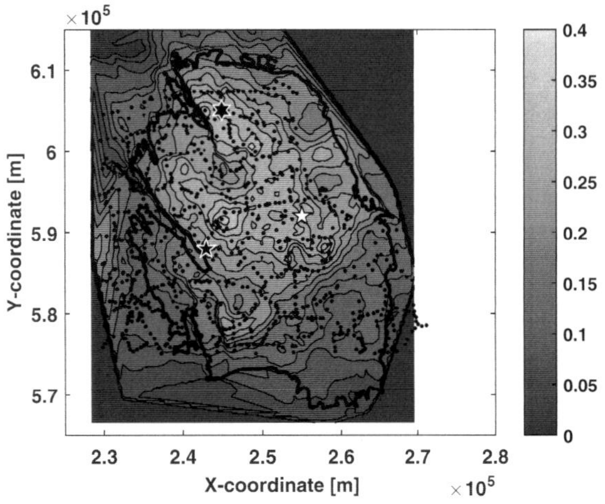

Figure 1. Prior estimate of the compaction field of the Groningen gas field in 2013 (color-coded), outline of the gas-bearing layers (solid line) and surface locations of the benchmarks used in the study (filled dots).

differences of the interpolated height maps to estimate surface movement. The procedure to obtain these differences includes the coupling to a reference benchmark or a set of reference benchmarks which are supposed to be stable, by integrating along the path of measurements to the stable benchmark. This procedure is sensitive to errors in the network and it accumulates the inaccuracy of all the measurements in the connecting path. The latter drawback can be addressed by providing the full covariance matrix of the resulting height estimates; this is, however, rarely done. Also reference benchmarks which, in hindsight, are not stable give rise to further inaccuracies. We have therefore chosen to use height difference measurements directly. The procedure to obtain double-difference estimates has been outlined in an earlier paper (Fokker and Van Thienen-Visser, 2015); it involves the determination of height differences between corresponding benchmark pairs in subsequent measurement campaigns, which have not necessarily been achieved in the same order.

Optical levelling campaigns have been performed many times in Groningen with different coverage. We had access to a total of 92 campaigns, dating from 1938 to 2012. Within a total of 7995 benchmarks, more than 26 000 height differences had been measured. In this set, 1572 benchmarks had been identified as stable ones in the resulting optical levelling database. We have constructed differences between stable benchmarks only, using the measurement paths along the unstable ones, and used these to construct the double differences. Further, we discarded benchmarks west of the line with $x = 230\,000$ m and south of the line with $y = 575\,000$ m in the local coordinate system (RD) to exclude the influence of other sources of compaction in those areas (e.g. the depletion of the Annerveen gas field south of Groningen). Still, a total of 10 860 double differences could be constructed between 987 benchmarks. The locations of these benchmarks are shown in Fig. 1.

3 Forward model

Gas production causes reservoir compaction, which, in turn, results in surface movement. Compaction in the reservoir may also change certain reservoir parameters. For the current study, a one-way coupling suffices – the change in porosity due to compaction only affects the reservoir pressure negligibly. We employed a linear-elastic model for the subsurface response, with the compacting blocks in the reservoir as source terms (Fokker and Orlic, 2006). Using an influence function approach, the subsidence at any surface point then is a superposition of the contributions of all compacting reservoir blocks. For the elastic profile in the subsurface we used a homogeneous elastic modulus down to a rigid basement at a depth of 5000 m. The reservoir is located at a depth of 3000 m. The connection to the double differences measured with the optical leveling can be made by making the appropriate time differences combined with space differences.

The goal of the present study was to employ an inverse algorithm on the interpreted double differences to improve the history match of the reservoir model and the predictive capability of the model in terms of reservoir pressures and subsidence. We considered the compaction of the reservoir as the uncertain parameter – the reservoir pressures and the porosities underlying it would involve too large computational efforts for this assessment. To map the uncertainty of the reservoir compaction we employed a field of multiplication factors at a spacing of 3200 m in space and 4 years in time. Values at the actual grid and intermediate times were obtained by interpolation. The prior multiplication values were defined as a constant value of unity over the field. A standard deviation of 0.3 was assumed. The mathematics of development

4 Inverse model

For the inverse model we define the vector $\boldsymbol{m}$ as the collection of adjustable model parameters, the vector $\boldsymbol{d}$ as the collection of double-difference data, and the matrix $\mathbf{G}$, working on the model parameters, as the forward model. The inverse problem is then formulated as the task of estimating the vector $\hat{\boldsymbol{m}}$ for which $\mathbf{G}\hat{\boldsymbol{m}}$ approaches the data vector $\boldsymbol{d}$ best. With additional information present in the form of a prior model ($\boldsymbol{m}_0$) and covariance matrices of the measurements ($\mathbf{C}_d$) and of the prior model ($\mathbf{C}_m$), the conventional least-squares solution is obtained by maximizing the objective function J given by Tarantola (2005) (or by minimizing $-\log[J]$):

$$J = \exp\left[-\frac{1}{2}(\boldsymbol{m}-\boldsymbol{m}_0)^T \mathbf{C}_m^{-1}(\boldsymbol{m}-\boldsymbol{m}_0) - \frac{1}{2}(\boldsymbol{d}-\mathbf{G}_m)^T \mathbf{C}_d^{-1}(\boldsymbol{d}-\mathbf{G}_m)\right]$$

For the linear problem at hand, the estimate and its covariance are given by

$$\hat{\boldsymbol{m}} = \boldsymbol{m}_0 + \mathbf{C}_m\mathbf{G}^T\left(\mathbf{G}\mathbf{C}_m\mathbf{G}^T + \mathbf{C}_d\right)^{-1}(\boldsymbol{d}-\mathbf{G}_{m_0})$$
$$= \boldsymbol{m}_0 + \left(\mathbf{G}^T\mathbf{C}_d^{-1}\mathbf{G} + \mathbf{C}_m^{-1}\right)^{-1}\mathbf{G}^T\mathbf{C}_d^{-1}(\boldsymbol{d}-\mathbf{G}_{m_0})$$
$$\mathbf{C}_{\hat{m}} = \mathbf{C}_m - \mathbf{C}_m\mathbf{G}^T\left(\mathbf{G}\mathbf{C}_m\mathbf{G}^T + \mathbf{C}_d\right)^{-1}\mathbf{G}\mathbf{C}_m$$
$$= \left(\mathbf{G}^T\mathbf{C}_d^{-1}\mathbf{G} + \mathbf{C}_m^{-1}\right)^{-1}$$

in which the first or second line of both expressions can be chosen according to the number of data points and model parameters (Tarantola, 2005). A smoothness constraint was added by extending the data vector with a number of elements equal to the number of multipliers in the model parameters, and by assigning the Laplacian working on $\boldsymbol{m}$ as the forward model for those elements. Furthermore, an independent constant vertical velocity for every benchmark was used as an additional unknown parameter to allow for movement not caused by the depletion of the gas field.

5 Results

The inversion exercise yielded an update of the fields of multiplication values and values for the autonomous movement of the benchmarks. With the original unit values and with the expected values of the multiplication factors, the forward model was rerun. The quality of the fit, indicated by $\chi^2 = \frac{1}{N}(\mathbf{G}_m - \boldsymbol{d})^2/\sigma_d^2$, improved from 8.8 to 5.9 – the first and second number being calculated with the prior and the estimated model parameters, respectively. The remaining value around 6, much larger than an optimal value around 1, is presumably related to a remaining instability in the selected benchmarks, however it could also mean that the standard deviation of the height difference is too optimistic. The average of the background movement of the benchmark is zero; the standard deviation is 0.5 mm year^{-1}.

There is a clear effect on the compaction fields. Examples of prior and updated compaction fields are given in Fig. 2. They show that around some areas the compaction levels must be adjusted to explain the measurements. These areas consistently return, independent of variations of the amount of smoothing or the precise form of the influence function in the forward model. More compaction than assumed in the prior model seems to have taken place around Ten Boer [$(x_{RD}; y_{RD}) = (243\,000; 588\,000)$]; less around Delfzijl [$(x_{RD}; y_{RD}) = (255\,000; 592\,000)$ and less around Uithuizen [$(x_{RD}; y_{RD}) = (245\,000; 605\,000)$]. The improvement of the double difference estimates and the effect on the subsidence estimates benchmarks is represented in Figs. 3 and 4.

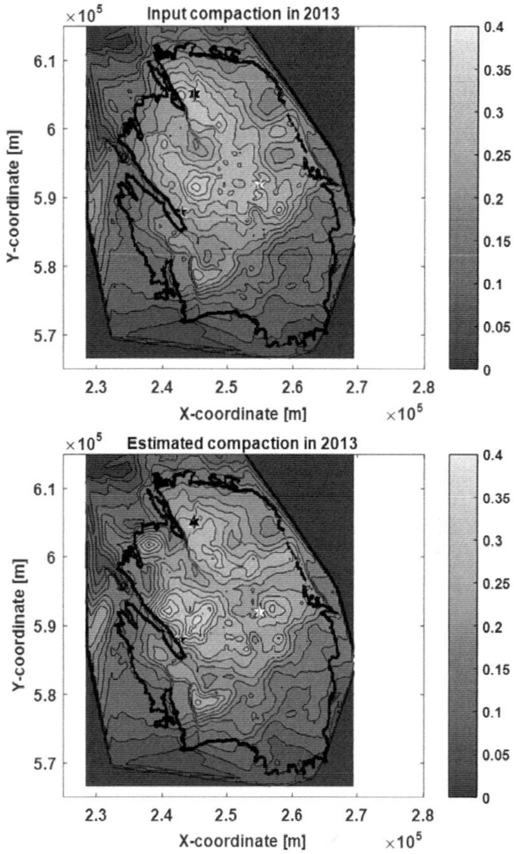

Figure 2. Prior compaction fields and estimated compaction fields in 2013.

6 Discussion

The correlation between measured and predicted double differences is much better for the estimated values of the multiplication factors than for the prior values. Still, the scatter remains considerable and there are many points with estimated value around zero that show comparatively large measured double differences. In view of this, it is remarkable that the inversion results in a consistent increase of compaction around Ten Boer and consistent decreases around Delfzijl and Uithuizen. This result was even apparent when no background movement was taken into account and the resulting correlation between measured and predicted double difference values was even worse. We assume that instabilities of individual benchmarks will cause deviations of double differences connected to them which are compensated with deviations with opposite sign for double differences starting from them.

Independent support for the updated compaction field has been found in a separate study (Van Thienen-Visser and Bre-

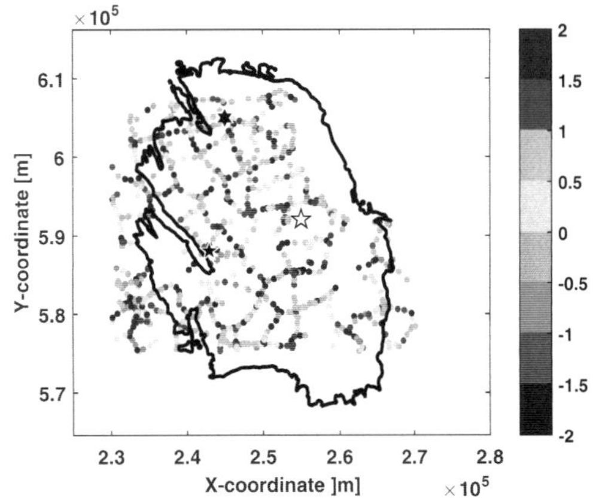

Figure 3. Improvement of the fit of double differences measured for two out of the 92 campaigns towards 1993 (starting times are variable for the different points). The color code indicates the ratio between prior and posterior offset: $(\mathbf{G}m_{\mathrm{E}} - \boldsymbol{d})/(\mathbf{G}m_0 - \boldsymbol{d})$. Absolute values of this number smaller than 1 (yellow or green) indicate improvement.

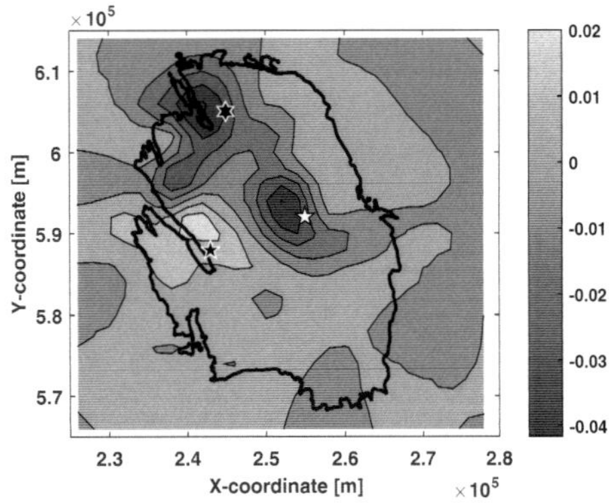

Figure 4. Effect of the compaction adjustment on the calculated subsidence in 2013.

unese, 2015). In that study, a different forward compaction model was employed and the predicted surface subsidence was compared to differences of interpreted heights at stable benchmarks and PS-InSAR measurement of the surface movement velocity. The areas that we found here were also identified in that study, and an additional effort was already recommended there to improve the subsurface model in those areas as it pointed towards inaccuracies of the porosity model and the assumed aquifer activity.

7 Conclusions

The present study proves the possibility of using double differences of optical levelling between stable benchmarks for the determination of reservoir parameters by its application on the Groningen gas field. The inverse study that we performed yielded a consistent update of the compaction of Groningen gas field during the lifetime of the field. The area around Ten Boer is compacting more than in the prior compaction model; the areas around Delfzijl and Uithuizen less. This is consistent with independent results obtained from comparing predicted subsidence with temporal differences of interpreted benchmark elevations. A renewed effort of reservoir modelling is required to improve the understanding of the reservoir in these areas.

Acknowledgements. The geological model and measured height differences were provided by NAM.

References

Fokker, P. A. and Orlic, B.: Semi-analytic modelling of subsidence, Math. Geol., 38 , 565–589, 2006.

Fokker, P. A. and Van Thienen-Visser, K.:. On the use of double differences in inversion of surface movement measurements, Paper ARMA 15-096, presented at the 49th US Rock Mechanics/Geomechanics Symposium, San Francisco, CA, USA, 28 June–1 July 2015.

Tarantola, A.: Inverse Problem Theory and Methods for Model Parameter Estimation, SIAM, Paris, France, 64–68, 2005.

Van Thienen-Visser, K. and Breunese, J. N.: Induced seismicity of the Groningen gas field: history and recent developments, The Leading Edge, special issue Injection Induced Seismicity, 34, 664–671, 2015.

Van Thienen-Visser, K., Pruiksma, J., and Breunese, J. N.: Compaction and subsidence of the Groningen gas field in the Netherlands, Proc. IAHS, in this volume, 2015.

Proc. IAHS, 372, 379–384, 2015
proc-iahs.net/372/379/2015/
doi:10.5194/piahs-372-379-2015

Open Access
PIAHS

Subsidence monitoring network: an Italian example aimed at a sustainable hydrocarbon E&P activity

M. C. Dacome[1], R. Miandro[1], M. Vettorel[1], and G. Roncari[2]

[1]ENI SpA – Geodynamics department, via del Marchesato 13, 48122 Marina di Ravenna, Italy
[2]ENI SpA – Geodynamics department, via F. Maritano 26, 20097 S. Donato M.se, Italy

Correspondence to: M. C. Dacome (maria.dacome@eni.com)

Published: 12 November 2015

Abstract. According to the Italian law in order to start-up any new hydrocarbon exploitation activity, an Environmental Impact Assessment study has to be presented, including a monitoring plan, addressed to foresee, measure and analyze in real time any possible impact of the project on the coastal areas and on those ones in the close inland located.

The occurrence of subsidence, that could partly be related to hydrocarbon production, both on-shore and off-shore, can generate great concern in those areas where its occurrence may have impacts on the local environment.

ENI, following the international scientific community recommendations on the matter, since the beginning of 90's years, implemented a cutting-edge monitoring network, with the aim to prevent, mitigate and control geodynamics phenomena generated in the activity areas, with a particular attention to conservation and protection of environmental and territorial equilibrium, taking care of what is known as "sustainable development".

The current ENI implemented monitoring surveys can be divided as:

- Shallow monitoring: spirit levelling surveys, continuous GPS surveys in permanent stations, SAR surveys, assestimeter subsurface compaction monitoring, ground water level monitoring, LiDAR surveys, bathymetrical surveys.
- Deep monitoring: reservoir deep compaction trough radioactive markers, reservoir static (bottom hole) pressure monitoring.

All the information, gathered through the monitoring network, allow:

1. to verify if the produced subsidence is evolving accordingly with the simulated forecast.
2. to provide data to revise and adjust the prediction compaction models
3. to put in place the remedial actions if the impact exceeds the threshold magnitude originally agreed among the involved parties.

ENI monitoring plan to measure and monitor the subsidence process, during field production and also after the field closure, is therefore intended to support a sustainable field development and an acceptable exploitation programme in which the actual risk connected with the field production is evaluated in advance, shared and agreed among all the involved subjects: oil company, stakeholders and local community (with interests in the affected area).

1 Introduction

A large part of ENI extraction activities in Italy is located in the central-northern Adriatic offshore, this area being characterised by shallow water depth and by a naturally subsiding sedimentary basin (Teatini et al., 2005). The coast is characterized by the presence of lagoons, deltaic areas and marshes with the average lands elevation around 2 m a.m.s.l. and some areas even below, up to −5 m. Most of the domestic offshore ENI exploited fields are located in this environmentally sensitive area, at a variable distance from the coastal line ranging from 3 km up to over 50 km. In this situation the risk of subsidence propagation toward the coast, especially for the field located closer to the shoreline, has to be carefully managed.

ENI, as any other Oil Company operating in Italy, to carry out its hydrocarbon extraction activities and in particular before starting any oil or gas production, especially for the offshore fields located in the central-northern Adriatic sea, is obliged, by authorities, to define a monitoring plan able to foresee, measure and analyse any possible impact, direct or indirect, of the production project on the coastal areas and on those ones in the close inland located.

Therefore ENI with the aim of a sustainable development of its activities and to safeguard the integrity of a territory, characterised by the presence of a high number of very important naturalistic and artistic sites some of which with unique features as Venice and Ravenna, implemented, since 90's years, a cutting edge geodynamics monitoring network. For its design the recommendations set out and presented to the *Seventh International Symposium on Land Subsidence* in Shanghai (Gambolati et al., 2005) have been largely followed. The network has been implemented, during the years, with innovative technologies in order to ensure an accurate understanding of the altimetrical variation of the land surface in every components of the movement with high resolution (sub-centimetric resolution in z component).

Inside the Company dedicated departments have been set up with the mission to prevent, mitigate and control all the geodynamic territory phenomena, with particular attention to the protection of the territorial and environmental equilibrium.

In the ENI subsidence prediction and control programme three are the main intervention steps followed:

- Subsidence prediction, in the phenomenon interested area, through the use of numerical Finite Element models
- Design of a monitoring plan
- Prevention of the anthropic subsidence expected or mitigation of the subsidence recorded during hydrocarbon production.

The information gathered through the monitoring network allow:

- To rapidly and continuously ascertain if subsidence phenomena are evolving according to initial forecast and if necessary to put in action all the remedial activities to mitigate subsidence effects to safeguard the territory's environment and economy.
- To provide data for the periodical calibration of the numerical forward models for subsidence prevision and control, according to the following sequence:
  - Monitoring → check → model calibration

2 The geodynamic monitoring network

Geodynamic monitoring showed in the last decade a very high technological improvement that allows nowadays to define, together with the reservoir developing program, an effective monitoring plan of the environmental effects due to subsidence.

In particular the following measurements are regularly acquired:

- Formation pressure both in the gas bearing layers and aquifer
- Gas bearing layers deep compaction
- Ground vertical displacements (subsidence/rebound)

To distinguish between anthropic subsidence induced by hydrocarbon extraction and natural subsidence and/or due to fresh water withdrawal from shallow aquifers, the measurements of soil vertical displacements are performed on a larger area than reservoir dimension and the shallow compaction of layers subjected to fresh water extraction it is also measured.

The beginning of the monitoring phase always starts before the hydrocarbon production, it continues during the reservoir production life and it is extended for several years (roughly between 5 to 10 years) after the end of production, to record residual subsidence and/or partial ground rebound due to layers pressure re-balance (Gambolati et al., 2007).

The monitoring technologies used are a set of complementary technologies, each one of them is not exhaustive on its own, but only if used in conjunction with all the others. In such a way it is possible to reach the desired measurement precision and a general data reliability in order to obtain in the following step a trustfully synthesis and interpretation data.

The current ENI implemented monitoring survey can be divided as:

- Shallow monitoring:
  - High precision levelling surveys
  - Continuous GPS surveys in permanent stations
  - SAR surveys
  - Assestimeter subsurface compaction monitoring

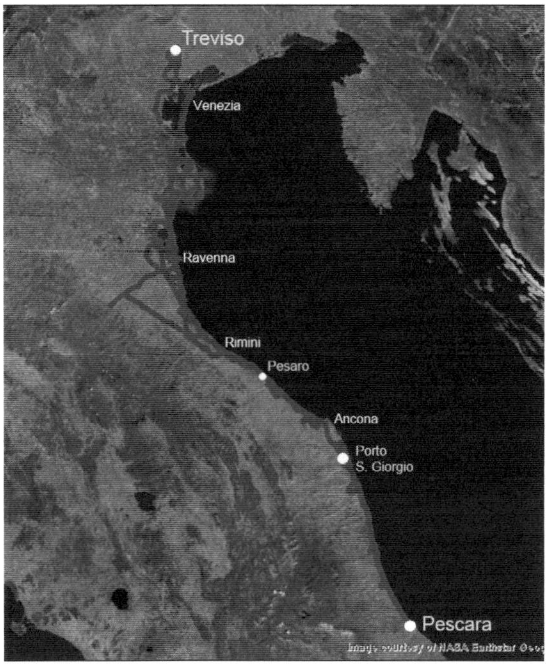

Figure 1. The Adriatic ridge of the levelling network.

Figure 2. CGPS network, yellow dot onshore stations, red dot offshore stations. SSU locations are also indicated.

  - Ground fresh water level monitoring
  - LiDAR surveys
  - bathymetrical surveys
- Deep monitoring:
  - reservoir deep compaction trough radioactive markers
  - reservoir static (bottom hole) pressure monitoring

The high precision levelling surveys is nowadays the technology for subsidence monitoring with the theoretical highest precision. The ENI implemented network (Fig. 1) consists of 1822 km and it allows to monitor all the central-northern Adriatic coast facing the main offshore gas fields. All the network is linked to stable external benchmarks with nil movements with respect to those ones present in monitored area.

The entire network is completely surveyed, respecting an high precision standard, every year since 2002 till 2009, every three years since 2011.

The Continuous GPS monitoring network currently consists of 65 permanent recording stations of which 48 located on production platform offshore and 17 onshore (Fig. 2). Any CGPS station of the network is linked to the international reference network EUREF, the acquired data is processed and solved according to the ITRF2008-IGB08 network. Among the onshore stations, 6 are in a configuration called by ENI: Satellite Survey Unit (SSU) (Fig. 3) where a GPS antenna, a SAR totem corner reflector and a levelling benchmark are mounted on a concrete base so that to have the same soil movement recorded simultaneously by the three different systems (SAR, GPS and levelling). The SSU stations allow to calibrate SAR data, linking them to a point with known absolute movements (GPS antenna/levelling) (Werner et al., 2003).

The Synthetic Aperture Radar (SAR Interpherometry) data (Fig. 4) provide measurements of surface deformation (mean trend and instantaneous variations) by radar interpherometric studies (Strozzi et al., 2001). The interferometric analysis consists in the radar signal phase evolution study between two different (in time) radar surveys. One of the causes of phase signal variation is the soil movements. Because the method provides only relative movements the SAR data are linked, as said above, to GPS station and/or levelling benchmarks.

To monitor and identify the amount of the contribution to the total subsidence of an area, provided by the shallow multi-aquifer system, assestimeters are used. They provide the low-depth compaction assessment to compute the water extraction contribution to subsidence (Bonsignore et al., 2010).

Figure 3. SSU station. Unique configuration assembling in a only one position three measuring methods.

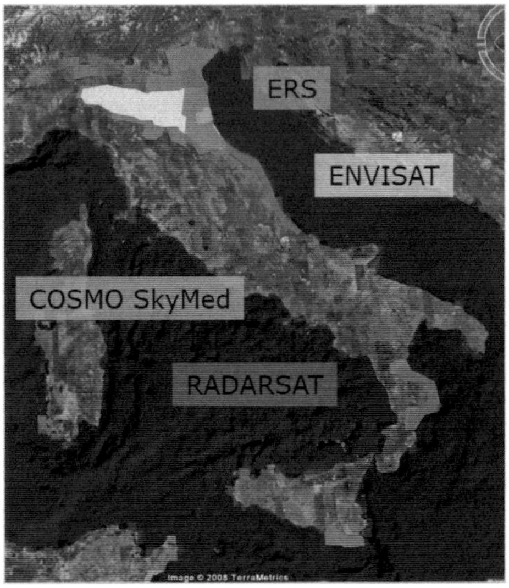

Figure 4. SAR Image coverage yearly processed. Cosmo SkyMed: 1200 km^2, ERS: 1500 km^2, ENVISAT: 6300 km^2, RadarSAT: 26 000 km^2.

The assestimeter tools have been installed by Eni in 6 stations along the Adriatic coast (Fig. 5).

The stations are in a configuration named by ENI: EPSU (Extensometric Piezometric Survey Unit) where the assestimeters are associated to one or more piezometers, monitoring the main exploited fresh water aquifers of the area. The deepest assestimeters installed by Eni, reaching the depth of about 370 m, are in the north, whereas the southward installed tools reach an average depth of 45 m. All the assestimeter installations are anchored below the deepest known drained aquifer for water supply.

Further monitoring technologies, useful in a sustainable development of hydrocarbon resources, are altimetrical and bathymetrical surveys (LiDAR, LADS, single/multi beam surveys), for fields located in offshore or close to the coast. Although these surveys don't reach the requested precision for subsidence monitoring, unless when the soil movements are very high, their use is important to fulfil the knowledge picture of the area subjected to hydrocarbon exploitation. ENI, depending on field location (close or faraway from the coast), always acquires the most appropriate survey before the start of every project, to have a "white" picture of the sea bottom morphology. Then the survey will be repeated over time in accordance with expected changes to constantly follow its evolution and to monitor any possible subsidence bowl formation.

A reliable subsidence prevision, using numerical modelling, of the expected subsidence due to a field exploitation requires a regular "in situ" monitoring of two important parameters: Compaction of gas/oil bearing layers and pressure of the layers fluids. These information are obtained by ENI with the use of the radioactive markers method to obtain reservoir/aquifer layers deep compaction and with the reservoir static (bottom hole) pressure monitoring (Cassiani and Zoccatelli, 2000). Both the radioactive and the bottom hole pressure surveys are acquired every year (Fig. 6).

3 Conclusions

In Italy, since 1997, for each new hydrocarbon development project an EIA (Environmental Impact Assessment, in Italian VIA) has to be submitted to the Ministry of Environment (in Italy MATTM) in order to start-up activities.

All the VIA decrees contains rules about subsidence monitoring that have to be respected through the whole life of the field and even after the end of the exploitation.

The monitoring network set up over the years in order to satisfy the assigned rules has gradually increased and now it is likely to be one of the widest and most complex monitoring network worldwide.

It allows to combine data acquired through different technologies and to integrate them with the double goal to (1) continuously update the simulation model and (2) be aware of the evolution of the phenomenon in order to be able to implement remedial actions if necessary (Verdecchia et al., 2010).

The implementation of the subsidence monitoring network described here represents a piece concurring to the puzzle of a sustainable development that is the result of a sharing of goals, risks and remediation activities between ENI and its stakeholders.

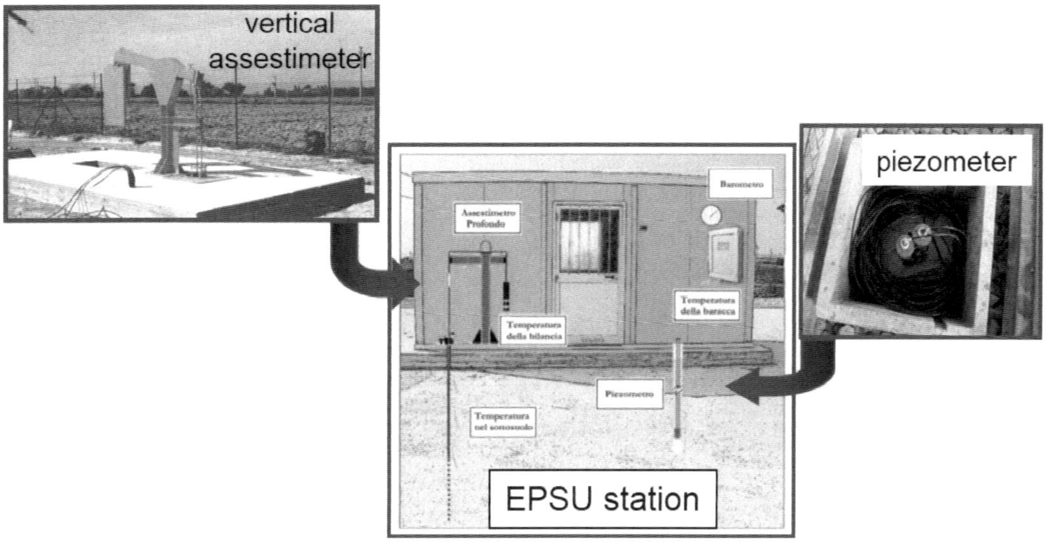

Figure 5. EPSU station, provide shallow compaction assessment, to compute the water extraction contribution to subsidence.

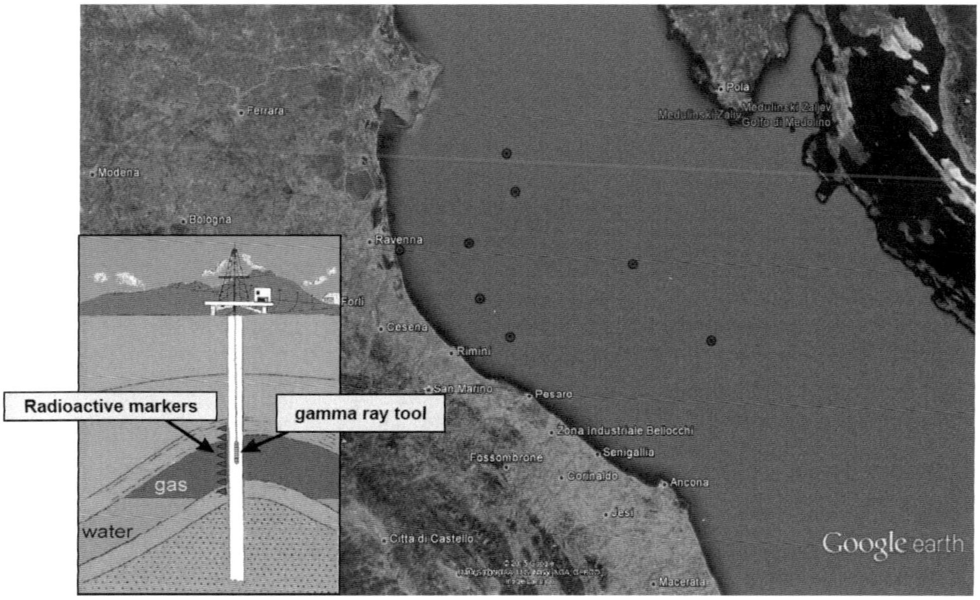

Figure 6. Wells instrumented with radioactive markers.

References

Bonsignore, F., Bitelli, G., Chahoud, A., Macini, P., Mesini, E., Severi, P., Villani, B., and Vittuari, L.: Recent extensometric data for the monitoring of subsidence in Bologna (Italy), in: Land Subsidence, Associated Hazard (Proc. Eigth Int. Symp. Land Subsidence Queretaro, Mexico, October 2010), edited by: Carreon-Freyre, D., Cerca, M., and Galloway, D. L., 333–338, IAHS publication 339, 2010.

Cassiani, G. and Zoccatelli, C.: Towards reconciliation between laboratory and in-situ measurements of soil and rock compressibilità, in: Proceedings of the Sixth International Symposium on Land Subsidence (24–29 September), edited by: Carbognin, L., Gambolati, G., and Johnson, A. I., vol. 2 CNR, Ravenna, Italy, 3–15, 2000.

Gambolati, G., Teatini, P., and Ferronato, M.: Basic conceptual guidelines for a sustainable gas/oil development, in: Land Subsidence (Proc. Seventh Int. Symp. Land Subsidence Shanghai, Pr China, October 2005), edited by: Zhang, A., Gong, S., Carbognin, L., and Johnson, A. I., II, 919–924, Shanghai Scient. & Tech. Publ. Shanghai China, 2005.

Gambolati, G., Teatini, P., and Ferronato, M.: Linee guida per lo studio dei fenomeni di subsidenza nell'ambito dei progetti di sviluppo sostenibile dei campi ad olio e gas, Technical report 2007 Department of Mathematical Model and Methods for Applied Sciences, University of Padova Italy, 2007.

Strozzi, T., Wegmuller, U., Tosi, L., Bitelli, G., and Spreckels, V.: Land subsidence monitoring with differential SAR interferometry, Photogramm. Eng. Remote Sens., 67, 1261–1270, 2001.

Teatini, P., Ferronato, M., Gambolati, G., Bertoni, W., and Gonella, M.: A Century of Land Subsidence in Ravenna, Italy, Environ. Geol., 47, 831–846, 2005.

Verdecchia, F., Zoccatelli, C., Norelli, E., and Miandro, R.: Integrated monitoring network for surface deformation in Capo Colonna archaeological area, Crotone, Italy, in: Land Subsidence, Associated Hazard (Proc. Eigth Int. Symp. Land Subsidence Queretaro, Mexico, October 2010), edited by: Carreon-Freyre, D., Cerca, M., and Galloway, D. L., 345–351, IAHS publication 339, 2010.

Werner, C., Wegmuller, U., Strozzi, T., and Wiesmann, A.: Interferometric point target analysis for deformation mapping, in: IGARSS 2003, VII, 4362–4364, IEEE, 2003.

Proc. IAHS, 372, 385–391, 2015
proc-iahs.net/372/385/2015/
doi:10.5194/piahs-372-385-2015

Open Access
PIAHS

Multiple sensors applied to monitorland subsidence in Central Taiwan

W.-C. Hung[1], C. Wang[2], C. Hwang[3], Y.-A. Chen[4], H.-C. Chiu[4], and S.-H. Lin[4]
[1]General Manager, Green Environmental Engineering Consultant Co. LTD
[2]Deputy General Manager, Green Environmental Engineering Consultant Co. LTD
[3]Department of Civil Engineering, National Chiao Tung University
[4]Green Environmental Engineering Consultant Co. LTD

Correspondence to: W.-C. Hung (khung@itrige.com.tw)

Published: 12 November 2015

Abstract. During 1992–2013, pumping of groundwater caused large-scale aquifer-system compaction and land subsidence in the Choshui River Alluvial Fan (CRAF) in Taiwan. The subsidence has already endangered the operation of Taiwan High Speed Rail (THSR). In this paper, we introduce the multiple sensors monitoring system to study the extent of subsidence in CRAF and its mechanism, including GPS (Global Positioning System), PSI (Persistent Scatterer Interferometry), leveling and multi-layer compaction monitoring well. These sensors complement each other in spatial and temporal resolutions.

1 Introduction

Due to continuous growth of the population and economy, more water resources are needed. The use of the underground water is inevitable when the surface water sources become insufficient. Inappropriate excessive exploitation of the underground water may result in land subsidence, which leads to land sinking and other disasters such as long-term water logging. The ocean water invades the water-bearing stratum and results in soil salinization. The utilization value of the soil may be degraded, and it will be difficult to introduce industries and the vital functions are downgraded. The underground water resources will no longer be utilized. These problems will result in the increase of the social cost. The above cases can be found in Table 1.

Taiwan is located in the subtropical monsoon region, and the climate of Taiwan is oceanic tropical and subtropical. The rainfall is plentiful; the annual rainfall reaches up to 2500 mm, 2.5 times the average world rainfall. However, Taiwan is a country suffering from severe dehydration due to various reasons. Taiwan is also highly populated. On this island, mountains are high and torrents are swift; therefore, most of the rainfall rapidly flows into the ocean. Taiwan strives for high economic development and the total water consumption rises continuously. As a result, underground water has become a major water source. Excessive exploitation of the underground water is common in the country areas. Therefore, critical land subsidence has occurred in the southwestern coastal areas, especially Choushui River Alluvial Fan (CRAF) in recent years.

CRAF is the most important agricultural area in western, central Taiwan, with elevations ranging from 0 to 100 m. CRAF covers a total area of 2000 km^2 and is bounded by Wu River (north), Pekang Creek (south), Douliu Mound (east) and Taiwan Strait (west). Figure 1 shows the geographical location of CRAF, which is centered at 24° N and 120.5° E.

Choshui River is the longest river in Taiwan. The sediments in CRAF originate from rock formations in the upstream watershed of Choshui River, including slate, metamorphic quartzite, shale, sandstone, and mudstone (Fig. 2). Sediment loads composed of weathered rock fragments of different sizes gradually settled on the riverbed, floodplain, and seabed to form CRAF. The head of CRAF contains mainly gravel and coarse sand, and the toe is a delta covered by fine sand. Because the upstream watershed area of Choshui River is wide and infiltrative, surface water in the head of the fan will penetrate the ground to recharge the subsurface aquifers.

Table 1. A list of world countries experiencing subsidence problems.

Area	Maxima cumulative subsidence (m)	Period	Reference
China	3.0	1921–2006	Gong and Yang (2008)
Mexico	10	2007–2011	Chaussard et al. (2014)
Spain	0.15	1993–2007	Tomás et al. (2010)
Taiwan	0.14	2006–2008	Hung et al. (2011)
CA, US	9	1925–1977	Galloway and Sneed (2013)
Australia	0.35	2006–2011	Ng et al. (2015)
Italy	0.5	1897–1937	Tosi et al. (2007)
Japan	0.4	2011–2014	ElGharbawi and Tamura (2014)

Figure 1. Geographical location of Choushui River Alluvial Fan (CRAF).

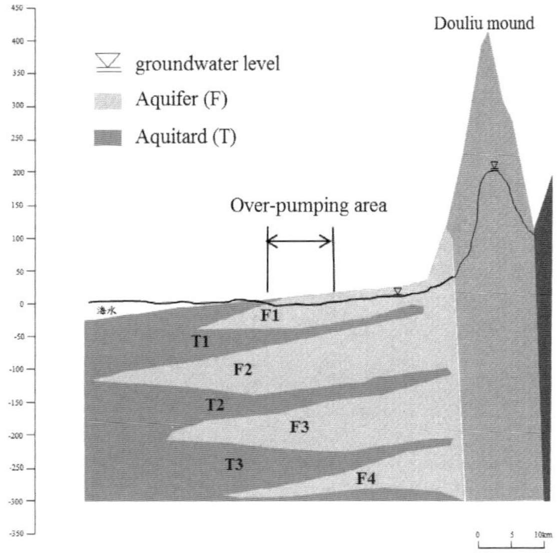

Figure 2. Geological settings of Choshui River Alluvial Fan (modified from Central Geological Survey of Taiwan, http://www.moeacgs.gov.tw/).

Due to the lack of surface water and the flat terrain that prohibits dam construction, the water for civilian and agricultural use in CRAF comes largely from underground. According to the statistics of Water Resource Agency of Taiwan, the extracted groundwater in Yunlin is 94 metric tons a day, totaling 305 million tons a year. Withdrawing groundwater without recharge here will decrease the water level and consequently reduce the porous pressure and increase the effective stress, inevitably leading to land subsidence. The Taiwan High Speed Rail (THSR) passes through Yunlin, where subsidence poses a serious threat to its operation (Hung et al., 2011).

2 Synergy of monitoring sensors

In order to further investigate the mechanism of land subsidence, the multi-sensors monitoring system including InSAR, GPS, leveling, multi-layer compaction monitoring well and Piezometer are used to monitor land subsidence in CRAF. Figure 3 is the sample picture of the multi-sensor monitoring system.

Figure 4 shows the distributions of leveling networks, Continuously Operating Reference Stations (CORS) and multi-layer compaction monitoring well used in this paper.

Table 2 summarizes the resolutions and accuracies of these sensors. With these sensors, we may inter-compare different results and complement each other in the spatial and temporal resolutions. With a good spatial correlation and proper environmental corrections, TCPInSAR with ALOS-PALSAR images will deliver areal displacements accurate to few cm with a 25 m spatial resolution at a time scale equivalent to the satellite repeat period. At a continuous GPS station, the accuracy of vertical changes can be determined to 1 cm on a daily

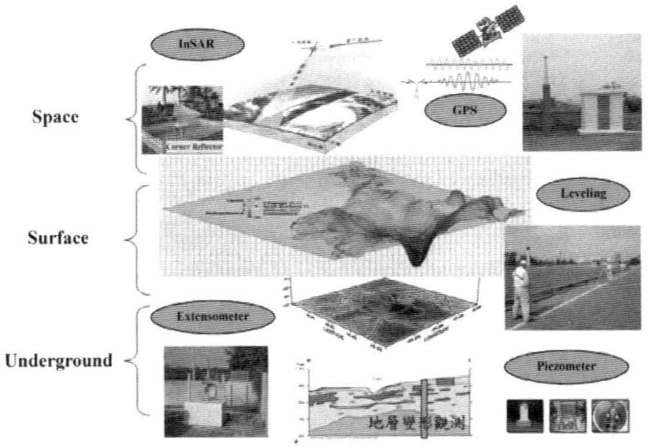

Figure 3. Sample pictures of Multi-sensor monitoring system.

solution. With a precision level and good correction models, leveling can deliver point-wise vertical displacements accurate to few mm along leveling routes.

In this study, the leveling and INSAR results used to detect the land subsidence area and calculate land subsidence rate in CRAF. In the subsidence center, the Continuously Operating Reference Stations (CORS) and multi-layer compaction monitoring well are used to analyze the mechanism of land subsidence.

3 Monitoring Result of CRAF

3.1 Leveling

In addition to the THSR leveling network, a leveling network west of THSR was established by Water Resource Agency (WRA) of Taiwan to monitor surface subsidence in Yunlin County.

The length of the leveling network in CRAF is 830 km. The accuracy requirement for the leveling is a $3\sqrt{K}$ mm misclosure in any double run, where K is the distance between two neighboring benchmarks in km. At each leveling setup, the distances to the foresight and backsight were measured by an electronic distance measurement (EDM) device to ensure the two are identical. Corrections for collimation error, atmospheric refraction and the earth's curvature were applied to the differential heights (Wolf and Ghilani, 2005). Because the cumulative differences between foresight and backsight distances do not exceed 10 m, residual collimation errors, the earth's curvature and atmospheric effects were negligible.

The corrected height differences from leveling were least-squares adjusted to determine the heights. To ensure the quality of least-squares solution, we screen the observables for outliers using Baada's data snooping technique (Koch, 1987; Caspary, 1988). First the standardized residual of an observable is computed as

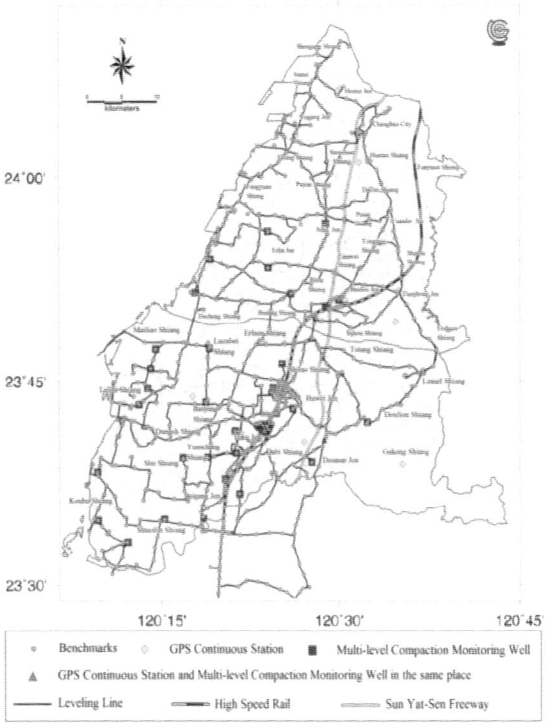

Figure 4. Distributions of leveling benchmarks, monitoring wells and GPS stations in CRAF, the names of townships in Yunlin County and Changhua County are shown here.

$$W_i = \frac{V_i}{\delta V_i}, \quad (1)$$

where δV_i is the standard error of residual V_i, which is the squared root of a diagonal element of matrix $\mathbf{Q}_v$ as follows

$$\mathbf{Q}_v = P^{-1} - A(A^T P A)^{-1} A^T. \quad (2)$$

With a significance level of 0.01, the critical value for data snooping in (4) is 3.3. That is, if $W_i > 3.3$, the observable (height difference) associated with V_i is regarded as an outlier and is removed from the input to the adjustment. A second adjustment is then performed without these outliers. Because all misclosures are smaller than $3\sqrt{K}$ mm and all raw data are carefully inspected, we did not find any outliers in the data from the three campaigns.

The cumulate subsidence is defined as change of height in 1992–2003. Contours of subsidence are computed using Kriging and are shown in Fig. 5. In the past two decades, the maximum cumulative subsidence is over 150 cm and shows a basin-like subsidence pattern centering at Huwei, Tuku, and Yuanchang Township. The serious subsidence are in central Youlin County, i.e. Tuku, Huwei and Yuanchang township.

Table 2. Multi-sensor comparison table.

	Leveling	Continuous GPS	Monitoring Well	InSAR
Spatial Resolution	1.5–2 km	10–15 km	5–10 km	25 m
Survey frequency	1 year	1 day	1 month	35day (C band)
Vertical Accuracy	0.5–1 cm	0.5–1 cm	0.1–0.5 cm	0.5–1 cm

THSR passes through series subsidence area, and land subsidence will result in a major safety concern.

3.2 Continuously operating reference stations

The used GPS network, presented in Fig. 6a, in this study consists of 36 GPS stations belong to the governmental agencies in Taiwan including Water Resource Agency (WRA), Central Geological Survey (CGS), Ministry of Interior (MOI), Central Weather Bureau (CWB), and Industry Technology Research Institute (ITRI). The Bernese V5.0 developed by Bern University was adopted to obtain the change of horizontal and vertical coordinates via the automatic data processing strategy. In addition to the used Continuously Operating Reference Stations (CORS), the non-continuously campaign GPS surveying with the fixed-centering, fixed-leveling, and fixed-heighting GPS Monumentation (Patent Number: 182781) was also utilized in this investigation on the observation of land surface change. The GPS pilar has been conformed to National Geodetic Survey (NGS), which indicate the fixed-centering, fixed-leveling, and fixed-heighting are necessary for the campaign GPS surveying with 2cm accuracy campaign (Zilkoski et al., 1997).

According to the results from the GPS CORSs observed from April 2010 to April 2014, presented in Fig. 6b, the region along the high speed rail runs through the main subsidence area in the Yunlin County, especially around the Huwei Township (GFES GPS station located), Tuku Township (TKJS GPS station located), and Yuanchang Township (KTES GPS station located). The rate of subsidence can reach 6 cm per year. The agreement between the results from GPS CORS and Leveling survey can be obtained. Because the advantage of continually observing from GPS CORS, the deformation of land surface caused by the seasonal variation and the change of groundwater in wet/dry season can be detected via the Principal Component Analysis (PCA) strategy.

3.3 Multi-layer compaction monitoring well

As mentioned previously, the subsurface deposits constituted the aquifer system in CRAF are not only heterogeneous but also with different hydraulic and mechanical properties in each layer. In order to well understand the mechanism of subsidence, the multi-layer compaction monitoring well technique is applied to measure compaction at different stratigraphic intervals within the aquifer system (Fig. 7).

The basic principle of multi-layer compaction monitoring well is to measure the depths of magnetic rings precisely and to analyze the depth differences of all the rings. The depth difference of two adjacent rings indicates the deformation of the stratigraphic section between the two rings. Hence the compaction amount of each layer to the total subsidence can be determined. In general, 20 to 26 magnetic rings are anchored in each well based on the different stratigraphic types, and the depths of the rings are measured at a one-month interval. Moreover, the advantage of multi-layer compaction monitoring well is the capability of monitoring multi-layer compaction with high accuracy (about 1 mm) and stability (Hung et al., 2012). Currently there are 31 multi-layer compaction monitoring wells installed in CRAF (Fig. 4).

Figure 8 shows the cumulative compactions at the TKJS monitoring well which is located at the most serious subsidence area along the rail of THSR from 2004 to 2011. As the result of the aquifer system which contains highly compressible alternations of sand and clay, the majority and the secondary compaction occur at depths deeper than 200 m and 50 to 200 m, respectively. Moreover, based on the compactions at different strata, the monitoring well result can illuminate the main source of water pumping as well. For example, if the major compaction occurs at a deeper stratum, the municipal and industrial water are the main sources of water use. On the contrary, if the major compaction occurs at a shallow stratum, the agricultural water is the main source of water use. Therefore, the compaction monitoring well is not only to understand the strain source effectively but also to mitigate the land subsidence problem usefully.

3.4 Areal subsidence by multi temporal InSAR, MT-InSAR

In the past decades, synthetic aperture radar interferometry (InSAR) has been proved to be a powerful and effective geodetic technique, with high spatial resolution, sub-centimeter accuracy and useful observation cadence, for measuring surface deformation. Nevertheless, due to the different surface features with time and the variation of water vapor stratification between satellite passes would easily cause spatial decorrelation in conventional SAR interferometry, and lead to a degraded accuracy in the deformation measurements. Thus, the multi-temporal InSAR (MT-InSAR) techniques are developed to overcome the limitation of DInSAR technology. Currently, there are three categories

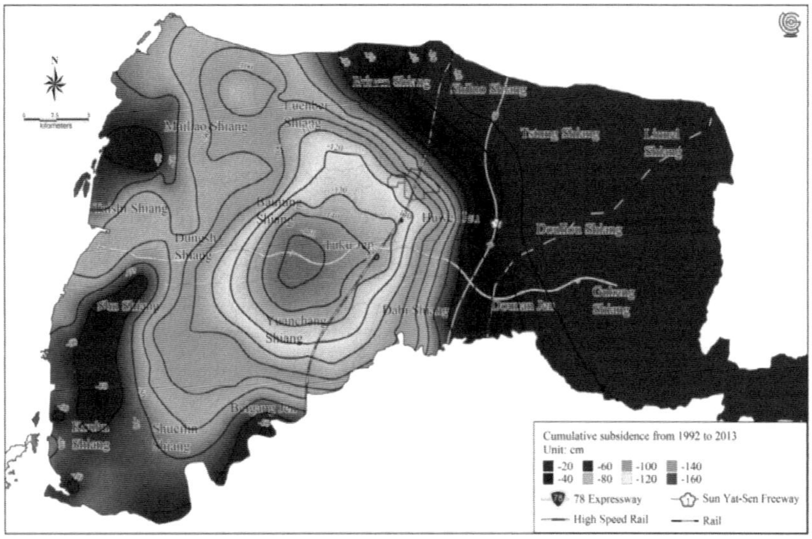

Figure 5. Cumulative subsidence derive from leveling over 1992–2013.

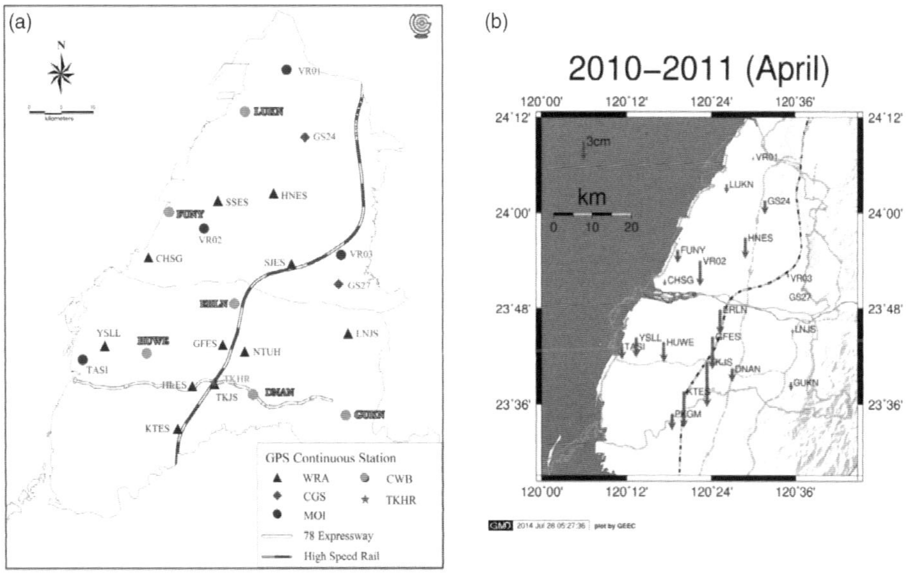

Figure 6. (a) Distributions of continuous GPS stations in CRAF (b) Vertical displacement rate from GPS over April 2010–April 2011.

of MT-InSAR approaches used to monitor land subsidence in Taiwan, such as persistent scatterersInSAR (PSI) (Hooper et al., 2004), Stanford method for persistent scatterers/multi-temporal InSAR (StaMPS/MTI) (Hooper, 2008) and temporarily coherence point InSAR (TCPInSAR) (Zhang et al., 2011).

In this study, we utilize TCPInSAR approach to measure the ground deformation in CRAF from 2007 to 2011 and to compare the TCPInSAR result with precise leveling data. From the comparison result, the TCP pixel density over CRAF is 346 pixels km^{-2} which is much higher than the leveling benchmarks (0.2 points km^{-2}). Moreover, the vertical displacements inferred from TCPInSAR show good consistency with the leveling result to 10 mm $year^{-1}$ (RMS), and provides a higher spatial resolution of subsidence than the latter. Besides, we also develop a data fusion method that considers the high-precision leveling data as a smoothed correction to the PSI measurement (Hung et al., 2011). Figure 9

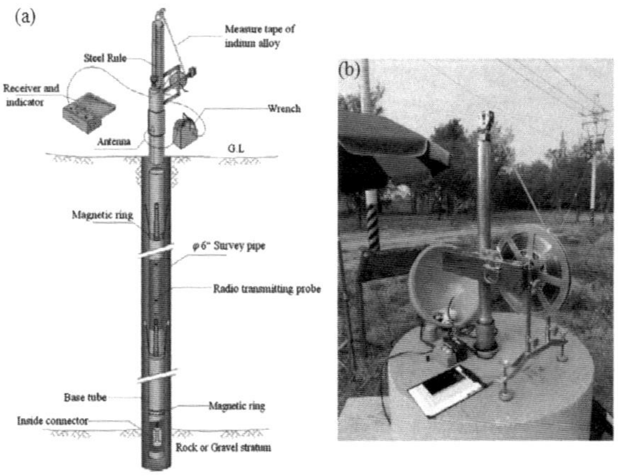

Figure 7. Sample pictures of multi-layer compaction monitoring well.

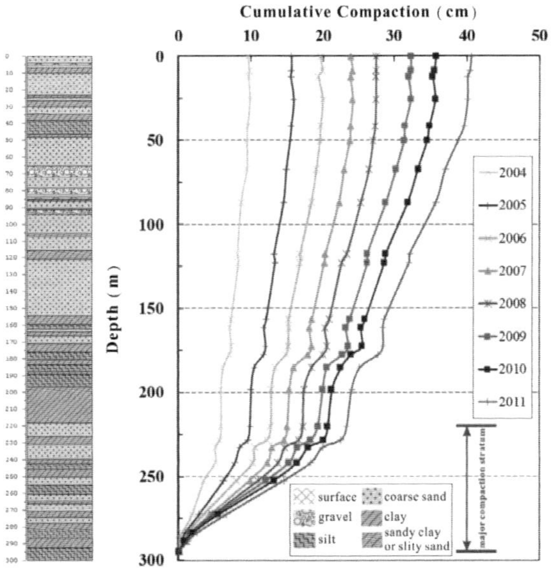

Figure 8. Layers of sediments and cumulative compactions from 2004 to 2011 at TKJS.

shows the vertical displacement rate after data fusion processing during the period between 2007 and 2011 in CRAF, and the deformation pattern is quite consistent with the leveling result. The major subsidence areas in CRAF are located in western areas of Sun Yat-Sen Freeway, e.g. middle fan and distal fan areas (see Fig. 1). The combined displacement field is more representative of overall deformation characteristics than the TCPInSAR-only or leveling-only field, and it is better suited for the assessment of the impact of subsidence over CRAF.

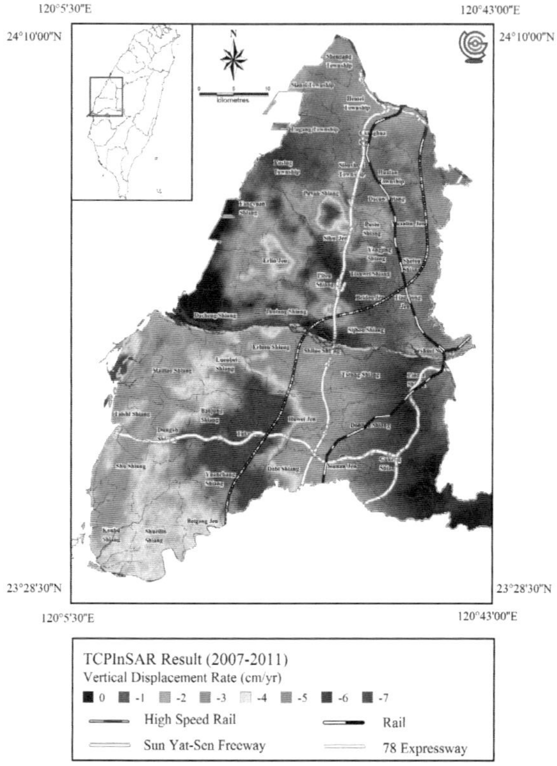

Figure 9. Vertical displacement rate from TCPINSAR over 2007–2011.

4 Discussion and conclusions

The security of country's territory has been affected by the long term land subsidence. The multi-sensors system developed by Taiwanese government including GPS, PSI, Leveling, monitoring well, and groundwater well can monitor the land subsidence from space, ground, and underground. Not only does it control the deformation of land surface efficiently and understand the magnitude and major source of stress for the shallow and deeper land, but also goes with policy of industry. The rate and area of land subsidence can be reduced efficiently. In the future, monitoring action will be conducted by extending the area and executing automatically. 400 GPS CORSs around Taiwan will be included for the understanding of the variation of three dimensions. Meanwhile, combining with airborne PSI, GPS, and data fusion technology, the affected range and change of land subsidence can be controlled quickly and precisely.

Acknowledgements. This study is supported by Water Resource Agency, Dept. of Economics, Taiwan, R.O.C.

References

Caspary, W. F.: Concepts of Network Adjustment and Deformation Analysis. Monograph 11, Scholl of Surveying, University of New South Wales, Kensington, NSW, Australia, 183, 1988.

Chaussard, E., Wdowinski, S., Cabral-Cano, E., and Amelung, F.: Land subsidence in central Mexico detected by ALOS InSAR time-series, Remote Sens. Environ., 140, 94–106, 2014.

El Gharbawi, T. and Tamura, M.: Measuring deformations using SAR interferometry and GPS observables with geodetic accuracy: Application to Tokyo, Japan, ISPRS J. Photogramm. Remote Sens., 88, 156–165, 2014.

Galloway, D. L. and Sneed, M.: Analysis and simulation of regional subsidence accompanying groundwater abstraction and compaction of susceptible aquifer systems in the USA, Bol. Soc. Geol. Mex., 65, 123–134, 2013.

Gong, S. L. and Yang, S. L.: Effect of land subsidence on urban flood prevention engineering in Shanghai, The Key Magazine of China Technology, 39, 1–20, 2008.

Koch, K. R.: Parameter Estimation and Hypothesis Testing in Linear Models, Springer, Berlin, 378, 1987.

Hooper, A.: A multitemporal InSAR method incorporating both persistent scatterer and small baseline approaches, Geophys. Res. Lett., 35, 2008.

Hooper, A., Zebker, H., Segall, P., and Kampes, B.: A new method for measuring deformation on volcanoes and other natural terrains using InSAR persistent scatterers, Geophys. Res. Lett., 31, 2004.

Hung, W.-C., Hwang, C., Chen, Y.-A., Chang, C.-P., Yen, J.-Y., Hooper, A., and Yang, C.-Y.: Surface deformation from persistent scatterers SAR interferometry and fusion with leveling data: A case study over the Choushui River Alluvial Fan, Taiwan, Remote Sens. Environ., 115, 957–967, 2011.

Hung, W.-C., Hwang, C., Liou, J.-C., Lin, Y.-S., and Yang, H.-L.: Modeling aquifer-system compaction and predicting land subsidence in central Taiwan, Eng. Geol., 147, 78–90, 2012.

Ng, A. H. M., Ge, L., and Li, X.: Assessments of land subsidence in the Gippsland Basin of Australia using ALOS PALSAR data, Remote Sens. Environ., 159, 86–101, 2015.

Tosi, L., Teatini, P., Carbognin, L., and Frankenfield, J.: A new project to monitor land subsidence in the northern Venice coastland (Italy), Environ. Geol., 52, 889–898, 2007.

Tomás, R., Herrera, G., Delgado, J., Lopez-Sanchez, J. M., Mallorquí, J. J., and Mulas, J.: A ground subsidence study based on DInSAR data: Calibration of soil parameters and subsidence prediction in Murcia City (Spain), Environ. Geol., 111, 19–30, 2010.

Wolf, P. R. and Ghilani, C. D.: Elementary surveying: an introduction to geomatics, Pearson Prentice Hall, 2005.

Zhang, L., Ding, X., and Lu, Z.: Ground settlement monitoring based on temporarily coherent points between two SAR acquisitions, ISPRS J. Photogram. Remote Sens., 66, 146–152, 2011.

Zilkoski, D. B., D'Onofrio, J. D., and Frakes, S. J.: Guidelines for Establishing GPS-derived Ellipsoid Heights (standards, 2 Cm and 5 Cm), Version 4.3: US Department of Commerce, National Oceanic and Atmospheric Administration, National Ocean Service, National Geodetic Survey, 1997.

Proc. IAHS, 372, 393–394, 2015
proc-iahs.net/372/393/2015/
doi:10.5194/piahs-372-393-2015

In-situ formation compaction monitoring in deep reservoirs by use of fiber optics

H. Ikeda[1], S. Kunisue[1], D. Nohara[2], K. Ooba[3], and T. Kokubo[4]

[1]Kanto Natural Gas Development Co., Ltd., Mobara-shi, Chiba-ken, Japan
[2]Asahi Glass Co., Ltd., Oamishirasato-shi, Chiba-ken, Japan
[3]Godo Shigen Co., Ltd., Chosei-mura, Chiba-ken, Japan
[4]Asano Taiseikiso Engineering Co., Ltd., Uenohara-shi, Yamanashi-ken, Japan

Correspondence to: H. Ikeda (h.ikeda@k-and-o-energy.co.jp)

Published: 12 November 2015

Abstract. We have devised a new in situ monitoring method for the amount of stratified compaction in borehole drilled several hundred meters underground. This newly developed epoch-making monitoring system differs from conventional monitoring methods for land subsidence in that it is designed to continuously monitor the amounts of displacement in several intervals separately, using optical fibers fitted in the sensor assembly. This report presents results from a deep observation well. This is a continued report from the previous one on EISOLS 2010.

1 Introduction

Gas fields distributed in the Southern Kanto region in Japan – collectively known as the Minami kanto natural gas field – are ranked among the world's largest for natural gas dissolved in water. In particular, the Kujukuri region of Chiba Prefecture not only has abundant reserves of natural gas dissolved in water but also a high concentration of iodine found in brine (in which natural gas is dissolved) that represent precious domestically-produced resources for a resource-poor country like Japan (Fig. 1). The problems posed by land subsidence in this area were manifested in the late 1960s. Although land subsidence in recent years has been reduced in many areas, some places still experience land subsidence at levels exceeding 10 mm per annum. The eight private corporations currently engaged in development in this area are making efforts to address environmental problems, including land subsidence, by organizing the Environment Committee of the Japan Natural Gas Association and Keiyo Natural Gas Association.

2 In situ formation compaction monitoring

Levelling is one thing most traditional methods. In recent years, GPS or SAR (synthetic aperture radar) surveys using artificial earth satellites measurements are also increasing gradually. However, these methods observe relative changes in the ground surface. A purpose of this study is to understand the actual in situ formation compaction in a certain underground section (Fig. 2).

The monitoring site is located the area where has an annual ground deformation rate of at least 10 mm by levelling every year. In this area, the natural gas dissolved in water is produced by pumping up brine from formations several hundred meters to 1000 m in depth. The gas reservoirs of the Minami kanto gas field consist of alternation of unconsolidated sandstone and mudstone.

The monitoring system basically consists of the settlement gauges and rods used in displacement transmission (Fig. 3). Features of this system are applied to the optical fibers as settlement gauges and transmission lines, with an eye on deformation measuring technology through optical fiber. The use of optical fiber is advantageous in terms of high accuracy as settlement gauges, which will most likely enable measurement of underground deformations. And the optical fibers make long distance communication possible.

In the measurement principle of this system, settlement gauges are installed at several positions inside the borehole at a certain interval to monitor changes within the interval length. The settlement gauges are interconnected with rods.

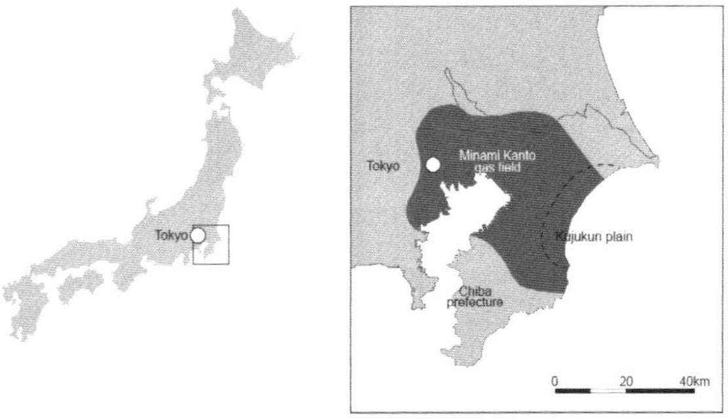

Figure 1. Minami kanto gas field.

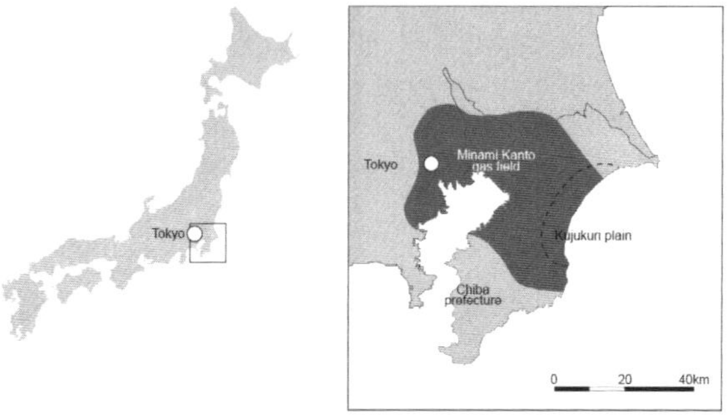

Figure 3. In situ formation compaction monitoring method.

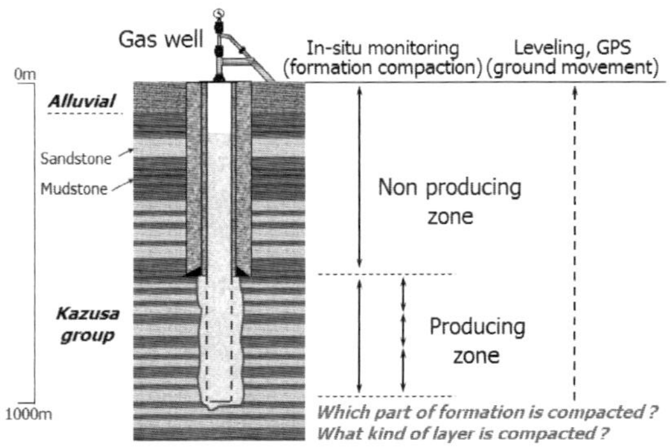

Figure 2. Concept of in situ formation compaction monitoring.

Each rod is fixed with an anchor at the top end; its lower end is directly connected with an optical fiber sensor inside the lower settlement gauge. The optical fiber sensor (made of stainless plate) is designed to bend when pushed down by the rod. The lower part of the settlement gauge is equipped with an anchor mechanism that fixes the gauge directly to borehole. When the settlement gauge drops to the predetermined depth, the metal pad will expand in three directions as a result of the water pressure and be pressed against the side wall to be anchored. The pad is mechanically designed to prevent it from shrinking after being expanded.

3 Observation system and results

In order to verify the reliability of the observation system, and sorted out problems on the construction of measurement system applying to the strata of the actual test well depth of 80 m equipped prototype tool, conducted demonstration tests. It was confirmed that in a pilot test well off the 10 m water well and pumping artificial contraction stratum, layer compaction can be observed properly. Then 800 m deep borehole (the observation well) was drilled at monitoring site and, later, the settlement gauge were installed in the borehole on June 2012. Continuous monitoring was launched in September 2012. Monitoring in five natural gas production zone has been conducting. Settlement gauges installed in the hole observed during the nearly two and a half years that working properly except for the part. Displacement data are measured in each settlement is in optical fiber signals to Earth, 30 min in every recorded in the recording equipment. The result shows that trends of formation compaction are not same among observation intervals. Also rate of formation compaction is not constant. The observation will be continued to collect data in the future. On the other hand, it has turned out that the observation equipment on the ground is affected by changes in temperature. Because it is difficult to constant temperature in the data recording equipment, relationship between temperature and change in wavelength of receiving light has been adopted to correct data.

At the time of the April 2015, total compaction of all observation intervals is 1.2 mm for about two and a half years.

4 Conclusions

The observation system based on optical fiber sensing technology has been developed and installed in the natural gas reservoir. It has monitored and recorded formation compaction data for about two and a half years from the observation starting without big trouble. This system has been running and we will continue observation. However, we consider that the examination about the correction of the observed data is required.

Acknowledgements. This research was carried out by the support for application of new technologies and technical studies program which Japan Oil, Gas and Metals National Corporation (JOGMEC) undertook.

Proc. IAHS, 372, 395–398, 2015
proc-iahs.net/372/395/2015/
doi:10.5194/piahs-372-395-2015

Numerical simulation of earth fissures due to groundwater withdrawal

Z. Wang[1], Y. Zhang[1], J. Wu[1], J. Yu[2], and X. Gong[2]

[1]School of Earth Sciences and Engineering, Nanjing University, Nanjing, 210093, China
[2]Key Laboratory of Earth Fissures Geological Disaster, Ministry of Land and Resources of China, Geological Survey of Jiangsu Province, Nanjing, 210049, China

Correspondence to: Y. Zhang (zhangyun@nju.edu.cn)

Published: 12 November 2015

Abstract. Excessive groundwater withdrawal can cause land subsidence and earth fissures. The initiation and propagation of earth fissures are related to tensile failure and crack propagation in soils. Based on fracture mechanics, the crack band model (CBM), one of the smear crack models which is relatively easy to construct and convenient to be integrated into standard finite element codes is used in this paper. The calculated results of CBM are less dependent on the sizes of finite elements. The model was applied to simulate the formation and propagation of earth fissures in the hydrostratigraphic units with a bedrock ridge. The simulated positions and patterns of earth fissures coincide with field observations, suggesting that the modeling approach is adept to simulate the initiation and propagation of earth fissures due to groundwater withdrawal.

1 Introduction

Discontinuous problems may arise when performing numerical simulation analysis on earth fissures. It is rather inefficient to treat a discontinuity as an internal boundary when applying a standard finite element method (FEM), especially when formation and propagation of fissures are concerned. This approach usually requires intense work in order to deal with the constantly changing shape and topology of the discontinuity. New methods have been presented aiming to solve such discontinuous problems, such as the extended finite element method (XFEM) (Khoei et al., 2014), the element-free Galerkin method (EFGM) (Belytschko et al., 1995), the discrete element method (DEM) (Camones et al., 2013), the displacement discontinuous method (DDM) (Detournay and Gordeliy, 2011). Most of them behave reasonably well in some particular circumstances but they show limitations in other cases because of their different design nature. On the other hand, it would be quite convenient if a similar approach would be compatible with existing standard finite element methods. The smeared cracking model (SCM) was presented by Rashid (1968), aiming to adapt the standard finite element analysis where instances of discontinuous problems are present. It is relatively easy to build from an existing standard FEM and has been proved successful in describing mode-I fracture behavior. In addition, the method is widely-used in many fracture mechanics field such as concrete fracturing (Li et al., 2007). Early SCM implementations used stress as a failure criterion, which leads to a serious mesh-dependency (Cedolin and Bazant, 1980). One substantial modification has been made since the energy-based criterion from fracture mechanism was considered. Softening process now has been incorporated instead of a pure brittle failure. One typical case is the Crack Band Model (CBM) proposed by Bazant and Oh (1983), which will be applied to simulate earth fissures in this paper.

2 Crack band model

Cracking within an element is considered as a three-stage process in the crack band model: intact stage, micro-fissure stage and cracked stage (Fig. 1). The critical strains between these three stages are denoted as ε_0 and ε_f, which are determined by Eqs. (1) and (2).

Published by Copernicus Publications on behalf of the International Association of Hydrological Sciences.

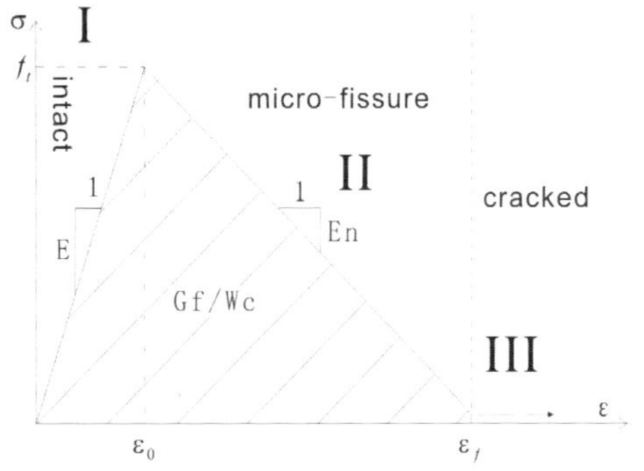

Figure 1. Constitutive relation in CBM.

$$\varepsilon_0 = \frac{f_t}{E} \tag{1}$$

$$\varepsilon_f = \frac{2G_f}{f_t w_c} \tag{2}$$

where E is the tensile modulus, kPa; f_t is the tensile strength, kPa; G_f is the release energy rate, J m^{-2}; and w_c is the element size, m.

When an element is deemed as a cracked or micro-fissure element, this element is treated as mechanical anisotropy. A reduced modulus perpendicular to crack face is then used in order to decrease the stress with an increasing strain. Based on addition of strain which means that the total strain is the sum of the crack strain and the material strain, the constitutive relation is modified from

$$\Delta\boldsymbol{\sigma} = \left[\mathbf{D}^{\mathrm{co}}\right]\Delta\boldsymbol{\varepsilon} \tag{3}$$

to (Rots, 1991)

$$\Delta\boldsymbol{\sigma} = \left[\mathbf{D}^{\mathrm{co}} - \mathbf{D}^{\mathrm{co}}\mathbf{N}\left(\mathbf{D}^{\mathrm{cr}} + \mathbf{N}^T\mathbf{D}^{\mathrm{co}}\mathbf{N}\right)^{-1}\mathbf{N}^T\mathbf{D}^{\mathrm{co}}\right]\Delta\boldsymbol{\varepsilon} \tag{4}$$

where $\Delta\boldsymbol{\sigma}$ is the stress increment; $\Delta\boldsymbol{\varepsilon}$ is the strain increment; $\mathbf{N}$ is the coordinate-transformation matrix from global coordinate system to local coordinate system in front of a crack tip; $\mathbf{D}^{\mathrm{co}}$ is the modulus matrix of intact material under the global coordinate system; $\mathbf{D}^{\mathrm{cr}}$ is the modulus matrix under the local coordinate system with its x coordinate coincident with the crack surface, which contains an element modulus normal to the crack surface and a modulus tangent to the crack surface, and it is given as

$$\mathbf{D}^{\mathrm{cr}} = \begin{bmatrix} E_n & 0 \\ 0 & E_s \end{bmatrix} \tag{5}$$

where E_n is the modulus normal to the crack face, and E_s is the modulus tangent to the crack face.

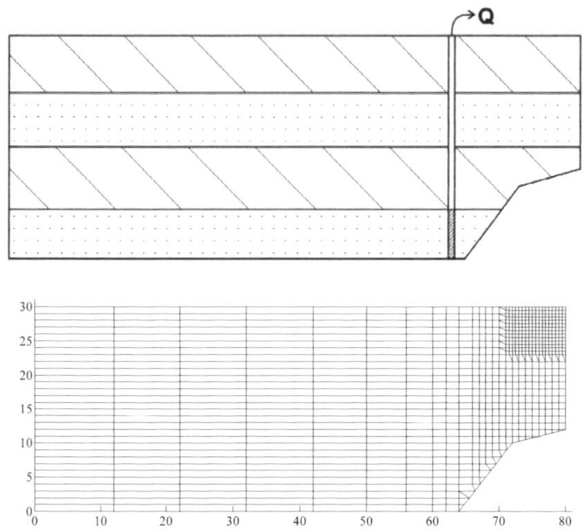

Figure 2. Conceptual model (top panel) and the finite element mesh (bottom panel) of the earth fissure example.

3 Simulation on pumping-reduced earth fissure formation

The non-tectonic formation and propagation of earth fissures are often considered to be closely related to groundwater extraction activities. This is clearly the case in Suzhou–Wuxi–Changzhou (SXC) area in Jiangsu Province, China (Zhang et al., 2008; Wu et al., 2003). A simulation is performed based on the crack band model, aiming to find the mechanism of earth fissure formation and propagation under the aquifer system with a bedrock ridge.

According to the local geological conditions and the current developmental state of earth fissures in the SXC area (Zhang et al., 2008), we can then propose a conceptual model (Fig. 2) assuming a plane strain problem. The depth of the whole Quaternary strata is 30 m. Four layers are considered of alternating clay soil and sandy soil beds. More specifically, the first layer (on the top surface) and the third one are clay soil; and the other two beds correspond to the sandy soil. The irregular shape on the lower right corner of the conceptual model represents the bedrock ridge, which is cutting off the third and the fourth layers at depth (Fig. 2, top panel). The left boundary is set as a boundary of fixed water level with no excess pore pressure and no horizontal displacement. Finally, other boundaries are assumed impervious. Pumping activities are assumed to lie on both sides of the bedrock ridge at the fourth layer with a constant flow rate (100 m^3 day^{-1}). The phreatic level is 5 m below the ground surface and remains unchanged during the whole simulation because it is unexploited. The upper part of strata where earth fissures usually occur lies above the phreatic level. It can be considered as a linear elastic material when subjected to a tensile load because the soil in the upper part is often over-consolidated

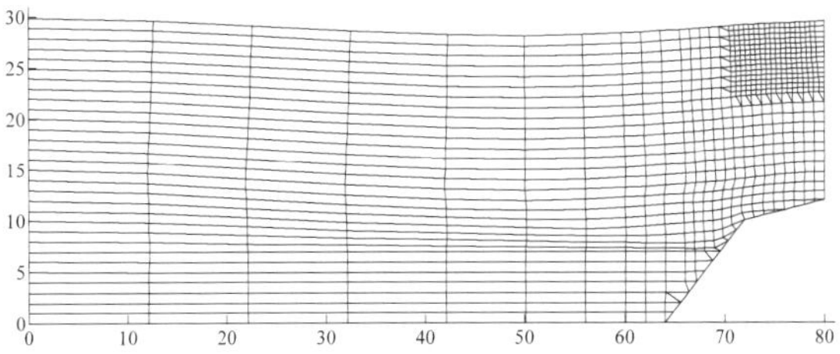

Figure 3. Deformation after consolidation.

with low confining pressure and unsaturation. Therefore, the constitutive relation is chosen to be the Duncan-Chang model under compressive conditions and linear-elasticity under tensile conditions.

Considering its symmetry, the simulation is conducted on the left part, as shown in the conceptual model in Fig. 2 (top panel). Calculation is carried out following the Biot's consolidation equations. Since fluid-solid coupling is concerned, it is notable that the pore pressure should be calculated in an incremental form instead of total form as given in Eq. (6)

$$\begin{bmatrix} \overline{\mathbf{K}} & \mathbf{K}' \\ \mathbf{K}'^T & \widetilde{\mathbf{K}} \end{bmatrix} \begin{Bmatrix} \Delta\boldsymbol{\delta} \\ \Delta\boldsymbol{\beta} \end{Bmatrix} = \begin{Bmatrix} \boldsymbol{R} - \boldsymbol{R}_t - \mathbf{K}'\boldsymbol{\beta}_\mathrm{t} \\ \Delta t\,\boldsymbol{Q} - \widetilde{\mathbf{K}}\boldsymbol{\beta}_\mathrm{t} \end{Bmatrix} \qquad (6)$$

where $\overline{\mathbf{K}}$ is the element stiffness matrix; $\widetilde{\mathbf{K}}$ is the seepage matrix; $\mathbf{K}'$ is the coupled matrix; $\Delta\boldsymbol{\delta}$ is the incremental displacement; $\Delta\beta$ is the incremental pore pressure; $\boldsymbol{R}$ is the load; $\boldsymbol{R}_t$ is the load already balanced by stress; Δt is the time interval; $\boldsymbol{Q}$ is the flow at nodes; $\boldsymbol{\beta}_\mathrm{t}$ is the total pore pressure in the last time interval which equals zero in the first time step.

The strata deformation after pumping is indicated in Fig. 3. When the pumping started, pore pressure at depth decreased, forming a zone with negative excess pore pressure, which was centered at the pumping point. Effective stress increased accordingly, which has caused a land subsidence bowl on top of the pumping well. It should be noted that the main deformation occurs at the deep clay soil.

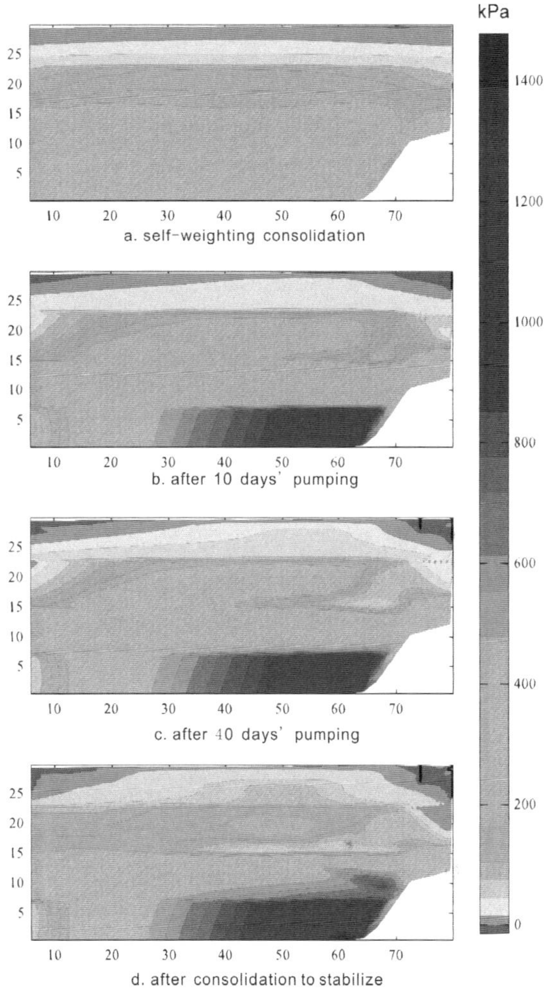

Figure 4. Distribution of minor principal stress during pumping.

Figure 4 shows the minor principal stress distribution as well as earth fissure formation and propagation at different time during pumping. The red zone represents the tensile zone and the black lines within represent the earth fissure simulated by the crack band model. Firstly, a self-weighting consolidation simulation is performed to determine the initial stress distribution instead of simply assuming a k_0 consolidation state, as in Fig. 4a. The self-weight in all elements is not applied simultaneously but activated layer by layer from bottom to top, aiming to represent the deposition process. It can be observed that the stress distribution yielded due to the self-weighting consolidation is slightly different from a simple k_0 consolidation state. The bedrock ridge induces a decrease in the upper layers in the minor principal stress. When pumping activity is initiated, the minor principal stress in the soil at depth increases significantly. Moreover, maximum change in stresses occurs near the pumping well. Meanwhile, the minor principal stress above the bedrock ridge begin to decrease, the tensile zone emerges and extends constantly. After 10 days of pumping, an earth fissure has occurred at the very top of the bedrock ridge, which was propagating downward (Fig. 4b). After a 40-day pumping period, a second crack parallel to the main crack was formed (Fig. 4c). It is noted that after each earth fissure formation or propagation, the tensile zone shrinks towards the active crack horizontally but grows deeper at the front of the active crack tip because of the release of unbalanced stresses. This becomes clearer when the crack reaches a larger length. So activities of each crack are tending to impede nearby its crack and promote it (i.e. the crack). After the pumping period is ended, the tensile zone became even narrower but deeper than that at the beginning of the pumping period. It can be observed that, in the second layer, a tensile zone is also formed (Fig. 4d). Minor principal stress in this area is even smaller than the adjacent area above and is formed because of the discrepancy of displacement between the two different layers with contrasting mechanical characteristics. The simulation results are in good agreement with the field observations, indicating that it is suitable to use the crack band model to simulate earth fissure formation and propagation.

4 Conclusions

This paper discusses the simulation of earth fissuring using the crack band model. The simulation results indicate that the crack band model is useful to model earth fissures generation. An example of earth fissure formation and propagation is simulated based on the crack band model under the usual geological conditions with bedrock ridges in the Su–Xi–Chang area of Jiangsu Province, China.

Acknowledgements. This paper was financially supported by the Ministry of Land and Resources of China (grant no. 201411096-03), the National Nature Science Foundation of China (grant no. 41572250), and UNESCO and IUGS (no. 641).

References

Bazant, Z. P. and Oh, B. H.: Crack band theory for fracture of concrete, Mater. Struct., 16, 155–177, 1983.

Belytschko, T., Lu, Y. Y., and Gu, L.: Crack propagation by element-free Galerkin methods, Eng. Fract. Mech., 51, 295–315, 1995.

Camones, L. A. M., Vargas Jr., E. A., Figueiredo, R. P., and Velloso, R. Q.: Application of the discrete element method for modeling of rock crack propagation and coalescence in the step-path failure mechanism, Eng. Geol., 153, 80–94, 2013.

Cedolin, L. and Bazant, Z. P.: Effect of finite element choice in blunt crack band analysis, Comput. Meth. Appl. Mech. Eng., 24, 305–316, 1980.

Detourna, E. and Gordeliy, E.: Displacement discontinuity method for modeling axisymmetric cracks in an elastic half-space, Int. J. Solids Struct., 48, 2614–2629, 2011.

Khoei, A. R., Vahab, M., Haghighat, E., and Moallemi, S.: A mesh-independent finite element formulation for modeling crack growth in saturated porous media based on an enriched-FEM technique, Int. J. Fract., 188, 79–108, 2014.

Li, Q., Zhang, B., Yu, Y., and Wang, J.: Numerical simulation of the process of hydraulic fracturing in earth and rockfill dams, Chinese J. Geotech. Eng., 29, 212–217, 2007.

Rashid, Y. R.: Ultimate strength analysis of pre-stressed concrete pressure vessels, Nucl. Eng. Design, 7, 334–344, 1968.

Rots, J. G.: Smeared and discrete representations of localized fracture, Int. J. Fract., 51, 45–59, 1991.

Wu, Z., Yu, Q., and Zhang, Y.: Forming process of earth fissure hazard in the Suzhou-Wuxi-Changzhou area, Hydrogeol. Eng. Geol., 30, 67–71, 2003.

Zhang, Y., Xue, Y., Wu, J., Yu, J., Wei, Z., and Li, Q.: Land subsidence and earth fissures due to groundwater withdrawal in the Southern Yangtse Delta, China, Environ. Geol., 55, 751–762, 2008.

Proc. IAHS, 372, 399–401, 2015
proc-iahs.net/372/399/2015/
doi:10.5194/piahs-372-399-2015

Open Access
PIAHS

Research on the deformation of a confined aquifer based on Cosserat continuum mechanics

Y. S. Xu, N. Zhang, Y. Yuan, and S. L. Shen

State Key Laboratory of Ocean Engineering, Department of Civil Engineering, Shanghai Jiao Tong University, Shanghai, China

Correspondence to: S. L. Shen (slshen@sjtu.edu.cn)

Published: 12 November 2015

Abstract. Recent monitoring of land subsidence and soil deformation indicates a new phenomenon where excessive and continuous deformation occurs in the sandy aquifers in Shanghai and the Su-Xi-Chang region of China. It is hard to explain factors contributing to this phenomenon with traditional Cauchy continuum mechanics in which low normal stress in the ground could not cause such large deformation. Steep hydraulic gradient would be formed in the aquifer if groundwater is pumped from densely distributed wells, and shear stresses would develop then. Accumulated shear stress could then lead to deformation of the aquifer or even land subsidence. Accumulated shear stress due to the drawdown of groundwater level is one of the main factors that contribute to deformation within an aquifer. Traditional Cauchy continuum mechanics cannot consider this shear stress because of the hypothesis of equal shear stress in the aquifer unit. Cosserat continuum mechanics can be applied to analyse the mechanism of aquifer deformation controlled by accumulated shear stress by considering the scale effect and the asymmetric distribution of shear stress in the aquifer unit.

1 Introduction

Land subsidence due to groundwater exploitation is of global concern. Recent research presents some new characteristics of aquifer deformation including excessive deformation which occurs in aquifer systems, and delayed deformation compared with groundwater decline in coastal areas in China (e.g. in Shanghai and the Su-Xi-Chang region). These new characteristics are significant for investigating the continuing deformation of sandy aquifers (Gong et al., 2009; Wang et al., 2010; Zhang et al., 2006). Some researchers have discussed factors contributing to such deformation in terms of aquifer properties (Wang et al., 2010; Zhang et al., 2006), however the yield mechanism of the sandy aquifer under lower vertical stress has not previously been addressed.

The hydraulic gradient due to groundwater withdrawal can contribute to shear stress development in a sandy aquifer. Some research demonstrates that the sand gets to a critical state under the high shear stress level (Cai, 2004). The accumulated shear stress due to the hydraulic gradient can result in the deformation of the aquifer or even land subsidence. Therefore, the accumulated shear stress in the aquifer may be one of the main factors contributing to the deformation of the aquifer sand. Traditional methods based on Cauchy mechanics does not consider the shear and rotational deformation caused by accumulated shear stress (Budhu and Adiyaman, 2010). It is therefore important to investigate a new prediction model which considers the asymmetric distribution of shear stress in aquifers.

2 Budhu's approach based on Cosserat continuum mechanics

Cosserat continuum mechanics was developed to analyse the deformation of a continuous medium. It simultaneously considers the isotropic effective stress and the torsion on the unit cell in the equilibrium equation. Budhu and Adiyaman (2010) proposed an approach to predict land subsidence in an unconfined aquifer based on Cosserat mechanics. Figure 1 shows the stress distribution in an unconfined aquifer with a variation in groundwater level. Based on Budhu's approach, the changes in effective stress due to the decreasing groundwater level are broken down into two parts. Part I is an

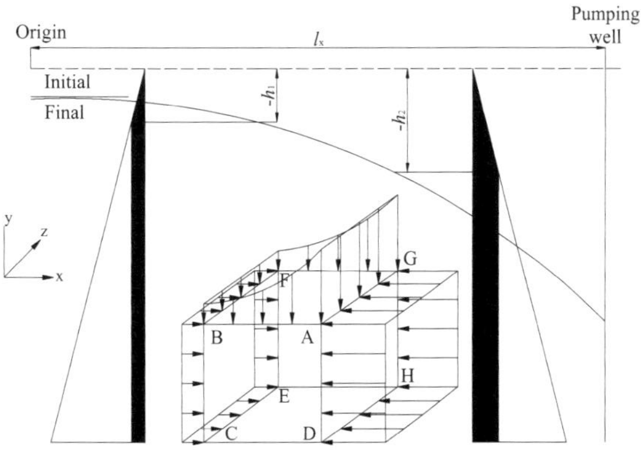

Figure 1. Stress distribution with a change in groundwater level (Budhu and Adiyaman, 2010).

isotropic effective stress, which contributes to isotropic compression. Part II is a vertical effective stress varying with lateral distance, which causes simple shear and rotation. The isotropic deformation due to compression stress is analysed using Cauchy mechanics, and the deformation due to the shear stress and torsion is simulated using Cosserat continuum mechanics. The hypotheses of Budhu's approach are as follows: (1) the aquifer is an isotropic elastic continuous medium without any cracks; (2) there is no replenishment from other aquifers to the unconfined aquifer. The stress in the unconfined aquifer remains unchanged when the groundwater level changes; (3) the groundwater level in the initial state is the same in the horizontal direction; (4) the effective stress in the aquifer changes instantaneously when the groundwater level changes.

3 Shen's approach based on Cosserat continuum mechanics

3.1 Hypotheses for Shen's approach

Budhu's approach is not appropriate for confined aquifers because it was proposed based on the unconfined aquifers in Arizona. In order to apply Budhu's approach in confined aquifers, the following additional hypotheses for confined aquifers (Shen et al., 2013) are presented: (1) the confined aquifer is an isotropic, uniform continuous medium without cracks; (2) the overburdened load from an aquitard above an aquifer remains unchanged during groundwater pumping from the aquifer; (3) variation in groundwater replenishment from the overburdened aquitard depends on the consolation rate of the overburdened aquitard; (4) the groundwater level in the initial state remains the same in the horizontal direction; (5) the groundwater level in the aquifer declines slowly with pumping time.

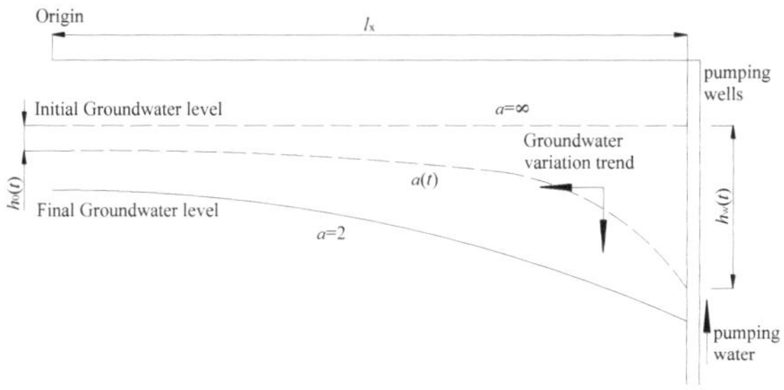

Figure 2. Variation tendency of groundwater level in aquifer.

3.2 Shen et al. (2013)'s equation

To simulate the rate of the replenishment from upper overburdened aquitards, μ' is introduced, and is included in Shen et al. (2013)'s equation as follows:

$$\overline{\sigma}'_{ij} = \mu' \left(\frac{1}{V}\right) \int\limits_V \sigma'_{ij} \mathrm{d}V, \qquad (1)$$

where σ'_{ij} is the effective stress tensor, $\overline{\sigma}'_{ij}$ is the average effective stress tensor, and V is the volume of the soil unit. At the start of pumping, when there is a large hydraulic gradient between the aquifer and the aquitard, groundwater replenishment is high and the effective stress in the aquifer increases slowly so that μ' equals 0. After a period of time, once groundwater motion becomes a stable seepage and the soil completes consolidation in the aquitard, the effective stress in the aquifer reaches its maximum value; in this case, μ' equals 1.

Figure 2 shows the variation tendency of the groundwater level in the aquifer. The groundwater in the initial state is assumed to be in the horizontal direction. When the groundwater is pumped from the aquifer, the groundwater level declines gradually with time. To consider the impact of time on groundwater level, the groundwater level equation is as follows:

$$h_{\mathrm{w}}(t) = -\left(\alpha_i x^a + h_0(t)\right), \qquad (2)$$

where $h_{\mathrm{w}}(t)$ is the drawdown of the groundwater head with time from the initial state at the pumping well, $h_0(t)$ is the drawdown of the groundwater head with time from the initial state at an arbitrary selected origin, α_i is a soil layer constant, x is the distance from the origin, and a is the time factor, which can be calculated using the following equation:

$$a = \lambda' - \ln t \qquad (3)$$

where t is time, and λ' is a constant time factor.

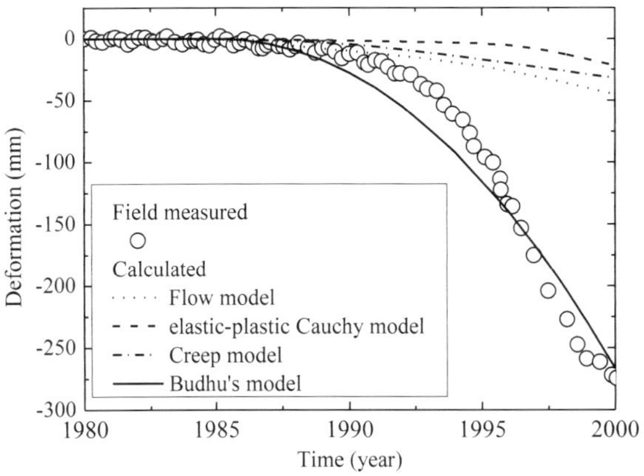

Figure 3. Comparison of the deformation of Aquifer IV at Laodong Park.

4 Discussion

Shen et al. (2013) used the modified approach to simulate the deformation of Aquifer IV in Shanghai. Figure 3 depicts the measured and calculated curves using four different models: (1) groundwater flow model, (2) elasto-plastic model based on Cauchy mechanics, (3) model considering the creep behaviour of sandy soil, and (4) Shen et al. (2013)'s approach based on Cosserat mechanics. The results predicted by the classic models (groundwater flow model, elasto-plastic model, and creep model) do not fit the measured data well. However, Shen et al. (2013)'s approach based on Cosserat mechanics can reasonably predict the variation in deformation of Aquifer IV in Shanghai.

5 Conclusions

1. Budhu's approach based on Cosserat mechanics can be used to predict the deformation of an unconfined aquifer. As the consolidation effect of the overburdened aquitards is not considered, Budhu's approach cannot directly predict the deformation of a confined aquifer.

2. Shen et al. (2013) modified Budhu's approach to simulate the deformation of a confined aquifer by considering the effect of an overburdened aquitard and the impact of time on groundwater level.

Acknowledgements. The research work described herein was funded by the National Nature Science Foundation of China (Grant No. 41472252) and partially funded by the National Basic Research Program of China (973 Program: 2015CB057806). This financial support is gratefully acknowledged.

References

Budhu, M. and Adiyaman, I. B.: Mechanics of land subsidence due to groundwater pumping, Int. J. Numer. Anal. Meth., 34, 1459–1478, 2010.

Cai, Z. Y.: The deformation behavior of sand, Yellow River Conservancy Press, China, 172 pp., 2004.

Gong, S. L., Li., P. C., Wu, J. H., and Li, X. G.: Analysis of groundwater seepage field and stratum stress field of land subsidence in Shanghai, Journal of Liaoning Technical University: Natural Science, 28, 232–235, 2009 (in Chinese).

Shen, S. L., Ma, L., Xu, Y. S., and Yin, Z. Y.: Interpretation of increased deformation rate in aquifer IV due to groundwater pumping in Shanghai, Can. Geotech. J., 50, 1129–1142, 2013.

Wang, F., Miao, L. C., and Lv, W. H.: Study on the new deformation characteristics of the pumped aquifers in Su-Xi-Chang area, China, in: Geoenvironmental Engineering and Geotechnics (GSP 204), edited by: He, Q. and Shen, S. L., ASCE Press, Reston, Virginia, 57–62, 2010.

Zhang, Y., Xue, Y. Q., Wu, J. C., and Li, Q. F.: Characteristics and parameters of sand strata deformation due to groundwater pumping in Shanghai, Journal of Hydraulic Engineering, 37, 560–566, 2006 (in Chinese).

Proc. IAHS, 372, 403–409, 2015
proc-iahs.net/372/403/2015/
doi:10.5194/piahs-372-403-2015

Open Access
PIAHS

The spatiotempora variations rules of Songzao coal mining subsidence based on numerical simulation

J. Lu[1,2], Y. Li[1,2], H. Cheng[3], and Z. Tang[1,2]

[1]Key Laboratory of Exogenic Mineralization and Mine Environment, Institute of Geology and Mineral Resources, Chongqing 400042, China
[2]Research Center of State Key Laboratory of Coal Resources and Safe Mining, Chongqing 400042, China
[3]Energy College, Chongqing 400041, China

Correspondence to: J. Lu (lujiahao09@mails.ucas.ac.cn)

Published: 12 November 2015

Abstract. With the increasing demand of coal, coal mining at Songzao makes the area of land subsidence growing larger. Land subsidence in coal mining area not only made large subsided farmland out of production and caused the enormous loss to local agricultural production, but also brought a number of serious problems to the local social economy and ecology Environment. To use Probability-integral Method based on numerical simulation of Songzao Mine, its subsidence simulation data from 1999 to 2009 was obtained. Hence, overlay analysis between Goaf data and the simulation data in 2009, and between field investigation and the simulation data in 2009 were carried out. After the coal mining underground was identified as the crucial cause of surface subsidence. Therefore, the accuracy and feasibility of the simulation data had been verified, and the spatial pattern and spatiotemporal variations conforming to the actual values have been obtained. The results show five main findings. The first indicated that the surface subsidence is mostly located at the top of the Goaf, where the overlap areas between Goaf data and subsidence simulation data have accounted for 93.05 % of Goaf and 65.19 % of subsidence simulation data respectively. The second finding indicated that by end of 2009, the mining subsidence extent had reached about 5087.50 hm^2. This area accounts for about 40 % of total of the mining area. The third finding indicated that within 10 years from 1999 to 2009, the influence range of subsidence has expanded about 2340.54 hm^2, and the coal mining subsidence rate in Songzao Mine has increased gradually with time. Moreover, average increasing speed of the extent area in the second five years was larger than the first five years (about 75.08 $hm^2\,yr^{-1}$ more). The fourth finding indicated that maximum subsidence has increased from 2.0 m in 1999 to 2.5 m in 2004, and then 3.0m in 2009 with subsidence rate of about 0.1 $m\,yr^{-1}$. At the same time, the area affected by the subsidence 2.0 m in 1999 was more than twice the area affected by subsidence in 2004. This in return, it was more than 7 times larger than the area affected by subsidence in 2009 of the one affected by subsidence in 2004. Extent of the area affected by the 2.5 m subsidence has also enlarged rapidly. This area has expanded by about 40 times in 2009 than its value in 2004. In addition, the area of subsidence of value 3.0 m has reached about 0.44 hm^2 in 2009 from zero value. Finally, the fifth finding indicated that the overall extend of the mining subsidence was much more serious in southern than in northern side of the Songzao Mine. Moreover, it was indicated that the increasing rate of mining subsidence in the western side of the study area was as bigger as in the eastern side between 1999 and 2009. The spatiotemporal variations rules of songzao coal mining subsidence based on numerical simulation could provide reference for the subsequent subsidence prevention and land consolidation.

1 Introduction

Coal is currently the most important source of energy in China. It accounts for more than 70 % of the total energy production and consumption. Development of the coal industry not only promoted the development of other industries, but also caused a series of environmental problems, especially due to land subsidence caused by coal mining, that is causing damage to the regional ecological environment and ecological structure (e.g. He et al., 1994; Luo and Chen, 1994; Wu et al., 1997; Holzer, 1985; Guerrero et al., 1999; Sheoey et al., 2000; Hou and Zhang, 2004; Xia et al., 2008). According to incomplete statistics, when coal mining was about 1.0×10^4 t, the average land subsidence area was about 0.3 hm^2. The land subsidence area caused by national coal mining was more than 7.5×10^5 hm^2 (e.g. Chen et al., 2002). Moreover, this area is increasing by an annual rate of 2.7×10^4 hm^2 (Lin and Wang, 2005). This, in fact, may lead to imbalance between the current coal exploitation and ecological environment protection. Although mining subsidence governance work is being highly valued, the effect of subsidence governance increases is not obvious because the lack of overall planning and investment management. In this paper, subsidence due to coal mining is investigated using numerical simulation for Songzao Mine. In addition, overlay analysis between simulation data and Goaf data was performed to prove that the coal mining was the main cause for subsidence. This is done after accuracy and feasibility checks of the simulation results were verified through the site under investigation. Then, the spatial patterns and temporal variation of Coal Mining Subsidence was obtained, which could provide reference for the subsequent subsidence.

2 General situation of Songzao Mine

Songzao Mine (Fig. 1) is located in Qijiang District of Chongqing City, China. According to the plan in 2005, its mine planning production capacity is 10 800 kt yr^{-1}, with basic reserves 1 685 768 kt, and recoverable reserves of 931 282 kt. With continuous exploitation, mining subsidence continues. In addition, the Goaf becomes larger during the long process of mining exploitation. This, in fact, is causing serious damages and local scale problems (e.g., crack or subsidence in surface, water dewatering, drying up farmland, infrastructure damage, secondary geological disasters, soil erosion). This is also leading to ecological environmental deterioration as well as seriously negative impacts affecting the mining economic development and social stability. For example, the collapse basin had a huge impact on land especially arable land in its formation process. The chief influence is cracks or caves in houses and roads, and slopes, cracks, puddles, uneven and so on in arable land. Then under the combined effects of rain, wind and other factors, the phenomenon of soil erosion such as soil damage, move, deposition and so on show up day by day, accelerating surface erosion, resulting in cultivated soil degradation, and finally declining in the quality of cultivated soil (e.g. Chen et al., 1999). A survey shows that Mining Goaf in Songzao mine had damaged farmland about 3466.74 hm^2, farming production in dry land yields decreased 20–30 %, and farming production in paddy fields turned into dry land reduced by more than 50 %. Parts of the badly damaged fields forced to be fallow. This seriously damaged the land resources. At the end of 2008, about 1500 hm^2 of paddy fields converted to dry land in Songzao mining subsidence, and the indirectly affecting area was actually much larger than the data (Yang and Li, 2011). It requires that the spatiotemporal variations rules of coal mining subsidence should be found and then an effective long-term mechanism to the protection of farmland and damaged farmland reclamation should be established.

3 Numerical simulation of mining subsidence

At present, the widely used numerical method of China's mining subsidence areas is Probability-integral Method, which has advantages as parameters easy to determine, serviceability and so on (e.g. Ling et al., 1999; Gu and Hong, 2012). For example, Gu and Hong (2012) applied it based on MATLAB to do research about Ningtiaota Mine in the mountains, and achieved a multi-parameter nonlinear curve fitting expression which accurately describes the collapse rules in the surface after shallow coal seam mining. Gu and Gao (2011) used it to carry out the simulation and prediction of Xiangyang Coal mining subsidence in the hilly area, analyzed the dynamic collapse process for mining and provides reference in decision-making for post-mining subsidence prevention, remediation and recovery. Tan et al. (2006) also utilized it to research the uneven subsidence in the surface, tilt and curvature change by taking the geological conditions in a coal mine in Chongqing as the background condition, and then better achieved the ground subsidence influence evaluation on the building and construction. This paper also used it based on the soft MSPS independently developed by Mining Damage and Protection Research Institute of China University of Mining and Technology to achieve numerical simulation of Subsidence in Songzao, and then the result was used to analyze the spatiotemporal variations of Songzao Coal Mining Subsidence.

3.1 Probability-integral method

Probability-integral Method is a numerical method based on Stochastic Medium Theory. It has been widely used in the research field of mining subsidence in China, and its model (e.g. He et al., 1994) is as follows,

$$W_{ex} = \frac{1}{r} e^{-\pi \frac{x^2}{y^2}}, \tag{1}$$

where r is the radius of the main impact area.

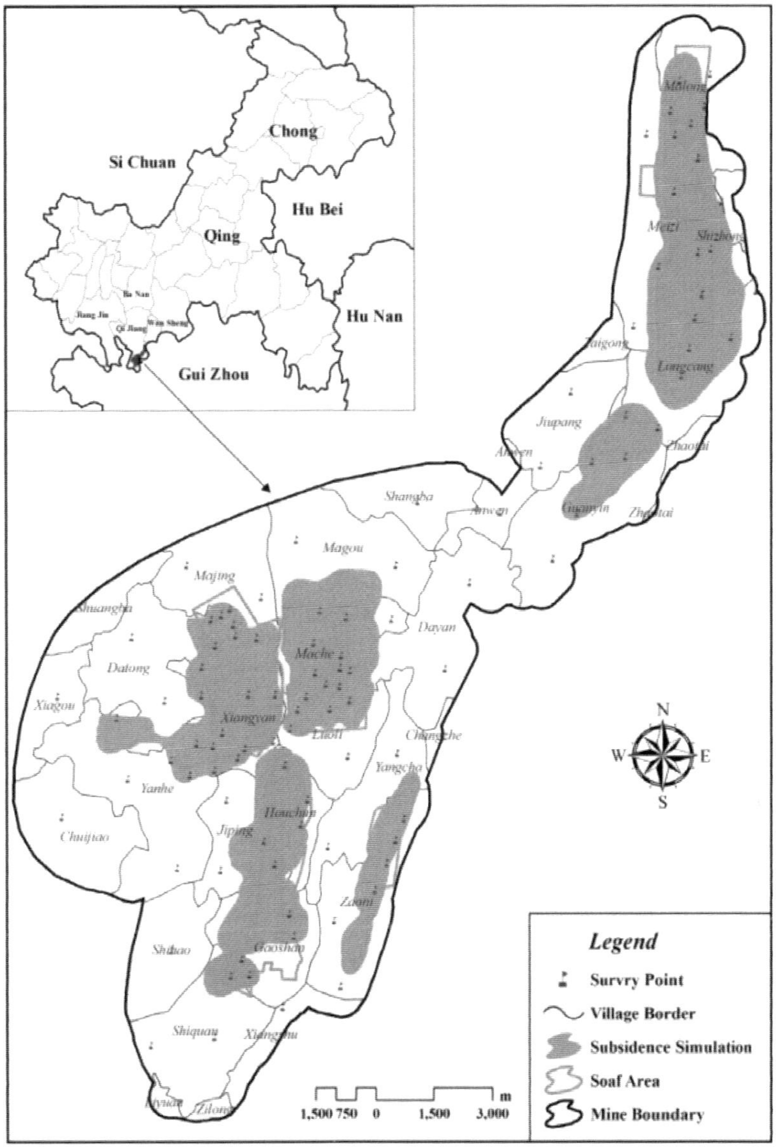

Figure 1. Location and overlay analysis map.

Table 1. The statistics of overlay analysis.

Goaf (hm^2)	Mining subsidence numerical simulation (hm^2)	Overlap area (hm^2)	Overlap area/mining subsidence numerical simulation (%)	Overlap area/mining subsidence numerical simulation (%)
3554.93	5073.80	3307.83	65.19	93.05

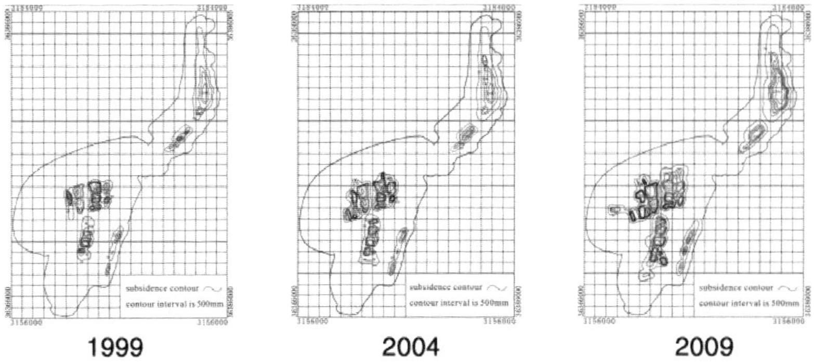

Figure 2. The subsidence contour of Songzao coal mining area.

1. $W(x, y)$ is the subsidence value of any surface point (x, y), and its computational formula shows as follows:

$$W(x,y) = \frac{1}{W_0} W^0(x) W^0(y), \tag{2}$$

$$W^0(x) = \frac{W_0}{2}\left[\mathrm{erf}\left[\frac{\sqrt{\pi}}{r}x\right] - \mathrm{erf}\left[\frac{\sqrt{\pi}}{r}(x-L)\right]\right], \tag{3}$$

$$W^0(y) = \frac{W_0}{2}\left[\mathrm{erf}\left[\frac{\sqrt{\pi}}{r_1}y\right] - \mathrm{erf}\left[\frac{\sqrt{\pi}}{r_2}(y-L)\right]\right], \tag{4}$$

$$W_0 = mq\cos\alpha, \tag{5}$$

where $W^0(x)$ is the subsidence value of the point whose abscissa is x on the movement/trend main sectional plane when critical extraction took place in the incline direction. $W^0(y)$ is the subsidence value of the point whose ordinate is y on the incline main sectional plane when critical extraction took place in the trend direction. m is the mining thickness. q is the subsidence coefficient. α is dip angle of coal seam. L is the length calculation of trend and tendency of limited mining under considering the displacement distance of inflection point. r, r_1, r_2 respectively is the radius of the main impact area of trend, downhill and uphill. erf is Gaussian error function.

2. The incline value, the curvature value, horizontal movement, horizontal displacement of any surface point along the tilt direction φ should be respectively calculated through formulas as follows:

$$i(x,y,\phi) = \frac{1}{W_0}\left[i^0(x)W^0(y)\sin\phi + i^0(y)W^0(x)\sin\phi\right], \tag{6}$$

$$K(x,y,\phi) = \frac{1}{W_0}\left\{\left[k^0(x)W^0(y) + k^0(y)W^0(x)\right]\sin^2\phi + i^0(x)i^0(y)\sin^2\phi\right\}, \tag{7}$$

$$U(x,y,\phi) = \frac{1}{W_0}\left[U^0(x)W^0(y)\cos\phi + U^0(y)W^0(x)\sin\phi\right], \tag{8}$$

$$\varepsilon(x,y,\phi) = \frac{1}{W_0}\left\{\varepsilon^0(x)W^0(y)\cos^2\phi + \varepsilon^0(y)W^0(x)\sin^2\phi + \left[U^0(x)i^0(y) + U^0(y)i^0(x)\right]\sin\phi\cos\phi\right\}, \tag{9}$$

where $i^0(x)$, $k^0(x)$, $U^0(x)$, $\varepsilon^0(x)$ is respectively the incline value, the curvature value, horizontal movement, and horizontal displacement on the trend main sectional plane of any surface point (x, y) when the critical extraction along the φ direction. At the same time, $i^0(y)$, $k^0(y)$, $U^0(y)$, $\varepsilon^0(y)$ is respectively the incline value, the curvature value, horizontal movement, and horizontal displacement on the incline main sectional plane of the same point (x, y).

3. When subsidence was caused by more than one workplane, the total influence of those should be considered to overlay analysis in the simulation process.

4. Three known values should be used to solve coordinate conversion in the Numerical simulation process. They are corresponding data that had been converted through Probability-integral Method from the point's coordinate

(X_0, Y_0) in the bottom left corner of work-plane and the trend direction azimuth α. When one point's coordinate value (X, Y) in geodetic coordinate system (left-handed coordinate system) or other independent coordinate system (right-handed coordinate system) required be converted to the calculating (right-handed coordinate system) as (X_j, Y_j), the formula as follows should be used:

$$\begin{bmatrix} X_j \\ Y_j \end{bmatrix} = \begin{bmatrix} \cos\alpha & \sin\alpha \\ \sin\alpha & -\cos\alpha \end{bmatrix} \begin{bmatrix} X - X_0 & Y - Y_0 \end{bmatrix}. \tag{10}$$

3.2 Numerical simulation and analysis of its result

According to Regulations of buildings, water, railway and main well lane leaving coal pillar and press coal mining, the 10 mm subsidence contour was taken as boundary line in the study area. And then after numerical simulation of mining subsidence of the study area, the subsidence contour map (Fig. 2) and the corresponding statistical data (Table 2) in 1999, 2004 and 2009 was respectively obtained.

3.2.1 Overlay analysis and proof

Through summarizing and arranging the data of geology and mineral maps, engineering plane maps and surface-underground contrast plans, the Goaf data was obtained with the area about 3554.93 hm^2 in 2009. Then the overlay analysis was respectively carried out between Goaf data and subsidence simulation results in 2009, field investigation and subsidence simulation results in 2009. The results (Fig. 1 and Table 1) showed that

1. the 70 field investigation points located in the extent of coal mining subsidence numerical simulation data had showed the same subsidence phenomenon as the simulation data, and these points also displayed the location as at the top of or close to the Coal Goaf. And the other 48 points unlocated in the simulation had not obviously appeared subsidence phenomenon in 2009. It could primely prove subsidence simulation data accord with the actual and the accuracy and feasibility of the simulation results was verified.

2. The surface subsidence mostly located at the top of the Goaf area.

3. The overlap areas between Goaf data and subsidence simulation results were totally about 3307.83 hm^2, accounting respectively for 93.05 % of mining Goaf and 65.19 % of subsidence simulation results. In short, the results inferred that coal mining was the basic and crucial driven factor for surface subsidence.

Table 2. The area statistics of mining subsidence numerical simulation.

Subsidence value (mm)	1999 (m^2)	2004 (m^2)	2009 (m^2)
10	27 469 616	37 295 259	50 875 011
500	11 779 468	17 304 094	26 613 164
1000	6 085 552	9 925 673	18 444 527
1500	2 818 673	4 863 881	11 717 092
2000	517 273	1 155 490	7 818 101
2500	0	58 014	2 234 698
3000	0	0	4375

3.2.2 Analysis of simulation result

Through comparative analysis of the three periods of numerical simulation data and field investigation, the results (Fig. 2 and Table 2) were obtained as follows:

1. The extent of Coal Mining Subsidence in Songzao from 1999 to 2004 expanded about 982.56 hm^2. When the extent was expanding in every direction and overall impact to the surrounding, the trend in north and southwest was more prominent.

2. The extent of Coal Mining Subsidence in Songzao from 2004 to 2009 expanded about 1357.98 hm^2. In the five years, each subsidence extent expansion accelerated significantly, especially the area of sink depth 2.0 m was fastest in the horizontal direction, and the west was fastest in the horizontal direction. And the subsidence foreign expansion feed is significantly higher, especially from the space dimension, the fastest subsidence foreign expansion feed distributed in the subsidence contour 2.0 m. When the extent was expanding in every direction, the trend in west was most prominent in the five years.

3. The extent of Coal Mining Subsidence in Songzao from 1999 to 2009 expanded about 2340.54 hm^2. The maximum subsidence depth changes from 2.0 m in 1999 to 2.5 m in 2004, and then having increased to 3.0 m in 2009. At the same time, the extent of 2.0 m in 2004 expanded by about 2 times than that in 1999, and then it is 7 times in 2009 to 2004. The extent of 2.5 m had enlarged rapidly, and its area expanded by about 40 times in 2009 than that in 2004. The extent of 3.0 m had grown out of nothing, and its area was about 0.44 hm^2.

4 Conclusions and discussion

Songzao mine is rich in coal resources, but it is located in exposed bedrock mountainous with complex terrain. The surface subsidence caused by coal mining is greatly different from the plains, such as it has mountain surface moving characteristics and bedrock bare mountain surface particularity movement. Through comparative analysis of the three periods of numerical simulation data and overlay analysis data, the realistic simulation results were obtained and then the mining subsidence spatiotemporal variations rule was drawn as follows:

1. The subsidence simulation data was realistic and the accuracy and feasibility of the simulation results was verified. Because the 70 field investigation points located in the extent of coal mining subsidence numerical simulation data had showed the same subsidence phenomenon as the simulation data, and these points also displayed the location as at the top of or close to the Coal Soaf. And the other 48 points unlocated in the simulation had not obviouslly appeared subsidence phenomenon in 2009.

2. The surface subsidence mostly located at the top of the Goaf, because the overlap areas between Goaf data and subsidence simulation data accounted respectively for 93.05 % of Goaf and 65.19 % of subsidence simulation data. All of the field investigation points especially those located in subsidence simulation areas could well prove subsidence simulation results was realistic. In short, it can be inferred that coal mining is the basic and crucial driven factor for surface subsidence.

3. Overall, the mining subsidence extent showed as that was much more serious in southern than northern, and the increasing rate of mining subsidence unfolded as much bigger in western than eastern from 1999 to 2009.

4. The coal mining subsidence rate increased gradually with time, and the increasing rate was significantly bigger than the previous post. The impact of coal mining subsidence range expanded from 2746.96 hm^2 in 1999 to 3729.52 hm^2 in 2004, and then increased to 5087.50 hm^2 in 2009 accounting for about 40 % of total area of the mining area. At the same time, the average increasing speed of extent area was bigger in the second five years than the first five years about 75.08 $hm^2\,yr^{-1}$.

5. The maximum subsidence depth of Songzao Mine increased with time, and the extent of each subsidence depth also expanded. In addition, the expanding speed of extent area was gradually accelerated over time. The maximum subsidence depth turned from 2.0 m in 1999 to 3.0 m in 2009, with subsidence depth increasing rate about 0.1 $m\,yr^{-1}$.

6. In the underground mining process of Songzao Mine, surface is prone to concentrate deformation, and then deformation bands appeared damage. The maximum exploitation of ground movement caused by coal mining is bigger than it in plains under the same conditions, and the range of surface movement is also relatively larger. But due to Songzao Mine located in Hilly area with complex geology and varied topography, the same surface subsidence deformation is not obvious as in plains. Coupled with the some uncertainty of the nature and mechanism of underground rock, the surface subsidence of coal mining is a constantly changing tectonic processes. Therefore, more and better research of Goaf subsidence rules should be developed with long-term tracking and monitoring, and investment in research subsidence area of governance should also be increased in order to achieve sustainable use of land resources in subsidence areas. As a result, the greatest harmony and unity among economic, social and ecological benefits could be created.

Acknowledgements. This study was supported by the Non-profit Industry Financial Program of Ministry of Land and Resources of China(Grant200911015-04)and Science and Technology Planning Project of Chongqing Municipal Bureau of Land Resources and Housing Management (Grant CQGT-KJ-2014004).

References

Chen, L., Deng, K., and Zhao, Z.: Space variation rules of physical characteristics about farmland soil due to mining subsidence, J. China Coal Soc., 24, 586–590, 1999.

Chen, L., Guo, D., Xu, S., Li, J., and Zhang, Y.: A study on state and comprehensive control ways of subsided land of Yanzhou coal mining area in China, J. Nat. Resour., 17, 504–508, 2002.

Gu, J. and Gao, Z.: Application of Probability-integral Method in the Prediction of Mining Subsidence, Mine Surveying, 2, 47–49, 2011.

Gu, S. and Hong, X.: Application of probability integral surface movement prediction method in mountainous area with shallow coal seam, J. Xi'an Univ. Sci. Technol., 1, 45–50, 2012.

Guerrero, O., Rudolph, D. L., and Cherry, J. A.: Analysis of long-term land subsidence near Mexico City: field investigations and predicative modeling, Water Resour. Res., 30, 3327–3341, 1999.

He, G., Yang, L., Ling, G., Jia, F., and Hong, D.: Mining subsidence Theories, China Mining Univesity Press, Xuzhou, China, 12–156, 1994.

Holzer, T. L.: Land Subsidence: Its Impacts And Costs In The U.S., Underground Space, US Geological Survey (USGS), Reston, VA, USA, 9, 1985.

Hou, Z. and Zhang, Y.: Movement rules of coal mining subsidence surface ground in Datong Mining Area, Coal Sci. Technol., 32, 50–53, 2004.

Lin, Z. and Wang, G.: Remediation of Subsided Land and Creation of Constructed Wetland in Mining Area – A case of Remediation of Subsided Land in Xuzhou Mining Area, J. Nat. Resour., 20, 790–795, 2005.

Ling, G., Wu, K., and Zhang, Z.: A Research on Analysis Method of Mining Subsidence Parameters for Space Problem and Its Application, J. China Univ. Mining Technol., 23, 64–69, 1999.

Luo, Z. and Chen, Q.: Research on Density Stability of Air Dense Medium Fluidized Bed, J. China Univ. Mining Technol., 4, 62–70, 1994.

Sheoey, P. R., Loui, J. P., Singh, K. B., and Singh, S. K.: Ground subsidence observation and a modified influence function method for complete subsidence prediction, Int. J. Rock Mech. Min., 37, 801–811, 2000.

Tan, Z., Hu, P., Cheng, W., Wang, X., Wang, Y., and Fu, S.: Safety evaluation of the ground surface subsidence under the condition of the seam group expbitation, J. Safety Sci. Technol., 2, 101–104, 2006.

Wu, K., Hu, Z., and Chang, J.: Distribution Rules of Ground Crack Induced by Coal Mining, J. China Univ. Mining Technol., 26, 56–59, 1997.

Xia, Y., Du, R., and Sun X.: Relative analysis and regressive estimation of mining subsidence in Tongchuan Mining Area, Coal Sci. Technol., 36, 89–92, 2008.

Yang, G. and Li, Y.: Effects of Coal Mining Subsidence on Environment in Songzao Mining Area and Land Reclamation, J. Anhui Agricult., 39, 20487–20489, 2001.

Proc. IAHS, 372, 411–416, 2015
proc-iahs.net/372/411/2015/
doi:10.5194/piahs-372-411-2015

Inverse modeling using PS-InSAR for improved calibration of hydraulic parameters and prediction of future subsidence for Las Vegas Valley, USA

T. J. Burbey[1] **and M. Zhang**[2]

[1]Virginia Tech, Blacksburg, VA, USA
[2]Duke University, Durham, NC, USA

Correspondence to: T. J. Burbey (tjburbey@vt.edu) and M. Zhang (meijing.zhang@duke.edu)

Published: 12 November 2015

Abstract. Las Vegas Valley has had a long history of surface deformation due to groundwater pumping that began in the early 20th century. After nearly 80 years of pumping, PS-InSAR interferograms have revealed detailed and complex spatial patterns of subsidence in the Las Vegas Valley area that do not coincide with major pumping regions. High spatial and temporal resolution subsidence observations from InSAR and hydraulic head data were used to inversely calibrate transmissivities (T), elastic and inelastic skeletal storage coefficients (S_{ke} and S_{kv}) of the developed-zone aquifer and conductance (CR) of the basin-fill faults for the entire Las Vegas basin. The results indicate that the subsidence observations from PS-InSAR are extremely beneficial for accurately quantifying hydraulic parameters, and the model calibration results are far more accurate than when using only water-levels as observations, and just a few random subsidence observations. Future predictions of land subsidence to year 2030 were made on the basis of existing pumping patterns and rates. Simulation results suggests that subsidence will continue in northwest subsidence bowl area, which is expected to undergo an additional 11.3 cm of subsidence. Even mitigation measures that include artificial recharge and reduced pumping do not significantly reduce the compaction in the northwest subsidence bowl. This is due to the slow draining of thick confining units in the region. However, a small amount of uplift of 0.4 cm is expected in the North and Central bowl areas over the next 20 years.

1 Introduction

Las Vegas Valley (Fig. 1), encompassing an area of 4150 km^2 in southern Nevada, has experienced a long history of rapid population growth and consequent growing rates of groundwater pumping that have led to large water-level declines of as much as 90 m (Burbey, 1995) and subsequent extensive and complex subsidence patterns with four regional subsidence bowls, with the Northwest bowl experiencing more than 1.5 m of land subsidence (Bell et al., 2002) in this structural basin filled with more than a thousand meters of unconsolidated to semi-consolidated heterogeneous sediments (Fig. 1b) (Amelung et al., 1999). In recent decades, mitigation strategies including aquifer storage and recovery (ASR) (Brothers and Katzer, 1990) practices have led to widespread recovery of water levels and substantial but uneven rebound of the land surface in some localities, while other localities continue to experience subsidence (Hoffmann et al., 2001). PS-InSAR reveals that from 2002 to 2010 much of the valley has experienced uplift or at least stabilization of the land surface due to water-level recoveries, but the Northwest and Southern bowls continue to subside (Fig. 2) even though the major pumping centers do not coincide with these subsiding areas. Thus the complex relation between pumping locations and rates, water-level changes, and surface deformation patterns are a clear reflection of the highly heterogeneous hydrogeologic units underlying the valley, which include basin fill faults and their poromechanical properties.

Past numerical models of Las Vegas Valley (Jeng, 1998; Morgan and Dettinger, 1996) have not been able to adequately capture these complex subsidence patterns and therefore their effectiveness as groundwater management tools

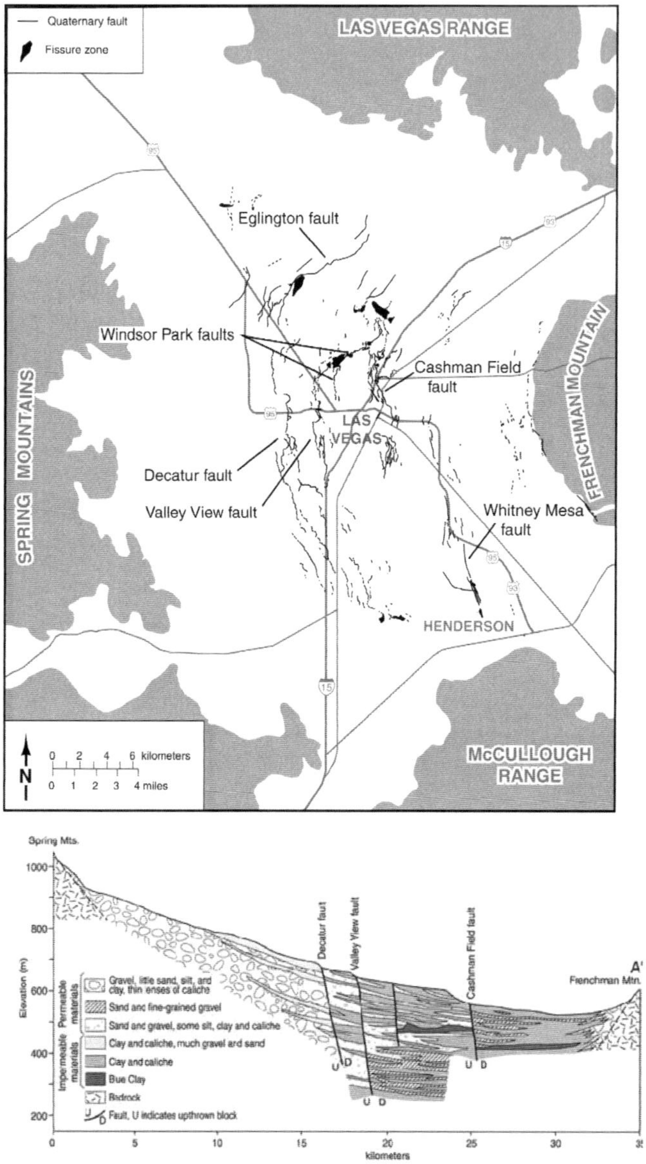

Figure 1. Site map showing location of basin fill faults (top) and A–A′ cross section (bottom) through Las Vegas Valley.

comes into question. The reason for the shortcomings of previous models includes: (1) coarse model discretization, (2) limited parameterization of transmissivity and elastic and inelastic storage coefficients, (3) omitting the effects of basin-fill faults, and (4) absence of temporal InSAR interferogram observations for calibration. Yan (2008) has recently included seasonal InSAR data from 1993 to 2002 into the original Las Vegas model (coarse grid). These subsidence data produced a more accurate and robust model, but no efforts were made to reparametrize (new parameter zonations) or rediscretize the model.

In this investigation we use newly acquired basin-wide temporal PS-InSAR and water-level data from 2002 to 2010 as observations to inversely estimate aquifer parameters from a more finely discretized groundwater flow model, which extends the previous modeling efforts of Yan (2008) and Yan and Burbey (2008). The large number of model cells and the computational requirements of the coupled forward and inverse models necessitated the use of the high-speed com-

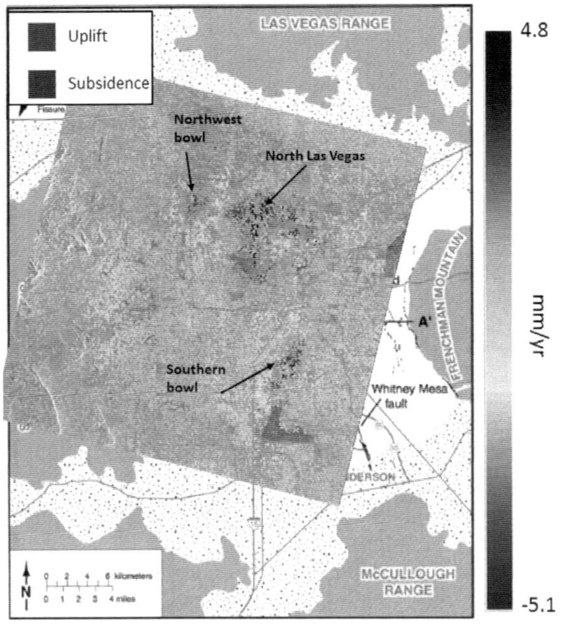

Figure 2. PS-InSAR interferogram showing rate of surface uplift and subsidence for the period 2002–2010.

puting facilities at Virginia Tech. The goal is to create a robust groundwater management model for the entire Las Vegas basin that accurately reflects the subsidence and recovery patterns observed during the past 100 years of aquifer development.

2 Numerical model for Las Vegas Valley

For this study we modify and extend the existing Las Vegas model developed by Jeng (Jeng, 1998) and later by Yan (Yan, 2008; Yan and Burbey, 2008) to account for data through 2010. The model simulates groundwater flow using MODFLOW_2005 (Harbaugh, 2005) and land subsidence using the SUB package (Hoffmann et al., 2003) for MODFLOW and considers the influence of basin-fill faults that are simulated using the HFB package (Hsieh and Freckleton, 1993) for the entire Las Vegas Valley using a 600 m × 600 m grid cell size. A four-layer model is used to represent the hydrogeologic conditions of the valley where the uppermost two layers represent the near-surface aquifer, layer three is the developed (principal) zone aquifer that extends 200–300 m below land surface and which is the primary focus of the calibration effort in this investigation because this is the layer from which all groundwater is pumped and virtually all subsidence occurs. Layer four is the deep-zone aquifer and extends to the basement rocks beneath the basin-fill sediments. The lateral boundaries are all prescribed as no-flow conditions because the basin fill is truncated by the surrounding carbonate and volcanic mountains that are considered to be of much lower permeability than the basin-fill sediments. A total of 87 stress periods is used to simulate the hydrogeologic conditions in the basin that includes nearly 100 years of aquifer development, from March 1912 to October 2010.

3 Initial conditions and observations

The key parameters that affect both water-level changes and land subsidence distributions throughout the valley and which are calibrated using inverse modeling are the transmissivities (T), elastic (S_{ke}) and inelastic (S_{kv}) skeletal storage coefficients and fault conductances of the developed-zone (or principal) aquifer (layer 3), which are believed to be largely responsible for the observed deformations and water level distributions. A total of 72 T parameters (zones) and six S_{ke} and S_{kv} parameters (zones) from the Yan model (Yan, 2008) were used as initial parameter estimates for this investigation. The zones and parameter values for all the other layers were kept the same as the Yan model. Fault conductance values were initially set to be equivalent to the adjacent aquifer conductance (or transmissivity) values. The large discrepancy between the number of transmissivity and storage zones was based on the limited clay thickness and compressibility data necessary for estimating storage values.

Water levels from both domestic and municipal wells were provided by the Las Vegas Valley Water District and the U.S. Geological Survey covering more than 50 years are used as observations for both the forward and inverse calibration procedures. New land surface deformations acquired from InSAR and PS-InSAR data processing were obtained from 58 ENVISAT satellite acquisitions (2002–2010) in descending track mode and were processed using ROI_PAC software (Rosen et al., 2004) to supplement the existing InSAR data from previous investigators (Bell et al., 2002). These new data (Fig. 2) were processed to obtain deformation phase data for each permanent scatterer (PS) to detect the average velocity fields from the time-series data. The PS-InSAR data were then filtered to remove topographic errors, atmospheric errors, the phase noise introduced by the filtering operation, to correct for elevation-dependent atmospheric effects and the orbit tilts. The final deformations were land-truthed with GPS data over a seven-year span. A 40 m resolution is used in both the azimuth and range directions to describe the subsidence distribution for Las Vegas Valley. These 40 m pixels were then averaged over each model grid cell to provide a widespread set of temporal and spatial subsidence (or uplift) values used as observations for the inverse model. Detailed monthly InSAR data are used to investigate how the seasonal variations in water levels are reflected in subsidence and rebound patterns. These elastic stress-strain signals are used to estimate the S_{ke}, which depends more heavily on the seasonal response of the developed zone aquifer that includes clay interbeds.

4 Inverse modeling and calibration

Iterative parameter zone reconstruction, which consists of either coarsening or refining the existing parameter zone number and distributions, was accomplished during the calibration process using an adaptive multi-scale algorithm (Ameur et al., 2002). Parameter inversion was accomplished using UCODE_2005 (Poeter et al., 2005). Prior information was used for parameters T and S_{se} and S_{sv} at some localities obtained from Morgan and Dettinger (1996). All simulations were conducted at Virginia Tech's ARC High Performance Computing Center due to both the size and scale of the iterative problem being solved.

Transmissivity zone distributions from Yan's model were refined through 15 iterations of successive approximations during the inverse procedure to calibrate the final T values for layer 3. Computed CSS (composite scaled sensitivity) values are all larger than 0.021 times the largest CSS, indicating that the model and the observed data provide sufficient information to estimate the transmissivities. Similarly, the S_{ke} and S_{kv} of layer 3 were calibrated at the basin scale using the multi-scale algorithm, which was accomplished through five iterations of successively refining and/or coarsening the zone domains. CSS values were all found to be larger than 0.11 times the largest CSS with S_{kv} and suggests that the inelastic skeletal storage coefficient is highly sensitive to the observations and perhaps the most important parameter for achieving accurate calibration of the model.

Fault conductance values near the Eglington Fault in NW Las Vegas Valley (Fig. 1) were calibrated using eight zones and assuming a 100 m wide fault zone. The calibrated transmissivity values ranged over two orders of magnitude (from 1 to 100 $m^2\,day^{-1}$) in this region, indicating that the fault zones can possess different hydraulic characteristics than the adjacent aquifers.

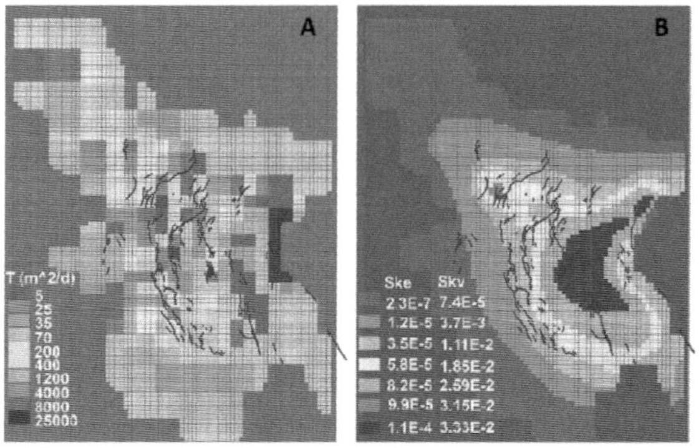

Figure 3. Calibrated **(a)** transmissivity and **(b)** storage zones.

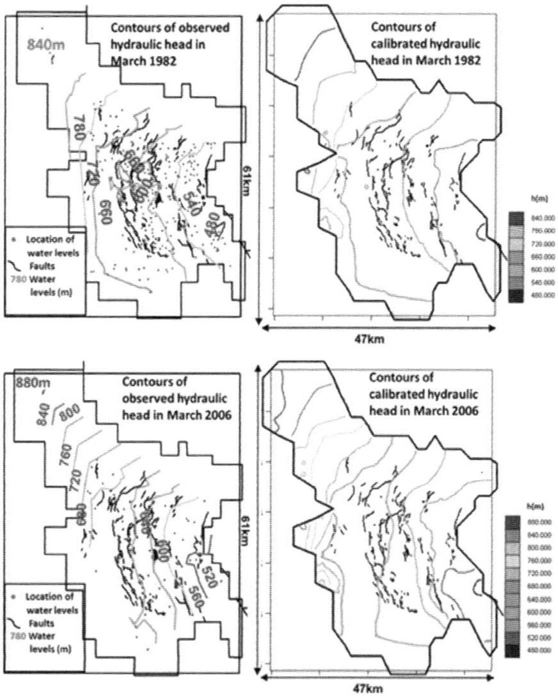

Figure 4. Observed (left) vs simulated (right) water levels after 1982 (top) and 2006 (bottom).

5 Results and predictions

The final calibrated transmissivity and storage zones and values are provided in Fig. 3. Final simulated water levels are compared with observed values across the entire Las Vegas Valley in Fig. 4 after the years 1982 and 2006. The Nash–Sutcliffe efficiency index (NS) (Nash and Sutcliffe, 1970), mass balance error (m) and the Schulz criterion (D) (Schulz et al., 1999) are used to evaluate model fit. These criteria suggest that the model fit is heads is excellent. The final simulated land subsidence distributions are shown in Fig. 5, which shows a similar pattern as developed by Bell (2002) based on observed values, except in the eastern part of the basin where larger simulated subsidence values occur. The NS and m criteria indicates that the fit between the simulated and observed land subsidence is very good.

Basin fill faults are shown to have an influence on the hydrogeologic setting of the basin. All four main subsidence bowls (Northwest, North Las Vegas, Central and Southern bowls) are bounded by basin-fill faults and are significantly offset from the center of artificial recharge and center of municipal pumping. For example, on the upthrown block of the Eglington fault, calibrated S_{ke}andS_{kv} values are higher than those from Yan (2008), so it is inferred that thicker interbeds (clay) may exist at this locality and may represent the primary factor controlling the formation of the Northwest subsidence bowl. Calibrated transmissivity values for the fault barriers indicate that a conductance barrier may exist along the upthrown block of the Eglington fault as suggested by

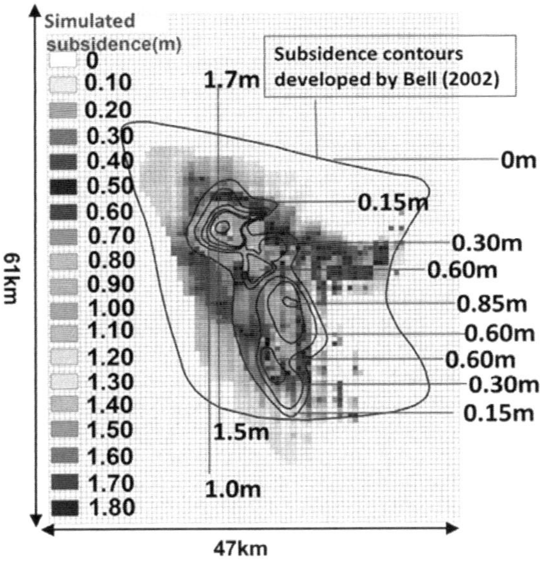

Figure 5. Simulated subsidence after 2002 compared with observed subsidence (solid lines) from Bell (2002).

Bell (2002), who implied that fault gouge or secondary carbonate cementation of the fault zone may exist, or perhaps some other mineralization may be occurring along the fault to cause it to act as a subsidence barrier. However, no significant hydraulic head discontinuity is found to exist across the fault indicating that the influence of the fault as a hydraulic barrier may not be significant or some more complex process such as strain partitioning may be contributing to the large subsidence gradients adjacent to several subsidence bowls.

The North bowl located not far from the pumping/recharge center has recently undergone land uplift and hydraulic head recovery (Fig. 2). At the Central and Southern bowls, groundwater levels have generally recovered, so the occurrence of land subsidence in the southern bowl suggests the existence of residual compaction as a result of slowly draining clay interbeds. The lack of significant subsidence/uplift at or near the center of active pumping/recharge could be explained by the absence of thick clay interbeds. The interbed thickness is large in the eastern part of the valley. The lack of significant subsidence in this area can be explained by the general lack of appreciable water level decline and by the fact that it juxtaposes a bedrock boundary.

Future trends in both water level and land subsidence were investigated by extending the current pumping and artificial recharge rates and patterns from 2011 to 2030. Secondary recharge rates, which have grown substantially since 1972, are based on a single average projected growth rate for 2011–2030 by extrapolating the expected rate from the 1972–2010 rates. Figure 5a shows the map of the predicted groundwater levels changes for the period 2011–2030, which generally

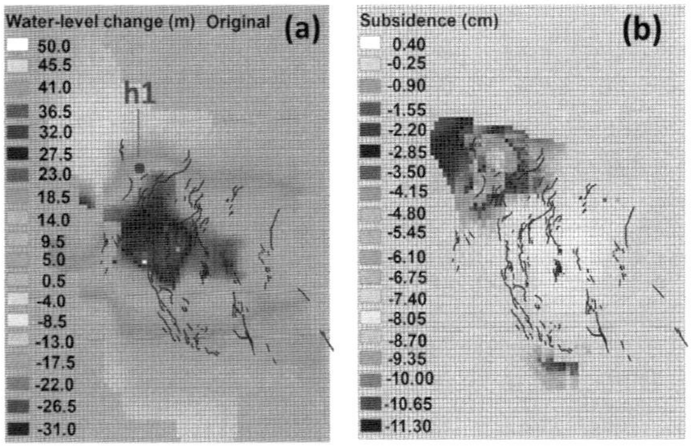

Figure 6. (a) is the predicted water level from 2011 to 2030 and (b) represents the predicted subsidence or uplift for the period 2011–2030.

shows an overall recovery of water levels of as much as 50 m. Even the water levels in the Northwest subsidence bowl show an increase in water levels. For example, location h1 (Fig. 6a) undergoes a head recovery of 14 m during this 20-year period. Figure 6b shows the predicted subsidence changes for this same 20-year period, which suggests that residual compaction from slowly draining confining units and clay interbeds will continue into the future leading to continued predicted subsidence of 11.3 cm in the Northwest subsidence bowl in spite of the water-level recovery in this area. However, uplift of approximately 0.4 cm will occur in the North and Central bowls. At h1 (Fig. 6a) the subsidence is estimated to be 6.3 cm in spite of the large recovery.

6 Conclusions

An updated numerical groundwater and subsidence model was created by updating observations of water level and land subsidence through the processing of new InSAR scenes for 2002–2010 and refined discretization. Early InSAR data revealed that subsidence was widespread across the region into the mid-90's and two subsidence bowls were shown to be controlled by basin fill faults and perhaps differences in interbed thickness. Recovery of water levels due to decreased pumping and increased artificial recharge are responsible for the observed uplift in recent years. However, residual compaction is occurring in areas where thick clay units are still expelling their water, such as the in the Northwest subsidence bowl (Fig. 2). Calibrated aquifer parameters of T, S_{ke} and S_{kv} are similar to Yan's model except near basin-fill faults where skeletal storage coefficients are generally lower than those of Yan. This may be due to abrupt changes in clay thickness or transmissivity differences along the fault zone.

Overall, we find that high resolution temporal and spatial subsidence observations from PS-InSAR are extremely beneficial for quantifying the complex patterns and values of aquifer properties in the Las Vegas basin, which provide considerably more information than water levels alone and therefore results in a more robustly calibrated model. Water levels are expected to recover into the future as are land surface elevations except in the Northwest subsidence bowl area where the draining of thick clay layers will continue to result in compaction in spite of notable water level recovery in the area.

Acknowledgements. We would like to thank David Donavan of the Las Vegas Valley Water District for supplying water-level data and Youquan Zhang of Capital Normal University, China who processed the InSAR data.

References

Amelung, F., Galloway, D. L., Bell, J .W., Zebker, H. A., and Laczniak, R. J.: Sensing the ups and downs of Las Vegas: InSAR reveals structural control of land subsidence and aquifer-system deformation, Geology, 27, 483–486, 1999.

Ameur, H. B., Chavent, G., and Jaffre, J.: Refinement and coarsening indicators for adaptie parameterization: application to the estimation of hydraulic transmsisivities, Inverse Probl., 18, 775–794, 2002.

Bell, J. W., Amelung, F., Ramelli, A. R., and Blewitt, G.: Land subsidence in Las Vegas, Nevada, 1935-2000: New geodetic data show evolution, revised spatial patters, and reduced rates, Environ. Eng. Geosci., 8, 155–174, 2002.

Brothers, K. and Katzer, T.: Water banking through artificial recharge, Las Vegas Valley, Clark County, Nevada, J. Hydrol., 115, 77–103, 1990.

Burbey, T. J.: Pumpage and water-level change in the principal aquifer of Las Vegas Valley, Nevada, 1980-90, U.S. Geological Survey, Water-Resources Information Report 34, 224 pp., 1995.

Harbaugh, A. W.: MODFLOW-2005, The U.S. Geological survey modular ground-water model–the ground-water flow process, U.S. Geological Survey Techniques and Methods 6-A16, 2005.

Hoffmann, J., Zebker, H. A., Galloway, D. L., and Amelung, F.: Seasonal subsidence and rebound in Las Vegas Valley, Nevada, observed by synthetic aperture radar interferometry, Water Resour. Res., 37, 1551–1566, 2001.

Hoffmann, J., Leake, S. A., Galloway, D. L., and Wilson, A. M.: MODFLOW-2000 Ground-Water Model–User guide to the subsidence and aquifer-system compaction (SUB) package, U.S. Geological Survey, Reston, Open-File Report 03-233, 44 pp., 2003.

Hsieh, P. A. and Freckleton, J. R.: Documentation of a computer program to simulate horizontal-flow barriers using the U.S. Geological Survey's modular three-dimensional finite-difference ground-water flow model, U.S. Geological Survey, Open-File Report 92-477, 32 pp., 1993.

Jeng, D. L.: A ground-water flow and land subsidence model of Las Vegas Valley, Nevada–A converted Modflow model, University of Nevada–Reno, Unpublished MS thesis, 123 pp., 1998.

Morgan, D. S. and Dettinger, M. D.: Ground-water conditions in Las Vegas Valley, Clark County, Nevada, U.S. Geological Survey, Water Supply Paper 2320-B, 124 pp., 1996.

Nash, J. and Sutcliffe, J.: River flow forecasting through conceptual models part 1–A discussion of principles, J. Hydrology, 10, 282–290, 1970.

Poeter, E. P., Hill, M. C., Banta, E. R., Mehl, S., and Christensen, S.: UCODE_2005 and six other computer codes for universal sensitivity analysis, calibration, and uncertainty evaluation, U.S. Geological Survey Techniques and Methods 6-A11, 299 pp., 2005.

Rosen, P. A., Hensley, S., Peltzer, G., and Simons, M.: Updated repeat orbit interferometry package released, Eos T. Am. Geophys. Un., 85, 47–47, doi:10.1029/2004EO050004, 2004.

Schulz, K., Beven, K., and Huwe, B.: Equifanality and the problem of robust calibration in nitrogen budget simulations, Soil Sci. Soc. Am. J., 63, 1934–1941, 1999.

Yan, T.: Effects of delayed drainage on subsidence modeling and parameter estimation, Virginia Tech, Blacksburg, VA, 88 pp., 2008.

Yan, T. and Burbey, T. J.: Effects of delayed drainage in an updated groundwater flow and subsidence model for Las Vegas Valley, Nevada, AGU 2008 Fall Meeting, San Francisco, CA, 15–19 December 2008, Eos T. Am. Geophys. Un., 89(53), Fall Meet. Suppl., Abstract H21D-0856, 2008.

Proc. IAHS, 372, 417–419, 2015
proc-iahs.net/372/417/2015/
doi:10.5194/piahs-372-417-2015

Open Access
PIAHS

Modifying the dissolved-in-water type natural gas field simulation model based on the distribution of estimated Young's modulus for the Kujukuri region, Japan

T. Nakagawa[1], R. Matsuyama[1], M. Adachi[2,a], S. Kuroshima[3], T. Ogatsu[1], and R. Adachi[2]

[1]Kanto Natural Gas Development Co., Ltd., Mobara, Chiba, Japan
[2]INPEX Corporation, Nigata, Japan
[3]Godo Shigen Co., Ltd., Chiba Office, Chiba, Japan
[a]now at: Teikoku Oil & Gas Venezuela, CA, Venezuela

Correspondence to: T. Nakagawa (t.nakagawa@k-and-o-energy.co.jp)

Published: 12 November 2015

Abstract. A simulation model, which covers the part of Southern-Kanto natural gas field in Chiba prefecture, was developed to perform studies and make predictions of land subsidence. However, because large differences between simulated and measured subsidence occurred in the northern modeled area of the gas field, the model was modified with an estimated Young's modulus distribution. This distribution was estimated by the yield value distribution and the correlation of yield value with Young's modulus. Consequently, the simulated subsidence in the north area was improved to some extent.

1 Introduction

The Southern-Kanto natural gas field is the largest dissolved-in-water type gas field in Japan. The natural gas is mainly produced in the Kujukuri region of Chiba prefecture for about 80 years. There is a high concentration of iodine found in the formation water. The Southern-Kanto natural gas field contributes about 30 % of total world iodine production. Therefore iodine is a precious resource for a resource-poor country like Japan.

The reservoir of the Southern-Kanto natural gas field is called the Kazusa group. The Kazusa group consists of alternate layers of mudstone and unconsolidated sand. The strata are several decimeters to several tens of meters thick. The alternate layers of the Kazusa group form a homocline with an inclination of approximately 5°.

Almost gas wells are vertical and are completed with perforated-pipe of several hundred meters in length. Natural gas and formation water are produced with a submersible pump or by gas lift from many strata simultaneously.

Land subsidence in this area was first observed in the late 1960s. The eight companies producing natural gas in this area have been making efforts to control land subsidence in accordance with the subsidence prevention agreement concluded with the local government. Although land subsidence in recent years has been reduced in many areas, subsidence of more than 10 mm per annum is still observed in some places. The eight companies currently operating in this area are addressing environmental problems including land subsidence through the Environment Committee of the Japan Natural Gas Association Keiyo Natural Gas Association (hereinafter referred to as the Environment Committee).

2 Simulation model

A simulation model, which covers the part of Southern-Kanto natural gas field in Chiba prefecture, was developed by the Environment Committee to perform studies and make predictions of land subsidence. The JARAS/3-D simulator was used for this simulation model. This simulator was developed by the former Japan National Oil Corporation and the eight companies (Nakagawa et al., 2010).

The effective stress of this reservoir until present has not yet reached the critical yield criterion that differentiates elastic and plastic deformation of the reservoir based on the re-

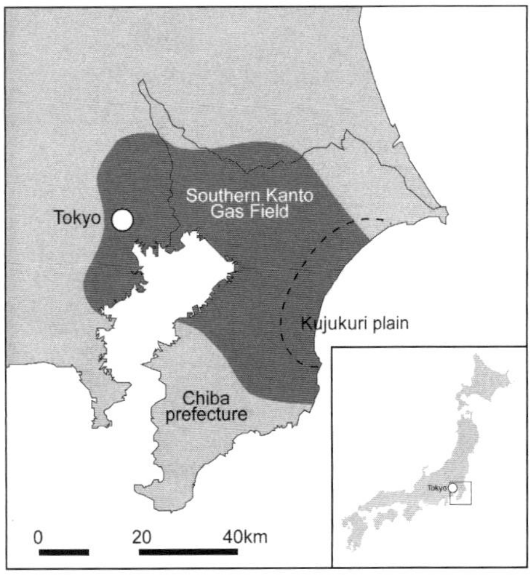

Figure 1. The Southern-Kanto natural gas field.

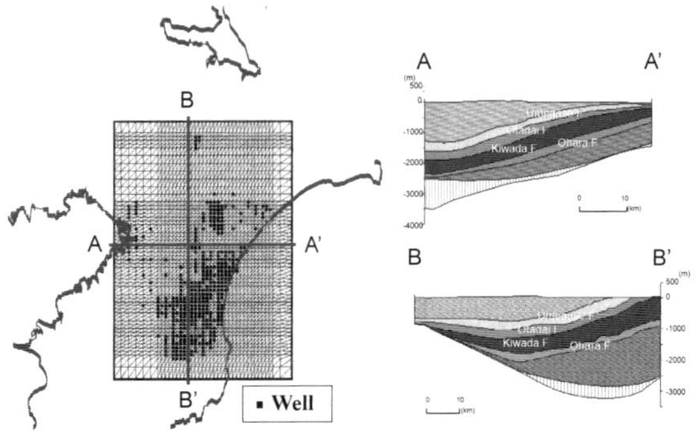

Figure 2. Simulation model.

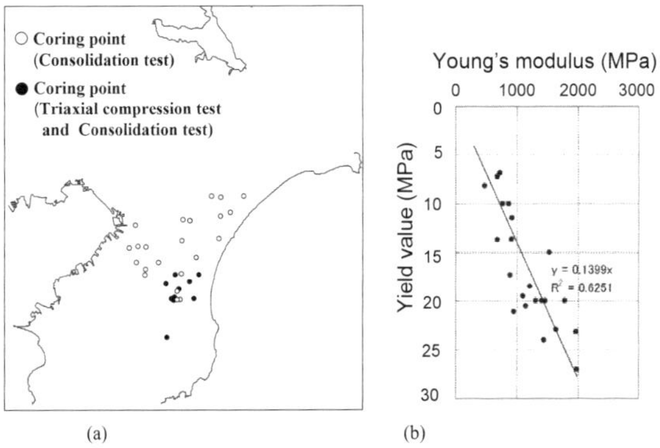

Figure 3. Core data used for estimation Young's modulus distribution: **(a)** coring point of consolidation tests and triaxial compression test and **(b)** correlation between yield value and Young's modulus.

sults of consolidation tests. Therefore only elastic deformation was simulated in this model. The model area (Fig. 2) is about 67 km (N–S direction) by about 48 km (E-W direction).

The model was calibrated through history matching of simulated and measured pore fluid pressures. Simulated land subsidence using the calibrated model was compared to subsidence measured using leveling data. Large differences between simulated and measured land subsidence occurred in the northern portion of the modeled area.

3 Improved subsidence simulation using the estimated Young's modulus distribution

Values of Young's modulus specified in the model were suspected as one of causes of the large differences between simulated and measured subsidence. Values of Young's modulus were determined from triaxial compression tests on cores collected only from the southern portion of model area (Fig. 3a) which indicated that the values depended on depth. The depth trend of Young's modulus was specified in the model uniformly throughout the modeled area. In order to improve estimates of Young's modulus in the northern portion of the modeled area, the yield value data from consolidation tests of cores collected throughout the modeled area (Fig. 3a) were used to estimate Young's modulus in the northern area.

First, the correlation of yield value with Young's modulus was determined for cores collected in the southern modeled area (Fig. 3b). Several of cores collected at each different depth on same coring point in Fig. 3a. Therefore it appears there are more points shown in Fig. 3b than the coring points for triaxial compression tests shown in Fig. 3a. Then, estimates of Young's modulus for the northern modeled area were made based on the yield values for cores collected in the northern modeled area and the correlation of yield value with Young's modulus determined for cores tested in the southern modeled area (Fig. 4). The results show that the depth dependent Young's modulus values decrease along a northwest direction. The simulation model was modified with this estimated Young's modulus distribution. Consequently, simulated subsidence in the north area was improved to some extent (Fig. 5).

4 Conclusions

Conclusions are as follows:

a. The gas field simulation model created by the Environment Committee showed large differences between simulated and measured land subsidence in the northern area of the simulation model.

b. A new distribution of Young's modulus was determined in an attempt to resolve the discrepancy between sim-

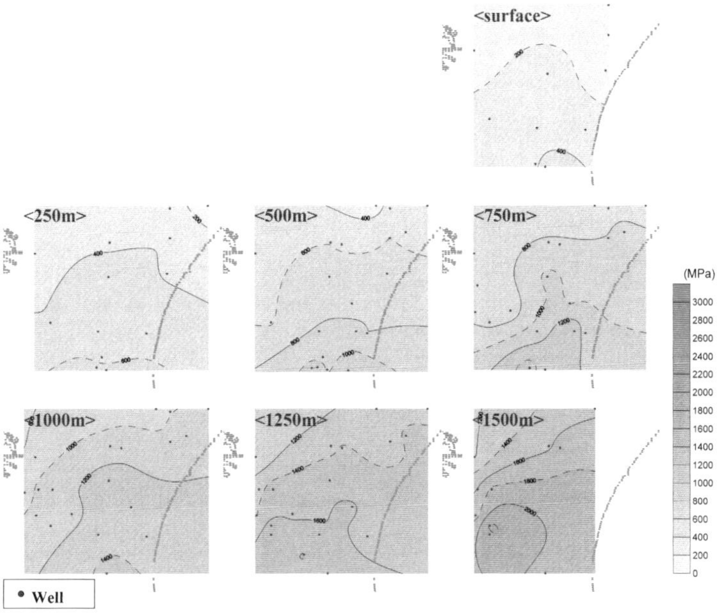

Figure 4. Estimated Young's modulus in each 250 m depth interval of reservoir.

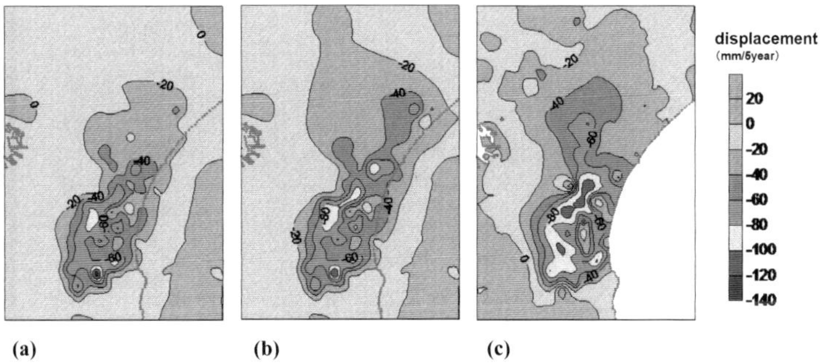

Figure 5. Comparison simulated and measured land subsidence from 1990 to 1994, **(a)** simulated result without modified Young's modulus distribution, **(b)** simulated result with modified Young's modulus distribution and **(c)** Leveling result.

ulated and measured subsidence. The revised Young's modulus distribution was estimated using the distribution of yield value determined from consolidation tests on cores collected throughout the modeled area and the correlation of yield value and Young's modulus determined from triaxial compression tests on cores collected in the southern area of the modeled area. The results show that the revised depth dependent Young's modulus values decrease along a north-west direction.

c. The simulation model was modified with the estimated Young's modulus distribution. Consequently, simulated subsidence in the northern modeled area was improved to some extent.

Acknowledgements. The content of this paper is based on the results of a joint study conducted by the Environment Committee of the Japan Natural Gas Association and Keiyo Natural Gas Association. The Environment Committee kindly permitted us to publish their study results in this paper.

References

Nakagawa, T., Suzuki, I., Nojo, M., Ogatsu, T., and Higuchi, T.: Introduction of the JARAS/3D simulator for natural gas dissolved in water, Land subsidence, Associated Hazards and the Role of Natural Resources Development, IAHS Publ., 339, 140–143, 2010.

Proc. IAHS, 372, 421–424, 2015
proc-iahs.net/372/421/2015/
doi:10.5194/piahs-372-421-2015

Open Access
PIAHS

Simulation study of the in-situ formation deformation behavior of a shallow formation in the Southern Kanto Natural gas field, Chiba Prefecture, Japan

M. Adachi[1,a], R. Matsuyama[2], T. Nakagawa[2], S. Kuroshima[3], T. Ogatsu[2], and R. Adachi[1]

[1]INPEX CORPORATION, Nigata, Japan
[2]Kanto Natural Gas Development Co., Ltd, Chiba, Japan
[3]GODO SHIGEN CO., LTD, CHIBA OFFICE, Chiba, Japan
[a]now at: Teikoku Oil & Gas Venezuela, C.A., Venezuela

Correspondence to: R. Matsuyama (r.matsuyama@k-and-o-energy.co.jp)

Published: 12 November 2015

Abstract. In 2010, eight companies which are exploiting natural gas and brine water in the Southern Kanto natural gas field, Chiba prefecture, Japan constructed an in-situ formation deformation monitoring well with a depth of approximately 80 m, and in-situ formation deformation was measured on a trial basis.

After this field test, by conducting the simulation study, we verified whether the deformation behavior at the monitoring well was perfectly elastic or not. In addition, we compared in-situ rock properties like Young's modulus and Poisson's ratio which were estimated by the simulation study with those determined from a triaxial compression test.

1 Introduction

In the Kanto region of Japan, a large amount of natural gas is widely dissolved in brine water. This natural gas field is known as "the Southern Kanto natural gas field" (Fig. 1). In particular, the Kujukuri plain of Chiba Prefecture has not only reserves of natural gas dissolved in brine water but a high concentration of iodine in brine water. These resources have been produced commercially for about 80 years in this area.

In the Kujukuri plain, land subsidence has been observed since the late 1960s. Recently, land subsidence has reduced in almost all areas, but land subsidence of more than 10 mm per annum has been observed in some places. So the eight companies exploiting natural gas in this area are making efforts to address environmental problems through the Environment Committee of the Japan Natural Gas Association Keiyo Natural Gas Association (hereinafter referred to as the Environment Committee).

2 In-situ formation deformation monitoring well

The Kazusa Group formation contains the natural gas deposits and consists of alternate layers of unconsolidated sandstone and mudstone. The thickness of the stratum is from several decimeters to several meters.

For measuring the in-situ formation deformaiton of each stratum, the eight natural gas companies constructed a monitoring well with a depth of approximately 80 m in 2010. This well is open to formation from 30 to 80 m and installed with settlement gauges, rods, optical fiber etc. By measuring displacement of settlement gauges, this well can measure the in-situ formation deformation of each stratum within a part of Kazusa Group continuously. In addition, water level was measured in this well.

At a distance of 10 m from the monitoring well, a production well was installed. This production well was screened from 45 to 59 m. Permeability of sandstone was estimated 2000 md from a pumping test and vertival permeability of mudstone was estimated 0.02 md from a core test. However, we did not use these values in the model because we could not match the measured drawdown and the simulated draw-

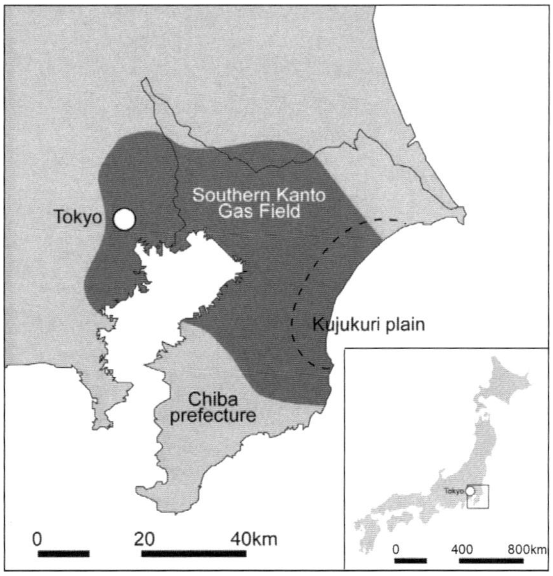

Figure 1. The Southern Kanto gas field.

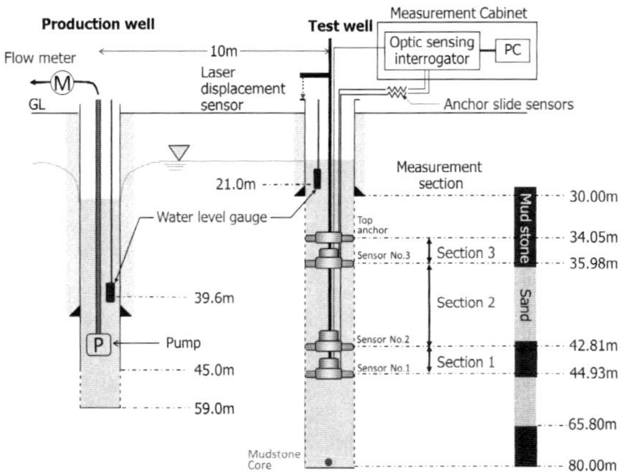

Figure 2. Formation deformation monitoring system (revised Murai et al., 2013).

down. Groundwater was pumped from 10 November 2010 to 16 December 2010 while vertical deformation of the formation was monitored. Figure 2 shows formation deformation monitoring system. Figure 3 shows the pump flow rate and bottom hole pressure at 55 m of each well.

3 Purposes of simulation study

We verified formation deformation behavior measured at the in-situ formation deformation monitoring well with a simulator. Purposes of this study are as follows:

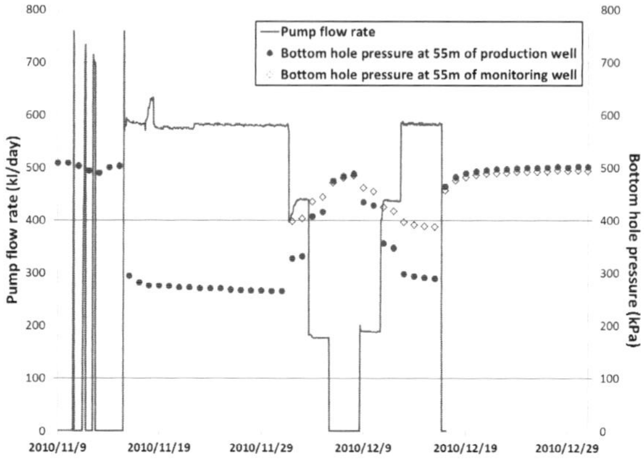

Figure 3. Pump flow rate and bottom hole pressure of each well.

- To confirm whether formation deformation behavior measured at the monitoring well can be explained by elasitc deformation simulated with a perfectly elastic model.
- To determine whether deformation properties estimated from the simulation and properties measured by a core test are consistent.

4 Settings of simulation model

Formation deformation behavior can be simulated with STARS provided by Computer Modelling Group Ltd. (CMG). STARS is a 3-D oil simulator using FDM in flow analysis and FEM in stress-strain analysis. Flow-deformation analysis is coupled. And it can simulate elastic deformation and elasto-plasticity deformation.

One production well and one monitoring well are located at the center of the simulation model (Fig. 4). The distance between two wells is about 10 m. In addition, the aquifer system in Kazusa Group is simulated as confined flow in this model. Figure 5 shows the grid setting for z axis. This model has 9 layers. Water is produced only from the aquifer represented by layer 9 in the model. The intervals of in-situ deformation monitoring are represented by layer 3 $\sim$ layer 7 in the model. Table 1 gives settings for this simulation. Young's modulus and Poisson's ratio are history matching parameters.

5 Results of simulation study

Figure 6 shows measured and simulated drawdown of each well. Figure 7 shows the measured and simulated vertical deformation of each layer. Red circles highlight mechanical errors in measurements. In-situ deformation behavior measured at the monitoring well was reasonably simulated using a perfectly elastic model. However, Young's modulus values

Table 1. Simulation settings.

Number of wells	Production well	1	
	Monitoring well	1	
Model size	X	10 000 m	
	Y	10 000 m	
	Z	65.8 m	
Porosity	Mudstone	0.45	From core test at this well
	Sandstone	0.40	Estimate from other wells
Permeability	Mudstone	10 md	From history matching
	Sandstone	1700 md	From history matching

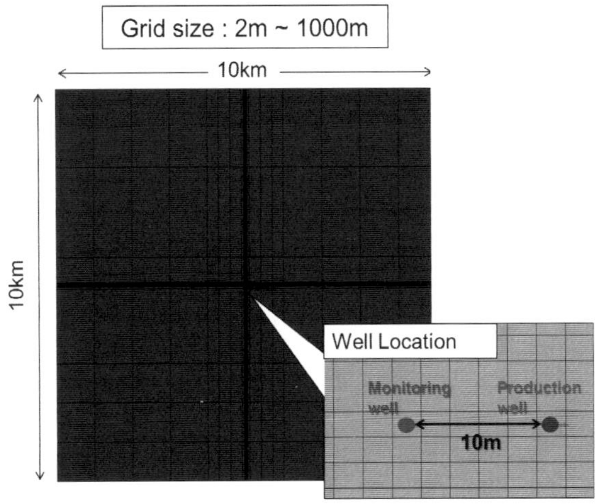

Figure 4. Model gridding (x-y plane).

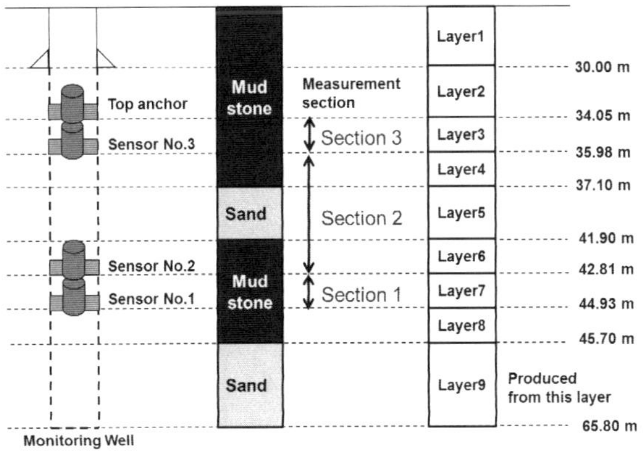

Figure 5. Model gridding (z axis).

estimated from the simulation were smaller than that determined from a triaxial compression test. On the other hand, estimated values of Poisson's ratios were fairly consistent with

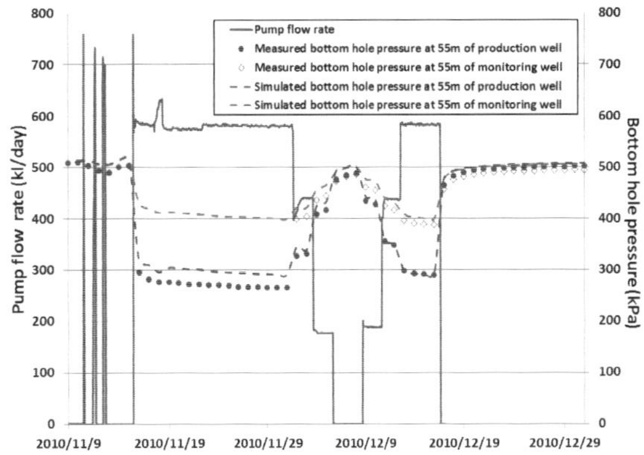

Figure 6. Simulated and measured drawdown.

the value determined from the core test. Because the in-situ drainage condition is strongly affected by the stress-strain response. The differences between the simulated response and the measured response from the core test are attributed to differences in drainage conditions.

Previous core tests show Young's modulus of sandstones and mudstones are equivalent. But the core test in this study was done on a core sampled from mudstone because the sandstones in the shallow formation were too fragile. Therefore, Young's modulus of sandstones and mudstones were assumed to be equivalent in this simulation model. In order to more accurately simulate the in-situ deformation behavior, given that one of either the Young's modulus or Poisson's ratio parameters are equivalent for sandstone and mudstone, then values of the other parameter should be different for sandstones and mudstones.

Measurements of these parameters (Young's modulus and Poisson's ratio) are needed for both sandstones and mudstones.

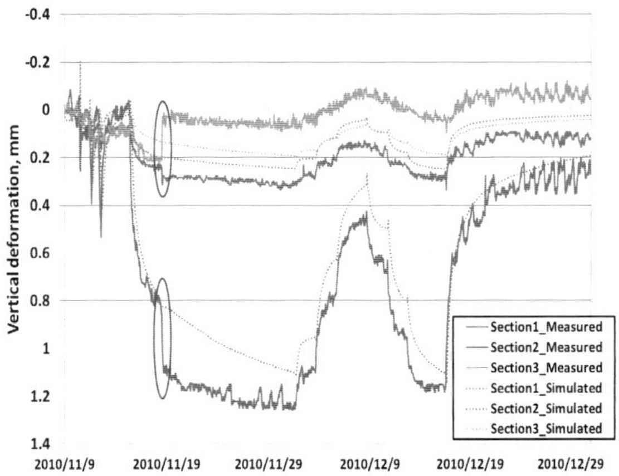

Figure 7. Simulated and measured vertical deformation.

6 Conclusions

- In-situ deformation behavior measured at the monitoring well was simulated reasonably well using a perfectly elastic model.
- Simulated Young's modulus values were assumed to be equivalent for sandstones and mudstones, and the estimated values from the simulation was smaller than that determined from a core test. Poisson's ratios estimated from the simulation were fairly consistent with that determined from a core test.
- It is possible that Poisson's ratio of sandstones and mudstones is different. Core tests for measuring static Poisson's ratios of sandstones and mudstones are required. Some core tests are planned in 2015 to validate these results.

Acknowledgements. The content of this paper is based on the results of a joint study conducted by the Environment Committee of the Japan Natural Gas Association and Keiyo Natural Gas Association. The Environment Committee kindly permitted us to publish their study results in this paper.

References

Murai, D., Kinisue, S., and Kokubo, T.: In-situ formation compaction monitoring in deep reservoirs by use of fiber optics, EGU General Assembly 2013, EGU2013-3860, 2013.

Proc. IAHS, 372, 425–429, 2015
proc-iahs.net/372/425/2015/
doi:10.5194/piahs-372-425-2015

Regional subsidence modelling in Murcia city (SE Spain) using 1-D vertical finite element analysis and 2-D interpolation of ground surface displacements

S. Tessitore[1,2,3], J. A. Fernández-Merodo[2,4,5], G. Herrera[2,4,5], R. Tomás[2,4,5,6], M. Ramondini[3], M. Sanabria[2,4,5], J. Duro[7], J. Mulas[2,4,5], and D. Calcaterra[1]

[1]Department of Earth Sciences, Environment and Resources, Federico II University of Naples, Naples, Italy
[2]Geohazards InSAR laboratory (InSARlab), Geoscience Research dept., Geological Survey of Spain (IGME), C/. Alenza 1, 28003 Madrid, Spain
[3]Department of Civil, Building and Environmental Engineering, Federico II University of Naples, Naples, Italy
[4]Grupo español de trabajo en subsidencia del terreno (SUBTER) de la UNESCO
[5]UNIRAD: Unidad Asociada de Interferometría Radar y Movimientos del Terreno de la Universidad de Alicante y el Instituto Geológico y Minero de España, Spain
[6]Departamento de Ingeniería Civil, Escuela Politécnica Superior, Universidad de Alicante, P.O. Box 99, 03080 Alicante, Spain
[7]Altamira Information, C/. Còrsega 381-387, 08037 Barcelona, Spain

Correspondence to: S. Tessitore (serena.tessitore@unina.it)

Published: 12 November 2015

Abstract. Subsidence is a hazard that may have natural or anthropogenic origin causing important economic losses. The area of Murcia city (SE Spain) has been affected by subsidence due to groundwater overexploitation since the year 1992. The main observed historical piezometric level declines occurred in the periods 1982–1984, 1992–1995 and 2004–2008 and showed a close correlation with the temporal evolution of ground displacements. Since 2008, the pressure recovery in the aquifer has led to an uplift of the ground surface that has been detected by the extensometers. In the present work an elastic hydro-mechanical finite element code has been used to compute the subsidence time series for 24 geotechnical boreholes, prescribing the measured groundwater table evolution. The achieved results have been compared with the displacements estimated through an advanced DInSAR technique and measured by the extensometers. These spatio-temporal comparisons have showed that, in spite of the limited geomechanical data available, the model has turned out to satisfactorily reproduce the subsidence phenomenon affecting Murcia City. The model will allow the prediction of future induced deformations and the consequences of any piezometric level variation in the study area.

1 Introduction

Subsidence due to ground water withdrawal could affect wide areas provoking severe damage to structures and infrastructures. More than 150 cities in the world are affected by this phenomenon (e.g. Venice, Bologna Ravenna, Mexico city, Santa Clara, San Joaquin Valleys, Antelope, Bangkok, Madrid, Guadalentín Valley, Almería, Granada and Sant Feliu de Llobregat, Orihulela city, Barcelona, ...) mainly due to an inefficient aquifer management. Among these, Murcia city (SE Spain) is located in the flood plain of the Segura River, (also known as the Vega Media of the Segura River, VMSR), whose aquifer was strongly exploited in the periods 1992–1995 and 2004–2008. This caused a moderate land subsidence in the whole VMSR which provoked damages to more than 150 buildings with an estimated cost of about EUR 50 million (Mulas et al., 2003). Consequently, the local authorities required detailed studies (e.g. Aragón et al., 2006; Herrera, et al., 2009; Mulas et al., 2003, 2010; Tomás, 2009) to understand the cause-effect mechanism of the phenomenon to improve the exploitation management. To this aim, an extensometric monitoring network was installed in 2001 and implemented in 2008 to cover the whole metropolitan area. Moreover, the land subsidence in Murcia City has been monitored by an innovative Differential SAR Interferometry (DInSAR) technique (Herrera, et al., 2009; Tomás, 2009). This technique allows measuring the ground surface displacements by calculating the differences between satellite SAR images acquired at different times and from different orbital positions. A-DInSAR techniques provide the displacement estimates, the velocity map and the displacements time series for the whole observation period along the satellite Line of Sight (LOS). These techniques have been effectively used to the subsidence monitoring as shown in many published works (Bru et al., 2013; Cascini et al., 2013; Ezquerro et al., 2014; Modoni et al., 2013; Sanabria et al., 2014; Tomás et al., 2014). Specifically, in the present work, the DInSAR processing of C-band SAR images achieved through the Stable Point Network (SPN) technique (Arnaud et al., 2003) for the period 1995–2008 (Terrafirma, 2008) has been used. In the present work, this information has been complemented with a subsidence finite element numerical model implemented by Tessitore et al. (2015), useful to analyse the cause-effect mechanism. This model allowed simulating the hydro-mechanical response of soils to water table variation. Specifically, the model was applied for the 24 columns where extensometric measurements are available (see their location in Fig. 1). Then, the 1-D subsidence results have been interpolated to obtain 2-D maps of deformation for the stable period (2001–2004), the drought period (2004–2008) and the recovery period (2008–2012). These maps have been compared with that achieved for the same periods through DInSAR and extensometric measurements.

2 Description of the study area

Murcia city (SE Spain) is located in the Vega Media of the Segura River Valley. The regional geological setting is characterized by the presence of a Permo-Triassic basement, mainly made up of limestones, dolomites and metamorphic rocks belonging to the Internal Zones of the Betic Cordillera. Three units represent the basin filling, ranging from Upper Miocene to Quaternary in age. The lowermost unit is given by Upper Miocene marls; the intermediate unit is constituted by a continental clastic succession with thick 250 m of marls and clays, intercalated with several levels of gravels and sands, dated from Pliocene-Quaternary to the present (Aragón et al., 2004). At its top recent sediments (mainly silts, clays and sands), with an overall thickness varying from 3 to 30 m (Rodríguez-Jurado et al., 2000) can be found. The geological settings generate a hydrogeological system constituted by (Aragón et al., 2004): (a) a deep aquifer which corresponds to a Pleistocene to Pliocene sequence of gravels and sands; (b) a surficial aquitard constituted by the first 5

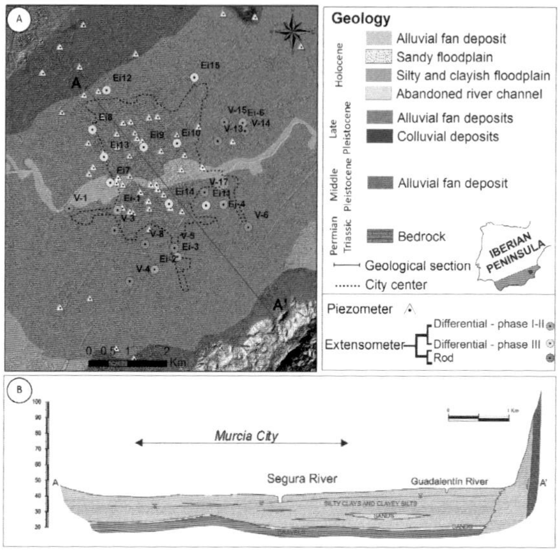

Figure 1. Geological Map **(a)**, Geological cross section of the study area **(b)**, (extracted from Mulas et al., 2003; modified after Tessitore et al., 2015).

to 30 m of recent clays, silts and sands (Fig. 1). The groundwater overexploitation interested the gravel layer located in the upper part of the deep aquifer. This provoked a water flow from the surficial aquitard to the gravel layer which triggered the consolidation of the surficial fine-grain soils. A geotechnical and hydrogeological characterization for the VMSR materials has been performed by Mulas et al. (2010) and used in the present work (Table 1). The parameters found in this study were obtained from hydrogeological tests, in situ (SPT, CPT-CPTU) and laboratory tests (triaxial tests, oedometric tests and uniaxial compressive strength tests) and stratigraphic records (ITGE and CPTOP, 2000).

3 Monitoring data

3.1 Piezometers

The piezometric network of the VMSR was installed in 1974, and extended during the years. Specifically, before 1994, only three piezometers were placed around the city of Murcia; in 1994, 46 new piezometers were drilled. In the following years (1998, 2000, 2006, 2008) the piezometric network reached the 111 units. Actually, 7 piezometers are currently available in the urban area of Murcia. The analysis of the measured groundwater table height shows three important piezometric level declines in the periods: 1982–1984, 1992–1995 and 2004–2008 respectively of 3.4, 8.3 and 12 m. After the last drought period, a recovery period has been observed which has triggered to a ground surface uplift since the 2008.

Table 1. Summary of the geotechnical properties of the soils from the metropolitan area of Murcia city (from Mulas et al., 2010 and Tessitore et al., 2015). γ_{sat}: specific saturated weight [$KN\,m^{-3}$]; k_y: vertical permeability [$m\,s^{-1}$]; E: Young's modulus [$KN\,m^{-2}$]; ν: Poisson's coefficient.

	City center				Sector II				Sector III			
	E	γ_{sat}	k_y	ν	E	γ_{sat}	k_y	ν	E	γ_{sat}	k_y	ν
Clay	($KN\,m^{-2}$)	($KN\,m^{-3}$)	($m\,s^{-1}$)		($KN\,m^{-2}$)	($KN\,m^{-3}$)	($m\,s^{-1}$)		($KN\,m^{-2}$)	($KN\,m^{-3}$)	($m\,s^{-1}$)	
Clay	7500	20-20.3	$0.7–5.6 \times 10^{-10}$	0.35	10 000	20	6.5×10^{-9}	0.35	13 000	20	3.87×10^{-9}	0.35
Sand	15 000–20 000	20–20.8	$2.8–28 \times 10^{-9}$	0.3–0.35	15 000–20 000	20–20.8	$2.8–28 \times 10^{-9}$	0.3–0.35	15 000–20 000	20–20.8	$2.8–28 \times 10^{-9}$	0.3–0.35
Gravel	80 000	21	2.8×10^{-6}	0.3	80 000	21	2.8×10^{-6}	0.3	80 000	21	2.8×10^{-6}	0.3

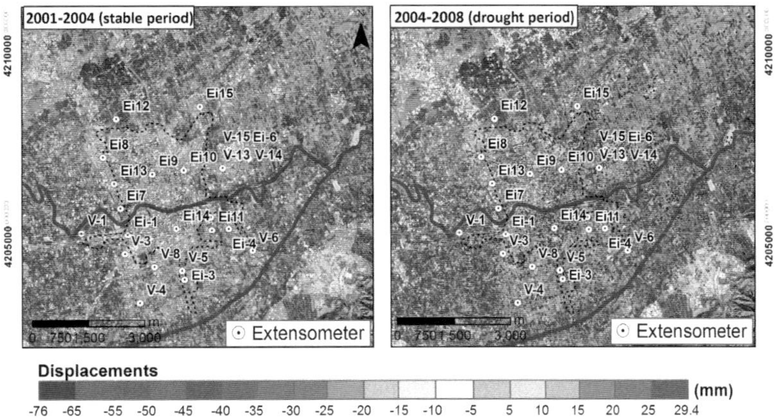

Figure 2. Cumulated DInSAR vertical displacement over two time periods: 2001–2004 (stable) and 2004–2008 (drought).

3.2 Extensometers

In Murcia city, a monitoring network of differential and rod extensometers was installed in February 2001 (phases 1 and 2) and extended in July 2008 (phase 3). In the first two phases 22 extensometers were drilled. In the 3rd phase, 9 additional incremental extensometers were installed, including a piezometric sensor. However, only 17 extensometers (from the 31 installed) were operational in 2012. The extensometers measure the deformation of the firsts 10–20 m depth with accuracy of 0.1 mm for both the available types (Mulas et al., 2010).

3.3 DInSAR data

In the present work, 129 SAR images (ERS and ENVISAT) acquired from descending orbits between July 1995 and December 2008 were processed through the Stable Point Network (SPN) technique (Arnaud et al., 2003). Two different time periods were analyzed: 1995–2005 and 2004–2008 (Terrafirma, 2008). In particular, both datasets have been integrated in order to provide a continuous time series of the displacements along the period 2001–2008, which coincides with the period monitored by the extensometer network.

Figure 2 shows the spatial development of the subsidence of Murcia city. Specifically, taking into account the drought period, the cumulated displacements turned out to be more intense in the southern side of the river and in the eastern part of the city.

4 Numerical model

In Murcia city subsidence was provoked by the groundwater overexploitation which triggered the consolidation of compressible soils (Herrera et al., 2009; Tomás, 2009). The Biot's equation, which states a time dependent coupling behaviour between the solid matrix and the interstitial fluids of soils, has been solved using a Finite Element Model (FEM). In particular, the finite element code GeHoMadrid (Fernandez-Merodo, 2001; Mira, 2001), developed by the M_{2i} group (Mathematical Models in Engineering) belonging to the Civil Engineering School of Madrid (Universidad Politécnica de Madrid) has been used.

In the present work, the 24 extensometric boreholes (see location in Fig. 1) have been modelled considering three main materials: gravel, sand and clay. Below the surficial gravel layer (where the pumping takes place) there are layers of consolidated hard clays and marls which have been considered as the incompressible and impervious bedrock for the displacement calculation. Therefore, only the deformations

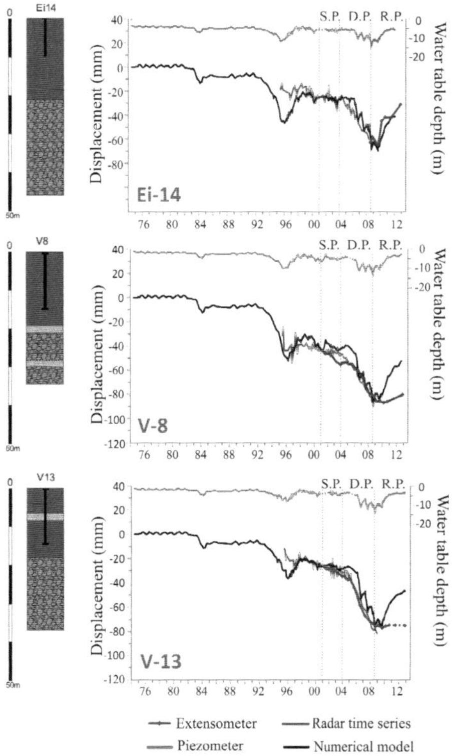

Figure 3. Modelled columns (clay: brown, sand: yellow, gravel: grey); computed and measured subsidence time series (extensometers and Radar); piezometric time series. (S.P.: stable period; D.P.: drought period; R.P.: recovery), redrawn after Tessitore et al. (2015).

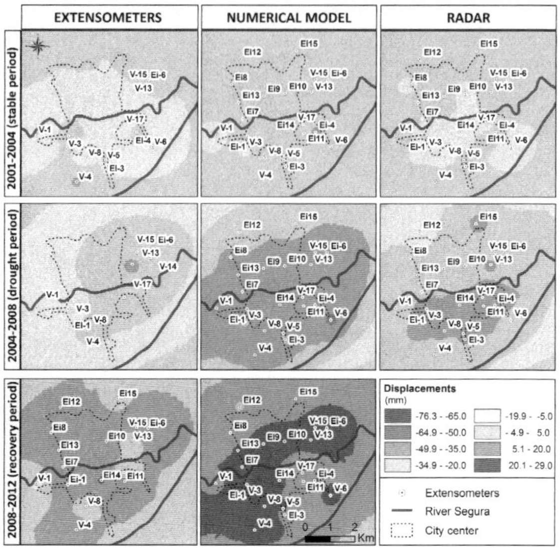

Figure 4. Spatial interpolation of the measured and computed displacements over the three considered periods: 2001–2004 (stable period), 2004–2008 (drought period), 2008–2012 (recovery period); redrawn after Tessitore et al. (2015).

occurred between the ground surface and the gravel layer bottom has been computed, considering an elastic behaviour for the materials and assuming a fully saturation condition (see Table 1).

For the boundary conditions, the gravel bottom has been fixed, the pore pressure variations have been prescribed on the base and no flux has been considered on the ground surface. Figure 3 shows some of the achieved results.

The 1-D model results have been interpolated to achieve the 2-D maps of deformations which have been compared with the ground-based and DInSAR-based ones (Fig. 4).

5 Discussion and conclusions

The comparisons between extensometric, radar data and numerical model results have allowed to analyse and justify the spatio-temporal differences among them. Specifically, the main differences found are referred to the drought period (2004–2008) where the extensometers registered a lower displacement than that measured through radar and the model ones. This can be due to the fact that the extensometer network is only measuring the deformations occurred in the firsts 10–20 m depth, while the radar measures the subsidence over the whole aquifer system. Concerning the recovery period, only the extensometers and the numerical model can be compared (Fig. 3). Columns V-8 and V-13 show that in some cases, the numerical model does not reproduce the measured deformation patterns. The numerical model predicts an uplift of the ground surface, due to the assumed elastic behaviour of the materials, whereas the extensometers exhibit no deformation, probably due to the fact that the measured displacement during the previous drought period (2004–2008) are plastic, producing irrecoverable deformation.

Thus, the implementation of a 1-D FEM model has permitted to effectively analyse the cause-effect mechanism of the subsidence phenomenon in Murcia city. In spite of the limited geotechnical and hydrogeological data available, the model has allowed to qualitatively and "quantitatively" reproduce the subsidence measured in Murcia city. Modelling is useful to the risk assessment because permit to predict the deformations induced by any piezometric level variation. This allows a better management of the ground water exploitation.

Acknowledgements. This work has been supported by the Spanish Ministry of Economy and Competitiveness and EU FEDER funds under projects ESP2013-47780-C2-2-R and TEC2011-28201-C02-02 and by the project 15224/PI/10 from the Regional Agency of Science and Technology in Murcia. The European Space Agency (ESA) Terrafirma project funded all the SAR data processing with the SPN technique.

References

Aragón, R., Lambán, J., García-Aróstegui, J. L., Hornero, J., and Fernández-Grillo, A. I.: Efectos de la explotación intensiva de aguas subterráneas en la ciudad de Murcia (España) en épocas de sequía: orientaciones para una explotación sostenible, Boletín Geológico Y Minero, 117, 389–400, 2006.

Arnaud, A., Adam, N., Hanssen, R., Inglada, J., Duro, J., Closa, J., and Eineder, M.: ASAR ERS interferometric phase continuityGeoscience and Remote Sensing Symposium, 2003 IGARSS '03 Proceedings 2003 IEEE International, 1133–1135, vol.1132, 2003.

Bru, G., Herrera, G., Tomás, R., Duro, J., De la Vega, R., and Mulas, J.: Control of deformation of buildings affected by subsidence using persistent scatterer interferometry, Struct. Infrastruct. E., 9, 188–200, doi:10.1080/15732479.2010.519710, 2013.

Cascini, L., Peduto, D., Reale, D., Arena, L., Ferlisi, S., Verde, S., and Fornaro, G.: Detection and monitoring of facilities exposed to subsidence phenomena via past and current generation SAR sensors, J. Geophys. Eng., 10, 064001, doi:10.1088/1742-2132/10/6/064001, 2013.

Ezquerro, P., Herrera, G., Marchamalo, M., Tomás, R., Béjar-Pizarro, M., and Martínez, R.: A quasi-elastic aquifer deformational behavior: Madrid aquifer case study, J. Hydrol., 519, 1192–1204, doi:10.1016/j.jhydrol.2014.08.040, 2014.

Fernandez Merodo, J. A.: Une approche a la modelisation des glissements et des effondrements de terrains: Initiation et propagation (Thesis), Ecole Centrale Paris, 2001.

Herrera, G., Fernández, J. A., Tomás, R., Cooksley, G., and Mulas, J.: Advanced interpretation of subsidence in Murcia (SE Spain) using A-DInSAR data – modelling and validation, Nat. Hazards Earth Syst. Sci., 9, 647–661, doi:10.5194/nhess-9-647-2009, 2009.

ITGE & CPTOP: Estudio Geotécnico para el análisis prevención y correción de la patología derivada de los cambios en el subsuelo de la Ciudad de Murcia, Informe inédito, 208 pp., 2000.

Mira, P.: Analisis por elementos finitos de problemas de rotura de geomateriales (Thesis), Universidad Politecnica de Madrid, ETSICCP, Madrid (Spain), 2001.

Modoni, G., Darini, G., Spacagna, R. L., Saroli, M., Russo, G., and Croce, P.: Spatial analysis of land subsidence induced by groundwater withdrawal, Eng. Geol., 167, 59–71, doi:10.1016/j.enggeo.2013.10.014, 2013.

Mulas, J., Aragón, R., and Clemente, A. A.: La subsidencia del terreno en la ciudad y área metropolitana de Murcia: Modelización, seguimiento y control (coord. Madrid: Instituto Geológico y Minero de España y Consejería de Obras Públicas y Ordenación del Territorio de la Comunidad Autónoma de la Región de Murcia), 2010.

Mulas, J., Aragón, R., Martínez, M., Lambán, J., García-Aróstegui, J. L., Fernández-Grillo, A. I., Hornero, J., Rodríguez, J., and Rodríguez, J. M.: Geotechnical and hydrogeological analysis of land subsidence in Murcia (Spain), Materials and Geoenvironment, 50, 249–252, 2003.

Sanabria, M. P., Guardiola-Albert, C., Tomás, R., Herrera, G., Prieto, A., Sánchez, H., and Tessitore, S.: Subsidence activity maps derived from DInSAR data: Orihuela case study, Nat. Hazards Earth Syst. Sci., 14, 1341–1360, doi:10.5194/nhess-14-1341-2014, 2014.

Tessitore, S., Fernández-Merodo, J. A., Herrera, G., Tomás, R., Ramondini, M., Sanabria, M., Duro, J., Mulas, J., and Calcaterra, D.: Comparison of water-level, extensometric, DInSAR and simulation data for quantification of subsidence in Murcia City (SE Spain), Hydrogeol. J., (HJ-2015-3677), submitted, 2015.

Tomás, R., Romero, R., Mulas, J., Marturià, J. J., Mallorquí, J. J., Lopez-Sanchez, J. M., Herrera, G., Gutiérrez, F., González, P. J., Fernández, J., Duque, S., Concha-Dimas, A., Cocksley, G., Castañeda, C., Carrasco, D., and Blanco, P.: Radar interferometry techniques for the study of ground subsidence phenomena: a review of practical issues through cases in Spain, Environmental Earth Sciences, 71, 163–181, doi:10.1007/s12665-013-2422-z, 2014.

Tomás, R.: Estudio de la subsidencia de la ciudad de Murcia Mediante Interferometría SAR diferencial avanzada, Universidad de Alicante, 2009.

Proc. IAHS, 372, 431–435, 2015
proc-iahs.net/372/431/2015/
doi:10.5194/piahs-372-431-2015

Open Access
PIAHS

Poroelastic modeling to assess the effect of water injection for land subsidence mitigation

M. Aichi and T. Tokunaga

Graduate School of Frontier Sciences, University of Tokyo, 5-1-5 Kashiwanoha, Kashiwa-shi, Chiba, 277-8563, Japan

Correspondence to: M. Aichi (aichi@k.u-tokyo.ac.jp)

Published: 12 November 2015

Abstract. The possible effect of water injection to mitigate land subsidence was studied through numerical simulations based on the theory of poroelasticity. The Kujukuri Plain, Japan, was chosen as a study area. The effect of past injection was evaluated by comparing a model with injection and the one without injection. The calculated results suggested that the past injection played a significant role to reduce land subsidence. For achieving more effective mitigation practices in the future, we proposed to install injection wells in shallower formations. The effect of proposed injection method to mitigate land subsidence from 2014 to 2030 was also investigated. The calculated results show that the proposed method can work similarly by lesser water injection than the past method. The results also indicate that the upper limit of injection rate should be carefully determined to control the pore pressure build-up in the formation to be small enough to avoid formation failure.

1 Introduction

The production of natural gas and iodine dissolved in groundwater has continued for more than 50 years at the Kujukuri Plain, Japan, and it has caused land subsidence problem in the area. To mitigate land subsidence, water injection has been conducted. Several studies (e.g., Horiguchi, 1998) reported that the rate of land subsidence became slower after the start of water injection. However, the impact of water injection has not yet been evaluated properly because other countermeasures such as the reduction of groundwater abstraction were also implemented at the same time, and it was difficult to discriminate the effect of each method from the measured rate of land subsidence. In addition, the design of spatial distribution of injection wells was not optimised because the injection wells were mainly converted from abandoned production wells. This approach also has caused a risk of reservoir deterioration because the formations into which water has been injected are the same as those from which natural gas and iodine are produced.

Thus, it is important to examine the design of water injection wells, both in spatial and injection scheme points of views, to make the mitigation of land subsidence more effective. In this study, water injection into shallower formations is proposed and its effect is quantitatively examined. There exist two expected advantages in the proposed method compared to the current method. One is that the shallower formations are more compliant than reservoirs and another is that the contribution of the deformation of shallower formation is larger to the displacement of land surface (Fig. 1).

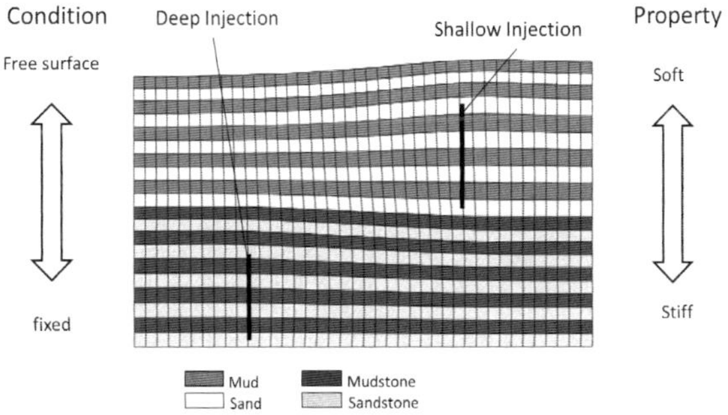

Figure 1. Schematic figure showing the deformation behaviour of the formation by injection. Note that water injection in shallower formations contributes more to the surface displacement compared to that in deeper formations.

Table 1. The formations and screen intervals in the model.

Formation	Depth (m)	Sand ratio	Production wells	Shallow injection
Soil	0			
Kokumoto	−1		0.30	
Upper Umegase	−150 to −80			
Middle Umegase	−200 to −150	0.30		screen
Lower Umegase Upper Ohtadai	−455 to −400	0.08–0.44	screen	
Lower Ohtadai				
Kiwada				
Ohara	−1250	0.17–0.36		
Namihana	−1500			
Miura Group	−3000	0.30		

To quantitatively discuss the effect of water injection into the shallower formations on land subsidence mitigation, a numerical modelling which can calculate the coupled process of both groundwater flow and deformation in subsurface formations is one of the approaches which we can use. There have been several numerical models applied to this area to study the processes around production wells (e.g., Tazaki, 1988, 1994) and regional land subsidence (Nakagawa et al., 2010). However, the effects of injection wells have not been carefully studied. In this study, the effect of past and possible furture water injection is investigated through poroelastic simulations at K-district (Fig. 2, Table 1).

2 Methodology

2.1 Poroelastic modelling

The set of governing equations used in this study is written as:

$$\frac{\partial}{\partial x_j}\left[G\left(\frac{\partial u_j}{\partial x_i}+\frac{\partial u_i}{\partial x_j}\right)+\frac{2G\nu}{1-2\nu}\frac{\partial u_k}{\partial x_k}\delta_{ij}-\alpha p\delta_{ij}\right] + \rho g_i = 0, \quad (1)$$

$$\frac{\partial}{\partial t}\left[\alpha\frac{\partial u_k}{\partial x_k}+S_\varepsilon p\right]=\nabla\left(\frac{k}{\mu}\nabla p\right)+Q, \quad (2)$$

which is based on the theory of poroelasticity (Biot, 1941), where G is shear modulus, ν is drained Poisson's ratio, α is Biot-Willis coefficient, ρ is bulk density of porous media, S_ε is specific storage under constant volumetric strain condition, k is intrinsic permeability, μ is viscosity of fluid, u_i is displacement vector, g_i is gravity acceleration vector, p is excess pore pressure and Q is source term where the injection is taken to be positive. To solve the Eqs. (1) and (2) simultaneously, ABAQUS (Dassault Systèmes) was used in this study.

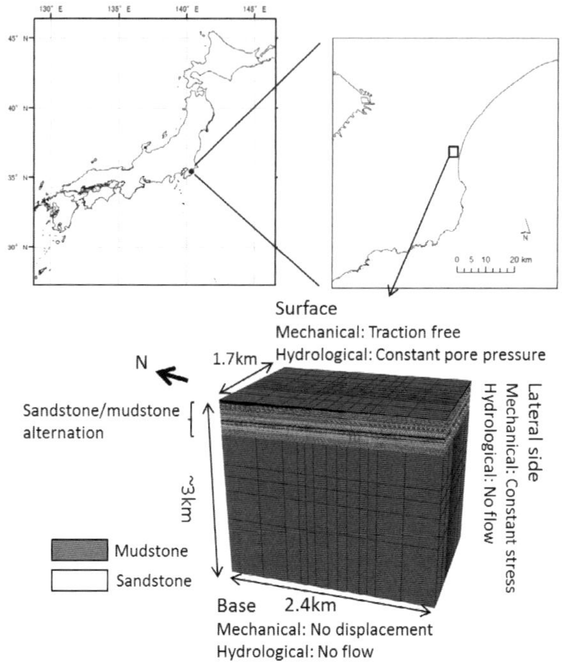

Figure 2. Location of K-district and overview of the model with boundary conditions used in this study.

2.2 Geological model and physical properties

The distribution of subsurface formations and physical parameters in the model were based on the JARAS/3D model (Nakagawa, 2010). The overview of the developed model is shown in Fig. 2. The drained Young's moduli were set to be dependent on depth, and were calculated from the following relation:

$$E = (350 - z)/0.64 \text{ [MPa]}, \quad (3)$$

where z is the elevation. The drained Poisson's ratio was assumed to be $\nu = 0.3$. The Biot-Willis coefficient and specific storage under constant volumetric strain condition were calculated by assuming that the unjacketed bulk modulus is 36 GPa which is a typical value in Wang (2000). Porosity was assumed to be 0.43 for mudstone and 0.37 for sandstone, respectively, after Tazaki (1988). Permeability of sandstone was set to be 1.5×10^{-14} m^2 and that of mudstone was 7.0×10^{-18} m^2.

2.3 Initial and boundary conditions

The extensive groundwater abstraction for methane and iodine production started in 1963 (Horiguchi, 1998). Because the known excess pore pressure at the initial condition, as reported by Tazaki (1994), was negligibly small compared to the pore pressure changes caused by production, the ini-

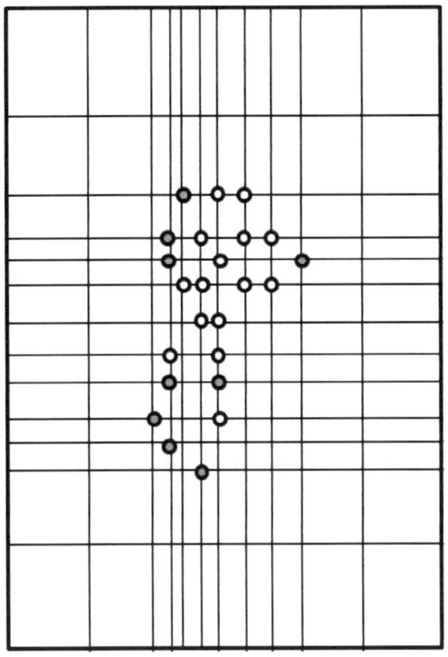

Figure 3. Horizontal view of mesh and the location of wells on the model surface. Open circles indicate the location of production wells or injection wells installed in reservoir. Grey-colored circles indicate the location of both production wells and injection wells installed in shallow formations.

tial condition for the model was set to be hydrostatic in this study.

Boundary conditions are summarized in Fig. 2. The bottom of the model was assumed to be hydrogeological basement and was set to be no flow and no displacement boundary. At the lateral sides, the boundary condition was set to be no flow (Tazaki, 1994) and constant stress. Boundary condition at land surface was set to be constant pore pressure and traction free.

2.4 Water abstraction and injection

The spatial distribution of the wells was set after Tazaki (1988) (Fig. 3). The rates of water abstraction and injection at each well were the same order of magnitude and ranged from several thousand to ten thousand cubic meters per month depending on seasons and years. Before 2010, the rates were specified based on the actual production and injection record. After 2010, the rates were assumed to be as same as that in 2009. The screens of production and injection wells converted from the production ones were installed in the lower Umegase and upper Ohtadai Formations (Table 1). The screens of shallow injection wells were installed in the upper Umegase Formation in the model (Table 1).

Three cases were calculated in this study. In the first case, only water production was set in the model and the simulation covered the period from 1963 to 2013. In the second case, both the water production and injection were set in the model and the simulation covered the period from 1963 to 2030. In the third case, the injection in shallow formations was added to case 2 from 2014 to 2030. The effect of past injection was evaluated by comparing cases 1 and 2. The effect of injection in shallow formations to mitigate land subsidence from 2014 to 2030 was discussed by comparing cases 2 and 3.

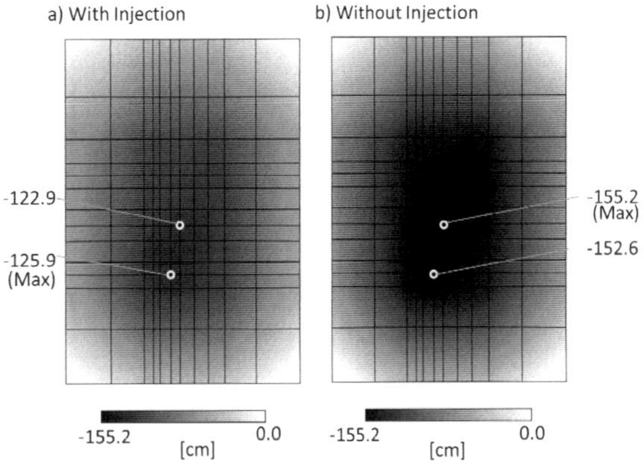

Figure 4. Calculated cumulative land subsidence from 1963 to 2013. **(a)** Model with injection in reservoir. **(b)** Model without any injection.

3 Results and discussions

3.1 The effect of past injection

The results of cases 1 and 2 are compared in Fig. 3. The contribution of past water injection to the reduction of land subsidence was found to be about 30 cm (20 %) at the point of maximum subsidence. This suggests that the past injection played a significant role to reduce land subsidence.

3.2 The effect of injection in shallower formations

The results of cases 2 and 3 are compared in Fig. 4. The calculated results showed that injecting 5 % of abstracted groundwater reduced about 2 cm at the site of maximum cumulative subsidence from 2014 to 2030. An injection of 10 % of the abstracted groundwater into shallow formations reduced 4 cm (25 %) of the maximum possible land subsidence (Fig. 5). The results indicate that the proposed method provides an effect similar to deep injection into reservoir formations with less water injection.

The pore pressure build-up due to injection is shown in Fig. 6. Until injecting 10 % of abstracted groundwater, the excess pore pressure does not appear. However, if 50 % of abstracted groundwater is injected into shallow formations, the excess pore pressure becomes almost as large as the overburden stress while the land uplift is about 13.5 cm at the maxi-

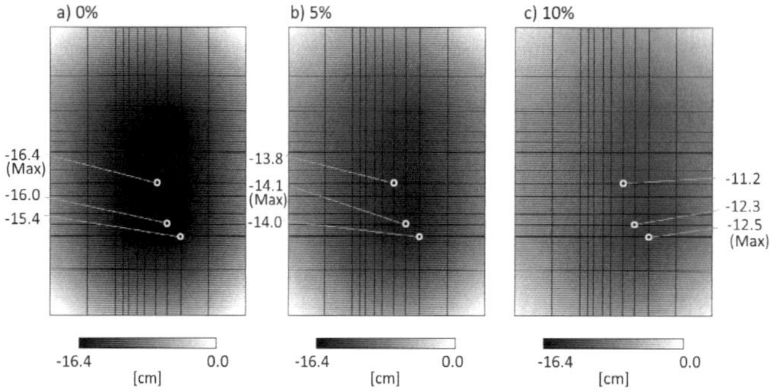

Figure 5. Calculated cumulative land subsidence from 2014 to 2030 in case 3. **(a)** No injection (case 2 for reference). **(b)** Injection of 5 % of abstracted water. **(c)** Injection of 10 % of abstracted water.

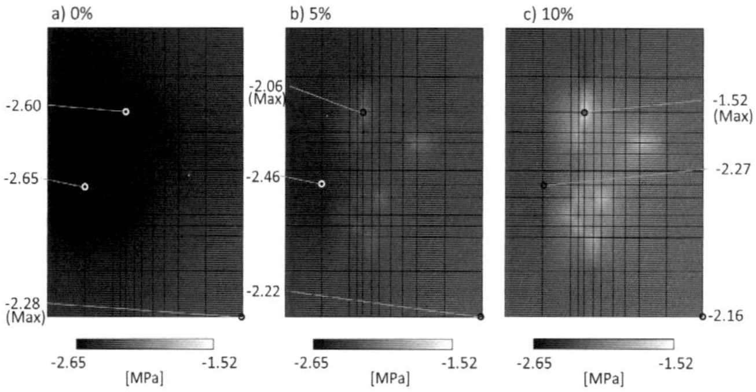

Figure 6. Calculated excess pore pressure at the Middle Umegase Formation in 2030 caused by water injection into shallow formations. **(a)** No injection in shallow formations (case 2 for reference). **(b)** Injection of 5 % of abstracted water. **(c)** Injection of 10 % of abstracted water.

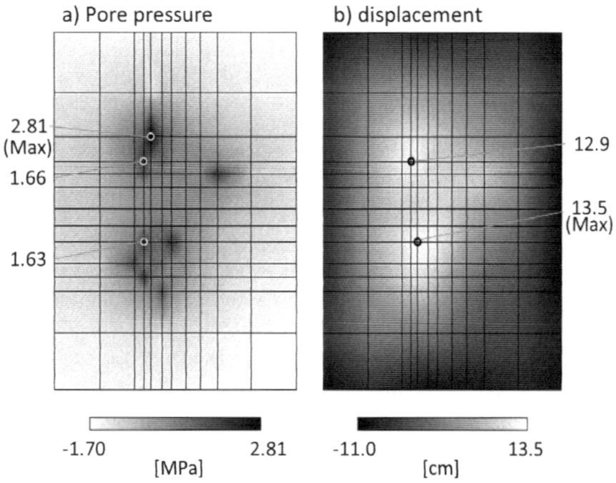

Figure 7. Calculated cumulative land subsidence from 2014 to 2030 and excess pore pressure at the Middle Umegase Formation in 2030 with the water injection of 50 % of abstracted water in shallow formations.

mum (Fig. 7) which is far less than the past cumulative land subsidence (Fig. 3). It indicates that the upper limit of injection rate should be carefully determined to keep the build-up of the pore pressure small enough to avoid formation failure, and it is difficult to totally cancel out the accumulated land subsidence only by the proposed method.

4 Conclusions

To quantitatively evaluate the effect of water injection on mitigating land subsidence, numerical simulations were conducted at the Kujukuri Plain, Japan. The past water injection was found to reduce about 25 % of the land subsidence. For the possible effective mitigation in the future, water injection into shallow formations are proposed. The proposed injection method reduced 25 % of total land subsidence from 2014 to 2030 by injecting only 10 % of abstracted groundwater. However, the calculated results also suggested that the upper limit of injection rate should be carefully determined to keep the build-up of pore pressure small enough not to induce the formation failure.

Acknowledgements. This research was done as a part of Kujukuri Project II and the authors are grateful to the Kanto Natural Gas Development Co. Ltd. and other members of the Project who support this study.

References

Biot, M. A.: General theory of three-dimensional consolidation, J. Appl. Phys., 12, 155–164, 1941.

Horiguchi, K.: Outline of water-dissolved natural gas field in Chiba Prefecture, J. Japanese Assoc. Petrol. Technol., 63, 475–484, 1998 (in Japanese).

Nakagawa, T., Suzuki, I., Nojo, M., Ogatsu, T., and Higuchi, T.: Introduction of the JARAS/3D simulator for natural gas dissolved in water, Proceedings of international symposium on land subsidence No.8, Santiago de Queretaro, Mexico, IAHS Publications, 339, 140–143, 2010

Tazaki, Y.: Natural gas deposits of dissolved-in-water type and sand/silt system, J. Japanese Assoc. Petrol. Technol., 53, 256–264, 1988 (in Japanese).

Tazaki, Y.: Reservoir engineering for water-dissolved natural gas reservoir in the Kanto sedimentary basin, Japan, Doctoral dissertation, Waseda University, 1994 (in Japanese).

Wang, H. F.: Theory of poroelasticity. Princeton University Press, Princeton, 2000.

Proc. IAHS, 372, 437–442, 2015
proc-iahs.net/372/437/2015/
doi:10.5194/piahs-372-437-2015

Open Access
PIAHS

3-D land subsidence simulation using the NDIS package for MODFLOW

D. H. Kang and J. Li

Department of Civil Engineering, Morgan State University, Baltimore, Maryland, USA

Correspondence to: J. Li (jiang.li@morgan.edu)

Published: 12 November 2015

Abstract. The standard subsidence package for MODFLOW, MODFLOW-SUB simulates aquifer-system compaction and subsidence assuming that only 1-D-vertical displacement of the aquifer system occurs in response to applied stresses such as drawdowns accompanying groundwater extraction. In the present paper, 3-D movement of an aquifer system in responses to one or more pumping wells is considered using the new aquifer-system deformation package for MODFLOW, NDIS. The simulation of aquifer- system 3-D movement using NDIS was conducted with a stress or hydraulic head dependent specific storage coefficient to simulate nonlinear deformation behavior of aquifer-system sedimentary materials. NDIS's numerical simulation for aquifer horizontal movement is consistent with an analytic solution for horizontal motion in response to pumping from a leaky confined aquifer (Li, 2007). For purposes of comparison, vertical subsidence of the aquifer system in response to groundwater pumping is simulated by the both the NDIS and MODFLOW-SUB models. The results of the simulations show that land subsidence simulated by MODFLOW-SUB is significantly larger and less sensitive to pumping rate and time than that simulated by NDIS. The NDIS simulations also suggest that if the total pumpage is the same, pumping from a single well may induce more land subsidence than pumping from multiple wells.

1 Introduction

Land subsidence has been attributed to the compaction (compression) of aquifer systems caused by ground-water withdrawal. Widespread land subsidence caused by the overexploitation of groundwater is a global problem. Over the past decades, many experts have attempted to develop models to evaluate transient aquifer-system compaction and subsidence accurately and efficiently using MODFLOW (Leake and Prudic, 1991; Hoffmann et al., 2003; Leake and Galloway, 2007). Most of the developed models only account for the 1-D-vertical displacement that is known to be associated with changes in aquifer-system storage, attributed principally to aquitards-fine-grained, low permeability interbeds and confining units. However, the occurrence of horizontal movement caused by ground water withdrawal indicates that aquifer-system deformation and subsidence result from 3-D movement of the aquifer system. The NDIS numerical model presented here simulates 3-D displacement of the solid matrix caused by groundwater extraction, and is based on the principle of effective stress for the case of a constant geostatic load.

The purpose of this paper is to present the NDIS numerical model for 3-D aquifer-system movement due to groundwater withdrawal, and to evaluate NDIS with respect to the simulated aquifer- system vertical movement (subsidence) by comparing simulation results from NDIS with those from MODFLOW-SUB. One of the advantages of the approach presented in this paper is that one can simplify the processes of solving for 3-D aquifer-system deformation with stress-dependent aquifer-system specific storage coefficients that change due to compaction driven by hydraulic force loading on the aquifer-system skeleton.

2 Theoretical background for the NDIS model

2.1 Basic equation

A governing equation for groundwater flow through a saturated anisotropic and hetergenous porous material is expressed by (Bear, 1972):

$$\nabla \cdot (\mathbf{K}\nabla h) + W = S_\mathrm{s}\frac{\partial h}{\partial t}, \quad (1)$$

where h is hydraulic head; **K** is hydraulic conductivity, a diagonal matrix of rank two; W denotes a volumetric flux per unit volume and represents sources and/or sinks of water mass within an aquifer system; and S_s ($= S_\mathrm{sk} + S_\mathrm{sw}$) is the specific storage coefficient, where S_sk and S_sw are the specific storage coefficients of the soil skeleton and pore water, respectively, and S_sk can be either a constant for linear elastic materials or a stress-dependent variable for inelastic materials. The individual solid particles are assumed to be incompressible. The term W denotes a volumetric flux per unit volume and represents sources and/or sinks of water mass within an aquifer system. For an aquifer system that has steady flow without the term W, Eq. (1) can be further simplified to the Laplace equation. The US Geological Survey (USGS) developed the 3-D program MODFLOW that numerically simulates steady and transient groundwater flow. To simulate land subsidence in response to groundwater withdrawal, USGS further developed the MODFLOW-SUB package simulating 1-D aquifer-system compaction and subsidence based on vertical stress and strain of the aquifer system. In contrast, a new 3-D model, NDIS developed from DIS (Zhang, 2009) similarly uses the parent program MODFLOW-96 (McDonald and Harbaugh, 1988) and the bulk flux relation through permeable, deformable and saturated porous materials. The bulk flux relation is defined by Helm (1979, 1984):

$$\boldsymbol{q}_\mathrm{b}(x,y,z,t) = \boldsymbol{v}_\mathrm{s}(x,y,z,t) + \boldsymbol{q}(x,y,z,t), \quad (2)$$

where $\boldsymbol{q}_\mathrm{b}$ is the bulk flux; $\boldsymbol{v}_\mathrm{s}$ is the velocity of solids; $\boldsymbol{q}$ is the transient specific discharge that is defined by $\boldsymbol{q} = n(\boldsymbol{v}_\mathrm{s} - \boldsymbol{v}_\mathrm{w})$; and $\boldsymbol{v}_\mathrm{w}$ is the fluid (water) velocity. Note that a bolded variable with an arrow represents a vector throughout this paper.

For steady flow, Eq. (2) can be expressed by the following form:

$$\boldsymbol{q}_{\mathrm{b-st}}(x,y,z) = \boldsymbol{v}_{\mathrm{s-st}}(x,y,z) + \boldsymbol{q}_{\mathrm{st}}(x,y,z), \quad (3)$$

where the subscripts st denote the steady flow. If no aquifer movement is assumed once an equilibrium condition (or steady flow) is reached, i.e., $\boldsymbol{v}_{\mathrm{s-st}}(x,y,z) = 0$, and the bulk flux is assumed to be constant, i.e., $\boldsymbol{q}_\mathrm{b}(x,y,z,t) = \boldsymbol{q}_{b-st}(x,y,z)$, combining Eqs. (2) and (3) results in:

$$\boldsymbol{v}_\mathrm{s}(x,y,z,t) = -[\boldsymbol{q}(x,y,z,t) - \boldsymbol{q}_{\mathrm{st}}(x,y,z)]. \quad (4)$$

The expression (4) indicates that aquifer velocity can be written in terms of the specific discharge for transient and steady flow. Moreover, on the right side of Eq. (4), the terms $\boldsymbol{q}(x,y,z,t)$ and $\boldsymbol{q}_{\mathrm{st}}(x,y,z)$ can be found by running MODFLOW for transient flow and steady flow, respectively. It should be pointed out that the assumption of $\boldsymbol{v}_{\mathrm{s-st}}(x,y,z) = 0$ suggests that deformation due to secondary consolidation is negligible and the assumption of the constant bulk flux implies a constant pumping rate. The former means that the aquifer system has no further displacement or deformation once the unsteady flow becomes steady one that is under the equilibrium condition, and the latter can be proved to be true for a constant pumpage (Li and Helm, 2010). Therefore, the velocity field of solids $\boldsymbol{v}_\mathrm{s}$ for transient flow can be computed as a function of space and time from Eq. (4). Using Eq. (4), the displacement field of solids, $\boldsymbol{u}_\mathrm{s}$ can be further computed through integrating the velocity field of solids over the time as follows:

$$\boldsymbol{u}_\mathrm{s} = \int_{t_0}^{t} \boldsymbol{v}_\mathrm{s}\mathrm{d}t + \boldsymbol{u}_{\mathrm{s}0}, \quad (5)$$

where subscript 0 stands for the initial value of displacement and time. The velocity and displacement fields of the solid skeleton vary with time.

2.2 Stress-dependent specific storage coefficient

Based on an exponential relation between the effective volume stress and volume strain of the soil skeleton, the specific storage coefficient can be found to be a function of volume stress written in the following form (Li and Ding, 2013):

$$S_\mathrm{sk}(\sigma'_\mathrm{v}) \triangleq \alpha\gamma_\mathrm{w} = S_{\mathrm{sk}0}\frac{\sigma'_{\mathrm{v}0}}{\sigma'_\mathrm{v}}, \quad (6)$$

where α is the compressibility of the soil skeleton; $S_{\mathrm{sk}0}$($= \gamma_\mathrm{w}\alpha_0$) is the initial specific storage coefficient, where the symbol $\triangleq$ stands for "by definition", γ_w is the water unit weight and α_0 is the initial compressibility of the aquifer system; $\sigma'_{\mathrm{v}0}$ is the initial effective volume stress; σ'_v is the current effective volume stress and the subscript 0 denotes the initial value of a variable throughout this paper. Because the effective volume stress is equal or larger than its initial value ($S_\mathrm{sk} \leq S_{\mathrm{sk}0}$ at $t \geq 0$), suggests that in response to changes in hydraulic head due to groundwater extraction, an aquifer system with elastic materials may deform larger than that with inelastic materials because of the strain-hardening effect. Equation (6) can be written in terms of transient hydraulic head, h through Terzaghi's principle (1925) of effective stress: $\sigma_\mathrm{v} = \sigma'_\mathrm{v} + \mathrm{p}$ and Hubbert's potential (1940): $h = z + p/\gamma_\mathrm{w}$ in terms of hydraulic head in the following form:

$$S_\mathrm{sk}(h) = S_{\mathrm{sk}0}\frac{\sigma'_{\mathrm{v}0}}{\sigma_\mathrm{v} - \gamma_\mathrm{w}(h-z)}, \quad (7)$$

where σ_v is the total volume stress and is normally assumed to be constant; z denotes the elevation from a chosen datum; p is the pore water pressure.

Note that $S_\mathrm{sk}(h)$ in Eq. (7) can be used in Eq. (1) because MODFLOW-96 for NDIS has been modified to solve the nonlinear partial differential equation (PED) for hydraulic head. For simplicity, the constant S_{ske0} is applied to Eq. (1) for elasticity when $\sigma_{\mathrm{v}}' \leq \sigma_{\mathrm{v-pre}}'$ and the head-dependent $S_{\mathrm{skv}}(h) = S_{\mathrm{skv0}}\sigma_{\mathrm{v0}}'/\sigma_{\mathrm{v}}'$ for inelasticity when $\sigma_{\mathrm{v}}' > \sigma_{\mathrm{v-pre}}'$ that is preconsolidation stress.

3 Numerical modeling

The finite-difference numerical simulation is based on a conceptual model that consists of five thin aquitards (thickness is 1 m for each) and two aquifers (thickness is 30 m for each). The top layer (Layer 1) is an unconfined aquifer and the bottom layer (Layer 7) is a confined aquifer. Layers 2–6 are aquitards that serve as a hydraulic separator between two aquifers. The schematic of the seven layer aquifer system is shown in Fig. 1 in which Q is the pumping rate, and H is the total thickness (65 m) of the aquifer system. The option of "no delay" in MODFLOW-SUB is chosen for simulation of the aquifer-aquitard system.

The lateral boundaries of the conceptual model are specified as constant-head boundaries. The initial head is assumed to be zero. Summary properties of the aquifer system are given in Table 1.

To simulate 3-D land movement at a basin-wide scale in an efficient and accurate way, the seven-layer aquifer system is discretized horizontally, and each layer is represented by a grid of 37 rows by 37 columns. The total number of elements (model cells) equals 9583.

Two cases are simulated for purposes of comparison. For one case, a single well is simulated at row 19 and column 19 in the confined aquifer in Layer 7, and for the second case of three pumping wells are also simulated. In the three-well case, the centre well is located at the same location as for the one-well case, and the other two wells are located on a line passing through the centre well. As indicated in Eq. (7), S_{skv} requires specification of the four parameters S_{skv0}, σ_{v}, σ_{v0}' and z, respectively. The parameter S_{skv0} is given in Table 1 and the values of the other three parameters in each layer are calculated and shown in Fig. 1, where the stress parameters are evaluated using the weight of the overburden material (geostatic load) and the initial head (h_0) where $\sigma_{\mathrm{v0}}' = \sigma_{\mathrm{v}} - \gamma_{\mathrm{w}}(h_0 - z)$. The soil unit weight is $21.0\,\mathrm{kN\,m^{-3}}$ for the aquifer material and $18.5\,\mathrm{kN\,m^{-3}}$ for the aquitard material. The empirical formulae are introduced to find K_0, the coefficient of earth pressure at rest, so that the volume stress can be evaluated from the vertical stress. In this paper, $K_0 = 0.19 + 0.233\log(\mathrm{PI})$ for clay and $K_0 = 1 - \sin\varphi$ for sand are used, respectively, where PI and φ are the plastic index and the internal friction angle, respectively. The total volume stress can be further calculated using the relation $\sigma_{\mathrm{v}} = \sigma_z(1 + 2K_0)/3$. For simplicity, the preconsolidation head is not introduced to each model layer although NDIS is capable to compute both elastic and inelastic deformation.

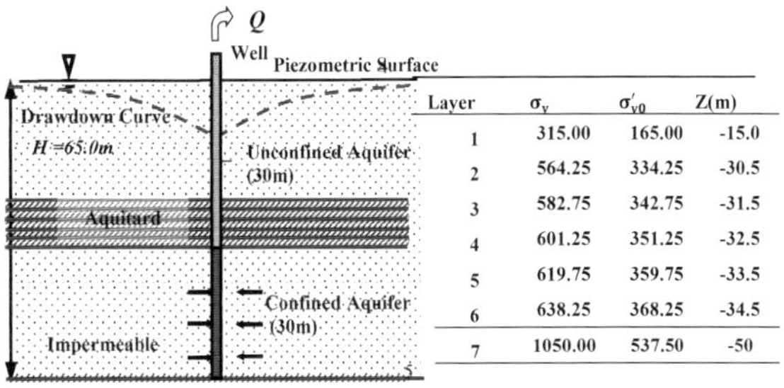

Figure 1. A conceptual model of an aquifer-aquitard system and the values of σ_{v} and σ_{v0}' at different depths.

Table 1. Properties of the conceptual model.

	Layer		
Properties	1	2 to 6	7
K_{h} $(\mathrm{m\,d^{-1}})$*	1.00×10^1	1.00×10^{-2}	1.00×10^1
K_{v} $(\mathrm{m\,d^{-1}})$*	1.00	1.00×10^{-3}	1.00
S_{skv0}	1.00×10^{-4}	1.00×10^{-2}	5.00×10^{-3}
S_{ske0}	1.00×10^{-5}	1.00×10^{-3}	5.00×10^{-4}
Thickness (m)	3.00×10^1	1.00	3.00×10^1
Layer	σ_{v} (kPa)	σ_{v0}' (kPa)	Z(m)
1	315.00	165.00	−15.0
2	564.25	334.25	−30.5
3	582.75	342.75	−31.5
4	601.25	351.25	−32.5
5	619.75	359.75	−33.5
6	638.25	368.25	−34.5
7	1050.00	537.50	−50

* K_{h}, K_{v} – horizontal and vertical hydraulic conductivities, respectively.

Four criteria specified in the numerical solver (Strongly Implicit Procedure package) in MODFLOW are: (1) the error criterion = 0.0001; (2) the acceleration parameter = 1.0; (3) the maximum number of iterations = 500; (4) seed = 0.001 (specified for calculation of iteration parameter). The computation starts by running MODLFOW for steady flow and then proceeds for transient flow with MODFLOW again according to Eq. (4).

Three different values of hydraulic diffusivity of the confining layer, $c_{\mathrm{v}} = K/S_{\mathrm{s}}$ (1000, 2000, and $4000\,\mathrm{m^2\,day^{-1}}$) are chosen along with three applied pumping rates (100, 1000, and $10\,000\,\mathrm{m^3\,day^{-1}}$). Three pumping periods (10, 100, and 1000 days) are simulated to evaluate the displacement sensitivity to various parameters.

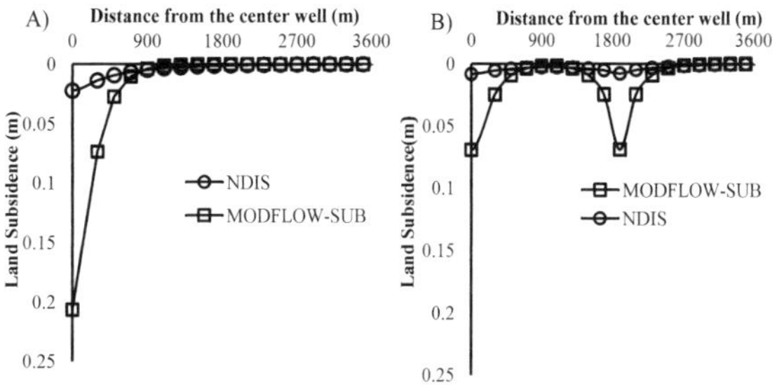

Figure 2. Comparison of land subsidence predicted by the NDIS and MODFLOW-SUB after 100 days pumping; **(a)** Land subsidence for a single well, **(b)** Land subsidence for three wells after 100 days pumping.

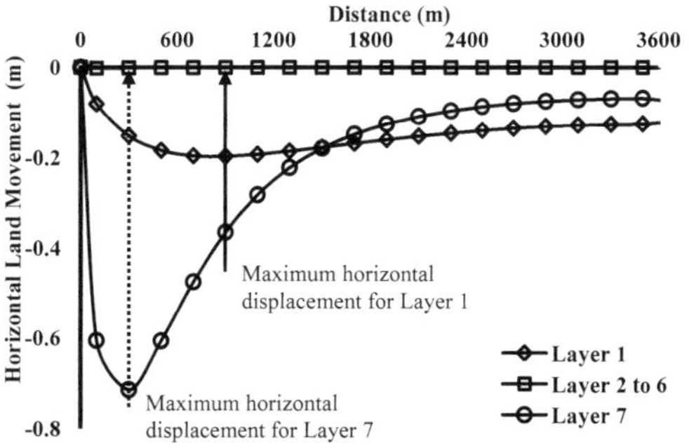

Figure 3. Horizontal subsidence after 100 day pumping simulated by NDIS.

4 Resulsts and analysis

Subsidence (cumulative model layer compaction) simulated by NDIS and MODFLOW-SUB is shown in Fig. 2a and b. A pumping rate of $1000\,m^3\,day^{-1}$ and a pumping period 100 days for both the single-well and three-well are simulated to evaluate the displacements. In the three-well case, the center well is located at the same location as for the one-well case, and the other two wells are located on each side of the center well along a line passing through it. The results show that the maximum subsidence for both models occurs near the well. Subsidence decreases rapidly with distance from the pumping well similar to the nature of the drawdown. The results indicate that the subsidence simulated by MODFLOW-SUB is significantly larger than that simulated by NDIS (Fig. 2). It is interesting to note that even though the total pumping rate in both the one-well and three-well cases are the same (i.e., $1000\,m^3\,day^{-1}$), land subsidence in response to three pumping wells is much less than that due to a single well. However, for the three-well case at the location between two pumping wells subsidence simulated by NDIS is slightly larger than that by MODFLOW-SUB (Fig. 2b).

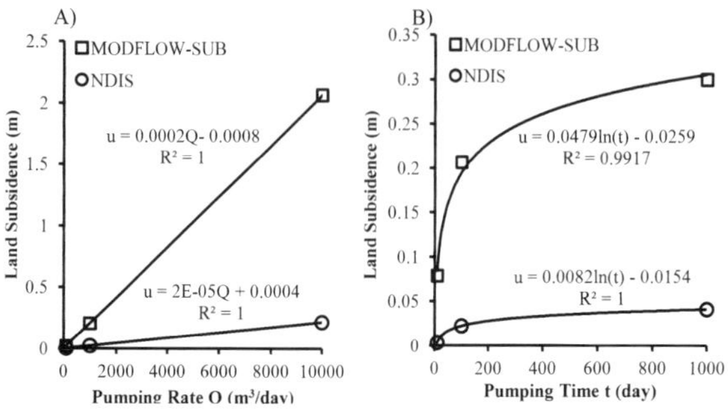

Figure 4. Sensitivity comparison using NDIS and MODFFLOW-SUB models. **(a)** Land subsidence vs. pumping rate. **(b)** Land subsidence vs. pumping time.

For horizontal movement (Figs. 3 and 5), the numerical solutions for the aquifer system with a leaky confining unit are consistent with the analytic solution developed by Li (2007). Namely, the aquifer movement has two zones. One is a compression zone and the other is an extension zone. The position of the boundary between the two zones is a function of time. This is because in the vicinity of the pumping well, the displacement of the solid matrix approaches zero as the well screen is assumed to be permeable only to water. The inward moving grains tend to accumulate in the surrounding area of the well and form a compression zone around the well that extends outward from the well as pumping continues.

The aquifers (Layers 1 and 7) show significant horizontal movement compared with the aquitards where there is negligible horizontal motion. Large radial strain occurs under compression near the well and decreases to zero at a transient point of the maximum radial displacement (Fig. 3). Similar to the relation between vertical displacement and distance from the pumping well, for horizontal displacement beyond the point of maximum horizontal motion the horizontal displacement gradually decreases along the radial direction. The displacement curves for all layers eventually approach the constant head boundary where strain becomes zero. The peak horizontal displacements in the confined (Layer 7) and unconfined aquifer (Layer 1) occur at about 300 and 700 m from the pumping well, respectively (Fig. 3).

Sensitivity analyses were conducted for different pumping rates and discharge periods to evaluate movement sensitivity using the NDIS and MODFLOW-SUB models. The results show that increasing pumping rate and time cause increases in displacements. Figure 4a and b show that subsidence simulated by NDIS is less sensitive than that by MODFLOW-SUB to changes in the pumping rate and time.

Horizontal land movement can result in the formation of earth fissures. Infrastructure associated with human development located in area affected by land-surface movements can result in a variety of potential problems. Figure 5 illustrates horizontal movement (displacement) and strain in the con-

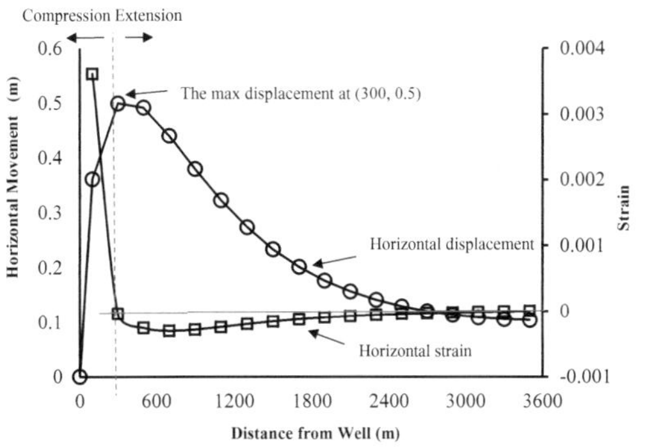

Figure 5. Horizontal movement and strain in Layer 7 after 100 days pumping.

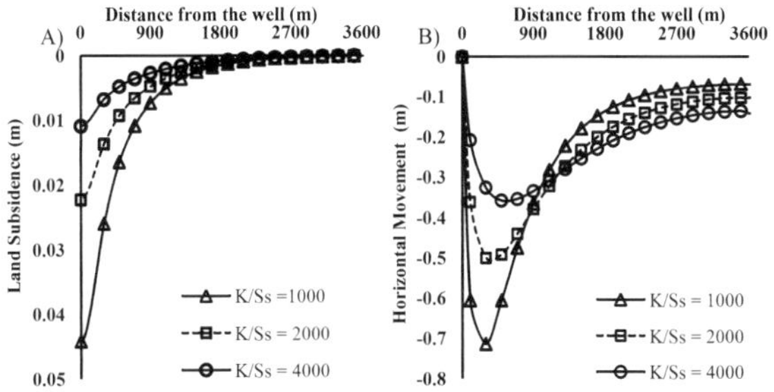

Figure 6. Sensitivity of displacement to hydraulic diffusivity (m^2/day) using the NDIS model **(a)** Subsidence, **(b)** Horizontal movement (Layer 7) after 100 days pumping.

fined aquifer (Layer 7). The peak value of the horizontal land subsidence in Layer 7 is located 300 m from the center of the well. This location is also a transition point between extensional and compressional strain.

Figure 6 shows the impact of hydraulic diffusivity on the vertical and horizontal movement (Layer 7) using the NDIS model. It is interesting to note that the vertical displacement at any location is decreasing along with increasing hydraulic diffusivity. However, this is true for horizontal displacement only when horizontal distance is less than 1000 m. When the horizontal distance larger than 1000 m, the horizontal displacement is increasing with increasing hydraulic diffusivity.

The plots in Fig. 7 compare the sensitivity of vertical displacement to hydraulic diffusivity for the NDIS and MODLFOW-SUB models. In general, subsidence decreases with increasing hydraulic diffusivity for the both models. However, the results show subsidence simulated by NDIS is less sensitive to hydraulics diffusivity than that simulated by MODFLOW-SUB.

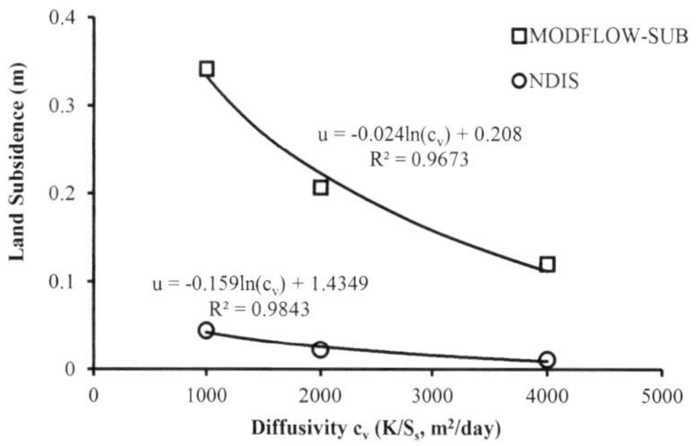

Figure 7. Comparison between NDIS and MODFLOW-SUB models for subsidence sensitivity to hydraulic diffusivity, c_V is diffusivity ($= K/S_S$, $m^2\,day^{-1}$).

5 Summary and conclusions

The summary and conclusions of this paper are given below:

- A new NDIS module for 3-D aquifer movement has been applied. The NDIS model is incorporated to the parent program MODFLOW-96 that was modified to solve a nonlinear partial differential equation for hydraulic head.

- Subsidence based on a hypothetical conceptual model is simulated using both the NDIS and MODFLOW-SUB models for purposes of comparison. The results indicate land subsidence simulated by MODFLOW-SUB is significantly larger than that simulated by NDIS.

- Two cases are considered for the conceptual model, one case has a single pumping well and the other case has three pumping wells, with two of the wells on opposite ends of a line intersecting, and at equal distances from, a center pumping well. The two cases are simulated with both the NDIS and MODLFOW-SUB models. The results show that for the same total discharge rate, the maximum land subsidence in response to a single pumping well is larger than that from the three pumping wells distributed as described above.

- Given the prescribed hydromechanical properties of the aquifers and aquitards in the conceptual model, the predominant horizontal land movement occurs in the aquifers, particularly the pumped aquifer. Horizontal motion in the aquitard's is relatively insignificant compared with the aquifer's. The characteristics of horizontal movement are consistent with the analytic solution of horizontal movement for an aquifer system with a leaky aquitard (Li, 2007). Namely, groundwater discharge radially induces two zones of horizontal deformation, compression and extension zones. The locations of the peak displacement are different in each of the two model layers representing the pumped and non-pumped aquifers and are a function of time.

- Sensitivity analysis using NDIS indicates that vertical displacement at a location reduces with increasing diffusivity at any location but this is true for horizontal displacement only within a 1000 m radius. Farther than 1000 m from the pumping well, the horizontal displacement increases with increasing hydraulic diffusivity.

- Comparison of sensitivity analysis between the NDIS and MODLFOW-SUB models shows vertical movement decreases with increasing of pumping rates and time for both the models. However, land subsidence simulated by NDIS is less sensitive to both the pumping rate and time than that simulated by MODFLOW-SUB.

Acknowledgements. The research and writing of this paper were supported by US Department of Energy under the contract DE-NA0000720 through the Samuel P. Massie Chairs of Excellence Program.

References

Bear, J.: Dynamics of fluids in porous media, Dover Publications, Inc., NY, 800 pp., 1972.

Helm, D. C.: A postulated relation between granular subsidence and Darcy's law for transient flow, in: Evaluation and Prediction of Subsidence, edited by: Saxena, S. K., ASCE, NY, 417–440, 1979.

Helm, D. C.: Analysis of sedimentary skeletal deformation in a confined aquifer and the resulting drawdown, in: Groundwater Hydraulic, edited by: Rosenshein, J. S. and Bennett, G. D., American Geophysical Union, Washington DC, 29–82, 1984.

Hubbert, K. M.: The theory of groundwater motion, J. Geology, 48, 745–944, 1940.

Hoffmann, J., Leake, S. A., Galloway, D. L., and Wilson, A. M.: MODFLOW-2000 ground-water model–user guide to the subsidence and aquifer-system compaction (SUB) package, USGS Open-File Report 03–233 Tucson, Arizona, 44 pp., 2003.

Leake, S. A. and Galloway, D. L.: MODFLOW ground-water model–user guide to the subsidence and aquifer-system compaction package (SUB-WT) for water-table aquifers, US Geological Survey, Techniques and Methods, 6–A23, Reston, VA, 42 pp., 2007.

Leake, S. A. and Prudic, D. E.: Documentation of a computer program to simulate aquifer-system compaction using the modular finite-difference ground-water flow model: US Geological Survey Techniques of Water- Resources Investigations, Book 6 Modeling Techniques, Chapter A2, US Geological Survey Open-File Report, Reston, VA, 88–482, 1991.

Li, J.: Transient radial subsidence of a confined leaky aquifer due to variable well flow rate, J. Hydrology, 333, 542–553, 2007.

Li, J. and Ding, D.: Modeling 3D land subsidence due to groundwater pumping with a variable parameter, Global View of Engineering Geology and the Environment, Beijing, China, CRC Press, Taylor & Francis Group, New York, 457–462, 2013.

Li, J. and Helm, D. C.: A theory of three-dimensional land motion in terms of its velocity field, Land Subsidence, the Proceedings of the 8th International Symposium on Land Subsidence, 17–22 October 2010, Queretaro, Mexico, 475–483, 2010.

McDonald, M. G. and Harbaugh, A. W.: A modular three-dimensional finite-difference ground-water flow model, Techniques of Water-Resources Investigations of the USGS: Chapter A1, Book 6, USGS, USGS, Reston, VA, 586 pp., 1988.

Terzaghi, K.: Settlement and consolidation of clay, Eng. News, Rec., 95, 874–878, 1925.

Zhang, L.: A model for transient three-dimensional underground deformation in response to groundwater pumpage, Doctoral Dissertation of Morgan State University, UMI/ProQest, Ann Arbor, MI, 182 pp., 2009.

Proc. IAHS, 372, 443–448, 2015
proc-iahs.net/372/443/2015/
doi:10.5194/piahs-372-443-2015

Open Access
PIAHS

Three dimensional numerical modeling of land subsidence in Shanghai

S. Ye[1], Y. Luo[1], J. Wu[1], P. Teatini[2], H. Wang[3], and X. Jiao[3]

[1]School of Earth Sciences and Engineering, Nanjing University, Nanjing, China
[2]Department of Civil, Environmental and Architectural Engineering, University of Padova, Padova, Italy
[3]Key Laboratory of Land Subsidence, Shanghai Institute of Geological Survey, Shanghai, China

Correspondence to: S. Ye (sjye@nju.edu.cn)

Published: 12 November 2015

Abstract. Shanghai city has been suffering land subsidence caused by overly exploitation of ground water since 1921, which is a serious problem for this coastal city with altitude of 2.2–4.8 m above mean sea level. The largest cumulative land subsidence amounted to 2.6 m in the downtown area. Measures to decrease the ground water exploitation, change the pumping aquifers, and increase aquifer artificial recharge have been used to mitigate land subsidence since 1961. It is necessary to develop a proper numerical model to simulate and predict land subsidence. In this study, a decoupled three-dimensional (3-D) finite element land subsidence model including a 3-D ground water flow model and a 3-D geo-mechanical model was developed to simulate the 3-D deformation of the aquifer systems in the center area of Shanghai. The area of downtown Shanghai is 660 km^2, with 10 million inhabitants, dense high buildings, and 11 metro lines. The simulation spans the period from 1979 to 1995. Two different assumptions have been tested on the side boundary, i.e., precluding the three components of the displacement, or assuming a free-displacement condition. The distribution of calculated land subsidence and horizontal displacements in different aquifers was analyzed. The computed vertical displacement fitted well with the available observations. It has been verified that the two different assumptions on the lateral boundaries in the geo-mechanical model caused different results just limited on nodes close to boundary. The developed 3-D land subsidence model is reasonable and can be used to simulate and predict 3-D movement of aquifer systems in the center area of Shanghai, which could provide scientific support to local government in controlling land subsidence and differential movements of the land surface.

1 Introduction

Anthropogenic land subsidence in Shanghai, which is induced by excessive groundwater withdrawal, has been observed over the last 94 years, i.e. since the year 1921 (Zhang and Wei, 2002). Shanghai government started to set up extensive monitoring networks of groundwater levels and land subsidence from the middle 1980s. Some earlier developed models (Gu et al., 1990, 1991, 1993) were focused on the shallow sedimentary layers, i.e. the upper 100 m depth above the second confined aquifer, and did not consider the whole aquifer system. Soil deformation was assumed one dimensional (1-D). The researchers from the Nanjing University, China, and the Shanghai Institute of Geological Survey have conducted comprehensive studies on numerical modeling of land subsidence since 2000. A regional land subsidence model, composed of a three dimensional (3-D) groundwater flow module and a 1-D subsidence module embodying complex visco-elasto-plastic deformation characteristics with variable parameters, was developed and applied to simulate and predict land subsidence in Shanghai (Ye, 2004; Ye et al., 2005; Xue et al., 2008; Shi et al., 2008; Wu et al., 2010; Ye et al., 2011, 2012). The 3-D groundwater flow model is used to calculate the time-dependent hydraulic heads in the sedimentary sequence, and the 1-D subsidence model is used to compute the time-dependent compaction of the sedimentary layers and the cumulative land subsidence.

All the land subsidence models mentioned above are based on the assumption of a 1-D deformation field in the vertical direction. However, the actual soil deformation caused

Published by Copernicus Publications on behalf of the International Association of Hydrological Sciences.

by groundwater withdrawal is fully three dimensional. The deformation of an aquifer system caused by fluid withdrawal is theoretically described by the 3-D fully coupled poroelasticity model originally developed by Biot (1941). Because of ill-conditioning and large CPU-time consume of coupled models when applied to regional land subsidence problems, uncoupling between the flow and the strain fields is usually assumed in the classical groundwater hydrology. In an uncoupled formulation, also known as a two-step or explicitly coupled approach (Teatini et al., 2010), the groundwater flow equation is first solved and then land subsidence is computed solving the equilibrium equations with the groundwater changes specified as an external distributed source of strength within the porous medium.

A 3-D uncoupled land subsidence model is developed in this study to simulate 3-D displacements caused by ground water exploitation from the multiaquifer system in the center area of Shanghai. Downtown Shanghai is bounded by the outer ring road as shown in Fig. 1, with an area of 660 km^2 and the largest cumulative land subsidence of 2 m since 1921. This zone is the most important area in Shanghai, with the densest 10 million population and concentration in high-rise buildings, and crossed by 11 metro lines. Therefore, it is important to calculate the 3-D deformation of the aquifer system caused by ground water exploitation to better evaluate the possible impact on the underground infrastructures such as metro lines, tunnels etc.

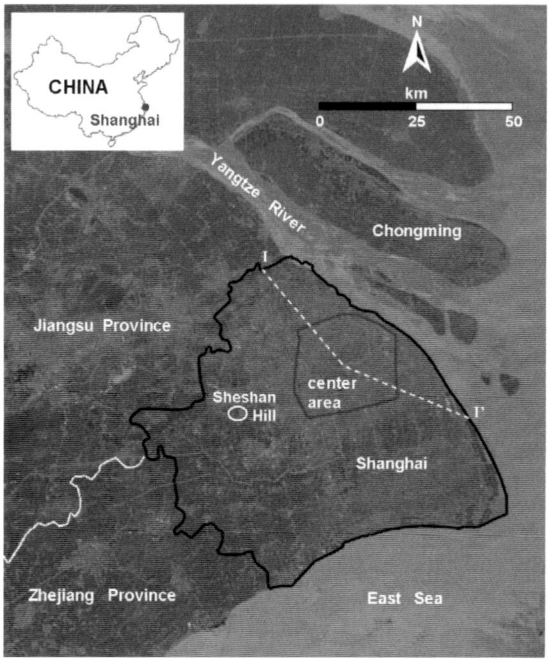

Figure 1. Location of the Shanghai City and the center area addressed in the presentmodelling study.

2 Model setup

In geomechanical investigations at regional scale, uncoupled pressure – displacements solution can be safely used in predicting land subsidence in compacting sedimentary on any time scale of practical interest (Gambolati et al., 2000; Teatini et al., 2006, 2010). In this study, an uncoupled approach is thus followed, with the flow field initially computed at each time step and the deformation then calculated using the pore pressure gradient as a known source of strength in the geomechanical model. The 3-D uncoupled land subsidence model is expressed as (Verruijt, 1969):

$$\frac{E}{2(1+\nu)}\left(\frac{1}{1-2\nu}\frac{\partial\varepsilon}{\partial x}+\nabla^2 u_x\right)=\frac{\partial P}{\partial x} \tag{1a}$$

$$\frac{E}{2(1+\nu)}\left(\frac{1}{1-2\nu}\frac{\partial\varepsilon}{\partial y}+\nabla^2 u_y\right)=\frac{\partial P}{\partial y} \tag{1b}$$

$$\frac{E}{2(1+\nu)}\left(\frac{1}{1-2\nu}\frac{\partial\varepsilon}{\partial z}+\nabla^2 u_z\right)=\frac{\partial P}{\partial z} \tag{1c}$$

$$\frac{\partial}{\partial x_i}\left(K_{ij}\frac{\partial H}{\partial x_j}\right)=S_s\frac{\partial H}{\partial t}+qi,\, j=x,y,z \tag{1d}$$

where E is the Young modulus, ν is the Poisson ratio, u_x, u_y and u_z are the displacement components along the coordinate axes x, y and z, respectively, ε is the incremental volume strain $\varepsilon=\frac{\partial u_x}{\partial x}+\frac{\partial u_y}{\partial y}+\frac{\partial u_z}{\partial z}$, P is the incremental pore water pressure, which is equals to $\gamma_w \cdot H$ with γ_w the specific weight of water and H the incremental hydraulic head, and ∇^2 is the Laplace operator. In Eq. (1d), $K_{i,j}$ is the hydraulic conductivity tensor, t is time, q is the source/sink term, and S_s is the specific storage $S_s=\gamma_w(\alpha+n\beta)$, with α the vertical oedometric compressibility, which is related to the Young modulus and Poisson ratio through $\alpha=\frac{(1+\nu)(1-2\nu)}{E(1-\nu)}$, n is the medium porosity, and β the water volumetric compressibility. The parameter α is not constant and takes the elastic value α_{ke} when the effective stress is less than the preconsolidation stress, otherwise the plastic value α_{kv} (Ye et al., 2011, 2012). The specific storage is denoted by S_{ske} and S_{skv}, accordingly. Special care must be taken in the explicitly coupled approach to use S_s values in the groundwater flow Eq. (1d) consistent with those used for the geomechanical parameters E and ν in the equilibrium Eqs. (1a)–(1c).

The simulated domain, which is the center area of Shanghai (Fig. 1), is comprised between the ground surface and a rigid 250–300 m deep basement (Fig. 2). A 3-D mesh consisting of 19 524 nodes and 34 309 triangular prisms is developed to accurately represent the multiaquifer system including six aquifers and five aquitards (Fig. 3). Zero flux on the basement and Dirichlet conditions on the top and outer boundaries are prescribed in the flow model. The known hydraulic heads on the top and outer boundaries are obtained from the ground flow model developed for the whole Shanghai city from 1961 to 2005 (Xue et al., 2008). A Dirichlet

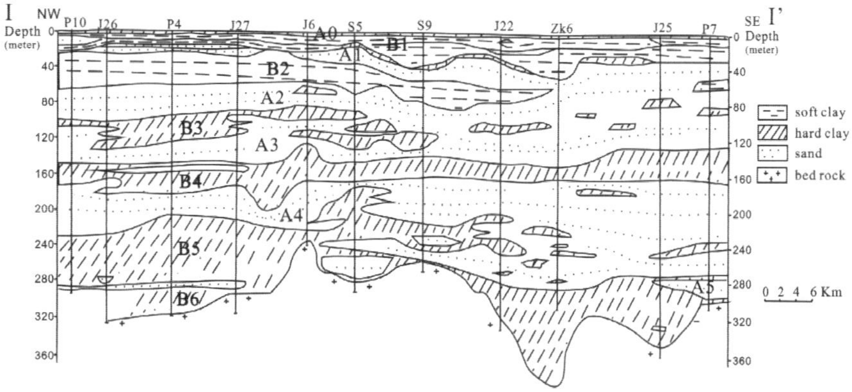

Figure 2. Schematic stratigraphic section in Shanghai (after Ye et al., 2011).

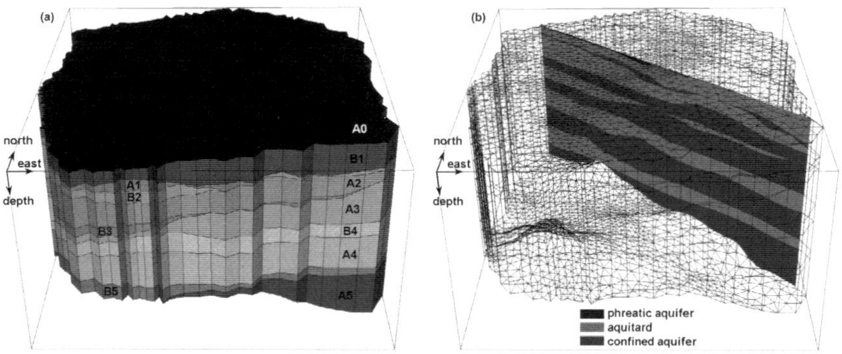

Figure 3. Perspective views of the 3-D finite-element mesh of the center area of Shanghai. The various geologic layers are distinguished in **(a)**. In **(b)** the hydro-stratigraphic units are highlighted on the center part of the vertical section passing along the I-I' profile traced in Fig. 2.

condition with fixed bottom boundary and a traction©/free land surface are assumed in the geomechanical model. Two different boundary conditions are tested on the outer boundary in the geomechanical model. In the first case a Dirichlet condition with fixed boundary is prescribed, and in the other a traction©/free outer boundary is used.

The simulations span the interval from 30 December 1979 to 30 December 1995. There are many recharge wells and a few discharge wells in the second confined aquifer, and many discharge wells with high flow rates and some recharge wells with low flow rates in the fourth confined aquifer. The 16-year long simulation period is divided into 64 time steps 3-month long each. The initial groundwater levels in aquifer A2 ranged from -1.5 to 2 m above mean sea level (m.s.l.) with a flow generally from west to east, and that in aquifer A4 ranged from -22 to -10 m above msl with a flow generally from east to west.

3 Modeling results and discussion

3.1 Parameter subzones and parameter values

The whole studied aquifer system is finally divided into 103 subzones according to the difference in lithology, hydraulic, and mechanical properties of sediments. Due to lack of specific information, the Poisson ratio ν is set to 0.3 which is a typical values for recent sedimentary deposits. Hydraulic conductivities, elastic specific storage, and plastic specific storage of aquifers and aquitards are provided in Table 1.

3.2 Ground water levels

Figure 4 compares the model results and the measured head in a few observation wells. The simulated heads agree generally well with the observed heads.

Figures 5 shows the groundwater head maps in the second confined aquifer on 30 December 1995, in which the simulation outcomes agree well with the maps derived by interpolating the available records.

Table 1. Hydro-geo-mechanical Parameters Used in the Final Simulation of the 3-D Groundwater Flow and Geomechanical Models.

Aquifers/Aquitards	K_{xx} (m/d)	K_{yy} (m/d)	K_{zz} (m/d)	S_{ske} (1/m)	S_{skv} (1/m)
A1	1–10	1–10	0.1–1	10^{-6}–10^{-5}	10^{-6}–10^{-5}
A2	2–50	2–50	0.2–5	10^{-6}–10^{-5}	10^{-6}–10^{-5}
A3	7–30	7–30	0.7–3	10^{-6}–10^{-5}	10^{-6}–10^{-5}
A4	1–40	1–40	0.1–4	10^{-6}–10^{-5}	10^{-6}–10^{-4}
A5	1–20	1–20	0.1–2	10^{-6}–10^{-5}	10^{-6}–10^{-5}
B1–B5	10^{-2}–10^{-5}	10^{-2}–10^{-5}	10^{-3}–10^{-6}	5×10^{-6}–5×10^{-5}	9×10^{-6}–8×10^{-4}

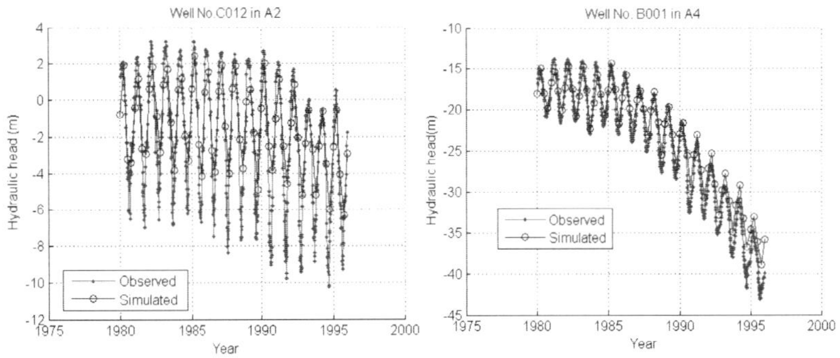

Figure 4. Calculated vs. measured groundwater levels in two boreholes.

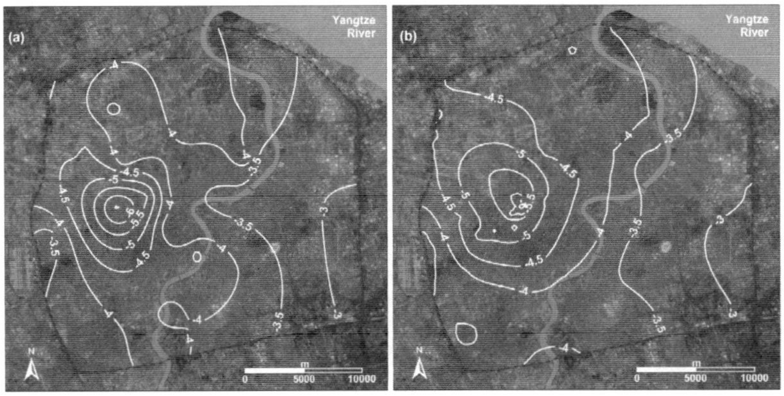

Figure 5. Computed **(a)** and measured **(b)** groundwater level in aquifer A2 on 30 December 1995, i.e. at the end of the simulation period.

3.3 Deformation and land displacements

Figure 6 compares the model results in term of vertical compaction/expansion and the observed deformation in a few depth intervals monitored at the F10 extensometric station. Both the computed long-term compaction as well as the seasonal compaction/expansion are in good agreement with the measurements.

Figure 7 shows the vector maps of the cumulative horizontal displacement of the bottom of the fourth confined aquifer. The maximum values are calculated in the range between 4 and 8 mm. The largest horizontal displacements are obtained in the north-eastern part of the map, which corresponds to the zone with the largest changes in groundwater levels and also the highest compressibility.

3.4 Impacts of boundary conditions of 3-D geomechanical model

As reported above, two different outer boundary conditions are tested in the 3-D geomechanical model. One is a fixed boundary, and the other one is a traction-free boundary. Fig-

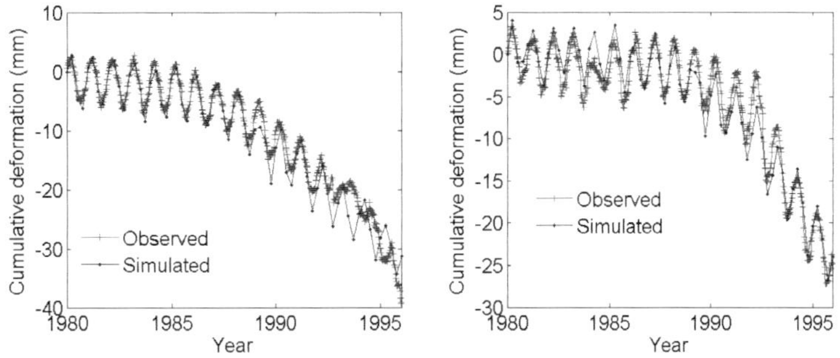

Figure 6. Simulated and measured vertical deformation (negative values mean compaction) for two depth intervals at the F10 extensometer. **(a)** 165–294 m depth range; **(b)** 26–78 m depth range.

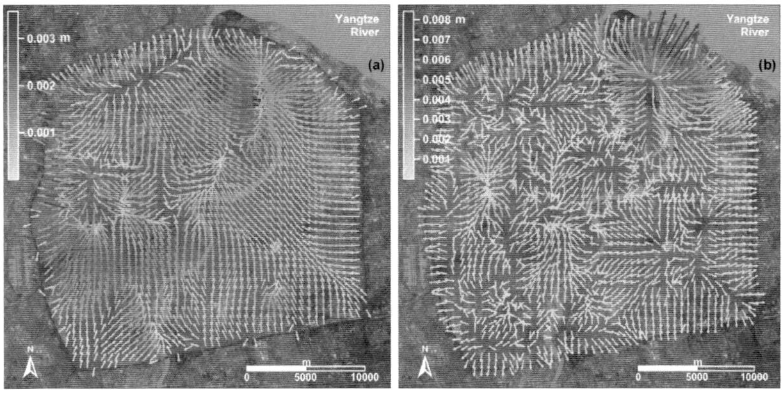

Figure 7. Cumulative horizontal displacements at 30 December 1995, as provided by the 3-D geomechanical model in aquifer A4.

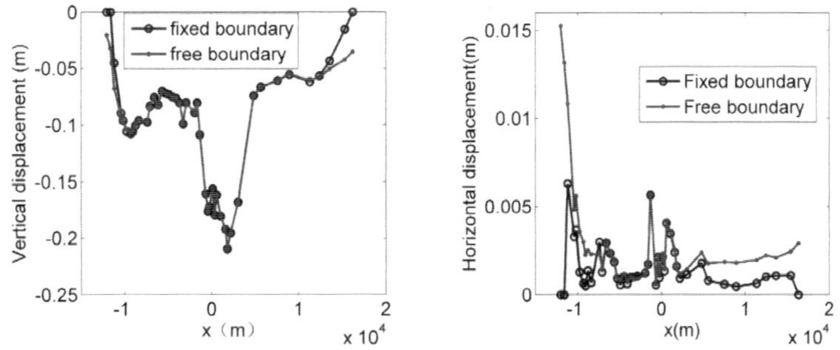

Figure 8. Comparison between the simulated vertical **(a)** and horizontal **(b)** displacements provided by the geomechanical model with different boundary conditions along a west-east profile crossing the downtown area.

ure 8 shows the simulated vertical and horizontal displacements as obtained implementing the two outer boundary conditions along a west-east profile through the center of the study area. The differences between the results are significant only in a relatively narrow strip along the domain boundary.

4 Conclusions

A 3-D one-way coupled geomechanical model is developed to simulate the groundwater level changes and 3-D displacements in the aquifer system at the center area of Shanghai due to groundwater withdrawal and recharge between 30 December 1979 and 30 December 1995. Elastic and plastic compressibility values are appropriately considered in the groundwater flow and geomechanical models depending on the actual effective stress with respect to the preconsolidation stress. The model outcome satisfactorily agrees with the available observations. The cumulative land subsidence reaches 250 mm in the downtown area and the cumulative horizontal displacement peaks 11 mm on the ground surface and is smaller than 8.5 mm at depth.

Fixed and traction-free conditions are tested on the outer boundary of the geomechanical model. The differences that remain limited to a relatively narrow strip close to the model frontier. The horizontal displacements appear more influenced by the boundary conditions than land subsidence. Although no measurement is available for the model validation in term of horizontal displacements, their magnitudes properly resemble the distribution of the hydraulic head changes and the compressibility of aquifers and aquitards. Hence, it can be concluded that the 3-D uncoupled geomechanical model can well simulate the three-dimensional displacement field of the aquifer system in the center area of Shanghai.

Acknowledgements. Funding supported by KLLSMP No. 201401, NSFC No. 41272259 and NSF of Jiangsu Province No. BK2012730 is appreciated. Pietro Teatini was partially supported by the University of Padova, Italy, within the 2014 International Cooperation Programme.

References

Biot, M. A.: General theory of three-dimensional consolidation, J. Appl. Phys., 12, 155–164, 1941.

Gambolati, G., Teatini, P., Bau, D., and Ferronato, M.: Importance of poroelastic coupling in dynamically active aquifers of the Po river basin, Italy, Water Resour. Res., 36, 2443–2459, 2000.

Gu, X. Y., Gong, S. L., and Huang, H. C.: Quantitative study of land subsidence in Shanghai, Proc. VI International Congress of IAEG, 1990.

Gu, X. Y., Tsien, S. I., Huang H. C., Liu, Y.: Analysis of Shanghai Land Subsidence, Proc. IV International Symposium on Land subsidence, Publ. 200, 603–612, 1991.

Gu, X. Y., Deng, W., Xu, D. N., and Liu, Y.: Computation of land subsidence in Shanghai with secondary consolidation effect, Proc. International conference on soft soil engineering, 65–70, 1993.

Shi, X., Wu, J., Ye, S., Zhang, Y., Xue, Y. Q., Wei, Z. X., Li, Q. F., and Yu, J.: Regional land subsidence simulation in Su-Xi-Chang area and Shanghai City, China. Eng. Geol., 100, 27–42, 2008.

Teatini, P., Ferronato, M., Gambolati, G., and Gonella, M.: Groundwater pumping and land subsidence in the Emilia-Romagna coastland, Italy: Modeling the past occurrence and the future trend, Water Resour. Res., 42, W01406, doi:10.1029/2005WR004242, 2006.

Teatini, P., Ferronato, M., Gambolati, G., Baù, D., and Putti, M.: Anthropogenic Venice uplift by seawater pumping into a heterogeneous aquifer system, Water Resour. Res., 46, W11547, doi:10.1029/2010WR009161, 2010.

Verruijt, A.: Elastic storage of aquifers, in: Flow through porous media, Academic Press, New York, 331–375, 1969.

Wu, J., Shi, X., Ye, S. J., Xue, Y. Q., Zhang, Y., Wei, Z. X., and Fang, Z.: Numerical Simulation of Viscoelastoplastic Land Subsidence due to Groundwater Overdrafting in Shanghai, China. J. Hydrologic Eng., 15, 223–236, 2010.

Xue, Y, Wu, J., Zhang, Y., Ye, S., Shi, X. Q., Wei, Z.X., Li, Q. F., and Yu, J.: Simulation of regional land subsidence in the southern Yangtze Delta, Science in China Series D-Earth Sciences, 51, 808–825, 2008.

Ye, S.: Study on the Regional Land Subsidence Model and Its Application, Ph.D Thesis, Nanjing University, China, 2004.

Ye, S., Xue, Y., Wu, J., Zhang, Y., Wei, Z. X., and Li, Q. F.: Regional land subsidence model embodying complex deformation, ICE-Water Management, 164, WM10, 519–531, 2011.

Ye, S., Xue, Y., and Wu, J. C., and Li, Q. F.: Modeling visco-elastic-plastic deformation of soil with Modified Merchant Model, Environ. Earth Sci., 66, 1497–1504, 2012.

Ye, S. J., Xue, Y. Q., Wu, J. C., Zhang, Y., Wu, J. F., and Li, Q.: Study on the groundwater flow model for land subsidence modeling in Shanghai, Proc of the Seventh International Symposium on Land Subsidence, Shanghai, China, 2005.

Zhang, A. G. and Wei, Z. X.: Past, present and future research on land subsidence in Shanghai (in Chinese), Hydrogeol. Eng. Geol., 33, 72–75, 2002.

Proc. IAHS, 372, 449–453, 2015
proc-iahs.net/372/449/2015/
doi:10.5194/piahs-372-449-2015

Open Access
PIAHS

Examples of deformation-dependent flow simulations of conjunctive use with MF-OWHM

R. T. Hanson, J. Traum, S. E. Boyce, W. Schmid, and J. D. Hughes
U.S. Geological Survey, 4165 Spruance Rd. Suite 200, San Diego, CA 92101, USA
Correspondence to: R. T. Hanson (rthanson@usgs.gov)

Published: 12 November 2015

Abstract. The dependency of surface- and groundwater flows and aquifer hydraulic properties on deformation induced by changes in aquifer head is not accounted for in the standard version of MODFLOW. A new USGS integrated hydrologic model, MODFLOW-OWHM, incorporates this dependency by linking subsidence and mesh deformation with changes in aquifer transmissivity and storage coefficient, and with flows that also depend on aquifer characteristics and land-surface geometry. This new deformation-dependent approach is being used for the further development of the integrated Central Valley hydrologic model (CVHM) in California. Preliminary results from this application and from hypothetical test cases of similar systems show that changes in canal flows, stream seepage, and evapotranspiration from groundwater (ET_{gw}) are sensitive to deformation. Deformation feedback has been shown to also have an indirect effect on conjunctive surface- and groundwater use components with increased stream seepage and streamflows influencing surface-water deliveries and return flows. In the Central Valley model, land subsidence may significantly degrade the ability of the major canals to deliver surface water from the Delta to the San Joaquin and Tulare basins. Subsidence can also affect irrigation demand and ET_{gw}, which, along with altered surface-water supplies, causes a feedback response resulting in changed estimates of groundwater pumping for irrigation. This modeling feature also may improve the impact assessment of dewatering-induced land subsidence/uplift (following irrigation pumping or coal-seam gas extraction) on surface receptors, inter-basin transfers, and surface infrastructure integrity.

1 Introduction and background

With conjunctive use of surface water and groundwater, excessive groundwater withdrawals in alluvial aquifers can lead to a physical reconfiguration of hydrostratigraphy, aquifer properties, surface infrastructure, and conjunctive water-use management owing to the effects of land subsidence. This has occurred in many of the major alluvial aquifers throughout the world. For example, in California's Central Valley, groundwater pumping resulting from the growing irrigation demand has led to significant land subsidence in the areas of the western San Joaquin Valley (Faunt et al., 2009, 2011) and, to the south, in the Tulare Basin, where historical declines in groundwater levels were greatest and are predicted to continue with climate change (Hanson et al., 2012). Subsidence and differential subsidence also are affecting the integrity and efficiency of the Delta Mendota Canal and California Aqueduct along the west side of the valley (Sneed et al., 2013).

Conjunctive use is the combined use of surface water and groundwater (UNESCO, 2010). The evolution of conjunctive-use management has come to include a framework of both natural and anthropogenic components of surface water and groundwater resources. The engineered components of supply and demand have become more sophisticated and ubiquitous in the urban and agricultural landscape. Many of these components such as canal operation, pumping of multiple-aquifer wells, aquifer storage and recovery, and reuse have an effect on conjunctive-use strategies. The connection to potential land subsidence has not been systemically linked to key processes and hydraulic properties of groundwater flow in order to understand the effects on conjunctive use. The feedback from these and other components that are linked to conjunctive use may require combined as-

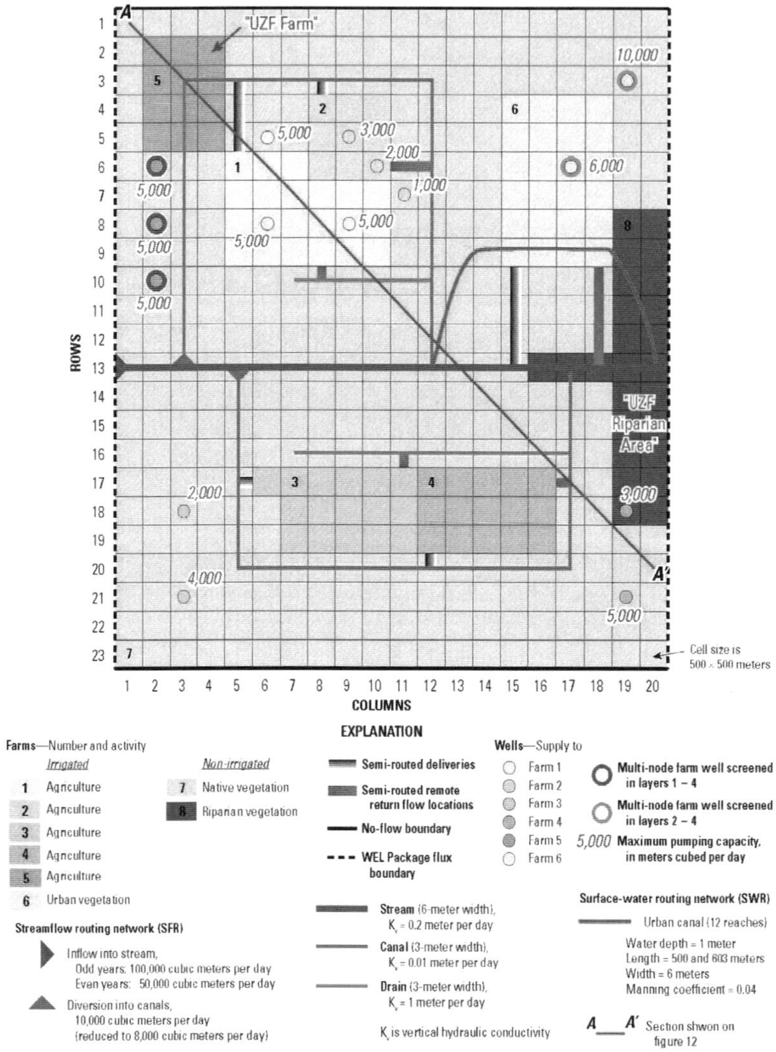

Figure 1. Plan view of model domain, grid resolution, boundary conditions, distribution of farms and farm wells, and streamflow routing network with points of diversion to farms and points of return flow from farms and surface-water canal traversing urban area (Schmid et al., 2014).

sessment within an integrated physically-based and supply-and-demand modeling structure.

Linking the effects of subsidence to changes in aquifer properties and other components of the hydrologic landscape is critical to conjunctive-use management. This type of analysis helps to identify and quantify the flow capacities and limits that these components can bring to conjunctive use. Flow capacities and limits may depend on aquifer head, other flow components, or deformation, and, hence, cannot be directly quantified through simple water allocation models. One of the current issues for regions within which large amounts of surface water are transported via canal systems is the potential effects on infrastructure from land subsidence. In particular, reduced surface-water supplies owing to dry climate and environmental constraints are being offset by additional groundwater pumpage that can cause additional land subsidence. The feedback from this cause-and-effect linkage within a conjunctive-use system is an impairment of the conveyance and integrity of the canal systems as well as changing related streamflows and runoff. Land subsidence and differential land subsidence can affect elevations and slopes across the landscape, which can result in reduction in freeboard, reduced or reversed slopes, and even fractures in the canals. Similarly, the ability to convey natu-

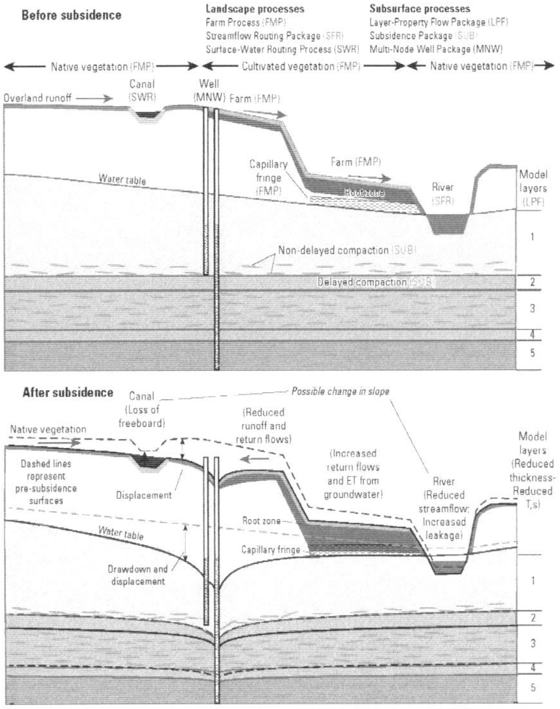

Figure 2. Diagram showing the relation between surface and subsurface processes with linkage to land subsidence (Schmid et al., 2014; Hanson et al., 2014).

ral streamflow, deliver water for irrigation, return landscape runoff to streams, or reuse such water is directly and indirectly affected by land subsidence. Finally, the effects of land subsidence are also occurring in the subsurface where compaction is releasing water from fine-grained interbed storage and, as a result, aquifer systems suffer reduced storage and transmission properties associated with the compaction.

MODFLOW (MF) uses a modular approach to represent surface processes that can interact with groundwater. In contrast to simulations that include only head-dependent flows, the assessment of conjunctive-use problems requires simulations with additional connectivity of flows of one stress term that directly depend on flows of other stress terms. For the analysis of deformation-dependent flows, a version of MODFLOW (MF) called MODFLOW-One Water Hydrologic Model (MF-OWHM, Hanson and Schmid, 2013) has been developed to include linkages between deformation, aquifer properties, and affected flow terms.

2 Coupled land subsidence

In MF-OWHM (Hanson et al., 2014), new linkages were provided for the Subsidence Package (SUB) (Hoffmann et al., 2003) with other packages and processes in a fully integrated hydrologic model, which calculate flow rates that are affected by the deformation of the land surface or hydrostratigraphic boundaries (Fig. 1). The new capability allows MF-OWHM to adjust elevations of the streambeds, canal and surface-water control-structure elevations, the land surface used by various agricultural, drain, and riparian evapotranspiration, and the tops and bottoms of aquifer layers. These changes also affect secondary calculations, such as the slope of a streambed, slope-related runoff in farms, proximity of the capillary fringe to the root zone, and changing hydraulic properties (transmissivity and storage coefficient) of aquifer layers (Fig. 2). The effects of these deformation-dependent flows were first analyzed with respect to a hypothetical problem that had typical properties and features (Schmid et al., 2014). This analysis demonstrated significant effects from the direct feedback of deformation onto physical properties of the land surface and subsurface. The indirect feedback of deformation on conjunctive use is facilitated by connecting the effects of layer deformation and associated changed transmissivities to changes in groundwater, landscape, and surface water flows. These, in turn, may affect changes in surface-water and groundwater deliveries to farms. One might also consider a third level of feedback of changes of surface-water and groundwater deliveries back to further changes in subsidence. That is, for instance, a reduction in surface-water deliveries might contribute to increased groundwater pumping and thus increased subsidence.

A realistic hypothetical example (Fig. 1) was used to explore and demonstrate the effects of the subsidence linkage within a fully coupled hydrologic model (Schmid et al., 2014). The coupling also affects the differential deformation of each layer (Fig. 3a). The supply-and-demand timing and amount of surface water available for conjunctive use is affected within the valley, along the urban canal, and as an inflow to the downstream valley (Fig. 3b). Finally, the stage (and related freeboard) and flows in the urban canal are also adversely affected by land subsidence (Fig. 3b). There are reductions in outflow of as much as 8 % during the summer months and a reduction in stage elevation in the urban canal of as much as 0.62 m that is potentially a reduction in freeboard simulated with the Surface Water Routing Process (SWR). The reductions in flows and stage affect the ability to deliver water through the urban canal and ultimately may also affect inter-basin transfers as deformation affects the integrity of the physical structures. The coupled deformation also affects water use and movement across the landscape related to irrigation and urban deliveries.

Substantial relative temporal and overall differences in streamflow occur at diversions to (Fig. 3c) and return flows (Fig. 3c) from farms that affect the amounts of surface water that is available to downstream farms for irrigation. There are also increased deliveries and reduced return flows during periods of subsidence. Thus, the amounts and sources of irrigation water are altered and the amount of water that is potentially divertible again for downstream farms is affected.

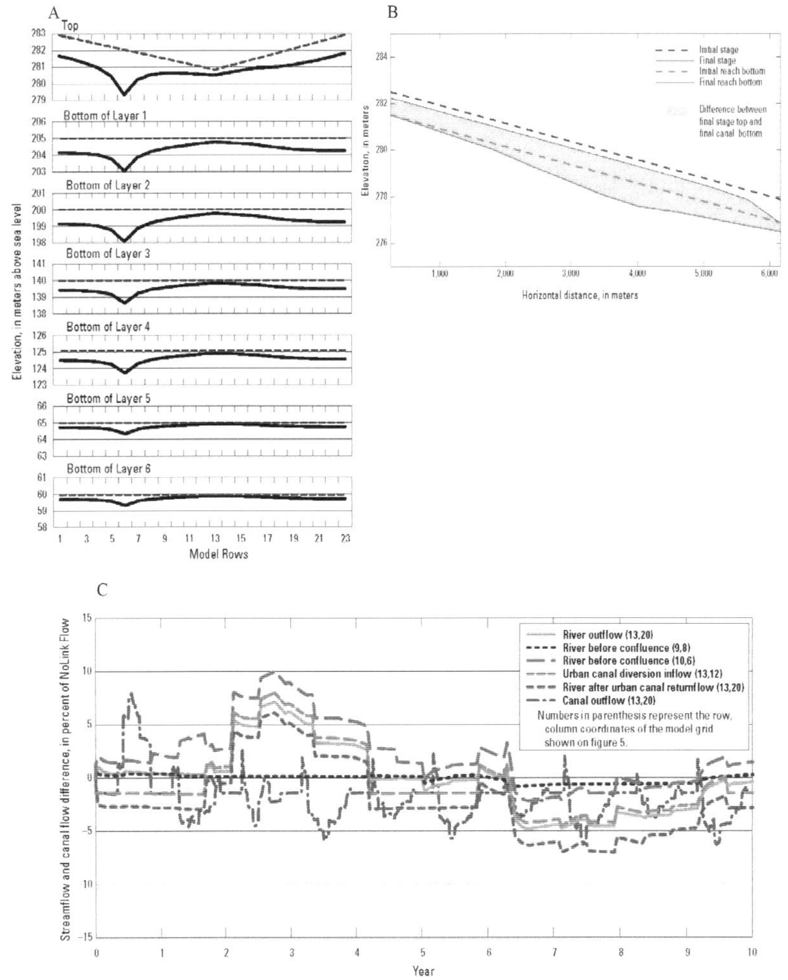

Figure 3. Example of simulated changes in bed altitudes **(a)** along model column 17 showing the final displacement of land surface and layer bottoms using SUB-Link (dashed line initial surface and solid line final surface), **(b)** with the deformable mesh using the Surface-water Routing Process of MF-OWHM along the urban canal, and **(c)** Difference in streamflow between using SUB-Link and using just SUB with no linkage (Schmid et al., 2014).

A real world example is under development for the Central Valley of California, where deformation of Delta Mendota Canal will be simulated by using the updated and refined Central Valley Model (Faunt et al., 2011). Examples of this development suggest that there are significant effects on the gradient of the canal and the bottom elevation, as well as differential elevation along the canal.

3 Summary and conclusions

The vertical displacement of the land surface and model-layer tops and bottoms calculated by the MODFLOW Subsidence Package has been linked to other MODFLOW packages and processes that are affected by the altitude and slope of the land surface as well as by the model layer thicknesses. In a model of a hypothetical aquifer system, these feedback effects of land surface and mesh deformation were found to be significant. The effects were large enough spatially and temporally to affect the supply-and-demand components of conjunctive water use. Thus, modeling deformation-dependent flows and hydraulic properties may be necessary to investigate the rate limiting aspects of conjunctive water use for a variety of flows in subsidence prone regions. In addition, the modeled feedback of deformation-dependent flows resulted in increased or decreased supplies and reduced or increased groundwater pumpage. That is, in some regions changes in land surface could either enhance or impede the use and movement of water.

The effects of feedback onto the land surface and aquifers from mesh deformation with MF-OWHM flow terms were found to be significant when compared to simulations without the subsidence linkage. The inclusion of land subsidence resulted in an even larger difference when compared to simulations that did not consider land subsidence for a hypothetical alluvial aquifer that is subject to sustained agricultural and urban demand that drive related secondary effects such as land subsidence. Therefore, these linkages are best suited for evaluating conjunctive water use where the vertical displacements or differential displacements may affect the sources of water, the proportions of multiple sources of water, as well as their use and movement across the landscape, the performance of conveyance through canals and rivers across the landscape, and the flow to and from the aquifers. Future work will involve applying methods described here to a real-world model of the Central Valley of California, USA.

References

Faunt, C. C., Hanson, R. T., Belitz, K., Schmid, W., Predmore, S. P., Rewis, D. L., and McPherson, K. R.: Numerical model of the hydrologic landscape and groundwater flow in California's Central Valley (Chapter C), in: Groundwater Availability of the Central Valley Aquifer, edited by: Faunt, C. C., California, U.S. Geological Survey Professional Paper 1776, 121e212, 2009.

Faunt, C., Hanson, R., Martin, P., and Schmid, W.: Planned Updates and Refinements to the Central Valley Hydrologic Model with an Emphasis on Improving the Simulation of Land Subsidence in the San Joaquin Valley, World Environmental and Water Resources Congress 2011, 864e870, 2011.

Hanson, R. T. and Schmid, W.: Economic resilience through "One-Water" Management, U.S. Geological Survey Open-File Report 2013–1175, p. 2, 2013.

Hanson, R. T., Flint, L. E., Flint, A. L., Dettinger, M. D., Faunt, C. C., Cayan, D., and Schmid, W.: A method for physically based model analysis of conjunctive use in response to potential climate changes, Water Resour. Res., 48, 23, doi:10.1029/2011WR010774, 2012.

Hanson, R. T., Boyce, S. E., Schmid, W., Hughes, J. D., Mehl, S. M., Leake, S. A., Maddock III, T., and Niswonger, R. G.: MODFLOW-One-Water Hydrologic Flow Model (OWHM), U.S. Geological Survey Techniques and Methods 6-A51, p. 122 (http://pubs.usgs.gov/tm/tm6a51/) 2014.

Hoffmann, J., Leake, S. A., Galloway, D. L., and Wilson, A. M.: MODFLOW-2000 Groundwater Model–user Guide to the Subsidence and Aquifer-system Compaction (SUB) Package, U.S. Geological Survey Open-File Report, 03e233, 46 pp., 2003.

Schmid, W., Hanson, R. T., Hughes, J., Leake, S. A., and Niswonger, R.: Feedback of land subsidence on the movement and conjunctive use of water resources, Environ. Modell. Softw., 62, 253–270, 2014.

Sneed, M., Brandt, J., and Solt, M.: Land subsidence along the Delta-Mendota Canal in the northern part of the San Joaquin Valley, California, 2003–10, U.S. Geological Survey Scientific Investigations Report 2013–5142, p. 87, doi:10.3133/sir20135142, 2013.

UNESCO: United Nations Organization for Education, Science and Culture (UNESCO) Glossary, http://webworld.unesco.org/water/ihp/db/glossary/glu/EN/GF0237EN.HTM (last access: 1 August 2014) 2010.

Proc. IAHS, 372, 455–462, 2015
proc-iahs.net/372/455/2015/
doi:10.5194/piahs-372-455-2015

Open Access PIAHS

Numerical and experimental study of strata behavior and land subsidence in an underground coal gasification project

N. N. Sirdesai[1], R. Singh[1], T. N. Singh[1], and P. G. Ranjith[2]

[1]Department of Earth Sciences, Indian Institute of Technology Bombay, Mumbai, India
[2]Deep Earth Energy Lab, Monash University, Clayton, VIC, 3800, Australia

Correspondence to: N. N. Sirdesai (nikhil.sirdesai@gmail.com)

Published: 12 November 2015

Abstract. Underground Coal Gasification, with enhanced knowledge of hydrogeological, geomechanical and environmental aspects, can be an alternative technique to exploit the existing unmineable reserves of coal. During the gasification process, petro-physical and geomechanical properties undergo a drastic change due to heating to elevated temperatures. These changes, caused due to the thermal anisotropy of various minerals, result in the generation of thermal stresses; thereby developing new fracture pattern. These fractures cause the overhead rock strata to cave and fill the gasification chamber thereby causing subsidence. The degree of subsidence, change in fluid transport and geomechanical properties of the rock strata, in and around the subsidence zone, can affect the groundwater flow. This study aims to predict the thermo-geomechanical response of the strata during UCG. Petro-physical and geomechanical properties are incorporated in the numerical modelling software COMSOL Multiphysics and an analytical strength model is developed to validate and further study the mechanical response and heat conduction of the host rock around the gasification chamber. Once the problems are investigated and solved, the enhanced efficiency and the economic exploitation of gasification process would help meet country's energy demand.

1 Introduction

Most of the energy requirement is fulfilled by conventional fuels such as coal, oil and gas with a minor share contributed by nuclear fuel and unconventional energy sources such as solar, wind, tidal and geothermal energy. It has become imperative to address the growing energy demand by using unconventional energy sources and by increasing the efficiency of the conventional sources with the application of novel techniques of fuel recovery. One such novel technique is Underground Coal Gasification (UCG). This technique involves the conversion of coal into fuel gas, also known as syngas, within the earth. The coal undergoes partial combustion by reacting with the mixture of air and steam, which is injected into the seam through one well. The product of the combustion is syngas, which is recovered through the production well/s. The result of the UCG process is a large cavity which is left behind upon the burning of the coal seam.

Due to the combustion of coal, the temperatures inside the reactor/cavity reach up to 1200 °C (Tian et al., 2011). The rocks around the cavity get heated for a duration of several months. The continuous exposure at such high temperatures for elongated periods of time causes melting, which induces several changes in the physical, mineral and mechanical properties of rocks (Kratzsch, 1983). A huge amount of laboratory research has been carried out to ascertain the effects of high temperatures on the physical (Wu and Liu, 2008; You and Kang 2009; Tian et al., 2012) and the mechanical (Rao et al., 2007; Vishal et al., 2011; Ranjith et al., 2012; Tian et al., 2015) properties of rock. The microstructure of the rock undergoes significant change, thereby affecting the physical and the mineralogical properties (Somerton 1992; Hajpál and Török, 2004). The constituent materials of the rock experience thermal expansion which leads to the creation of new micro fractures and the widening of the inherent micro-cracks (Den'gina et al., 1994). The results obtained from these tests indicate that heating, effects change in the physical and the thermomechanical properties. However, the effect varies with the change in rock type and the conditions prior to thermal treatment. The change in properties due to heating and the extent of the cavity induces instability in the overlying layer, tending it to fail and fill in the cavity (Kratzsch, 1983; Reddish and Whittaker, 1989). This heavy movement of the overlying area results in subsidence. Cracks formed from the burnt coal seam to the overlying strata, may result in a leak and cause groundwater contamination (Whittaker and Reddish, 1989; Burton et al., 2007). The scale of a commercialized UCG operation will be bigger than that of the pilot and field tests. This would result into notable ground subsidence caused by the large underground cavities. Success of a UCG project depends upon the amount of associated ground movement, thus making it an important factor to evaluate.

The shape of the UCG cavity governs the mechanical stability, the extent of ground subsidence and the economic feasibility of the operation (Shirsat and Riggs 1986; Luo et al., 2009). During UCG, the cavity formed underground spreads out both radially and axially along the gasification channel (Cena et al., 1987). The shape of the cavity depends on the type of UCG process used. As seen in Figs. 1 and 2, the cavity is the largest around the vicinity of the injection well in Linked Vertical Well (LVW) method (Shirsat, 1989). Whereas in the Controlled Retraction Injection Point (CRIP) method, due to constant retraction of injection point, the cavity has a constant cross section (Bell et al., 2011). The CRIP chamber created is basically not more than 20 m in width and 20 m in length (Tian, 2013).

In this research programme, a numerical model of the overlying strata was simulated in COMSOL Multiphysics in order to study the behaviour of mechanical parameters. The dimensions and the depth at which the cavity exists were altered to study the strata behavior and to analyse the subsidence caused on the surface. The aim of the ongoing research at the Department of Earth Sciences, Indian Institute of Technology Bombay, is to add to the available pool of scientific knowledge regarding subsidence during a UCG project and to propose possible mitigation techniques so as to minimize the damage caused to human life, infrastructure and economy.

2 Numerical modelling

In order to understand the effect of UCG on the overlying strata, a numerical model was prepared using the Finite Element Analysis (FEA) simulation software COMSOL Multiphysics. The process of UCG has been simulated for three coal seam condition. Shallow seated, medium seated and deep seated coal seams have been simulated at a depth of 100, 500 and 1000 m respectively (Fig. 3). This study would help us in determining the effect of cavity depth on subsidence. Similarly, the effect of the cavity width on subsidence has been determined by simulating widths ranging from 20 to 200 m (Fig. 4). As stated earlier, the cavity created does not exceed 20 m in width. But, in the multiple inlets and single outlet method, the size of the cavity near the injection points would be large.

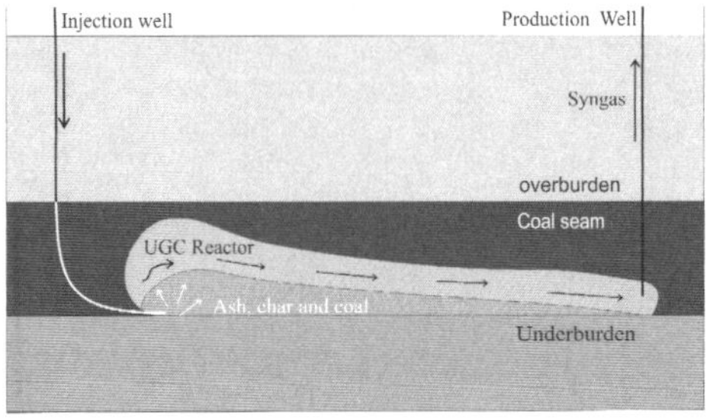

Figure 1. Linked Vertical Well Method (Modified after Shirsat, 1989).

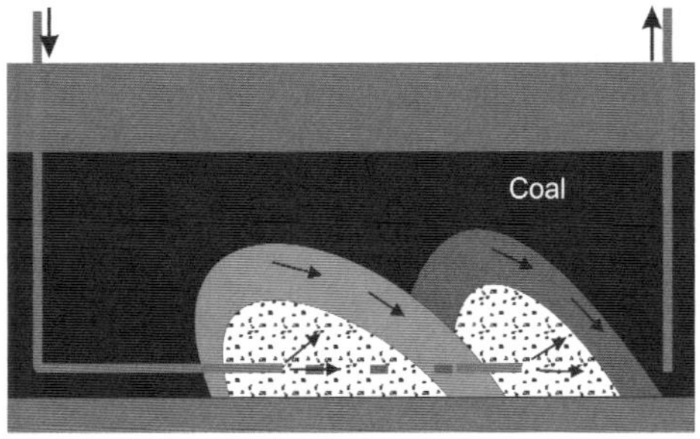

Figure 2. Controlled Retraction Injection Point Method (Modified after Bell et al., 2011).

The cavity has been visualized to be seated in a host consisting of fine grained sandstone. The mechanical properties of the host rock are shown in Table 1. The host rock is considered to be homogenous in nature. The limitation is to simulate geological discontinuities in the modelling. Thus, the Mohr-Coulomb Failure Criterion has been used. The criterion states that failure occurs when the shear stress and the normal stress acting on any element in the material satisfy the equation

$$|\tau| + \sigma \cdot \tan\varphi - c = 0 \tag{1}$$

here, "τ" is the shear stress, c the cohesion, and "ϕ" denotes the angle of internal friction.

With the help of Mohr's circle, this criterion can be written as

$$\frac{1}{2}(\sigma_1 - \sigma_3) + \frac{1}{2}(\sigma_1 + \sigma_3) \cdot \sin\varphi - c \cdot \cos\varphi = 0. \tag{2}$$

Initially, the model was simulated without the presence of cavity and gravity was applied to the entire system. Later

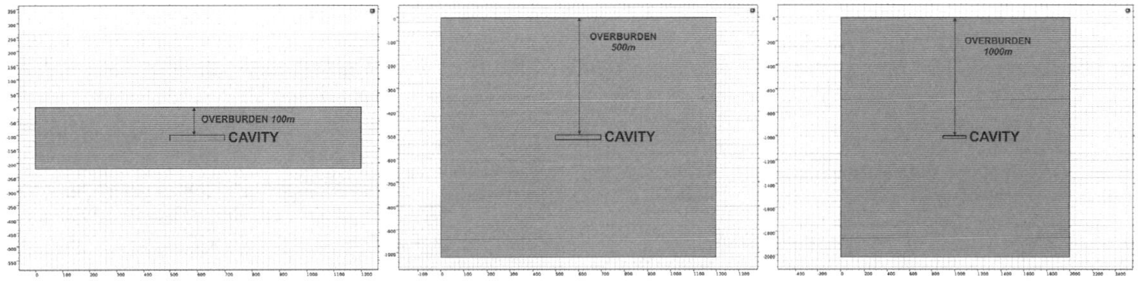

Figure 3. COMSOL Models for Shallow (100 m), Medium (500 m) and Deep (1000 m) seated cavity.

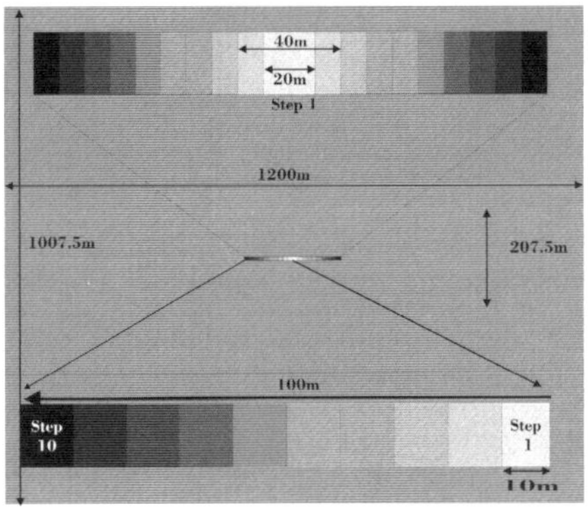

Figure 4. Simulated cavity widths.

Table 1. Mechanical Property of Host Rock.

Mechanical Property	Value
Density	2231.04 [kg (m^{-3})]
Young's Modulus	25.45×10^9 [Pa]
Poisson's Ratio	0.24
Cohesion	10×10^6 [Pa]
Angle of Internal Friction	50.19 [deg]
Dilation Angle	0 [deg]

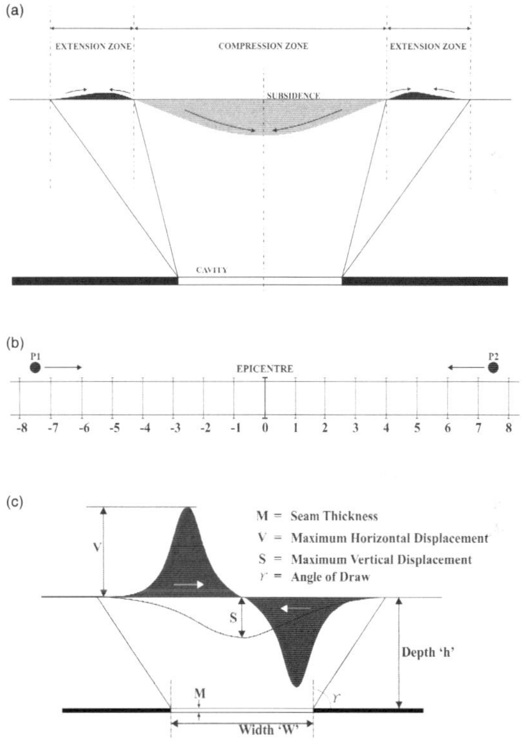

Figure 5. (a) Compression and Extension zones during Subsidence (NCB, 1975; Whittaker and Reddish, 1989). **(b)** Movement of materials around the epicentre of subsidence. **(c)** Empirical Subsidence Model (NCB, 1975; Whittaker and Reddish, 1989).

the cavity was modelled and the vertical and the horizontal movement of the upper boundary was observed to study the displacement caused at the surface. Parametric sweep was applied to observe the effect of various cavity widths on the behaviour of strata.

The effect of high temperatures on the strata and the rate of heat transfer was observed by applying high temperature element (1000 °C) to the cavity boundaries. The thermal properties of the host rock are shown in Table 2. Equation (3) was used to calculate the heat transfer through solids.

$$\rho \cdot C_p \cdot \frac{\partial T}{\partial t} + \rho \cdot C_p \cdot u \cdot \nabla T + \nabla \cdot q = Q. \quad (3)$$

For a steady-state problem the temperature does not change with time and the first term disappears. The equation includes the following material properties: density "ρ", heat capacity "C_p", velocity field "u" and a heat source (or sink) "Q".

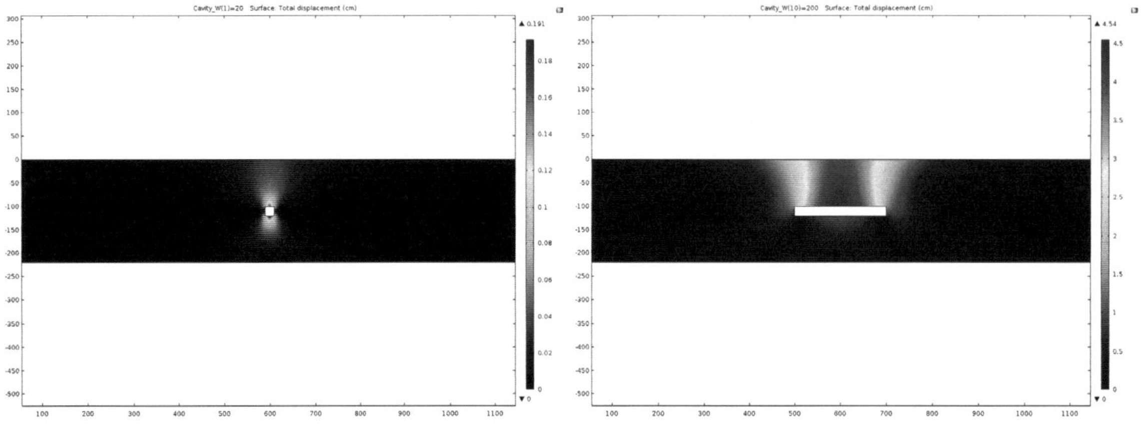

Figure 6. Total strata displacement around a shallow seated cavity.

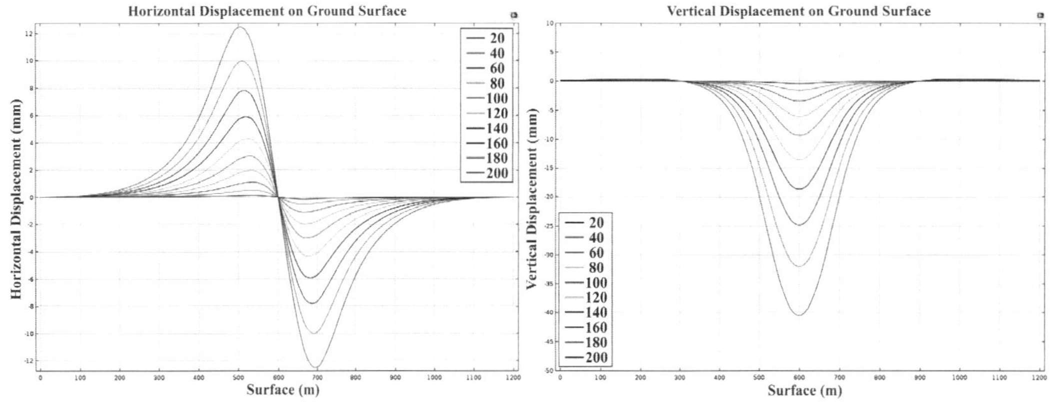

Figure 7. Horizontal and Vertical displacement of surface induced by a shallow seated cavity.

Table 2. Thermal Property of Host Rock.

Thermal Property	Value
Thermal Conductivity	2.348 $[\mathrm{W\,(m \cdot K)^{-1}}]$
Heat Capacity at Constant Pressure	752.72 $[\mathrm{J\,(kg \cdot K)^{-1}}]$
Coefficient of Thermal Expansion	11×10^{-6} $[\mathrm{K^{-1}}]$

3 Results and discussion

During the period of 1950–1965, 200 coalfields were observed for subsidence and an empirical model was formed. The model includes the change in surface profile, zones of extensions and compression and the direction of ground movement (NCB, 1975; Whittaker and Reddish, 1989). A schematic view of the subsidence model is provided in Figs. 5a and 5c. The subsided surface is categorized in two zones namely, Extension Zone and Compression Zone. Subsidence causes the material, near the subsided zone, to move towards the epicentre, thereby creating a compression zone. As the material moves away from the sides, an extension zone develops (Fig. 5a). If the epicentre is considered to be the origin i.e. "0", the material on the negative x axis move towards the origin, as seen in Fig. 5b. Due the surficial movement, particle P1 is drawn towards the epicentre. This causes its value to increase. Similarly, particle P2 is also drawn towards the centre, thereby causing a decrease in its value. If the movement of all such particles are plotted, the resulting graph of the total horizontal displacement would be as seen in Fig. 5c.

Figure 6 shows the total displacement caused on the surface due to the smallest and the largest cavity. The 20 m wide cavity does not affect the overlying strata as much as the 200 m wide cavity. A cavity induces instability in the overlying strata. This affects the equilibrium of the rock directly above the cavity causing it to break away and fill in the cavity. This process continues until an equilibrium is attained.

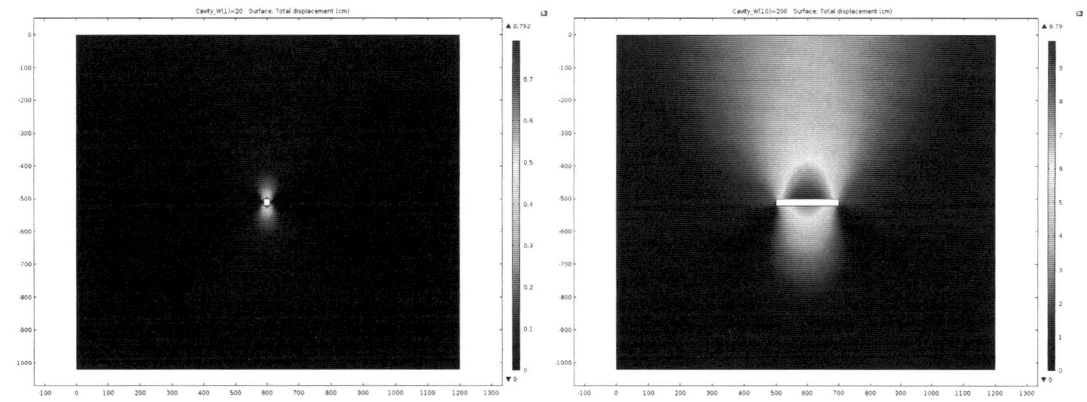

Figure 8. Total strata displacement around a medium seated cavity.

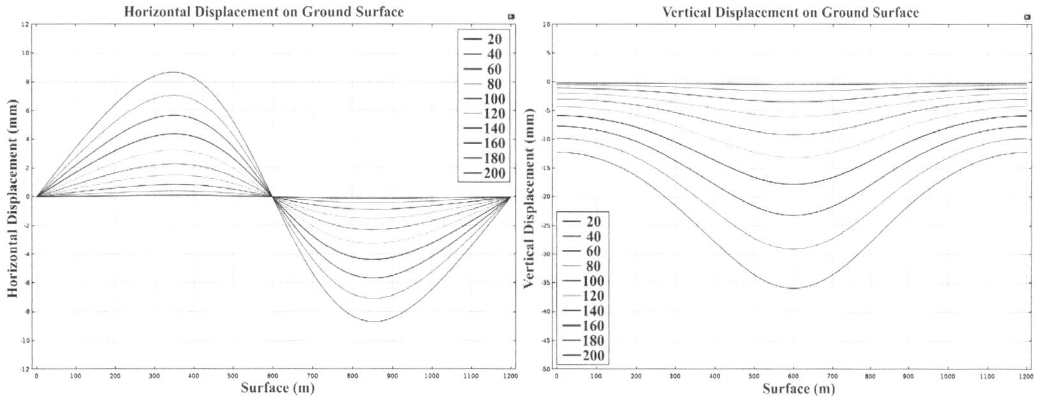

Figure 9. Horizontal and Vertical displacement of surface induced by a medium seated cavity.

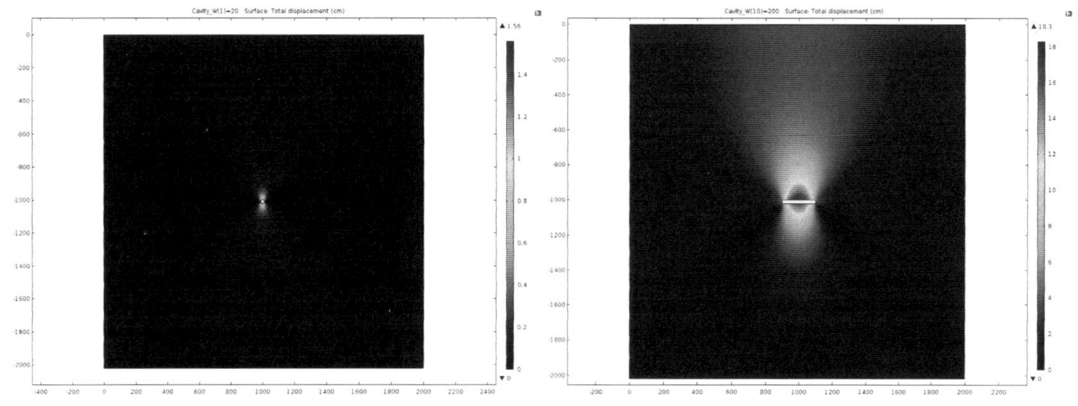

Figure 10. Total strata displacement around a deep seated cavity.

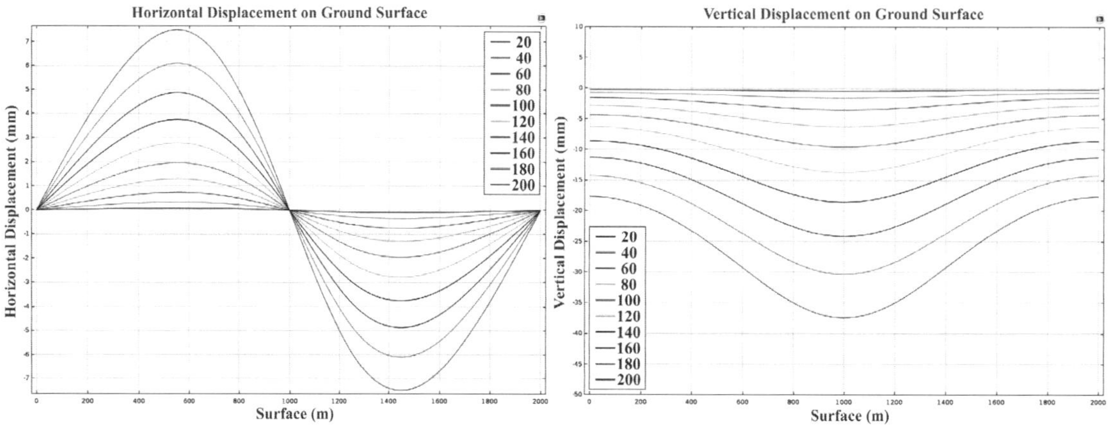

Figure 11. Horizontal and Vertical displacement of surface induced by a deep seated cavity.

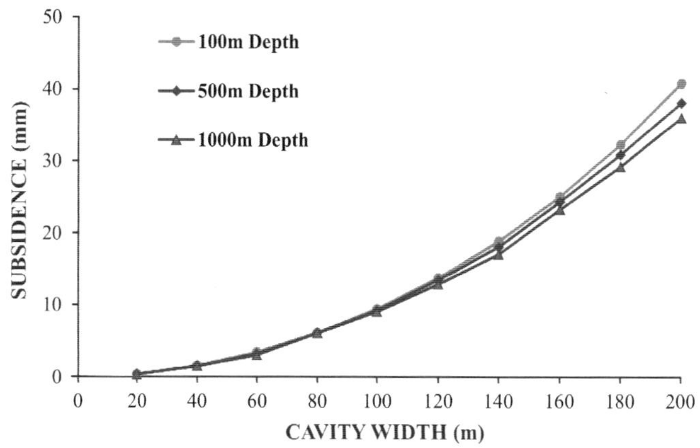

Figure 12. Effect of cavity width and depth on subsidence.

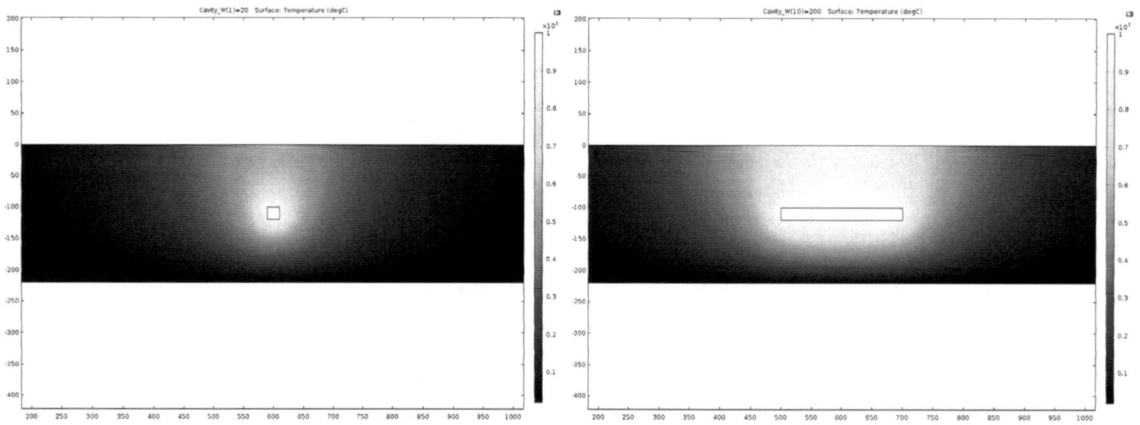

Figure 13. Temperature distribution around a shallow seated cavity.

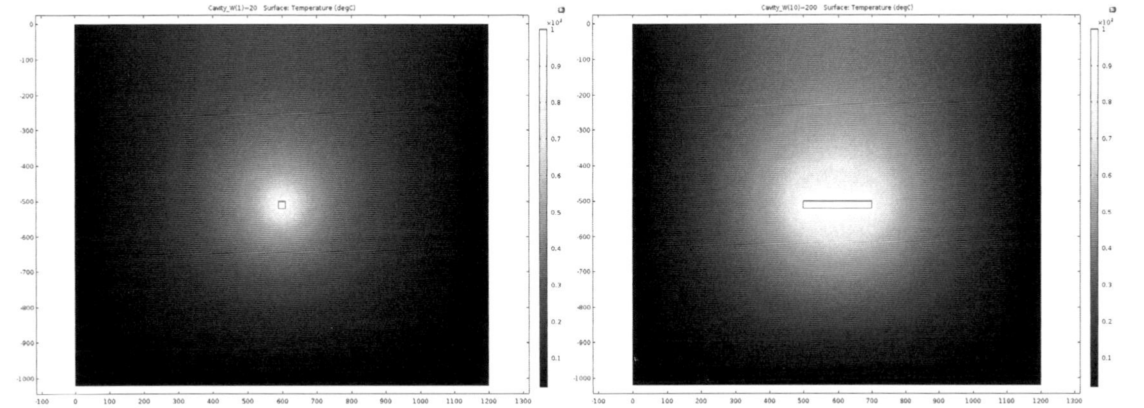

Figure 14. Temperature distribution around a medium seated cavity.

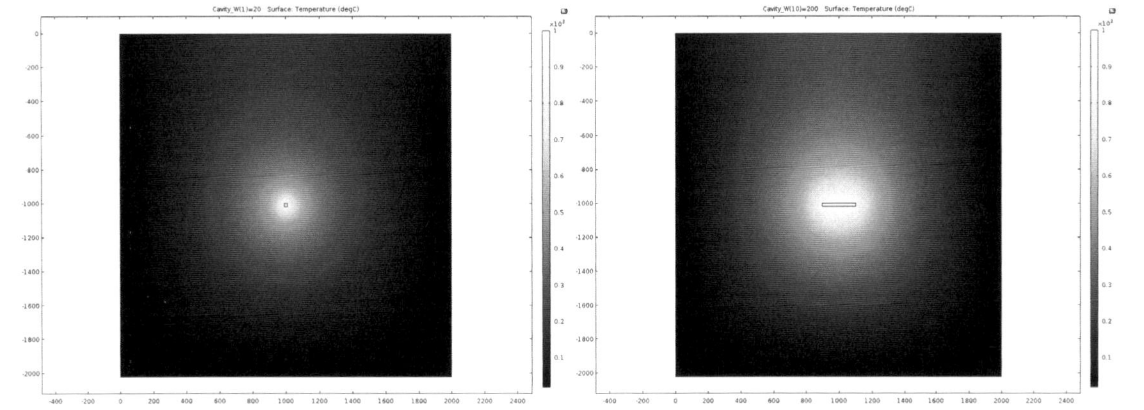

Figure 15. Temperature distribution around a deep seated cavity.

The amount of material required to fill in the 20 m cavity is far less than the 200 m cavity. A small opening thus causes less disturbance to the strata as compared to a large opening. The horizontal and the vertical displacements of the surface is shown in Fig. 7. Point 600 on the x axis of both the graphs represent the midpoint of the cavity below the ground. A subsidence of approximately 40 mm is observed when the cavity width is 200 m. Cavities created at a depth of 500 m induced less disturbance to the overlying strata as compared to the shallow seam model. The area just above the cavity displays the highest amount of displacement (Fig. 8). The effect of the cavity on the surface can be observed in Fig. 9. The surficial movements are less than those observed in shallow seam model. With the increase in the depth, the subsidence influence zone decreases. Maximum subsidence of 38 mm is observed when the cavity width is 200 m. As the depth increases, the effect on the surface reduces. This can be seen in Fig. 10, where the cavity was created at a depth of 1000 m. The effect on the surface is the least as compared to the shallow and the medium depth models (Fig. 11). The 200 m wide cavity induced the highest vertical displacement of 36 mm. The graphs obtained for the horizontal displacement in Figs. 7, 9 and 11 follow the empirical model. A comparative analysis of vertical displacements obtained for cavities at various depths is shown in Fig. 12. It can be inferred that cavities having width below 120 m induce subsidence of similar magnitude. Only after 120 m, do we observe a distinct variance in the magnitude of subsidence.

Variations in the temperature field around the cavities are shown in Figs. 13, 14 and 15. The region near the cavity are heated up to 1000 °C which causes the rock to undergo several changes. These changes further affect the strength of the rocks. A shallow seated (100 m) UCG process generates a very high temperature zone over the ground surface as compared to medium (500 m) and deep seated (1000 m) processes. Figure 16 illustrates the temperatures observed for the point on the ground surface lying exactly over the centre of the cavity.

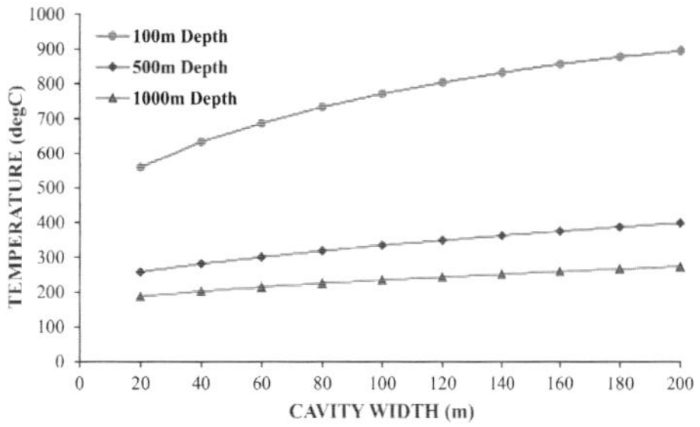

Figure 16. Effect of cavity width and depth on temperature.

4 Conclusions

The cavities less than 100 m in width induce low and similar magnitudes of subsidence on the surface. High magnitudes of subsidence are observed when the chamber is near to the ground surface. The magnitude of subsidence is inversely proportional to the depth at which the cavity is situated. The thermal studies of the cavity and the strata exhibit a similar relationship. It should be noted that the geomechanical and thermal studies were conducted for a homogenous strata consisting of single type of rock. The actual scenario observed in the field could vary owing to the presence of various strata and various geological discontinuities. The magnitude of subsidence and the heat distribution depends on the type of litho-units. Rocks such as siltstone, claystone and shale exhibit higher swelling properties as compared to sandstone. The presence of such rock types in the strata may lower the subsidence effect on the surface. Moreover, presence of ground water and geological discontinuities would affect the temperature distribution in the strata. Thus it is imperative to carry out a detailed geological exploration of the area before the commencement of UCG process. A cavity seated at a higher depth has minimal effect on the strata. Therefore, a deep seated coal seam is conducive for the application of UCG process.

Acknowledgements. The authors thank the anonymous reviewers for their careful review and contributions, which led to the improvement of the manuscript. The first author is thankful to the other three authors for their constant support and invaluable advice during the experimental and numerical analysis.

References

Bell, D. A., Towler, B. F., and Fan, M. H.: Coal gasification and its applications, William Andrew/Elsevier, Oxford, 101–111, 2011.

Burton, E., Friedmann, J., and Upadhye, R.: Best practices in Underground Coal Gasification, Lawrence Livermore National Laboratory, Livermore, California, 13–21, 2007.

Cena, R. J., Britten, J. A., and Thorsness, C. B.: Excavation of the partial seam CRIP underground coal gasification test site, Proceedings of the 13th Annual Underground Coal Gasification Symposium, 24 August 1987, Laramie, Wyoming, USA, 1987.

Den'gina, N. I., Kazak, V. N., and Pristash, V. V.: Changes in rocks at high temperatures, J. Min. Sci., 29, 472–477, 1994.

Hajpál, M. and Török, Á.: Mineralogical and colour changes of quartz sandstones by heat, Environ. Geol., 46, 311–322, 2004.

Kratzsch, H.: Mining subsidence engineering, Springer-Verlag, Berlin, 30–31, 1983.

Luo, Y. G., Coertzen, M., and Dumble, S.: Comparison of UCG cavity growth with CFD model predictions, Proceedings of the 7th International Conference on CFD in the minerals and process industries, 9–11 December 2009, CSIRO, Melbourne, Australia, 1–5, 2009.

NCB (National Coal Board): Subsidence Engineers' Handbook, NCB Publications, Hobart House, London, 90–105, 1975.

Ranjith P. G., Viete, D. R., Chen, B. J., and Perera, M. S. A.: Transformation plasticity and the effect of temperature on the mechanical behaviour of Hawkesbury sandstone at atmospheric pressure, Eng. Geol., 151, 20–127, 2012.

Rao Q. H., Wang Z., Xie H. F., and Xie, Q.: Experimental study of mechanical properties of sandstone at high temperature, J. Cent. South Univ. T., 15, 478–483, 2007.

Reddish, D. J. and Whittaker, B. N.: Subsidence: Occurrence, Prediction and Control, Elsevier, Amsterdam, 359–376, 1989.

Shirsat, V. A.: Modeling of cavity growth in underground coal gasification, MS Thesis, Chemical engineering, Texas Tech University, 16–48, 1989.

Shirsat, V. A. and Riggs, J. B.: UCG cavity growth model, Proceedings of the 12th Annual Underground Coal Gasification Symposium, 21–28 August 1986, Saarbrucken, Germany, 297–304, 1986.

Somerton, W. H.: Thermal properties and temperature-related behavior of rock/fluid systems, Elsevier, Amsterdam, 148–166, 1992.

Tian, H.: Development of a Thermo-Mechanical Model for Rocks Exposed to High Temperatures during Underground Coal Gasification, PhD Thesis, Rheinisch-Westfälischen Technischen Hochschule (RWTH) Aachen University, Aachen, Germany, p. 10, 2013.

Tian, H., Kempka, T., Feinendegen, M., and Ziegler, M.: Study on ground subsidence development during and after underground coal gasification, EGU General Assembly, Vienna, Austria, 3–8 April 2011, EGU2011-1277-2, 2011.

Tian, H., Kempka, T., Neng-Xiong, X., and Ziegler, M.: Physical Properties of Sandstones After High Temperature Treatment, Rock Mech. Rock Eng., 45, 1113–1117, 2012.

Tian, H., Kempka, T., Yu, S., and Ziegler, M.: Mechanical Properties of Sandstones Exposed to High Temperature, Rock Mech. Rock Eng., Springer Verlag, Wien, 1–7, 2015.

Vishal, V., Pradhan, S. P., and Singh, T.: Tensile Strength of Rock Under Elevated Temperatures, Geotechnical and Geological Engineering, 29, 1127–1133, 2011.

Wu, G. and Liu, S.: Research on ultrasonic characteristics of sandstone after heating to high temperature, Boundaries of Rock Mechanics, Taylor & Francis Group, London, 207–211, 2008.

You, L. J. and Kang, Y. L.: Effects of thermal treatment on physical property of tight rocks, Progress in Geophysics, 2015, 1850–1854, 2009.

Proc. IAHS, 372, 463–469, 2015
proc-iahs.net/372/463/2015/
doi:10.5194/piahs-372-463-2015

Open Access
PIAHS

A fusion model used in subsidence prediction in Taiwan

S.-J. Wang[1,a], K.-C. Hsu[1,2], and C.-H. Lee[1,2]

[1]Georesources Research Center, National Cheng Kung University, Tainan, Taiwan
[2]Department of Resources Engineering, National Cheng Kung University, Tainan, Taiwan
[a]now at: Georesources Research Center, NCKU, Tainan, Taiwan

Correspondence to: S.-J. Wang (sjwang1230@gmail.com)

Published: 12 November 2015

Abstract. The Taiwan Water Resources Agency uses four techniques to monitor subsidence in Taiwan, namely data from leveling, global positioning system (GPS), multi-level compaction monitoring wells (MCMWs), and interferometry synthetic aperture radar (InSAR). Each data type has advantages and disadvantages and is suitable for different analysis tools. Only MCMW data provide compaction information at different depths in an aquifer system, thus they are adopted in this study. However, the cost of MCMW is high and the number of MCMW is relatively low. Leveling data are thus also adopted due to its high resolution and accuracy. MCMW data provide compaction information at different depths and the experimental data from the wells provide the physical properties. These data are suitable for a physical model. Leveling data have high monitoring density in spatial domain but lack in temporal domain due to the heavy field work. These data are suitable for a black- or grey-box model. Poroelastic theory, which is known to be more conscientious than Terzaghi's consolidation theory, is adopted in this study with the use of MCMW data. Grey theory, which is a widely used grey-box model, is adopted in this study with the use of leveling data. A fusion technique is developed to combine the subsidence predicted results from poroelastic and grey models to obtain a spatially and temporally connected two-dimensional subsidence distribution. The fusion model is successfully applied to subsidence predictions in Changhua, Yunlin, Tainan, and Kaohsiung of Taiwan and obtains good results. A good subsidence model can help the government to make the accurate strategies for land and groundwater resource management.

1 Introduction

Subsidence is a worldwide hazard (Galloway et al., 1999; Chen and Rybczyk, 2005; Galloway and Burbey, 2011). The subsidence problem in Taiwan is suspected to be due to groundwater over pumping for aquaculture and irrigation along its coastal plain. The Taiwan Water Resources Agency (TWRA) uses four types of monitoring system to investigate land subsidence in Taiwan, namely data from leveling, global positioning system (GPS), multi-level compaction monitoring wells (MCMWs), and interferometry synthetic aperture radar (InSAR) (Hung et al., 2010). Leveling and MCMW data are adopted in this study due to their relatively long monitoring periods. MCMWs have several magnetic rings anchored at various depths. Variations are detected using the magnetic rings about once a month to measure the compaction in individual formations. The accuracy of the MCMW data is 0.1–0.5 cm. Leveling is a commonly used method for measuring height variations on a land surface by using backsight and foresight readings from leveling rods. The accuracy of leveling is 0.5–1.0 cm. Although accuracy is high, measurement efficiency is low. The measurement period is about one to two years in Taiwan.

There are three types of model named white box, grey box, and black box models. White box models are purely theoretical, black box models have no model form, and grey box models are between white box and black box models. A physical model (white box model) avoids the incorrect behavior of a grey box model, whereas a grey box model avoids the assumptions and uncertainty of a physical model (white box model). Therefore, a technique that can fuse the results of physics-based and grey-box-based models would make the prediction of land subsidence more reasonable and reliable (Wang, 2015). Wang (2015) proposed a technique that can fuse two types of time series for subsidence prediction results in both the spatial and temporal domains and applied to subsidence fusion in Jhuoshuei River Alluvial Fan, Taiwan. The fusion method uses a slope criterion and a distance weighting factor to obtain a spatially and temporally connected two-dimensional subsidence distribution by upgrading two one-dimensional models. Wang et al. (2015) adopted this fusion technique to combine the physics-based nonlinear poroelastic model (NPM) and the grey-box-based grey system model (GSM) to evaluate the subsidence under the climate change effect in Tainan, Taiwan and got good results. The authors have also applied the fusion technique to predict the subsidence in Kaohsiung, Taiwan, which is not yet published.

2 Background

2.1 Study area

The fusion technique used for subsidence prediction was successfully applied to Changhua and Yunlin counties and Tainan and Kaohsiung cities in Taiwan. In this study, only the results for Jhuoshuei River Alluvial Fan (including Changhua and Yunlin counties) and Tainan City are shown (Wang, 2015; Wang et al., 2015). There are less data in Kaohsiung and they are not discussed here.

Jhuoshuei River Alluvial Fan, located in central Taiwan (see Fig. 1), is the largest alluvial fan in Taiwan. Wu River and Beigang River flow through the north and south areas, respectively, and Jhuoshuei River flows across the middle of the alluvial fan. Changhua and Yunlin counties are located in the north and south divisions of Jhuoshuei River Alluvial Fan, respectively. The western boundary is the Taiwan Strait and the eastern boundary is Bagua and Douliu foothills, where are the main areas for groundwater recharge.

Tainan is located in southwestern Taiwan. The Tainan area is in the south central division of the Chianan Plain. Tainan is bounded by the Bajhang River in the north, the Erren River in the south, the Chiayi and Xinhua foothills in the east, and the Taiwan Strait in the west. The main geography is the Tainan Plain in the western part and some foothills and mountainous areas in the eastern part, as shown in Fig. 2.

2.2 Subsidence data

There were 29 MCMWs and 708 leveling points in 2011 in Jhuoshuei River Alluvial Fan, Taiwan. The measurement frequencies are monthly for the MCMWs and yearly for leveling in this area. The land subsidence data from 2007 to 2011 are adopted, and thus the subsidence quantities shown in this study are relative to the level in 2007. Only 26 MCMWs and 452 leveling points are used in this study due to a lack of monitoring data. The monitoring depth is 200 to 330 m. From the 26 monitoring wells, the subsidence quantity at the ground surface is the largest at Yuanchang well (= 0.252 m) and the smallest is at Jiaxing well (= 0 m) in 2011 based on

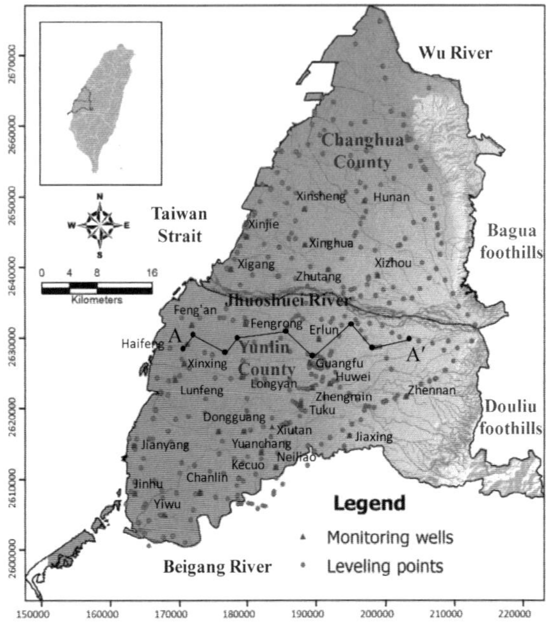

Figure 1. Study area of Jhuoshuei River Alluvial Fan, Taiwan, and distribution of leveling points and MCMWs (From Wang, 2015).

2007. However, the compaction quantity at different depths can be different.

There are only four MCMWs in Tainan due to the subsidence being relatively low. The depths of the MCMWs are all at 300 m. Since the wells were installed in different years, their monitoring periods are different. The compaction at the ground surface continuously increases at the Xiaying, Xuejia, and Chengda wells. The land surface at the Iian well shows an initial decrease and then recovery. The recovery behavior is due to an increase in groundwater level around this area and shows an elastic response behavior. Since this study focuses on subsidence evaluation, the subsidence of the Iian well is set to zero in the entire area estimation.

3 Methodology

3.1 Nonlinear poroelastic model

The governing equations for the one-dimensional poroelastic model can be written as (Wang and Hsu, 2009a):

$$\begin{cases} a^{-1}\dfrac{\partial^2 w(z,t)}{\partial z^2} - \alpha\dfrac{\partial P(z,t)}{\partial z} = 0 \\ \alpha\dfrac{\partial^2 w(z,t)}{\partial t\partial z} + Q^{-1}\dfrac{\partial P(z,t)}{\partial t} - \kappa\dfrac{\partial^2 P(z,t)}{\partial z^2} = 0 \end{cases} \tag{1}$$

where w is the vertical displacement; P is the change in pore water pressure; z is the vertical direction; t is time; a is called the final compressibility with $a^{-1} = \lambda + 2\mu$, where

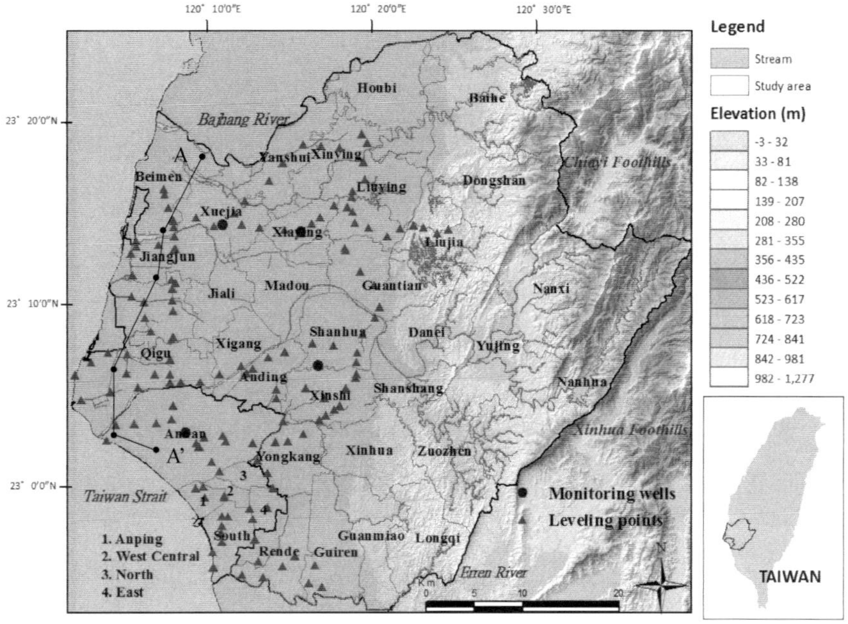

Figure 2. River system and geographic map with distribution of districts and monitoring points in Tainan City (From Wang et al., 2015).

λ and μ are Lame's constants ($\mu = E/2(1+\nu)$, and $\lambda = E\nu/(1+\nu)(1-2\nu)$, where E is Young's modulus, and ν is Poisson's ratio); α is the dimensionless coefficient of the effective stress; Q^{-1} is the compressibility introduced by Biot; and κ is the Darcy conductivity, which is related to hydraulic conductivity K by $K = \gamma_w \kappa$, where γ_w is the unit weight of water ($= 9810\,\mathrm{N\,m^{-3}}$).

Following Wang and Hsu (2009b), the porosity (n), hydraulic conductivity (K), and Young's modulus (E) under the deformation effect can be written, respectively, as:

$$n_1 = \frac{n_0 + \varepsilon}{1+\varepsilon} \tag{2}$$

$$K_1 = K_0 \frac{(1+\varepsilon/n_0)^3}{1+\varepsilon} \tag{3}$$

$$E_1 = E_0 \left(\frac{1}{1+\varepsilon}\right)^2 \tag{4}$$

where ε is the volume strain, and the subscripts 0 and 1 indicate before and after the deformation effect, respectively. The nonlinear parameters are embedded in the calculation during the modeling period.

3.2 Grey system model

Deng (1989) proposed grey system prediction theory. The GSM with one first-order variable is called the GM (1,1) model. The grey differential equation of the GM (1,1) model can be written as:

$$\frac{\mathrm{d}x^{(1)}}{\mathrm{d}t} + a x^{(1)} = b \tag{5}$$

where a and b are coefficients often evaluated using the least squares error method by fitting the theory and observation data and $x^{(1)}$ is a canonical series written as:

$$x^{(1)} = \left(\sum_{k=1}^{1} x^{(0)}(k), \sum_{k=1}^{2} x^{(0)}(k), \cdots \sum_{k=1}^{N} x^{(0)}(k) \right) \tag{6}$$

where $x^{(0)}$ is the original data series (i.e., leveling data series in this study) and N is the data number. Substituting the evaluated a and b into Eq. (5) and letting the initial condition be $x^{(1)}(1) = x^{(0)}(1)$ yields:

$$\widehat{x}^{(1)}(n+1) = \left[x^{0}(1) - \frac{b}{a} \right] e^{-an} + \frac{b}{a} \tag{7}$$

where $\widehat{x}^{(1)}(n+1)$ is the predicted value for $x^{(1)}$ at the $n+1$ point. Using back differential for $\widehat{x}^{(1)}$ yields:

$$\widehat{x}^{(0)}(n+1) = \widehat{x}^{(1)}(n+1) - \widehat{x}^{(0)}(n) \tag{8}$$

The predicted value $\widehat{x}^{(0)}(n+1)$ is thus obtained. Equation (8) is the basic equation for the GM(1,1) model.

3.3 Fusion method

A fusing method that can combine two types of trend series and obtain a spatial-temporal relative subsidence distribution

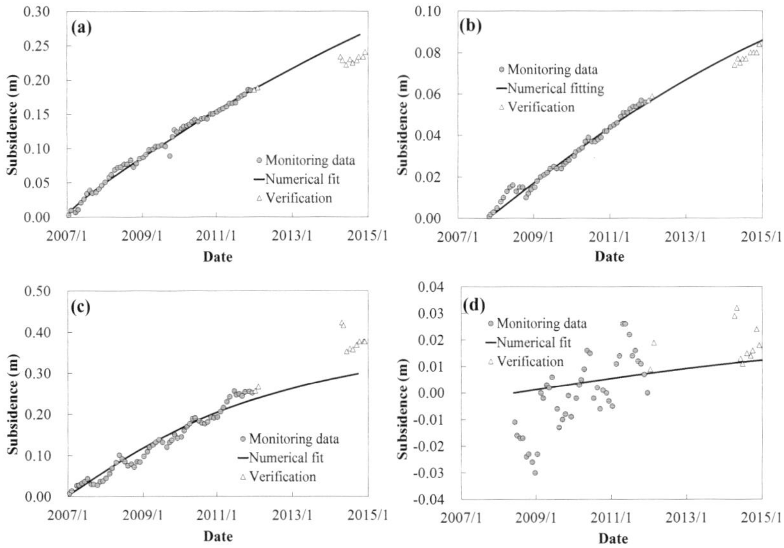

Figure 3. Four MCMWs for demonstrating fitting results. **(a)** Xinghua, **(b)** Xizhou, **(c)** Yunchang, and **(d)** Jiaxing wells. White triangles show data used for model verification (From Wang, 2015).

was proposed by the present author (Wang, 2015). The fusion concept is to use the results of NPM to re-evaluate the results of GSM under a slope criterion of the subsidence trend in the temporal domain and a weighting factor between the monitoring points in the spatial domain.

Considering a distance weighting factor between the locations of GSM_i and NPM_j, the mean slope of the subsidence trend for NPM_j with respect to GSM_i can be written as:

$$\bar{S}_{\mathrm{NPM}_{j=1,m}}(t,D)=\sum_{j=1}^{m}\left\{\frac{d\left(\mathrm{NPM_j}\right)}{\mathrm{d}t}\times\frac{1}{D_{ij}}\right\}\Bigg/\left(\sum_{j=1}^{m}\frac{1}{D_{ij}}\right) \quad (9)$$

where D_{ij} is the distance from GSM_i to NPM_j, m is the number of monitoring points for NPM (i.e., the number of MCMWs), and $\bar{S}_{\mathrm{NPM}_{j=1,m}}(t,D)$ is the mean slope of NPM_j with respect to GSM_i in the time interval $\mathrm{d}t$. Assuming that the subsidence trends obtained from the NPM results and the initial subsidence quantities obtained from the GSM results are correct, the initial subsidence quantity of GSM is kept, but the slope of the subsidence trend obtained from the GSM is adjusted using the mean slope calculated from NPM. If the slope of the subsidence trend obtained from GSM_i in the time interval $\mathrm{d}t$ is larger than the mean slope of NPM_j in the same time interval, the GSM_i slope is substituted with the mean NPM_j slope. The slope criterion can be written as:

$$\begin{cases} S_{\mathrm{GSM}_i}(t)=S_{\mathrm{GSM}_i}(t) \\ \quad \text{if} \quad S_{\mathrm{GSM}_i}(t)\le\bar{S}_{\mathrm{NPM}_{j=1,m}}(t,D) \\ S_{\mathrm{GSM}_i}(t)=\bar{S}_{\mathrm{NPM}_{j=1,m}}(t,D) \\ \quad \text{if} \quad S_{\mathrm{GSM}_i}(t)>\bar{S}_{\mathrm{NPM}_{j=1,m}}(t,D) \end{cases} \quad (10)$$

where S_{GSM_i} and S_{NPM_j} represent the slopes of the subsidence trend for GSM_i and NPM_j, respectively.

The criterion is used to evaluate the slope of the subsidence trend in each time interval for each leveling point with respect to all MCMWs by considering the distance weighting factor. Both the spatial and temporal results are combined in the fusion technique. The independent one-dimensional evaluations in the NPM and GSM are then upgraded to a spatially and temporally connected two-dimensional distribution.

4 Results and discussion

4.1 Results for NPM

The MCMWs with maximum and minimum subsidence in Changhua and Yunlin, respectively, are chosen as examples for illustrating the fitting situation, as shown in Fig. 3. The Xinghua and Xizhou MCMWs are in Changhua County and the Yunchang and Jiaxing MCMWs are in Yunlin County. The subsidence fluctuation is more dominant in Yunlin County (Fig. 3c and d) than in Changhau County (Fig. 3a and b) due to the large fluctuation of groundwater level in Yunlin County. The fitting for Jiaxing well is poor due to the negative subsidence and the adopted compaction NPM, as shown in Fig. 3d. Negative subsidence indicates an uplift situation, which is not considered in the NPM in this study. From the fitting results, the quantity of discharge is 0 to $2.58\times10^{-9}\,\mathrm{m\,s^{-1}}$ and that of loading is 0 to $1.07\times10^{4}\,\mathrm{N\,m^{-2}}$. Two and 15 MCMWs show zero discharge and zero loading, respectively. The main driving force for the land subsidence in Jhuoshuei River Alluvial Fan

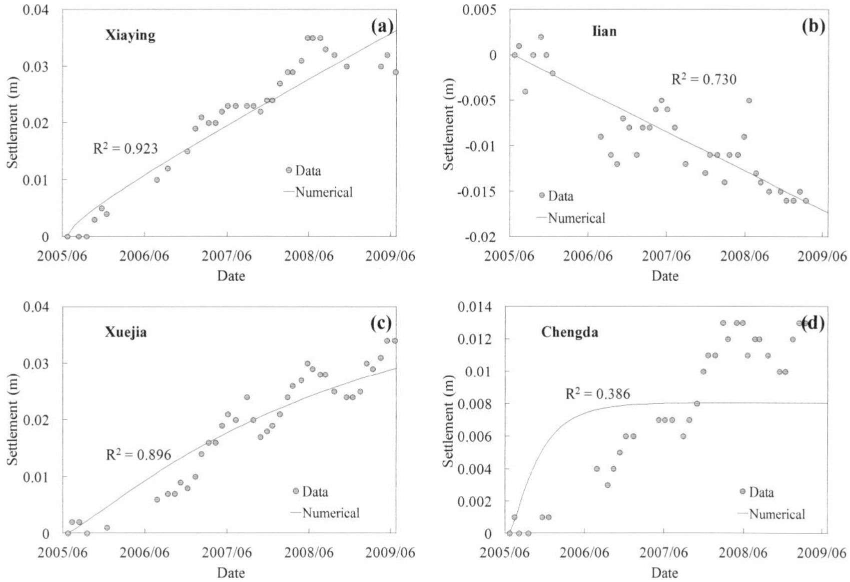

Figure 4. Fitting results for **(a)** Xiaying, **(b)** Iian, **(c)** Xuejia, and **(d)** Chengda MCMWs obtained using the NPM (From Wang et al., 2015).

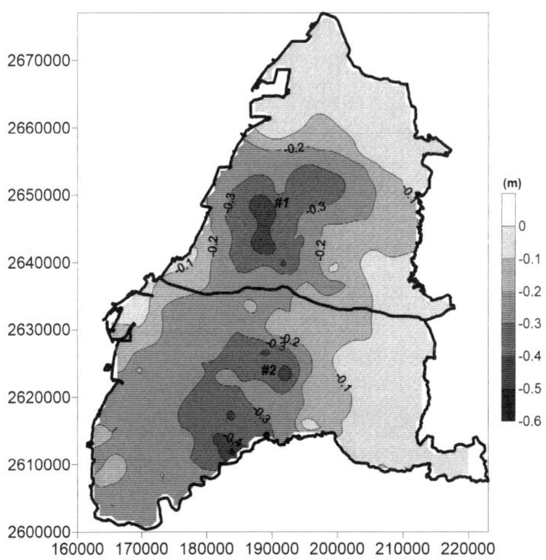

Figure 5. Subsidence distribution in Jhuoshuei River Alluvial Fan in 2039 obtained using NPM and GSM. Triangle symbols numbered #1 and #2 are locations of demonstration points used for fusion results in Fig. 6 (From Wang, 2015).

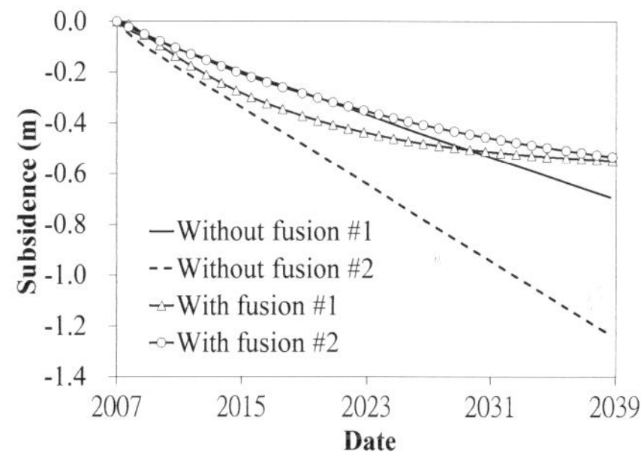

Figure 6. Comparison of results obtained with and without fusion for two leveling points marked in Fig. 5 (From Wang, 2015).

is groundwater pumping, which is expressed as discharge (pumping rate in unit area). The influence of top loading is relatively small, occurring only at 11 MCMWs. The quantities of loading are also small, lower than atmospheric pressure ($= 1.013 \times 10^5\,\mathrm{N\,m^{-2}}$).

The best fits for the compaction data from the wells in Tainan and the NPM are shown in Fig. 4. The figures show that the fits are excellent at the Xiaying and Xuejia wells with correlation coefficients (R-squared) of about 0.9, good at the Iian well, but poor at the Chengda well due to both the hydraulic conductivity and Young's modulus being large. The additional loading is from 0 to $2.03 \times 10^3\,\mathrm{N\,m^{-2}}$, and the additional discharge is from -1.33×10^{-10} to $8.23 \times 10^{-10}\,\mathrm{m\,s^{-1}}$. The negative discharge indicates a recharge condition due to the uplift of the ground surface near the Iian well. The evaluated additional discharge can be used to evaluate the additional pumping quantity around the MCMW by considering an influence area. The maximum additional loading of $2.03 \times 10^3\,\mathrm{N\,m^{-2}}$ is in Xiaying District; it is only about 2 % of the barometric pressure ($1.01 \times 10^5\,\mathrm{N\,m^{-2}}$). Considering the influence radius of 1.0 km from the monitoring well, the additional

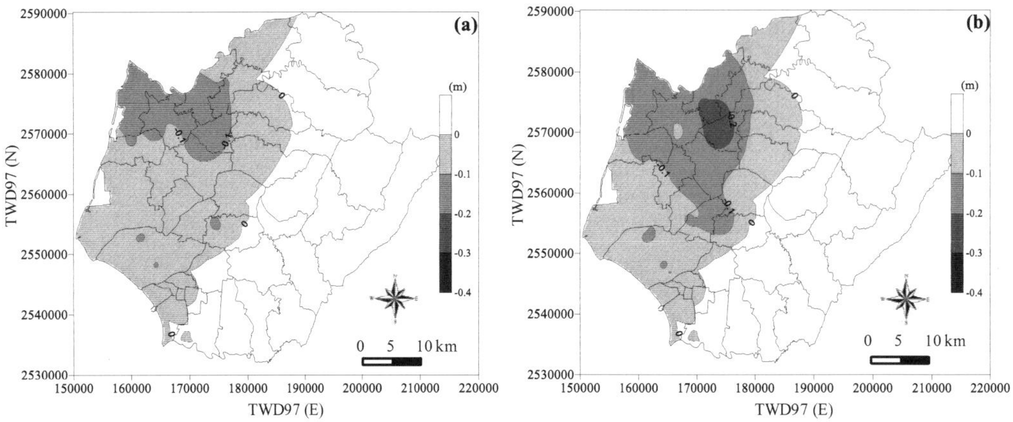

Figure 7. Subsidence distribution in Tainan predicted using combination of the NPM and GSM for the target years **(a)** 2020 and **(b)** 2039 (From Wang et al., 2015).

pumping rate can be calculated by multiplying the additional discharge by the influence area. The quantities of additional pumping rate around the monitoring wells in Tainan are from -36 to $223\,\mathrm{m}^3\,\mathrm{day}^{-1}$. However, the evaluated additional pumping rate is affected by the influence area and could be different in the field.

4.2 Subsidence prediction

The slope criterion for the subsidence trend series is adopted to re-calculate the results of the GSM by considering the mean slope calculated from the NPM with a distance weighting factor. The fusion results for subsidence distribution in Jhuoshuei River Alluvial Fan for 2039 are shown in Fig. 5. The maximum subsidence decreases from 2.31 to 0.59 m. The areas with subsidence quantities larger than 0.3 m are mainly located in the south-west areas of Changhua County and south-central areas of Yunlin County. Comparisons of the results obtained with and without fusion for two leveling points are shown in Fig. 6. The locations of the two leveling points are marked in Fig. 5. The subsidence predicted using the GSM for the leveling points shows a linear increasing trend. The subsidence quantities for points #1 and #2 are 0.71 and 1.23 m, respectively, for 2039. After using the fusion technique, both subsidence curves show convergence trends. The subsidence quantities are 0.55 and 0.53 m, respectively, and the differences are 0.16 and 0.70 m, respectively. The decreases vary with leveling point. The subsidence behavior becomes reasonable for both the subsidence quantity and subsidence rate.

The subsidence distributions obtained by combining the NPM and GMS in the years 2020 and 2039 in Tainan are shown in Fig. 7. The results show that the main areas with large subsidence move from Xuejia and Beimen Districts to Yanshui and Xiaying Districts, as evidenced by the large settlement increment at the Xiaying well. The pattern in the subsidence areas obtained by combining the NPM and GSM is similar to that obtained using only the GSM. However, the largest subsidence areas are different, and the subsidence quantity is much lower for the former due to the constraint of the physics-based NPM. The similar pattern might be due to the limited number of MCMWs (four) in Tainan, and thus there is relatively low influence exhibited in the combination. However, with more MCMWs, the combination could produce better results.

5 Conclusion

A physics-based NPM and a grey-box-based GSM were used to investigate land subsidence in Changuha, Yunlin, and Tainan, Taiwan. A technique was proposed to fuse the subsidence results from NPM and GSM in the spatial and temporal domains using a slope criterion and a distance weighting factor. The linear increasing trend of subsidence predicted using a grey-box-based GSM was corrected to a convergence trend with a physics-based NPM. The fusion results show dramatic decreases of subsidence quantity and subsidence rate. The subsidence behavior became physically reasonable due to the convergence trend of subsidence under the assumption of constant discharge of groundwater and top loading. The proposed fusion technique upgrades the independent one-dimensional evaluations to a spatially and temporally connected two-dimensional distribution.

The parameters in the NPM were taken from laboratory experiments and the data were insufficient. Field or more laboratory experiments will provide the typical values and improve the modeling results. There are few MCMWs installed in Taiwan. More available monitoring wells will provide better predicted subsidence results from the combination of the NPM and GSM. The recent NPM assumed a soil column with uncertainty embedded for subsidence evaluation. Layer system or complex concept could be used to develop a detail model for subsidence evaluation. Therefore, there are uncertainties embedded in the model. More data and further investigations can improve the estimations.

Author contributions. Shih-Jung Wang developed the model, analyzed the results, and wrote the manuscript. Kuo-Chin Hsu advised the numerical model and revised the manuscript. Cheng-Haw Lee reviewed the manuscript and provided correction comments and suggestions.

Acknowledgements. This study was supported by Ministry of Science and Technology, Taiwan, ROC, under grant NSC 101-2221-E006-194-MY3 and the Water Resources Agency, Taiwan, ROC, under grants MOEAWRA 1020314 and MOEAWRA 1030239.

References

Chen, Z. and Rybczyk, J.: Coastal Subsidence, in: Encyclopedia of Coastal Science, edited by: Schwartz, M., Kluwer Academic Publishers, Dordrecht, the Netherlands, 2005.

Deng, J. L.: Introduction to Grey System Theory, The J. Grey Syst., 1, 1–24, 1989.

Galloway, D. and Burbey, T.: Review: Regional land subsidence accompanying groundwater extraction, Hydrogeol. J., 19, 1459–1486, 2011.

Galloway, D. L., Jones, D. R., and Ingebritsen, S. E.: Land Subsidence in the United States: US Geological Survey Circular 1182, 175 pp.; US Department of the Interior, US Geological Survey, Washington, DC, USA, 1999.

Hung, W. C., Hwang, C. W., Chang, C. P., Yen, J. Y., Liu, C. H., and Yang, W. H.: Monitoring severe aquifer-system compaction and land subsidence in Taiwan using multiple sensors: Yunlin, the southern Choushui River Alluvial Fan, Environ. Earth Sci. 59, 1535–1548, doi:10.1007/s12665-009-0139-9, 2010.

Wang, S. J.: Dimensional upgrade approach for spatial-temporal fusion of trend series in subsidence evaluation, Entropy, 17, 3035–3052, doi:10.3390/e17053035, 2015.

Wang, S. J. and Hsu, K. C.: The application of the first-order second-moment method to analyze poroelastic problems in heterogeneous porous media, J. Hydrol., 369, 209–221, doi:10.1016/j.jhydrol.2009.02.049, 2009a.

Wang, S. J. and Hsu, K. C.: Dynamics of deformation and water flow in heterogeneous porous media and its impact on soil properties, Hydrol. Process., 23, 3569–3582, doi:10.1002/hyp.7458, 2009b.

Wang, S. J., Lee, C. H., Chen, J. W., and Hsu, K. C.: Combining gray system and poroelastic models to investigate subsidence problem in Tainan, Taiwan, Environ. Earth Sci., 73, 7237–7253, doi:10.1007/s12665-014-3902-5, 2015.

Proc. IAHS, 372, 471–474, 2015
proc-iahs.net/372/471/2015/
doi:10.5194/piahs-372-471-2015

The interaction between land subsidence and urban development in China

Y. Yang[1,2], R. Wang[1], Y. Zhou[1,2], Y. Jiang[1], and X. Wang[1]

[1]Beijing Institute of Hydrogeology and Engineering Geology, Beijing, China
[2]School of Earth and Space Sciences, Peking University, Beijing, China

Correspondence to: Y. Yang (alovelystair@163.com)

Published: 12 November 2015

Abstract. The Yangtze River Delta and North China Plain are experiencing serious land subsidence development and are also the areas that have undergone the fastest urbanization. Rapid urban development inevitably requires more water resources. However, China is a country with small per capita water resources, nonuniform distribution of water resources, and over-exploitation of groundwater – all of which are critical factors contributing to the potential for a land subsidence disaster. In addition, land subsidence has brought about elevation loss, damaged buildings, decreased safety of rail transit projects, lowered land value, and other huge economic losses and potential safety hazards in China. In this paper, Beijing, a typical northern Chinese city deficient in water, is taken as an example to explore (a) the problems of urban development, utilization of water resources, and land subsidence development; (b) the harm and influence of land subsidence hazards on urban construction; and (c) the relationship between urban development and land subsidence. Based on the results, the author has predicted the trend of urban development and land subsidence in Beijing and puts forward her viewpoints and suggestions.

1 Introduction

Since the 20th century, the issue of land subsidence attracted increasing, global attention in the urban development and construction industries. This attention has continued to the present, and a lot of work has been done in monitoring and studying land subsidence disasters and their prevention (Zhang and Yang, 2010).

The monitoring and study of land subsidence disasters in China began in the 1920s, and revealed that the distribution and development of subsidence is largely influenced by urban construction (Zhang and Wei, 2005). Beijing, the capital of China, is representative of the Chinese cities that have the fastest rates of land subsidence development in the country. It is also representative of other cities in North China that are deficient in water. From 2007 to the present, the maximum annual settlement rate in the Beijing Plain was more than 100 mm a^{-1}, and land subsidence may develop rapidly. Therefore, the damage and other impacts inflicted by land subsidence disasters must be seriously considered for urban development in Beijing and elsewhere in China.

2 Urban development in China

The formation of urban areas in China has been ongoing for several thousand years; however, China's urbanization was relatively slow until the last decades of the 20th century. The urbanization level in 1949 was only 10.6 % (Zhuang et al., 2012), but accelerated in the 1980s. By the end of 2013, urban residents in China comprised 53.73 % of the total population. It is predicted that in 2050, urban residents will comprise about 75 % of the total population and the urban population may exceed to 1 billion (Gu, 2010).

With the rapid development of urbanization and the sharp increase in the urban population, deficiencies in water and land resources have been some of the most important factors influencing the sustainability of the urbanization process. More than 600 cities in China (nearly two-thirds of the country's cities) are deficient in water, and approximately 110 other cities face serious water deficiencies. To satisfy the demand for water for urban development and population increases, the majority of cities have no choice but to mine for groundwater. Land resources are also a concern. Given eleva-

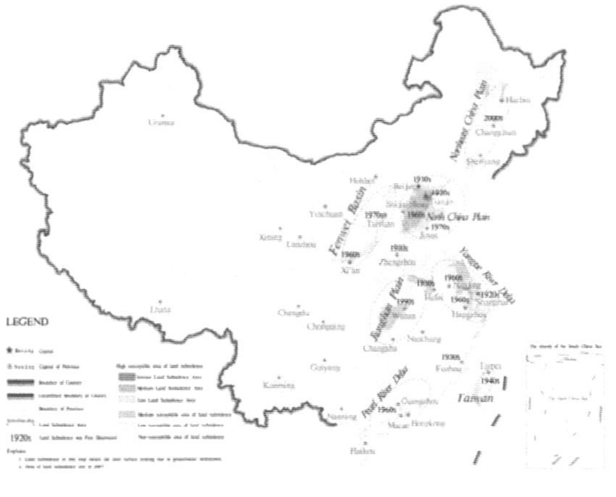

Figure 1. Main regions of land subsidence in China, with years subsidence was discovered.

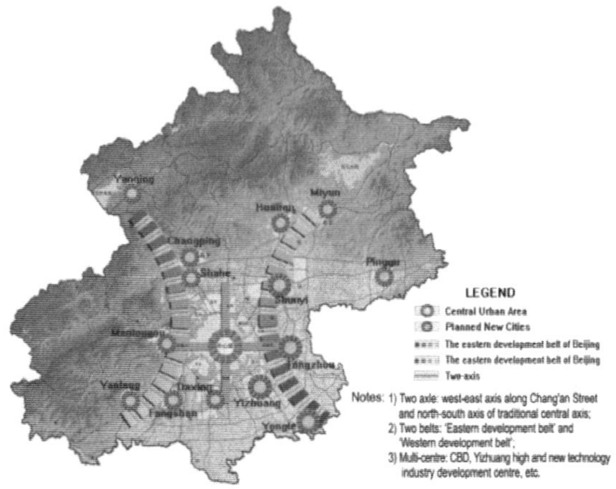

Figure 2. Beijing's planned spatial structure (2004–2020).

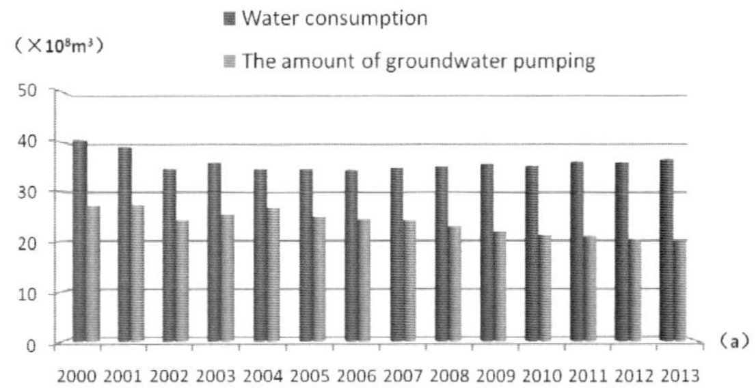

Figure 3. Water consumption and amount of groundwater pumping in Beijing (2000–2013).

tion, temperature, rainfall, and geological and other factors, the area of land suitable for modern habitation is less than 50 % of China's total area. An even smaller percentage of the land is suitable for urban development and construction. Therefore, the loss of land resources caused by land subsidence has received much attention.

3 Land subsidence development in China

The Yangtze River Delta and North China Plain experienced China's earliest urban development and fastest urbanization. They are also the regions with the most obvious land subsidence. By the 1930s, Shanghai was already the largest financial center in the Far East (Wang, 2014), with large building density and early mining of groundwater. It was the first area in China to show land subsidence. Tianjin and Beijing in the North China Plain and other provinces and cities in the Yangtze River Delta and Fen-wei Basin, successively experienced land subsidence (see Fig. 1).

4 Impact of urban development on land subsidence

4.1 Analysis of urban development and water resources

Beijing is a city with a long history. Since 1949, Beijing has operated under six different urban plans and has grown continuously. The current vision of the city as a national political and cultural city follows the city's early development as an industrial base and then a center of science and technology. Beijing's urban planning has been constantly updated to reflect these changes. For example, its current "two axle – two belts – multicenter" spatial layout (shown in Fig. 2) is significantly different from the early single-center layout.

Beijing is also extremely deficient in water resources, as noted previously. In 2014, the water resource per capita in Beijing was only 154.8 m^3 per person, approximately 1/14 of the national per capita level and 1/44 of the global per capita level (Cao, 2014). Groundwater consumption accounts for approximately two-thirds of the total urban water consumption (shown in Fig. 3).

To relieve the water resource burden caused by urban development and reduce groundwater pumping, Beijing Municipal has strengthened its water resource management, actively introduced external water resources (such as the South-to-North Water Diversion Project), enhanced the utilization of reclaimed water, optimized the water consumption structure, and taken other measures that have yielded remarkable results (see Fig. 4).

4.2 Analysis of urban development and land subsidence

The factors influencing land subsidence development in Beijing mainly include (1) active fault and structure development and geologic features and a thicker quaternary system of a compressible clay layer; and (2) over-exploitation of groundwater. The latter is the primary cause of fast development of land subsidence. Figure 5 shows the relationship be-

Table 1. Urban development and land subsidence development in Beijing.

Year	Urban development	Land subsidence development condition
Before 1930s	The urban area was < 100 km^2 and the scale of modern industry was very small.	Land subsidence was not discovered.
1930s–1940s	Beijing's downtown began to expand, and exploitation of groundwater began on a small scale.	Slow development of Xidan–Dongdan belt began.
1950s	Heavy industry began, groundwater mining increased, and the city area was expanded to the third ring road.	In East Bali Village and Jiuxianqiao, the local settlement rate was less than 5 mm a^{-1}.
1960s 1970s	Steel, machinery, and other heavy industry rose rapidly, and Beijing became one of the most important industrial bases in China. A groundwater depression cone formed.	Local subsidence was seen at East Balizhuang, Dajiaoting, and other areas. The subsidence area was larger than 400 km^2. The maximum settlement rate reached 28.2 mm a^{-1}.
1980s 1990s	The city adjusted its industrial structure and light industry dominated. Shunyi, Changping, and other suburban counties developed, and the urban area expanded to the fourth ring road. Exploited groundwater accounted for 50 % of the city's water supply.	Shunyi, Changping, and several other districts and counties discovered land subsidence, and the subsidence area was larger than 2800 km^2. The maximum sedimentation rate was 34.5 mm a^{-1}, and the maximum accumulative settlement was 722 mm.
2000s 2010s	The new town and satellite town were developed and were dominated by the third industry. Large numbers of rail transit projects and other large projects were constructed. Groundwater was continuously over-exploited and accounted for 70 % of total water consumption.	Approximately 10 districts and counties discovered land subsidence, and the subsidence area took up approximately two-thirds of the total Beijing Plain, approximately 4300 km^2. The maximum settlement rate was greater than 150 mm a^{-1}, and the maximum accumulative settlement was greater than 1.6 m.

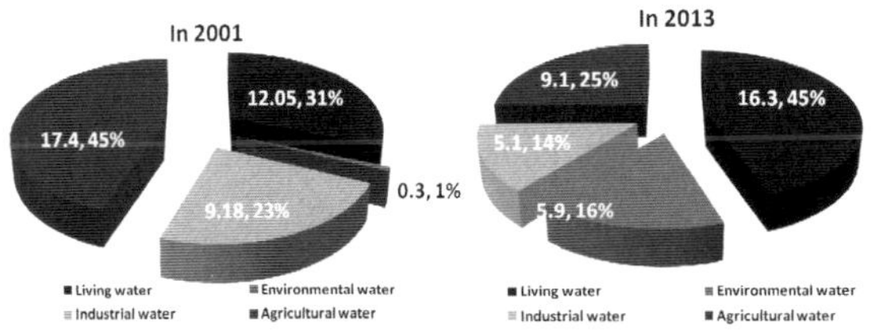

(Data source: Beijing Water Supplies Bureau)

Figure 4. Contrast analysis diagram of water consumption structure in 2001 and 2013 in Beijing.

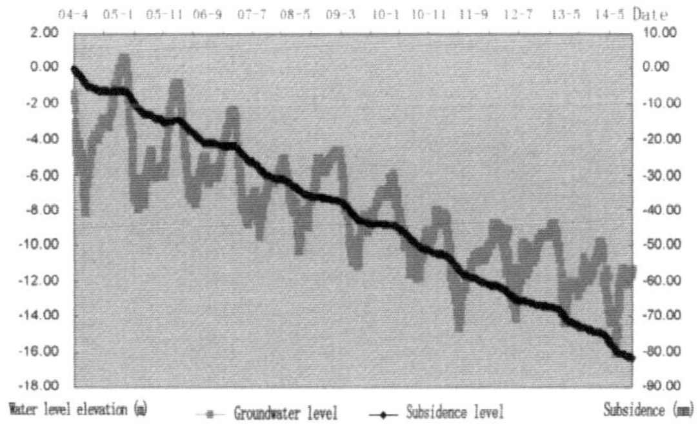

Figure 5. Groundwater and land subsidence levels in Beijing.

tween groundwater and land subsidence, as revealed through dynamic monitoring of groundwater and land subsidence research. The land subsidence disaster development condition caused by over-exploitation of groundwater is shown in Table 1.

4.3 Predictive analysis of urban development and land subsidence development in Beijing

The planning of new city zone construction, traffic and infrastructure construction, ecological and environmental protection, and conservation of the historic cultural city are the key components of urban development in Beijing. The carrying capacity of resources and the geological and environmental safety of the Beijing–Tianjin–Hebei region are also issues that must be considered.

Beijing has a serious deficiency of water resources, a high rate of urban construction and population growth, and a particular regional geological background. Given these factors and the hysteresis of land subsidence development, it is predicted that land subsidence around Beijing will continue to develop at a rapid rate. In 5 to 10 years, the settlement rate of the local settlement area will still be greater than $100\,\mathrm{mm\,a^{-1}}$. Therefore, balancing the exploitation and supplementation of groundwater is the key for guaranteeing sustainable urban development and preventing a land subsidence disaster in Beijing.

4.4 Impact of urban development on land subsidence

The negative impact of urban development and construction on land subsidence reflects in elevation loss, inwelling, acute storm surges, failure of floodwalls, and other serious consequences (e.g. failure of municipal facilities, roads, bridges, underground utilities, levelling signs, and well casings) (Yang, 2015). Subsidence has caused damage to buildings in Shanghai, Tianjin, and other coastal cities and has decreased the safety of rail transit projects. Moreover, subsidence has triggered and aggravated ground fissures and other urban geological disasters and has decreased comprehensive land use value. China will bear the resulting economic loss, which amounts to thousands of billions of renminbi (Ye et al., 2006).

Although influencing urban construction, land subsidence has also restricted future urban development planning.

5 Conclusions and suggestions

1. The ecological and environmental carrying capacity is the key factor restricting China's urban development. Managing the relationships among economic construction, population growth, resource utilization, and ecological and environmental protection is the key for realizing sustainable urban development.
2. Over-exploitation of groundwater is the critical factor influencing the development of land subsidence disasters in Chinese cities, and controlling the mining of groundwater is the most effective approach to prevent land subsidence disasters.
3. Stratum structure unit and groundwater flow-field distribution characteristics should be taken as the standard for control zoning; these are required to comprehensively and rapidly promote the prevention and control of land subsidence disasters in China in order to achieve the aim of serving urban development.

References

Cao, H.: The crisis of water resources need to be solved in Beijing, China Econ, Times, 8, 6, 2014.

Gu, C.: China's urbanization space and its formation mechanism, School of Architecture, Tsinghua University, Beijing, 2010.

Wang, D.: Urban planning history of China, Southeast University Press, Nanjing, 2014.

Yang, Y.: Land Subsidence Disaster Prevention and Cure in Beijing-Tianjin-Hebei Area, China, Urban Geol., 1, 1–7, 2015.

Ye, X., He, Q., Li, S., et al.: Economic loss assessment of land subsidence in the north China plain, China Ocean University Press, Beijing, 2006.

Zhang, A. and Wei, Z.: Land subsidence, Shanghai Scientific and Technical Publishers, Shanghai, 2005.

Zhang, A. and Yang, T.: Latest progress review of research on international land subsidence, Shanghai Geol., 4, 57–63, 2010.

Zhuang, D., Jiang, D., Xu, X., et al.: China's urban space layout of suitability evaluation, Science Press, Beijing, China, 2012.

Proc. IAHS, 372, 475–479, 2015
proc-iahs.net/372/475/2015/
doi:10.5194/piahs-372-475-2015

Settlement characteristics of major infrastructures in Shanghai

X. Jiao[1,2], **X. X. Yan**[1,2], **and H. M. Wang**[1,2]

[1]Shanghai Institute of Geological Survey, Shanghai, China
[2]Key Laboratory of Land Subsidence Monitoring and Prevention, Ministry of Land and Resources of China, Shanghai, China

Correspondence to: X. Jiao (jiaoxun83@126.com)

Published: 12 November 2015

Abstract. Critical infrastructures in Shanghai have undergone uneven settlement since their operation, which plays an important role in affecting the security of Shanghai. This paper, taking rail transportation as example, investigates settlement characteristics and influencing factors of this linear engineering, based on long-term settlement monitoring data. Results show that rail settlement is related to geological conditions, regional ground subsidence, surrounding construction activities and structural differences in the rail systems. In order to effectively decrease the impact of regional ground subsidence, a monitoring and early-warning mechanism for critical infrastructure is established by the administrative department and engineering operators, including monitoring network construction, settlement monitoring, information sharing, settlement warning, and so on.

1 Introduction

Since the opening of Metro Line 1 in January 1990, a total of 11 rail lines have been put into operation in Shanghai, and the mileage has increased to 410 km in 2010. These subway systems play a great role in relieving traffic congestion on the ground, which bring good social, economic and environmental benefits. However, the uneven settlement, caused by rail construction and operation, emerged in several lines, and resulted in negative impacts on rail safety and management(Chen and Zhan, 2000; Wang, 2009; Wang and Hang, 2004). In this paper, we surveyed plenty of long-term monitoring data for tract bed settlement, mostly coming from three lines (Lines A, B, C in the text), and drew up the settling characteristics and affecting factors for subway tunnels. Also the management for rail-settlement controls in Shanghai were introduced.

2 Laws for rail settlement

Of the three lines, Line A was the first to be put into operation, followed by Lines B and C.

From the monitoring results, Line A has a maximum cumulative settlement, of about −111.1 mm from 1995 to 2010. Figure 1 shows that two sedimentation bowls exist in Line A, from Site 260 to Site 340 and Site 560 to Site 780. While Line B had an average cumulative settlement of −57.9 mm from 1999 to 2010, and two intervals, from Site 480 to Site 1680 and Site 1860 to Site 2220 in Fig. 2, underwent larger settlement. However Line C exhibits a smooth trend in overall settlement, with a cumulative settlement of −4.4 mm from 2008 to 2010 (Fig. 3).

According to the curves of Lines A and B (Figs. 2, 3), the tunnel sedimentation rate was much larger in the first few years, particularly in the bowl area whose settlement rate exceeded 20 mm a^{-1}, then the rate gradually reduced to a smaller value and eventually transitioned to a stable state.

We compared the monitoring results of rail settlement and extensometer data, and found that the subsidence trend was consistent from the tunnel to the bedrock, and the soil compression below the tunnel floor played a controlling role in tunnel settlement. For example, results of Line A Site 782 shows a high level correlation with the settlement at the nearby extensometer from 1999 to 2010 (Fig. 4). At the end of 2010, Site 782 shows a cumulative settlement of about

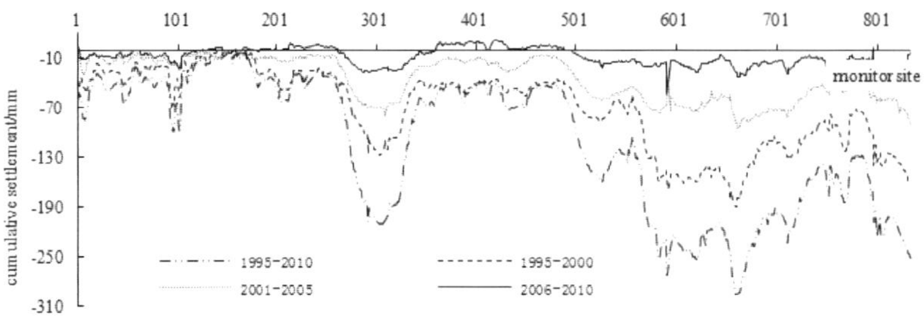

Figure 1. Graph of cumulative settlement of A in different periods

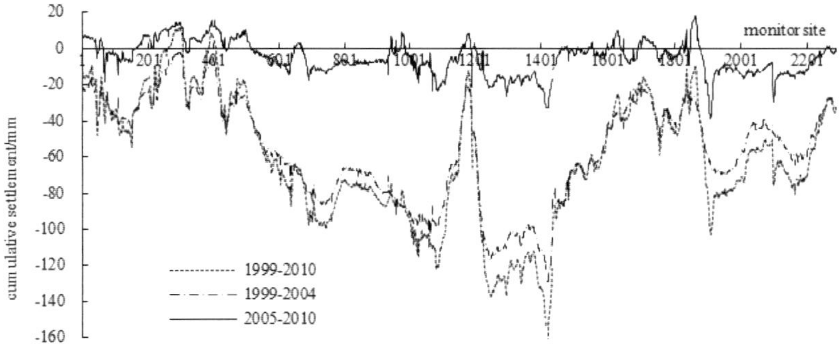

Figure 2. Graph of cumulative settlement of B in different periods.

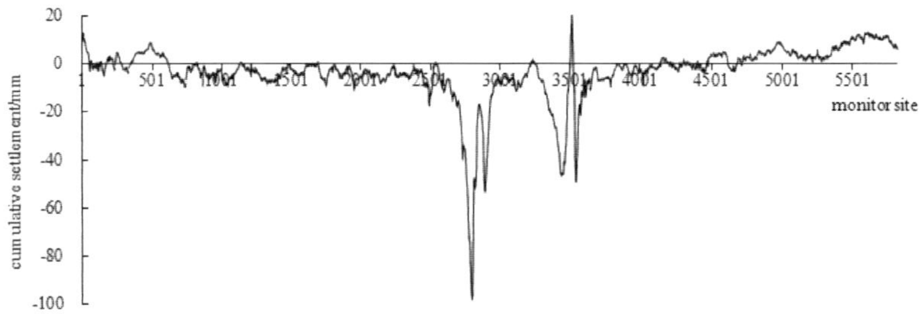

Figure 3. Graph of cumulative settlement of C from 2008 to 2010.

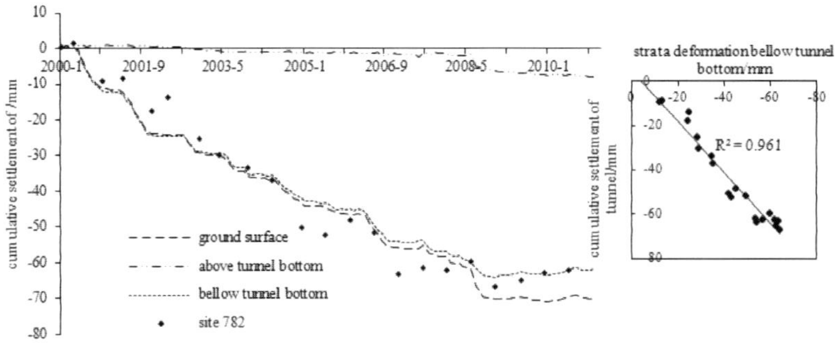

Figure 4. Contrast the monitoring results of point 782 with the ones of extensometer.

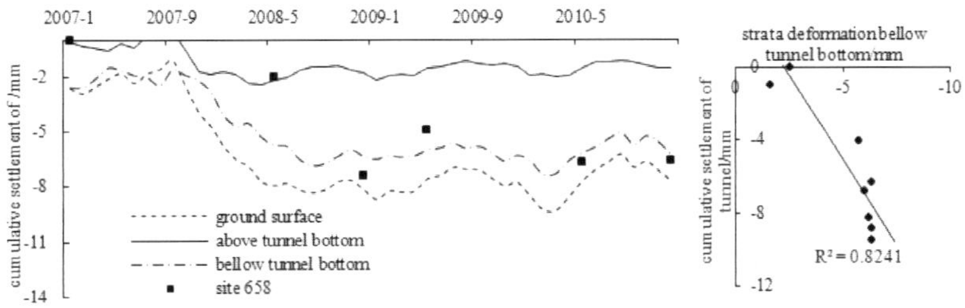

Figure 5. Contrast the monitoring results of point 658 with the ones of extensometer.

60 mm, while the soil compression below the tunnel floor was about 62 mm. Also Site 658 had the similar compression, which produced $-2.6 \sim -7.3$ mm of settlement while compression in the adjacent soils was $-1.5 \sim -7.4$ mm (Fig. 5).

3 Factors for rail settlement

3.1 Geological differences

Tunneling and structure differences of underlying soil are important factors influencing the settlement of the rail way (Shi, 2010; Yan and Shi, 2006). By analyzing the subsidence characteristics of operating lines, we found that tunnels, which passed through shallow sand and ancient river formations, often had large and obvious settlement. In addition, if there are thick soft clays underlying the rail tunnel, it would grow into subsidence bowl along the entire line. For example, Line C passes through a length of sandy soil, whose thickness is about 10–15 m. When trains pass, pore water in the sandy soil is prone to leakage from soil to tunnel. Subsequently, settlement in this area arises.

3.2 Regional land subsidence

As an underground structure in urban environment, subway tunnel is linear throughout the soft soil below the surface. So geological changes caused by urban activities would inevitably affect the subway tunnel. The regional land subsidence, resulting from groundwater exploitation and construction activities, forced the passing tunnel to give similar deformation features.

The cumulateve settlement curve of the subway and surface leveling-points along the Line are compared in Figs. 6 and 7. It is apparent that the two settlement characteristics are in agreement. When the amount of surface subsidence is large, the settlement of subway tunnel is also large. Thus, regional land subsidence along the subway directly affects the overall trend of tunnel settlement, which has become one of important factors for rail risk.

3.3 Surrounding construction

The trend of surrounding construction activities has a negative impact on rail settlement. For example, the excavation and contruction of a mallbegan on 28 September 2008, close to Line C, while the dewatering well associatied with the development commenced pumping on 22 April 2009. On 27 December 2009, the foundation pit construction of the main building was completed. Monitoring results in Fig. 8 show that, from June 2009 to December 2009 tunnel sedimentation rate suddenly increased, which produced a total settlement of -60 to -80 mm. This is closely related to the mall's dewatering activities. When pit construction was completed, the rail sedimentation rate was drastically reduced.

Nearby pit construction would cause the outside soil to release stress during soil excavation, and the tunnel would produce horizontal and vertical displacements. In the process of reliving and dewatering constructions, the groundwater seepage changed, and the aquifer pressure was reduced. Then the pore water pressure in clayey soil layers decreased, which caused further consolidation of clay layers. The joint effect resulted in soil deformation and caused rail settlement. Therefore, this region should be focused as a high-risk zone when large construction activities take place near the tunnel.

3.4 Structural differences

Due to long-term monitoring of the operating lines, it was found that structural differences also have contributed to zone settlement. These special sections, for example, aisles connecting the tunnels and transition-section between underground and above-ground lines, are structuraly different with regard to their connecting pieces, which creates large changes of stress in the soil layer or tunnel body. After a period of operation, these sections grew up with dislocation and partial leakage, followed by longitudinal settlement.

4 Management for prevention of rail settlement

In order to ensure safe operation of Shanghai rail system, the Shanghai administrative department collaborated with major engineering and management units, and established a moni-

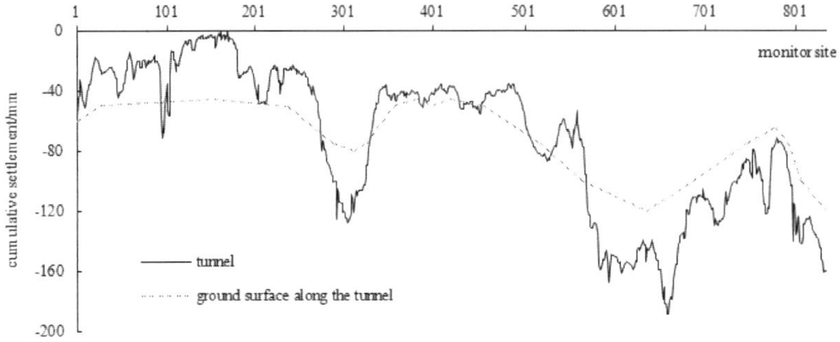

Figure 6. Graph of tunnel settlement and land subsidence along A from 1995 to 2000.

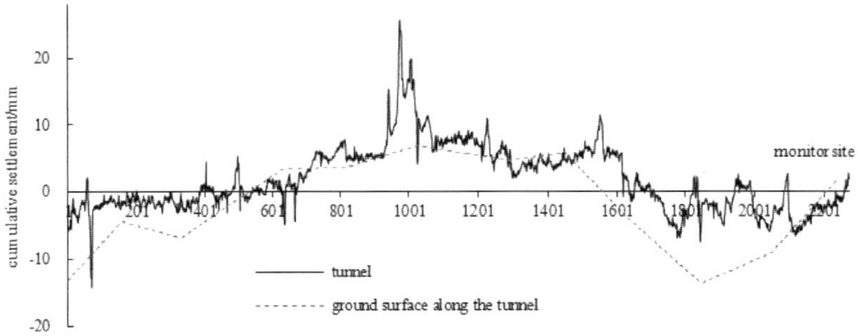

Figure 7. Graph of tunnel settlement and land subsidence along B in 2009.

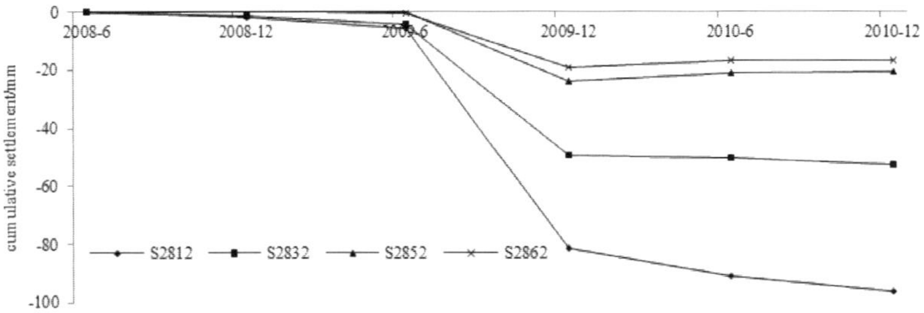

Figure 8. Duration curve of cumulative settlement in typical C sites nearby the pit.

toring and early-warning mechanism for major facilities' settlement. It included the following:

1. *Regular meeting*: regular meetings would be convened to discuss the dynamic data of major facilities' settlement, draw up monitoring plans, and discuss information sharing and other related matters.

2. *Network construction*: the two entities define their duties regarding-network construction, according to Shanghai Plan for Controlling and Management of Land Subsidence. This includes improving the monitoring network and optimizing the monitoring sites combined with municipal spatial layout of all major facilities.

3. *Settlement monitoring*: the Shanghai administrative department oversees the monitoring of regional and other settlements that occur along the major facilities. The frequency of investigation is not less than once a year. However, the engineering or management units should monitor their responsible projects, and report results to the administrative department. Monitoring frequency is made by each unit in accordance with safe operation of major facilities.

4. *Information sharing*: a data system is established. The two units create a uniform standard for data collection for major facilities' settlement, and all of the data will be put into the system, and shared by each unit.

5. *Early-warning mechanism*: Shanghai administrative departments collect monitoring results, analyze and work on a regular summary report, which will be offered to facility-operating units in a timely manner. When the monitoring values of land subsidence along major facilities exceeds the control requirements, the administrative department will immediately inform the operational unit and share each data to find the problems and take steps toward mitigation.

5 Conclusions

Shanghai has operated metro lines for over 20 years. By means of continuous learning and research, some rules of rail settlement have been discovered. In this paper, we briefly illustrate the contributing factors, including operating time, geological conditions, structural differences, and external construction. In addition, a monitoring and early-warning system was established, which aims to reduce negative impacts on the rail transit system and brings greater security to the public.

Acknowledgements. The authors thank their colleagues for continuing support.

References

Chen, J. W. and Zhan, L. X.: Deformation measuring of the metro tunnel and deformation data analysis of shanghai metro line, Shanghai Geology, 21, 51–56, 2000.

Shi, Y. J.: Engineering geological conditions and major geological problems during construction of metro tunnels in Shanghai area, Journal of Engineering Geology, 18, 774–780, 2010.

Yan, X. X. and Shi, Y. J.: Structure characteristic of engineering geology in Shanghai, Shanghai Geology, 26, 51–56, 2006.

Wang, R. L.: Factors influencing deformation of shanghai soft soil metro tunnel and deformation analysis, Underground Engineering and Tunnels, 1–6, 2009.

Wang, Y. and Hang, H. W.: Hierarchy-Fuzzy Comprehensive Judgment for Safety Evaluation of Metro Running Tunnel, Underground Space, 24, 301–305, 2004.

Proc. IAHS, 372, 481–484, 2015
proc-iahs.net/372/481/2015/
doi:10.5194/piahs-372-481-2015

Characterization of land subsidence induced by groundwater withdrawals in Wenyu River alluvial fan, Beijing, China

R. Wang[1], **Y. Luo**[1], **Y. Yang**[1,2], **F. Tian**[1], **Y. Zhou**[1,2], **and M.-Z. Tian**[1]

[1]Beijing Institute of Hydrogeology and Engineering Geology, Beijing, China
[2]School of Earth & Space Sciences,Peking University, Beijing, China

Correspondence to: R. Wang (happywangrong@126.com)

Published: 12 November 2015

Abstract. The Beijing plain area has suffered from severe land subsidence owing to groundwater overdraft. A major example is the Wenyu River alluvial fan in the Beijing plain area. This area has experienced as much as 10 m of land subsidence through 2000s. An integrated subsidence-monitoring program, including borehole extensometer and multilayer monitoring of groundwater, has been designed to meet the needs of monitoring land subsidence in this region. This work has allowed us to characterize land subsidence and understand the mechanical properties of the strata. The analysis results show the development of the land subsidence in this area is consistent with water-level change. The major strata contributing to compression deformation are Mid-Pleistocene stratum which contributed around 70 % of total subsidence. The shallow stratum and deep stratum show elastic mechanical behavior the intermediate stratum exhibit elastic-plastic mechanical behavior.

1 Introduction

Land subsidence, as a non-compensable loss of environment and resources, is regional geological disaster caused by the damage of geological environment system. It can induce a series of other environmental hazards to form disaster chains (Bawden et al., 2001). China also suffers serious land subsidence. A long sequences of land subsidence monitoring data was used to analyze the deformation behaviour of soil in Shanghai (Zhang et al., 2006), Tianjin (Dong et al., 2008), Suzhou, Xi'an and Changzhou (Shi et al., 2006), China. The deformation behavior of compression layer is also analysed at the rapidly development stage in Beijing (Tian et al., 2012), and, the study of the relationship between land subsidence and groundwater is also carried out at the same time (Yang et al., 2013). The previous research is insufficiency in some respects such as the different layer characters of land subsidence and the relationship between land subsidence and the groundwater level change with the same monitoring data in the same area, and the reason of the compression characteristic emerged through multiple factors and multi-angle analysis. However, it is critical for the regulation of land subsidence and the optimization planning of groundwater pumping.

The paper takes Wenyu Alluvial Fan as the study area, which contains the Baxianzhuang subsidence region. It has complicated compression types and varied influence factors, and acts as one of the most serious and representative subsidence regions in Beijing currently.

2 Development characteristics of land subsidence

2.1 Delamination development characteristics

Since 2004, seven land subsidence monitoring station has been built to obtain the compression in different layers and the groundwater level change in Beijing plain area. The groundwater level monitoring wells and borehole extensometer can provide the compression of aquifer system and the change of pore pressure, which is useful for understanding the hydrology and mechanics characteristics of the hydrogeological unit in different depth. There are ten borehole extensometer wells in Baxianzhuang monitoring station. The

Table 1. Monitoring land subsidence list in Baxianzhuang.

ID	Layer (m)
D4-1	304–333
D4-2	259–278
D4-3	211–235
D4-4	147–195
D4-5	60–107
D4-6	13.5–23
F4-1	below 304
F4-2	262.00 ~ 304
F4-3	235.00 ~ 262
F4-4	197.00 ~ 235
F4-5	144.00 ~ 197
F4-6	108.00 ~ 144
F4-7	83.00 ~ 108
F4-8	59.00 ~ 83
F4-9	24.00 ~ 59
F4-10	2.00 ~ 24

Table 2. Stratums compaction of different geologic age.

Stratigraphic age	The depth of bottom boundary	Compression proportion
Holocene, Late Pleistocene	37	< 1 %
Middle Pleistocene	135	70 %
Early Pleistocene	350	29 %

monitoring layers of borehole extensometer and groundwater level are showed in Table 1.

The borehole extensometer result showed that the major compression layer is the Middle Pleistocene strata, which accounts for about 70 %. The most dominant compression layer has a depth of 108–144 m, which can account for 34.62 % of the total, followed by the depth of 24–59, 59–83, 83–108, 144–197, and 197–235 m with 12.40, 7.66, 15.31, 14.52, and 11.02 % compression amount respectively. The compression of other strata is only 2 %.

Different geological age has great impact on the Sedimentary thickness and consolidation status. According the Paleomagnetic, pollen and C^{14} of engineering geology hole in Baxianzhuang, the interface of Quaternary chronostratigraphic can be determined. Combined with the measured subsidence of different layer, the bottom boundary of Holocene and Late Pleistocene has a depth of 37 m below the surface, whose contribution is small for the subsidence. The bottom boundary of Middle Pleistocene has a depth of 135 m, and it is the major contribution layer to the subsidence which can account for about 70 %. The bottom boundary of Early Pleistocene has a depth of 350 m, and the subsidence can account for about 29 %.

2.2 Groundwater withdrawal and land subsidence

There are five aquifer groups in the Quaternary aquifer system of the study area (Liu et al., 2007). Due to the low-quality of groundwater, the groundwater withdrawal of first aquifer group is small and the groundwater level drops slowly, and in some areas the groundwater level even rises. The second aquifer group has a large exploitation quantity due to the agricultural irrigation, and the groundwater level shows seasonal changes. The water of third aquifer group is mainly used as domestic and industrial water, and the groundwater level drops quickly owing to the large amount of groundwater exploitation. The forth pumping is mainly for domestic water and partly for industrial water. There is little groundwater pumping in the fifth group.

The main pumping strata is the third layer, which can account for 75 %, and the corresponding compression is 50 % of the total. The second layer accounts for 21.64 % of the total pumping, and the compression of this strata is 35.36 %. Therefore, groundwater exploitation is an important factor for land subsidence. In the same area, large water pumping in the different layer can cause large land subsidence, however, they are not directly related. There are other factors will certainly influence the compression volume and compression character.

2.3 Groundwater regime and land subsidence

Long-term over-exploitation of groundwater can cause significant decline of groundwater level, then lead to the decline of the pore-pressure of the overlaying soil. As a result, the water loss, consolidation and compression of soil would happen. At the same time, soil particles of aquifer layer become more closely aligned to form land subsidence.

With long series land subsidence monitoring, the groundwater level change of aquifer in different layers is compared with the trend of land subsidence of different strata. The result shows that the groundwater level is consistent with development of land subsidence. The land subsidence has an increasing state when the groundwater level has a cyclical variation. The rate of subsidence repeating is in accordance with the seasonal change of groundwater level. The accumulative subsidence is increasing fast when the groundwater level drops, and it has a delay and the rate is slow down when the groundwater level rises (shown in Fig. 1).

3 Deformation characteristics of soil water release

Based on the monitoring of the groundwater level and borehole extensometer, the stress-strain map is used to analyze the character of compression in different layer. It can be used to select appropriate constitutive equation and provide a scientific basis for land subsidence prediction (Burbey, 2001). The result shows that the strata in different depth has disparate compression and deformation character of soil. The

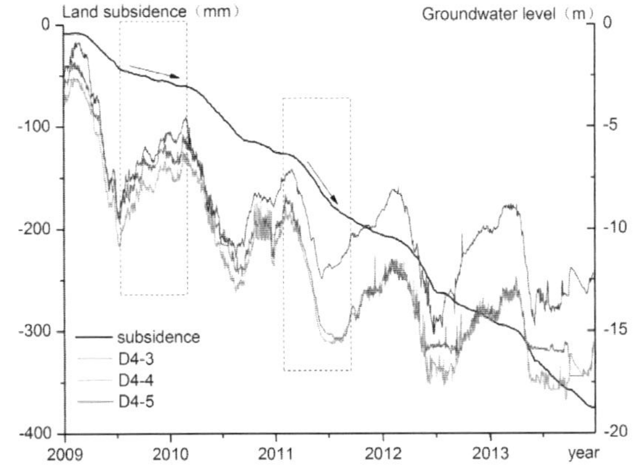

Figure 1. Relation of land subsidence and groundwater level in monitoring station.

compression of soil has elastic and elastic-plastic characteristics.

3.1 Elastic deformation

3.1.1 Deformation characteristics of shallow strata

The monitoring data of Baxianzhuang station shows that the compression of shallow strata (0–24 m) is small. From 2009 to 2013, the deformation is from −1 to 1 mm and the groundwater level is rising (Fig. 2). The result is that the level is rising slowly and there is no compression in this layer.

3.1.2 Deep strata

The compression of strata below 235 m is small in Baxianzhuang which accounts for 3.6 % of the total land subsidence. The monitoring result shows that the deformation of this strata is well consistent with the groundwater level change. Along with cyclical fluctuations in the water table, the deformation also shows seasonal deformation characteristics. Since 2009, the groundwater level shows significantly periodic characteristics, and the compression period of soil also shows almost the same trend.

The confined water show a circular and cyclic quality in such depth. However, the rise value is usually less than the decline value in each cycle, which cause the continuous declination of groundwater level (Fig. 3a). Accordingly, the soil is also suffered from repeated loading and unloading effect. The stress increased is greater than the stress reduction, and the effective stress of the overall soil continued to increase. When the water level drops, the compression amount is relatively large. When the water level rises, the compression amount slowed a significantly decreasing tendency, and a slight rebound phenomenon (Fig. 3b). The stress-strain analysis indicates that there is smaller hysteresis loop appeared within the repeated loading-unloading process (Fig. 3c). The strata shows significant elastic deformation characteristics.

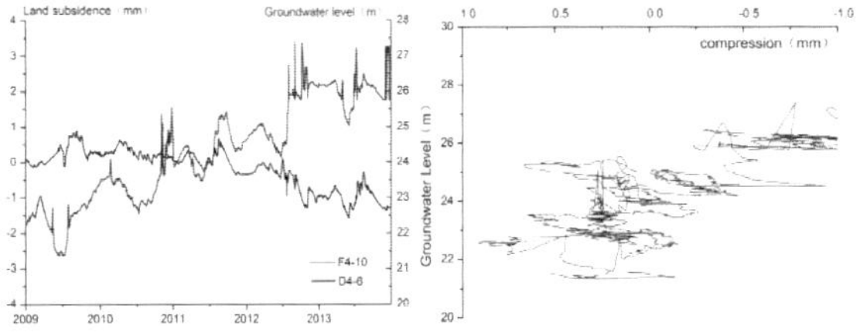

Figure 2. Soil deformation and water level of F4-10.

3.2 Elastic-plasticity deformation

The middle-deep strata (24 ~ 235 m) is the main layers studied for groundwater exploitation, the groundwater level of the aquifer system at such strata has overall decline trend. The groundwater level has significant seasonal fluctuation with the maximum amplitude up to 4 m. At the same time, the strata is the major layer contributes to the formation of surface deformation, whose compression amount accounts for more than 95 % of the total compression amount. Since 2009, the groundwater level periodically reciprocating movements, in each cycle, the water level declined more than rise and showed an overall decline trend. It means that the effective stress of soil continues to increase, and the soil sustained to compress (Fig. 4a). Whenever the water level rises, soil compression slows down. The residual compression is greater than the rebound, and the soil shows a sustained compress trend (Fig. 4b). Stress-strain analysis showed (Fig. 4c), the water level overall continued to decline with dynamic changes, but still higher than the lowest level. There are six borehole extensometers at the depth within 24–235 m in the Immortals village land subsidence monitoring station With the analysis of layered subsidence and the corresponding water level monitoring, each layer has basically the same stress–strain characteristics, which is elastic-plastic deformation that dominants by the plasticity characteristics.

For normally consolidated or unconsolidated formations, its deformation mechanism can be explained on the basis of the principle of effective stress. It can be described as that drawdown increases, the pore pressure decreases, the effective stress of soil skeleton increases and land subsidence occurs. For over-consolidated formations, the driving conditions of ground subsidence is the key to study its deformation mechanism. In over-consolidated formations, the occurrence of land subsidence needs a driver such as a initial aditional stress more than the over-consolidation stress, or a initial drawdown greater than the of the critical water level.

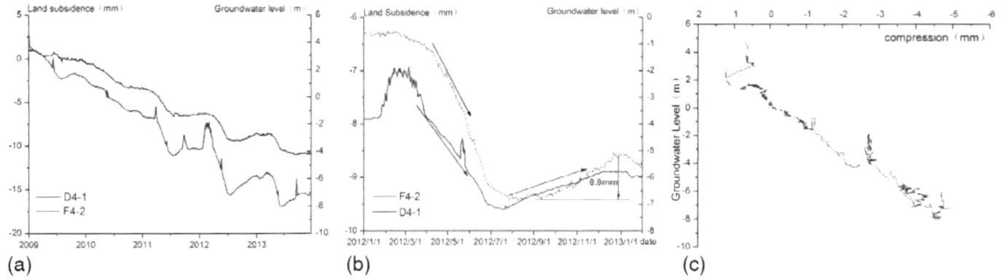

Figure 3. Soil deformation and water level of F4-2.

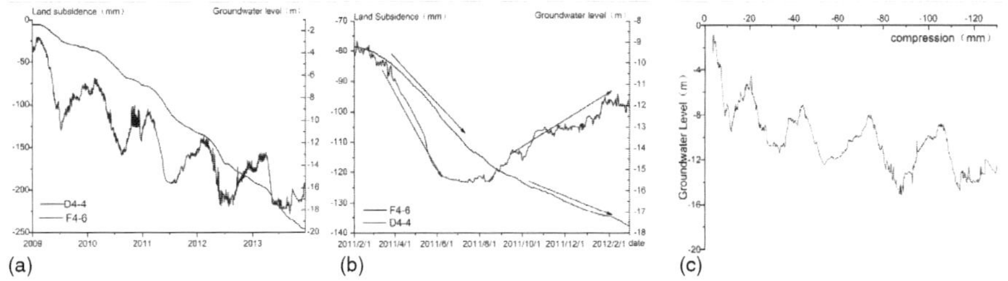

Figure 4. Soil deformation and water level of F4-6.

4 Conclusion

This paper focuses on the land subsidence of Baxianzhuang, and analyzes the compression characteristics of different strata and the influence factors with monitoring data of different layers. According to the geotechnical test, the deformation characteristics of aquifer system when releasing water can be acquired. The results are obtained as below.

The borehole extensometer result showed that the major compression layer is the Middle pleistocene strata, which accounts for about 70 %.

The compression of early Pleistocene strata is low, and the compression of strata below 200 m can account for 29 % of the total land subsidence. Most soil is at normal consolidated or under-consolidated state and the thickness of clay is small. With the same exploitation amount, the compression of this strata is small. Therefore, it is the priority layer to pump groundwater in order to regulate land subsidence.

Acknowledgements. We thank Beijing Geology Minerals Investigation Development Bureau, China Institute of Geo-Environment for supporting our work and data processing tasks that aided our research. We are especially grateful to Ye Chao, Liu Wenchen, Liu Jiurong and our peers who have been working tirelessly on the land subsidence monitoring and prevention in Beijing Plain.

References

Bawden, G. W., Thatcher, W., Stein, R. S., Hudnut, K. W., and Peltze, G.: Tectonic constraction across Los Angeles after removal of groundwater pumping effects, Nature, 412, 812–815, 2001.

Burbey, T. J.: Stress-strain analyses for aquifer-system characterization, Groundwater, 39, 128–136, 2001.

Dong, K. G., Wang, W., Yu, Q., Zhou, J., Lu, Y., Yi, Ch. R., and Zheng, Y. P.: Ratio of soil loss to groundwater-exploitation in the Tianjin land subsidence area, Hydrogeology and Engineering Geology, 5, 76–80+91, 2008.

Liu, Y., Ye, C., and Jia, S.-M.: Division of water-bearing zones and compressible layers in Beijing's land subsidence areas, City Geology, 1, 10–15, 2007.

Shi, X. Q., Xue, Y. Q., Wu, J. Ch., Zhang, Y., Yu, J., and Zhu, J. Q.: A study of soil deformation properties of the groundwater system in the Changzhou area, Hydrogeology and Engineering Geology, 3, 1–6, 2006.

Tian, F., Guo, M., Luo, Y., Zhou, Y., and Jia, S. M.: The deformation behavior of soil mass in the subsidence area of Beijing, Geology in China, 39, 236–242, 2012.

Yang, Y., Zheng, F. D., Liu, L. C., Wang, S. F., and Wang, R.: Study on the correlation between groundwater level and ground subsidence in Beijing plain areas, Geotechical Investigation & Surveying, 8, 44–48, 2013.

Zhang, Y., Xue, Y. Q., Ye, Sh.j., Wu, J. Ch., and Li, Q. F.: Analysis of deformation of sand strata and land subsidence based on modes of groundwater level changes in Shanghai city, The Chinese Journal of Geological Hazard and Control, 3, 103–109, 2006.

Proc. IAHS, 372, 485–491, 2015
proc-iahs.net/372/485/2015/
doi:10.5194/piahs-372-485-2015

An integrated assessment framework for land subsidence in delta cities

T. H. M. Bucx, C. J. M. van Ruiten, G. Erkens, and G. de Lange
Deltares, Delft, the Netherlands

Correspondence to: T. H. M. Bucx (tom.bucx@deltares.nl)

Published: 12 November 2015

Abstract. In many delta cities land subsidence exceeds absolute sea level rise up to a factor of ten by excessive groundwater extraction related to rapid urbanization and population growth. Without change, parts of Jakarta, Ho Chi Minh City, Bangkok and numerous other delta (and coastal) cities will sink below sea level. Increased flooding and also other widespread impacts of land subsidence result already in damage of billions of dollars per year.

In order to gain insight in the complex, multi-sectoral aspects of subsidence, to raise awareness and to support decision making on appropriate adaptation strategies and measures, an Integrated Assessment Framework (IAF) for subsidence is introduced, illustrated by several (delta) case studies. Based on that a list of 10 generic key issues and possible solutions is presented in order to further develop and support a (generic) approach how to deal with subsidence in current and future subsidence-prone areas. For exchange of experiences and knowledge development.on subsidence in deltas the Delta Alliance, a knowledge network of deltas worldwide, can be supportive.

1 Introduction

In many coastal and delta cities land subsidence exceeds absolute sea level rise up to a factor of ten. A major cause of this severe land subsidence is excessive groundwater extraction due to rapid urbanization and population growth. Current global mean absolute sea level rise is around 3 mm year^{-1} and projections until 2100 based on IPCC scenarios expect a global mean absolute sea level rise in a range of 3–10 mm year^{-1} (Slangen et al., 2012, 2014). However current observed subsidence rates in coastal megacities are in the range of 6–100 mm year^{-1} and projections till 2025 expect similar subsidence rates, depending on future measures. This is illustrated in Fig. 1. Without action, parts of Jakarta, Ho Chi Minh City, Bangkok and numerous other coastal cities will sink below sea level.

Land subsidence causes increased flooding and widespread impacts on roads, embankments, subsurface infrastructure and housing, resulting already in damage of billions of dollars per year. In order to gain insight in the complex, multi-sectoral aspects of subsidence, to raise awareness and to support decision making on appropriate adaptation strategies and measures, an Integrated Assessment Framework for subsidence is introduced, illustrated by several (delta) case studies addressing causes, impacts and responses.

2 Case study results

In the Greater Jakarta area, metropolitan authorities and technical agencies are advocating the reduction of groundwater extraction in vulnerable areas by completely phasing out the use of groundwater and taxing groundwater consumption. This would include the development of alternative water supply for large industrial users or relocation of large groundwater users, outside the "critical zones". The number of "unregistered" users is still a problem. Ongoing economic development and city expansion, lead to the filling of low-lying and flood prone lands with mineral aggregates and often waste materials. To some extent, spatial planning measures were applied to avoid subsidence-prone areas, but fast growth of informal settlements has made many of these plans obsolete (Dam, 2012).

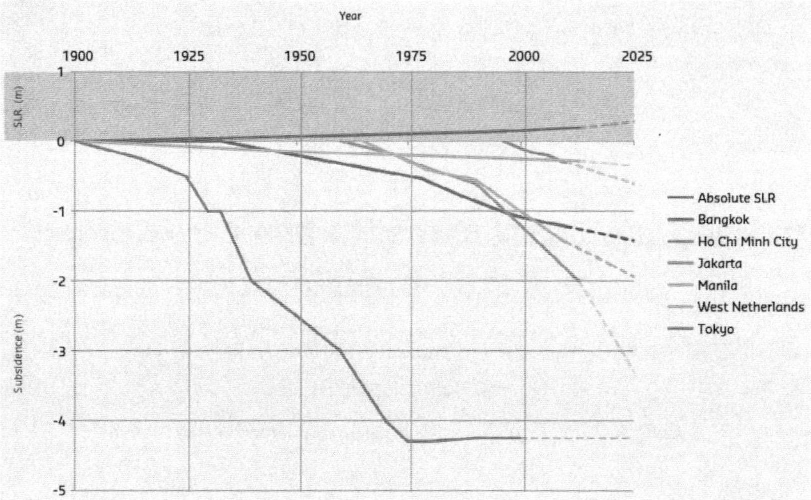

Subsidence in Sinking Cities

	Mean cumulative subsidence in period 1900 - 2013 (mm)	Mean current subsidence rate (mm/yr)	Maximum subsidence rate (mm/yr)	Estimated additional mean cumulative subsidence until 2025 (mm)
Bangkok	1250	20-30	120	190
Ho Chi Minh City	300	Up to 80	80	200
Jakarta	2000	75-100	179	1800
Manila	1500	Up to 45	45	400
New Orleans	1130	6	26	> 200
Tokyo	4250	Around 0	239	0
West Netherlands	275	2-10	> 17	70

Sea level rise (SLR)

	Cumulative mean SLR in period 1900 - 2013 (mm)	Current rate (mm/yr)	Maximum rate (mm/yr)	Possible additional future SLR until 2025 (mm)
Worldwide mean	195	3	-	86

Sources: Bangkok: MoNRE-DGR 2012; Ho Chi Min City: Le Van Trung & Ho Tong Minh Dinh, 2009; Jakarta: Bakr 2011; Manila: Eco et al 2011; West Nederlands; Van de Ven 1993; Tokio: Kaneko & Toyota 2011; Church and White, 2011, Slangen et al., 2013

Figure 1. Global sea level rise (SLR) and average land subsidence for several coastal cities (please note that subsidence can differ considerably within a city area, depending on groundwater level and subsurface characteristics).

Recently the Jakarta Coastal Defense Strategy (JCDS) programme integrated the results of various subsidence studies and tried to obtain a reliable number for current and future subsidence trends. This subsidence prognosis is regarded as an extremely vital component for an integrated flood management and coastal defense strategy, but so far a sufficiently detailed hydrogeological model is lacking to reliably predict long term scenarios. This is now under construction (Aobpaet, 2013; Phisut, 2012).

In Bangkok extreme land subsidence by groundwater extraction was successfully reduced by regulations and restrictions for groundwater extraction. A specific law (Groundwater Act) was enacted in 1977. Most severely affected areas were designated as Critical Zones with more control over private and public groundwater activities. Groundwater Use Charges were first implemented in 1985 and gradually increased. In Bangkok currently about 10 % of the total water use is from groundwater extraction, however subsidence is still ongoing but at a much slower pace than before.

Although land subsidence in Ho Chi Minh City has been observed since 1997, there is still considerable discussion about its causes and impacts (Bakr et al., 2013). This is partly due to poor monitoring data on land subsidence and on groundwater extraction. Restrictions on groundwater extraction have been initiated but it is too early to observe effects. Besides the registered exploitation, which draws mainly from the deeper aquifers there is a large amount of unregistered extraction for domestic water supply. Hence, the total drawdown rate shows no sign of decreasing. Possibly, this can also be attributed to the fact that recharge is hindered by the reduction in infiltration area due to urbanization.

In New Orleans and the Mississippi delta there is as yet no coordinated strategy on mitigating subsidence. The extraction of oil and gas is of great economic importance for the region and will be stimulated, rather than limited. The debate on groundwater use in New Orleans has only recently started, as its contribution to subsidence is so far unknown (Dokka, 2011). The recently published water management strategy for New Orleans, however, suggests raising water levels in areas with organic rich soils, reducing oxidation of organic matter and mitigating subsidence. The Mississippi delta is starved of sediment because of constructions of dams and erosion-prevention measures upstream in the catchment. In the Coastal Masterplan for the Mississippi Delta plans are included to reintroduce sediment-loaded floodwaters to the delta once more.

In Tokyo, after taking regulatory measures on restriction of groundwater use in the early sixties, the groundwater levels increased again and after around 10 years the subsidence was stopped (see Fig. 1).

With increasing insight in the relationship between groundwater extraction and land subsidence, techniques were developed in Shanghai to restore groundwater levels with active or passive recharge. Although this reduced the further lowering of groundwater tables and limited subsidence, it did not solve immediate problems, notably the effect of subsidence on infrastructure, roads and buildings. The subsidence rate was 39.1 mm year^{-1} during the period of 1921–1965 as a result of the increasing utilization of groundwater for industry development. Since the government began to control extraction and to recharge groundwater after 1965, the rate of land subsidence decreased to 6.2 mm year^{-1} on average until 2007 (Shi et al., 2008). According to Gong (2008), the primary cause of land subsidence in downtown is construction of high-rise buildings (60 %) after 1990 while in suburbs 70 % is attributed to withdrawal of groundwater.

In Dhaka, increasing problems with flooding and water supply are resulting in more attention to excessive groundwater extraction and subsidence. In this rapidly expanding city many areas are subsidence-prone but there is currently a lack of data on subsidence and its impacts (Steckler et al., 2012). At present around 87 % of the supplied water is from ground water extraction and it has been acknowledged that the focus has to be changed to using surface water instead. However, treating surface water is much more technically complex and expensive than using groundwater, also because the large rivers nearest to Dhaka city, are polluted among others by the economically important textile industry.

While subsidence in large urban environments in deltaic regions can usually be associated with groundwater extraction, in the case of the Netherlands, where deep extractions are rare, centuries of drainage of surface water, often from former peat marshes has resulted in the widespread lowering of the landsurface by several metres. This is exacerbated by the historic exploitation of peat for fuel, leaving large areas below sea level. These areas were reclaimed and converted in fertile clay polders, but require constant drainage of upwelling groundwater from the underlying confined aquifers. Although the recent upconing of brackish water in these polders poses a water management challenge to adapt to these changing environmental and agricultural circumstances, the area where the surficial and shallow peat remained poses the greatest problems (Cuenca et al., 2010). Not only does the oxidation of drained peat produce a significant amount of green house gas, the continuing subsidence in these areas leads to larger areas where the volumes of upwelling water become a significant economical cost factor. Here also the risk of upconing of brackish groundwater is looming in the background. The greatest cost factor however has been shown to be the gradual deterioration of foundations and infrastructure.

In recent years studies have been performed on how to mitigate this situation. In the agricultural areas keeping the drainage level as high as possible has reduced the rate of subsidence considerably. In urban areas in the peat areas, this would only worsen the situation as the settling surface of streets and gardens leaves insufficient freeboard to accommodate heavy rains. In 2014 the summer rains were especially heavy and widespread flooding has triggered a research effort to systematically mitigate the settlement in urban areas on soft soils. Results from analyses combining InSAR monitoring with geological information indicate that damage in urban areas is dominated by the settlement caused by the use of surcharge in the urban developments. Subsidence rates of 10 mm year^{-1} over a decade are not uncommon in these areas, which is comparable to the rates incurred by peat oxidation. Research efforts will therefore concentrate on ways to reduce the load on the soft soils while at the same time improve the drainage capacity of the urban environment.

Costs of subsidence

Subsidence leads to direct and indirect damage. Direct effects are for instance loss of functionality or integrity of structures like buildings, roads and underground utility networks (critical infrastructure). The most common indirect effects of damage are related to changes in relative water levels, both groundwater levels and surface water levels. The estimation of associated costs is very complex. Moreover due to economic and urban development the potential damage costs will increase considerably, especially in subsidence prone areas such as flood plains.

In practice operational and maintenance costs are considered in several short and long term policies and budgeting. The costs appear on financial sheets as ad hoc investments or planned maintenance schemes, but not as damage costs related to subsidence.

In China the average total economic loss due to subsidence is estimated at around USD 1.5 billion year^{-1} of which 80–90 % are indirect losses. In Shanghai, over the period 2001–2010, the total loss cumulates to approximately USD 2 billion. In Bangkok many private and public buildings, roads, pavements, levees and underground infrastructure (sewerage, drainage) are severely damaged by subsidence, but proper estimates of the costs of damage are not available.

In 2006, the total costs of subsidence related damage in the Netherlands was estimated at over EUR 3.5 billion year^{-1}. The majority of these costs will not be recognized directly as damage due to subsidence. Also the construction site preparation and construction costs in soft-soil areas should be considered as subsidence-related costs, as these are mainly incurred to prevent consolidation. The greater cost of maintenance of existing structures and buildings as well as the extra costs involved in ground improvement in new developments compared to areas without peat in the subsurface has led to a national governmental fund for municipalities in this situation. A recent inventory of the extra costs incurred in the Netherlands in urban areas and on infrastructure estimated at least EUR 220 million year^{-1}. This is excluding damage to historic wooden foundations. A maximum total amount of EUR 40 billion was estimated as cost incurred by soft soil conditions, including projected effects of climate change, when no mitigation measures are taken to counteract present and future adverse conditions.

3 Integrated assessment framework

Land subsidence is often literally a "hidden" issue. Because of its complex, cross-sectoral nature it not only takes place out-of-sight, but it is usually not fully recognized (or acknowledged), especially in the domain of governance and institutional mandates and responsibilities (i.e. lack of a responsible problem owner). The detrimental effects of subsidence are ignored until they become a serious and costly issue. If proper attention is paid to developing the required technical, administrative and institutional capabilities, the negative impacts of land subsidence can be mitigated and the process largely stopped. Therefore a comprehensive and integrated approach is needed. Deltares has developed an Integrated Assessment Framework addressing all aspects of subsidence, which can be applied on any subsidence case. It is based on the DPSIR approach (Driving forces, Pressures, State, Impacts and Responses) and Spatial Layer model. The DPSIR aspects cover the cause-effect-response chain being elaborated for three spatial layers: the Occupation layer (land and water use), Network layer (infrastructure) and Base layer (natural resources subsurface). The main steps and elements of this approach are elaborated in Table 1.

Every city that wants to tackle its subsidence problem has to deal with questions such as: what are the main causes, how much is the current subsidence rate and what are future scenarios, what are the impacts and risks, how can adverse impacts be mitigated or compensated for, and who is involved and responsible to act? In order to answer these technical and governance questions, an integrated approach supports the (policy) development path that cities should follow, from problem identification to planning and implementation of solutions and their evaluation. Every subsiding city is somewhere along this development path (see Table), ranging from being in an early analysis stage (for instance Dhaka, Bangladesh) to cities that more or less seem to have solved the problem at the other end of the spectrum (for instance Tokyo, Japan).

Delta Alliance – knowledge platform

To address subsidence in delta cities through an integrated approach, the Delta Alliance can be supportive in information gathering, exchange of experiences and knowledge development (joint research initiatives), building on the Delta Alliance publication "Comparative assessment of the vulnerability and resilience of deltas – Extended version with 14 deltas, Synthesis report" (see www.delta-alliance.org).

Delta Alliance is an international knowledge-driven network organization with the mission of improving the resilience of the world's deltas. With increasing pressure from population growth, industrialization, changing climate and subsidence, it is more important than ever that these valuable and vulnerable locations increase their resilience to changing conditions. Delta Alliance brings stakeholders involved in delta management together in order to benefit from each other's experience and expertise and as such contribute to an increased resilience of their delta region, building on scientific research and practical experiences.

Delta Alliance has sixteen network wings where activities are focused: California Bay (USA), Ciliwung and Mahakam (Indonesia), Mekong (Vietnam), Rhine-Meuse (the Netherlands), Nile (Egypt), Pantanal (Brazil), Ganges-Brahmaputra (Bangladesh), Mississippi (USA), Yangtze (China), Parana (Argentina), Taiwan, (Zambezi) Mozambique, (Tana) Kenya, Ghana, (Ebro) Spain and (Ayeyarwadi) Myanmar. Some other deltas have shown keen interest to join the Delta Alliance network, notably from Senegal, Thailand, Japan, Romania, Benin and Nigeria.

Research and investment in addressing the unique challenges of river delta regions are happening in many projects and programs around the world. Delta Alliance provides a framework to support the symbiosis of these countless activ-

Table 1. The main steps and elements of the Integrated Assessment Framework for subsidence.

Steps	Questions	Technical aspects	Governance aspects	City example (state of development)
1. Problem analysis	– How much subsidence is there? – What are the causes? – Who is involved and responsible?	– Measurement data collection – Data analyses to disentangle subsidence causes – (Inverse) modeling to make predictions	– Awareness raising – Stakeholder analysis/ identification of problem owners	– Dhaka – Manila – New Orleans – Jakarta
2. Planning	– How much future subsidence is predicted? – What are the current and future impacts (monetised)? – What are most vulnerable areas? – What are possible solutions?	– Scenario constructions – Modeling/forecasting – Damage assessments – Vulnerability and risk assessments – Decision support systems (see 4.6) – Cost-benefit analyses/multi-criteria analysis – Selection of structural measures in an integrated multi-sectoral perspective	– Capacity building/education – Multi-sectoral planning, participation, stakeholder engagement and commitment (see 4.4 and 4.5) – Political action, development of policy, strategy and legal instruments – Planning and design of buildings and infrastructure, incl. building codes (see 4.8) – Decision-making on implementation (see 4.5)	– HCMC
3. Implementation	– What will be done, how and when and by whom?	– Installing monitoring systems (see 4.7) – Setting up pilot projects – Proposing innovative (alternative) solutions (see 4.3) – Implementation of structural mitigating and/or adapting measures (see 4.1, 4.2, 4.3) – Exchange of knowledge and best practices (see 4.10)	– Multi-sectoral cooperation/organisational structure – Implementation of non-structural measures (see 4.1) – Legal framework/operational procedures/guidelines – Enforcement of laws and regulations – Financing mechanisms/asset management (see 4.9)	– Bangkok
4. Evaluation	– Is the problem under control?	– Monitoring, re-modeling – Compliance checking – Assessment and outlook	– Stakeholder evaluations – Public hearing	– Tokyo – Shanghai

ities ongoing in delta regions, reducing unnecessary overlap and identifying gaps in efforts. Delta Alliance is a vehicle for increased cooperation between parties in river deltas worldwide, with the ultimate aim of improving the resilience of the world's deltas through more integrated and effective efforts, building on scientific research and knowledge exchange.

The members of Delta Alliance together develop and share knowledge to:

- envision resilient deltas
- measure, monitor and report resilience in deltas, and
- provide inspiration and support for improved resilience in deltas

In pursuing these goals, members strive to:

- share knowledge, expertise and best practices between deltas worldwide
- co-create and integrate knowledge across deltas, sectors and disciplines
- connect expertise and parties from throughout the watershed
- incorporate local knowledge in research and other knowledge-building activities

- provide knowledge and insight to support any use, including education, policy development, investments and disaster response.

The Delta Alliance has many linkages with other networks and initiatives such as the Global Water Partnership strategy 2014–2018, the Belmont Forum Deltas project, the GEF Transboundary Waters Assessment Programme, the Alliance for Global Water Adaptation, and the SIWI/UNEP initiated From Source to Sea partnership. This is further supported by the Sustainable Deltas Initiative 2015 as endorsed by ICSU.

One of the most important research activities of the Delta Alliance in the past years was a comparative assessment of the vulnerability and resilience of 14 deltas, addressing various aspects of delta development and management through an integrated approach. The following deltas were involved: Nile, Tana, Incomati, Zambezi, Ganges- Brahmaputra-Meghna, Yangtze, Ciliwung, Ayeyarwady, Mekong, Rhine-Meuse, Danube, San Francisco Bay and Delta, Mississippi River and Parana. In this study subsidence was recognized as an important issue to be dealt with in many deltas.

4 Conclusions and the way forward

As yet, insufficiently are taken into account natural resource management, regional (urban) development and strategic spatial planning, and in particular urban flood management, infrastructure design and infrastructure maintenance.

In the framework of an integrated approach the following 10 generic key issues and possible solutions should be further elaborated on a case by case basis:

4.1 Restriction of groundwater extraction

This measure is very important in most cases to counteract human induced subsidence. In vulnerable areas extraction of groundwater should be reduced or completely phased out.

The following regulatory measures can be considered:

- appropriate legislation and consistent implementation and enforcement of regulations
- designation of groundwater regions and critical zones
- restricted licensing and compliance checking for groundwater well drilling
- universal groundwater use metering and charges for groundwater use

4.2 Development of alternative water supply (instead of groundwater)

In order to fulfill the increasing (urban) water demand, an alternative water supply for industry and domestic users is required. This process should include water demand assessments (water footprint) and cost/benefit assessments. Addressing and reducing surface water pollution is vital, to develop a sustainable alternative water supply.

4.3 Natural and artificial recharge of aquifers

When addressed consistently and effectively, the reduction of groundwater mining can eliminate one of the primary causes of land subsidence. However, the prolonged effects of settlement, possibly taking up to 10 years, are not immediately solved. Natural and/or controlled groundwater recharge may be applied to speed up recovery, as well as controlled aquifer storage and recovery (ASR) as currently being developed and implemented in Shanghai and Bangkok.

4.4 Integrated (urban) flood water management

Improved groundwater management and subsidence studies should be part of an integrated urban water (resources) management strategy including the whole water-subsurface system. Water resources management should be linked to flood mitigation. Ultimately, land subsidence is intricately linked to integrated land and water management, including surface as well as subsurface resources and constraints.

4.5 Improving governance and decision-making

In many cases governance is inadequate to address subsidence in an integrated multi-sectoral approach and to develop sustainable short and long term solutions. This involves (public) awareness, encouraging (public) participation, cooperation and coordination between stakeholders at different scales and levels, and also enabling good decision-making supported by decision support models and tools.

4.6 Decision support models and tools

Development of models and tools are important to support good decision-making. This involves among others analysis of the relationship between groundwater level and subsidence, development of modeling and forecasting capabilities by development and implementation of an integrated groundwater–subsidence monitoring and analytical model. Moreover, it is essential that local agencies have the expertise and tools to conduct studies, and are engaged in ongoing capacity building, training and knowledge exchange.

4.7 Appropriate monitoring and database system

Ongoing studies show that the dependence on reliable ground truth data of observed subsidence is an alarming "Achilles heel" in the battle against subsidence and related flood risk. Therefore, it is necessary to develop and maintain geodetic monitoring networks throughout the metropolitan areas, with stable, precisely calibrated benchmarks levels and periodic leveling surveys, in order to build a good database with long time measurements of subsidence.

4.8 Integrating geotechnical aspects in planning and design of buildings and infrastructure

In the planning and design of (heavy) buildings and road infrastructure, geotechnical research and modelling of the subsoil should be taken into account in order to avoid subsidence problems including differential settlements, in the short or long term. This will save considerable damage and high maintenance costs of infrastructure and buildings (foundations). During underground construction activities (deep parking lots, metro-stations, tunneling), the effects of dewatering should be minimized and, if necessary monitored and/or mitigated.

4.9 Asset management, financing and public-private-partnerships (PPP)

To minimize damage caused by subsidence, the main financial risks associated with investments and maintenance of assets (buildings, infrastructure) should be assessed. This will lead to improved design options, programming and prioritization of investments, making use of Real Options Theory and Asset Management. This involves determining performance indicators, functional specifications, risk mitigation measures and bonus/malus in (innovative) contracts. Moreover PPP and Private Financing should be explored building on sustainable business models.

4.10 Exchange of knowledge and best practices

Through organization of international conferences, workshops, expert meetings and courses knowledge and best practices can be exchanged to extend the common knowledge base efficiently and effectively, especially regarding lessons learned from higly build-up urban areas and rural argicultural areas. This can be further supported by development of collaborative research projects preferably in the framework of international (research) networks and initiatives such as the UNESCO, CCOP and the Delta Alliance.

References

Aobpaet, A., Cuenca, M. C., Hooper, A., and Trisirisatayawong, I.: InSAR time series analysis of land subsidence in Bangkok, Thailand, Int. J. Remote Sens., 2012.

Bakr, M.: Land subsidence in North Jakarta – preliminary analysis results, part of Jakarta Coastal Defence Strategy (JCDS) study, JCDS Atlas, 2011.

Bakr, M., de Lange, G., and Toan, D. N.: Ho Chi Minh City Flood and Inundation Management – Final Report, Volume 2, IFRM Strategy, Annex 3, Land Subsidence, Royal Haskoning-DHV and Deltares, 2013, HCMC FIM, 9T4178/HCMC/R/0042/MB, 2013.

Church, J. A. and White, N. J.: Sea-Level Rise from the Late 19th to the Early 21st Century, Surv. Geophys., 32, 585–602, doi:10.1007/s10712-011-9119-1, 2011.

Cuenca, M. C., van Leijen, F. J., and Hanssen, R. F.: Shallow subsidence in the Dutch wetlands estimated by satellite radar interferometry, Delft University of Technology, First International Conference on Frontiers in Shallow Subsurface Technology, Session: Shallow Subsurface Characterization, Imaging and Monitoring Techniques I, 2010.

Dam, R.: Jakarta Coastal Defence Strategy (JCDS) study, Activity Report, Land subsidence and adaptation/mitigation strategies, JCDS Bridging Phase, Deltares, 2012.

Dokka, R. K.: The role of deep processes in late 20th century subsidence of New Orleans and coastal areas of southern Louisiana and Mississippi, J. Geophys. Res., 116, B06403, doi:10.1029/2010JB008008, 2011.

Eco, R. C., Lagmay, A. A., Bato, M. P.: Investigating ground deformation and subsidence in northern Metro Manila, Philippines using Persistent Scatterer Interferometric Synthetic Aperture Radar (PSInSAR), American Geophysical Union, Fall Meeting, 2011.

Gong, S.: Synthesis analysis of impact factors for land subsidence and its systematic adjustment in Shanghai, East China Normal University, 2008.

Kaneko, S. and Toyota, T.: Long-Term Urbanization and Land Subsidence in Asian Megacities: An Indicators System Approach, In Groundwater and Subsurface Environments: Human Impacts in Asian Coastal Cities, doi:10.1007/978-4-431-53904-9_13, 2011.

Le Van Trung and Ho Tong Minh Dinh: Monitoring Land Deformation Using Permanent Scatterer INSAR Techniques (case study: Ho Chi Minh City), 7th FIG Regional Conference Spatial Data Serving People: Land Governance and the Environment – Building the Capacity Hanoi, Vietnam, 19–22 October 2009.

Shi, X., Wu, J., Ye, S., Zhang, Y., Xue, Y., Wei, Z., Li, Q., and Yu, J.: Regional land subsidence simulation in Su-Xi-Chang area and Shanghai City, China, Eng. Geol., 100, 27–42, doi:10.1016/j.enggeo.2008.02.011, 2008.

Slangen, A. B. A., Katsman, C. A., van de Wal, R. S. W., Vermeersen, L. L. A., and Riva, R. E. M.: Towards regional projections of twenty-first century sea-level change based on IPCC SRES scenarios, Clim. Dyn., 38, 1191–1209, doi:10.1007/s00382-011-1057-6, 2012.

Slangen, A. B. A., Carson, M., Katsman, C. A., van de Wal, R. S. W., Köhl, A., Vermeersen, L. L. A., and Stammer, D.: Projecting twenty-first century regional sea-level Changes, Climatic Change May 2014, 124, 317–332, doi:10.1007/s10584-014-1080-9, 2014.

Steckler, M. S., Akhter, S. H., Seeber, L., Bilham, R. G., Kogan, M. G., Masson, F., Maurin, T., Mondal, D., Piana Agostinetti, N., Rangin, C., and Saha, P.: GPS Velocities and Structures across the Burma Accretionary Prism and Shillong Anticline in Bangladesh, Fall Meeting, 2012.

Phisut: Technology: The study of systematic land subsidence monitoring on critical groundwater used area project – executive summary (report in Thai), Ministry of Natural Resources and Environment, Department of Groundwater Resources, Bangkok, Thailand, 2012.

van der Ven, G. P.: Man-made lowlands, history of water management and land reclamation in the Netherlands, uitgeverij Matrijs, Utrecht, ISBN-13: 978-9053450307, 1993.

Proc. IAHS, 372, 493–497, 2015
proc-iahs.net/372/493/2015/
doi:10.5194/piahs-372-493-2015

Open Access
PIAHS

Numerical analysis to determine the impact of land subsidence on high-speed railway routes in Beijing, China

C. Ye[1], Y. Yang[1,2], F. Tian[1], Y. Luo[1,3], and Y. Zhou[1,2]

[1]Beijing Institute of Hydrogeology and Engineering Geology, Beijing, China
[2]School of Earth and Space Sciences, Peking University, Beijing, China
[3]Institute of Geology and Geophysics, Chinese Academy of Sciences, Beijing, China

Correspondence to: C. Ye (wildhorse898@126.com)

Published: 12 November 2015

Abstract. More than 10 high-speed railway routes with top speeds of $300\,km\,h^{-1}$ are expected to be operational from Beijing by the year 2020. However, the safety of these routes is affected by the occurrence of land subsidence. This paper focuses on the Beijing–Tianjin Intercity High-Speed Railway (BTR), the first high-speed railway in China, to analyze the operational safety of high-speed railway routes by analyzing both regional land subsidence and local differential subsidence caused by groundwater drawing. The Beijing construction stratum is mainly composed of cohesive soil, and the BTR has a maximum accumulative subsidence of > 800 mm and a maximum subsidence rate of $> 80\,mm\,a^{-1}$. In this paper, finite-element software ABAQUS is used to analyze groundwater drawdown and land subsidence caused by local water drawing, and its effect on the bearing capacity of railway bridge pile foundations and the orbit concrete supporting course. The analysis provides a technical basis for developing prevention and control engineering measures against land subsidence so as to guarantee the safe operation of these high-speed railway routes.

1 Introduction

A large number of urban rail transit projects are under construction in Beijing. By 2020, the city will have > 10 high-speed railway routes in operation, offering top speeds of $300\,km\,h^{-1}$. However, it is necessary to consider regional geological and environmental conditions in relation to the development trends along these high-speed railway routs. These geological and environmental conditions, such as differences in the intensity of groundwater exploitation, the degree of urban development/construction, and the degree of differential subsidence, influence land subsidence and ultimately lead to disasters.

Three main phenomena are generally displayed where land subsidence occurs in relation to high-speed railway routes: uneven subsidence resulting in the changing of orbit comfort, a worsening of the structure of the subgrade and its stability, and a change in the bridge and track orbit comfort (Li et al., 2008). Uneven subsidence of the ground surface affects the stability of bridges, orbit comfort, and produces a gradually increasing influence that eventually causes the rail bed to undulate in a series of waves (Yang et al., 2013). This not only results in increased maintenance costs but also affects the safety of the railway operation performance, potentially causing adverse effects to the high-speed trains, resulting in a threat to passenger safety, lives, and property.

2 Land subsidence along the route of the high-speed railway in Beijing

In comparison with ordinary railways, high-speed railway routes demand higher reliability, comfort, stability, and durability throughout the structure along the entire line (Hwang, 2008), and their sensitivity to uneven subsidence is also higher. Beijing is rapidly developing areas where land subsidence occurs. In addition, there are currently several high-speed railway routes that are either in operation or are under construction, which traverse areas where land subsidence oc-

Table 1. Main parameters used in the model.

Type	Elastic modulus (E/MPa)	Poisson's ratio (μ)	Cohesion (c/kPa)	Internal friction angle ($\varphi/°$)	Dilatancy angle ($\psi/°$)
Pile	2.8×10^4	0.167	–	–	–
Reinforced area	2–50	0.35	10	20	0.1
Bearing stratum	100–300	0.3	0.1	35	10

curs, such as the Beijing–Tianjin inter-city (BTR) and the Beijing–Shanghai railway. Such areas are affected in different degrees by the damage caused by differential subsidence (Yang, 2008).

For example, BTR has operated since 2008, while InSAR monitoring and land leveling results show that there have been significant regional land subsidence along the development path, with a largest amount of cumulative subsidence of > 800 mm and a maximum subsidence rate of > 80 mm a^{-1}. Long-wave irregularity of the rail bed with local fluctuations of > 10 mm/150 m and short-wave irregularity with surface fluctuations of > 2 mm/5 m have also been reported. According to Code for Design of High Speed Railway, the designed value tolerance for uneven subsidence is 5 mm between bridge piers, but the uneven subsidence has already occurred over 5mm between parts of the bridge pier. However, the bridge structure is currently stable and has not yet been affected.

Bridge girders are affected by land subsidence in two main ways: first, by the level difference caused by the uneven subsidence between the bridge pile foundations, and second, by the response change of the bridge single-pile-bearing capacity under a decline in water levels, causing additional stress changes. Using the BTR as an example, this paper focuses on an analysis of the latter case. In this respect, ABAQUS is used as the finite-element software to analyze the different pile lengths and single-pile-bearing capacity characteristics under different water-level changes (Martin Hernandez-Marin, 2010).

3 Stability influence: a numerical analysis of land subsidence on a high-speed railway bridge

3.1 Analytical model

Within this model, a linear elastic model is used to analyses the pile, and the Mohr–Coulomb model is used to analyses the soil body. The model is developed using axial symmetry, and the main parameters are shown in Table 1. The boundary conditions for the model include a zero-displacement constraint (vertical and radial) along the bottom, and a radial displacement constraint along the lateral margins of the model. An interface coulomb friction model is established to describe the contact relationship between the pile and soil. The friction angle is defined as:

$$\delta = \tan - 1(\sin\varphi \times \cos\varphi/(1+\sin 2\varphi)). \quad (1)$$

As φ is the friction angle of the soil body, δ is therefore 0.33.

Most of the pile lengths used in the bridge pile foundation for BTR transit are between 20 and 50 m, and the design bearing capacity of a single pile is between 2000–5000 kN. We therefore used pile lengths of 20 and 50 m in this model, with diameters of 0.75 and 1.0 m, respectively. The depth of the model is 2 times of pile, and the length is equal to the length of pile.

Three steps were used in the analytical process: first, the balance of the initial stress field was identified; second, the deformation of the design load conditions under loading on the pile cap were obtained; and third, deformation under changing water levels or moduli, was determined. The calculations used for the third step were also used to determine the proportion of loading in the second step.

3.2 Single-pile-bearing capacity characteristics

Single pile test curves for piles lengths of both 20 and 40 m, under a bearing layer modulus of 2000 MPa and a modulus for the reinforced area of 2, 5, 10, 20, 30, 40, and 50 MPa, are shown in Fig. 1. The results show the bearing capacity characteristic values for both pile lengths to be > 3000 kN.

The subsidence curve in the single pile test load is linear, and is maintained under the imposed load (Li et al., 2007). In addition, it is less than the bearing capacity characteristic value, which ensures that the total deformation is within the elastic range. This is contrary to the results relating to deformation caused by water-level changes, which account for a proportion of the deformation under the above-mentioned load conditions.

3.3 The influence of water level changes

Variations in the bulk density occur in relation to ground water level changes. We analyzed water level changes from the ground surface and at depths of 2, 3, 4, 5, 6, 8, 10, and 15 m, where different moduli in the reinforced area were used (the moduli of the bearing stratum were 300 and 100 MPa, respectively). The ratio of the deformation caused by the water

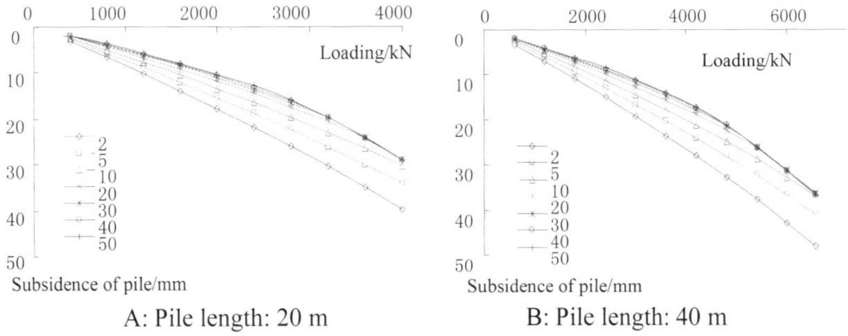

Figure 1. Single pile experimental curves.

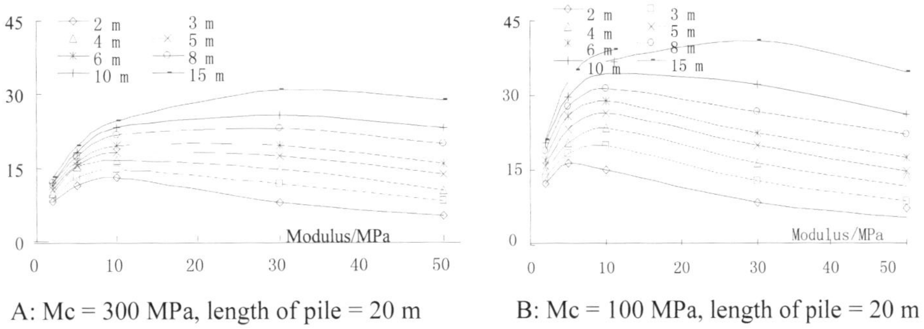

Figure 2. Changes in water level and resulting pile top deformation ratio.

level changes were found to account for a proportion of the vertical load deformation (for a pile length of 20 m, the load is 2000 kN) as shown in Fig. 2. The results show that as the depth of water falls, a greater proportion of the deformation was found to be caused by the change in water level.

Results show that when the water level falls within the upper 2 m, there is a corresponding deformation ratio for $M_c = 300$ MPa and $M_c = 100$ MPa of 12.9 and 16.2 %, respectively. However, when the water levels falls in the upper 10 m, the corresponding maximum deformation ratio reaches 25.8 and 25.8 %, respectively. In addition, the deformation percentage increases with the modulus of the reinforced area, but there is no increase in this value after > 10 MPa. The deformation proportion for a pile length of 40 m has a virtually similar trend, The deformation proportion for a pile length of 40 m also reaches 9.7 %, when groundwater falls 2m and using the same parameters.

Comprehensive analysis shows that the pile-bearing capacity is improved by an increase in subsoil modulus within the same area of drawdown, and that this decreases subsidence to a great degree. Therefore, proper reinforcement applied around the pile soil body could effectively mitigate the effect of land subsidence.

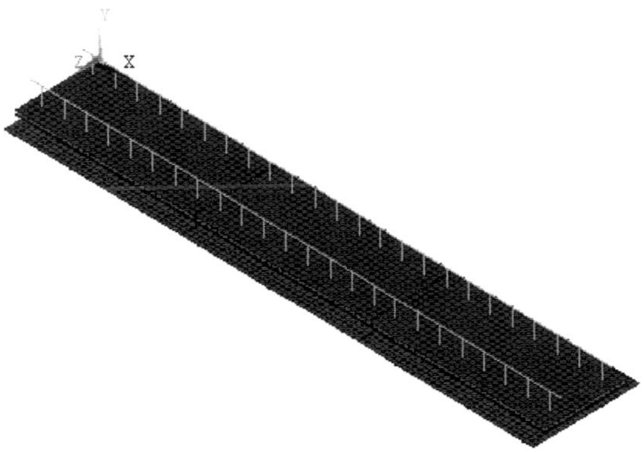

Figure 3. Beam–plate model.

4 Numerical analysis of the impact of track stability in relation to land subsidence on a high-speed railway

We used the finite-element method (FEM) to establish the beam–plate model (Fig. 3), where the rail uses an elastic-point-supported beam model, the fastener adopts a linear spring simulation, the track plate and concrete supporting layer adopt a plate/shell element simulation, the mortar cush-

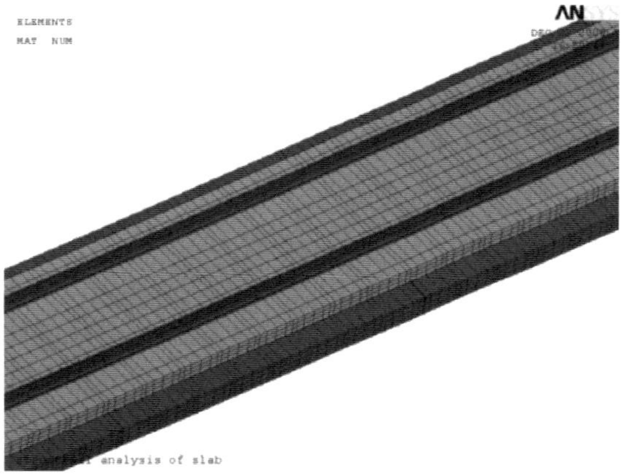

Figure 4. Solid model.

ion uses a solid unit simulation, and the roadbed uses a spring element simulation (Liu, 2011; Yang, 2012; Qi, 2009).

Uneven roadbed subsidence adopts the method of cosine curve simulation by and the curve chord assumes 20 m. According to the amplitude range of the deformation curve, an analysis of the track structure under forces of uneven subsidence was made using measurements of 5, 8, 9, 10, 15, 20, 30, 40, 50, 60, 70, 100 mm. The results of these calculations show that the tensile stress of the concrete supporting layer is greater than the tensile strength, showing that the structure would easily break under uneven roadbed subsidence of ≤ 20 mm and the maximum subsidence is > 10 mm.

FEM was applied to establish a solid model (Fig. 4), where the parameter selection used is the same as that in the beam–plate model (the calculation results are shown in Fig. 5).

According to the calculation results of the solid model, for an uneven subsidence wavelength ranging over 20 m, the amplitude should not be > 15 mm.

In conclusion, according to analysis and calculation for both models, the amount of uneven subsidence subsidence should be controlled less than 10–15 mm between the distance of 20 m.

If the amount of subsidence increases, the fragility of the bearing layer structure increases and is liable to break.

According to the effective stiffness iterative algorithm, under a serviceability limit state and bearing capacity limit state, and a dominant condition of temperature control and load control, calculates the deformation quantity able to withstand in continuous base plat bearing. In addition, the results show that a uniform subsidence of 60 mm over a range of 20 m will not destroy the continuous base plate or the track board structure, and it will not create a risk for the safety of the track structure.

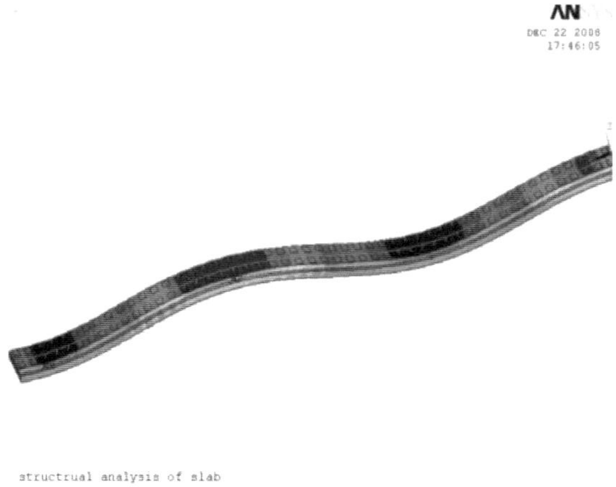

Figure 5. Simulation of differential subgrade subsidence.

5 Conclusions

1. ABAQUS finite-element software was used to analyze the bridge pile foundation bearing capacity, and it is known that the bigger the amount of water level drawdown, the larger deformation proportion in relation to the design load. When the groundwater level drops to 2 m from the ground surface, the single-pile-bearing capacity is decreased to 12.9 and 9.7 % of the design load corresponding to a modulus of the bearing stratum, with pile lengths of 20 and 40 m, respectively;

2. Using the FEM to calculate the effect of land subsidence on the concrete bearing layers of the tracks, subsidence is in the range of 20 mm when the subgrade is uneven, and the maximum amount of subsidence should be controlled within the range of 10–15 mm. Therefore, subsidence can be easily exceeded in the bearing layer structure.

References

Li, G., Xu, Z., Sun, S., et al.: The influence of surface subsidence on constrution of High-speed railway in North China Plain and its counterm easures, J. Railway Eng., 8, 7–12, 2007.

Li, G. H., Sun, S. L., Xu, Z. L., et al.: Ground subsidence impacting on high-speed railway bridge Construction and countermeasures, J. Railway Eng., 4, 37–41, 2008.

Liu, F. S.: SXC regional land subsidence trends and their impaction the analysis of the Beijing-Shanghai high-speed rail, J. Railway Invest. Surv., 1, 33–37, 2011.

Martin Hernandez-Marin, T. J.: Control on initiation and propagation of pumping-induced earth fissures: insights from numerical simulations, Hydrogeol. J., 18, 1773–1785, 2010.

Qi, B.: Analysis of land subsidence impacting the Beijing-Shanghai high-speed railway (Beijing-Jinan section), SI CHUAN: Earth

Science and Environment Engineering, Southwest Jiao tong University, 2009.

Yang, Y., Jia, S. M., and Wang, H. G.: Land subsidence of Beijing plain area situation and development trend analysis, J. Shanghai Geol., 31, 23–28, 2010.

Yang, Y., Zheng, F. D., Liu, L. C., et al.: The relations between the plain of Beijing's groundwater level and land subsidence, J. Eng. Surv., 8, 44–48, 2013.

Yang, Y. Z. H.: SXC land subsidence impact on the Beijing-Shanghai high-speed railway, SI CHUAN: Earth Science and Environment Engineering, Southwest Jiao tong University, 2012.

Proc. IAHS, 372, 499–502, 2015
proc-iahs.net/372/499/2015/
doi:10.5194/piahs-372-499-2015

Open Access
PIAHS

The *Kabu-ido* system: a pioneering solution for uncoordinated groundwater pumping in Japan

T. Endo

College of Sustainable System Sciences, Osaka Prefecture University, Sakai city, Osaka Prefecture, Japan

Correspondence to: T. Endo (tte23042@osakafu-u.ac.jp)

Published: 12 November 2015

Abstract. The *Kabu-ido* system was a customary institution for groundwater management in a ring levee area of the Tokai region in Japan. It consists of three programs, a permit system for groundwater pumping, groundwater pricing, and economic compensation. The purpose of this paper is to clarify characteristics of the *Kabu-ido* as a groundwater management institution.

1 Introduction

The *Kabu-ido* system was a customary institution for groundwater management in a ring levee area of the Tokai region in Japan. The ring levees have developed at the confluence of three large rivers, the Ibi, Nagara, and Kiso. Many villages and towns are surrounded by successive levees to prevent against flooding. Such ring levees are often locally called *waju* in Japanese.

Early in the 19th century, a conflict caused by groundwater pumping took place in the ring levee area. The conflict was mitigated by implementation of three programs, a permit system for groundwater pumping, groundwater pricing, and economic compensation. These programs were gradually developed and later called the *Kabu-ido* system, which was created in the early 19th century and disappeared around the 1930s. The *Kabu-ido* system at least existed in the Fukuzuka, Takasu, and Shimogasa ring levees. This study focuses on the Fukuzuka ring levee, which pioneered the *Kabu-ido* system (Fig. 1).

The *Kabu-ido* system has been investigated since the 1930s. Bekki (1932), Nakazawa et al. (1936), Katano (1941), Mori (1964), and Matsubara (1968) clarified the *Kabu-ido* systems in the Takasu and Fukuzuka ring levees based on a wide review on historical documents. These studies were followed by the more recent works of Itoh and Aoki (1987), Kubota (2008), and Shimizu (2014). These works ably clarified the history of the *Kabu-ido* system, but did not deal with institutional aspects of the system. The purpose of this paper is to clarify characteristics of the *Kabu-ido* system as an institution for groundwater management.

2 The *Kabu-ido* system in the Fukuzuka ring levee

2.1 Water issues of the Fukuzuka ring levee

Although the Fukuzuka ring levee is flat, the altitude of its northeast part is higher than that in the southeast. While villages located in higher places are called upper villages, the ones in the lower places are called lower villages. There were ten upper villages and six lower ones within the ring levee. The location and name of those villages are showin in Fig. 2.

Some of the upper villages diverted water from the Ibi River, but began to use groundwater after the advent of pumping techniques in the late 18th century. Accumulated drainage from artesian wells in the upper villages increased flood risk and hindered agricultural activities in the lower villages (Editorial committee of a history of Wanouchi town, 1981).

2.2 Internal agreement on artesian wells in Fukuzuka ring levee of 1812 (hereafter the 1812 agreement)

There was a major dispute between the upper and lower villages in 1801. After this trouble, an internal agreement on artesian wells was reached between the upper and lower villages in July 1812 (Internal agreement on artesian wells in Fukuzuka ring levee of 1812; Sakurai, 1934). The agreement includes the following articles.

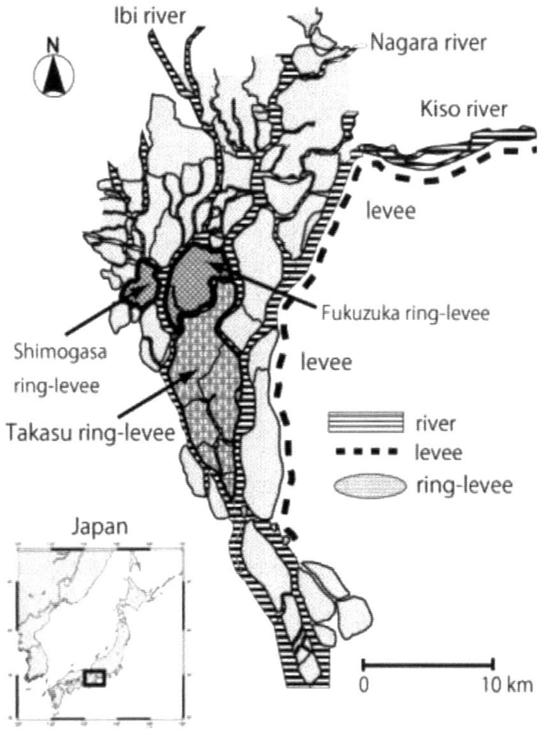

Figure 1. Ring levees around 1870 (based on a map in Nanno town, 1978).

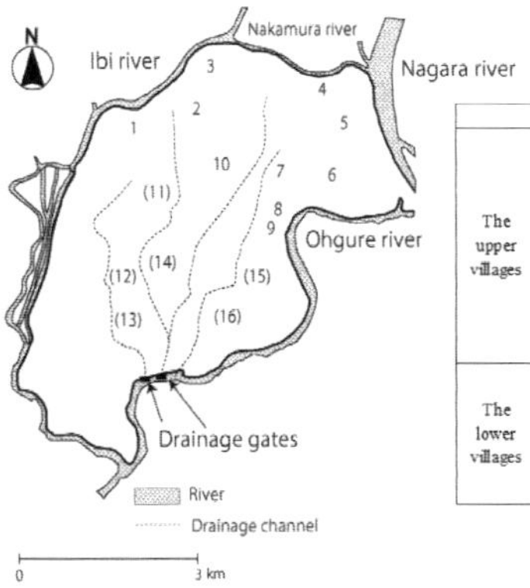

	Number	Village name
The upper villages	1	Fukuzuka
	2	Sato
	3	Nanba
	4	Nishijo
	5	Niremata
	6	Ohyabu
	7	Niremata-Shinden
	8	Kami-Ohgure
	9	Gotango
	10	Nakagoh
The lower villages	(11)	Fukuzuka-Shinden
	(12)	Moike-Shinden
	(13)	Miru-Shinden
	(14)	Nakagoh-Shinden
	(15)	Gotango-Shinden
	(16)	Shimo-Ohgure-Shinden

Figure 2. Lower villages are shown by numbers with parentheses. (Based on a map provided by an Editorial committee of a history of Wanouchi town, 1996.)

- Article 1: the number of artesian wells is confined to 45. No additional wells are permitted. If any upper village breaches this article, all existing wells in the village must be eliminated.
- Article 2: four people (two from the upper villages and two from the lower ones) organize well operators. A well must be opened in the presence of these well operators.
- Article 3: the annual salary of a well manager is RYO 1 (a unit of money at the time). The upper villages pay this salary.
- Article 4: each well must be locked. The key is deposited with well managers of the upper villages in a key box. The key of the key box is deposited with well managers of the lower villages.
- Article 5: a broken well must be repaired to prevent unexpected drainage.
- Article 6: those who break the lock to secretly open wells at night must be strictly warned.
- Article 7: a well whose owner intentionally lets groundwater flow under the guise of a broken well should be buried in the presence of well managers from the lower villages.
- Article 8: a well inside a house must be buried.
- Article 9: if an irrigation well does not function adequately, the owner can dig another well in the presence of a well manager from the lower villages.
- Article 10: when drainage gates in the lower villages are closed, wells must not be opened even in case of drought.
- Article 11: if water flows out from a buried well, the well must be completely buried in the presence of the well managers. If residents of the lower villages find that upper village residents secretly use the well without reporting its use, the former residents should consult the local magistrate office.

2.3 Agreement on artesian wells and changes in drainage channels of 1877 (hereafter the 1877 agreement)

In 1877, the agreement described above was revised. The new agreement consisted of 19 articles. The articles addressing groundwater management are the following.

- Article 4: the number of wells is set at 395 in the upper villages (including 45 existing wells) and 100 in the lower villages. If additional wells should be necessary, KOKU 7 of rice must be paid per well (rice functioned as currency at the time; the KOKU was a unit of volume). However, a well can be sold and bought with the permission of well managers of the upper and lower villages. After its sale, the well must be completely buried.
- Article 5: well managers consist of two people from the upper villages and two from the lower villages. The annual salary is YEN 5 per person per year.

- Article 6: numbered certificate poles must be established at all wells. Well managers and assistants must check all wells once per year. Broken wells should be replaced in the presence of well managers. Letting water flow out of a well without a certificate pole is strictly prohibited.
- Article 12: a penalty of 10 *koku* of rice will be imposed for an uncertified well, and the well must be buried. Half of this fee goes to the person who discovers the uncertified well.
- Article 16: the penalty fee for uncertified wells and revenue from newly permitted wells are used for public works of the ring levee.
- Article 19: a survey of existing (including uncertified) wells must be done and a report submitted no later than 28 February 1878.

In Japanese, *kabu* means "the privilege to do business" (Miyamoto, 1977) and *ido* means "well"; therefore, *Kabu-ido* can be interpreted as meaning "privileged well" or "the special right to dig wells". Whereas the word *ido* is found in the 1812 agreement, the word *Kabu-ido* is not. Sakurai pointed out that development of the *Kabu-ido* system was completed in the 1877 agreement because the word had never been used before (Sakurai, 1934).

3 Characteristics of the *Kabu-ido*

3.1 Correction of negative externality

The main function of the *Kabu-ido* was to remedy the negative externality. Groundwater pumping by an actor often causes negative impacts on another actor without any compensation. They include not only drainage problems but also seawater intrusion, decline of groundwater table and land subsidence. These effects are called as "negative externalities". Whatever forms these may take, they suggest excessive use of groundwater (Scitovsky, 1954).

The *Kabu-ido* system had the function of reducing negative externality. The permit system decreased drainage volume by limiting the total number of wells (Articles 1, 2 4 and 9 of the 1812 agreement; Article 4 of the 1877 agreement). Groundwater pricing encouraged groundwater users to take negative externality into consideration (Article 4 of the 1877 agreement). This increased the cost of groundwater pumping and promoted internalization. The revenue related to groundwater pricing was used for the construction of drainage gates that further mitigated the negative externality (Article 16 of 1877 agreement).

3.2 Market-based approach

The *Kabu-ido* was a groundwater management institution that used economic tools. Article 4 of the 1877 agreement gave the well manager the ability to permit well transfers. This rule was not found in the 1812 agreement. This transfer rule had a strong relationship with the limitation on total numbers of wells. Given this limitation, those who wanted an additional well had no choice but to buy an existing one from somebody else. Article 4 of the 1877 agreement anticipated the possibility of such a well transaction. In other words, the *Kabu-ido* introduced a cap-and-trade for wells, which is a type of tradable permit. Cap-and-trade prevents excessive use of resources by introducing a usage limit and promotes their efficient use by permitting transactions.

A document on well transaction includes the phrase "under permission of village mayor and well managers". This shows that a well was transacted under the authorization of well managers as described in the 1877 agreement. This transaction was made on 27 February 1884 and the well price was YEN 15. How widely such a transaction occurred and how well prices changed remain to be proven (Bill of sale on *Ido-kabu*, 1884).

3.3 The *Kabu-ido* as a self-organized institution

The *Kabu-ido* was a self-organized institution that was operated by local groundwater users themselves. The primary issue of self-organized collective action is how to monitor and penalize those who breach local rules without the involvement of external government. It is because voluntary monitoring tends to be hindered by the free-rider problem (Ostrom, 1990). For example, if a lower village conducts such monitoring activities, other villages can enjoy the benefit without contribution. In that situation, no village will have an incentive to monitor and penalize, expecting other villages to do so.

Article 3 of the 1812 agreement solved this problem. The lower villages made upper ones pay the cost of monitoring activities to avoid the free-rider problem. The 1877 agreement also provided other solutions. These were well numbering (Article 6) and a reward system (Article 12). Well numbering not only imposed a cap on the total number of wells but also reduced monitoring cost. It enabled patrolmen from the lower villages to quickly distinguish between an authorized and unauthorized well, because they could easily identify wells without a number and those with an unauthorized duplicate number. In addition, the personal reward system for well detection increased the personal benefit of monitoring activities, towards the voluntary provision of a public good.

4 Conclusion

The purpose of this paper was to clarify characteristics of the *Kabu-ido* as a groundwater management institution.

The first characteristic of the *Kabu-ido* system is its function. It works to reduce the negative externality related to groundwater pumping. The second characteristic regards economic tools. The 1877 agreement had already introduced a cap-and-trade system for groundwater. The final characteristic is that the *Kabu-ido* was a self-organized institution.

The *Kabu-ido* was initiated about 200 years ago, but has disappeared. After the 1877 agreement, the government subsidized the installation of electric drainage pumps in the Fukuzuka ring levee. This solved the drainage problem so effectively there was no need for well regulations. In this sense, the *Kabu-ido* can be regarded as an institution of undeveloped drainage techniques.

Acknowledgements. This work was financially supported by the Asahi Glass Foundation, Societal Adaptation to Climate Change: Integrating Paleoclimatological Data with Historical and Archaeological Evidences (Research Institute for Humanity and Nature Project) and the Japan Society for the Promotion of Science KAKENHI (Grant-in-Aid for Challenging Exploratory Research) grant number 26550101. All historical documents on the *Kabu-ido* cited (i.e., Agreement on artesian wells and changes in drainage channels of 1877, Bill of sale on *Ido-kabu*, 1884, and Internal agreement on artesian wells in Fukuzuka ring levee of 1812), are available in the Gifu Prefecture historical document center. The views presented are those of the author and should in no way be attributed to Gifu Prefecture. Responsibility for the text (including any errors) rests entirely with the author.

References

Agreement on artesian wells and changes in drainage channels, Tanahashi family documents, Gifu Prefecture historical document center, 1877 (in Japanese).

Bekki, A.: A geographical study of ring levees in Seino plain, Collections of essays on geography, 1, 230–282, 1932 (in Japanese).

Bil of Sale on *Ido-kabu*, Tanahashi family documents, Gifu Prefecture historical document center, 1884 (in Japanese).

Editorial committee of a history of Wanouchi town: A history of Wanouchi town, Dai-ichi Houki Shuppan, Tokyo, 1981 (in Japanse).

Editorial committee of growing Wanouchi: Growing Wanouchi, Wanouchi town, Gifu, 1996 (in Japanse).

Internal agreement on artesian wells in Fukuzuka ring levee, Nakajima family documents, Gifu Prefecture historical document center, 1812 (in Japanese).

Itoh, Y. and Aoki, N.: Ring levee, Gakuseisha, Tokyo, 1987 (in Japanese).

Katano, A.: The *Kabu-ido* in Fukuzuka and Takasu ring levees, Noubi culture, 1, 47–77, 1941 (in Japanese).

Kubota, M.: Living with rivers: stories of Nagara, Ibi rivers, Fubaisha, 2008.

Matsubara, Y.: On the development of "The *Kabu-ido*", authorized artesian wells for irrigation in Tkasu-wadju of Nobi Plain, Geographical Review of Japan, 41, 491–504, 1968 (in Japanese).

Miyamoto, M.: A study on Kabu group, Koudansha, Tokyo, 1977 (in Japanese).

Mori, G.: A history of Hirata town, the second volume, Rinsen Book Company, Kyoto, 1964 (in Japanese).

Nakazawa, B., Akiyama, K., and Yamada, K.: A history and geography of ring levee villages, Research Institute of Japanese Agricultural Villages, Tokyo, 1936 (in Japanese).

Nanno town: A history of Nanno town, Nanno town, Gifu, 1978 (in Japanese).

Ostrom, E.: Governing the commons, the evolution of institutions for collective action, Cambridge University Press, New York, 1990.

Sakurai, I.: On the Kabu-ido in Fukuzuka Ring levee, Keiseikaku Documents, Ohgaki City Library, 1934 (in Japanese).

Scitovsky, T.: Two concepts of external economy, J. Polit. Econ., 62, 143–151, 1954.

Shimizu, S.: Fukuzuka ring levee and the *Kabu-ido*, Wanouchi studies, 3, 2–10, 2014 (in Japanese).

Proc. IAHS, 372, 503–506, 2015
proc-iahs.net/372/503/2015/
doi:10.5194/piahs-372-503-2015

Land subsidence of coastal areas of Jiangsu Province, China: historical review and present situation

J. Q. Zhu[1], Y. Yang[2], J. Yu[1], and X. L. Gong[1]
[1]Key Laboratory of Earth Fissures Geological Disaster, Ministry of Land and Resources, Geological Survey of Jiangsu Province, Nanjing, 210018, China
[2]School of Earth Sciences and Engineering, Hohai University, Nanjing, 210098, China
Correspondence to: Y. Yang (yy_hhu@hhu.edu.cn)

Published: 12 November 2015

Abstract. Surface faults related to land subsidence have been observed in coastal cities, such as Nantong, Yancheng, and Lian Yungang, in Jiangsu Province (CAJS) since the early 1970s. Nowadays, increases flooding and rising sea levels are attributed to subsidence caused by groundwater pumping. In this work we present a brief description of land subsidence in CAJS, we examine the mechanisms of land subsidence induced mainly by groundwater pumping and its evolution and economic implications as well as the implemented measures by the local government to prevent new damage.

1 History and current status of land subsidence in CAJS

1.1 Development history of land subsidence

Before the 1970s, the coastal areas of Jiangsu Province (CAJS) experienced very slight land subsidence (1–2 mm annually), dominated by natural consolidation and settlement of loose soil and rocks (Hu et al., 2004). However, entering the 1980s, large-scale groundwater exploitation began in the cities of Yancheng, Dafeng, and Nantong of Jiangsu Province, leading to a rapid decline of the groundwater table in coastal areas. In the 1990s, as the intensity of groundwater exploitation in the urban districts rose, the issue of land subsidence began to affect the entire region with greater severity (Xue et al., 2005).

In the past decade, from 2005 to 2013, land subsidence of coastal areas began to extend to the north of Yancheng City and to the south of Lianyungang City (Fig. 1). In addition to the surrounding areas of Yancheng City and Dafeng City, the cumulative subsidence of the counties of Guanyun, Guannan, Xiangshui, and Funing exceeded 200 mm, with an annual average subsidence rate of 25 mm a^{-1}. On the basis of monitoring data from 2005 to 2013, it was found that coastal regions with cumulative subsidence exceeding 200 mm covered an area of 1.06×10^4 m^2. A subsidence cone was formed centered around Nantong City, Haimen City, and the urban districts of Dafeng, Yancheng, and Rudong City as shown in Fig. 1.

In addition, land subsidence has reached a critical level in the north coastal areas of Dafeng City. The cumulative subsidence in all these regions has exceeded 250 mm, with the exceptions of the southwest areas of Guanyun County, northwest areas of Guannan County, east of Xiangshui County, east of Binhai County, northeast of Sheyang County, and south of Dafeng City. Several land subsidence cones have formed in the urban districts of Nantong City, Haimen City, and Matang Town of Rudong County, which is located in Dongtai City, and the coastal areas to the south of Dongtai. In some of these regions, the cumulative subsidence has also exceeded 250 mm.

1.2 Current status of land subsidence

In 2013, both the magnitude and area of subsidence increased at accelerated rate in the northern coastal areas of Jiangsu Province (Fig. 2). The annual subsidence levels in the long, narrow strips on the both banks of the abandoned Yellow River, Guan River, and North Jiangsu Irrigation Canal exceeded 30 mm. The annual subsidence of Yanweigang Town in Lianyungang City at the border between Lianyungang and Yancheng City exceeded 40 mm. The confined aquifer III

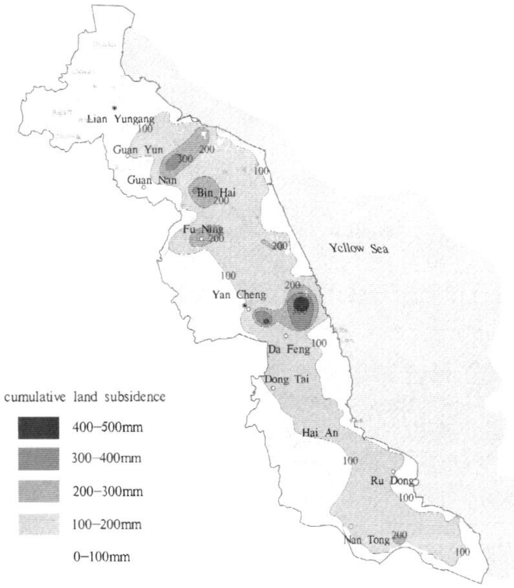

Figure 1. Cumulative land subsidence in CAJS from 2005 to 2013.

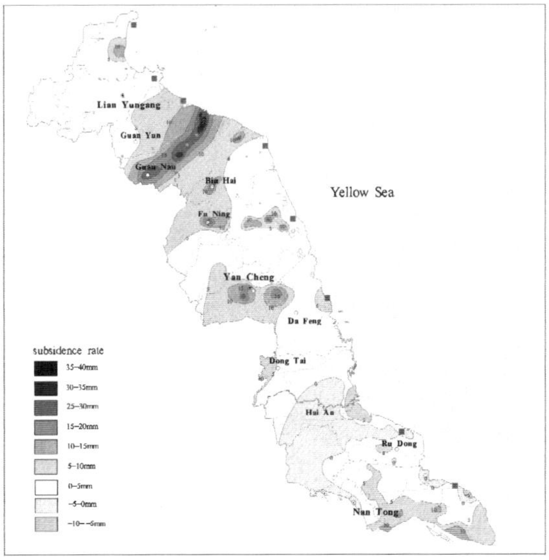

Figure 2. Subsidence rates in CAJS at the end of 2013.

was mainly exploited in this region, where the artesian head is above 30 m. This region also suffered from the most serious land subsidence in Jiangsu Province.

Dafeng City and the urban district of Yancheng City showed much earlier signs of land subsidence. Because of effective countermeasures adopted by local water authorities, the decline in the artesian head was reversed to some extent. The water table in the urban district began to be restored, and the land subsidence rate decreased. However, due to the long history of regional subsidence, the subsidence rate in 2013 was still over 15 mm.

2 Mechanisms of land subsidence development

Through long-term monitoring and analysis, land subsidence has been found to have the following characteristics: slow progression, persistency, hysteresis, resilience, irreversibility, non-uniformity, and complexity. The factors influencing land subsidence are very complicated, and the degree of land subsidence and the intensity of influential factors vary spatially. The main influential factors are the thickness of loose overlying stratum, aquifer development, groundwater exploitation dynamics, and engineering and construction activities (Xu et al., 2008).

2.1 Geological conditions as endogenous factors of land subsidence

Coastal areas usually have thick superficial loose sediment layers, which mainly consist of marine or interactive marine and terrestrial deposits. Sullage-containing, soft-soil layers of varying thickness generally have not finished self-consolidation and may be associated with land subsidence (Zhang et al., 2007). The burial depth of bedrock in the urban district of central Yancheng City and south Dongtai County has exceeded 1000 m. The compressible loose layers are thick and are easily susceptible to subsidence due to groundwater exploitation.

Neotectonic movement has resulted in pre-Quaternary topography featured by depression and uplift. The loose Quaternary deposits in the depressed areas are thicker than 300 m, whereas the thickness of loose deposits is smaller in the uplifted region. Since the Quaternary, the frequent alternation of marine and terrestrial environments has led to the complex genesis of Quaternary deposits and interbedding of sand layers and clayey soil. Moreover, an obvious rhythmic deposition pattern has been noted, providing the material basis for land subsidence.

The thickness of loose layers in the middle and southern coastal areas of Jiangsu Province is greater than 1000 m, and can even exceed 1400 m in the mid-regions and 1800 m in the southern regions (Fig. 3). Generally, four confined aquifers have been developed, indicating abundant groundwater resources. However, land subsidence is likely to occur due to groundwater overexploitation. For example, in Dafeng City and the urban district of Yancheng City, cumulative subsidence is greater than 700 mm, mainly due to the thick loose layers. Although the loose deposits in northern coastal areas are less thick, the evidence still indicates that the land subsidence was due to the highly developed soft soil layers in the offshore zone.

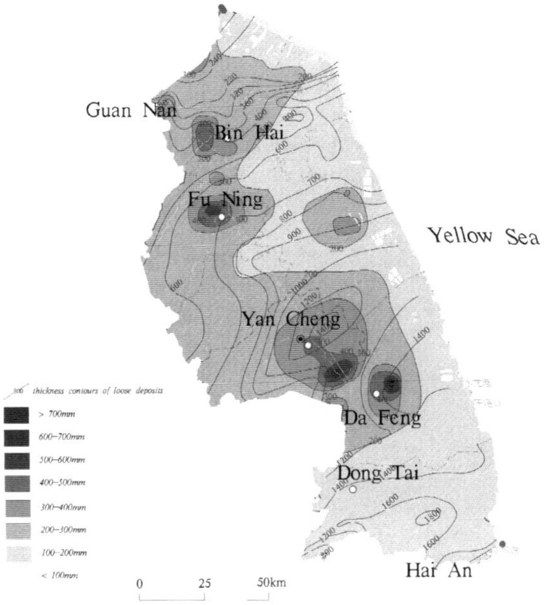

Figure 3. Relationships between cumulative land subsidence and the thickness of loose deposits from 1985 to 2013.

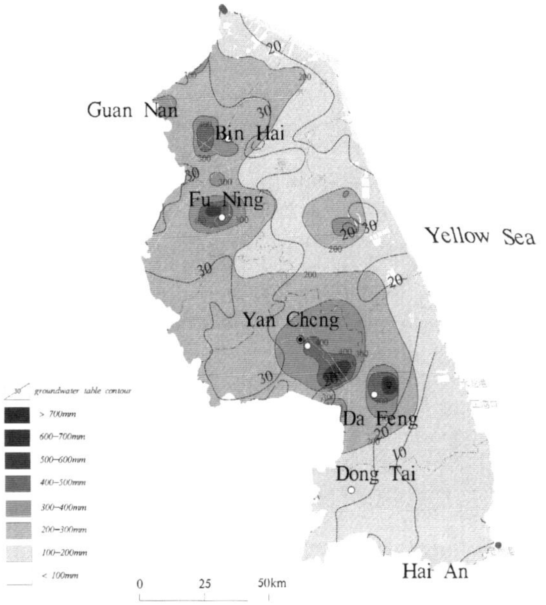

Figure 4. Relationships between subsidence rates and groundwater depression cone in 2013.

2.2 Groundwater overexploitation as the main cause of land subsidence

Land subsidence in the coastal areas of Jiangsu Province is closely related to groundwater exploitation with a high correspondence in time and space. Taking Nantong City as an example, before the 1980s, groundwater exploitation mainly occurred in Nantong City and the surrounding areas, leading to land subsidence in the urban district. After the 1990s, large-scale groundwater exploitation caused land subsidence throughout the entire region. The groundwater table has continuously declined along with the increase in exploitation, and the subsidence rate has risen as well.

The land subsidence cone basically overlaps the groundwater depression cone (Fig. 4). For example, along the banks of Guan River and in the urban districts of Yancheng and Nantong City, close spatial and temporal correlations between land subsidence and groundwater exploitations are clearly visible.

Land subsidence is jointly influenced by the compaction of the sand layer due to excess swabbing of the confined aquifer and deformation of the upper-confined bed and above due to water release by clayey soil. Land subsidence in coastal areas is controlled by the intensity of confined water exploitation. Earlier and more intensive exploitation corresponds with larger land subsidence. The subsidence rate is determined by the burial depth of the exploited groundwater layer and the physical properties of the aquifer and overlying aquitard.

However, land subsidence is not completely coupled to the depression cone of confined water, and the water table depth has to be considered. It has been found that water table depth directly corresponds with subsidence. The intensification of groundwater exploitation increases the stress on the soil layers. The aquifer bears the largest stress, followed by the overlying confining units, with the disrupted equilibrium lying between upward water stress and soil-layer stress. Due to the higher compactness of the water-bearing sand layer compared with clayey soil, the sand layer has a smaller subsidence under compaction, whereas clayey soil has a greater subsidence. The intensification of groundwater exploitation will cause more strata to lose the equilibrium between water and soil stress. As a result, land subsidence will increase, and the relationship between the water table and subsidence will not be in equilibrium.

A relationship exists between the annual cumulative subsidence of the confined aquifer III (the main exploitation layer) and the overlying geological strata and changes in the water table of the confined aquifer III. It has been found that subsidence is higher under a continuously declining water table. When the amount of exploitation is under control (indicated by a rise in the water table), land subsidence decreases.

2.3 Engineering and construction activities increase land subsidence

Under the progress of coastal development and the construction of coastal cities and ports, the impacts of engineering facilities on land subsidence are increasing (e.g. high-rise buildings, expressways, high-speed railways, ports, water usage, and drainage pipes). Foundation pits are usually excavated either openly or underground, leading to obvious subsidence. This is particularly prominent in the large-scale tunnel excavation.

Land subsidence caused by engineering activities usually occurs locally in shallow strata affected by frequent and intense engineering activities. This is considerably distinguishable from the subsidence of regional and deep strata caused by groundwater exploitation. However, the contribution of engineering activities to total subsidence is increasing and can no longer be neglected.

Given the geological conditions of the coastal areas of Jiangsu Province, intensive groundwater exploitation activities have led to the formation of several large subsidence cones. Holocene soft soil in shallow strata is likely to consolidate and deform due to engineering activities. Thus, these coastal areas are facing the grim challenges of land subsidence due to multiple factors.

3 Suggestions for the prevention and control of land subsidence

3.1 Establishment of a land-subsidence monitoring network

In contrast to the cities of Suzhou, Wuxi, and Changshu in Jiangsu Province, the coastal areas have less intense engineering activities and more complicated geological conditions such as neotectonic activities, thick loose deposits, and more groundwater exploitation layers. Therefore, the mechanisms of land subsidence are more complicated. Due to the combined factors of low elevation and strong storm tides, land subsidence may pose greater danger. Therefore, it is necessary to build a land-subsidence monitoring network for coastal areas that includes GPS, bedrock markers, layerwise markers, dynamic groundwater monitoring, automated monitoring systems, and a land subsidence surveillance center.

3.2 Establishment of multi-department linkage mechanisms

Authorities in water conservancy, municipal administration, and construction agencies should coordinate their activities to enhance groundwater exploitation management and reverse the trend of increasing land subsidence. Municipal engineering and construction and land-use planning should fully consider the impacts of land subsidence. Caution should be exercised to strictly control land subsidence due to engineering in densely built areas (e.g. dewatering in deep foundation pits). Geotechnical investigations and foundation treatments should be properly conducted in soft-soil regions.

References

Hu, R. L., Yue, Z. Q., and Wang, L. C.: Review on current status and challenging issues of land subsidence in China, Eng. Geol., 76, 65–77, 2004.

Xu, Y. S., Shen, S. L., and Cai, Z. Y.: The state of land subsidence and prediction approaches due to groundwater withdrawal in China, Nat. Hazards, 45, 123–135, 2008.

Xue, Y. Q., Zhang, Y., and Ye, S. J.: Land subsidence in China, Environ. Geol., 48, 713–720, 2005.

Zhang, Y., Xue, Y. Q., Wu, J. C., Ye, S. J., Wei, Z. X., Li, Q. F., and Yu, J.: Characteristics of aquifer system deformation in the Southern Yangtze Delta, China, Eng. Geol., 90, 160–173, 2007.

Proc. IAHS, 372, 507–510, 2015
proc-iahs.net/372/507/2015/
doi:10.5194/piahs-372-507-2015

Research on critical groundwater level under the threshold value of land subsidence in the typical region of Beijing

Y. Jiang[1], J.-R. Liu[2], Y. Luo[2], Y. Yang[2], F. Tian[2], and K.-C. Lei[2]

[1]Beijing geology prospecting & developing bureau, Beijing, 100195, China
[2]Hydrogeology and Engineering Geology Team of Beijing, Beijing, 100195, China

Correspondence to: Y. Jiang (yuan.12.3@163.com)

Published: 12 November 2015

Abstract. Groundwater in Beijing has been excessively exploited in a long time, causing the groundwater level continued to declining and land subsidence areas expanding, which restrained the economic and social sustainable development. Long years of study show good time-space corresponding relationship between groundwater level and land subsidence. To providing scientific basis for the following land subsidence prevention and treatment, quantitative research between groundwater level and settlement is necessary. Multi-linear regression models are set up by long series factual monitoring data about layered water table and settlement in the Tianzhu monitoring station. The results show that: layered settlement is closely related to water table, water level variation and amplitude, especially the water table. Finally, according to the threshold value in the land subsidence prevention and control plan of China (45, 30, 25 mm), the minimum allowable layered water level in this region while settlement achieving the threshold value is calculated between −18.448 and −10.082 m. The results provide a reasonable and operable control target of groundwater level for rational adjustment of groundwater exploited horizon in the future.

1 Introduction

Beijing suffered successive droughts since 1999, the precipitation decreased 19 % in recent 12 years. The groundwater has been over exploited in order to ensure the safety of water supply for the capital, the average groundwater depth in Beijing plain decreased from 11.9 m (1998) to 24.3 m (2012), with an average annual decline of 0.89 m, which resulted in intensified development of ground subsidence. In the future, the contradiction between supply and demand of water resources is difficult to solve, the land subsidence will be in sustained and rapid development (Yang et al., 2010).

Years of research shows that, the main causes of land subsidence in Beijing plain is groundwater overexploitation. The characteristic of land subsidence development is closely related to the intensity and depth of groundwater exploitation. The speed and magnitude of the groundwater level change directly affect the speed and amount of ground subsidence.

2 Correlation between land subsidence and groundwater level

The relationship between groundwater level and land subsidence in Beijing plain has the following characteristics: on the plane, the range and distribution of land subsidence and groundwater depression are similar. The groundwater depression area has been expanding continuously from 1971, now the area is more than 1000 km^2 (Fig. 1). In general, the area of land subsidence is consistent with the distribution of groundwater depression regional.

In vertical, through the comparison regional land subsidence trend, layered water level and settlement, there is a good consistency between groundwater level variety and land subsidence trend. The periodic change of the settlement rate is consistent with the seasonal fluctuation of groundwater level (Yang et al., 2013) (Fig. 2).

Published by Copernicus Publications on behalf of the International Association of Hydrological Sciences.

Table 1. Layer water level and settlement in Tianzhu station.

Layered water level	D3–2	D3–3	D3–4	D3–5	D3–6
Depth	210–218	120–146.8	85.7–91.3	59.3–63.4	27.5–31.0
Layered settlement	F3–3	F3–4	F3–6	F3–8	F3–10
Depth	148.49–218.89	117–148.49	82.3–102	48.5–64.5	2.4–35.43
Lithology	silt fine sand	silt coarse sand	Powder stick fine sand	Fine sand coarse sand	silt

Table 2. Results of multiple regression analysis.

Monitoring layer	Regression equation	R	R^2	F
D3–2/F3–3	$y = 1.750x_1 + 0.108$	0.933	0.871	58.727
D3–3/F3–4	$y = 2.095x_1 + 0.662x_2 + 2.038x_3 + 0.199$	0.926	0.858	52.227
D3–4/F3–6	$y = 1.219x_1 + 0.657x_2 + 1.918x_3 - 5.240$	0.942	0.887	67.900
D3–5/F3–8	$y = 0.248x_1 + 0.092x_2 + 0.128x_3 - 0.662$	0.894	0.800	34.673
D3–6/F3–10	$y = 2.190x_1 + 0.378x_2 + 2.840x_3 - 37.338$	0.510	0.260	3.044

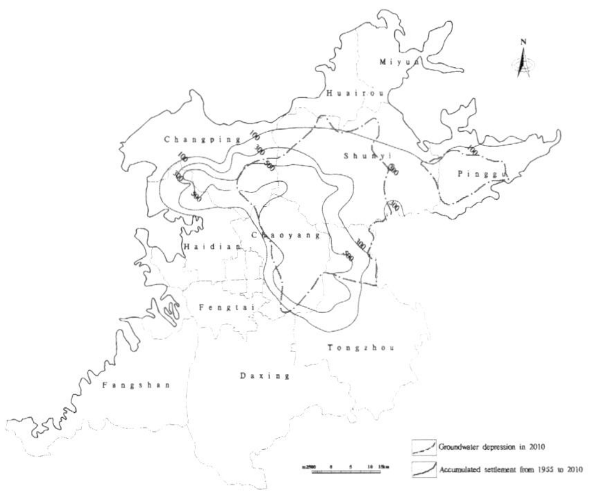

Figure 1. Groundwater depression in 2010 and accumulated settlement from 1955 to 2010.

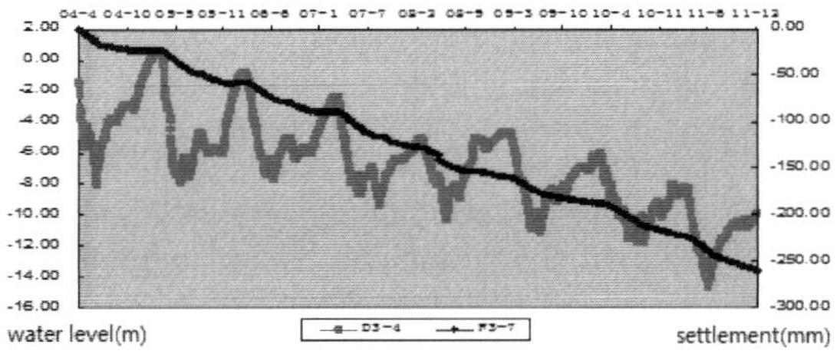

Figure 2. F3–7 (64.5–82.3 m) settlement with the water level elevation of D3–4 in Tianzhu station.

3 Correlation research on water level and settlement

According to the sediment regulation of the quaternary stratum, the underground water layer structure, exploitation utilization of groundwater and other factors, the aquifer in Beijing area can be divided into four groups: the first aquifer group, mainly phreatic water and confined water whose buried depth is less than 50 m shallow; the bottom boundary depth of second aquifer group is 80–120 m, while the third aquifer group is 150–180 m and the fourth one is 300 m.

The first phase project of Land Subsidence Monitoring Network and Warning System in Beijing has been built in 2004, three land subsidence monitoring stations were set up. There are marked and layered bedrock standard group, groundwater level and pore water pressure monitoring wells in every station (Table 1). Monitoring data of layered water table and settlement in long time series provide a solid basis on the further research between them.

3.1 Establishment of the regression equation

For further identification of the intrinsic relationship between the two, multi-linear regression models are set up by long series factual monitoring data about layered water table and settlement in the Tianzhu monitoring station based on quarterly average water table, water level variation and amplitude (Eq. 1).

$$y = \beta_1 x_1 + \beta_2 x_2 + \beta_3 x_3 + \varepsilon \quad (1)$$

In the equation: y – quarterly accumulated settlement ($\mathrm{mm\,a^{-1}}$); x_1 – quarterly average water table (m); x_2 – quarterly water level variation (m); x_3 – quarterly water level amplitude (m); β_1, β_2, β_3, ε – the coefficients of regression equation.

Table 3. The minimum permissible level of groundwater in different control of settlement velocity.

Monitoring layer	Control targets (mm) 45	30	25	Water level variation (m)	Control targets (mm) 45	30	25
	Layered control targets (mm)				The minimum permissible water level (m)		
D3–2/F3–3	3.843	2.562	2.135	0.952	−18.448	−17.716	−17.472
D3–3/F3–4	4.437	2.958	2.465	2.114	−16.806	−16.270	−16.091
D3–4/F3–6	2.705	1.803	1.503	2.081	−14.502	−14.021	−13.861
D3–5/F3–8	0.342	0.228	0.190	2.208	−10.529	−10.194	−10.082

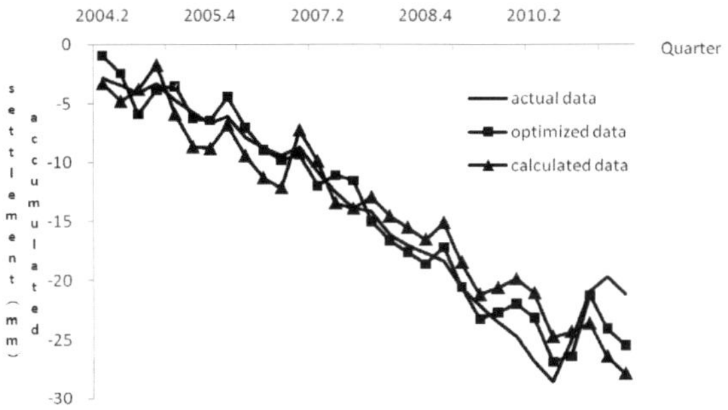

Figure 3. The calculated, actual and optimized value of F3–3.

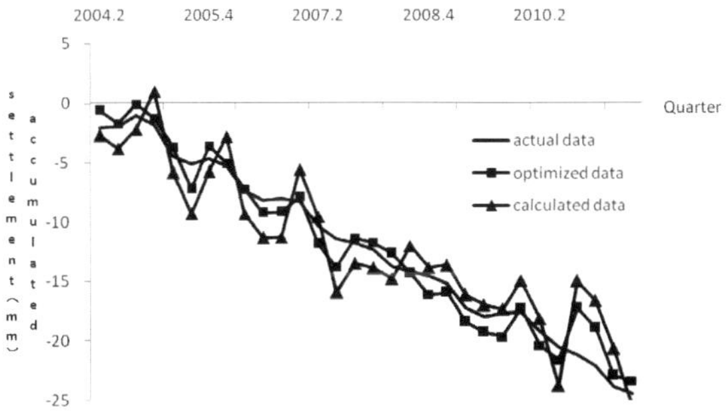

Figure 4. The calculated, actual and optimized value of F3–4.

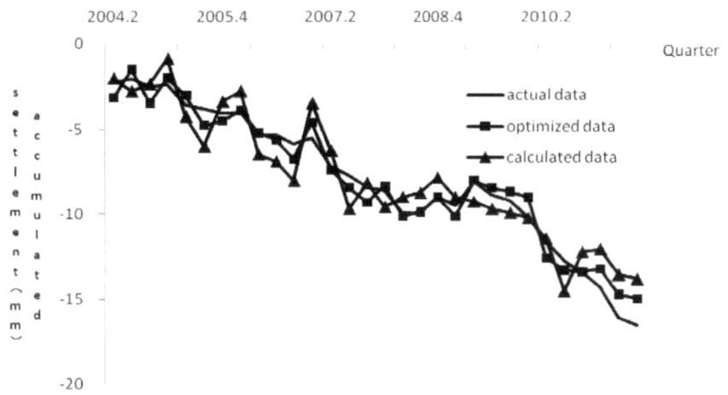

Figure 5. The calculated, actual and optimized value of F3–6.

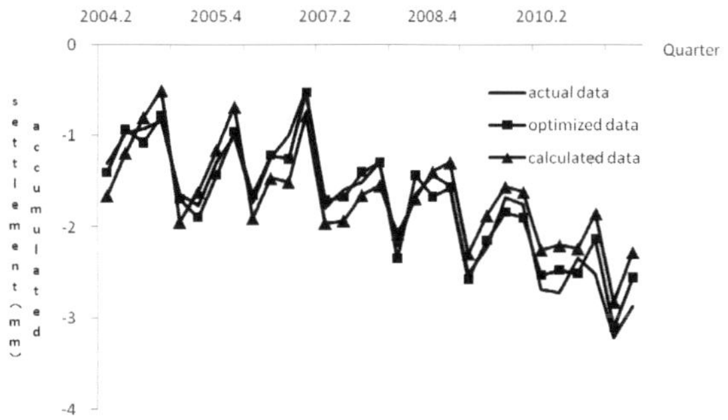

Figure 6. The calculated, actual and optimized value of F3–8.

3.2 Significant test

The correlation coefficient of R describes the varying degree of relationship between variable X and Y, which can test the significant of the equation. The correlation coefficient is larger, means the sample points are closer to the regression line, the degree of linear correlation is greater (Jia et al., 2003).

The analysis results show that (Table 2), in addition to D3–6/F3–10 monitoring layer, the correlation coefficients of the other layers are greater than 0.8, the ground ettlement has highly positive correlation with the above factors (Dong, 2006; Hao et al., 2011). F0.05 (1, 28) = 4.196, so that at the significance level of 0.05, in addition to D3–6/F3–10, the other regression equations are significant and practical.

3.3 Analysis and optimization of regression equation

The coefficient of quarterly average water table variable β_1 is positive, means the water level is positively correlated with the accumulated settlement (settlement is negative); The coefficient of quarterly water level variation variable β_2 is positive, means the greater water level decline, the greater the amount the cumulative settlement. In addition, from the coefficient of the regression equation, quarterly average water table has the strongest impact on accumulated settlement (Figs. 3–6).

4 Allowable layered water level under the threshold value of land subsidence

According to the annual control targets of land subsidence in Beijing, the threshold values are set as 45, 30, 25 mm.The layered control target of F3–3, F3–4, F3–6 and F3–8 are calculeted by the historical data of settlement.The minimum allowable layered water level in this region while settlement achieving the threshold value can be calculated (Table 3).

5 Conclusions

1. Ground subsidence has high degree correlation with the groundwater level in Beijing plain area. On the plane, the range and distribution of land subsidence and groundwater depression are similar. In vertical, there is good consistency between groundwater level variety and land subsidence trend.

2. The ground settlement has highly positive correlation with the average water table, variation and amplitude. Quarterly average water table has the strongest impact on accumulated settlement.

3. According to the threshold value in the land subsidence prevention and control plan of China (40, 30, 25 mm),the minimum allowable layered water level in this region while settlement achieving the threshold value is calculated between −18.448 and −10.082 m. The results provide a reasonable and operable control target of groundwater level for rational adjustment of groundwater exploited horizon in the future.

References

Hao, Y., Xu, J., and Zhang, Y.: Research on Application of settlement monitoring model in the calculation of multi factors based on the city ground, J. Railway Eng. Soc., 3, 20–23, 2011.

Jia, Y., Huang, Z., and Zhang, M.: Application of multiple regression analysis and prediction in ground subsidence with water table, J. Geotech. Invest. Surv., 1, 77–79, 2003.

Dong, F.: Prediction model and application of land subsidence, Tianjin University, Tianjin, 2006.

Yang, Y., Jia, S.-M., and Wang, H.-G.: The development of land subsidence in Beijing Plain, J. Shanghai Geol., 31, 23–28, 2010.

Yang, Y., Zheng, F., and Liu, L.: Study on the correlation between groundwater level and ground subsidence in Beijing plain areas, J. Geotech. Invest. Surv., 8, 44–48, 2013.

Proc. IAHS, 372, 511–514, 2015
proc-iahs.net/372/511/2015/
doi:10.5194/piahs-372-511-2015

Effects of groundwater regulation on aquifer-system compaction and subsidence in the Houston-Galveston Region, Texas, USA

M. J. Turco and A. Petrov

Harris-Galveston Subsidence District, Friendswood, Texas, USA

Correspondence to: M. J. Turco (mturco@subsidence.org)

Published: 12 November 2015

Abstract. Subsidence has been a primary concern in the Houston area for many years. Since 1906, about 4 m of subsidence has occurred in the coastal areas, with a broad area of about 2 m of subsidence existing today throughout most of the Houston Area. In 1975, as a result of area residents and local governments becoming increasingly alarmed by the continued impact of subsidence on economic growth and quality of life in the region, the Harris-Galveston Subsidence District was created by the 64th Texas Legislature as an underground water conservation district with the mission to provide for the regulation of the withdrawal of groundwater to control subsidence. Over its nearly 40 years of existence, the District has developed substantial data sets providing the foundation for its regulatory plan. The District's regulatory plan allows groundwater users an amount of annual groundwater based on a percentage of an individual permitee's total water demand. The management of the groundwater resources within the District has involved significant coordination with regional ground and surface water suppliers; ongoing interaction with other state and local regulatory bodies; analysis of accurate and up to date predictions on water usage; the enforcement of disincentives to those who rely too heavily on groundwater and a commitment to practicing and promoting water conservation. Areas nearest the Gulf of Mexico coast have completed the conversion to alternative (other than groundwater) water sources, consequently the effective stress on the aquifer has decreased and subsidence rates have been reduced. Areas within the District further inland are currently proceeding through the conversion process, and subsidence has continued in those areas as development of the groundwater resources has continued.

1 Introduction

Subsidence has been a concern in the Houston, Texas area throughout most of recent history. The U.S. Geological Survey first documented the correlation between shallow fluid withdrawal and subsidence in 1929 at the Goose Creek Oilfield near Baytown, Texas just southeast of Houston (Pratt and Johnson, 1926). Following that revelation, the area continued to develop on groundwater, water-levels in aquifer declined, and subsidence continued.

The occurance of subsidence in a coastal area prone to tropical weather can have serious consequences including loss of property and loss of life. About 4 m of subsidence has occurred along parts of the coastal community near Houston and Galveston, TX. These areas have seen entire developments inundated by the Gulf of Mexico. There has also been evidence of infrastructure damage as a result of subsidence in the area including damage to homes and water infrastructure.

2 Hydrogeology

There are three primary aquifers comprising the Gulf Coast aquifer system best categorized in relative terms as "shallow and younger" sediments and "deep and older" sediments. The uppermost Chicot and underlying Evangeline aquifers are hydraulically connected and comprise the younger and shallow sediments of the system. The regionally persistent Burkeville Confining unit is stratigraphically positioned between the Evangeline and the deeper and older Jasper aquifer (Kasmarek et al., 2014). The lowermost unit of the system is the Miocene-age Catahoula confining system which contains the Catahoula Sandstone (Baker Jr., 1979). Through time, geologic and hydrologic processes created these accretionary sediment wedges of clay and other fine-grained clastic material more than 2300 m thick at the coast (Chowdhury and Turco, 2006; Baker Jr., 1979). The shallow and young sediments of the Chicot and Evangeline aquifer are the most concern when considering the potential for clay-compaction and resulting land surface subsidence.

3 Water use

Prior to the creation of the Harris-Galveston Subsidence District, the Houston region had relied mostly on groundwater for municipal, industrial and agricultural water supply. This reliance on groundwater coupled with the rapid development and population growth of the city of Houston led to substantial groundwater level declines and subsidence. The subsidence that occurred as a result of the increased groundwater use led to significant outcry from many in the local community and led to the creation of the HGSD by the State of Texas in 1975.

Regulating the use of groundwater is an effective measure to minimize active subsidence and prevent areas from subsiding in the future. The HGSD regulates groundwater as a percentage of the areas total water demand which allows continued growth of the area while still combating the issue of subsidence. The District began immediately focusing on those areas, currently referred to as Area 1, nearest the Gulf of Mexico, in perilous proximity to frequent tropical storms, hurricanes, and storm surge, requiring those areas to convert to 90 % alternative water as soon as possible. Subsequently, conversion to 80 % alternative supply occurred in Area 2 and in 2035 Area 3 will reach a similar conversion ratio. Beginning in 1976 and continuing through 2014, the percentage of groundwater used with respect to the total water demand of the area has decreased. In 1976 about 63 000 hectare meter (ha m) of water were sourced from groundwater with a total water demand at that time of 103 000 ha m. Whereas in 2014 about 26 500 ha m of water were sourced from groundwater with a total water demand of about 129 000 ha m (Fig. 1). In 2014, groundwater accounts for about 21 % of the total water demand within the District, a dramatic change over the 40-year history of groundwater regulation in the area.

4 Impact of changes in water use on subsidence

As water use has changed in the Houston area, groundwater-levels have risen in areas one and two and subsidence rates have slowed considerably. The HGSD operates a network of about 80 GPS subsidence monitoring stations (Fig. 2) throughout the area along with a network of 13 extensometers operated by the USGS to monitor the impact of changes in water use on subsidence in the area. The HGSD network

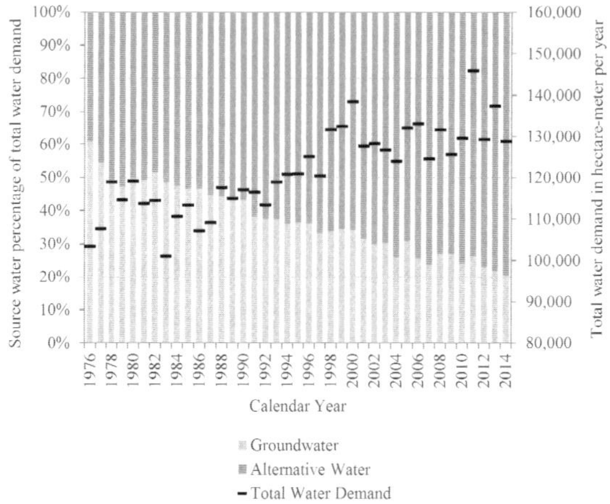

Figure 1. Groundwater and alternative source water percentage and total water demand, 1976–2014, Harris-Galveston Subsidence District, Texas, USA.

Figure 2. Subsidence monitoring station located near Texas City, TX, USA.

consists of continuously operating and periodically operating GPS stations equipped with real-time global positioning system receivers that are post-processed, determining a relative change in ellipsoid height of the monitoring station refer-

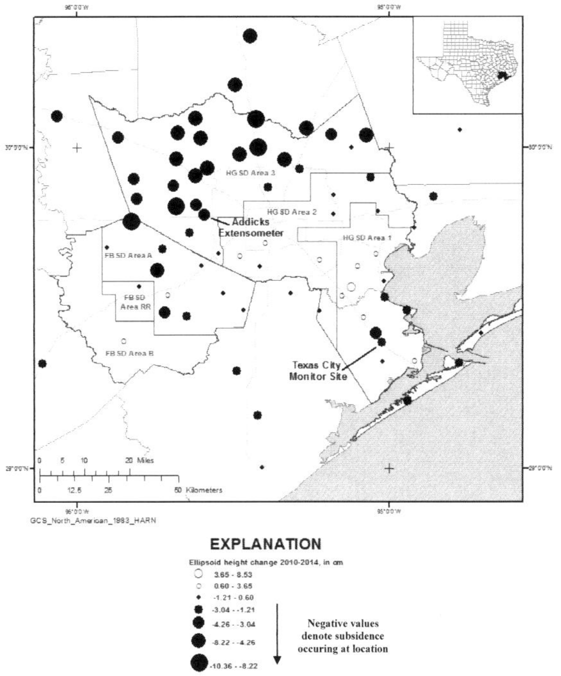

Figure 3. Observed change in ellipsoidal height calculated (Zilkoski et al., 2003) at GPS monitoring stations from 2010 to 2014, Houston area, Texas.

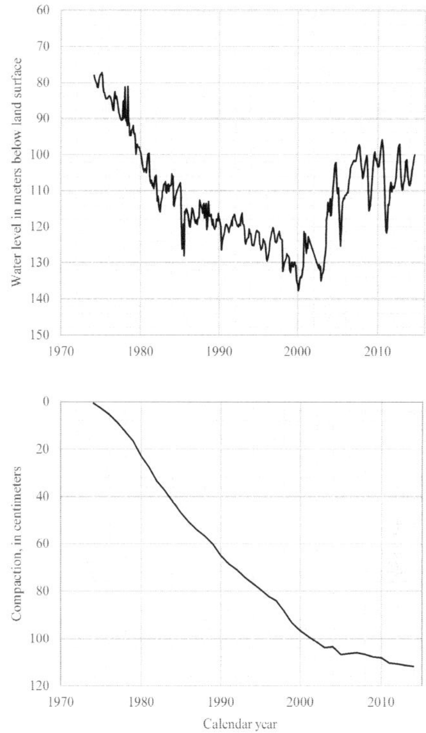

Figure 4. Water-level and cumulative compaction at the Addicks extensometer located in Harris Galveston Subsidence District Regulatory Area 3, Houston, Texas, USA, 1974–2015.

enced to one of the stable extensometer benchmarks at depth (Zilkoski et al., 2003).

Generally, the network shows a good correlation between measured water-levels and observed subsidence. Figure 3 shows distribution of ellipsoidal height change from 2010 to 2014, generally showing the area of highest occurrence and magnitude of subsidence occurring in Area 3, currently not fully converted to alternative water.

The Addicks extensometer is located in Area 3 and is currently the most active extensometer. Since 1975, nearly 115 cm of clay-compaction has occurred (Fig. 4). From late in 2010 through 2012, the Houston area experienced a prolonged drought, resulting in the highest total water demand for the area on record. Consequently, subsidence rates increased in areas one, which had not seen appreciable subsidence occur in recent history (Fig. 3).

5 Conclusions

The Gulf Coast aquifer system in the Houston area is uniquely susceptible to clay-compaction and subsidence due to its aquifer composition and large amounts of historical groundwater withdrawal. Changes in water use due to groundwater regulation and the conversion of groundwater users from groundwater to alternative source waters has resulted in groundwater levels above the historical minimum and minimized subsidence in coastal areas. Local collaboration between the regulatory and regulated communities is absolutely necessary in coordinating the conversion from one source water to another. Considering the variability in water demand as it relates to climatic change, on-going monitoring of water-level and land surface subsidence is required to adequately gage the aquifer's response to changes in water-use.

Acknowledgements. The authors wish to thank Robert Thompson, Kevin Ebrom, Mike Chrismer, and Greg Lackey of the Harris-Galveston Subsidence District for their assistance in the preparation of this paper and the USGS Texas Water Science Center for their continued cooperation in the groundwater and subsidence research in the Houston area. Additionally, this information would not be available without the collaboration of numerous regional water providers such as the City of Houston and the North and West Harris County Regional Water Authorities.

References

Baker Jr., E. T.: Stratigraphic and hydrogeologic framework of part of the Coastal Plain of Texas, Texas Department of Water Resources, USA, Report 236, 43 pp., 1979.

Chowdhury, A. H. and Turco, M. J.: Geology of the Gulf Coast aquifer, Texas, chap. 2, in: Aquifers of the Gulf Coast of Texas, edited by: Mace, R. E., Davidson, S. C., Angle, E. S., and Mullican, W. F., Texas Department of Water Resources, USA, Report 289, 101 pp., 2006.

Kasmarek, M. R., Johnson, M. R., and Ramage, J. K.: Water-level altitudes 2014 and water-level changes in the Chicot, Evangeline, and Jasper aquifers and compaction 1973–2013 in the Chicot and Evangeline aquifers, Houston-Galveston region, Texas, U.S. Geological Survey Scientific Investigations Map 3308, pamphlet, 16 sheets, scale 1 : 100 000, available at: http://dx.doi.org/10.3133/sim3308, last access: 14 July 2015, 2014.

Pratt, W. E. and Johnson D. E.: Local subsidence of the Goose Creek oil field, The University of Chicago Press, The Journal of Geology, 34, 577–590, 1926.

Zilkoski, D. B., Hall, L. W., Mitchell, G. J., Kammula, V., Singh, A., Chrismer, W. M., and Neighbors, R. J.: The Harris-Galveston Coastal Subsidence District/National Geodetic Survey Automated GPS Subsidence Monitoring Project, in: U.S. Geological Survey Subsidence Interest Group Conference, Proceedings of the Technical Meeting, Galveston, Texas, 27–29 November 2001, edited by: Prince, K. R. and Galloway, D. L., U.S. Geological Survey, Open-File Report 03-308, 162 pp., 2003.

Proc. IAHS, 372, 515–518, 2015
proc-iahs.net/372/515/2015/
doi:10.5194/piahs-372-515-2015

A model, describing the influence of water management alternatives on dike stability

J. W. M. Lambert, E. Vastenburg, and F. J. Roelofsen
Deltares, Delft, the Netherlands
Correspondence to: J. W. M. Lambert (john.lambert@deltares.nl)

Published: 12 November 2015

Abstract. The awareness is rising that economic effects of Land Subsidence are high. Nevertheless, quantifying these economic losses is difficult and, as far as known, not yet done in a sophisticated way. Also, to be able to decide about future strategies, for example to avoid or decrease subsidence, it is necessary to know the financial consequences of measures and possible solutions. As a first step to quantify these economic effects, a MODFLOW-SCR (coupled MODFLOW-Settlements model) is coupled with the model DAM. Based on the local stratigraphy, the shape and composition of the existing dike or levee, the level of the surface water and the surface level, macro-stability of the dike is calculated and – if the dike does not meet the required stability – adaptions are proposed. The model enables to separate effects that are caused by sea-level rise and the effects of subsidence. Coupling the DAM model with an economic model to calculate costs of these adaptions is under construction.

1 Introduction

Dikes and levees are designed and constructed to safeguard people and economic interests from flooding. The minimum height of a dike is derived from the maximum expected water height and may be reduced by accepting a certain frequency of overflow. In design, a future increase of the water height, due to climate change, is taken into account. Besides the criterion of required height, criteria of stability and piping play an important role. In this paper mechanisms that play a role in dike-safety, will be discussed and a model is presented that enables to assess the long term influence of water management and climate change effects on dike stability. A case-study for Delft in the Netherlands has been elaborated. This coupled groundwater – subsidence – dike stability model can be considered as a first step and a show case in the future development of a water management support system.

2 Case-study Delft

In Delft, industrial groundwater extraction of about 1200 $m^3\,h^{-1}$ by DSM (an industrial site) takes place since the early 20th century. The groundwater has been used for cooling purposes and caused seasonal fluctuations in the piezometric levels in the Pleistocene aquifers in the order of 6 m and could be observed in a radius of about 20 km from the extraction. The groundwater abstraction also caused subsidence with a maximum of about 0.50 m.

After DSM decided to stop groundwater abstraction, the possible effects in terms of piezometric levels were predicted with a groundwater model and – parallel – a subsidence model was developed to enable forecasting of possible resulting land raise.

As the groundwater abstraction has lasted many decades and it was never expected it would ever stop, possible effects of a rising water table were never taken into account in design and construction of buildings and infrastructure. These effects have been investigated one by one and it was concluded that they are so uncertain that the municipality of Delft, the Province of South Holland and the waterboard Delfland decided to maintain the abstraction until 2017, taking mitigating measures in the meantime and install a monitoring network to support stepwise decrease of the extraction rate during the years after 2017.

3 Failure mechanisms of dikes and levees

Each dike is assessed on a regular base to establish its functionality under possible – future – extreme circumstances. Controlled mechanisms are: e.g. overtopping and piping.

More complicated mechanisms are outward and inward stability of the dike. The driving force in instability is the weight of the dike body, taking into account the difference of wet and dry volumetric weight and the shape of the dike. Under extreme low (river-level) conditions outward instability should be controlled and under extreme high conditions the inward stability should be guaranteed. As indicated; for these mechanisms, the weight of the dike plays an important role, but also – in case of inward instability – the water pressure at the outside and the groundwater conditions in and under the dike. These mentioned forces have to be resisted by friction forces in the soil. The allowable forces are strongly dependent from the soil, but also water pressures play an important role. This is especially the case if piezometric levels in the (sandy) subsoil are more or less in balance with the (clay or peaty) covering layers on top. In such cases, which are rather common in the Netherlands, the critical slip circle may change because – at least theoretically – the resistance against slip becomes almost frictionless where the piezometric level is in balance with the weight of the covering layer.

Climate change and water management strategies cause boundary conditions for dike safety change in time. The developed model is able to predict the influence of different strategies on dike safety and such it can support decision making.

Figure 1. 3-D Geological model on a scale of $100 \times 100 \times 0.5$ ($L \times W \times H$) based on all available boreholes and CPTs. Colors indicate variation in lithology and are user-defined. The picture indicates the heterogeneity of the subsoil (pink = peat, green and blue are different clays, yellow are Holocene sands).

4 The models

4.1 iMOD

iMOD (Vermeulen, 2006) is an easy to use Graphical User Interface and an accelerated version of the well-known MODFLOW 3-D Groundwatermodel. Modflow was developed by USGS in the eighties of the past century; the iMOD version has been developed by Deltares in co-operation with USGS. The main feature of iMOD is that it allows gathering all available input data at its available resolution, where iMOD enables up- and down-scaling towards the model area and grid resolution that is required for a specific user-defined problem. So, different sources of information can easily be combined and applied in whatever resolution as required. Shifting the area of interest or changing the resolution can easily be done.

4.2 SCR-module

SCR (Settlement Creep) is an extension on MODFLOW that is recently developed by Deltares in co-operation with USGS; the SCR-module enables to predict the influence of changing groundwater levels and piezometric levels on land subsidence. The new module simulates groundwater storage changes and compaction in discontinuous interbeds or in extensive confining units, accounting for temporal and spatial variability of geostatic and effective stresses. It implements two methods of compaction (settlement) calculation accounting for secondary compaction (i.e. creep). The NEN-Bjerrum and the Isotache methods are included in the new package. The first method adapts a linear strain relationship while the second adapts a natural strain relationship to calculate compaction/settlement. Several features of the package make it useful for application in shallow, unconfined flow systems.

4.3 DAM

DAM is a Dike Analysis Module that evaluates the safety of levees and dams against instability under present and under decisive (e.g. high tide) circumstances. Surface levels, the phreatic levels and the piezometric levels are important input for this analysis, besides possible change in the decisive level, as a result of climate change. DAM analysis and presents dike-stability according to Bishop theory, possible uplift and seepage erosion. Coupling of DAM to iMOD-SCR allows to predict the effect of possible water management decisions on dike safety.

5 Results of the case-study

In 2008, a preliminary model-study was carried out to predict the effects of stopping the groundwater extraction in terms of piezometric and phreatic groundwater levels. These results were exported from MODFLOW and imported in DSettle (a 2-D-settlement model) to predict changes in surface level in specified cross-sections. Separately, the predicted phreatic

Figure 2. Cross-sections in DAM.

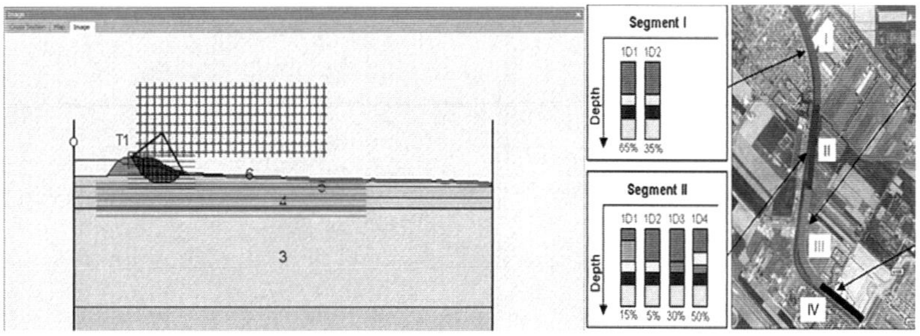

Figure 3. Input and results DAM-calculations. The figure shows one cross-section, representing Segment II for which – statistically – a distribution of possible geological settings is estimated. The grid represents all possible slip circlesof which the most critical is established by calculating them all.

and piezometric changes were imported in DStab to predict effects on levees and dikes. Also, the effects of stopping groundwater extraction on buildings, infrastructure and other effects had to be assessed. All effects had to be assessed more or less manually, and were therefore very time-consuming. Therefore, an optimization of scenarios never took place and it was decided to install a monitoring network, take already a number of mitigating measures and – after 2017 – decrease the extraction rate step-wise.

Using the integrated iMOD-SCR-DAM model, the effects of stopping the groundwater extraction on dike stability, were repeated, which gave a much more detailed insight in the effects of the management decisions and also giving the opportunity to compare the effects of different strategies.

Figure 1 shows the geological model that is based on an interpretation of all available CPT and borehole data and in Fig. 2 the calculated DAM locations are given, as well as a typical cross-section.

Figure 3 shows the input of the DAM-model as it is derived from iMOD and – following – all possible slip-circles through the dam body are calculated simultaneously, selecting the most critical of them and calculating the safety factor against instability.

The results of this calculation are given in Fig. 4, in terms of a decrease of the safety factor.

6 Conclusions

Although groundwater abstraction has many negative effects, decrease or even stopping abstraction can also cause negative effects, such as possible instability of dikes and levees. In this paper these negative consequences on dike stability have been elaborated.

Decision makers need insight in the consequences of their decisions. The quicker this insight can be given; the better an optimum can be proposed. For this reason, coupling of already existing models that describe the effects of alterna-

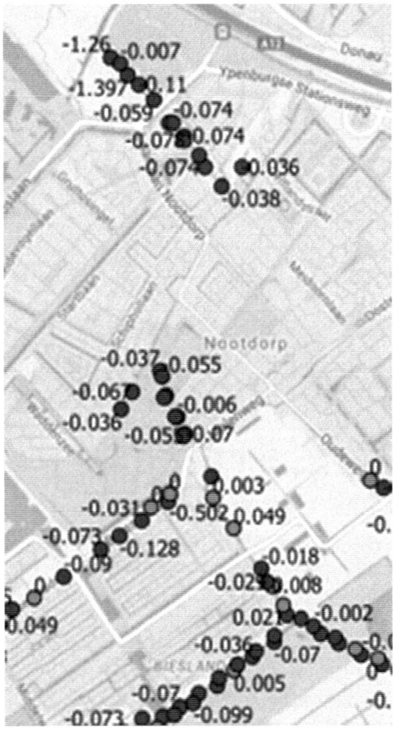

Figure 4. Calculated change in stability factor. Green dots mean an increase of the safety factor while the red ones indicate a decrease. The indicated values are the calculated decrases and increases.

tives is important. Presented is a model that describes the effects on dike stability of alternatives in water management and – from the results – costs can be derived rather easily. The results of the calculations give a quick insight in the locations that need to be adapted; alternative scenarios can be elaborated quickly, such that advantages and disadvantages of different decisions become clear and an optimum can be chosen.

7 Future developments

Given the complex connections between cause and effects of alternative water management decisions, development of models that quantify these effects are urgently required. The described interconnection between a groundwater – subsidence model and a model that controls dam stability is an example how existing models can easily be coupled and give information on the effects of different decisions. This shown example should be seen as a first step in coupling the groundwater – subsidence model to other models that describe effects, such as effects on foundations, effects on infrastructure and consequences on flooding. Future developments will also contain coupling with economic models and adding a module that enables to predict the effects of loading the subsoil. Although this seems (and is) a huge amount of work, it should be considered that – in common – existing models are already available and can be integrated rather easily.

Acknowledgements. The activities for this development were funded by Deltares.

References

Bakr, M.: Coupled Modeling of Groundwater Flow and Land Subsidence With Creep (MODFLOW/SUB-CR), 2012.

Vermeulen, P. T. M.: Model-Reduced Inverse Modeling, PhD thesis, Delft University of Technology, Delft, 2006.

Proc. IAHS, 372, 519–523, 2015
proc-iahs.net/372/519/2015/
doi:10.5194/piahs-372-519-2015

On the possible contribution of clayey inter-layers to delayed land subsidence above producing aquifers

G. Isotton, M. Ferronato, G. Gambolati, and P. Teatini
Università degli Studi di Padova, Padova, Italy
Correspondence to: G. Isotton (g.isotton@m3eweb.it)

Published: 12 November 2015

Abstract. In recent years, measurements of land subsidence above pumped aquifers by permanent GPS and InSAR have exhibited some delay relative to drawdown ranging from months to years. The current modeling approaches accounting for water fluid dynamics and porous medium geomechanics may fail to predict such a delay and may underestimate the land settlement after the well shutdown. In the present communication, an investigation is made on the residual compaction of the intervening clayey formations as a possible contribution to retarded land subsidence. The pore pressure variation within the aquifer and its propagation in the clay are simulated by a finite element flow model, with the resulting pore pressure decline used as input data in a hypoplastic geomechanical model. A proper sensitivity analysis on (i) aquifer depth, (ii) ratio between the sandy and the clayey layers thickness and hydraulic conductivity, (iii) oedometric compressibility in first and second loading cycles, is performed for a typical geology of a Quaternary sedimentary basin. The results show that a certain fraction, up to 20 % of the overall land subsidence, can take place after the shutdown of the producing wells depending on actual basin, litho-stratigraphy and parameter values.

1 Introduction

A major consequence of groundwater pumping is anthropogenic land subsidence. This can be a matter of concern if the affected areas are highly urbanized, with the loss in ground elevation generating possible structural problems to the buildings. If the aquifers underlie the coastland, the environmental problems are the exposure to flooding during high tides and severe sea storms. Hence, land subsidence must be investigated, monitored and reliably predicted to implement adequate remedial measures.

Extensometric records of soil deformation combined with land subsidence measurements by permanent GPS stations and InSAR techniques have provided evidence of some delay between the well shutdown or a significant reduction of the pumping rates and the resulting land settlement (Hettema et al., 2002; Teatini et al., 2006; Wu et al., 2010; Galloway and Sneed, 2013; Chang et al., 2014). While pore pressure in the aquifers progressively recovers, the subsidence may still continue, with a delay ranging from a few months to a few years. This suggests that the clayey layers confining the aquifer may keep on depressurizing and depleting after the shutdown of the producing wells. In the present study an investigation is made on the possible correlation between the delayed subsidence and the depletion of the clayey inter-layers of a multi-aquifer system.

The current standard modelling approach accounting for water fluid-dynamics and porous medium geomechanics usually does not accurately address these inter-layers (or aquitards) and tend to underestimate the land settlement after field abandonment. Before using a more complex modelling approach, such as visco-elastic or visco-plastic (Wu et al., 2010; Chang et al., 2014), a sensitivity analysis is performed with the aid of a 1-way coupled hysteretic hypoplastic model (Gambolati et al., 2001) in which the pore pressure variation propagates in the clayey inter-layer. The sensitivity analysis takes into account the main parameters controlling land subsidence: aquifer depth, ratio between the sandy and the clayey layers thickness and hydraulic conductivity, and oedometric compressibility in the first loading and second unloading-loading cycle.

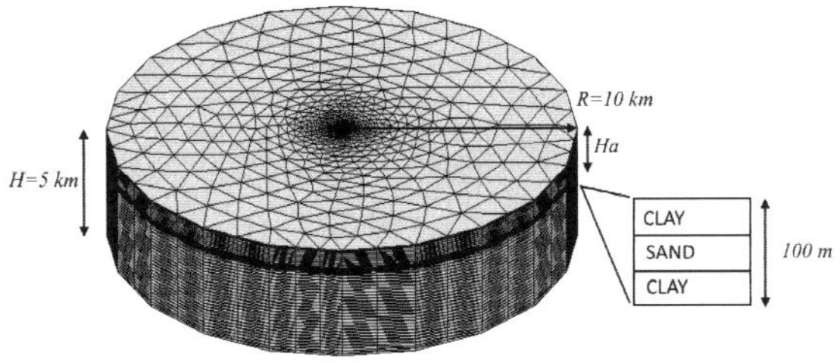

Figure 1. Geomechanical model used in the numerical simulations.

2 Mathematical model

Aquifer dynamics and land subsidence caused by water withdrawal is theoretically described by the coupled process involving mechanics and flow in porous media (Biot, 1941). Uncoupling the flow field from the stress field is a usual assumption in groundwater hydrology. Gambolati et al. (2000) have shown that uncoupling has generally no measurable influence over any timescale of practical interest. Hence, in the present analysis a 1-way coupled approach is followed where the pore pressure variation is first simulated by a 3-D groundwater flow model, and then the land settlement is predicted by a 3-D geomechanical model with the pore pressure change used as an external distributed source of strength.

The aquifer hydrodynamics relies on the classical groundwater flow equation:

$$\nabla(k\nabla h) = S_s \frac{\partial h}{\partial t} + q \tag{1}$$

where k is the hydraulic permeability tensor, h is the hydraulic head change, t is time, q the source/sink and S_s the specific elastic storage defined as $S_s = \gamma(c_M + \varphi\beta)$ with γ the specific weight of water, c_M the oedometric compressibility of the porous medium, φ the porosity and β the volumetric fluid compressibility. A nonlinearity is present in Eq. (1) since S_s is a function of the vertical effective stress σ_z, hence h, via c_M. Equation (1) is solved in space by the finite element method using linear tetrahedral elements and in time by an implicit finite difference scheme, with the nonlinearity addressed by a Newton-like iterative method.

The incremental pore pressure variation $p = h \cdot \gamma$ induced by water pumping is implemented into the geomechanical model to compute the medium displacements. The equilibrium equations along the coordinate directions x, y and z in an isotropic medium can be written in terms of incremental soil displacement u (Verruijt, 1969):

Table 1. The values of the four parameters used in the simulations: depth H_a and permeability of the producing aquifer, thickness of the sandy (ΔS_s) and clayey (ΔS_c) layer, ratio between the oedometric compressibility in first and second loading cycle and ratio between the sandy and the clayey hydraulic conductivity.

H_a (m)	500	1000	2000
Permeability (mD)	200	100	50
ΔS_s (m)	10	50	90
ΔS_c (m)	45	25	5
C_r	3	5	8
R_k	1×10^{-2}	1×10^{-4}	1×10^{-6}

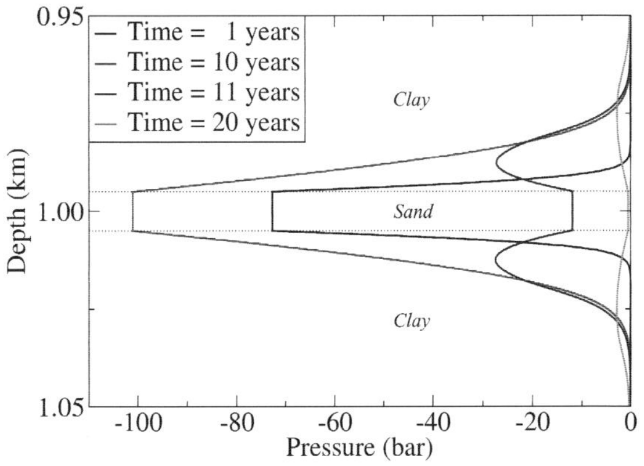

Figure 2. Pressure change predicted by the groundwater flow model along the symmetrical axis in the case with $H_a = 1000$ m, $\Delta S_c = 45$ m, $R_k = 1 \times 10^{-6}$ and $C_r = 8$.

$$\begin{aligned} G\nabla^2 u_x + (\lambda + G)\frac{\partial \varepsilon}{\partial x} &= \frac{\partial p}{\partial x} \\ G\nabla^2 u_y + (\lambda + G)\frac{\partial \varepsilon}{\partial y} &= \frac{\partial p}{\partial y} \\ G\nabla^2 u_z + (\lambda + G)\frac{\partial \varepsilon}{\partial z} &= \frac{\partial p}{\partial z} \end{aligned} \tag{2}$$

where u_x, u_y and u_z are the components of u along the coordinate directions, λ and G are the Lamé constant and shear modulus of porous medium, respectively, and ε is the volume strain. A nonlinearity is introduced in Eq. (2) since λ and G are functions of σ_z, and hence u, via c_M as $\lambda = \nu/[(1-\nu)c_M]$ and $G = (1-2\nu)/[2(1-\nu)c_M]$ with ν the Poisson ratio. Using data based on field and lab measurements the compressibility c_M is related to σ_z, with a different relationship in first and second loading cycle accounting for geomechanical hysteresis (Baù et al., 2002). Equation (2) is solved by the finite element method using the same tetrahedral elements as the groundwater flow model. The nonlinearity is solved by a Newton-like iterative method.

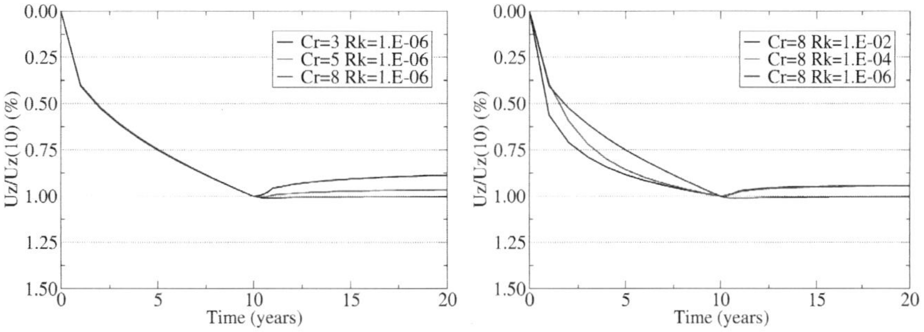

Figure 3. Maximum vertical displacement U_z normalized with respect to the displacement at the end of the pumping period (10 years) $U_z(10)$ in the case with $H_a = 500$ m and $\Delta S_s = 10$ m. In the left panel $R_k = 1 \times 10^{-6}$ and $C_r = 3$, 5 and 8. In the right panel $C_r = 8$ and $R_k = 1 \times 10^{-2}$, 1×10^{-4} and 1×10^{-6}.

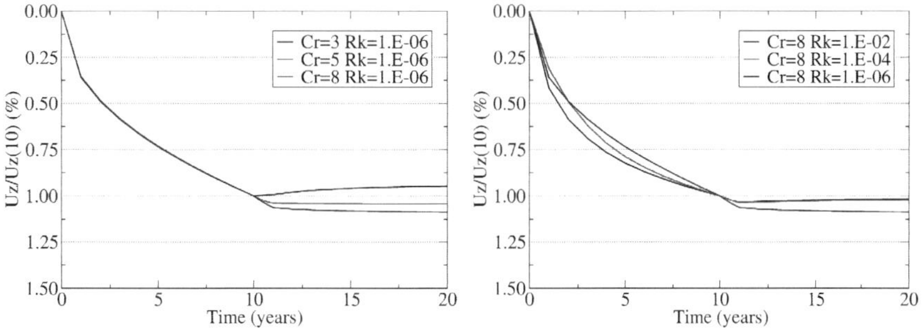

Figure 4. Maximum vertical displacement U_z normalized with respect to the displacement at the end of the pumping period (10 years) $U_z(10)$ in the case with $H_a = 1000$ m and $\Delta S_s = 10$ m. In the left panel $R_k = 1 \times 10^{-6}$ and $C_r = 3$, 5 and 8. In the right panel $C_r = 8$ and $R_k = 1 \times 10^{-2}$, 1×10^{-4} and 1×10^{-6}.

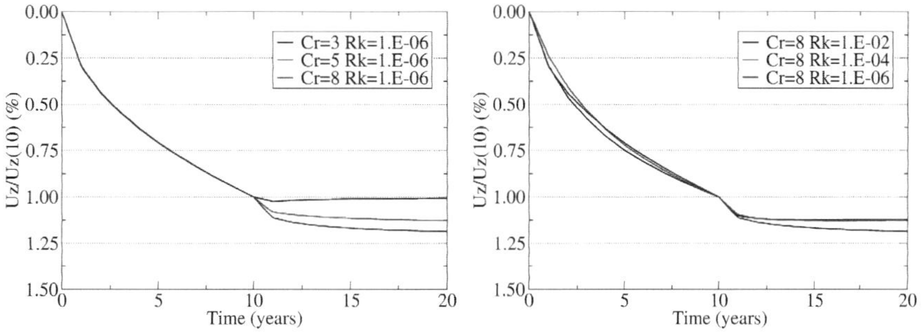

Figure 5. Maximum vertical displacement U_z normalized with respect to the displacement at the end of the pumping period (10 years) $U_z(10)$ in the case with $H_a = 2000$ m and $\Delta S_s = 10$ m. In the left panel $R_k = 1 \times 10^{-6}$ and $C_r = 3$, 5 and 8. In the right panel $C_r = 8$ and $R_k = 1 \times 10^{-2}$, 1×10^{-4} and 1×10^{-6}.

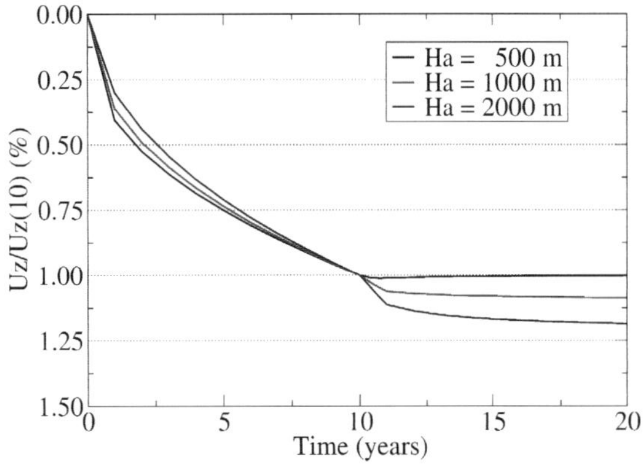

Figure 6. Maximum vertical displacement U_z normalized with respect to the displacement at the end of the pumping period (10 years) $U_z(10)$ in the case with $\Delta S_s = 10$ m $C_r = 8$, $R_k = 1 \times 10^{-6}$ and $H_a = 500$ m, 1000 m and 2000 m.

3 Modeling set-up

The geometry of the porous system used as test problem for the sensitivity analysis is shown in Fig. 1. The cylindrical domain of the geomechanical model is characterized by a radius $R = 10$ km and height $H = 5$ km, with zero displacements prescribed on the bottom and the outer boundaries. An aquifer is located at depth H_a and confined by two clayey layers. An accurate discretization of 1 m thick elements is used to simulate the pore pressure propagation at the sand-clay interface and the in less permeable units along the vertical direction. A zero pore pressure is prescribed on the side surface and the natural condition is prescribed on the top and bottom boundaries. The aquifer permeability decreases with depth as summarized in Table 1 with typical values of a Quaternary sedimentary basin.

A 10-year constant water withdrawal rate Q^* is prescribed from a 500 m-radius cylindrical aquifer volume to simulate the pore pressure decline experienced by several producing wells. The fluid-dynamic simulations last for 10 years after the cessation of the production. The value Q^* is calibrated to have a uniform 100 bar pressure decline in the produced volume at the end of the withdrawal period.

The sensitivity analysis takes into account four parameters:

- the depth H_a of the sandy layer;
- the ratio between the sandy and the clayey layer thickness, ΔS_s and ΔS_crespectively, with $\Delta S_s + 2\Delta S_c = 100$ m;
- the ratio C_r between the oedometric compressibility in first *e* second loading cycle;
- the ratio R_k between the sandy and the clayey layers permeability.

The values of the four parameters used in the simulations are summarized in Table 1.

4 Numerical results

Figure 2 shows the pore pressure variation along the symmetry axis for the case with $H_a = 1000$ m, $\Delta S_c = 45$ m, $R_k = 1 \times 10^{-6}$ and $C_r = 8$: the pore pressure recovery in the clay layers is significantly delayed relative to the aquifer. The pore pressure in the aquitards may still decline after the shutdown of the producing wells, possibly causing a delayed land subsidence.

The results of the geomechanical model are presented in terms of the maximum vertical land displacement normalized with the respect to the value at the end of the water abstraction ($t = 10$ years). A normalized displacement larger than 1 highlights a delay between the pore pressure in the aquifer and the land settlement. The results obtained for the investigated H_a values and $\Delta S_s = 10$ m are shown in Figs. 3–6.

Note that the sudden increase of the aquifer stiffness at the well shutdown generates an abrupt increase of land subsidence although the pore pressure recovers in the aquifer. This 3-D deformation process, which is mainly accounted for the Poisson ratio and, subordinately, the depth of the depleted layer, was already pointed out by Ferronato et al. (2001).

Small values of R_k and ΔS_s and large values of H_a and C_r enhance a delayed land settlement because the pore pressure propagates in the clay more slowly. For all the investigated H_a values the delay between the pore pressure recovery and the land settlement is maximum with $\Delta S_s = 10$ m, $C_r = 8$ and $R_k = 1 \times 10^{-6}$. With these configurations, a 9 and 20 % of the final land subsidence takes place after the cessation of withdrawal for $H_a = 1000$ and 2000 m, respectively. With $H_a = 500$ m subsidence is almost unvaried after the well closure.

5 Conclusions

A set of numerical simulations are performed using a 3-D 1-way coupled groundwater flow and geomechanical model to investigate the possible contribution of clayey inter-layers to the delayed land subsidence measured above deep pumped aquifers.

The parameters addressed by the analysis include: aquifer depth H_a, thickness of the sand ΔS_s and of the clay ΔS_c units, ratio between the oedometric compressibility in first *e* second loading cycle C_r, and ratio between the aquifer and the aquitard permeability R_k. The range of variability of these parameters is typical of a Quaternary sedimentary basin.

The results show that the delay between the pore pressure variation and land settlement can partly be accounted for by the delayed pressure propagation in the clayey layers confining the aquifer. The delay increases if:

- the aquifer depth H_a increases;
- the sandy layer thickness ΔS_s decreases relative to the aquitard thickness;
- the ratio between the oedometric compressibility in first and second loading cycle C_r rises;
- the ratio between the sandy and the clayey layers permeability R_k decreases.

Within the investigated configurations the maximum subsidence occurring after the well shutdown amounts to 20 % and therefore must be taken into account for reliable subsidence evaluation and prediction.

Coupling this process with a visco-plastic mechanical behaviour of the depleted layers might cause an enhancement of the land subsidence delay.

References

Baù, D., Ferronato, M., Gambolati, G., and Teatini, P.: Basin-scale compressibility of the Northern Adriatic by the radioactive marker technique, Geotechnique, 52, 605–616, 2002.

Biot, M. A.: General theory of three-dimensional consolidation, J. Appl. Phys., 12, 155–164, 1941.

Chang, C., Mallman, E., and Zoback, M.: Time-dependent subsidence associated with drainage-induced compaction in Gulf of Mexico shales bounding a severely depleted gas reservoir, Am. Assoc. Petr. Geol. B., 98, 1145–1159, 2014.

Ferronato, M., Gambolati, G., Teatini, P., and Baù, D.: Land surface uplift above overconsolidated compacting reservoirs, Int. J. Solids Struct., 38, 8155–8169, 2001.

Galloway, D. L. and Sneed, M.: Analysis and simulation of regional subsidence accompanying groundwater abstraction and compaction of susceptible aquifer systems in the USA, Boletin de la Sociedad Geologica Mexicana, 65, 123–136, 2013.

Gambolati, G., Teatini, P., Baù, D., and Ferronato, M.: Importance of poroelastic coupling in dynamically active aquifers of the Po river basin, Italy, Water Resour. Res., 36, 2443–2459, 2000.

Gambolati, G., Ferronato, M., Teatini, P., Deidda, R., and Lecca, G.: Finite element analysis of land subsidence above depleted reservoirs with the pore pressure gradient and total stress formulations, International Journal of Numerical and Analytical Method in Geomechanics, 25, 307–327, 2001.

Hettema, M., Papamichos, E., and Schutjens, P.: Subsidence Delay: Field Observations and Analysis, Oil Gas Sci. Technol., 57, 443–458, 2002.

Teatini, P., Ferronato, M., and Gambolati, G.: Grounwater pumpimg and land subsidence in the Emilia-Romagna coastland, Italy: Modeling the past occurrence and future trend, Water Resour. Res., 42, W01406, doi:10.1029/2005WR004242, 2006.

Verruijt, A.: Elastic storage of aquifers, in: Flow Through Porous Media, edited by: De Wiest, R., Elsevier, NY, 331–376, 1969.

Wu, J., Shi, X., Ye, S., Xue, Y., Zhang, Y., Wei, Z., and Fang, Z.: Numerical Simulation of Viscoelastoplastic Land Subsidence due to Groundwater verdrafting in Shanghai, China, J. Hydrol. Eng., 15, 223–236, 2010.

Proc. IAHS, 372, 525–528, 2015
proc-iahs.net/372/525/2015/
doi:10.5194/piahs-372-525-2015

Performance of pile foundation for the civil infrastructure of high speed rail in severe ground subsidence area

H. W. Yang

Taiwan High Speed Rail Corporation, Taiwan

Correspondence to: H. W. Yang (hw_yang@thsrc.com.tw)

Published: 12 November 2015

Abstract. In this study, the performance of pile foundation is assessed for the 30 km long viaduct bridge structure seating in the ground subsidence area in the central part of Taiwan. The focus of this paper is placed on the settlement behaviour of a continuous 3-span R. C. viaduct bridge supported on piles adjacent to highway embankment. Monitoring data accumulated over the last 12 years indicate that the observed pace of the settlement of the viaduct structure in other sections matches that of the ground. However, the bridge piers adjacent to the embankment have been suffering an additional approximately 1 cm settlement every year since the completion of the embankment construction. It was believed that the piers may suffer from enormous negative skin friction owing to the surcharge from the embankment and groundwater depression. This paper first summarizes the settlement analysis results to quantify the settlement of pile due to regional ground subsidence and the combined effects from ground water fluctuation and embankment surcharge loading. Accordingly, a loading path on P′-q stress space is formulated to illustrate the loading variation subject to the combined loading effects that can explain why the combined effect becoming significant on settlement control for civil infrastructure in ground subsidence area.

1 Introduction

Taiwan, like many other countries, has been suffering from land subsidence problems resulting from the over-extraction of groundwater over the last four decades. Around 30 km of the Taiwan High Speed Rail (HSR) crosses over the area with the most severe ground subsidence by viaduct bridge seating on pile foundations. Monitoring records have shown that the observed settlements of the viaduct structure match well with those of the ground. However, excluding the regional ground subsidence effect, one continuous 3-span viaduct bridge crossing over a highway embankment has been suffering at least an additional 1 cm annual settlement since the completion of its embankment construction in 2003.

The bridge in question is a continuous 3-span R.C. viaduct bridge supported on reverse circulation drilled (RCD) piles. It was built in 2002, and shortly after its completion, another earth-filled embankment for a highway was placed underneath its main span. Since 2003, the accumulated settlement of piers adjacent to the highway embankment is 575 mm, and the differential settlements between piers are near 30 mm. The pier settlement is considered attributable to the regional ground subsidence and surcharge effect from embankment. This study aims to quantify the pile settlement behaviours subject to dead load from the viaduct structure, surcharge from the adjacent embankment, seasonal fluctuation of groundwater level, and regional ground subsidence.

2 Site and structures

2.1 Soil and groundwater

The site is located at the toe of the southern Chuoshui River Alluvial Fan (CRAF). Based on the conceptual hydrogeological model (1), the sediments in CRAF contain unconsolidated sand, gravel, silt, and clay, with a total depth of over 2 km. The aquifers near the head of CRAF mainly contain gravel and coarse sand that become finer towards the toe of CRAF. The aquifers in the toe of the southern CRAF contain highly compressible granular material with loosely scattered clay (Yang et al., 2007).

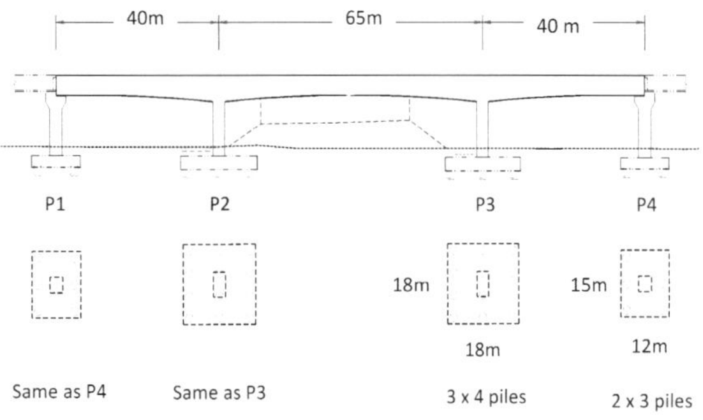

Figure 1. Cross section of viaduct and pile cap plan.

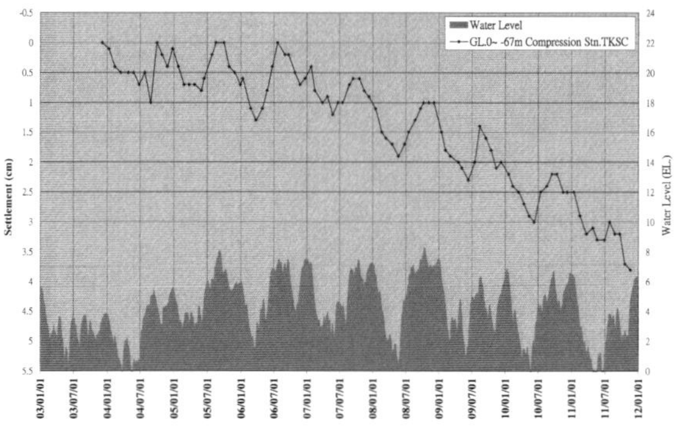

Figure 2. Groundwater level fluctuation and compression of soil.

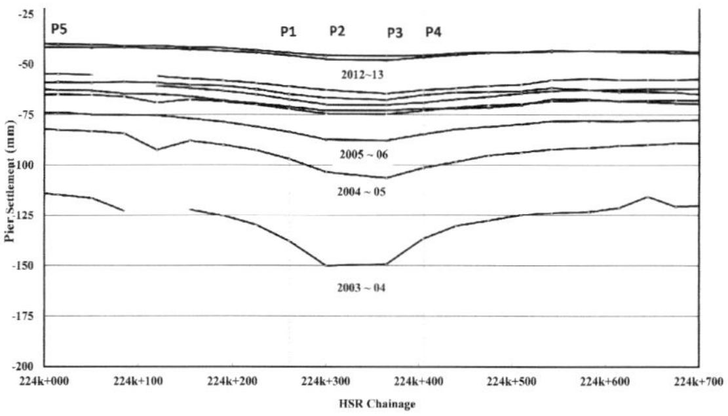

Figure 3. Settlement trough with the center of embankment.

The groundwater table at the site was 2–4 m below the ground surface. In the sandy layer at 50–160 m depth, called the "Second Aquifer," a lower piezometric head was observed owing to the over-extraction of groundwater from this aquifer. A higher piezometric head can be seen annually in the wet season from April to November. In the dry season, the groundwater pressure head declines by ~6 m.

2.2 Viaduct bridge & earth-filled embankment

The continuous 3-span viaduct bridge is a "π"-shaped structure. The central span of the bridge is 65 m and the two outer spans are 40 m. The superstructure was constructed using the balanced-cantilever method from the two main piers. The last and center segments of the bridge were completed in April 2002. Each bridge pier is supported by an individual pile group. Figure 1 shows the cross section of the bridge and the plan of the pile groups. RCD piles of 2 m diameter were used, with pile lengths of 63.5 m for P2 and P3 and a slightly shorter length of 53.4–59.5 m for P1 and P4. The remainder of the viaduct bridge extending from the two end piers P1 and P4 had simple supported beams with piers on similar piles to 62-m depth. All piles at this site are 100 % friction piles supporting only the design load, which is 3427 kN/pile for P2 and P3 and 4562 kN/pile for P1 and P4.

The earth-filled embankment is 6 ± 0.5 m high with 53.6 m width at the bottom, and it has slope angles of $36° \pm 2°$. The embankment was filled in two stages—first to a height of 2.55 m from 1998 to 2001 and then to its full height from January 2003 to November 2004. The bridge's pile foundations were constructed during the period between these two stages.

3 Characteristics of regional ground subsidence near site

The study site is located in the most severe ground subsidence area in CRAF in Yunlin County. The piezometric head in the aquifers usually drops to its lowest level and gradually rises to the peak in the following rainy season. The cyclic fluctuation of groundwater pressure changes the effective stress and inevitably leads to the compaction of the aquifer and, in turn, land subsidence. In 2003, the Stn. TKSC monitoring station was established by the government included a GPS station, a 300 m-long "sondex" tube type multilevel compaction monitoring well, groundwater monitoring well, and benchmark for leveling survey. Based on the data from Stn. TKSC, Yang (2013) summarized the characteristics of ground subsidence and groundwater variations, particularly within 67.4 m depth, as follows.

Annual average subsidence of ground below GL.-300 m was rather constant at 25 mm yr^{-1}.

Total ground surface settlement from 2003 to 2011 was 608 mm, of which 570 and 38 mm was from soil compression below and above 67.4 m depth, respectively.

As the groundwater pressure has been dropping progressively in recent years, the total compression of the ground became larger than its total upheaval, resulting in an accumulated compression of 38 mm in the 0–67.4 m soil layer from 2003 to 2011. Figure 2 demonstrates the variation of groundwater level and compression of soil layer.

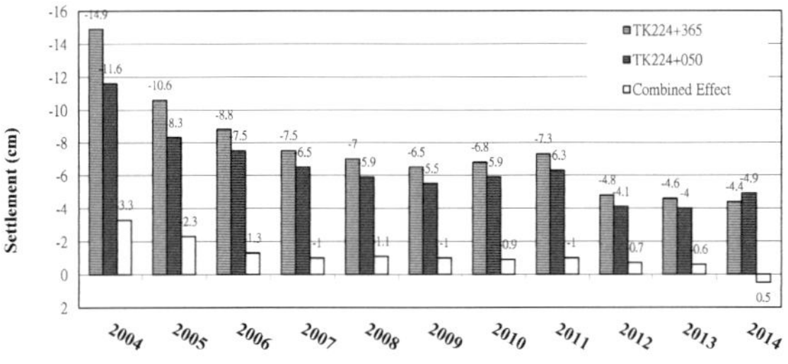

Figure 4. Pile settlment due to combined effect.

4 Settlement of viaduct bridge

Starting from October 2003, settlement markers fixed on all piers were monitored annually. These four piers as indicated in Fig. 1 are categorized as the main piers P2 and P3, which are adjacent to the embankment, and the outer piers P1 and P4. The settlement rates of the piers at each side of the embankment are plotted in Fig. 3. It is observed that a settlement trough with the center of the embankment has been formed since 2004. Five piers at each side of the embankment are within the area of this settlement trough. These ten piers appear to be affected by the surcharge loading from the embankment. The main piers suffer the highest settlement and simultaneously take the highest surcharge loading from the embankment. The surcharge loading from the embankment could have caused ground compression next to the foundation piles, resulting in downdrag force or negative skin friction acting on the foundation piles.

Two piers, P5 and P3, are selected to comparatively study the pile settlement behaviour. Both piers are located in an area with the same background of ground subsidence. However, P5 is 260 m distant from P3 and is located outside the settlement trough shown in Fig. 3, which can represent the typical pile foundation subject to the effect of ground subsidence only. The monitoring results of multilevel compaction in the monitoring well at Stn. TKSC represent the natural ground subsidence at various soil depths. Pier P5 settled 707 mm from 2003 to 2014. A comparison of the ground subsidence below the pile tip level and pier settlement showed that the latter was only 3 mm larger than the former. Similarly, under the same natural condition but with an additional surcharge load next to P3, P3 settled 132 mm more than the ground subsidence at pile tip level. Figure 4 shows a comparison of the settlement history of P3 and P5; both piers show similar annual settlement patterns, with larger differences in the first 3 years. Since 2007, a constant deviation rate of 10 mm yr^{-1} was observed. The behaviour of P3 can be attributed to the combined effect of surcharge load and cyclic groundwater lowering

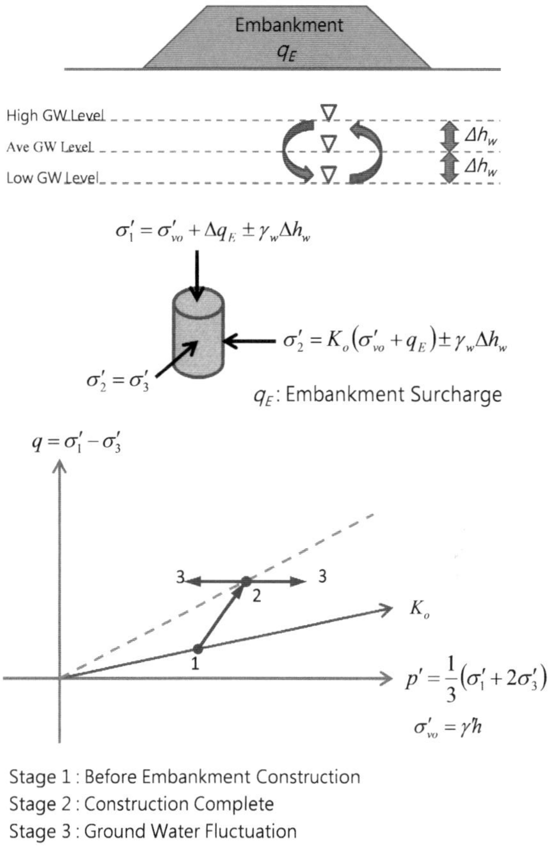

Figure 5. Stress path on P′-q space for loadings of combined effect.

5 Stress path of the loading for combined effect

Figure 5 is the stress path on P′-q space considering surcharge load from embankment and groundwater fluctuation It demonstrates the stress path of soil layers in three stages, namely prior to embankment construction (stage 1), embankment construction complete (stage 2), and ground subject to the existing embankment with seasonal ground water level fluctuation (stage 3). Within the aquifer, the principle effective sstress σ'_1, $\sigma'_{vo} + q_E \pm \gamma_w \Delta h_w$, changes with cyclic groundwater lowering. Combined with the surcharge loading increment q_E, there arises a zone with effective stress $\sigma'_{vo} + \Delta q_E \pm \gamma_w \Delta h_w$. In the dry season, while the groundwater level is relatively low, the combined effective stress $\sigma'_{vo} + \Delta q_E$ increases by $\gamma_w \Delta h_w$, and it decreases by the same amount in the wet season. Hence, within the zone of combined effects, the ground may settle more in the dry season and heave less in the wet season compared to natural ground without the surcharge effect. For soils at greater depth, where q_E is low, the soil heave in the wet season may be limited by the thick overburden. For soils at shallow depth, where

q_E is relatively high, ground settlement/upheaval will be predominantly governed by q_E. Yang et al. (2013) reported that at this site, fluctuation of ground movements was only observed in the second aquifer, particularly in the sand and silt layers. For soils below 230 m, the monitored settlement-time curve was smooth with little fluctuation, indicating that very limited soil heave occurred at this depth. Xi et al. (2012) recently performed a series of laboratory model tests to study the deformation characteristics of remolded silt specimens under the coupling effect of cyclic groundwater fluctuation and footing pressure. Larger settlement occurred when the groundwater was in the lowering stage with lighter footing pressure.

Based on the studies above, pier settlement of pier P5 is outside of the affected zone and free from the combined effect. On the other hand, the pier P3 and P2 is under the influence of surcharge loading from the embankment. The settlement difference between P3 and P5 should be resulted from the loading combined effects.

6 Conclusions

The study of this complicated problem is still on-going, and further measurements of ground settlement and pier settlement together with groundwater level changes at the site are being performed. Thus far, the following tentative conclusions can be derived from this study.

The regional ground subsidence occurred below the pile tip has changed the vertical alignment and elevation of the viaduct structure. Only the portion of ground subsidence contributed by soil compression above the pile tip could affect the performance of pile structure due to relative movement and possible negative skin friction on the working piles.

A combined effect occurred under the embankment surcharge and groundwater fluctuation and lowering is deemed to have increased the relative movement between ground and working pile structure.

The study demonstrates that a total of 12.4 cm settlement on piles for pier P3 during the period from 2003 to 2011 due to the combined effect from embankment surcharge and ground water fluctuation.

References

Yang, H. W. and Hseih, J.: Engineering Assessment of Ground Subsidence to Railway Infrastructure, Proc. of 7th International Symposium of Land Subsidence, Shanghai, China, 2007.

Yang, H. W., Kao, T., Yu, C., Ho, S. K., and Woo, S. M.: Settlement Behaviours of Structure and Fluctuation of Ground Water in Ground Subsidence Area, Proc. of 15th Conf. on Current Research in Geotechnical Engineering in Taiwan, 2013.

Xi, Y., Xuewen, L., Qingshan, M., and Yue, Q.: Test Study on Deformation Characteristics of Silt – Soil Foundation under Condition of Groundwater Fluctuation, Yangtze River, 43, 84–87, 2012.

Proc. IAHS, 372, 529–531, 2015
proc-iahs.net/372/529/2015/
doi:10.5194/piahs-372-529-2015

Open Access
PIAHS

Tools to support maintenance strategies under soft soil conditions

J. W. M. Lambert[1], J. J. van Meerten[1], M. P. Woning[1], M. J. Eijbersen[2], and M. Huber[1]

[1]Deltares, Delft, the Netherlands
[2]CROW, Ede, the Netherlands

Correspondence to: J. W. M. Lambert (john.lambert@deltares.nl)

Published: 12 November 2015

Abstract. Costs for maintenance of infrastructure in municipalities with soft soil underground conditions, are estimated to be almost 40 % higher than in others. As a result, these municipalities meet financial problems that cause overdue maintenance. In some cases municipalities are even afraid to be unable to offer a minimum service level in future.

In common, traditional practice, roads and sewerage systems have been constructed in trenches that consist of sandy material that replaces the upper meters of the soft soil. Under influence of its weight, this causes accelerated settlements of the construction. A number of alternative constructions have been developed, e.g. using light-weight materials to limit settlement velocity.

In order to limit future maintenance costs, improvement of maintenance strategies is desired. Tools have been and will be developed to support municipalities in improving their maintenance strategies and save money by doing that.

A model (BALANS) that weighs the attractiveness of alternative solutions under different soil, environmental and economic circumstances, will be presented.

1 Introduction

The presence of soft soils near the surface is common in many Delta areas in the world. These areas are attractive for agricultural activities because of their fertility, but also population concentrates in these zones because of their favourable transport situation near the sea and near the mouth of rivers.

As a result, the soils in Delta eras are commonly intensively used for agriculture, housing, industrial zones and transport infrastructure. These different activities require different, sometimes conflicting conditions. Especially water management to optimize agricultural production and groundwater abstraction in order to supply inhabitants with drinking water and for industrial water supply, may lead to subsidence and damage to other interests, such as to houses, buildings and line infrastructure such as railways, roads and pipelines.

In the Netherlands, soft soils cover a big part of the country and the mentioned activities concentrate in these soft soil areas. The thickness of the Holocene peat and clay layers is variable and may reach up to 15 m of extremely soft soil.

Municipalities in these regions suffer high costs for maintenance of their infrastructure. Alternative constructions have been developed and new alternatives will be developed. In this paper, a tool to support municipalities and other infrastructure managers in their design and maintenance management is described. The tool has been developed by Deltares, Delft Cluster (a consortium of knowledge institutes in the Netherlands) and ISMH (Community of Soft Soil Municipalities)

2 Traditional solution

Municipal roads and sewerage systems were originally constructed by digging a trench into the soft soil, filling the trench with sand and then cover the trench with a road construction. After the pavement and foundation have settled more than about 0.40 m a reconstruction takes place by lifting the surface with sand again (Fig. 1).

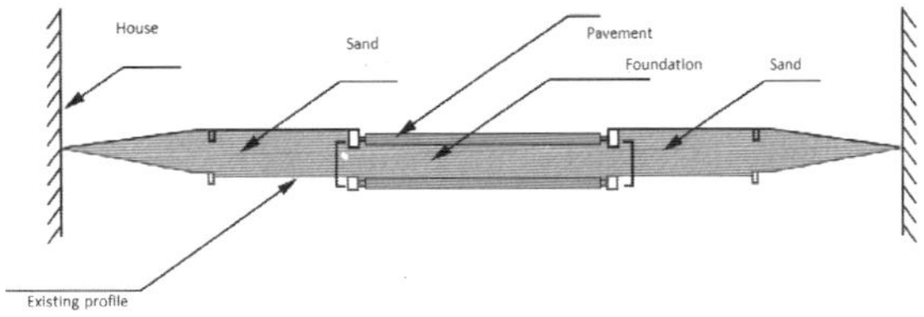

Figure 1. Traditional solution.

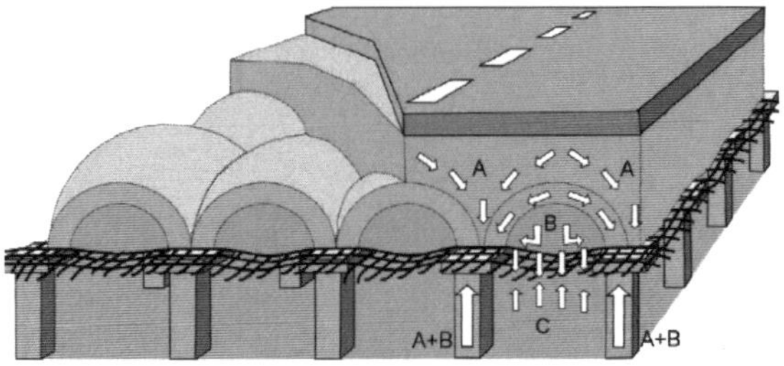

Figure 2. Load distribution in a reinforced piled embankment.

As the volumetric weight of sand (e.g. $19\,\mathrm{kN\,m^{-3}}$) is much higher than that of the excavated peat (e.g. $11\,\mathrm{kN\,m^{-3}}$) or clay, the underlying peat or clay layers are burdened and sink rapidly after road-construction and the infrastructure may need frequent maintenance and re-construction.

3 Alternative solutions

In the past decades many alternative road foundation constructions have been developed and have been applied. Especially light materials, such as Bims (volcanic porous rock), flugsand, and foamed glass have been applied, but also constructions consisting of light weight blocks or panels, such as EPS (polystyrene) and strengthened sludge. Also under piled constructions, such as the geosynthetic reinforced piled embankment, have been applied. A schematic cross-section of this solution is given in Fig. 2.

4 Tools for management support

Municipalities need support to design new roads (and sewerage systems) as well as long term maintenance and reconstruction strategies in order to save public money and to minimize environmental impact. A design tool (model BALANS) was developed to support municipalities to choose the system that fits best their circumstances and requirements. Also, a tool to support their management strategy is under construction. This tool will be based on InSAR-observations.

Besides this, a pilot project in the village Kamerik (Municipality Woerden) is under construction. In the pilot, four different foundation constructions will be constructed and monitored in order to develop objective methods to compare the performance of the different constructions. In this paper, the BALANS tool and the tool to support the management strategy are described

5 Balans

Balans (Dutch for: Balance) enables to assess the financial and environmental consequences of different alternative road constructions, taking into consideration geotechnical and hydrological circumstances, as well as the governing boundary conditions, such as cables, pipelines, sewerage systems, houses and their foundations, hindrance during construction, etc.

In the first place, the present geological, geotechnical (in terms of settlement parameters) and geohydrological (groundwater levels) must be imported in the model.

Then, the chosen construction has to be imported in terms of: excavation level, planned future construction level and volumetric weight of all construction materials. Balans is coupled with a material database that derives standard costs and environmental costs of the applied materials and proposes the volumetric weight to be applied (this can be adapted manually).

Given the proposed construction and materials, the program predicts the settlements and derives from these expected costs for maintenance and reconstruction during a user defined lifetime (e.g. 50 or 60 years). These costs are presented as Net Contant Value (NCV), taking into account a user defined interest percentage. Figure 3 gives an example of the calculated settlements and proposed maintenance periods.

BALANS is also coupled with a database, that contains LCA's (Life Cycle Analysis) of a number of materials; import of the amounts of the used materials enables to calculate the environmental costs of the proposed construction, also in terms of NCV.

Besides costs, other circumstances may influence the optimal choice. These circumstances are very much depending of local conditions, such as the presence of houses and their foundations, requirements in terms of availability of the road and other boundary conditions, such as the presence of sewerage systems, subsoil cables and pipelines, possible risks etc. The user may define the weight of each of these mentioned topics.

BALANS is able to compare two or more possible constructions at once and presents the results in a diagram. Figure 4 shows such a diagram, which gives insight how decisions may depend from the weight that is given to each of the mentioned criteria.

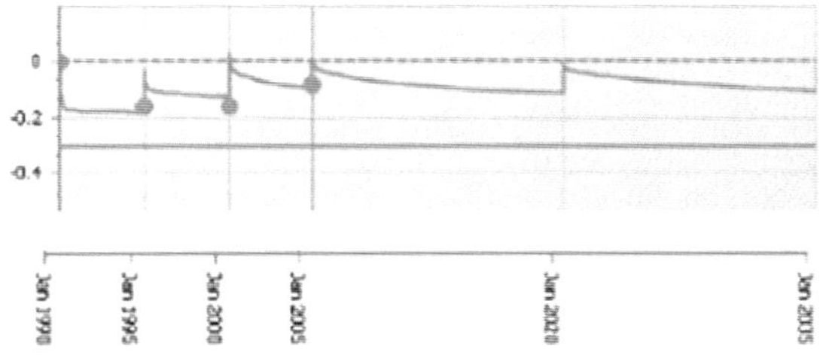

Figure 3. Predicted settlements and (green dots) predicted maintenance moments.

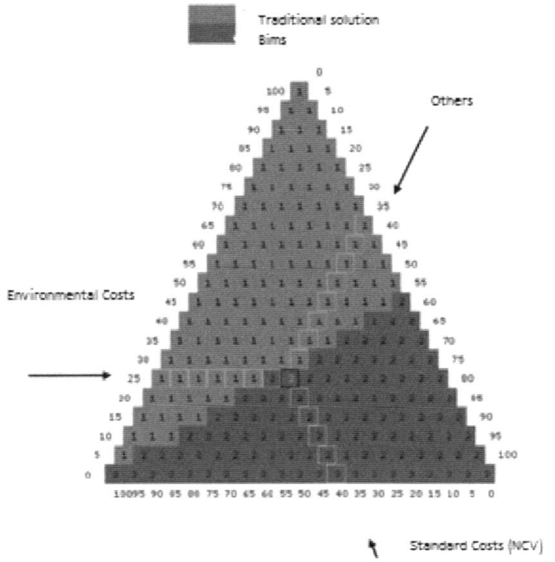

Figure 4. Output from BALANS, comparison of a traditional solution and a Bims construction.

6 Test site Woerden (Kockengen)

However many different alternative constructions and materials have been developed and applied in the Netherlands, there is still much misunderstanding and confusion about the applicability of the different solutions. The main reason is that no objective method exists to evaluate and, moreover, reconstructions after execution are rarely monitored.

At this moment, the municipality of Woerden is preparing the construction of a test site to compare foundations constructed with bims, foamed glass, sand and reinforced piled embankment.

The subsoil is consisting of very soft peat until a depth of about 6 m below surface. The roads were constructed in the seventies of the 20th century and have been settling about 0.50 m since then. As adjacent houses have been constructed on concrete piles, the difference in height between the houses and the road becomes unacceptable. Moreover, the road surface is approaching the groundwater level.

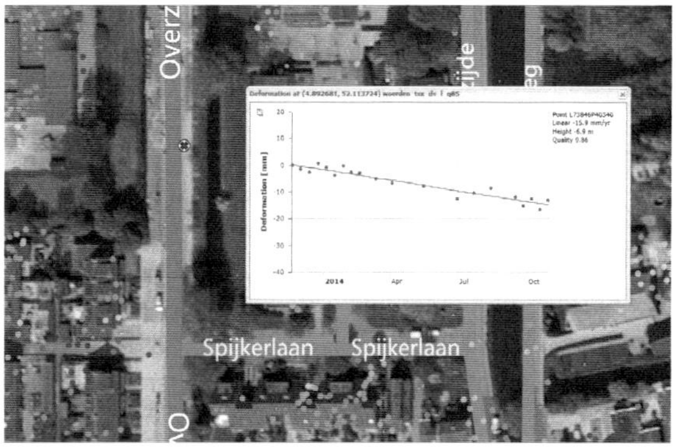

Figure 5. Example of present settlement observations by InSAR (SkyGEO).

In the test site, the mentioned foundation constructions will be applied. During construction all activities and used materials will accurately be monitored, while – after construction – the future settlements of each construction will be monitored by means of InSAR, as well as by traditional levelling methods.

Figure 5 shows an example of the present settlements, observed by means of InSAR near the test location.

7 Conclusions

A model has been developed to compare possible alternative foundation constructions for municipal roads under soft soil conditions. The model takes into account not only the present construction costs, but also accounts with the expected maintenance and reconstruction costs in future. Difference is made between real costs, environmental costs and the influence of other boundary conditions. A test site, to compare the real future performance of different methods is under preparation and will be executed in Kamerik (Municipality of Woerden); besides traditional levelling, InSAR techniques will be applied for this comparison.

In our opinion, the discussed model is a useful tool to evaluate alternatives for road foundations under soft soil conditions. It makes clear that the cheapest construction may be the most expensive on the long term.

References

Deltares: Leidraad Balans (only available in Dutch).

Proc. IAHS, 372, 533–538, 2015
proc-iahs.net/372/533/2015/
doi:10.5194/piahs-372-533-2015

Open Access
PIAHS

Land subsidence, seismicity and pore pressure monitoring: the new requirements for the future development of oil and gas fields in Italy

P. Macini[1], E. Mesini[1], L. Panei[2], and F. Terlizzese[2]

[1]University of Bologna, School of Engineering and Architecture, Bologna, Italy
[2]Italian Ministry for Economic Development, Rome, Italy

Correspondence to: P. Macini (paolo.macini@unibo.it)

Published: 12 November 2015

Abstract. The Emilia earthquake of 2012 (Italy) stimulated a controversial debate concerning the possibility that the event could have been induced or triggered by underground fluids production. The public discussion led the Italian Government to issue a protocol of guidelines for the monitoring of microseismic activity, ground deformation and reservoir pore pressure. The guidelines will be put into operation as soon as practicable when licensing is being considered, and all data provided by mining operators must be made available to the relevant Authorities. The implementation of an outreach and communication program to local residents and administrative authorities is prescribed, so that the civil and scientific community at large can gain confidence that operations are being managed optimally.

1 Introduction

After the well-known "Emilia" seismic sequence of May 2012 (Pizzi and Scisciani, 2012), located in the densely populated Emilia-Romagna Region (Italy), causing 27 casualties and relevant damages to buildings and industrial infrastructures, doubts have been raised on the possibility that the above seismic sequence could have been induced or triggered by underground fluids production or injection.

The area under investigation (and more extensively the whole Po Plain area) is well known for the large number of oil and gas fields, discovered since the early 1950's. These fields significantly boosted the Italian industrial reconstruction after World War II and the area was intensely explored until the late 1980's. Today, hydrocarbon exploration and production in this area is in its final stage. The Po plain is a mature petroleum province and it is becoming increasingly marginal, not the least because of the strong environmental limitations and the low social acceptability of management practices and conditions in the field of fossil fuels.

In particular, near the measured epicenters of the two main shocks (in the order of several tens of km) there are some oil and gas fields, an underground gas storage inside a depleted gas field and a closed-loop water-dominated medium-temperature geothermal field.

In the view of the above, the Italian Department of Civil Protection, following the request of the Government of Emilia-Romagna Region, appointed an independent International Commission for evaluating the possible relationships between hydrocarbon exploration and production and a marked increase of seismicity in the area hit by the May 2012 earthquakes. In particular, the Commission, named ICHESE (International Commission on Hydrocarbon Exploration and Seismicity in the Emilia region) reviewed and evaluated all available geological, geophysical and fluid production information in order to assess the likelihood of the above concerns. In particular, the final report (ICHESE, 2015), issued in April 2014, focused on three active production sites, located close to the epicenters of the seismic sequence: the geothermal field (Casaglia, geothermal production lease "Ferrara"), the underground gas storage field (Minerbio, gas storage lease "Minerbio Stoccaggio") and an oilfield in its final stage of production (Cavone, hydrocarbon production lease "Mirandola").

ICHESE report concluded: "it is highly unlikely that the activities of hydrocarbon exploitation and the geothermal activity have produced sufficient stress change to generate an "induced" seismic event. While it cannot constitute proof, the current state of knowledge and all the processed and interpreted information does not allow the ruling out of the possibility that *the actions involved in hydrocarbon exploitation in the oilfield* may have contributed to "trigger" the Emilia seismic activity" (author's italics). ICHESE report eventually suggests some recommendations to investigate the issue in more details. Amongst these, we recall the following: implementation of a coupled hydro-mechanical model of the area and monitoring of seismic activity, ground deformation and pore pressure. ICHESE report originated a sharp debate in the media, mining operators, regulating and administrative authorities, local residents and the scientific community. The oil and gas industry reacted very responsibly, granting a positive and transparent cooperation with the Authority responsible for the control (the Dept. of Energy and Mineral Resources of the Italian Ministry for Economic Development), the Regional Government, the scientific community and the public at large.

In particular, the Italian oil and gas industry (the supposedly guilty party indicated by the ICHESE report) reacted immediately, by promoting a collaboration agreement between the Ministry for Economic Development (MISE), the Emilia-Romagna Regional Government and the oil company "Società Padana Energia" (owner of the Mirandola lease and of Cavone oil plant), together with Assomineraria (the Italian Petroleum and Mining Industry Association). The aim of the project was the immediate development of monitoring and research at the "Cavone" oilfield, in order to provide a complete answer on the questions remained open and unresolved in the ICHESE report. The project is known as "Cavone Laboratory", whose activities are detailed in the last chapter of this paper. The results of the activities, still ongoing, are continuously updated and available online at http://labcavone.it/ and http://www.assominerariacavone.org/. Transparency and accountability characterize these activities. Cavone Laboratory fully reflects these principles and on request make public the entire database of monitoring and modeling of the reservoir. Historical data and recent ones are available to the academic world, students and scholars who wish to analyze them.

In parallel, an updated, complete and extensive analysis of geological and geophysical data (together with the construction of a novel geomechanical model) of the Cavone reservoir and surrounding formations has been prepared (Astiz et al., 2015; Jha et al., 2014). The first step of the above activities concluded: "there is no physical reason to suspect that pressure changes at the hypocenters associated with production or injection activities at Cavone oilfield triggered the seismic sequence of May 2012". The above project produced also an updated fully coupled hydro-mechanical model of the Cavone reservoir, as recommended by the ICHESE report.

In the same period, another study was published by the Italian National Institute for the Environmental Protection and Research (ISPRA, 2015), concerning the possible relationship between induced/triggered seismicity and anthropogenic activities in Italy.

Notwithstanding all the above, ICHESE's conclusions and recommendations prompted MISE to issue (MISE, 2015) a protocol of guidelines for the monitoring of land subsidence, pore pressure and seismic activity. In particular, existing and future hydrocarbon and geothermal activities must be supported by high technology monitoring networks aimed to study the time evolution of the three essential aspects: microseismic activity, ground deformation and pore pressure. These should be put into operation as soon as practicable when licensing is being considered. Finally, it is necessary that all the relevant data provided by mining operators are made available to the authorities responsible for the control. It is critically important to implement also an outreach and communication program to local residents and administrative authorities, so that the civil and scientific community at large can gain confidence that operations are being managed optimally.

This paper details and illustrates the above guidelines, and analyzes the preliminary results of experimental monitoring activities performed at selected pilot test sites.

2 The new rules of the game

The abovementioned document (MISE, 2015) is a first formulation of the technical "Guidelines" for the implementation, management and control of the monitoring, as well as the decision-making methods in case of possible changes of the monitored parameters. These guidelines will be firstly tested on some pilot sites, chosen among the onshore leases that operate the injection of the produced water and selected underground gas storage leases (in most of which, both seismic and ground deformations monitoring networks are already in operation). The monitoring results will be published and shared with MISE and the Government of the Regions concerned with such activities. It is hoped that that this document will be re-evaluated after two years, on the following basis:

- the experience gained in the pilot sites representing different case studies concerning, among the other, carbonatic reservoirs that include injection activities;
- a feasibility study, including construction and operative costs.

The Guidelines aim to establish procedures and monitoring protocols including methods for the analysis of spatial and temporal parameters related to seismicity, ground deformation and pore pressure. In particular, seismic monitoring is addressed to identify and locate the seismicity in a control volume surrounding the anthropic activities, also with the aim of eventually distinguishing the natural seismicity from the one possibly caused by such activities. Monitoring should allow to track the evolution of seismicity in space, time and magnitude in order, if necessary, to modify or, under certain circumstances, to halt the activities. Ground deformation monitoring aims to identify possible surface deformation (related to the anthropic activities) and to study its spatial and temporal evolution. Pore pressure monitoring is addressed to estimate reservoir pressure through downhole measurements, possibly also with the support of interference test with neighboring wells, in order to validate the hydrodynamic model of the reservoir and to evaluate potential pressure variations in space and time.

In general, these Guidelines (after an appropriate revision) might also be adopted in the future for industrial activities such as dam construction, geothermal fluids production (conventional and closed-loop), enhanced geothermal systems (at present not performed in Italy), CO_2 geological storage (at present not performed in Italy), underground mining and tunneling.

The Guidelines recommends that the new monitoring infrastructures are designed also taking into account the principle of "upgrading the existing" and, therefore, to provide, after due evaluation, a possible improvement of infrastructures already operating to obtain the required performances.

Aside from the specifications of the monitoring network (see below), it is recommended that the mining operators prepare and make available, even in the light of geological and seismotectonic studies previously carried out, the following data and information, useful for the design of monitoring networks:

a. detailed geological cross sections (three at least);

b. seismic profiles (possibly in 3-D);

c. 3-D stratigraphic-structural model;

d. estimation (log or cores) of primary and secondary porosity along a reference well profile;

e. evaluation of rock compaction and of expected subsidence;

f. simulation of fluid migration nearby the reservoir;

g. location of active faults adjacent (within 3 km) or neighboring (within 15 km) to the reservoir;

h. general kinematic framework of the area;

i. evaluation of lithostatic stress changes;

j. evaluation of the possible volume around an injection well through which the injected fluids can migrate, and planning of the monitoring actions;

k. implementation of a geomechanical model, and its periodical updating.

It is also recommended to construct surface piezometric wells (possibly equipped with an extensometer), and to perform geoelectric or magnetotelluric surveys in order to assess resistivity anomalies and possible variation in time.

3 Seismic monitoring

The Guidelines firstly specify the geometrical definition of two distinct control volumes (referred as "internal" and "extended" control volume), i.e. the size of subsoil potentially interested by specific anthropic activities.

The internal control volume circumscribes the area of the activities on a local scale, but it must be wide enough to include the phenomena of induced seismicity. In this control volume, the seismic monitoring network must have the maximum sensitivity, to allow the use of the most advanced techniques for the localization of earthquakes and the reconstruction of seismic velocity models and, if data are adequate, to track the possible migration of seismicity. The seismic monitoring network must strictly comply with the following requirements:

- in the internal control volume, detect and locate earthquakes in the range of local Magnitude (M_L) in between 0 and 1 ($0 \leq M_L \leq 1$) and with uncertainty in locating the hypocenter of some hundreds of meters;

- in the extended control volume, improve of about 1 unit the level of the magnitude of completeness of records kept by the National or Regional seismic monitoring networks implemented for purposes of civil protection at National or Regional level (the monitoring network must be able to locate the hypocenter with an uncertainty of about 1 km);

- Determine the acceleration and velocity of ground motion caused by any earthquake at the measurement point;

- Integrate accurately with the existing monitoring networks (National, Regional and Local) in order to improve the accuracy and completeness of the measurements.

Concerning the timing, it is recommended that the seismic monitoring will have to start at least one year before the start of production (or storage) operations, in order to verify and measure the natural background in "undisturbed" conditions; it will operate along all the production stage and continue for at least one year after the conclusion of the activities.

4 Ground deformation monitoring

The Guidelines recommend that the monitoring must cover the surface area corresponding to internal and extended control volume, complying with the following requirements:

- it must provide a detailed deformations framework, estimated by means of Earth observation satellite, i.e. interferometric (InSAR) measurements performed during the past 15 to 20 years (minimum during the last 10 years). The accuracy must be of the order of 5 to 10 mm (for InSAR LOS measurements) and about 1 to 2 mm yr^{-1} for the estimates of deformation rates.
- the above framework must be updated by InSAR measurements repeated 1 to 4 times a year and must continue for at least three years after the conclusion of production (or storage) operations.
- the deformation obtained by InSAR must be integrated and complemented with additional measurements acquired by a continuous GPS network.
- the Controller should evaluate the opportunity to integrate InSAR and GPS information with precision spirit leveling measurements. The precision spirit leveling must be carried out every 2 or 3 years.

The complete results of the ground deformation measurements will be described and analyzed in dedicated reports, to be published indicatively every 3 to 12 months (recommended: 6 months).

5 Pore pressure monitoring

The Guidelines suggest performing pore pressure monitoring only for future field activities. Concerning the ongoing activities, the monitoring should be implemented only in the cases in which the existing wells can be fitted with the equipment to perform this measurement. The same holds also for interference test, which will be performed only if a suitable set of neighboring wells already exists.

As far as new storage or injection wells are concerned (excluding the production wells), pore pressure (or reservoir pressure) will be measured by downhole tools ("surface readout") inserted in the well completion, which provide a continuous measurement in real time. In some existing wells, it is possible to use "memory gauges" temporarily run at the bottom of the well by a wireline unit. Memory gauges allow the pressure measurement (not in real time) for a limited time (a few weeks): in fact, data are stored in an internal digital memory and downloaded only when the tool is retrieved at the surface.

Periodic static pressure profile will be carried out on specific wells of the field. Pore pressure in the volume of rock surrounding the monitored wells, which however is difficult to measure and highly uncertain, can be estimated by performing correlations with other monitoring wells, also through the application of mathematical models. The results of pore pressure measurements will be described and analyzed in dedicated reports, to be published indicatively twice a year.

6 The responsible of the monitoring, data analysis and publication

Today in Italy there is not a structure with specific expertise to monitor and identify possible effects of seismicity induced by human activities. Therefore, it is necessary to identify a third-party highly qualified technical-scientific subject (the Responsible of the Monitoring, RM) to support the Control Authority (MISE) in the management, analysis and use of monitoring data. The RM will be chosen among of one or more Universities or Research Institutions of proven expertise, possibly in consortium, or even with private structures. The Operator, with the advice of the RM, designs and constructs the monitoring networks (i.e. purchase, install or modify the existing equipment, and is responsible for their maintenance and management). The RM examine and evaluate the monitoring project (that must be finally approved by MISE), and performs data processing, analysis and interpretation.

An important aspect concerns the analysis and the procedures to steer the actions to be taken on the basis of the results of the monitoring. The working group that drafted the Guidelines concluded that in Italy today there are not proven methods of physical or statistical analysis that allow correlating induced seismicity within a maximum of one or two days, a reasonable interval needed to take decisions through applicable procedures. For this reason, it was chosen a "traffic light" approach based on parameters measured only within a limited control volume around the well and specific ground deformation measurements.

It is proposed to adopt, on an experimental basis, a decision-making system defined through four levels of activation, based on the assessment of the geodynamic model of the area and the overall framework of a series of parameters monitored in the control volume, including:

- variation of the number, frequency, magnitude and spatial distribution of seismic events;
- peak values of acceleration and ground motion speed;
- change of ground deformation rates;
- variation of pore pressure.

The four levels of activation are the following: (a) ordinary (green), (b) attention (yellow), (c) reduction of field activities (orange), (d) interruption of field activities (red).

This traffic light system provides procedures for taking actions associated with different activation levels, based on threshold values of the monitored parameters. At present, it seems not appropriate to adopt a decision model with automatic actions linked to precise threshold values. In fact, given the variability of the geological contexts, the depth and the technologies used for fluid production operations, the natural background seismicity and its depth, in general it is not possible to determine these threshold values uniquely.

Therefore, the decision-making system establishes the transition from one level to the other via a combined evaluation of RM, MISE, Regional Government and Operator, each one within their respective responsibilities. This must be undertaken as soon as originates a framework of monitored parameters "out of the ordinary", and the recognition of a possible correlation between the abovementioned changes and the ongoing field activities is detected.

The guidelines prescribes to adopt a traffic-light system with automatic actions (i.e. linked to precise threshold values) exclusively for the activities involving the injection of produced fluids (Table 1). For any transition from one level to the other is sufficient that one of the parameters (measured in the internal control volume) exceed the indicated threshold.

In order to ensure the effectiveness and transparency of the activities, it is planned the construction of a dedicated website of the MISE, with a special area containing specific sections for individual fields and leases, dedicated to the dissemination of information on existing and ongoing activities and data acquired during the monitoring.

Table 1. Threshold values of seismic parameters measured in the internal control volume utilized to establish the levels of activation: Maximum Magnitude (M_{max}), Peak Ground Acceleration (PGA), Peak Ground Velocity (PGV). For any transition from one level to the other is sufficient that one of these parameters exceed the indicated threshold.

Activation level	Traffic-light	M_{max}	PGA (% g)	PGV ($cm\,s^{-2}$)
0	Green	$M_{max} \le 1.5$	–	–
1	Yellow	$1.5 < M_{max} \le 2.2$	0.5	0.4
2	Orange	$2.2 < M_{max} \le 3.0$	2.4	1.9
3	Red	$M_{max} > 3.0$	6.7	5.8

7 Cavone monitoring laboratory

The Cavone Monitoring Laboratory originated from an "Agreement" between MISE Ministry, Emilia-Romagna Regional Government, oil company "Società Padana Energia", and Assomineraria to set up a pioneering center for research, monitoring and dissemination of local data. Having obtained the opinion of the Local Government Bodies, the Ministry and the Emilia-Romagna Regional Government asked the Società Padana Energia to make the Cavone oilfield available for a 90-day time period, in order to perform monitoring and research activities in line with the recommendations set forth in the ICHESE report. The operations carried out were the following: (a) updating of the static and dynamic models of the Cavone reservoir; (b) planning of interference and injectivity tests at selected wells; (c) assessing feasibility of further monitoring for the first-time application of the Guidelines. The results of these operations were made available, with all due transparency and dissemination, by a dedicated information point, set up at the Cavone site, and through the website http://www.labcavone.it. After the 90-day period, all the parties met to evaluate the outcomes of the operations and to plan any further action to be taken.

The Cavone Laboratory project started on 18 April 2014, when the Work Group began to define the details and started the operations as set down in the abovementioned Agreement. At the same time, the local information point was set up, where the results of operations were updated daily. On 16 May 2014, the labcavone website was published. The website was designed to inform local communities and the community at large of the contents and daily progress in the study and monitoring phases, and has remained active after the conclusion of the Laboratory program, as a service provided to the community. By means of data provided by the Italian National Institute of Geophysics and Volcanology (INGV) and by local monitoring stations, it displays, within 24 h, the data relating to the seismicity of the entire lowland area of the Province of Modena and to a larger area of approximately 81 000 km^2. The website has a section dedicated to the answers to frequently asked questions, in order to make the relevant topics as clear as possible.

From 16 May to 16 June 2014, the field activity program was carried out, with the aim of collecting data. After its completion, approximately one million pressure and temperature values had been recorded, together with 5000 flow rate measurements. The advanced studies carried out during the 90 days of activity of the Laboratory led to identify the oilfield of Cavone as the first Italian testing site for the application of the Guidelines prepared by MISE. At present, based on a further agreement for the first application of the Guidelines in Italy (23 July 2014), at Cavone oilfield, study and research are still under way, aimed to the field implementation of the three main monitoring areas: seismic, pore pressure and ground deformation monitoring.

As of today (May 2015), the Cavone microseismic network consists of four stations, whose task is to record and localize the local seismicity within a range and depth of 10 km of the next station. This local network is linked to the Italian National seismic network, a complex structure including over 300 monitoring points. The large majority of these stations was installed and are managed by INGV, though the Italian National Earthquake Center, but a significant number of seismometers were installed and are maintained by other INGV Departments (Naples, Catania, Milan) or by other Italian or European research institutions.

Pore pressure monitoring is performed by means of specific instruments (memory gauges), which record pressure and temperature, temporarily located at the bottom of one or more existing wells. Periodical campaigns are carried out for the measurement of the static pressure of the field. The average pressure value in the volume surrounding the wells is estimated by correlation with other monitoring wells and by mathematical models.

Ground deformation monitoring, so far based on InSAR interferometry only, was tested at the Cavone oilfield. The Canadian satellite RADARSAT-1 recorded images of the concerned areas from 20 to 29 May 2012 (the "Emilia" seismic sequence of May 2012) and, based on such images, it was reconstructed an accurate map of the ground deformation induced by the abovementioned earthquake.

8 Conclusions

The Guidelines (MISE, 2015) are the first Italian document collecting technical specifications of monitoring networks, decision-making framework and related procedures in the field of underground fluid production and storage activities. The document highlights some elements for which there is a lack of operative experience; therefore, it points out that the Guidelines should be re-evaluated, according to direct experience, approximately two years after its first experimental application on pilot test sites.

One of the main concern is hydrocarbon production involving the injection of formation fluids. In order to minimize the potential impact, for these activities is recommended to preserve the original natural stress (maintaining the balance of fluids extracted and injected as close to zero as possible) and to keep the pressure of the injection as close as possible to the initial (natural) pressure.

One of the most analyzed point was the general organizational structure, for which it is necessary to assign the designing of the monitoring networks (together with data analysis and interpretation) to third-party public or private research agencies, the Responsible of the Monitoring (RM). These are controlled by the Ministry itself, in order to separate the Operator and the RM, to ensure the independence of scientific assessments and, finally, to establish a proper level of control over the monitoring.

Another issue raised was that of the definition of the areas to be monitored and how to implement effective decision-making procedures. Especially for the seismic monitoring, it was necessary to consider and synthesize the experiences derived from different productive activities. The unifying principle has been identified in the definition of two volumes, respectively "internal" and "extended" control volume based on the positions of the wells and the extent of the reservoir.

The last point is that concerning the decisions to be taken depending on the results of monitoring. It was chosen an approach based on four levels of activation. It was suggested the adoption, on an experimental basis, a strict traffic light system only for the fields involving injection activities. The traffic light was defined by threshold values based on magnitude, ground velocity and acceleration measured within the internal control volume. Aware of the inherent difficulties in the implementation of the system, it is believed that these Guidelines will have greater applicability in the future, when knowledge and scientific methodologies will be further established, capable of providing immediate and reliable answers.

References

Astiz, L., Dieterich, J. H., Frohlic, C., Hager, B. H., Juanes, R., and Shaw, J. H.: On the Potential for Induced Seismicity at the Cavone Oilfield: Analysis of Geological and Geophysical Data, and Geomechanical Modeling. Study prepared for the Italian Petroleum and Mining Industry Association, Assomineraria, 2014, http://www.assominerariacavone.org/uploads/tappe_lavori/11/attachment/cavonereportfinal_v03.pdf, last access: April 2015.

ICHESE – International Commission on Hydrocarbon Exploration and Seismicity in the Emilia Region: Report on the Hydrocarbon Exploration and Seismicity in Emilia Region, February 2014, http://geo.regione.emilia-romagna.it/gstatico/documenti/ICHESE/ICHESE_Report.pdf, with appendices: http://geo.regione.emilia-romagna.it/gstatico/documenti/ICHESE/Appendices.pdf, last access: April 2015.

ISPRA – National Institute for the Environmental Protection and Research: Possible relationship between induced/triggered seismicity and anthropogenic activities in Italy (Rapporto sullo stato delle conoscenze riguardo alle possibili relazioni tra attività antropiche e sismicità indotta/innescata in Italia, Jun. 27, 2014, in Italian), http://www.isprambiente.gov.it/files/notizie-ispra/notizia-2014/rapporto-sismicita-indotta-innescata-in-italia/Rapporto_sismicitaindottainnescata_initalia.pdf, last access: April 2015.

Jha, B., Plesch, A., Shaw, J. H., Hager, B. H., and Juanes, R.: Coupled Flow and Geomechanical Modeling of Fluid Production and Injection in the Cavone Oil Field, Northern Italy: an Assessment of the Potential for Induced Seismicity, AGU Fall Meeting, 15–19 December 2014, San Francisco, CA, 2014.

MISE – Italian Ministry for Economic Development: Oilfield Monitoring Guidelines (Indirizzi e linee guida per il monitoraggio della sismicità, delle deformazioni del suolo e delle pressioni di poro nell'ambito delle attività antropiche, Nov. 24, 2014, in Italian), http://unmig.sviluppoeconomico.gov.it/unmig/agenda/upload/85_238.pdf, last access: April 2015.

Pizzi, A. and Scisciani, V.: The May 2012 Emilia (Italy) earthquakes: preliminary interpretations on the seismogenic source and the origin of the coseismic ground effects, Ann. Geophys., 55, 751–757, 2012, http://www.ann-geophys.net/55/751/2012/.

Proc. IAHS, 372, 539–542, 2015
proc-iahs.net/372/539/2015/
doi:10.5194/piahs-372-539-2015

Open Access
PIAHS

The investigation of a new monitoring system using leveling and GPS

M. Nojo[1], F. Waki[2], M. Akaishi[3], and Y. Muramoto[4]

[1]Godo Shigen Co., Ltd
[2]Nihonn Tennen Gas Co., Ltd
[3]Ise Chemicals Co., Ltd
[4]Kanto Natural Gas Development Co., Ltd

Correspondence to: M. Nojo (m.nojo@godoshigen.co.jp)

Published: 26 November 2015

Abstract. We installed GPS stations for monitoring of land subsidence of Kujukuri Plain in Chiba prefecture in 2003, and have been monitoring since then. In this paper, we discussed a data processing method capable of reducing noise in data collected by the GPS stations. Also, we compare the accuracy of leveling and GPS, and collected basic data for establishing a new monitoring system based on the combination of leveling and GPS.

1 Background

In Chiba prefecture,Japan, eight companies have exploited natural gas in Southern Kanto Natural Gas Field, which is a deposit of natural gas dissolved in water. In particular, the Kujukuri district has good gas production ratio, therefore natural gas with formation water is mainly produced there.

Land subsidence in this area was found in the late 1960s. The eight companies in this area have been making efforts to control land subsidence and also have concluded the subsidence prevention agreement with the local government. Although land subsidence in recent years has been reduced in many areas, land subsidence of more than 10 mm per annum has been observed in some places. The eight companies currently operate addressing environmental problems including land subsidence through the Environment Committee of the Japan Natural Gas Association Keiyo Natural Gas Association (hereinafter referred to as the Environment Committee). The monitoring and observation working group works under the commitee.

2 Result summary

2.1 The installation of six GPS stations and subsequent monitoring

Natural gas with formation water is mainly produced in the Kujukuri district. The companies conduct leveling surveys twice a year. Although data from leveling surveys is accurate, it is only conducted perioducally. On the other hand, GPS system can collect location infomations of the GPS station continuously. The WG installed three GPS stations in and near the Kujukuri district in 2003, and an additional three stations in that area in 2008. Land subsidence has been monitored including GSI GPS (Fig. 1). To analize GPS data, the WG regards the Ohara station (GSI GPS) as the base station (refference point), because there is a reffernce point of leveling in Ohara near the Ohara station. The analysis software used Bernese ver 4.2. A monitoring data result in GPS stations which was established in WG is shown in Fig. 2.

2.2 The smoothing of the GPS data which used a trend model

Data observed in the GPS stations often includes abnormal values due to noise. The WG has smoothed the data with a 10-day moving average. However, even if the data includes abnormal values, the moving average system calculates an

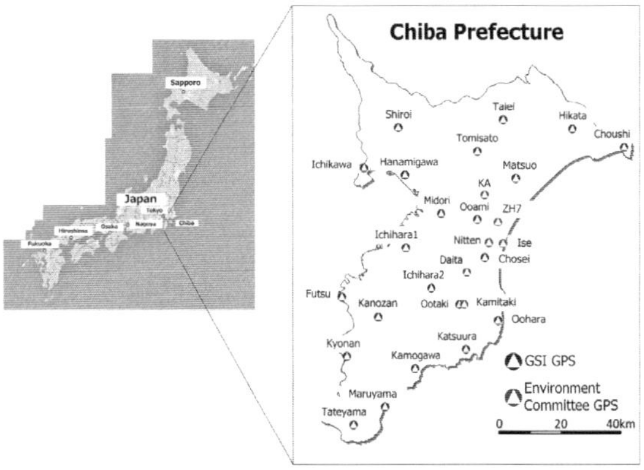

Figure 1. GPS station in Chiba Prefecture.

Figure 2. GPS observation result.

average of the whole data, and as a result, a calculated value may not represent the trend. To solve this problem, the WG smoothed the data with the trend model used for inversion analysis. It supposed that the time-series data followed an auto regression model and it implemented an anti-jaggy by Kalman filter. We consider the result is more reasonable than the observational data and the 10-day moving average.

2.3 The accuracy comparison of the leveling and the GPS data

To promote the utility of monitoring of land subsidence by the GPS stations, it is necessary to check accuracy of the observation data collected from the GPS stations. The WG conducted leveling survey between the two GPS stations every month for a total of six times between 1 June 2013 and 1 Norvember 2013, to check the accuracy by comparing leveling data with the data from the GPS stations.

Two fields inside the development area of natural gas were selected for the testing. Field 1 (Kamitaki district) is 12 km from the Pacific Ocean, and the relevant two GPS stations are separated by a distance of approximately 1.3 km in the east-west direction. Field 2 (Daita district) is located in the hill area 10 km north of the Field 1, and the relevant two GPS stations are separated by a distance of approximately 2.4 km in the north and south direction.

Table 1. GPS analysis conditions.

Analytical method	Static relative positioning
Baseline analysis software	Bernese ver 4.2
Base station	GEONET Ohara
Analysis range	GPS station in or near Chiba Prefecture
Ephemeris	IGS precise ephemeris

Figure 3. GPS smoothing result.

A comparable height difference between the two stations was calculated individually for GPS and leveling, based on the value observed on 1 June and the results were compared.

The comparison indicated that, the maximum difference between the GPS data and the leveling data in the Kamitaki district was comparably large at 5.7 mm. On the other hand, the data in the Daita district showed a maximum difference of 2.4 mm. Characteristically, Field 1 is flat near the ocean and a relatively large river and Field 2 is mountainous further from the ocean. There is a possibility that the influence of water vapor content in the surrounding areas generates the measurement discrepancies.

2.4 The examination of the leveling network which used a GPS station as a fixed point

The leveling data of the companies reflects the result of leveling conducted by the Chiba prefectural government. Since the Chiba prefectural government releases the leveling result more than six months after the leveling survey, there is a delay in fixing their observation values by the companies. Therefore, we discussed a method to acquire data in a short time by integrating the data collected at the GPS stations and the leveling results.

Each GPS antenna and an attached benchmark are installed on the same base. Therefore, we assume that variation measured with the GPS and the variation amount of the

Table 2. GPS analysis conditions.

Course of Traverse		Course of Traverse 1		Course of Traverse 2	
Observation Point		Kamitaki Station[a]	Ootaki Station[b]	Daita Station[a]	Jibiki Station[c]
Coordinates	E-W (m)	1333		986	
difference	S-N (m)	164		2200	
GPS distance	(km)	1343		2411	
Leveling distance	(km)	2237		2536	

[a] Envirpnment Committee GPS, [b] GSI GPS and [c] Kanto Natural Gas Development GPS.

Table 3. GPS observation result.

GPS station	The amount of oval Date 1 Jan 2014 [mm]	1 Jan 2015 [mm]	Quantity of change $[\mathrm{mm\,yr^{-1}}]$
Kamitaki GPS station	57 480.0	57 479.6	−0.4
Daita GPS station	69 399.9	69 387.8	−12.1

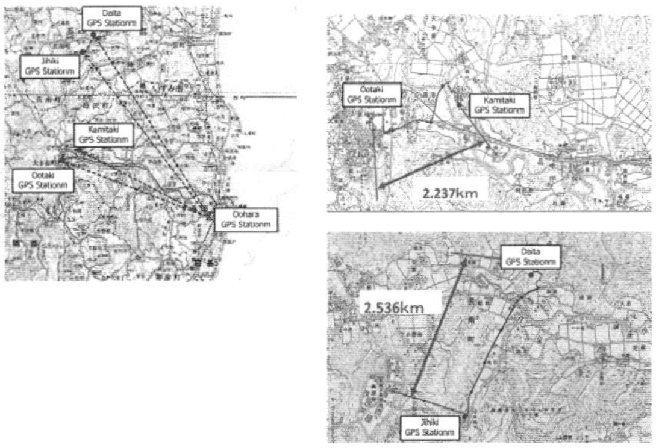

Figure 4. Course of traverse.

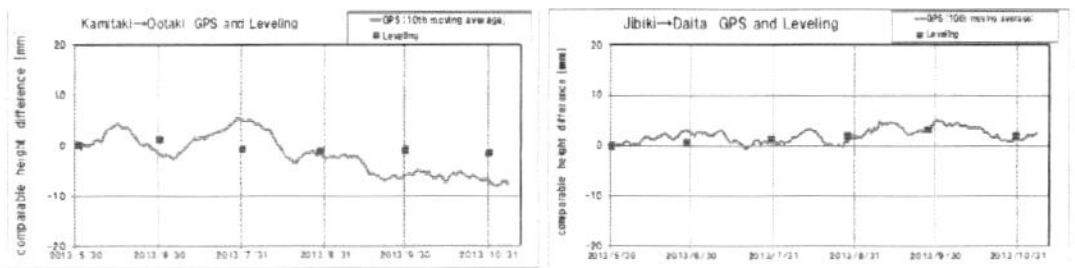

Figure 5. Comparable height difference GPS and Leveling.

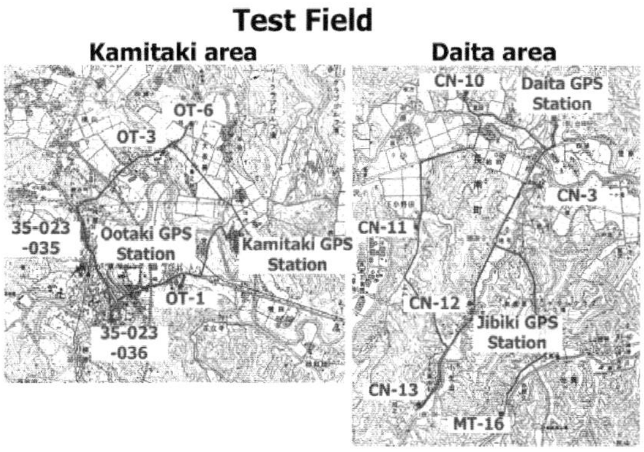

Figure 6. Test field.

attached benchmark is the same. When an elevation value of the attached benchmark installed in the GPS station is acquired at any time, an elevation value of the attached benchmark can be found based on the variation amount measured with the GPS. We consider an elevation value of a benchmark near the GPS station can be acquired by leveling from the GPS-attached benchmark.

We conducted a test in the Kamitaki district and Daita district, described as the two test fields in the previous section.

3 Test procedure

1. A leveling network containing each GPS station and its attached benchmark is set.
2. The WG carry out surveying on 1 January 2014.
3. Based on the results of levelling measurements that the WG conducted according to the result of leveling by the Chiba prefectural government, an elevation value is given to the GPS attached benchmark.
4. The WG carry out surveying on 1 January 2015.
5. An elevation value of the attached benchmark on 1 January 2015, is determined based on the annual difference amount of the GPS station.
6. According to the attached benchmark's elevation value, an elevation value of each benchmark is calculated from the leveling survey results from the 1 January 2015 survey.

Table 4. The change predicted value that put leveling and GPS together.

Area	Bench mark	Chiba prefecture leveling 1 Jam 2014 [mm]	WG leveling[a] 1 Jan 2014 [mm]	Altitube leve 1 Jan 2014 [mm]	GPS quantity of change 2014–2015 [mm yr^{-1}]	WG leveling[b] 1 Jan 2015 [mm]	Expected altitube level [mm]	Expected quantity of change [mm yr^{-1}]
Kamitaki area	Kamitaki GPS attached bench mark		−9072.1	22 253.9	−0.4	+0.0	22 253.5	−0.4
	OT-1	31 326.0	+0.0	31 326.0		+9071.9	31 325.4	−0.6
	Ootaki GPS attached bench mark		−5350.9	25 975.1		+3716.9	25 970.4	−4.7
	35-023-036		−5214.5	26 111.5		+3852.8	26 106.3	−5.2
	35-023-035		−7726.9	23 599.1		+1337.6	23 591.1	−8.0
	OT-3		−10 616.0	20 710.0		−1551.2	20 702.3	−7.7
	OT-6		−6702.9	24 623.1		+2360.1	24 613.6	−9.5
Daita area	CN-12	29 315.6	+0.0	29 315.6		−6759.7	29 303.4	−12.2
	Jibiki GPS attached bench mark		+209.0	29 524.6		−6550.5	29 512.6	−12.0
	CN-11		−104.0	29 211.6		−6862.5	29 200.6	−11.0
	CN-10		−6130.8	23 184.8		−12 889.0	23 174.1	−10.7
	Daita GPS attached bench mark		+6759.6	36 075.2	−12.1	+0.0	36 063.1	−12.1
	CN-3		−9037.1	20 278.5		−15 795.4	20 267.7	−10.8
	MT-16		+6323.2	35 638.8		−436.3	35 626.8	−12.0
	CN-13		+11 752.0	41 067.6		+4993.6	41 056.7	−10.9

[a] CN-12 and OT-1 besed and [b] GPS attached bench mark based.

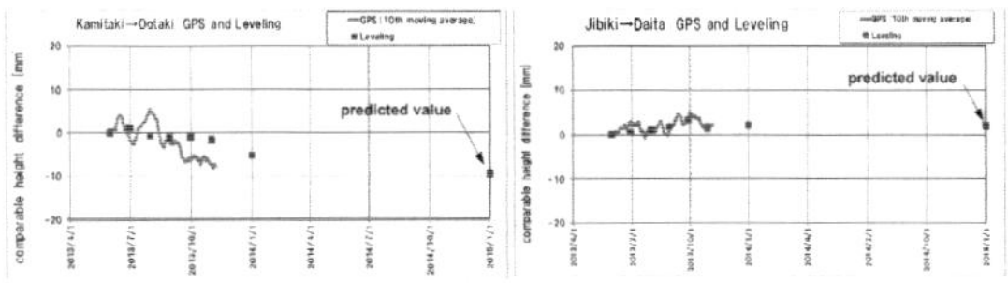

Figure 7. Comparable height difference Leveling (predicted value).

We compare the data acquired in the calculation above with the leveling results released by the Chiba prefectural government six months after the govemment's survey, and verify this test result. If the result is reasonable, we can shorten the period to fix the value after leveling from more than six months to a few months.

4 Conclusions

Establishment of the GPS analysis method and improvement of accuracy allow monitoring of land subsidence.

We can shorten the period to fix the value after leveling from more than six months to a few months, if the data using leveling network which used a GPS station as a fixed point are correct in comparison with the leveling results released by the Chiba prefectural government.

Proc. IAHS, 372, 543–553, 2015
proc-iahs.net/372/543/2015/
doi:10.5194/piahs-372-543-2015

Risk evaluation of land subsidence and its application to metro safety operation in Shanghai

J. Liu, H. Wang, and X. Yan
Shanghai Institute of Geological Survey, Shanghai 200072, China
Correspondence to: J. Liu (dabao_0323@163.com)

Published: 26 November 2015

Abstract. Based on sufficiently investigating characteristics and risk connotation of land subsidence, a risk evaluation index system for land subsidence disaster is established, which is combined with the sensitivity feature of the hazard bearing body to land subsidence. An appropriate evaluation method system is established by using an improved fuzzy analytic hierarchy process method. So risk evaluation is developed for providing theoretical basis and technical support for the regional management of land subsidence prevention and control. On this basis, as a case of Shanghai metro, firstly, the paper studies the identifying risk sources of the metro. According to metro linear characteristics, external indexes of representing subsidence risk are obtained. Studying the subsidence risk of the metro, relevant achievement has provided the technical basis for daily main monitoring, early warning and work arrangement.

1 Introduction

Shanghai will has been built to become the world's economic, financial, trade and shipping center, but it is inhibited by the concentration of population and poor resources. The combination of high-intensity exploitation of land and water resources and a susceptible geological environment makes this rigion prone to land subsidence problems. Specifically, the thick Quaternary overburden whose structure is very soft produces land subsidence by means of imperceptible compression deformation from the influence of natural and artificial factors. Land subsidence does serious harm to the sustainable development of the society and the economy. At present, the average cumulative subsidence in the inner city of Shanghai is generally greater than 0.6 meters, the maximum cumulative subsidence tends to be close to 3 meters, and the height of the inner city is lower than the average elevation of the entire city.

In recent years, Shanghai has further improved its mechanismsfor subsidence prevention and treatment. In particular, groundwater exploitation and artificial recharge management was continually strengthened, and land subsidence monitoring and control capabilities were continually improved. The magnitude of subsidence and its scope of influence has been gradually reduced, but the deep confined aquifer's large cone of depression caused by the long-term over exploitation has influenced land subsidence for a long time. Then after a superposition of other factors such as the deep foundation dewatering and drainage pit, the number of land subsidence cones have gradually increased, the resulting change of equal distance in land subsidence cone areas were more significant, and the nonuniformity of land subsidence has increased year by year. This brought new challenges for the management of land subsidence prevention and control.

This paper introduces the risk evaluation and management method, which is applied to the management of land subsidence prevention and control. Risk evaluation and management is a new subject in the study of geological disasters. Domestic and foreign literature regarding tunnel, landslide and debris flow risk evaluation is well established (Brabb, 1984; Woude and Jonker, 2003; Yoo Chungsik et al., 2006). However, only a few studies exist on the theoretical aspect of land subsidence risk evaluation (Hu et al., 2008; Yu et al., 2008; Chen et al., 2013, 2014).

2 Risk evaluation of land subsidence

2.1 Risk evaluation indexes of land subsidence

The risk degree of land subsidence mainly depends on two conditions. One is the dynamic conditions of land subsidence, including geological conditions (the distribution of aquifer, the thickness and structure of soft soil layer, etc), topography (landform, terrain elevation, etc), man-made geological dynamic activities (underground water exploitation, engineering construction, etc). Normally, the ampler dynamic conditions of land subsidence are, the intenser land subsidence disaster activities become. When the damages caused by land subsidence are more serious, the risk of land subsidence is greater. The other is the economic vulnerability of human society. That is the resistant and recoverable ability of life, property and various economic activities in the disaster areas, which includes the population density, the residential environment, the land-use type, the investment for disaster prevention and reduction, etc. Two aspects abovementioned are named for hazard and vulnerability respectively. They determine the risk degree of land subsidence.

The hazard of land subsidence refers to the occurrence probability and danger degree of land subsidence in a certain area and a certain period of time. It is evaluated quantificationally by using H. The evaluation factors of hazard include geological conditions, topography, geological dynamic activities, scale of geological hazard, the occurrence probability, growth rate, etc.

The vulnerability of land subsidence refers to the probability of suffered damage and the difficult of damage caused by land subsidence. This concept implies the ability to cope with the land subsidence disaster which the human society and the development level of economic technical reply. It is evaluated quantificationally by using V.

According to the hazard evaluation and the vulnerability evaluation of land subsidence, the final results of the risk evaluation of land subsidence provide basic data for monitoring, forecasting and controlling the land subsidence, and offer decision-making information for the government departments.

2.2 Risk evaluation method of land subsidence

According to the current basic data of land subsidence and social economic in Shanghai, the fuzzy comprehensive evaluation model is established by using the improved fuzzy analytic hierarchy process method (Csutora and Buckley, 2001; Feng, 2007). Compared to the traditional analytic hierarchy process, this method not only solves the problem of the consistency of judgment matrix but also overcomes the fuzziness and the distribution difficulty of the factor's weight. The specific calculation steps are as follows.

1. Decision on the evaluation factors. Based on the analysis of the influencing factors and the actual monitoring data, the evaluation factors are selected and formed the factor set.

   $$U = \{u_1, u_2, \ldots, u_n\} \tag{1}$$

2. Establishing evaluation collection. The land subsidence risk evaluation set is assumed. Its levels of values range are between 0 and 1. Specific division is five grades including very low, low, medium, high and very high.

   $$V = \{v_1, v_2, \ldots, v_m\} \tag{2}$$

3. Determining the single factor evaluation matrix.

   $$R = \left(r_{ij}\right)_{n\times m} (i = 1, 2, \ldots, n;\ j = 1, 2, \ldots, m) \tag{3}$$

   where r_{ij} is the membership grade that the index r_j was assessed v_j. For example, the factor "land subsidence susceptibility" may be considered to the "low" grade by twenty percent of experts and the "medium" grade by eighty percent of experts. Then twenty percent and eighty percent respectively refer to the membership grade of "low" and "medium".

4. Using the improved analytic hierarchy process (AHP) to determine the weight vector W.

   $$W = \{w_1, w_2, \ldots, w_n\} \tag{4}$$

   First, established the priority relation matrix.

   $f_{ij} = \begin{cases} 0.5, s(i) = s(j) \\ 1.0, s(i) > s(j), \\ 0.s(i) < s(j) \end{cases}$ s_i and s_j showed the relative importance degree of factor u_i and u_j. Fuzzy complementary matrix $F = \left(f_{ij}\right)_{n\times m}$ is obtained based on the analysis of the measured data of land subsidence and its influencing factors. Then the sum method of matrix rows $e_j = \sum\limits_{j=1}^{n} f_{ij}$, and the following mathematical manipulation are implemented, $e_{ij} = \frac{(e_i - e_j)}{2(n-1)} + 0.5$. The complementary judgment matrix $E = \left(e_{ij}\right)_{n\times n}$ is changed into a reciprocal matrix $D = \left(d_{ij}\right)_{n\times n}$ by using the transformation formula $d_{ij} = \frac{e_{ij}}{e_{ji}}$.

   The sum normalization method is used to obtain the ordering vector.

   $$W^{(0)} = \left(w^1, w^2, \ldots, w^n\right)^T = \left(\frac{\sum\limits_{j=1}^{n} d_{1j}}{\sum\limits_{i=1}^{n}\sum\limits_{j=1}^{n} d_{ij}}, \frac{\sum\limits_{j=1}^{n} d_{2j}}{\sum\limits_{i=1}^{n}\sum\limits_{j=1}^{n} d_{ij}}, \ldots, \frac{\sum\limits_{j=1}^{n} d_{nj}}{\sum\limits_{i=1}^{n}\sum\limits_{j=1}^{n} d_{ij}}\right)$$

   Regarded the ordering vector $W^{(0)}$ as an eigenvalue method to solve the ordering vector with higher accuracy $W^{(k)}$, the paper inputs the initial value $\alpha^{(0)} = W^{(0)} = (\alpha_{01}, \alpha_{02}, \cdots, \alpha_{0n})^T$, uses the iterative formula $\alpha^{k+1} = E\alpha^k$, seeks the eigenvalue α^{k+1}, and finds the infinite norm $\|\alpha^{k+1}\|_\infty$ of α^{k+1}. If $\left|\|\alpha^{k+1}\|_\infty - \|\alpha^k\|\right| < \varepsilon$, the $\|\alpha^{k+1}\|_\infty$ is the largest eigenvalue$\lambda_{\max}$, and the α^{k+1} is used for normaliczation process $\alpha^{k+1} = \left(\frac{\alpha_{k+1,1}}{\sum_{i=1}^{n}\alpha_{k+1,i}}, \frac{\alpha_{k+1,2}}{\sum_{i=1}^{n}\alpha_{k+1,i}}, \cdots, \frac{\alpha_{k+1,n}}{\sum_{i=1}^{n}\alpha_{k+1,i}}\right)$. The obtained vector $W^{(k)} = \alpha^{k+1}$ becomes the ordering vector and iterated is over. Otherwise, this method regardes the vector $\alpha^k = \alpha^{k+1}/\|\alpha^{k+1}\|_\infty$ as a new initial value, and iterates again, finally the obtained $W^{(k)}$ serves as the weight allocation vector W.

5. Fuzzy comprehensive evaluation.

$$V = W \times R \tag{5}$$

6. Fuzzy vector uniformization. The evaluation set $V = \{v_1, v_2, \ldots, v_m\}$ is assigned one to five, then the corresponding evaluation results are divided into five levels.

7. The hazard evaluation and the vulnerability evaluation are carried out through the fuzzy comprehensive evaluation, then the risk degree can be calculated by the next formula.

$$R\ (\text{risk}) = H\ (\text{hazard}) \times V\ (\text{vulnerability}) \tag{6}$$

2.3 Risk evaluation of land subsidence

1. Hazard evaluation of land subsidence

   Through analyzing the development mechanism, the incubation conditions, the monitoring results and other factors of land subsidence, the indexes including land subsidence susceptibility, historical disaster intensity, sedimentation rate and ground elevation are choosen to evaluate the hazard of land subsidence. The single factor evaluation map is compiled based on the land subsidence hazard evaluation of each single index evaluation. Then the basic properties of all kinds of influencing factors are extracted through the space intersection analysis. The above method is used for evaluating the hazard (Fig. 1).

   It can be seen from the evaluation results that land subsidence in Shanghai mainly locates in medium-hazard areas. High hazard and very high hazard areas are mainly distributed in the central city and the local area of Jiading district and Jinshan district which are adjacent to Jiangsu province and Zhejiang province. Low hazard and very low hazard areas are mainly distributed in the local area of Fengxian district, Qingpu district and other districts.

2. Vulnerability evaluation of land subsidence

   Due to the difficulty of acquiring the data of suffering disasters zone and the easiness of obtaining the regional social and economic data based on the administrative unit (district, county), the vulnerability evaluation indexes of land subsidence mainly consider the social and economic statistic (GDP, population density, proportion of land for construction), linear infrastructure influenced by land subsidence (flood control wall, metro, water supply network, elevated road), disaster prevention and reduction investment in the past (monitoring and control facilities). It can be seen from the evaluation results (Fig. 2) that the vulnerability of land subsidence reflects the comprehensive characteristics about the development level of regional economic and land subsidence.

3. Risk evaluation of land subsidence

   Based on hazard and vulnerability evaluation of land subsidence, risk zoning map of land subsidence in Shanghai is compiled after dividing risk grade (Fig. 3). The spatial distribution characteristics can be seen from the evaluation results. The risk of land subsidence shows three-level system characteristics which revolves around the inner city of Shanghai: the very high risk region of land subsidence mainly locates in the inner city, the high to medium risk region of land subsidence mainly locates in the suburban area and the low to very low risk region of land subsidence mainly locates in the far-suburban area.

3 Regional management of land subsidence prevention and control

3.1 Division of land subsidence prevention and control

According to the current development situation of land subsidence and the development trend of regional social and economic, the regional management of land subsidence prevention and control can be realized on the basis of the land subsidence risk evaluation. Combined two-level management system of land subsidence prevention and control, the unit is identified as the administrative region which are city and district in Shanghai. It also reads the urban and rural planning for reference. It can enhance the management level of land subsidence prevention and control (Fig. 4).

1. Emphasis prevention area of land subsidence (area I)

   It is divided into two subregions that are area I_1 and area I_2. Area I_1 is the inner city inside the outer ring road where the serious land subsidence disaster has happened. In this area, the situation of flood control is severe, inhomogeneous land subsidence is obvious, and safe operation of grave projects are affected greatly.

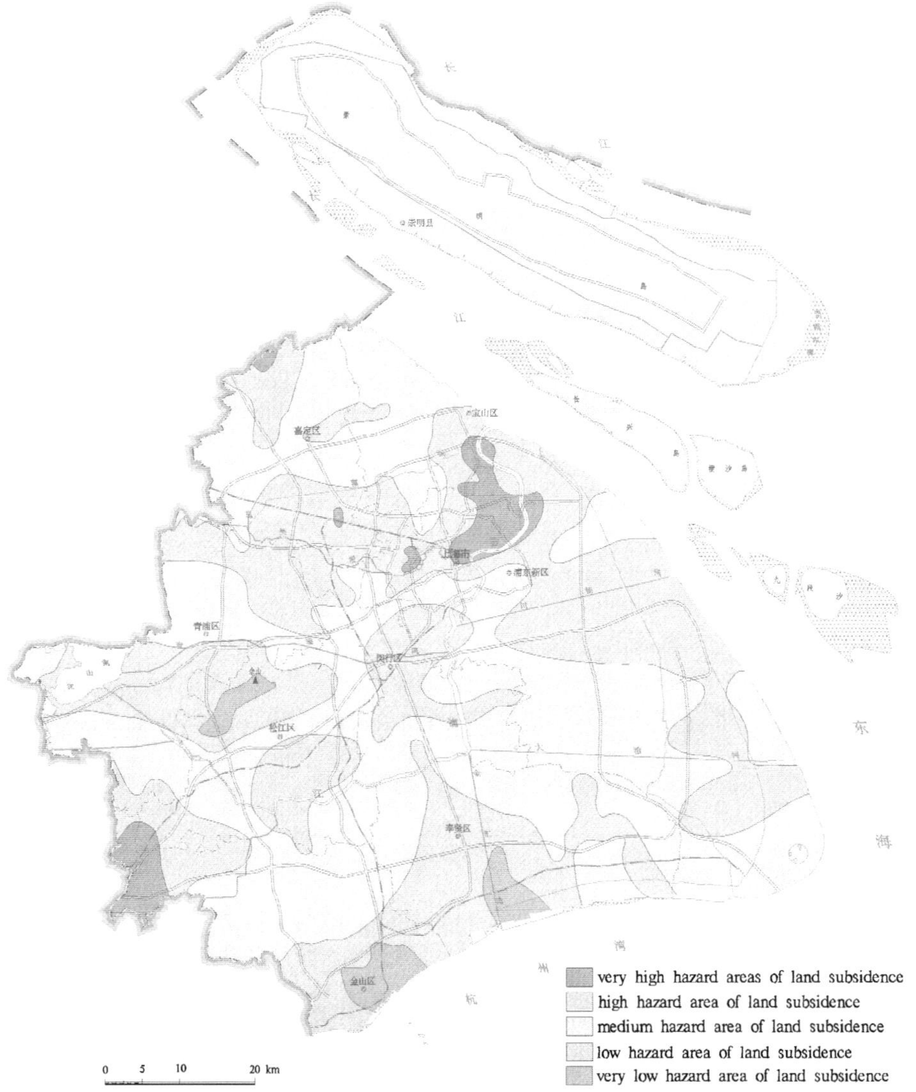

Figure 1. Hazard evaluation of land subsidence in Shanghai.

Area I_2 includes Pudong new district and Hongqiao area outside the outer ring road. With the adjustment and implementation of urban master planning, in recent evolution of land subsidence presents an aggravating tendency. Because of engineering construction, some land subsidence centers have formed, for instance, Hongqiao, Sanlin, etc. In area I_2 the phenomenon of inhomogeneous land subsidence is serious, which seriously influences infrastructure safety operation including metro, maglev trains, flood control wall, etc.

2. Second-emphasis prevention area of land subsidence (area II)

Area II includes Baoshan district, Jiading district and Minhang district except for area I. In this area land subsidence is medium, however, as the future development region of urban master planning, the influence of engineering for land subsidence should not be neglected. In history, underground water level of major exploiting aquifer (the fourth confined aquifer) in this erea is very low, and in some parts of area groundwater cones are due to the exploitation of underground water in neighboring province.

3. General prevention area of land subsidence (area III)

Figure 2. Vulnerability evaluation of land subsidence in Shanghai.

Area III Includes Fengxian district, Songjiang district, Jinshan district, Qingpu district and Chongming county. In this area the overall development intensity is low, and land subsidence is medium or low, but land subsidence is still outstanding in some areas because of the exploitation of underground water in neighboring province, especially, seawall in north shore of Hangzhou Bay.

3.2 Control objectives of land subsidence regional prevention and control

1. Control objectives of emphasis prevention area of land subsidence (area I)

   In emphasis prevention area of land subsidence (area I_1), by the end of 2015, the annual average amount of land subsidence will control within 7 mm, the groundwater table of the fourth confined aquifer groundwater table will return to -12 m and the impact of asymmetrical land subsidence will be further reduced. In emphasis prevention area of land subsidence (area I_2), by the end

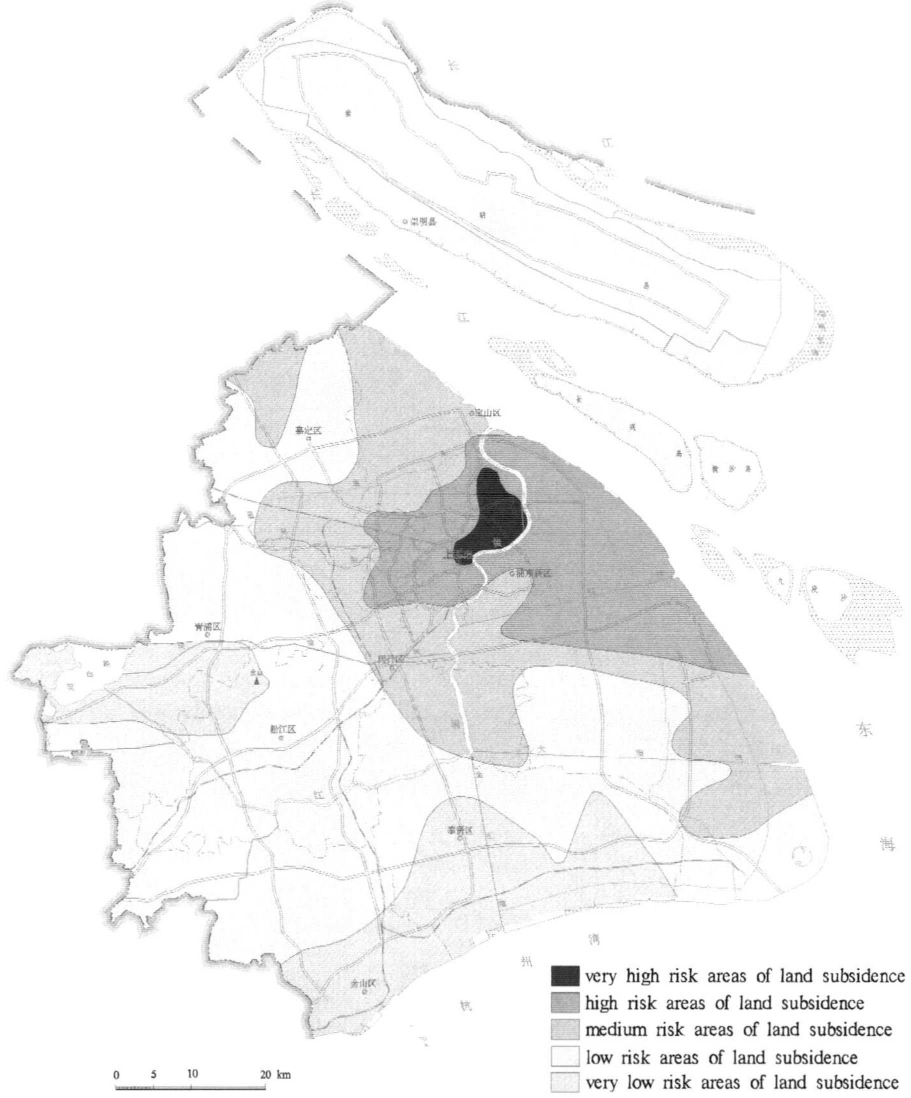

Figure 3. Risk evaluation of land subsidence in Shanghai.

of 2015, the annual average amount of land subsidence will control within 10 mm, the groundwater level will lift steadily and the impact of asymmetrical land subsidence will be relieved gradually.

2. Control objectives of second-emphasis prevention area of land subsidence (area II)

   By the end of 2015, the annual average amount of land subsidence will control within 6 mm and the confined aquifer groundwater funnel along the lifeline engineering will be eliminated basically.

3. Control objectives of general prevention area of land subsidence (area III)

   By the end of 2015, the annual average amount of land subsidence will control within 5 mm and the confined aquifer groundwater cone triggered by the exploitation of groundwater will be eliminated actively.

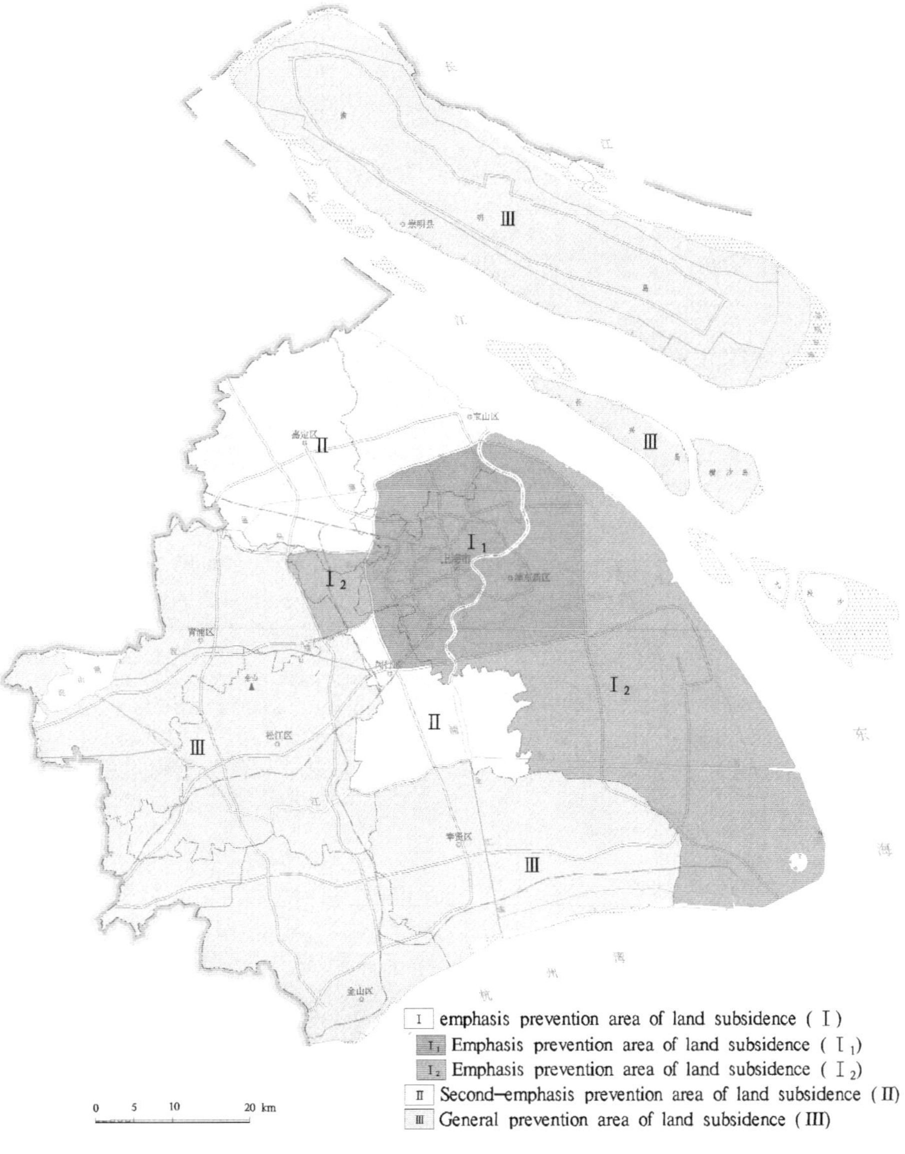

Figure 4. Land subsidence regional prevention and control management.

3.3 Measures of land subsidence regional prevention and control

1. Measures of emphasis prevention area of land subsidence (area I)

   In emphasis prevention area of land subsidence (area I_1), many measures will have been taken to prevent and control land subsidence, such as the stricter management of groundwater exploitation and recharge, the special recharge in shallow aquifer and the early warning mechanism for the track traffic, flood wall and other lifeline engineering. In emphasis prevention area of land subsidence (area I_2), many measures will have been taken to prevent and control land subsidence, such as the compression of groundwater exploitation, the distensible size of artificial recharge, the intensive monitoring and management of land subsidence caused by deep foundation pit and other construction activities, the enhanced subsidence monitoring of the metro, the maglev train, the sea wall, and other lifeline engineerings.

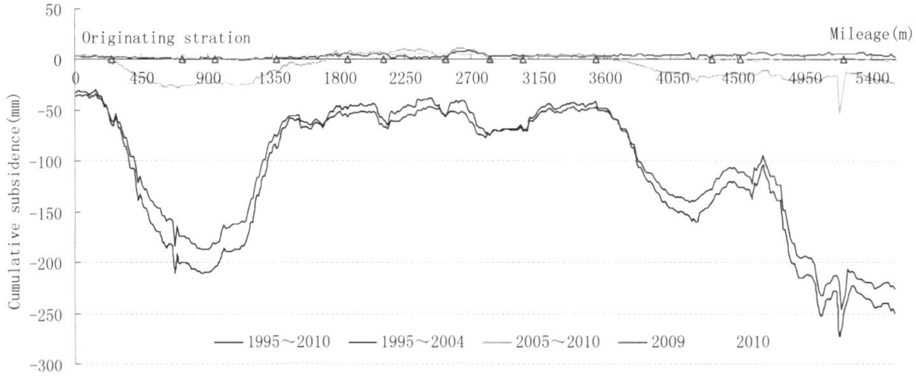

Figure 5. The cumulative subsidence curve of metro in each stage.

Table 1. Evaluation indexes system of metro subsidence.

First Grade	Second Grade
subsidence	cumulative subsidence quantity subsidence rate radius curvature
convergence	change rate cumulative convergence variable
geological environment	topography across the sand layer across the clay layer
structural percolating water	water mud sand ratio infiltration rate infiltration aquare

2. Measures of second-emphasis prevention area of land subsidence (area I)

   The existing arrangements of groundwater exploitation and recharge situation will have been optimized and the management of groundwater development and utilization along high speed railway will have been strengthened.

3. Measures of general prevention area of land subsidence (area III)

   The existing arrangements of groundwater exploitation and recharge situation will have been optimized, linkage mechanism of land subsidence will have been established in the Yangtze River delta and the construction of monitoring net for the metro and the sea wall will have been improved.

4 Application of land subsidence risk evaluation to metro safety operation

The thinking of the risk evaluation of land subsidence can be used to evaluate the risk of specific projects subsidence. Different projects have different characteristics and different sensitive degree of land subsidence, so it is difficult to establish the vulnerability evaluation of subsidence. This paper attempts to take the metro as the breakthrough point and makes risk evaluation of metro subsidence. The results provide a basis for the deployment of daily monitoring and early warning.

4.1 The identification of risk sources of metro

1. Regional land subsidence

   An comparison of metro subsidence and regional land subsidence shows that their overall trend is consistent (Fig. 5). The metro subsidence cone area is generally greater than regional land subsidence in a short time after the construction of metro, then metro subsidence and regional land subsidence are consistent. In recent years, metro subsidence has become alleviative along with a slow down in land subsidence growth. The soil deformation below the tunnel floor plays a crucial role in the metro tunnel subsidence, and the soil deformation above the tunnel floor has a little effect on the metro tunnel subsidence.

2. Geological structure difference

   In the metro areas, the difference of geological structure of soil layers is large. Different engineering geological characteristics and different disturbance extent of soil layers in construction stage can obviously cause different subsidence of the metro. The soft clay layer in the ancient river cutting areas which is easy to produce deformation by the static load and dynamic load of train is usually thicker than the normal areas. Some sections

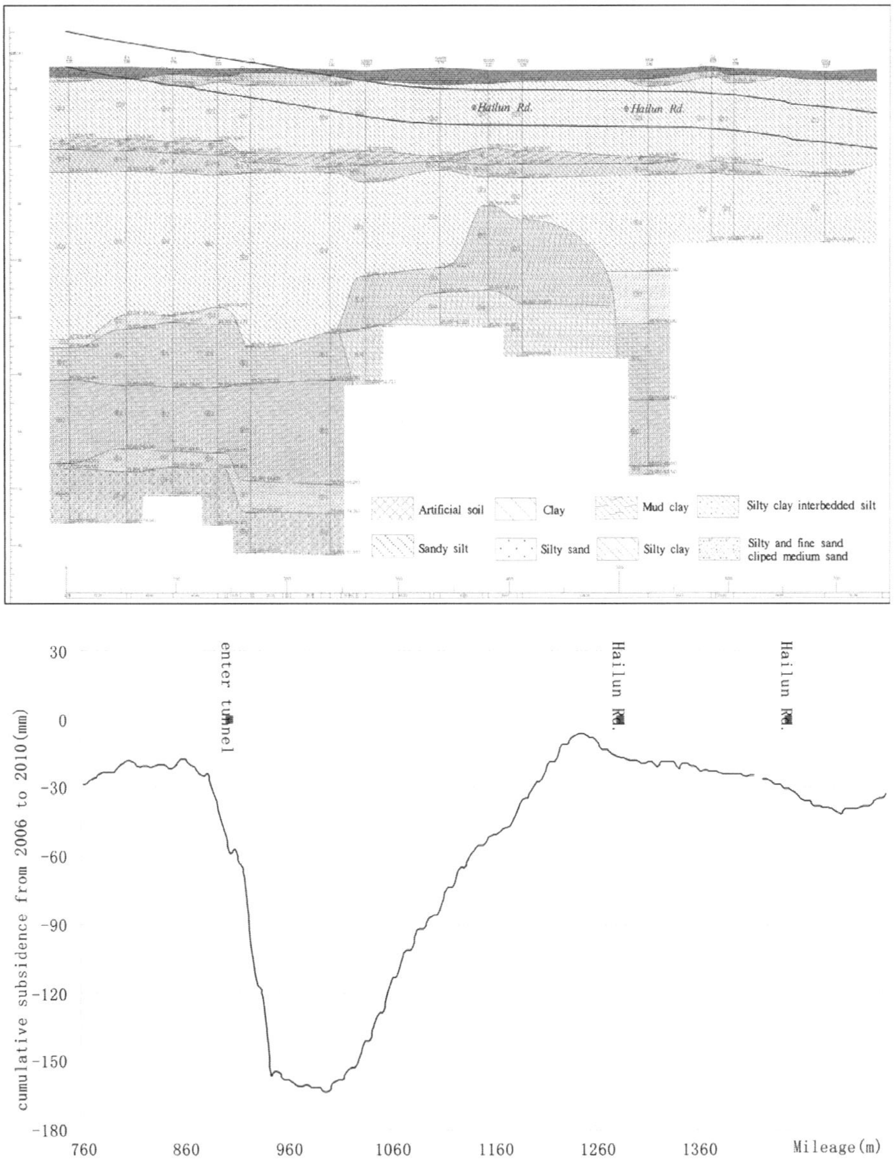

Figure 6. Engineering geological profile comparison chart of tunnel cumulative subsidence.

located in the thick sand layer can cause the large subsidence of metro because the structure of sand soil is obviously destroyed (Fig. 6).

3. Safety inspection to metro

The structural diseases can be found in a timely manner by rail tunnel daily inspections, such as seepage of underground structure, tube cracks, tube loss, and so on. When structural diseases are founded, they serve as the sources of risk and require more attention to the subsidence development.

4.2 Risk evaluation indexes of metro subsidence

Based on an thorough investigation into geological environmental characteristics and risk sources of the metro, appropriate and quantifiable indexes are selected combining with the characteristics of metro structure itself. Then the risk

Figure 7. Risk evaluation of metro subsidence in Shanghai.

evaluation indexes system of metro subsidence is established in Table 1.

4.3 Risk evaluation of Metro subsidence

Based on the analysis of above-mentioned risk sources, the risk evaluation method is used to develop the risk evaluation of metro subsidence, and provide the safety precaution of metro by the evaluation results. The evaluation results show that the high risk sections of metro subsidence mainly have located in the emphasis prevention area of land subsidence (area I), as illustrated in Fig. 7. The largest risk sections located at line 4 from Baoshan road station to Helen road station which is the transition section between surface and beneath. Specifically, the uneven subsidence at line 4 near Helen road station was serious and its risk was very high. It would cause the greater harm to the metro security. Suggestions were given on strengthening monitoring and maintenance, or doing maintenance in a wide range when necessary. The above conclusion provided the technical support for an overhaul on 22 January 2012, New Year's eve.

5 Conclusions

Conclusions can be drawn by the numerical experiments. (1) Using the improved fuzzy analytic hierarchy process, not only solves the problem of the consistency of judgment matrix, but also overcomes the fuzziness of the factor's weight.This method avoids the difference of risk evaluation results caused by human factors. (2) According to developmental characteristics and influence factors of land subsidence in Shanghai, the land subsidence risk evaluation is developed, the land subsidence prevention management areas are delimited and the corresponding prevention measures are adopted. The policy and social behavior of disaster management mode under the government guidance can be realized. (3) Based on the recognition of risk sources on the metro in Shanghai, the paper builds the risk evaluation index system of metro subsidence, and applies the risk evaluation results to the daily security operation of the metro in Shanghai. Thus it can provide the technical support for daily key monitoring and early warning in the management of metro operation.

Acknowledgements. The authors thank the finacial assistances given to this work by the Science and Technology Comission of Shanghai Municipality (Funding no. 12231200700).

References

Brabb, E. E.: Innovation approaches to landslide hazard and risk mapping, Proc. 4th ISL, Torronto, 307–323, 1984.

Chen, Y., Shu, L., and Burbey, T. J.: Composite subsidence vulnerability assessment based on an index model and index decomposition method, J. Human Ecol. Risk Assess., 19, 674–698, 2013.

Chen, Y., Shu, L., and Burbey, T. J.: An integrated risk-assessment model of township-scaled land subsidence based on an evidential reasoning algorithm and fuzzy set theory, Risk Analysis, 34, 656–669, 2014.

Csutora, R. and Buckley, J.: Fuzzy Hierarchical Analysis: the Lambda-Max Method, Fuzzy Set. Syst., 120, 181–195, 2001.

Feng, K.: A New Fuzzy Hierarchical Analysis Method Based on the Ideal Alternative. Proceedings of the Six International Conference On Machine Learning and Cybernetics, Hong Kong, 19–22, 2007.

Hu, B., Jiang, Y., and Zhou, J.: Assessment and zonation of land Subsidence disaster risk of Tianjin binhai area, Sci. Geogr. Sin., 28, 693–697, 2008.

Woude, S. V. and Jonker, J. H.: Risk management for the Betuweroute shield driven tunnels, (Re) Claiming the Underground Space, 1043–1049, 2003.

Yoo Chungsik, Y.-W. and Choi, B.-S.: IT-based tunneling risk management system (IT-TURISK) – Development and implementation, Tunneling and Underground Space Technology, 21, 190–202, 2006.

Yu, J., Wu, J., and Wang, X.: Preliminary research on risk of land subsidence, Geol. J. China Universit., 14, 450–454, 2008.

Proc. IAHS, 372, 555–557, 2015
proc-iahs.net/372/555/2015/
doi:10.5194/piahs-372-555-2015

Preface: Prevention and mitigation of natural and anthoropogenic hazards due to land subsidence

K. Daito[1] and D. L. Galloway[2]

[1]Daido University, Japan
[2]USGS, USA

Correspondence to: K. Daito (daito@daido-it.ac.jp)

Published: 16 December 2015

An awareness of problems related to land subsidence and its anthropogenic causes has been growing worldwide since the second half of the Twentieth century. The problem of land subsidence was first included by the UNESCO program of the International Hydrological Decade (IHD), 1965–74. In 1969 UNESCO convened the 1st International Symposium on Land Subsidence in Tokyo. In 1975 land subsidence was retained under the framework of the International Hydrological Program (IHP) as subproject 8.4 "Investigation of Land Subsidence due to Groundwater Exploitation", and UNESCO IHP formally codified the Working Group on Land Subsidence. In collaboration with UNESCO IHP, IAHS, and other scientific organizations, the UNESCEO–IHP Working Group on Land Subsidence convened eight more International Symposia on Land Subsidence in different countries in Asia, Europe and North America: Anaheim, USA (1976); Venice, Italy (1984); Houston, USA (1991); The Hague, Netherlands (1995); Ravenna, Italy (2000); Shanghai, China (2005); Queretaro, Mexico (2010); and Nagoya, Japan (2015).

Though subsidence is a global phenomenon, the consequences and their remediation generally are local. Land subsidence has become an important area of research in Japan, and globally. The proceedings of the Ninth International Symposium on Land Subsidence (NISOLS, Nagoya, 2015) represents a body of high quality and globally relevant scientific and technical information for scientists, engineers, and other stakeholders concerned about land subsidence and the associated hazards. The papers in this volume cover international science and social issues related to land subsidence and the role of natural resources development.

The main topics addressed by NISOLS includes Land Subsidence processes related to (1) Aquifer-system compaction and subsidence caused by groundwater withdrawal, (2) Risk management of subsidence related hazards, (3) Anthropogenic land subsidence in coastal regions, (4) Land subsidence related to soil oxidation, (5) Anthropogenic uplift (CO_2-sequestration included), (6) Land subsidence and liquefaction, due to the East Japan Great Earthquake of March 11, 2011, (7) Ground failure (fracturing, fault activation, fissuring), (8) Land surface displacement, measuring and monitoring, (9) Numerical modelling, (10) Social, cultural and economic influence of land subsidence, (11) Water management strategies, (12) Subsurface deformation due to shale gas production, (13) Settlements and geotechnical construction activities and (14) Seismic activity, triggered by fluid extraction and injections. Many presenters describe the development of new techniques for monitoring, analysis, interpretation and prediction of subsidence and (or) related fracturing processes – an emphasis of NISOLS.

This symposium and the proceedings are of particular interest for Japan because of the close relation between land subsidence, ground fracturing, and groundwater use and management. These phenomena provoke widespread problems in the rapidly growing urbanized areas of central Japan. The analysis of these phenomena requires a multidisciplinary approach to improve understanding of the triggering factors, failure modes, and propagation processes of fracturing. Resolution of the often conflicting socio-economic demands placed on our natural resources and the desires to protect and preserve our natural resources is an inevitable condition for the sustainable development in developed areas. To face this challenge in Japan and elsewhere we need to improve our understanding of Land Subsidence processes. An effort was made in NISOLS to integrate a broad spectrum of the subsidence-interest community; three peripheral technical meetings were convened to focus on subsidence-related topics: (1) Workshop on Subsidence Monitoring, (2) Workshop on MODFLOW Subsidence Simulation, and (3) Workshop on Geotechnical Databases.

On behalf of Inter Group Corporation, Daido University and the UNESCO–IHP Working Group on Land Subsidence we are pleased to present the proceedings of NISOLS 2015 and hope that you find much relevant and useful information in the contributed papers.

Acknowledgements

The Organizing Committee is grateful to the many people who gave their time, effort and knowledge to produce a successful program, productive workshops and enjoyable field trips. The Scientific Advisory Board diligently peer reviewed each of the contributed abstracts and papers. The Local Organizing Committee accomplished the overwhelming tasks of arranging, coordinating, directing and hosting the symposium and processing the contributed abstracts, presentations and papers.

The Editors are especially grateful for the sponsorship and endorsement provided for the symposium by the following organizations and companies:

Sponsoring organizations

- Daido University
- Daiko Foundation
- FORUM8 Co. Ltd.
- Japan Natural Gas Association/Keiyo Natural Gas Association
- Kajima Foundation
- Sakabe Environmental Technology Office
- The Association for Cavity Filling Technology
- Tokai Three-Prefecture Investigation Committee on Land Subsidence

Endorsing organizations

- UNESCO, United Nations Educational, Scientific and Cultural Organization
- Japanese National Commission for UNESCO
- Ministry of the Environment Government of Japan
- Aichi Prefecture
- City of Nagoya
- Tokai Three Prefecture Investigation Committee on Land Subsidence of Japan
- The Japanese Geotechnical Society
- Japanese Association of Groundwater Hydrology
- Japan Society of Civil Engineers
- Japan Society of Engineering Geology

Booth exhibition attending organizations

- UNESCO-IHP Working Group on Land Subsidence
- Japan National Commission for UNESCO
- Tokai Three-Prefecture Investigation Committee on Land Subsidence
- Aichi Prefectural Government, Department of Environment
- City of Nagoya, Department Regional Environment Policies Division, Environmental Affairs Bureau Regional Environment Policies
- Daido University, Department of Architecture, Civil Engineering and Environmental Design Course
- Japan Natural Gas Association/Keiyo Natural Gas Association, Environment Committee
- The Association for Cavity Filling Technology
- FORUM8 Co. Ltd.

Organizing committee

- Devin L. Galloway – Chairman, USGS, USA
- Dora Carreón-Freyre – Vice Chairman, UNAM, Mexico
- Kenji Daito, Daido University, Japan
- Pietro Teatini, University of Padova, Italy

Local organizing committee

- Kenji Daito – Chair, Daido University, Japan
- Takeshi Sato, Gifu University, Japan
- Takeshi Makinouchi, Meijo University, Japan
- Kano Ueshita, Nagoya University, Japan
- Ryuichi Sugisaki, Nagoya University, Japan
- Fusetsu Takagi, Nagoya University, Japan
- Mamoru Adachi, Nagoya University, Japan
- Katsuro Ogawa, Nagoya University, Japan
- Tsutomu Sugiura, Aichi University of Education, Japan
- Kazuki Mori, Mie University, Japan

Special thanks to Kengo Suzuki and Mizuru Suzuki of Inter Group Corporation for providing the technical-scientific secretarial support for the symposium and invaluable assistance in the publication of this proceedings volume.

AUTHOR INDEX

AUTHOR INDEX

AUTHOR INDEX